NUMERICAL PROBLEMS IN CHEMISTRY

NUMERICAL PROBLEMS IN CHEMISTRY

OBJECTIVE TYPE

(FOR ALL COMPETITIVE EXAMINATIONS)

- **WITH SOLUTIONS OF ALL PROBLEMS**
- **SIMPLE EXPLANATIONS**
- **PRACTICE PAPERS**

A.S. BHATIA
B. Tech (IITD)

DEEP & DEEP PUBLICATIONS PVT. LTD.
F-159, Rajouri Garden, New Delhi - 110 027

NUMERICAL PROBLEMS IN CHEMISTRY

ISBN 978-81-8450-203-9

Typeset by RAHUL COMPOSERS
358, Pocket-B, Phase-II, Sector-16B, Dwarka, New Delhi - 110 078

Printed in India at MAYUR ENTERPRISES
WZ Plot No. 3, Gujjar Market, Tihar Village, New Delhi - 110 018

Published by DEEP & DEEP PUBLICATIONS PVT. LTD.
F-159, Rajouri Garden, New Delhi - 110 027 • Phone : 25435369, 25440916
E-mail : ddpubs@gmail.com • ddpbooks@yahoo.co.in
Showroom :
2/13, Ansari Road, Daryaganj, New Delhi - 110 002 • Telefax : 23245122

Dedicated to
My Parents
Sardar Sarup Singh Bhatia
and
Sardarni Parkash Kaur Bhatia

CONTENTS

Part IV
SOLUTIONS

TABLES

PREFACE

I feel emmense pleasure in bringing out this book on Numerical Problems in Chemistry.

For as long as chemistry has been taught in schools, students have found this subject difficult to understand and learn. Despite the publication of hundreds of textbooks in this field, each one intended to provide an improvement over previous textbooks, students continue to remain perplexed, and the subject is often taken in class only to meet school/departmental requirements for a selected course of study.

This book can, be an invaluable aid to students in chemistry as a supplement to their textbooks.

The study of Chemistry is incomplete if students confine themselves only to the theory portion. Keeping this in view, I felt it necessary to write a book on Numerical Problems.

This book is intended to aid students in chemistry to overcome the difficulties in solving problems. Solutions to the problems are illustrated with detailed step by step explanations, to save the students the large amount of time that is often needed to fill in the gaps that are usually found between steps of illustrations in textbooks. To obtain maximum benefit from the book, students should familiarize themselves with the section, on formulae.

The theory portion as key points has been summarized in the beginning of each chapter in order to make the subject more understandable for solving numerical problems. A quick study chart is provided to refresh the important formulae and concepts while attempting to solve the problems.

The book is subdivided into chapters, each dealing with a separate topic. The subject matter is developed with fundamental gas laws, and sections on equilibrium, chemical kinetics, thermodynamics, and quantum chemistry have also been included.

S.I System of units has been used throughout. Students should practice to solve the numericals on stoichiometry; solutions; chemical kinetics & chemical equilibrium with proper steps using formula and correct units.

Solutions of problems helps better estimate the theory and contributes to its deeper understanding. I have tried to select problems covering virtually every aspect of the course in physical chemistry.

Each chapter begins with the fundamental equations and symbols, with are followed by problems. Solutions to the problems are provided at the end of the book.

In preparing this edition, I have taken into account the main changes in the physical chemistry course and the trends (M. K. system) in this field.

It is hoped that the present edition of the book will be appreciated by our learned colleagues and young readers.

The numerical problems appeared in the various board and competitive examinations have been solved up-to-date. Typical and important problems have been incorporated.

This book will also be helpful to the students preparing for various medical and engineering entrance examination.

Gratitude is also expressed to many persons involved in the difficult task of typing the manuscript with its endless changes, and to the staff who prepared the numerous detailed illustrations together with the layout and physical features of the book.

Finally, special thanks are due to unique talents to render those difficult border-line decisions and constructive suggestions related to the design and organizations of the book.

I shall appreciate constructive comments and suggestions for any improvement in the book.

Three test papers are provided to check the achievement of the student in chemistry. I expect them to be solved independently before checking their solutions.

A.S. BHATIA

NOTE TO STUDENTS

This Quick Study chart on following few pages is a outline of 'Chemistry.' Keep it handy as a quick reference source in the classroom, while doing homework and use it as a memory refresher when reviewing prior to exams. It is a durable and inexpensive study tool that can be repeatedly referred to during your school years and beyond.

MEASUREMENT AND UNITS

mass: kilogram (kg) = 100 g
length: meter (m) = 100 cm = 10^{10} A°
time: second (s); **temperature:** Kelvin, (K)
T(K) = T(°C) + 273.15

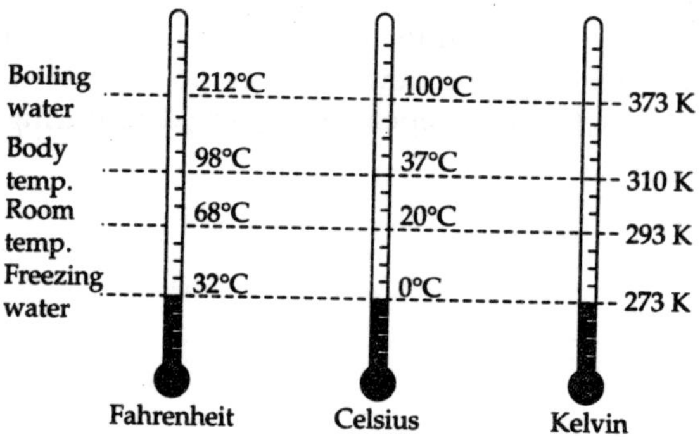

volume: liter (L) = 1000 mL
pressure: Pascal, Pa (N/m^2), 1 atm = 101,325 Pa
force: Newton, N (J/m); **charge:** coulomb, C
energy: Joule, J(kg m^2/s^2)
Prefixes:

tera T (10^{12})	**giga** G (10^9)	**mega** M (10^6)
kilo k (10^3)	**centi** c (10^{-2})	**milli** m (10^{-3})
micro μ (10^{-6})	**nano** n (10^{-9})	**pico** p (10^{-12})

FUNDAMENTAL CONSTANTS

R = 8.314 J $mole^{-1}$ K^{-1} (for energy calculation)
R = 0.082 atm $mole^{-1}$ K^{-1} (for gas property calculation)
Avogadro's Number: $N_A = 6.022 \times 10^{23}$ $mole^{-1}$
Bolzamann constant: k= $R/N_A = 1.381 \times 10^{-23}$ J $molecule^{-1}$ K^{-1}
Elementary charge of the electron e: 1.602×10^{-19} C
Faraday Constant: $\mathscr{F}$ charge of N_a electrons.
Mass of a proton $m_p = 1.673 \times 10^{-27}$ kg;
Mass of a neutron, $m_n = 1.675 \times 10^{-27}$ kg
Mass of an electron, $m_e = 9.110 \times 10^{-31}$ kg
Planck's Constant, h: 6.626×10^{-34} j s
Sped of light in a vacuum, c: 2.9979×10^8 m s^{-1}

QUICK STUDY CHART

ATOMIC STRUCTURE

Atomic Number Z: # of protons in the nucleus; for neutral atoms Z = # of electrons; for anion of charge –q, Z + q electrons; for carbon of charge +q, Z-q electrons.
Mass Number, A: A = Z_+ N, the number of neutrons in the nucleus. **Isotopes:** atoms with the same Z, different A. An element can have a number of isotopes, any sample contains a number of isotopes; in practical work use the average **Mass Number** or **Atomic Weight** (weighted average of natural isotopes for a given element).

ATOMIC QUANTUM NUMBERS AND ORBITALS

n; principle
l : angular momentum (orbital shape)
m_l: magnetic (orbital direction) m_s : electron spin
For each *n*, the possible *l* values are 0, 1... *n*–1, for each *l* the possible m_l, values are –*l*, 0, +*l*, m_s has two possible values +½ and –½ (spin up, spin down) Each value of *n* denotes a "shell" in the atomic structure, each *l* denotes a subshell
s-orbital: *l* = 0, 1 type
p-orbital: *l* = 1, 3 types
d-orbital: *l* = 3, 5 types
f-orbital: *l* = 3, 7 types

MANY ELECTRON ATOMS

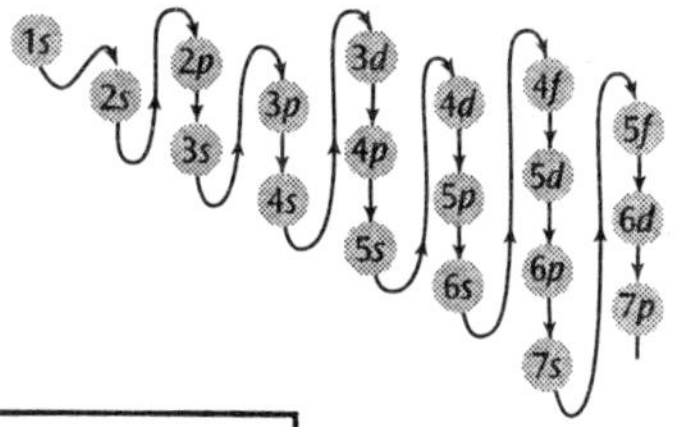

Orbitals and quantum numbers derived for the hydrogen atom are used to describe all many-electron atoms and ions. A more rigorous treatment determines exact energy levels and orbital properties. The main result is that sub-shells have different energies. The filling of levels is guided by the **Aufbau Principle**

n	*l*	m_i	m_s	*element*	
1	0	0	+½	H	Is-orbital
1	0	0	+½	He	spherical
2	0	0	+½	Li	2s-orbital
2	0	0	–½	Be	spherical
2	1	–1	+½	B	2p-orbitals
2	1	0	+½	C	
2	1	+1	+½	N	bimodal
2	1	–1	–½	O	$p_x\ p_y\ p_z$
2	1	0	–½	F	
2	1	+1	–½	Ne	

electron configuration atomic orbital occupancy
pauli Exclusion Principle. each electron has a unique set of quantum numbers. An orbital may hold up to two electrons, one with spin-up $m_s = +\frac{1}{2}$, and one with spin-down, $m_s = -\frac{1}{2}$.

Hund's Rule electrons fill *p*, *d* and *f* sub-shells in a manner to maximize the number of repaired electrons (half-filled orbitals)
ionization potential (IP): energy required to remove electron from an atom or ion. **First IP:** removal of first valence electron.
electronegativety: tendency of an atom to attract electrons in a chemical bond, 0–4 scale, 0 for rare gas.

FORMAL BONDING MODELS

Lewis Structure: visual representation of a molecule with the valence electrons (dots) allocated as bonding plus and non-bonding lone-pairs

H:H :C̤̈l:C̤̈l: :N:::N: Ö::Ö

H:C̤̈l: Ö::C::Ö H:Ö:H

Octel Rule: atoms surrounded by 8 electrons, forming 4 bonds (except H which will have 2 electrons and atoms with d-orbitals which can form 6 bonds e.g. SF_6).
multiple bonds: bonded atoms may share 1,2 or 3 electron-pairs.
bond order: the number of bonds for an atom divided by the number of neighbors. Single bond (order 1.0), double (2.0), triple (3.0), may be fractional. For a given bond-type, bond length decreases with increasing bond order.
resonance: if different Lewis structures are possible an average of them is the more accurate view of the bonding. Resonance delocalizes electrons and stabilizes energy. For example, Ozone (O_3) has two resonance structures.

:Ö:Ö::Ö: ⟷ :Ö::Ö:Ö:

formal charge: denotes that an atom has gained or lost valence electrons, relative to the free atom. Calculate as follows : (number of valence electrons in free atom) $-\frac{1}{2}$ (number of bonded electrons) – (# lone pair electrons). Other things equal structures with smaller formal charges are more stable place negative charge on more electronegative atom.

VALENCE BOND THEORY

A chemical bond forms by the overlap of singly-occupied valence atomic or hybrid orbitals. The resulting pair of electrons form the chemical bond

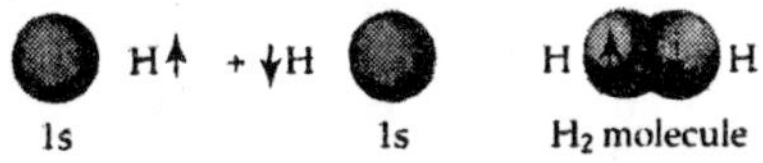
H↑ + ↓H
1s
1s
H H
H₂ molecule

CHEMICAL BONDING AND QUANTUM MECHANICS

Particle-wave Duality: electrons and light exhibit wave and particle character. The wavelength (λ) of light describes the "color", related to the frequency (v) and speed of light (c) by $\lambda v = c$, the energy of light is quantized in photons, hv (h, Planck's Constant) in simple form, the electron wave-property is described by the deBroglie model: the λ is related to its mass (m) and velocity (v) by $\lambda = h/mv$. Electrons in atoms occupy discrete energy levels and are described with quantum number and wavelike **atomic orbitals** (AO's). The Aufbau principle guides the filing of the atomic energy levels. The complete description of the wave character is termed the **wavefunction** ψ. The AO's give ψ for an atom.

MOLECULAR ORBITAL (MO) THEORY

MO's give the ψ for a molecule; each MO is linear combination (weighted sum) of AO's.

Overlap of atomic orbital lobs with the same sign gives constructive interference (**bonding interaction**); overlap of orbital lobes with opposite sings gives destructive interference; a less stable arrangement termed **antibonding**.

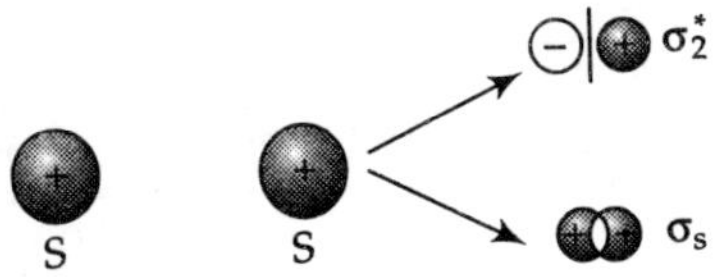

The MO's describe the electron density and energy. The bonding interaction produces an

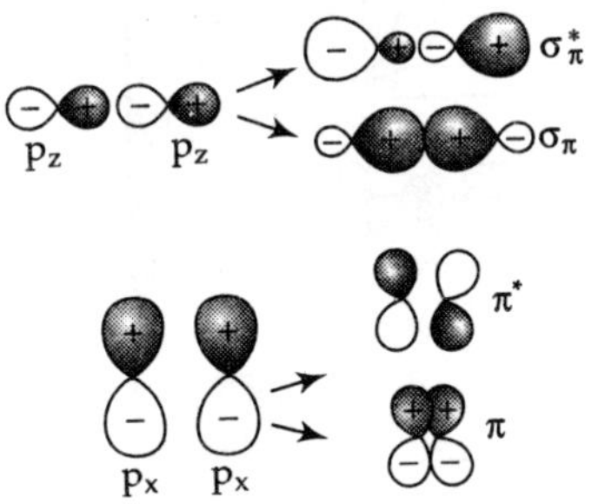

Interactions on the bond axis are termed σ; off axis are termed π. Antibonding interactions are noted π^* and σ^*

accumulation of electron density in the bonding region.

A diagram of MO's gives energy levels for electrons in the molecule. As for AO's each MO can have at most 2 electrons (with paired spin).

The **bond order** is given by (number of bonding electron pairs).– (number of antibonding electron pairs).

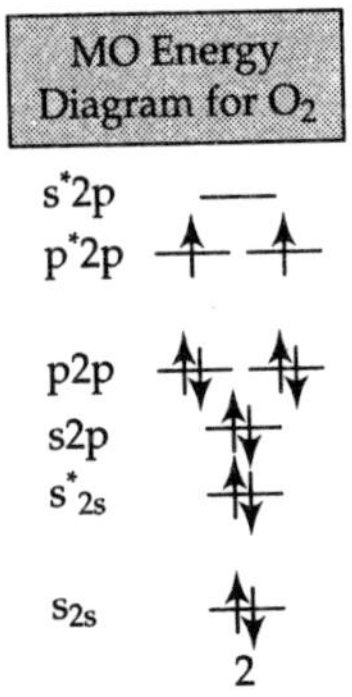

MOLECULAR PROPERTIES

Geometry: Valence Shell Electron Pair Repulsion Theory (VSEPR): optimum arrangement of bonded neighbors (X) and non-bonded or lone pairs (E) all structures are derived from ideal arrangement of X + E objects about central atom.

AX_2 $BeCl_2$	linear	$\mu = 0$	Hybrid = sp
AX_4 GH_4	trigonal planer	$\mu = 0$	Hybrid = sp^3
AX_2E GeF_2	bent	$\mu > 0$	
AX_4 GH_4	tetrahedral	$\mu = 0$	Hybrid = sp^3
AX_2E NH_3	trigonal pyramidal	$\mu > 0$	
AX_4E_2 H_2O	bent	$\mu = 0$	
AX_5 PF_5	trigonal bipyramidal	$\mu = 0$	Hybrid = sp^3d
AX_4E SF_4	seesaw	$\mu > 0$	
AX_3E_2 ICl_3	T-shaped	$\mu > 0$	
AX_2E_3 ICl_2^-	linear	$\mu = 0$	
AX_6 SF_6	octahedral	$\mu = 0$	Hybrid = sp^3d^2
AX_5E SbS_5^{2-}	square pyramidal	$\mu > 0$	
AX_4E IF_4	square planer	$\mu = 0$	

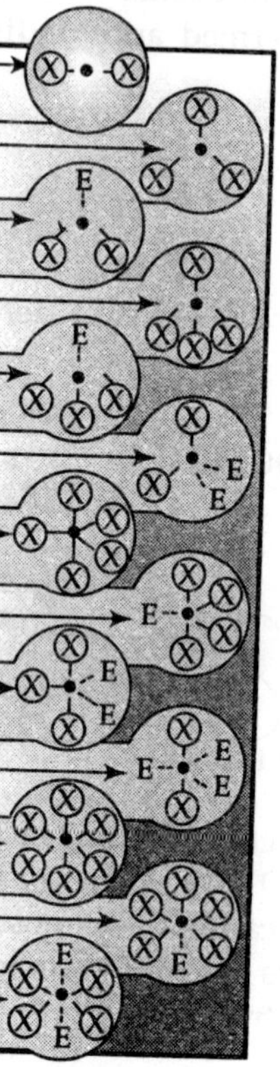

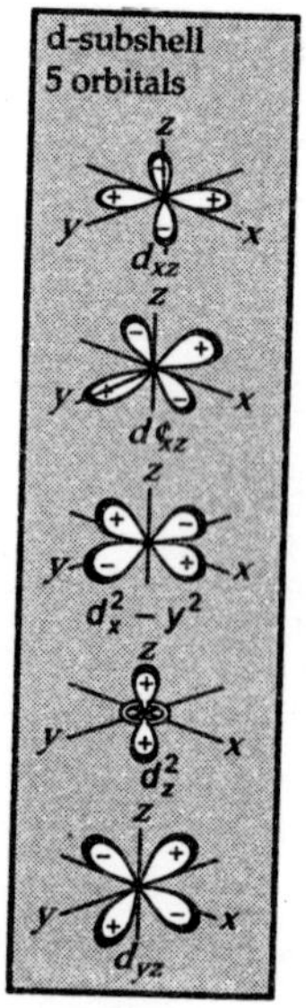

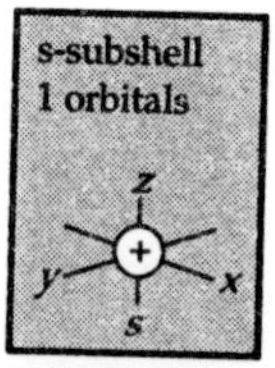

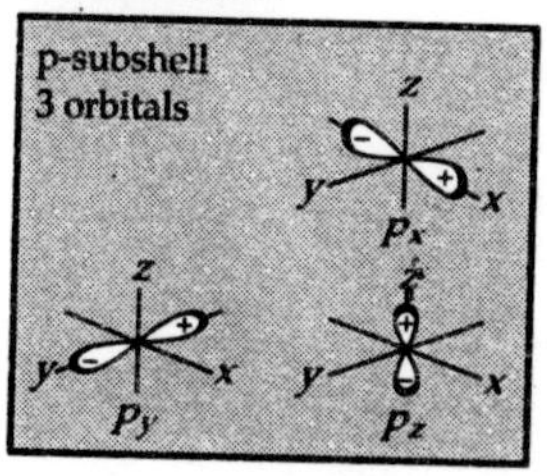

HYBRID ORBITALS

S, *p*, *d* and *f* AO's can mix or **hybridize** to form equivalent lone pairs and bonding paris orbitals. Supports VSEPR model. *S* and *p* can hybridize to *sp* (2) sp^2 (3) or sp^3 (4 equivalent orbitals), *d* orbitals expand the option to five ($sp^3 d$) and six bonded neighbors (sp^3d^2).

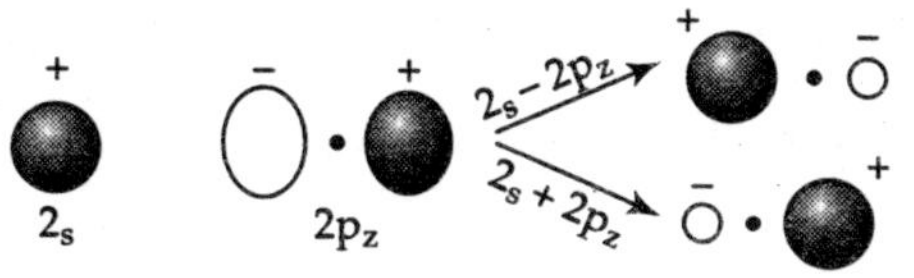

STOICHIOMETRY: MOLE/MASS RELATIONSHIPS

Reaction equation is a molecular equation; the masses of individual molecules are too small for routine use. The **mole** (Avogadro's number of atoms, particles or molecules) gives a more usable quantity.

The **molar mass** of a material is the mass (in grams) of 1 mole of the material. (This is termed the **formula weight** or **molecular weight**.). It is determine by summing the atomic weights for the elements comprising the material, weight by the formula coefficients (e.g. 1 mole of carbon 12 is 12.0 grams).

The balanced **molar equation** can also be viewed as a balanced molecular equation with the coefficients interpreted as **molar quantities.**

Theoretical yield: determine the mass of products from mass of reactants, molar masses, and balanced equation.

Limiting reagents: for two or more reactants, the one consumed first will limit the amount of product.

Solution calculation: the number of moles is determined by the product of solution volume (in liters) and the molarity (moles per liter) of the reagent.

BEHAVIOUR OF GASES

P: pressure is the force/area exerted on the container walls (in atm, mm Hg, Pa) **V: volume** of the gas sample (in liters)
T: temperature (in Kelvin) **n: moles** of gas particles
R: ideal Gas Constant STP (Standard T and P: 1 atm, 273.15 K, one mole of **Ideal Gas** occupies 22.414 liter.
Boyle's Law: $P \times V$ is a constant for fixed T; $P \propto 1/V$
Charles's Law: $V \propto T$ fixed P **Avogadro's Law** $V \propto n$

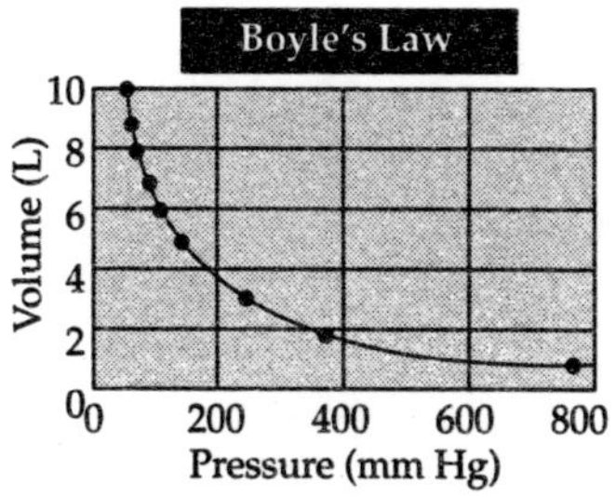

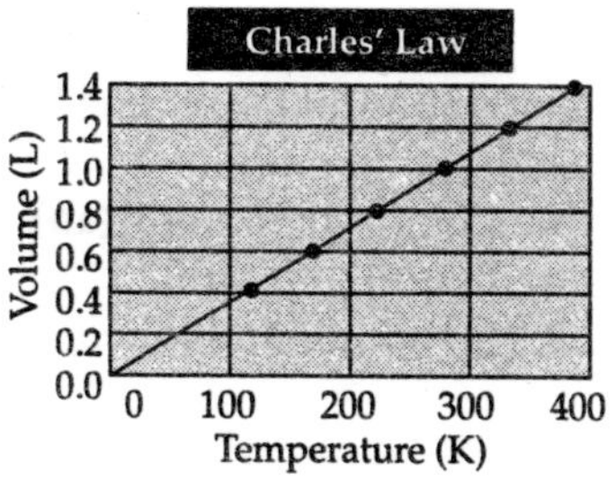

Ideal Gas Law: PV = nRT
Model relationship for describing properties of a gas sample. Non-interacting massless particles. Good for low P, high T, **velocity and energy:** the kinetic energy of 1 mole of Ideal Gas is 3/2 RT; the velocity (rms) for a gas with molar mass "M" at a given T is $(3RT/M)^{1/2}$ (the average speed of a typical gas molecule ranges from 400–1800 m/sec at STP)
Graham's Law: rate of effusion is $\propto 1/M^{1/2}$ (Effusion – gas expands into a vacuum); also fits diffusion for low P (gas mixes with other gases, with molecular collisions)

Gas Mixtures **Dalton's Law of Partical Pressure** (fixed V, T): $P_{total} = P_1 + P_2 + \ldots P_i$; total pressure equation the sum of gas component pressures; $P_i = x_i P_{total}$ (x_i = mole fraction gas "i")
Real gases violate the **Ideal Gas Law.**

- PVT data does not fit Ideal predictions
- Gases liquefy at low T and high P.
- Molecules have rotational and vibrational energy

The van der Waals equation improves the **Ideal Gas** model; adds terms for molecular volume and intermolecular attractions.

OXIDATION-REDUCTION REACTION

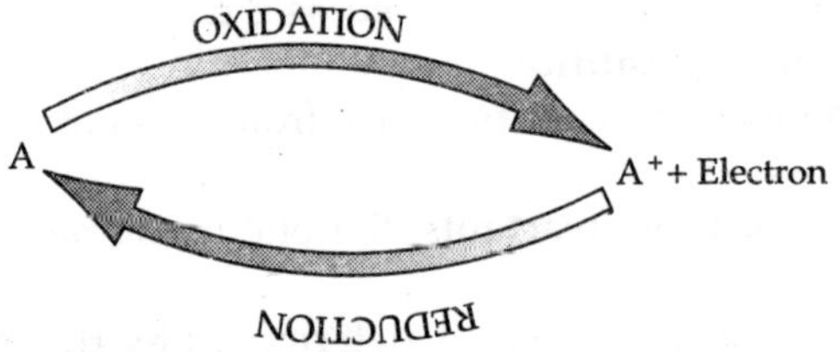

oxidation number: element (0), ionic substance (charge of the ion), covalent compound (charge on atom if all valence electrons shift to the more electronegative atom).
redox chemistry: electrons are exchanged in the reaction; oxidation numbers change.
View reaction "A + B ⇒ AB" in two steps **(half-reactions)**:
first **oxidation:** (A loses electrons), $A \Rightarrow A^+ + e^-$
then **reduction** (B gains electrons) $B + e^- \Rightarrow B^-$
$A^+ + B^-$ form the product AB
Balancing Redox Reactions: Two common approaches:

1. **Half-reaction Method:** balance the reduction and oxidation "half-reactions", Combine, with electron flow balanced.
2. **Oxidation-Number Method**: identify changes in elements valence; balance electron-exchange.

For acidic use H_3O^+ and H_2O t balance. For basic: use OH^- and H_2O.
Examples of redox reactions
battery/galyanic: $Zn(s) + Cu^{2+}(aq) \Rightarrow Zn^{2+}(aq) + Cu(s)$
electrolysis, $2\ H_2O(I) \Rightarrow 2H_2(g) + O_2(g)$
corrosion: $2\ Al\ (s) + 3O_2(g) \Rightarrow 2\ Al_2O_3(s)$

combustion: $C + O_2 \Rightarrow CO_2$
electrochemical cell: An external circuit connects two electrodes (**anode** – site of oxidation, and **cathode** – site of reduction) to facilitate the reaction.
cell EMF: electrical potential generated by the cell $\varepsilon > 0$ for spontaneous process
galvanic (voltaic): spontaneous reaction produces a flow of current; used to make batteries.
Diagram for Zn, Cu cell Zn(s) $[Zn^{2+|}(aq)$ || $Cu^{2+}(aq)$ | Cu(s)

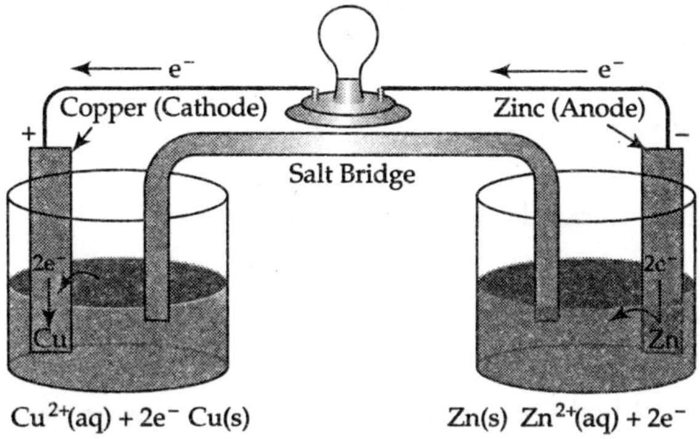

Electrolysis of molten NaCl
electrolytic: external current/voltage drives the reaction. Battery powered electrolytic process

For $\varepsilon < 0$ reverse is spontaneous. $\Delta G = -n \mathcal{F}\varepsilon$ ($\mathcal{F}\varepsilon$, Faraday constant, "n" moles of electrons); reverse reaction, change the sign of the potential.

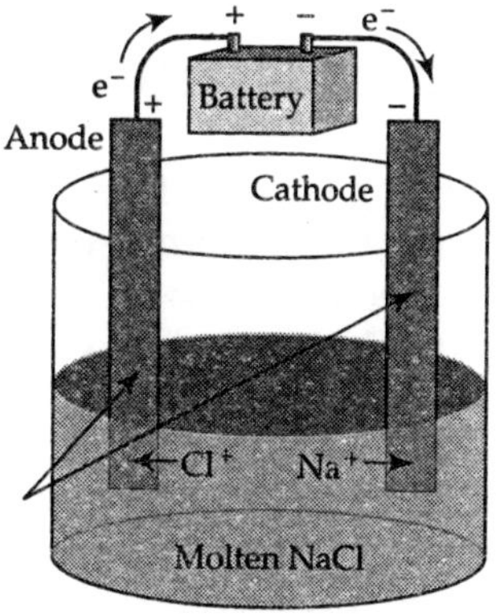

Standard potential

$$\varepsilon^o = \varepsilon^o_{anode} + \varepsilon^o_{cathode}$$

Reduction potentials: Inert standard tabulation of electrodes electrode half-reactions (written as reduction).

$$2Cl^-(l) \rightarrow Cl_2(g) + 2e^- \; 2\,Na^+(l) \; 2e^- \rightarrow 2Na(l)$$

Reference potential: H_2 electrode 0.00 V

- Non metals: $E_2 + 2\,e^- \Rightarrow 2E^-$
  A larger, positive number is evidence of a more reactive material.
  F_2 2.87 Cl_2 1.36 Br_2 1.09 I_2 0.54
- Metals: M^{x+} (aq) + x e^- $\Rightarrow$ Metal
  The positive value is evidence of a less reactive metal.

Ag(I) + 0.80	Cu(II) + 0.34	Pb(II) –0.13	Ni(II) –0.26	Fe(II) –0.45
Zn –0.76	Al –1.66	Mg –237	Na –2.71	Li –3.04

MIXTURES AND SOLUTIONS

Physical combination: solute (solid or lesser amount) dissolves in the solvent (liquid or larger amount)
Major Factor Promoting Mixing of Fluids: a mixture is more disordered than separate phases
Variable Factor: interaction between materials, "like-dissolves-like", polar materials mix

(miscible), as do nonpolar. Polar and nonpolar are usually **immiscible** (form **separate** liquid layers).

Solution units:

M: **molarity**—moles of solute dissolved in 1 liter of solution

m: **molarity**—moles of solute per kg of solvent

x: **mole fraction**—moles of solute divided by total moles

colligative properties: Depend only on the number of solute particles and the identity of the solvent. Ionic materials dissociate.

vapor pressure lowering:

$\Delta P = x_{\text{solute}}\ P^{\circ}$ (solvent vapor pressure).

freezing pt. depression:

$\Delta T = -m_{\text{solute}} k_{bp}$ (constant depends on solvent).

boiling pt. elevation:

$\Delta T = m_{\text{solute}} k_{bp}$ (constant depends on solvent).

osmotic pressure:

$\Pi = M_{\text{solute}}\ RT$, R is the Ideal Gas constant; account for turgor pressure in plants and shape in animal cells.

PHYSICAL PROCESSES

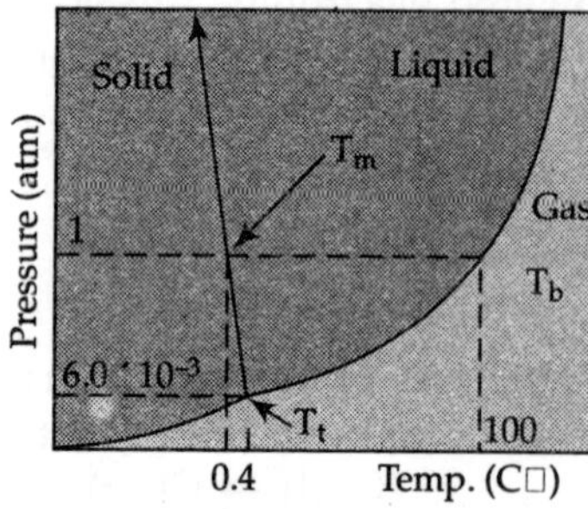

melting ($s \Rightarrow 1$); freezing ($1 \Rightarrow s$)
(at 1 atm: normal melting point "T_m")

evaporating ($1 \Rightarrow g$), condensation ($g \Rightarrow 1$)
(at 1 atm, normal boiling point "T_b")

sublimation ($s \Rightarrow g$)

solution formation mixing of gases liquids or solids.

distillation: Separate mixture by selective evaporation

triple point: s, 1, g in equilibrium "T_f"

vapor pressure: The partial pressure of the gas in equilibrium with liquid (depends on temperature); equals 1 atm at "T_b"

CHEMICAL INTERACTIONS

A. Electronic Properties:

Dipole moment (μ): position of the molecular electron density relative to the center of mass: imparts partial charge to the molecule: $\mu = 0$ **for symmetric (H_2)**

$\mu > 0$ **for asymmetric**

$^{\delta+}H—Cl^{\delta-}$

Polarizability: tendency of electron cloud to disort from equilibrium due to external electrostatics; **increases with atom size**

B. Intermolecular (between molecules)

London Forces (Dispersion): attraction of induced-dipole moments: stronger for more polarizable species; accounts for lique-faction of gases like argon and methane.

Dipole-diple: molecules with dipole moments experience attractive forces for certain relative orientations. Based electrostatic forces.

Stable Stable

Hydrogen-bonding: enhanced dipole-interaction between hydrogen of an –OH or –NH group and a nearby oxygen or nitrogen atom electrostatic: strong interaction between ions; attractive (oppostic charges) or repulsive (like charges). Inversely proportional to distance and dielectric constant of the medium. Water has a large dielectric constant, stabilizes ion formation in solution.

C. Chemical bonds (between atoms in molecules)

Valence electrons: the other electrons which form chemical bonds. The rest of the electrons (immer-shell) are inert core electrons. Bonding is described with three ideal models:

Covalent bond: electrons are shared; the polarity of the bond denotes unequal sharing and is determined by the electronegativity of the atoms; pure covalent: equal sharing

Ionic bond: electrostatic interactions between ions; created by the transfer of electrons between atoms to create ions with filled valene shells.

Metallic bond: electrons are delocalized: shared by a large number of metallic nuclei

Real bonds: partially covalent/partially ionic. Difference in electronegativity determines the % ionic character.

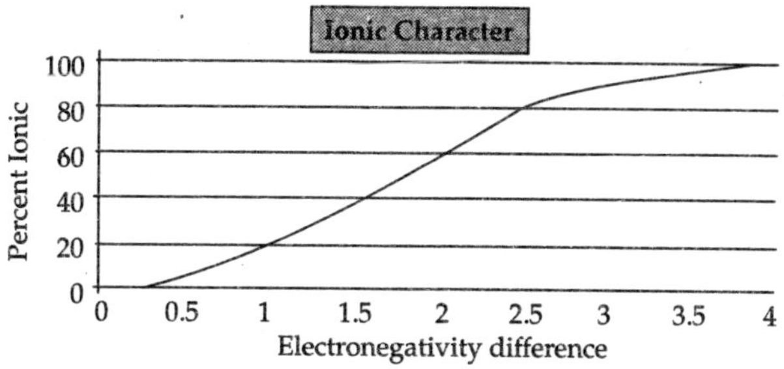

THERMODYNAMICS

The study of the heat and work associated with a **physical or chemical process.**

- **Types of Processes:**
  Reversible, the system is in a state of equilibrium
  Spontaneous (inversible), the system is moving towards a state of equilibrium
- **Laws of thermodynamics:**
  First Law—Conservation of energy (E): The heat (q) and work (w) associated with a process are interrelated: $\Delta E = q + w$.
  Any charge in the energy of the system must correspond to the interchange of heat or work with an external surrounding.

Second Law—Entropy, S, is conserved for a reversible process. The disorder of the system and thermal surrounding must increase for a spontaneous process.
Third Law—Entropy is zero for an ideal crystal at $T = 0$ K. The system is in its lowest possible energy state and most ordered arrangement.

- **Enthalpy (H):** ΔH is the heat absorbed or produced by a process under conditions of constant pressure (normal lab conditions)
  ΔH < 0 for an exothermic reaction and > 0 for endorthermic
  Enthalpies of Formation, ΔHP: the ΔH for the synthesis of the compound from standard elemental forms at 298K. Note: these quantities can be either positive or negative.
  ΔH = (sum of produce ΔHr°) – (sum of reactant ΔHr°)
- Entropy (S) thermodynamic disorder
  ΔS is the change in order in a system For s ⇒ 1 or 1 ⇒ g. ΔS is positive (the product in each case has more random motion).
  Standard entropy, S°, the entropy of a compound at 298K relative to 0K. **Note:** ***These quantities are always positive.***
  ΔS = (sum of product S°) 2 (sum of reactant S°)
- **Gibbs Free Energy** (G): $\Delta G = \Delta H - T\Delta S$.
  ΔG: the capacity of the system to perform work. $\Delta G = 0$ for equilibrium, $\Delta G < 0$ for spontaneous, if $\Delta G > 0$ the reverse process is spontaneous
- **Free energy of formation, ΔG_f** The ΔG for they synthesis for the compound from standard elemental forms at 298K
  ΔG = (sum of product ΔG_f) – (sum of reactant ΔG_f)
  The equilibrium constant K_{eq}, and ΔG are related by the equation: $\Delta G = -RT \ln (K_{eq})$

KINETICS

- **Rates of Chemical Process**
  For a generic reaction: $aA + bB \Rightarrow cC + dD$. the rate usually depends on $[A]$ and $[B]$.
- **First order rate law:** Rate = $k_1[A]$ or $k_i[B]$
  A graph of "ln [ ] vs. time" is linear, the slope is the rate constant k_1.
  half-life, $t_{1/2}$: The time repaired for the concentration to decrease by a factor of 2.
  $t_{1/2} = 0.693/k_1$; ex: radioactive decay.
- **Second order rate law:** Rate = $k_2[A]^2$ o $k_2[B]^2$
  A graph of "1/[ ] vs. time" linear slope is the rate constant k_2.
  Half-life changes during the reaction.
- **Zero order:** Rate = $k_0[A]^0 = k_0$
  The rate is independent of $[A]$: **Note $[A]^\circ = 1$**
- **Temperature-dependence** of rate constants
  Arrhenius Law: $k = A\, e^{E_a/RT}$. E_a, is the activation energy (energy barrier), a graph of "ln(k) vs. 1/T" is energy (energy barrier), a graph of "ln(k) vs. 1/T" is linear, slope is –Ea/R, and the intercept is in (A).
- **Kinetics and Thermodynamics**
  $A \Leftrightarrow B$: $K_{eq} = k_1/k_{-1}$ is forward-rate constant and k_{-1} is reverse-rate constant.

EQUILIBRIUM

- **Reactions to completion:** all reactants are converted to products
- **Equilibrium:** the reaction reaches a steady-state of forward and reverse reactions
- For: $aA \Leftrightarrow bB$: The equilibrium concentrations of reagents $[A]_{eq}$ and $[B]_{eq}$, are constrained by the relationship: $K_{eq} = [B]_{eq}^{b} / [A]_{eq}^{a}$. K_{eq} is a constant, characteristic of the reaction at a given temperature.
- **Solubility Product K_{sp}:** defines the equilibrium between a salt and its aqueous ions: For AX_2, the equilibrium is $AX_2(s) \Leftrightarrow A^{2+}(aq) + 2X^-(aq)$, and the $K_{sp} = [A^{2+}][X^-]^2$: small K_{sp} = low solubility
- LeChatelier's **Principle:** the equilibrium shifts in response to changes in temperature, pressure of reagent concentration. $A \Leftrightarrow B$, removing B or adding A shifts equilibrium towards the product.

  For increases in pressure, the equilibrium shifts to lower the total pressure. (Increasing the pressure rates the concentration). Most relevant for gas phase reactions.
- **Exothermic** reaction produces hear "$A \Rightarrow B$ + heat", lowering the temperature removes heat and shifts equilibrium towards the product. Raising the temperature has the opposite effect.
- **Endothermic** reaction absorbs heat: "Heat + $A \Leftrightarrow B$"; raising the temperature adds heat and shifts equilibrium towards the product. Lowering the temperature has the opposite effect.

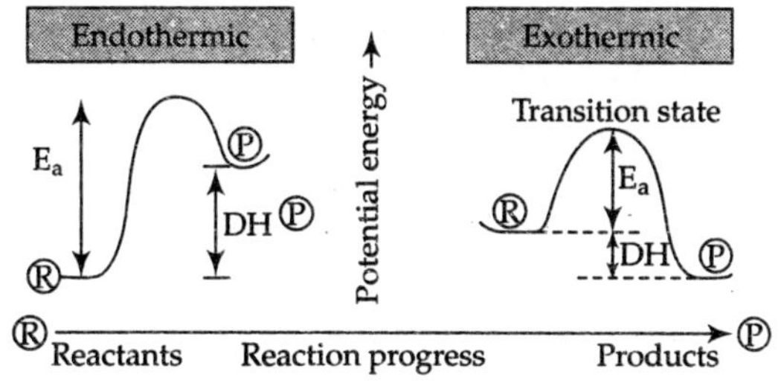

ACID-BASE REACTIONS

- Self-ionization of water. $2H_2O \Leftrightarrow OH^- + H_3O^+$

  $K_w = [OH^-][H_3O^+] = 1 \times 10^{-14}$ at 298K. For all aqueous solutions

  Natural solution $[OH^-] = [H_3O^+] = 1 \times 10^{-7}$

  $pH = -\log_{10}[H_3O^+]$: a measure of acidic strength: **Neutral solutions, pH = 7; acidic, pH < 7; basic, pH > 7.** (eg. a 0.01 M solution of H_3O^+ has a p^H of 2). p^{OH} ($p^{OH} = -\log_{10}[OH^-]$) can be used for basic solutions. $p^H + p^{OH} = 14$
- Acids (HA): $HA + H_2O \Leftrightarrow A^- + H_3O^+$

  A is the conjugate base of the acid HA

  Strong acid, total dissociation: HCl, HBr, HI, $HClO_4$, H_2SO_4 and HNO_3.

  Acid equilibrium is described by $K_a = [A^-][H_3O^+]/[HA]$

Weak acid have $K_a \ll 1$. $pK_a = -\log_{10}(K_a)$
Common weak acid (pK_a): acetic (4.19; HF (4.15); nitrous (3.35); carbonic, (6.37)
Lewis acid is an electron pair acceptor.

- Bases (B): $MOH \Leftrightarrow OH^- + M^+$
  $B + H_2O \Leftrightarrow HB^+ + OH^-$
  HB^+ is the conjugate acid of the base B
  Strong base, complete dissociation: NaOH, KOH and $Ba(OH)_2$
  Base equilibrium is dissociation: $K_b = [OH^-][HB^+]/[B]$
  Weak bases have $K_b \gg 1$. $pK_b = -\log_{10}(K_b)$. Common weak bases (pK_a): NH_3 (4.75); CN^- (4.70)
  Lewis base is an electron-pair donor.
- **Polyprotic acid:** A compound with more than one ionizable proton (eg. H_2SO_4, H_3PO_4)
- **Amphoteric substance.** A material which can react as an acid or a base
- **Hydrolysis:** water + aqueous ion ⇒ acidic or basic solution by forming OH^- or H_3O^+: $A^- + H_2O \Leftrightarrow HA + OH$ (basic, eg. F^- or acetate); $HX^+ + H_2O \Leftrightarrow X + H_3O^+$ (acidic, eg. N_4^+)
- **buffer:** a solution of weak acid and a salt of its conjugate base or a solution of a weak base and a salt of its conjugate acid. The mixture materials "constant" p^H, Henderson-Hasselbalch equation for an acid/salt buffer: $p^H = pK_s + \log_{10}([salt]/[acid])$.
- **acid-base titration:** react a known amount of acid with a basic solution of unknown concentration.
  At the equivalence point: **(a)** mole of acid = the moles of base. **(b)** For strong acid-strong base titration $p^H = 7$ **(c)** For weak acid-strong base titration $p^H > 7$; **(d)** For weak base strong acid titration $p^H < 7$ (Hydrolysis of the salt ions changes the p^H of the solution)
- **Reactions:** acid-rain: sulfur and nitrogen oxides react with water to give acids. Strong acid/bases react with metals. produce H_2 and salt. Strong acid/bases react with metals, produce H_2 and salt. Carbonates are decomposed by acid. Copper + nitric acid produces nitrogen oxides.

NUCLEAR CHEMISTRY

Nuclear reactions alter the nucleus; particles ejected or absorbed. Balance charge and mass-energy. Energy (E) changes to mass (m), follows the Einstein Equation $\mathbf{E = mc^2}$

Nuclide Symbol: for element X, nuclei is symbolized by ${}^{A}_{Z}X$ or X-A (eg. Deuterium, ${}^{2}_{1}H$ or H-2)

Types of processes:

Transmuttion: new elements made from particle collisions
Fusion (small atoms combirle): H-1 + H-2 ⇒ He-3
Fission (large atom splits)
a + U-235 ⇒ Ba-145 + Kr-91 + 3 n
β decay (energetic electron):
I-131 ⇒ Xe-131 + β
λ decay (photon): accompanied by α, β.
Characterized by half-life: time for half of the nuclei to decay.

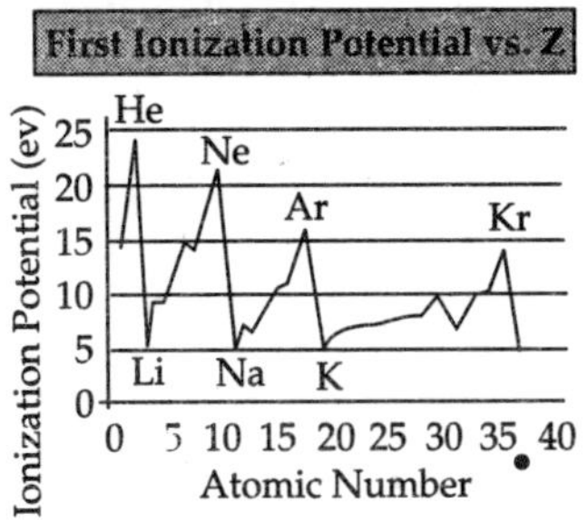

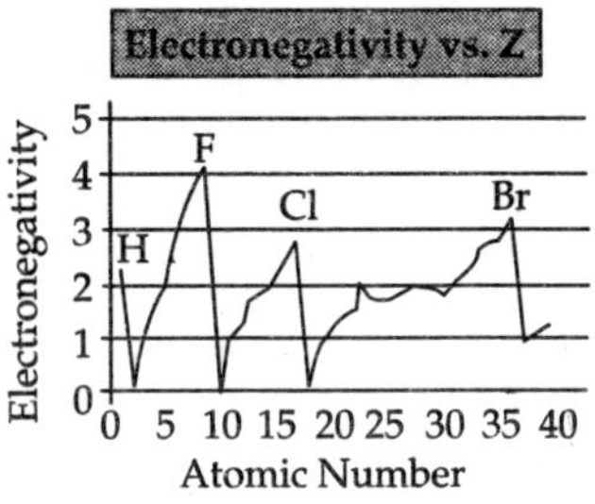

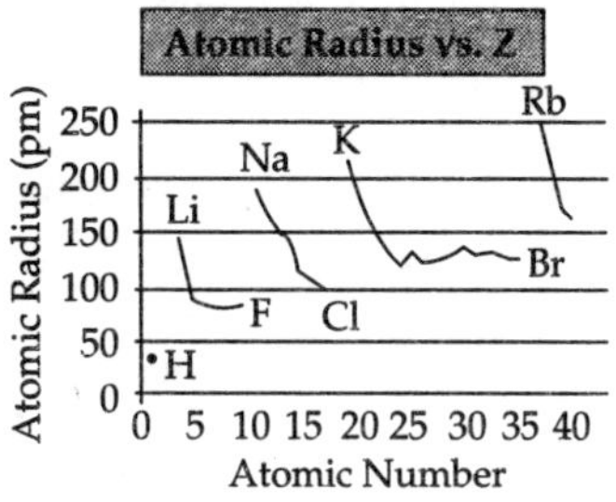

PROPERTIES OF INORGANIC SALTS

Aqueous solubility

	Acetate nitrate perchlorate chlorate	chloride bromide iodide	fluoride	sulfate	carbonate sulfide phosphate hydroxide chromate	oxide
alkali metals, ammonium	**Soluble**					form hydroxide
Ca, Mg			insol	s	insol	insol
Sr, Ba			insol	insol	insol	insol
Fe, Cu, Zn			insol	s	insol	insol
Pb(II)		insol	insol	insol	insol	insol
Hg(I)		insol	—	insol	insol	insol
Ag		insol	s	s	insol	insol

Flame test:

Character colors of ion in a flame:

violet	potassium, rubidium, cesium
blue:	copper (azure); lead, asremic, selenium
green:	copper (emerald), barium (yellowish), zinc (whitish)
yellow	sodium
red	lithium (carmine), strontium (scarlet), calcium (yellowish)

Chemical Formulae

- Cation symbol, followed by anion
- Subscripts denote relative composition
- Enclose polyatomic ions or molecules in parentheses
- Enclose polyatomic ions or molecules in parentheses
- **Molecular Formula:** Discrete molecule.
- **emperical Formula:** Relative molar ratio of elements for solids or molecules

Monatomic cations

Alkali metal (+1) alkaline earth (+2)

Transition metal: valence varies, give the valence in the name and formula, "ous" or "ic" ending. Roman-numerals are less confusing, more general

ferric, Fe (III) ferrous, Fe (II)
stamic, Sn(IV) stamous, Sn(II)
Plumbic, Pb(IV) plumbous, Pb(II)
cupric, Cu(II) Cuprous, Cu(I)
mercuric, Hg(II) mercurous, Hg(I)

(+1) **Polyatomic cations**

ammonium, NH_4^+
hydronium, H_3O^+; active form of acid in water
Monatomic anions
(-4) carbide, C^+, silicide, Si^+
(-3) nitride, N^{3-}, phosphide, P^{3-}, ansenide, As^{3-}
(-2) oxide, O^{2-}, sulfide, S^{2-}, selenide, Sc^{2-}, selenide, Sc^{2-}, telluride, Te^{2-}
(-1) hydride, H^- halides: fluoride F^-, chloride Cl^-, bromide, Br^-, iodide, I^- Acids: hydro-fluoric, -chloric, -bromic, -iodic

Polyatomic anions (and respective acids)

(-1)

acetate $C_2H_3O_2^-$ *acetic acid* $C_2H_4O_2$
nitrate, NO_2^- *nitric acid,* HNO_3
nitrite, NO_2^- *nitrous acid,* HNO_2
hypochlorite, ClO^- *hypochlorous acid,* $HClO$
Chlorite, ClO_3^- *chlorous acid,* $HClO_2$
perchlorate, ClO_4^- perchloric acid, $HClO_4$
cyanide, CN^- *hydrocyanic acid,* HCN
hydroxide, OH^- *formed by bases in water*
bicarbonate or hydrogen carbonate, HCO_3^-
bisulfate or hydrogen sulfate, HSO_4^-
dihydrogen phosphate, $H_2PO_4^-$
permanganate, Mn_4^-

(-2)

carbonate, CO_3^{2-} *carbonic acid,* H_2CO_3
sulfate, SO_4^{2-} *sulfuric acid,* H_2SO_4
sulfite, SO_3^{2-} *sulfurous acid,* H_2SO_3
chromate, CrO_4^{2-} *chromic acid,* H_2CO_4
peroxide O_2^{2-}

biphosphate or hydrogen phosphate, HPO_4^{2-}, dichromate, $Cr_2O_7^{2-}$ thiosulfate, $S_2O_3^{2-}$ (thio: S subst, for O atom) disulfide, S_2^{2-}

(-3)

phosphate, PO_4^{4-} *phosphoric acid* H_3PO_4

(-4)

silicate, SiO_4^{4-} *silicic acid,* H_4SiO_4

PERIODIC TABLE OF THE ELEMENTS

Key: 1 — ATOMIC NUMBER; H — SYMBOL; 1.008 — ATOMIC WEIGHT

H 1.01																	2 He 4.00
3 Li 6.94	4 Bo 9.01											5 B 10.81	6 C 12.01	7 N 14.01	8 O 16.00	9 F 19.00	10 Ne 20.18
19 Na 22.99	20 Mg 24.31											13 Al 26.98	14 Si 28.09	15 P 30.97	16 S 32.07	17 Cl 35.45	18 Ar 39.95
19 K 39.10	20 Ca 40.08	21 Sc 44.96	22 Tl 47.88	23 V 50.94	24 Cr 52.00	25 Mn 54.94	26 Fe 55.85	27 Co 58.93	28 Nl 58.69	29 Cu 63.55	30 Zn 65.39	31 Ga 69.72	32 Ge 72.61	33 As 74.92	34 Se 78.96	35 Br 79.90	36 Kr 83.80
37 Rb 65.47	38 Sr 87.62	39 Y 88.91	40 Zr 91.22	41 Nb 92.91	42 Mo 95.94	43 Tc 97.91	44 Ru 101.07	45 Rh 102.91	46 Pd 106.42	47 Ag 107.87	48 Cd 112.41	49 In 118.82	50 Sn 118.71	51 Sb 121.75	52 Te 127.60	53 I 126.90	54 Xe 131.29
55 Cs 132.91	56 Ba 137.32	57 La* 138.91	72 Hf 178.49	73 Ta 180.95	74 W 183.85	75 Re 186.21	76 Os 190.20	77 Ir 192.22	78 Pt 195.08	79 Au 196.97	80 Hg 200.59	81 Tl 204.38	82 Pb 207.20	83 Bi 208.98	84 Po 208.98	84 At 209.99	86 Rn 222.02
87 Fr 233.02	88 Ra 226.03	89 Ac* 227.03	104 Rf 261	105 Db 262	106 Sg 263	107 Bh 262	108 Hs 265	109 Mf 266	110 Disc 1994 269	111 Disc 1994 272	112 Disc 1996 277	113 Not Discovered	114 Disc 1999 289	115 Not Discovered	116 Di…99 retracted 28…	117 Not Discovered	118 Di…99 retracted 29…

58 Ce 140.12	59 Pr 140.91	60 Nd 144.24	61 Pm 144.91	62 Sm 150.36	63 Eu 151.97	64 Gd 157.25	65 Tb 158.93	66 Dy 162.50	67 Ho 164.93	68 Er 187.26	69 Tm 168.93	70 Yb 173.04	71 Lu 174.97
90 Th 232.04	91 Pa 231.04	92 U 238.05	93 Np 237.05	94 Pu 244.06	95 Am 243.06	96 Cm 247.07	97 Bk 247.07	98 Cr 251	99 Es 252.08	100 Fm 257.10	101 Md 258.10	102 No 259.10	103 Lr 260.11

Electronic Configurations of Elements in the Ground State

Atomic Number	Symbol	Atomic Weight	Name	Electron Config.	
1.	H	1.008	Hydrogen	$1s^1$	n=1
2.	He	4.003	Helium	$1s^2$	
3.	Li	6.941	Lithium	$1s^2\ 2s^1$	n=2
4.	Be	9.012	Beryllium	$1s^2\ 2s^2$	
5.	B	10.81	Boron	[He] $2s^2\ 2p^1$	
6.	C	12.01	Carbon	[He] $2s^2\ 2p^2$	
7.	N	14.01	Nitrogen	[He] $2s^2\ 2p^3$	
8.	O	16.00	Oxygen	[He] $2s^2\ 2p^4$	
9.	F	19.00	Fluorine	[He] $2s^2\ 2p^5$	
10.	Ne	20.18	Neon	[He] $2s^2\ 2p^6$	
11.	Na	22.99	Sodium	[Ne] $3s^1$	n=3
12.	Mg	24.31	Megnesium	[Ne] $3s^2$	
13.	Al	26.98	Aluminium	[Ne] $3s^2\ 3p^1$	
14.	Si	28.09	Silicon	[Ne] $3s^2\ 3p^2$	
15.	P	30.97	Phosphorus	[Ne] $3s^2\ 3p^3$	
16.	S	32.07	Sulfur	[Ne] $3s^2\ 3p^4$	
17.	Cl	35.45	Chlorine	[Ne] $3s^2\ 3p^5$	
18.	Ar	39.95	Argon	[Ne] $3s^2\ 3p^6$	
19.	K	39.10	Potassium	[Ar] $4s^1$	n=4
20.	Ca	40.08	Calcium	[Ar] $4s^2$	
21.	Sc	44.96	Scandium	[Ar] $3d^1\ 4s^2$	
22.	Ti	47.88	Titanium	[Ar] $3d^2\ 4s^2$	
23.	V	50.94	Vanadium	[Ar] $3d^3\ 4s^2$	
24.	Cr	52.00	Chromium	[Ar] $3d^5\ 4s^1$	
25.	Mn	54.94	Manganese	[Ar] $3d^5\ 4s^2$	
26.	Fe	55.85	Iron	[Ar] $3d^6\ 4s^2$	
27.	Co	58.93	Cobalt	[Ar] $3d^7\ 4s^2$	
28.	Ni	63.55	Nickel	[Ar] $3d^8\ 4s^2$	
29.	Cu	63.39	Copper	[Ar] $3d^{10}\ 4s^1$	
30.	Zn	65.39	Zinc	[Ar] $3d^{10}\ 4s^2$	
31.	Ga	69.72	Gallium	[Ar] $3d^{10}\ 4s^2\ 4p^1$	
32.	Ge	72.61	Germanium	[Ar] $3d^{10}\ 4s^2\ 4p^2$	
33.	As	74.92	Arsenic	[Ar] $3d^{10}\ 4s^2\ 4p^3$	
34.	Se	78.96	Selenium	[Ar] $3d^{10}\ 4s^2\ 4p^4$	
35.	Br	79.90	Bromine	[Ar] $3d^{10}\ 4s^2\ 4p^5$	
36.	Kr	83.80	Krypton	[Ar] $3d^{10}\ 4s^2\ 4p^6$	
37.	Rb	85.47	Rubidium	[Kr] $5s^1$	n=5
38.	Sr	87.62	Strontium	[Kr] $5s^2$	
39.	Y	88.91	Yttrium	[Kr] $4d^1\ 5s^2$	
40.	Zr	91.22	Zirconium	[Kr] $4d^2\ 5s^2$	
41.	Nb	92.91	Niobium	[Kr] $4d^4\ 5s^1$	
42.	Mo	95.94	Molybdenum	[Kr] $4d^5\ 5s^1$	
43.	Tc	98	Technetium	[Kr] $4d^5\ 5s^2$	
44.	Ru	101.1	Ruthenium	[Kr] $4d^7\ 5s^1$	
45.	Rh	102.9	Rhodium	[Kr] $4d^8\ 5s^1$	
46.	pd	106.4	Palladium	[Kr] $4d^{10}\ 5s^0$	
47.	Ag	107.9	Silver	[Kr] $4d^{10}\ 5s^1$	
48.	Cd	112.4	Cadmium	[Kr] $4d^{10}\ 5s^2$.	
49.	In	114.8	Indium	[Kr] $4d^{10}\ 5s^2\ 5p^1$	
50.	Sn	118.7	Tin	[Kr] $4d^{10}\ 5s^2\ 5p^2$	
51.	Sb	121.8	Antimony	[Kr] $4d^{10}\ 5s^2\ 5p^3$	
52.	Te	127.8	Tellurium	[Kr] $4d^{10}\ 5s^2\ 5p^4$	
53.	I	126.9	Iodine	[Kr] $4d^{10}\ 5s^2\ 5p^5$	
54.	Xe	131.3	Zenon	[Kr] $4d^{10}\ 5s^2\ 5p^6$	

Atomic Number	Symbol	Atomic Weight	Name	Electron Config.	
55.	Cs	132.9	Cesium	[Xe] $6s^1$	n=6
56.	Ba	137.3	Barium	[Xe] $6s^2$	
57.	La	138.9	Lanthanum	[Xe] $5d^1\ 6s^2$	
58.	Ce	140.1	Cerium	[Xe] $4f^1\ 5d^1\ 6s^2$	
59.	Pr	140.9	Praseodymium	[Xe] $4f^3\ 5d^0\ 6s^2$	
60.	Nd	144.2	Neodymium	[Xe] $4f^4\ 6s^2$	
61.	Pm	145	Promethium	[Xe] $4f^5\ 6s^2$	
62.	Sm	150.4	Samarium	[Xe] $4f^6\ 6s^2$	
63.	Eu	152.0	Europium	[Xe] $4f^7\ 6s^2$	
64.	Gd	157.3	Gadolinium	[Xe] $4f^7\ 5d^1\ 6s^2$	
65.	Th	158.9	Terbium	[Xe] $4f^9\ 6s^2$	
66.	Dy	162.5	Dysprosium	[Xe] $4f^{10}\ 6s^2$	
67.	Ho	164.9	Holdium	[Xe] $4f^{11}\ 6s^2$	
68.	Er	167.3	Erbium	[Xe] $4f^{12}\ 6s^2$	
69.	Tm	168.9	Thulium	[Xe] $4f^{13}\ 6s^2$	
70.	Yb	173.0	Ytterium	[Xe] $4f^{14}\ 6s^2$	
71.	Lu	175.0	Lutetium	[Xe] $4f^{14}\ 5d^1\ 6s^2$	
72.	Hf	178.5	Hafnium	[Xe] $4f^{14}\ 5d^2\ 6s^2$	
73.	Ta	180.9	Tantalum	[Xe] $4f^{14}\ 5d^3\ 6s^2$	
74.	W	183.8	Tungsten	[Xe] $4f^{14}\ 5d^4\ 6s^2$	
75.	Re	186.2	Rhenium	[Xe] $4f^{14}\ 5d^5\ 6s^2$	
76.	Os	190.2	Osmium	[Xe] $4f^{14}\ 5d^6\ 6s^2$	
77.	Ir	192.2	Iridium	[Xe] $4f^{14}\ 5d^7\ 6s^2$	
78.	Pr	195.1	Platinum	[Xe] $4f^{14}\ 5d^8\ 6s^2$	
79.	Au	197.0	Gold	[Xe] $4f^{14}\ 5d^{10}\ 6s^2$	
80.	Hg	200.6	Mercury	[Xe] $4f^{14}\ 5d^{10}\ 6s^2$	
81.	Tl	204.4	Thallium	[Xe] $4f^{14}\ 5d^{10}\ 6s^2\ 6p^1$	
82.	Pb	207.2	Lead	[Xe] $4f^{14}\ 5d^{10}\ 6s^2\ 6p^2$	
83.	Bi	209	Bismuch	[Xe] $4f^{14}\ 5d^{10}\ 6s^2\ 6p^3$	
84.	Po	209	Polonium	[Xe] $4f^{14}\ 5d^{10}\ 6s^2\ 6p^4$	
85.	At	210	Astatine	[Xe] $4f^{14}\ 5d^{10}\ 6s^2\ 6p^5$	
86.	Rn	222	Radon	[Xe] $4f^{14}\ 5d^{10}\ 6s^2\ 6p^6$	
87.	Fr	223	Francium	[Rn] $7s^1$	n=7
88.	Ra	226.0	Radium	[Rn] $7s^2$	
89.	Ac	227.0	Actiunium	[Rn] $6d^1\ 7s^2$	
90.	Th	232.0	Thorium	[Rn] $6d^2\ 7s^2$	
91.	Pa	231	Protactinium	[Rn] $5f^2\ 6d^1\ 7s^2$	
92.	U	238.0	Uranium	[Rn] $5f^3\ 6d^1\ 7s^2$	
93.	Np	237.0	Neptunium	[Rn] $5f^4\ 6d^1\ 7s^2$	
94.	Pu	244	Plutonium	[Rn] $5f^6\ 7s^2$	
95.	Am	243	Americium	[Rn] $5f^7\ 7s^2$	
96.	Cm	247	Curium	[Rn] $5f^7\ 6d^1\ 7s^2$	
97.	Bk	247	Berkelium	[Rn] $5f^9\ 7s^2$	
98.	Cf	251	Californium	[Rn] $5f^{10}\ 7s^2$	
99.	Es	252	Einsteinium	[Rn] $5f^{11}\ 7s^2$	
100.	Fm	257	Fermium	[Rn] $5f^{12}\ 7s^2$	
101.	Md	258	Mendelevium	[Rn] $5f^{13}\ 7s^2$	
102.	No	259	Nobelium	[Rn] $5f^{14}\ 7s^2$	
103.	Lr	260	Lawrencium	[Rn] $5f^{14}\ 6d^1\ 7s^2$	
104.	Ku	261	Rutherfordium	[Rn] $5f^2\ 6d^2\ 7s^2$	
105.	Ha	262	Dubuium	[Rn] $5f^2\ 6d^3\ 7s^2$	
106.	Sg	263	Seaborgium	[Rn] $5f^4\ 6d^4\ 7s^2$	

PART I

Chapter 1

ATOMIC STRUCTURE

KEY CONCEPTS

Equations and relationships that you need to know:

Postulates of the Quantum Theory

1. Atoms and molecules can exist only in discrete states, characterized by definite amounts of energy. When an atom or molecule changes state, it absorbs or emits just enough energy to bring it to another state.
2. When atoms or molecules absorb or emit light in moving from one energy state to another the wavelength of the light is related to the energies of the two states as follows.

$$E_{high} - E_{low} = \frac{h \cdot c}{\lambda}$$

3. The allowed energy states of atoms and molecules are described by quantum numbers.

$$h = \text{Planck's constant} = 6.626 \times 10^{-34} \ \frac{\text{J·s}}{\text{particle}}$$

$$= 6.626 \times 10^{-27} \ \text{erg} \cdot \text{sec}$$

$$c = \text{speed of light} = 2.998 \times 10^{10} \ \text{cm sec}^{-1}$$

$$\Delta E = \frac{1.196 \times 10^{5} \text{kJ} \cdot \text{nm}}{\lambda \cdot \text{mole}}$$

$$E = \frac{-1312 \text{kJ}}{n^2 \cdot \text{mole}} = h \cdot v = \frac{2.18 \times 10^{-11} \text{erg}}{n^2} = me^4$$

$$= \left(\frac{Z_{eff}}{n^2}\right) 1312 \text{kJ mole}^{-1}$$

$$Z_{eff} = Z - \sigma$$

where Z_{eff} = effective nuclear charge, Z = actual nuclear charge, and σ = shielding or screening constant

Rydberg-Ritz equation: $E_n = -R_H \left(\frac{1}{n^2}\right)$

$$\Delta E = \frac{h \cdot c}{\lambda} = h \cdot v = R_H \left(\frac{1}{n_i^2} - \frac{1}{n_f^2}\right)$$

n_i and n_f are quantum numbers

$R_H = 2.18 \times 10^{-18}$ J = 109,737 cm^{-1}

$$m.r.v = \frac{n \cdot h}{2\pi}$$

$$E_b - E_a = \frac{z^2 e^2}{2a_0}\left[\frac{1}{n_a^2} - \frac{1}{n_b^2}\right]$$

where n = quantum energy level

E = energy (at states a and b)

e = charge on electron

a_0 = Bohr radius

z = atomic radius

n = principal energy level

l = angular momentum = sublevel (s, p, d, and f)

m_l = magnetic quantum number = orientation of orbital

m_s = electron spin quantum number

Predicted Electron Configuration

$1s^2\ 2s^2\ 2p^6\ 3s^2\ 3p^6\ 4s^2\ 3d^{10}\ 4p^6\ 5s^2\ 4d^{10}\ 5p^6\ 6s^2\ 4f^{14}\ 5d^{10}\ 6p^6\ 7s^2\ 5f^{14}\ 6d^{10}$

Common Exceptions

chromium	$4s^1 3d^5$
copper	$4s^1 3d^{10}$
molybdenum	$5s^1 4d^{10}$
silver	$5s^1 4d^{10}$
gold	$6s^1 4f^{14} 5d^{10}$

PROBLEMS

Q. 1. Which of the following setups would be used to calculate the wavelength (cm) of a photon emitted by a hydrogen atom when the electron moves from the $n = 5$ state to the $n = 2$ state? (The Rydberg constant is $R_H = 2.18 \times 10^{18}$ J. Planck's constant is $h = 6.63 \times 10^{-34}$ J· sec. The speed of light = 3.00×10^{10} cm sec^{-1}).

(a) $(2.18\times10^{-18})\left(\frac{1}{5^2}-\frac{1}{2^2}\right)(6.63 \times 10^{-34})$

(b) $\dfrac{(6.63\times10^{-34})(3.00\times10^{10})}{(2.18\times10^{-18})\left(\frac{1}{5^2}-\frac{1}{2^2}\right)}$

(c) $\dfrac{(2.18\times10^{-18})(3.00\times10^{10})}{(6.63\times10^{-34})\left(\frac{1}{5^2}-\frac{1}{2^2}\right)}$

(d) $\dfrac{(2.18\times10^{-18})/(3.00\times10^{10})}{(6.63\times10^{-34})/\left(\frac{1}{5^2}-\frac{1}{2^2}\right)}$

Q. 2. A Co^{3+} ion has ______ unpaired electron(s) and is ______

(a) 1, diamagnetic
(b) 3, paramagnetic
(c) 3, diamagnetic
(d) 4, paramagnetic

Q. 3. Which of the following series of elements is listed in order of increasing atomic radius?

(a) Na, Mg, Al, Si
(b) C, N, O, F
(c) O, S, Se, Te
(d) I, Br, Cl, F

Q. 4. A characteristic that is unique to the alkali metals is:

(a) Their metallic character.
(b) The increase in atomic radius with increasing atomic number.
(c) The decrease in ionization energy with increasing atomic number.
(d) The noble gas electron configuration of the single charged positive ion.

Q. 5. Which of the following elements most readily shows the photoelectric effect?

(a) Noble gases
(b) Alkali metals
(c) Halogen elements
(d) Transition metals

Q. 6. The lithium ion and the hydride ion are isoelectronic. Which of the following statements is true of these two chemical species in the ground state?

(a) Li^+ is a better reducing agent than H^-.
(b) The H^- ion is several times larger than the Li^+ ion.
(c) It requires more energy to remove ab electron from H^- than from Li^+.
(d) The chemical properties of the two ions must be the same because they have the same electronic structure.

Q. 7. Which of the following configurations represents a neutral transition element?

(a) $1s^2\ 2s^2\ 2p^2$
(b) $1s^2\ 2s^2\ 2p^6\ 3s^2\ 3p^4$
(c) $1s^2\ 2s^2\ 3s^2$
(d) $1s^2\ 2s^2\ 2p^6\ 3s^2\ 3p^6\ 3d^8\ 4s^2$

Q. 8. The four quantum numbers (n, l, m_b and m_s) that describe the valence electron in the cesium atom are:

(a) 6, 0, –1, +1/2 **(b)** 6, 1, 1, +1/2

(c) 6, 0, 0, + 1/2 **(d)** 6, 1, 0, +1/2

Q. 9. When subjected to the flame test, a solution that contains K^+ ions produces the colour.

(A) yellow **(b)** violet

(c) crimson **(d)** green

Q. 10. Which of the following results indicates exothermic change in atoms?

(a) The production of line emission spectra.

(b) The appearance of dark lines in absorption spectra.

(c) The formation of ions from a metal in the gaseous state.

(d) The production of isotopic species of CO^+ and CO_2^+ when CO_2 is placed in a mass spectrometer.

Q. 11. A characteristic of the structure of metallic atoms is that:

(a) they tend to share their electrons with other atoms.

(b) their atoms are smaller and more compact than those of nonmetallic elements.

(c) their outermost orbital of electrons is nearly complete, and they attract electrons from other atoms.

(d) the small numbers of electrons in their outermost orbital are weakly held and easily lost.

Q. 12. Which of the following wave properties is proportional to energy for electromagnetic radiation?

(a) velocity **(b)** wave number

(c) wavelength **(d)** amplitude

Q. 13. Which of the following best describes the emission spectrum of atomic hydrogen?

(a) A discrete series of lines of equal intensity and equally spaced with respect to wavelength.

(b) A series of only four lines.

(c) A continuous emission of radiation of all frequencies.

(d) Several discrete series of lines with both intensity and spacings between lines decreasing as the wave number increases within each series.

Q. 14. Atomic emission spectra of an element CANNOT be used to

(a) Identify the element.

(b) Determine the mass number of the nucleus of the atom.

(c) Measure the difference in energy between pairs of stationary states of the atom.

(d) Calculate the ionization energy of the atom.

Q. 15. The ionization energy of gaseous Na atoms is 495.80 $kJ \cdot mol^{-1}$. The lowest possible frequency of light that can ionize a Na atom is

(a) $4.76 \times 10^{14}\ s^{-1}$ **(b)** $7.50 \times 10^{14}\ s^{-1}$

(c) $1.24 \times 10^{15}\ s^{-1}$ **(d)** $3.15 \times 10^{15}\ s^{-1}$

Q. 16. The cathode-ray tube experiments carried out by J.J. Thomson demonstrated that

(a) α-Particles are the nuclei of He atoms.

(b) The ratio of charge to mass for the particles of the cathode rays varies as different gases are placed in the tube.

(c) The mass of an atom is essentially all contained in its very small nucleus.

(d) The charge-to-mass ratio of electrons is about 1800 times larger than the charge to-mass ratio for a proton.

Q. 17. If the electron in a hydrogen atom drops from the $n = 6$ to the $n = 4$ level, the radiation emitted is in which series of lines in the spectrum of atomic hydrogen?

(a) Lyman (b) Balmer

(c) Paschen (d) Brackett

Q. 18. A line with wave number 1.028×10^{-2} nm^{-1} is emitted in the spectrum of atomic hydrogen. In what region of the electromagnetic spectrum does this line occur?

(a) far UV (b) near UV

(c) VIS (d) near IR

Q. 19. Which of the following statements about Millikan's oil drop experiment is TRUE?

(a) Some oil drops become positively charged and some become negatively charged after colliding with gaseous ions.

(b) The charge on each oil drop is the electronic charge.

(c) In the absence of the electric field, the speed with which the drop falls depends only upon the acceleration of gravity.

(d) Oil drops, rather than water drops, were used because oil is easier to see.

Q. 20. Because of the observation that some α-particles directed at a gold foil are scattered backwards at angles larger than 90°, Rutherford was able to conclude that

(a) All atoms are electrically neutral.

(b) The positively charged parts of atoms move with extremely large velocities.

(c) The positively charged parts of atoms occupy only a very small fraction of the volume of the atom.

(d) Negatively charged electrons are a part of all matter.

Q. 21. To move the electron in one H atom from the ground state to the second excited state, 12..084 eV are needed. How much energy is needed to cause 1 mol of H atoms to undergo this transition?

(a) 728 kJ (b) 984 kJ

(c) 1036 kJ (d) 1166 kJ

Q. 22. One of the lines in the spectrum of atomic hydrogen has wave number 5331.6 cm^{-1}. What is the frequency of this line?

(a) 5.623×10^{6} s^{-1} (b) 1.876×10^{11} s^{-1}

(c) 1.598×10^{12} s^{-1} (d) 1.598×10^{14} s^{-1}

Q. 23. The amount of energy required to remove the electron from a Li^{2+} ion in its ground state is how many times greater than the amount of energy needed to remove the electron from an H atom in its ground state?

(a) 12 (b) 3 (c) 9 (d) 6

Q. 24. A typical golf ball weighs 40.0 g. If it is moving with a velocity of 20.0 m. s^{-1}, its de Broglie wavelength is

(a) 1.66×10^{-34} nm **(b)** 8.28×10^{-32} nm

(c) 8.28×10^{-25} nm **(d)** 1.66×10^{-24} nm

Q. 25. What is the energy, in joules, of a photon of IR light with wavelength 4.0×10^3 nm?

(a) 5.0×10^{-20} **(b)** 7.5×10^{-20}

(c) 4.0×10^{-16} **(d)** 2.5×10^{-14}

Q. 26. The most intense line in the Brackett series of the spectrum of atomic hydrogen is the transition

(a) $n_H = \infty \rightarrow n_L = 1$ **(b)** $n_H = 8 \rightarrow n_L = 4$

(c) $n_H = 5 \rightarrow n_L = 4$ **(d)** $n_H = 4 \rightarrow n_L = 3$

Chapter 2

CHEMICAL BONDING

Equations and relationships that you need to know:

net dipole = charge × distance

Coulomb's law : $E = 2.31\ 5\ 10^{-19}\ \text{J·nm}\left(\frac{Q_1Q_2}{r}\right)$

lattice energy = $k\left(\frac{Q_1Q_2}{r}\right)$, k = constant

Q_1 and Q_2 — charges on the ions

r = shortest distance between centers of cations and anions

Equations and relationships that you need to know:

$$\text{bond order} = \frac{\text{number of bonding electrons} - \text{number of antibonding electrons}}{2}$$

Geometry and Hybridization Patterns

Number of Atoms Bonded to Central Atom X	*Number of Unshared Pairs on X*	*Hybridization*	*Geometry*	*Example*
(1)	*(2)*	*(3)*	*(4)*	*(5)*
2	0	*sp*	linear	$\underline{C}O_2$
2	1	*sp*^{2}	bent	$\underline{S}O_2$

(Contd.)

TABLE (Contd.)

(1)	(2)	(3)	(4)	(5)
2	2	sp^3	bent	$\underline{H_2O}$
2	3	sp^3d	linear	$\underline{Xe}F_2$
3	0	sp^2	trigonal planar	$\underline{B}Cl_3$
3	1	sp^3	trigonal pyramidal	$\underline{N}H_3$
3	2	sp^3d	T-shaped	$\underline{Cl}F_3$
4	0	sp^3	tetrahedral	$\underline{C}H_4$
4	1	sp^3d	distorted tetrahedral	$\underline{S}Cl_4$
4	2	sp^3d^2	square planar	$\underline{Xe}F_4$
5	0	sp^3d	triangular bipyramidal	$\underline{P}Cl_5$
5	1	sp^3d^2	square pyramidal	$\underline{Cl}F_5$
6	0	sp^3d^2	octahedral	$\underline{S}F_6$

Note : The underlined atom is the central atom x.

Character of Bonds

Electronegativity Difference	*Type of Bond*	*Example*
O → 0.2	nonpolar covalent	Br_2 HI, CH_4
0.2 → 1.7	polar covalent	NO, LiH
1.7 or greater	Ionic	LiBr, CuF

$$\text{formal charge} = \begin{pmatrix}\text{total number of}\\ \text{valence electrons}\\ \text{in the free atom}\end{pmatrix} - \begin{pmatrix}\text{total number of}\\ \text{nonbonding}\\ \text{electrons}\end{pmatrix} - 1/2\begin{pmatrix}\text{total number}\\ \text{of bonding}\\ \text{electrons}\end{pmatrix}$$

single bond = 1 sigma bond,
double bond = 1 sigma bond, 1 pi bond
triple bond = 1 sigma bond, 2 pi bonds

Homonuclear diatomic molecules of second-period elements, B_2, C_2, and N_2:

$$\sigma 1s < \sigma^* 1s < \sigma 2s < \sigma^* 2s < \pi 2p_y = \pi 2p_z < \sigma 2p_x < \pi^* 2p_y = \pi^* 2p_z < \sigma^* 2p_x$$

For O_2 and F_2:

$$\sigma 1s < \sigma^* 1s < \sigma 2s < \sigma^* 2s < \sigma 2p_x < \pi 2p_y = \pi 2p_z < \pi^* 2p_y = \pi^* 2p_z < \sigma^* 2p_x$$

PROBLEMS

Q. 1. Which of the following is the electron configuration for Ni^{2}?

(a) $1s^2\ 2s^2\ 2p^6\ 3s^2\ 3p^6\ 4s^2\ 3d^6$ **(b)** $1s^2\ 2s^2\ 2p^6\ 3s^2\ 3p^6\ 4s^2\ 3d^{10}$

(c) $1s^2\ 2s^2\ 2p^6\ 3s^2\ 3p^6\ 3d^8$ **(d)** $1s^2\ 2s^2\ 2p^6\ 3s^2\ 3p^6\ 4s^2\ 3d^8$

Q. 2. As the atomic number of the elements increases down a column.

(a) the atomic radius decreases

(b) the atomic mass decreases

(c) the elements become less metallic

(d) ionization energy decreases

Q. 3. What is the oxidation number of platinum in $PtCl_6^{2-}$

(a) –4 **(b)** –2 **(c)** –1 **(d)** +4

Q. 4. What type of bond would you except in CsI?

(a) Ionic **(b)** Covalent

(c) Hydrogen **(d)** Metallic

Q. 5. Arrange the following ions in order of increasing ionic radius: Mg^{2+}, F^-, and O^{2-}.

(a) O^{2-}, F^-, Mg^{2+} **(b)** Mg^{2+}, O^{2-}, F^-

(c) Mg^{2+}, F^-, O^{2-} **(d)** O^{2-}, Mg^{2+}, F^-

Q. 6. Which of the following is the correct Lewis structure for the ionic compound $Ca(ClO_2)_2$?

(a) 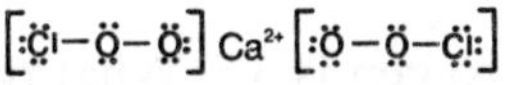

(b) 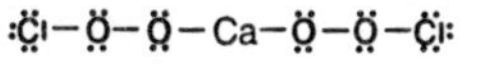

(c) 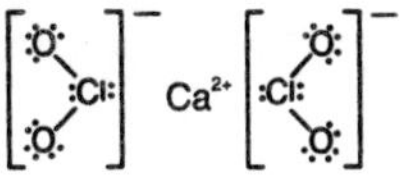

(d) 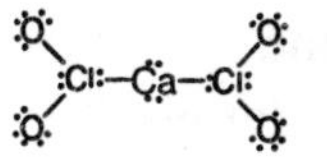

Q. 7. The compound expected when chlorine reacts with aluminum is

(a) Al_2Cl_3 **(b)** Al_3Cl_2

(c) $AlCl_3$ **(d)** Al_3Cl

Q. 8. What ions would you find in solution if potassium perchlorate was dissolved in water?

(a) KCl, O_2 **(b)** K^+, Cl^-, O^2

(c) KCl, O^{2-} **(d)** K^+, ClO_4^-

Q. 9. Which one of the following is correct?

(a) $KClO_3$ potassium perchlorate

(b) CuO copper oxide

(c) $Al_3(SO_2)_2$ aluminum sulfate

(d) $Na_2Cr_2O_7$ sodium dichromate

For Questions 10-14 use the following choices:

(a) Trigonal planar (b) Tetrahedral

(c) Pyramidal (d) Bent

(e) Linear

Q. 10. What is the geometry of S_2Cl_2?

Q. 11. What is the geometry of OPN?

Q. 12. What is the geometry of Seo_4^{2}?

Q. 13. What is the geometry of SiO_3^{2-}?

Q. 14. What is the geometry of the hydronium ion, H_3O^+?

Q. 15. Which one of the following is a nonpolar molecule with one or more polar bonds?

(a) H—Br (b) Cl—Be—Cl

(c) H—H (d) H—O—H

Q. 16. How many total sigma bonds are in the benzene molecule, C_6H_6?

(a) 6 (b) 9 (c) 12 (d) 14 (e) 18

Q. 17. Which one of the following does NOT exhibit resonance?

(a) SO_2 (b) SO_3 (c) HI (d) CO_3^{2-} (e) NO_3^-

Q. 18. What type of hybridization would you expect to find in BCl_3?

(a) *sp* (b) sp^2 (c) sp^3 (d) sp^3d^2

Q. 19. The electronegativity of carbon is 2.5, whereas that of oxygen is 3.5. What type of bond would you expect to find in carbon monoxide?

(a) Nonpolar covalent (b) Polar covalent

(c) Covalent network (d) Ionic

(e) Delta

Q. 20. Which one of the following contains a coordinate covalent bond?

(a) $N_2H_5^+$ (b) $BaCl_2$ (c) HCl (d) H_2O (e) NaCl

Q. 21. What is the formal number of pairs of unshared valence electrons in the NO_2^+ ion?

(a) 0 (b) 2 (c) 4 (d) 8 (e) 10

Q. 22. The bond energy of Br—Br is 192 kJ/mole, and that of Cl—Cl is 243 kJ/mole. What is the energy of the Cl—Br bond?

(a) 54.5 kJ/mole (b) 109 kJ/mole

(c) 218 kJ/mole (d) 435 kJ/mole

Q. 23. If the ionisation energy of hydrogen atom is 1.312×10^3 kJ mol^{-1}, then, the wave length of photon which will be emitted when the electron of hydrogen jumps from the second shell to the first shell is:

(Given $h = 6.6 \times 10^{-34}$ kg m^2 s^{-1}; c = 3×10^8 m s^{-1} and $N_A = 6 \times 10^{23}$ mol^{-1})

(a) 1207.3A° (b) 500A°

(c) 1000A° (d) 120.73A°

Q. 24. Iodine molecule dissociates into atoms after absorbing light of 4500Å. If one quantum of radiation is absorbed by each molecule, the kinetic energy of iodine atoms will be:
(a) 2.185×10^{-23} kJ (b) 21.85×10^{-23} kJ
(c) 2.185×10^{-20} kJ (d) 21.85×10^{-20} kJ

Q. 25. The wavelength of radiation emitted producing a line in Lyman series, when an electron falls from fourth stationary state in hydrogen atom. ($R_H = 1.1 \times 10^7$ m^{-1}) is:
(a) 9.5×10^{-8} m (b) 10×10^{-8} m
(c) 9.7×10^{-8} m (d) 9.9×10^{-8} m

Q. 26. If the density of mercury is 13.6 g cm^{-3}, then, approximately the diameter of an atom of mercury, assuming that each atom is occupying a cube of edge length equal to the diameter of the mercury atom will be:
(a) 29×10^{-10} m (b) 4.9×10^{-10} m
(c) 49×10^{-10} m (d) 2.9×10^{-10} m

Q. 27. In a hydrogen atom, an electron jumps from a third orbit to the first orbit. The frequency and wave length of the spectral line respectively will be:
(a) 1.925×10^{15} s^{-1}; 1025.4A° (b) 2.925×10^{15} s^{-1}; 1025.4A°
(c) 1025.4×10^{15} s^{-1}; 2925A° (d) 2.925×10^{15} s^{-1}; 2925A°

Q. 28. The wavelength and energy of radiation emitted for the electronic transition from infinity (∞) to stationary state one of the hydrogen atom will be:
(a) 9.18×10^{-6}m; 9.0×10^{-23} kJ (b) 2.17×10^{-6}m; 9.11×10^{-23} kJ
(c) 91.8×10^{-6}m; 21.8×10^{-23} kJ (d) 9.11×10^{-6}m; 217.9×10^{-23} kJ

Q. 29. Which one of the five oxides of chlorine is paramagnetic?
(a) Cl_2O (b) ClO_2 (c) Cl_2O_4 (d) Cl_2O_6

Q. 30. Pi (π) bonding occurs in each of the following EXCEP
(a) CO_2 (b) C_2H_4 (c) CH_4 (d) C_6H_6

Q. 31. Each of the following molecules has a nonzero dipole moment EXCEPT
(a) C_6H_6 (b) CO (c) SO_2 (d) NH_3

Q. 32. Of the elements Sr, Zr, Mo, Cd, and Sb, all of which are in the fifth period of the periodic table, the ones that are paramagnetic are
(a) Sr, Cd, and Sb only (b) Zr, Mo, and Cd only
(c) Sr, Zr, and Cd only (d) Zr, Mo, and Sb only

Q. 33. The geometrical structure of the PF_3 molecule is
(a) trigonal pyramidal (b) trigonal bipyramidal
(c) tetrahedral (d) trigonal planar

Q. 34. The total number of valence electrons in the phosphonium ion, PH_4^+ is
(a) 8 (b) 9 (c) 10 (d) 12

Q. 35. In the best Lewis octet structure for ICl_3 the formal charge on I is
(a) 0 (b) +1 (c) -I (d) +2

Q. 36. In which of the following molecules does the central atom use Sp^2 hybrid atomic orbitals in forming bonds?
(a) H_2S **(b)** SO_2 **(c)** Cl_2O **(d)** NH_3

Q. 37. The geometrical structure of the sulfite ion, SO_3^{2-}, is
(a) trigonal pyramidal **(b)** square planar
(C) trigonal planar **(d)** linear

Q. 38. In the best Lewis structure for perchlorate ion, ClO_4^-, the formal charge on Cl is
(a) –1 **(b)** 0 **(c)** +1 **(d)** +3

Q. 39. Which of the following molecules is nonplanar?
(a) C_6H_6 **(b)** SO_3 **(c)** CF_4 **(d)** XeF_4

Q. 40. Which hybrids can be used for bonding in a square-planar molecule or ion?
(a) sp^3 **(b)** dsp^2 **(c)** sp^2 **(d)** d^2sp^3

Q. 41. In one of the following triatomic molecules the observed bond angle is 116°49′. Which of these molecules would you expect to have a bond angle of about this magnitude?
(a) H_2O **(b)** OF_2 **(c)** CS_2 **(d)** O_3

Q. 42. The total number of valence electrons for the PO_4^{3-} ion is
(a) 32 **(b)** 28 **(c)** 29 **(d)** 30

Q. 43. In which of the following compounds does every atom have eight electrons in its valence shell?
(a) IF_5 **(b)** C_2H_4 **(c)** SiF_4 **(d)** NO_2

Q. 44. In which of the following compounds do the bonds have the largest percentage of ionic character?
(a) N_2O_4 **(b)** H_2O **(c)** HF **(d)** CO_2

Q. 45. There are both ionic and covalent bonds in each of the following compound EXCEPT for.
(a) $CaCO_3$ **(b)** Sr_3N_2 **(c)** $NaNO_3$ **(d)** K_2SO_4

Q. 46. Which of the following best describes the hybrids used by S in the sulfite ion, SO_3^{2-}?
(a) sp **(b)** sp^2 **(c)** sp^3 **(d)** dsp^2

Q. 47. Which of the following is an electron deficient compound?
(a) $NaBH_4$ **(b)** CO_2 **(c)** C_2H_6 **(d)** B_2H_6

Q. 48. Which of the following diatomic species do you expect to have the longest bond length?
(a) NO^+ **(b)** O_2^- **(c)** CO **(d)** O_2^+

Q. 49. Several resonance structures are used to describe the bonding in
(a) BF_3 **(b)** NO_3^- **(c)** $CHCI_3$ **(d)** NH_4^+

Questions 50 to 56. Select that one of the five molecules listed below that fits the description given
(a) CS_2 **(b)** $RaCl_2$ **(c)** HCN **(d)** F_2 **(e)** C_2Cl_4

Q. 50. Has only one pi (π) bond.

Q. 51. Has two double bonds.

Q. 52. Has one triple bond.

Q. 53. Exhibits ionic bonding.

Q. 54. Consists of discrete, polar molecules at room temperature.

Q. 55. Has a nonpolar single bond.

Q. 56. Is a solid at room temperature.

Q. 57. The geometry of the carbonate ion, CO_3^{2}, is
(a) square planar (b) trigonal pyramisal
(c) tetrahedral (d) trigonal planar

Q. 58. Which of the following is nonpolar, but contains polar bonds?
(a) HCl (b) H_2O (c) SO_3 (d) NO_2

Q. 59. Of the following molecules the only one which is NOT an exception to the octet rule is
(a) OF_2 (b) $SnCl_2$ (c) BeI_2 (d) ClO_2

Q. 60. Microwave ovens are becoming increasingly popular in kitchens around the world. They provide a rapid, energy-efficient means of preparing foods. Microwave ovens are also useful in the chemical laboratory, particularly in drying samples for chemical analysis. A typical microwave oven uses microwave radiation with frequency $v = 2.45 \times 10^9\ s^{-1}$.

(i) The wavelength of microwave radiation having a frequency of $2.45 \times 10^9 s^{-1}$ is
(a) 1220 cm (b) 1.22 cm (c) 122 cm (d) 12.2 cm

(ii) The energy associated with one photon of this microwave radiation will be:
(a) 16.2×10^{-24}J (b) 1.62×10^{-24}J (c) 0.162×10^{-24}J (d) 0.0162×10^{-24}J

(iii) The number of moles of this radiation that would contain sufficient energy to raise the temperature of 225 g of water from 297.5 to 372.5K
(a) 7.22×10^4 mol (b) 0.722×10^4 mol
(c) 72.2×10^4 mol (d) 0.0722×10^4 mol

Q. 61. An atom of phosphorus has 16 neutrons. The full symbol notation for this atom is :
(a) $^{31}_{15}P$ (b) $^{30}_{16}P$ (c) $^{32}_{14}P$ (d) $^{29}_{14}P$

Q. 62. The wavelength of light emitted when the electron in a hydrogen atom undergoes a transition from energy level n=4 to level n=2 is:
(a) 617 nm (b) 206 nm
(c) 486 nm (d) None of these

Q. 63. The wavelength of a red traffic light is centered on 700 nm. What is the frequency of this radiation?
(a) 2.33×10^{-17}Hz (b) 4.29×10^{-17} Hz
(c) 4.29×10^{14}Hz (d) 2.33×10^{14} Hz

Q. 64. The uncertainty in velocity of an electron, if uncertainty in its position is 10^{-10} m will be :

(Planck's constant h = 6.6×10^{-34} kg, m^2 s^{-1} and mass of electron is 9.1×10^{-31} kg.)

(a) 5.77×10^6 ms^{-1} (b) 0.577×10^6 ms^{-1}

(c) 23.08×10^6 ms^{-1} (d) 2.308×10^6 ms^{-1}

Q. 65. A moving electron has 5×10^{-25} joule of kinetic energy. Its de Broglie's wavelength will be :

(Given h = 6.6×10^{-34} kg m^2 s^{-1} and m = 9.1×10^{-31} kg).

(a) 6.9×10^{-4} m (b) 6.9×10^{-6} m

(c) 0.69×10^{-4} m (d) 0.69×10^{-6} m

Q. 66. An electron is moving with a kinetic energy of 4.55×10^{-25}J. Its de Brogile wavelength will be :

(a) 620Å (b) 520Å (c) 720Å (d) 820Å

Q. 67. The uncertainty in the position and velocity of a particle are : 10^{-10} m and 5.27×10^{-24} m s^{-1} respectively. The mass of the particle. (h = 6.625×10^{-34} Js), will be :

(a) 100 mg (b) 200 mg (c) 300 mg (d) 400 mg

Q. 68. The total number of electrons present in 1.6 g of methane, is :

(a) 6.023×10^{22} (b) 6.023×10^{23}

(c) 6.023×10^{21} (d) 6.023×10^{24}

Q. 69. The total number of neutrons and the total mass of neutrons in 7 mg of ^{14}C. (Assume that mass of a neutron = 1.675×10^{-27} kg.), will be :

(a) 24.096×10^{20}; 4.036×10^{-7} (b) 4.036×10^{23}, 24.096×10^{-7}

(c) 4.036×10^{20}; 24.096×10^{-7} (d) 24.096×10^{23}, 4.036×10^{-7}

Q. 70. The frequency and energy of the radiation corresponding to the spectral line of lowest frequency in Lyman series in the spectra of hydrogen atom respectively, will be:

(Given (R_H = 1.09678×10^7 m^{-1}, C = 3×10^8 ms^{-1}, h = 6.625×10^{-34} Js)

(a) 16.36×10^{15} s^{-1}; 2.47×10^{-19}J (b) 12.47×10^{15} s^{-1}, 16.36×10^{-19}J

(c) 2.47×10^{15} s^{-1}, 16.36×10^{-19}J (d) 16.36×10^{15} s^{-1}, 12.47×10^{-19}J

Q. 71. Electromagnetic radiation of wavelength 242 nm is just sufficient to ionise the sodium atom. The ionisation energy of sodium in kJ mol^{-1} will be :

(a) 49.45 KJ mol^{-1} (b) 4.945 KJ mol^{-1}

(c) 494.5 KJ mol^{-1} (d) 4945 KJ mol^{-1}

Q. 72. The wave number for the shortest wavelength transition in the Balmer series of atomic hydrogen is :

(a) 27419.25 cm^{-1} (b) 24719.25 cm^{-1}

(c) 29174.25 cm^{-1} (d) 21749.25 cm^{-1}

Q. 73. The number of waves made by a Bohr electron in one complete revolution in its 3rd orbit is :

(a) 1 (b) 2 (c) 4 (d) 3

Q. 74. The dipole moment of HBr is 2.60×10^{-30} C m. and the interatomic spacing is 1.41 Å. The percent ionic character of HBr is:

(a) 9.5% (b) 11.5% (c) 10.5% (d) 12.5%

Q. 75. NO^+ is more stable toward dissociation into its atoms than NO, whereas CO^+ is less sable than CO, because:

(a) NO^+ has lost an antibonding electron; CO^+ has lost a bonding electron.

(b) NO^+ has lost an antibonding electron; CO^+ has lost an antibonding electron.

(c) NO^+ has lost an bonding electron; CO^+ has lost a bonding electron.

(d) NO^+ has lost an bonding electron; CO^+ has lost a anti bonding electron.

Q. 76. The expected magnetic moment in the superoxide ion O_2^- is:

(a) 2.43 (B)M. (b) 1.0 B.M. (c) 1.73 B.M. (d) 3.46 B.M.

Q. 77. Compared to O_2 bond length, the peroxide ion O_2^{2-} will have a :

(a) Shorter bond length (b) Equal bond length

(c) Longer bond length (d) Stronger bond

Q. 78. Which of these molecules has the highest bond order—

(a) BN (b) CO (c) Ne_2 (d) F_2

Q. 79. The dipole moment of LiH is 1.964×10^{-29} C.m. and the interatomic distance between Li and H in this molecule is 1.596Å. The percent ionic character in LiH is :

(a) 84.8% (b) 80.8% (c) 72.8% (d) 76.8%

Q. 80. Assuming the additivity of covalent radii in the C—I bond, what would be the iodine-iodine distance in each of the three di-iodobenzenes respectively? Assume that the ring is a regular hexagon and that each C—I bond lies on a line through the center of the hexagon. The distance between adjacent carbons is 1.40Å.

(a) 3.5Å, 7.0Å, 6.06Å (b) 7.0Å, 3.5Å, 6.06Å

(c) 3.5Å, 6.06Å, 7.0Å (d) 6.06Å, 7.0Å, 3.5Å

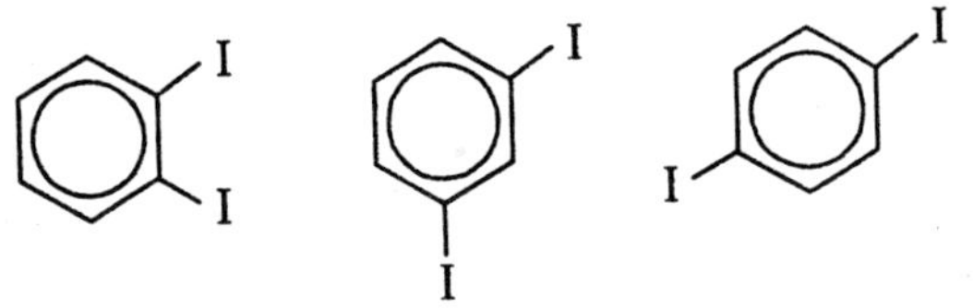

(i) (ii) (iii)

Q. 81. The length and width respectively of the carbon skeleton in naphthacene (Fig. below), assume hexagonal rings with equal carbon-carbon distance of 1.40Å is:

(a) 9.68Å, 2.8Å
(b) 6.89Å, 2.8Å
(c) 2.8Å, 9.68Å
(d) 6.89Å, 9.68Å

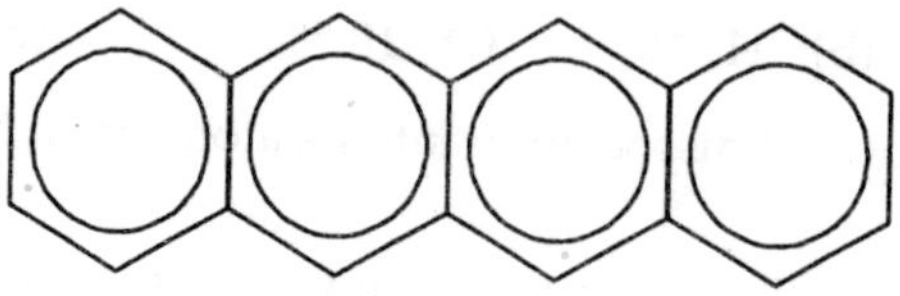

Naphthacene

Q. 82. In water , the H—0—H bond angle is 105°. Using the dipole moment of water 1.85D and the covalent radii of the atoms, 0.94Å, the magnitude of the charge on the oxygen atom in the water molecule will be:

(a) 1.6×10^{-10} e.s.u.
(b) 3.2×10^{-10} e.s.u.
(c) 2.4×10^{-10} e.s.u.
(d) 4.8×10^{-10} e.s.u.

Q. 83. The sets of hybridized orbitals which do not maximize the angles between electron pairs:

(a) sp (b) sp^2 (c) sp^3 (d) sp^3d

Q. 84. Which molecule, AX_3, AX_4, AX_5, AX_6, is most likely to have a trigonal bipyramidal structure and if the central atom, A, has no lone pairs, what type of hybridization will it have?

(a) AX_4, sp^3d
(b) AX_5, sp^3d
(c) AX_5, sp^3
(d) AX_6, sp^3

Q. 85. The bond order in NO is :

(a) 2.5 (b) 3.5 (c) 4.0 (d) 5.0

Chapter **3**

GAS LAWS

KEY CONCEPTS

Equations and relationships that you need to know :

- I atm = 760 mm Hg = 760 torr = 101.3 kPa = 14.7 ib/in^2
  K = °C + 273.15
- Avogadro's law: $V_1n_2 = V_2n_1$
  Boyle's law: $P_1V_1 = P_2V_2$
  Charles' law: $V_1T_2 = V_2T_1$
  combined gas law: $\frac{P_1V_1}{T_1} = \frac{P_1V_2}{T_2}$
  Dalton's law of partial pressures: $P_{total} = P_1 + P_2 + P_3 + ...$
  derivation: $P_1 = \frac{n_1}{n_{total}} \times P_{total}$
  density $= \frac{g}{V} = \frac{P \times MW}{R \cdot T}$
  Gay-Lussac's law: $P_1T_2 = P_2T_1$
  ideal gas law: $PV = nRT$
  R = 0.0821 liter · atm mole^{-1} · K^{-1} = 8.31 liter · kPa mole^{-1} · K^{-1}
  = 8.31 J mole · K^{-1} = 8.31 *V* · *C* mole^{-1} · K^{-1}
  = 8.31 × 10^{-7} g · cm^2 sec^{-2} · mole^{-1} · K^{-1}
  (for calculating the average speed of molecules)
  = 6.24 × 10^4 L · mm Hg mole^{-1} · K^{-1} = 1.99 cal mole^{-1} · K^{-1}

$$\text{Molecular weight} = \text{MW} = \frac{g \cdot R \cdot T}{P \cdot V}$$

Van der Waals (real gases): $(P+a/V^2)\ (V-b) = R \cdot T$

or

$$(P+n^2a/V^2)\ (V-nb) = n\ R\ T$$

Here a corrects for force of attraction between gas molecules, and b corrects for particle volume.

Graham's law of effusion:

$$\frac{r_1}{r_2} = \frac{\sqrt{d_2}}{\sqrt{d_1}} = \frac{\sqrt{MW_2}}{\sqrt{MW_1}} = \frac{t_2}{t_1} = \frac{u_1}{u_2}$$

where

r = rate of effusion; d = density

MW = molecular weight; t = time; u = average speed

Kinetic Molecular Theory

1. Gas are composed of tiny, invisible molecules that are widely separated from one another in empty space.
2. The molecules are in constant, continuous, random, and straight-line motion.
3. The molecules collide with one another, but the collisions are perfectly elastic (no net loss of energy).
4. The pressure of a gas is the result of collisions between the gas molecules and the walls of the container.
5. The average kinetic energy of all the molecules collectively is directly proportional to the absolute temperature of the gas. Equal numbers of molecules of any gas have the same average kinetic energy at the same temperature.

$$E_t = \frac{m \cdot u^2}{2} = cT$$

$$c = \frac{3R}{2N_A}$$

$$u^2 = \frac{3 \cdot R \cdot T}{m \cdot N_A} = \frac{3 \cdot R \cdot T}{\text{MW}}$$

where

E_t = average kinetic energy of translation; m = mass (of particle)

u = velocity (average speed); c = constant

N_A = Avogadro's number; R = 8.31×10^{-7} g · cm^2 sec^2 · mole^{-1} · K^{-1}

T = temperature in K

PROBLEMS

Q. 1. Which of the following would express the approximate density of carbon dioxide gas at 273K and 2.00 atm pressure (in grams per liter)?

(a) 2 g/L (b) 4 g/L (c) 6 g/L (d) 8 g/L

Q. 2. The combustion of carbon monoxide yields carbon dioxide. The volume of oxygen gas needed to produce 22 grams of carbon dioxide at STP is:

(a) 4.0 liters (b) 5.6 liters (c) 11 liters (d) 22 liters

Q. 3. If the average velocity of a methane molecule, CH_4 (MW = 16), is 5.00×10^4 cm sec^{-1} at 273K what s the average velocity of helium molecules at the same temperature and pressure conditions?

(a) 2.50×10^4 cm sec^{-1} (b) 5.00×10^4 cm sec^{-1}
(c) 1.00×10^5 cm sec^{-1} (d) 2.00×10^5 cm sec^{-1}

Q. 4. When 2.00 grams of a certain volatile liquid is heated, the volume of the resulting vapor is 821 mL at a temperature of 400K at standard pressure. The molecular weight of this substance is:

(a) 20.0 g $mole^{-1}$ (b) 40.0 g $mole^{-1}$
(c) 80.0 g $mole^{-1}$ (d) 120.0 g $mole^{-1}$

Q. 5. A sample of zinc metal reacts completely with excess by hydrochloric acid according to the following equation:

$$Zn_{(s)} + 2HCl_{(aq)} \rightarrow ZnCl_{2\,(aq)} + H_{2\,(g)}$$

8.00 liters of hydrogen gas at 720. mm Hg is collected over water at 313K (vapor pressure of water at 313K = 55 mm Hg). How much zinc was consumed by the reaction?

(a) $\dfrac{(720/760)\cdot 8.00}{(0.0821)\cdot 313}$ (b) $\dfrac{(760/720)\cdot 313}{(0.0821)\cdot 2}$

C. $\dfrac{(665/760)\cdot 8.00\cdot(65.39)}{0.0821\cdot 313}$ (d) $\dfrac{(665/760)\cdot 8.00}{(65.39)\cdot(0.0821)\cdot 313}$

Q. 6. What is the partial pressure of helium when 8.0 grams of helium and 16 grams of oxygen are in a container with a total pressure of 5.00 atm?

(a) 0.25 atm (b) 1.00 atm (c) 1.50 atm (d) 4.00 atm

Q. 7. For a substance that remains a gas under the conditions listed, deviation from the ideal gas law would be most pronounced at :

(a) 173K and 5.0 atm (b) 173K and 1.0 atm
(c) 273K and 1.0 atm (d) 373K and 1.0 atm

Q. 8. 100 grams of $O_2(g)$ and 100 gams of He(*g*) are in separate containers of equal volume. Both gases are at 373K. Which one of the following statements is true?

(a) The pressure of the He(*g*) would be greater than that of the $O_2(g)$.
(b) The average kinetic energy of the O_2 molecules is greater than that of the He molecules.

(c) The average kinetic energy of the He molecules is greater than that of the O_2 molecules.

(d) There are equal numbers of He molecules and O_2 molecules.

Q. 9. Which one of the manometers below represents a gas pressure of 750 mm Hg? (Atmospheric pressure is 760 mm Hg.)

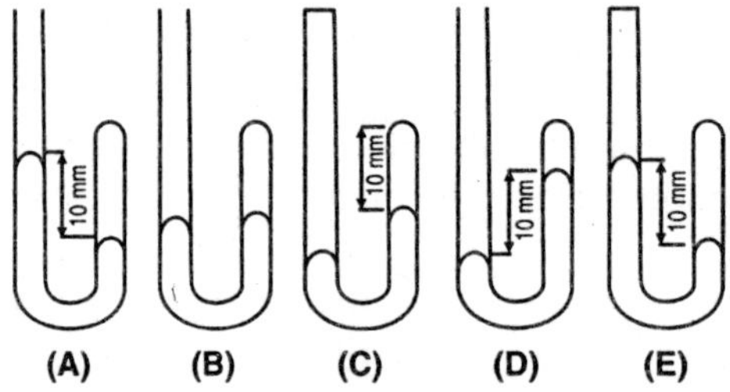

Q. 10. A mixture of gases at a total pressure of 1.0×10^5 Pa contains 3.4 mol of carbon dioxide, 4.5 mol of oxygen and 1.1 mol of nitrogen.

The partial pressure of each gas in the mixture respectively is:

(a) $(3.8, 5.0, 1.2) \times 10^4$ Pa
(b) $(0.38, 0.5, 0.12) \times 10^5$ Pa
(c) $(0.12, 05, 0.38) \times 10^4$ Pa
(d) $(3.8, 1.2, 5.0) \times 10^5$ Pa

Q. 11. Nitrogen and hydrogen are mixed in a mole ratio of 1:3 at 573K and in the presence of a catalyst. The reaction was allowed to reach equilibrium. At equilibrium, the total pressure of the equilibrium mixture is 800 kPa and the partial pressure of the ammonia is 75kPa. The numerical value for K_p under thee conditions is:

(a) 181×10^{-5} kPa
(b) 0.93×10^{-5} kPa
(c) 544×10^{-7} kPa
(d) 1.93×10^{-7} kPa

Q. 12. At 300K H_2 is leaked through a tiny hole into a vessels for 20 min. Ancther unknown gas, at the same temperature and pressure as that of H_2, is leaked through the same hole for 20 min. After the effusion of the gas the mixture exerts a pressure of 6 atm. The hydrogen content of the mixture is 0.7 mol. If the volume of the container is 3L, what is the molecular weight of the unknown gas?

(a) 1036
(b) 1033
(c) 1039
(d) 1030

Q. 13. The gas propane contains hydrogen and carbon only. When 100.0 cm^3 of it is passed into a syringe, the mass is found to be 0.178g. The pressure and temperature are 298 K and 100 kPa respectively. The M is:

(a) 0.441
(b) 4.41
(c) 441.0
(d) 44.1

Q. 14. A glass bulb contains 2.24L of H_2 and 1.12L of D_2 at S.T.P. It is connected to a fully evacuated bulb by a stopcock with a small opening. The stopcock is opened for some time and then closed. The first bulb now contains 0.10 g of D_2. Calculate the percentage composition by weight of the gases in the second bulb respectively:

(a) 58.8%, 41.2%
(b) 41.2%, 58.8%
(c) 100%, 0%
(d) 10%, 100%

Q. 15. 20 dm^2 of SO_2 diffuse through a porous partition in 60s. What volume of O_2 will diffuse under similar conditions in 30 s?

(a) 10.0 dm^3 (b) 14.1 dm^3 (c) 0 dm^3 (d) 7.05 dm^3

Q. 16. An open vessel at 300 K is heated until 3/5 parts of the air in it has been expelled. Assuming that the volume of the vessel remains constant, find the temperature to which the vessel has been heated.

(a) 750 K (b) 350 K (c) 550 K (d) 400 K

Q. 17. A 4 : 1 molar mixture of He and CH_4 is contained in a vessel at 20 bar pressure. Due to a hole in the vessel the gas mixture leaks out. What is the composition of the mixture effusing out initially :

(a) 4:1 (b) 8:1 (c) 2:1 (d) 1:1

Q. 18. A spherical balloon of 21 cm diameter is to be filled up with hydrogen at N.T.P. from a cylinder containing the gas at 20 atmosphere at 300 K. If the cylinder can hold 2.82 litres of water, calculate the number of balloons that can be filled up.

(a) 5 (b) 15 (c) 10 (d) 20

Q. 19. A balloon of diameter 20 m weights 100 kg. Calculate its payload if it is filled with He at 1.0 atm and 300 K. Density of air is 1.2 kg. m^{-3}. (R = 0.082 dm^3 atm K^{-} mol^{-1}).

(a) 3233.6 kg. (b) 2344.6 kg. (c) 4324.6 kg. (d) 4243.6 kg.

Q. 20. A 10 g. chunk of dry ice is placed in a 1 litre flask and tightly closed. What would be the final pressure in the bottle after all CO_2 has been evaporated and temperature reaches to 300 K?

(a) 5.59 (b) 4.59 (c) 7.59 (d) 6.59

Q. 21. At room temperature the following reactions proceed nearly to completion:

$$2NO + O_2 \rightarrow 2NO_2 \rightarrow N_2O_4$$

The dimer, N_2O_4, solidifies at 262 K. A 250 mL flask and a 100 mL flask are separated by a stopcock. At 300 K, the nitric oxide in the larger flask exerts a pressure of 1.053 atm and the smaller one contains oxygen at 0.789 atm. The gases are mixed by opening the stopcock and after the end of the reaction the flasks are cooled to 220 K. Neglecting the vapour pressure of the dimer, find out the pressure (and composition of the gas remaining) at 220 K. (Assume the gases to behave ideally).

(a) 0.42 atm (b) 0.32 atm (c) 0.52 atm (d) 0.22 atm

Q. 22. Expand the Vander Waals' equation (for 1 mol ga) into a cubic equation w.r.t the volume. At the critical temperature T_C, and the critical pressure P_C, the volume of 1 mol, is the critical volume V_C. *a* and *b* in terms of critical data respectively will be :

(a) $3V_c^2 P_C, 3V_C$ (b) $3V_c^2 P_c \frac{1}{3} V_C$

(c) $\frac{1}{3} V_c^2 P_c, \frac{1}{3} V_c$ (d) $\frac{1}{3} V_c^2 P_c, 3V_c$

Q. 23. A teacher enters in the classroom from the front door while a student from the back door. There are 13 equidistant rows of benches in the classroom. The teacher releases N_2O, the laughing gas from the first bench while the student releases the weeping gas $C_6H_{11}OBr$ from the last bench. The row at which the student start laughing and weeping simultaneously will be :

(a) 7th (b) 4th (c) 5th (d) 9th

Q. 24. A sample of gas occupies a volume of 430.0 mL at 301.2 K and 754.2 mm Hg. If the sample is cooled to 293 K, the pressure of the gas in millimeters of mercury is :

(a) 534.9 (b) 733.7 (c) 760.0 (d) 775.3

Q. 25. A 0.128-g sample of ethane gas, C_2H_6, is sealed in a glass tube at 297.4 K and a pressure of 765.3 mm Hg. The tube can withstand a maximum pressure of 2.50 atm without bursting. Maximum temperature the tube safely may be heated is:

(a) 339 K (b) 273 K (c) 611 K (d) 738.6 K

Q. 26. A sample of a pure gas has a density of 1.60 $g.L^{-1}$ at 299.5K and 680.2 mmH g. Which of the following could the sample be?

(a) CH_4 (b) C_2H_6 (c) CO_2 (d) Xe

Q. 27. A sample of butane gas at 296.8 K and 753.4 mm Hg, in a volume of 1.85 L, is completely burned in O_2 to yield only $H_2O(l)$ and $CO_2(g)$. $CO_2(g)$, measured at the same temperature and pressure, formed in this combustion L is:

(a) 7.40 L (b) 9.25 L (c) 12.0 L (d) 16.7 L

Q. 28. A 0.8763-g sample of sucrose, $C_{12}H_{22}O_{11}$, is completely burned with excess oxygen in a 3.40-L container that is maintained at a constant temperature of 298 K. The equation for the combustion is:

$$C_{12}H_{22}O_{11}(S) + 12\ O_2(g) \rightarrow 12CO_2(g) + 11H_2O(l)$$

The partial pressure, in millimeters of mercury, of CO_2 in the container after the combustion will be:

(a) 0.221 (b) 7.24 (c) 168 (d) 86.4

Q. 29. At room temperature mercury has a density of 13.6 $g{\cdot}cm^{-3}$, while liquid bromo-form, $CHBr_3$, has a density of 2.89 $g{\cdot}cm^{-3}$. How high a column of bromoform will be supported by a pressure that supports a column of mercury 200 mm high?

(a) 94.1 mm (b) 272.0 mm (c) 42.5 cm (d) 94.1 cm (e) 272.0 cm

Q. 30. A sample of gas at 308 K and 1.00 atm occupies a volume of 3.75 L. The temperature at which the gas should be kept, if it is desired to reduce the volume to 3.00 L at the same pressure?

(a) 299.6 K (b) 273 K (c) 276.9 K (d) 301 K

Q. 31. How many moles of helium gas occupy 22.4 L at 303 K and 1.00 atm?

(a) 0.110 (b) 0.900 (c) 1.00 (d) 1.11

Q. 32. If a sample of H_2 gas, collected over water at 299.4 K and 753.2 mm Hg, occupies a volume of 94.6 mL. The volume that H_2 gas occupy when it is dried, at 293 K and 100-atm pressure will be: (the vapor pressure of water at 299.4 K is 25.8 mm Hg).

(a) 68.6 mL. (b) 72.1 mL (c) 88.6 mL (d) 91.8 mL

Q. 33. The density (g/L) of Cl_2 gas at 323K and 1.60 atm is:
(a) 2.14 (b) 4.28 (c) 8.46 . (d) 13.8

Q. 34. A 0.4153-g sample of gas is collected over water at 296.5 K and 758.3 mmHg. The collected gas occupies 130 mL. The vapor pressure of water at 296.5 K is 21.7 mmHg. The molecular weight of this gas ($g \cdot mol^{-1}$) is :
(a) 70 (b) 80 (c) 90 (d) 100

Q. 35. At 300 K and 1.00 atm, the density of a gaseous hydrocarbon is 1.22 $g \cdot L^{-1}$. The hydrocarbon is :
(a) CH_4 (b) C_2H_4 (c) C_2H_6 (d) C_3H_8

Q. 36. A mixture of helium and argon contains 3 mol of He for every 2 mol of Ar. The partial pressure of argon is
(a) two thirds the total pressure (b) two fifths the total pressure
(c) three fifths the total pressure (d) one half the total pressure

Q. 37. A 1.7093-g sample of Al metal is reacted with excess aqueous hydrochloric acid to produce hydrogen gas:

$$2Al(s) + 6H+(aq) ® 3H_2(g) + 2Al^3+(aq)$$

What volume of H_2(g) is collected at a temperature of 297.8 K and a pressure of 738.5 mmHg?
(a) 88.45 mL (b) 199.0 mL (c) 1.063 L (d) 2.391 L

Q. 38. The molar volume of helium is 51.4 L at:
(a) 373 K and 1.00 atm (b) 298 K and 0.250 atm
(c) 313 K and 0.500 atm (d) 573 K and 1.00 atm

Q. 39. A mixture of 50.0 mL of ammonia and 60.0 mL of oxygen gas reacts according to the following equation:

$$4NH_3(g) + 5\ O_2(g) \rightarrow 4NO(g) + 6H_2O(g)$$

If all gases are at the same temperature and pressure, and the reaction continues until one of the gases is completely consumed, the volume of water vapor produced is:
(a) 48.0 mL (b) 60.0 mL (c) 72.0 mL (d) 75.0 mL

Q. 40. The partial pressure of SO_2 in millimeters of mercury, when 100 g of O_2 are mixed with 100 g of SO_2, and the total pressure is 600 mmHg will be:
(a) 500 (b) 400 (c) 300 (d) 200

Q. 41. A 0.20-mol sample of a: hydrocarbon C_xH_y yields, after complete combustion with excess O_2 gas, 0.80 mol of CO_2 and 1.0 mol of H_2O. The molecular formula of C_xH_y is :
(a) C_4H_{10} (b) C_4H_8 (c) C_4H_5 (d) C_8H_{16}

Q. 42. The average molecular speed is *greatest* in which of the following gas samples?
(a) 1.0 mol of N_2 at 560 K (b) 0.50 mol of Ne at 500 K
(c) 0.20 mol of CO_2 at 440 K (d) 2.0 mol of He at 140 K

Q. 43. A real gas most closely approaches the behavior of an ideal gas at :

(a) 15 atm and 200 K **(b)** 1 atm and 273 K

(c) 0.50 atm and 500 K **(d)** 15 atm and 500 K

Q. 44. If both gases are at the same temperature, the rate of effusion of O_2 is very close to

(a) 8 times that of He **(b)** 4 times that of He

(c) 2.8 times that of He **(d)** 0.35 times that of He

Q. 45. When the pressure of a sample of gas is increased from 0.50 to 100 atm at constant temperature, its volume decreases from 2.0 L to 13 mL. What could cause this deviation from Boyle's law?

(a) The average molecular speed has increased.

(b) The volume of the gas molecules is a significant fraction of the volume of the container at the higher pressure.

(c) The force of attraction between the gas molecules is greater when the pressure is higher.

(d) The molecules have dimerized at the higher pressure.

Q. 46. The temperature at which the rate of effusion of N_2 be 1.625 times the rate of effusion of SO_2 at 322 K will be:

(a) 87 K **(b)** 111 K **(c)** 230 K **(d)** 373 K

Q. 47. If a sample of an ideal gas in a sealed container of fixed volume is heated from 283 to 313 K, the value of which of the following quantities will remain constant?

(a) The average force of a molecular collision with the walls of the container.

(b) The density of the gas.

(c) The root-mean-square speed of the molecules.

(d) The total kinetic energy of the sample.

Q. 48. It takes 26 s for 10 mL of H_2 to effuse through a porous membrane. It takes 130 s for 10 mL of an unknown gas to effuse through the same membrane, when both gases are at the same temperature and pressure. The molecular weight of the unknown gas, in grams per mole will be:

(a) 100 **(b)** 80 **(c)** 50 **(d)** 40

Q. 49. The root-mean-square speed of the molecules in a sample of gas is proportional to :

(a) The square of the absolute temperature.

(b) The molecular weight of the gas.

(c) The square root of the absolute temperature.

(d) The reciprocal of the molecular weight of the gas.

Q. 50. A sample of gas in a closed container of fixed volume is at 250 K and 400-mmHg pressure. If the gas is heated to 375 K, its pressure increases to 600 mmHg. By what factor will the average speed of the molecules increase?

(a) 1.22 **(b)** 1.50 **(c)** 2.25 **(d)** 2.00

Q. 51. At 298 K, which of the following gases has the lowest average molecular speed?

(a) CO_2 at 0.20 atm **(b)** He at 0.40 atm

(c) Ne at 0.60 atm **(d)** CH_4 at 0.80 atm

Q. 52. The number of collusions of Ar atoms with the walls of their container per unit time:

(a) Increases when the temperature decreases.

(b) Remains the same when CO_2, is added to the container at constant temperature.

(c) Increases when CO_2 is added to the container at constant temperature.

(d) Decreases when the average kinetic energy per molecule increases.

Q. 53. At 320 K and 16.0 atm, the molar volume of ammonia gas (NH_3) is about 10% less than the molar volume of an ideal gas. The reason the actual volume is less than the ideal volume is that:

(a) NH_3 decomposes to N_2 and H_2 at 320 K.

(b) The force of attraction between NH_3 molecules is significant at this temperature and pressure.

(c) The volume occupied by the NH_3 molecules themselves is a significant fraction of the volume of the container at this pressure and temperature.

(d) NH_3 molecules move more slowly than predicted by the kinetic theory at this pressure and temperature.

Q. 54. If the rate of effusion of ammonia, NH_3, is 3.32 times faster than that of an unknown gas when both gases are at 350 K, what is the molecular weight of the unknown gas?

(a) 31.0 **(b)** 188 **(c)** 56.5 **(d)** 112

Q. 55. At what temperature will the molar kinetic energy of 0.30 mol of He be the same as the molar kinetic energy of 0.40 mol of Ar at 400 K?

(a) 533 K **(b)** 400 K **(c)** 346 K **(d)** 300 K

Q. 56. At what temperature will the total kinetic energy of 0.30 mol of He be the same as the total kinetic energy of 0.40 mol of Ar at 400 K?

(a) 533 K **(b)** 400 K **(c)** 346 K **(d)** 300 K

Q. 57. The total pressure in a 10 L cylinder which contains 0.4 g of helium, 1.6 g of oxygen and 1.4 g of nitrogen at 300 K is :

(a) 2.49 atm **(b)** 49.2 atm **(c)** 0.492 atm **(d)** 0.249 atm

Q. 58. Acetylene and hydrogen fluoride react to give difluoroethane when one mole of C_2H_2 and 5.0 mole of HF are reacted in a10 L flask. The pressure in the flask a 273 K when the reaction is completed is:

(a) 9.95 atm **(b)** 11.95 atm **(c)** 6.95 atm **(d)** 8.95 atm

Q. 59. An LPG (liquefied petroleum gas) cylinder weights 14.8 kg when empty. When full, it weights 29.0 kg. and shows a pressure of 2.5 atm. In the course of use at 300 K the weight of the full cylinder reduced to 23.2 kg. The volume of the gas in cubic meters used up at the normal usage conditions, and the final pressure inside the cylinder will be: (Assume LPG to be n-butane with normal boiling point of 273 K).

(a) 1.07 atm **(b)** 2.07 atm **(c)** 3.07 atm **(d)** 5.07 atm

Q. 60. At 300 K hydrogen is leaked through a tiny hole into a vessel for 20 minutes. Another unknown gas at the same temperature and pressure as that of H_2 is leaked through the same hole for 20 minutes. After the effusion of the gases the mixture exerts a pressure of 6 atmosphere. The hydrogen content of the mixture is 0.7 mole. If the volume of the container is 3 litres, the molecular mass of the unknown gas will be:

(a) 733 g mol^{-1} **(b)** 933 g mol^{-1} **(c)** 1033 g mol^{-1} **(d)** 1133 g mol^{-1}

Q. 61. When two moles of a gas are confined in a four litre flask exerts a pressure of 11.0 atm at a temperature of 300 K. If the value of 'b' is 0.05 L mol^{-1}, then the value of 'a' in vanderwaal's equation is:

(a) 6.46 atm L^2 mol^{-2} **(b)** 8.46 atm L^2 mol^{-2}

(c) 4.46 atm L^2 mol^{-2} **(d)** 10.46 atm L^2 mol^{-2}

Q. 62. A3 m long tube is provided with inlets a both the ends so that HCl and NH_3 gases can be admitted simultaneously. The distance from the HCl inlet end of the tube at which NH_4Cl will first appear if two gases are admitted at the same time, one from each end is :

(a) 1.19 m **(b)** 0.18 m **(c)** 0.81 m **(d)** 1.24 m

Q. 63. The temperature at which the root mean square velocity of hydrogen gas equals to 100 m s^{-1} is :

(a) 0.4 K **(b)** 2.4 K **(c)** 0.8 K **(d)** 1.6 K

Q. 64. The temperature at which the most probable velocity of oxygen equals that of hydrogen at 298 K is:

(a) 273 K **(b)** 4768 K **(c)** 328.4 K **(d)** 2384 K

Q. 65. The temperature at which N_2 molecules have the same average speed as He atoms at 330 K is:

(a) 231.0 K **(b)** 23.1 K **(c)** 2310 K **(d)** –231.0 K

Q. 66. The temperature at which the helium atoms have the same root mean square speed as hydrogen molecules have at 200 K is:

(a) 200 K **(b)** 400 K **(c)** 600 K **(d)** 800 K

Q. 67. The temperature at which the most probable speed of CO_2 molecules be twice that at 293 K is :

(a) 899 K **(b)** 117.2 K **(c)** 1172 K **(d).** 89.9 K

Q. 68. The ratio of the rates of effusion of neon gas to that of helium gas at the same temperature pressure is :

(a) $1:\sqrt{5}$ **(b)** $2:\sqrt{5}$ **(c)** $\sqrt{5}:1$ **(d)** $\sqrt{5}:2$

Q. 69. 10 moles of NH_3 occupy a volume of 25×10^{-3} m^{-3} at 25°C, pressure exerted using the Van der Waal's equation, given a = 0.417 n m^4 mol^{-2} and b = 0.037×10^3/mol. is :

(a) 18.8 5 10^5 N/m^2 **(b)** 9.4×10^5 N/m^2

(c) 18.8×10^7 N/m^2 **(d)** 9.4×10^7 N/m^2

Q. 70. Suppose water is used in a barometer instead of mercury. If the barometric pressure is 760.00 mmHg, The height of the water column in the barometer at 273 K will be:

(The densities of water and mercury at 273 K are 0.99987 g cm^{-3} and 13.596 g cm^{-3}, respectively).

(a) 760 mm of Hg (b) 103.34×10^2 mm
(c) 10.334×10^4 mm (d) 1.0334×10^4 mm

Q. 71. A volume of air occupying 12.0 dm^3 at 98.9 kPa is compressed to a pressure of 119.0 kPa. The temperature remains constant. The new volume will be:

(a) 0.997 dm^3 (b) 9.97 dm^3 (c) 1.994 dm^3 (d) 19.94 dm^3

Q. 72. How many grams of oxygen, O_2, are there in a 50.0-L tank at 294 K when the oxygen pressure is 15.7 atm?

(a) 1.04×10^3 g (b) 0.104×10^3g (c) 2.08×10^3 g (d) 10×10^3 g

Q. 73. What is the density of oxygen, O_2, in grams per liter at 278K and 0.850 atm?

(a) 11.1 g L^{-1} (b) 0.11 g L^{-1} (c) 1.11 g L^{-1} (d) None of these

Q. 74. You are given a quantity of a volatile (easily vaporized) liquid substance and asked to determine its molecular weight. To do that, a sample of the liquid is taken in a 200.0-mL, flask and submerge the flask in boiling water to vaporize all of the liquid. The vapor fills the flask and the excess escapes. Now cool the flask so that all the vapor in it condenses, and then weigh the flask and liquid. After subtracting the mass of the flask the mass of the substance in the flask is 0.970 g. The temperature of the boiling water was 372 K, the barometric pressure was 733 mmHg.

The molecular weight of the substance will be:

(a) 308 amu (b) 15.4 amu (c) 30.8 amu (d) 154 amu

Q. 75. Automobiles are being equipped with air bags that inflate on collision to protect the occupants from injury. Many such air bags are inflated with nitrogen, N_2, using the rapid reaction of sodium azide, NaN_3, and iron (III) oxide, Fe_2O_3, which is initiated by a spark. The overall reaction is :

$$6NaN_3(s) + Fe_2O_3(s) \rightarrow 3Na_sO(s) + 2Fe(s) + 9N_2(g)$$

The amount of sodium azide required to provide 75.0L of N_2 gas at 298 K and 748 mm Hg will be :

(a) 131 g. Na N_3 (b) 262g Na N_3
(c) 393 g. Na N_3 (d) 65 g Na N_3

Q. 76. Hydrogen gas is produced by the reaction of hydrochloric acid, HCl, on zinc metal.

$$2HCl(aq) + Zn(s) \rightarrow ZnCl_2(aq) + H_2(g)$$

The gas is collected over water. If 156 ml of gas is collected at 292 K and 769 mmHg total pressure; the mass of H_2 collected will be:

(a) 0.0064 g H_2 (b) 0.0258 g H_2 (c) 0.0129 g H_2 (d) 0.019.2 g H_2

Q. 77. 27 g of acetylene, C_2H_2, dissolves in 1 L of acetone at 1.0 atm pressures. If the partial pressure of acetylene is increased to 12 atm, its solubility in acetone will be:

(a) 320×10^2 g of acetylene **(b)** 0.32×10^2 g of acetylene

(c) 3.2×10^2 g of acetylene **(d)** None of these

Q. 78. If sulfur dioxide were an ideal gas, the pressure at 273 K exerted by 1.000 mole occupying 22.41 L would be 1.000 atm (22.41 L is the molar volume of an ideal gas at STP). The pressure of this volume of 1.000 mol SO_2 at 273 K will be : (values of a and b are 6.865 L^2 atm mol^{-2}, and 0.05679 L mol^{-1}. respectively.)

(a) 0.998 atm **(b)** 9.89 atm **(c)** 9.98 atm **(d)** 0.989 atm

Q. 79. The number of gas molecules left in a volume of 10^{-3} cm^3; if the gas is pumped out to give a vacuum of 10^{-5} mm Hg at 298 K is :

(a) 31.92×10^8 **(b)** 5.3×10^8 **(c)** 3.192×10^8 **(d)** 6.318×10^8

Q. 80. An iron tank of helium at 298 K contains the gas at 200 atmospheres. The tank will hold upto a pressure of about 700 atmospheres. If the tank is in a building that catches fire, will the tank blow up before it melts and the pressure exerted respectively is:
(The melting point of iron is 1808 K).

(a) No; 121.34 atm **(b)** Yes, 121.34 atm

(c) No; 1213.4 atm **(d)** Yes, 1213.4 atm

Q. 81. An equimolar mixture of N_2 and He is contained in a container. How many times faster will He initially leak through a pinhole in the container?

(a) 1.65 **(b)** 2.65 **(c)** 3.67 **(d)** 0.37

Q. 82. 5×10^{-3} kg of C_2H_6 are confined in a bulb of 10^{-3} m^3 capacity. The bulb is so weak that it will burst if the pressure exceeds 10 atm. At what temperature will the pressure of the gas each the bursting value?

(a) 121.8 K **(b)** 12.18 K **(c)** 73.09 K **(d)** 730.9 K

Q. 83. To determine the mol.wt. M of a volatile liquid, the end of a syringe is sealed with a self-sealing rubber cap through which a known mass of the liquid under investigation is injected. The liquid is volatile and has a low boiling point, the syringe is heated to above the boiling point and the volume of the gas produced is measured on the syringe.

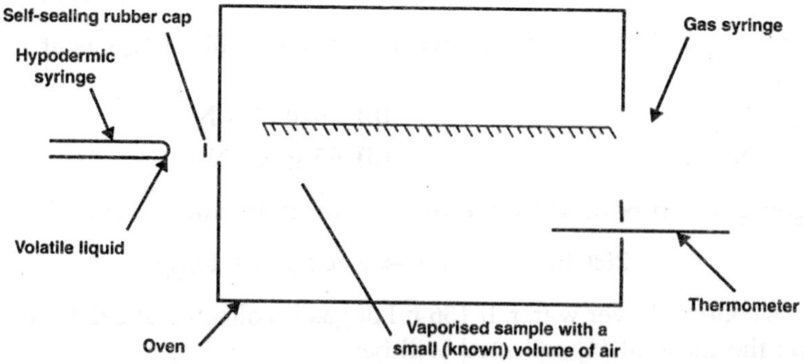

The apparatus used to measure the M of a volatile liquid

The following results were obtained from a sample of volatile liquid injected into a syringe :

Mass of liquid sample	0.180g
Volume of air	10 cm^3
Volume of vapour + air	100 cm^3
Temperature	353 K
Pressure	100600 Pa

The M of the sample is :
(a) 68 **(b)** 580 **(c)** 58 **(d)** 680

Q. 84. The vapour density of N_2O_4 at 333 K is 30.6. The percentage dissociation of N_2O_4 will be :
(a) 96% **(b)** 77.5% **(c)** 50.32% **(d)** 18.44%

Q. 85. The ratio of molecular velocities C_1/C_2 for an ideal gas when at constant pressure, temperature is doubled will be:
(a) 0.99 **(b)** 0.707 **(c)** 0.5 **(d)** 0.44

Q. 86. On P-V-T surface process AB shown in figure is:
(a) Polytropic **(b)** Isochoric **(c)** Isothermal **(d)** Isobaric

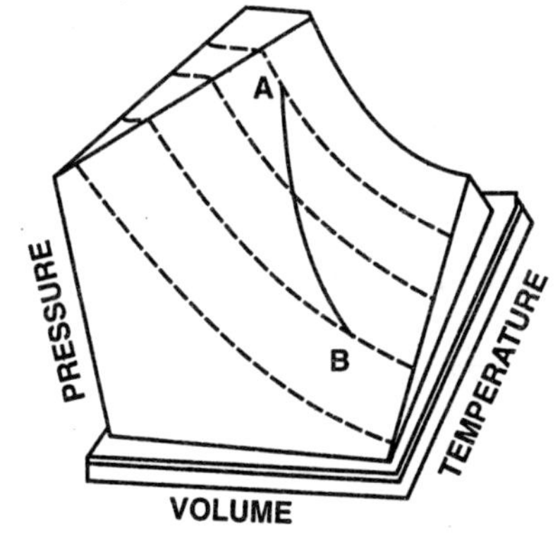

Chapter 4

STOICHIOMETRY

KEY CONCEPT

Equations and relationship that you need to know:

$$1 \text{ nm} = 1 \times 10^{-9} \text{ m} = 10 \text{ Å}$$

$$1 \text{ cm}^3 = 1 \text{ mL}$$

$$K = °C + 273.16$$

$$\text{density} = \frac{\text{mass}}{\text{volume}}$$

$$\text{Avogadro's number} = 6.02 \times 10^{23} = 1 \text{ mole}$$

$$\text{number of moles} = \frac{\text{mass}}{\text{molecular weight}}$$

$$\text{molecular weight of ideal gas} = \text{density} \times \text{molar volume}$$

$$\% \text{ yield} = \frac{\text{actual yield}}{\text{theoretical yield}} \times 100\%$$

$$\% \text{ composition} = \frac{\text{mass of element in compound}}{\text{mass of compound}} \times 100\%$$

$$\% \text{ error} = \frac{\text{observed value} - \text{expected value}}{\text{expected value}} \times 100\%$$

PROBLEMS

Q. 1. A popular bourbon whiskey is listed as being "92 Proof.". The liquor industry defines "Proof" as being twice the volume percentage of alcohol in a blend. Ethanol (drinking alcohol) as the structural formula CH_3CH_2OH (MW = 46 g.mol^{-1}). The density of ethanol is 0.79 g mL^{-1}. How many liters of whiskey must one have in order to have 50. moles of carbon?

(a) 0.80 liters (b) 1.6 liters (c) 3.2 liters (d) 4.0 liters

Q. 2. A sample of a pure compound was found to contain 1.201 grams of carbon, 0.202 grams of hydrogen and 7.090 grams of chlorine. What is the empirical formula for the compound?

(a) $CHCl_3$ (b) CH_2Cl (c) CH_2Cl_2 (d) CH_3Cl

Q. 3. Balance the following equation using the lowest possible whole number of coefficients:

$$NH_3 + CuO \rightarrow Cu + N_2 + H_2O$$

The sum of the coefficients is.

(a) 9 (b) 10 (c) 11 (d) 12

Q. 4. When 0.600 mole of $BaCl_{2(aq)}$ is mixed with 0.250 mole of $K_3AsO_{4(aq)}$, what is the maximum number of moles of solid $Ba_3(AsO_4)_2$ that could be formed?

(a) 0.125 mole (b) 0.200 mole (c) 0.250 mole (d) 0.375 mole

Q. 5. A test tube containing $CaCO_3$ is heated until *all* of the compound decomposes. If the test tube plus calcium carbonate originally weighed 30.08 grams and the loss of mass during the experiment was 4.400 grams, what was the mass of the empty test tube?

(a) 20.07 g (b) 21.00 g (c) 24.50 g (d) 25.08 g

Q. 6. 32.0 grams of oxygen gas, 32.0 grams of methane gas, and 32.0 grams of sulfur dioxide gas are mixed. What is the mole fraction of the oxygen gas?

(a) 0.143 (b) 0.286 (c) 0.333 (d) 0.572

Q. 7. Element X is found in two forms: 90.0% is an isotope that has a mass of 20.0 and 10.0% is an isotope that has a mass of 22.0. What is the atomic mass of element X?

(a) 20.1 (b) 20.2 (c) 20.8 (d) 21.2

Q. 8. What is the formula of a compound formed by combining 50 grams of element X (atomic weight = 100) and 32 grams of oxygen gas?

(a) XO_2 (b) XO_4 (c) X_4O (d) X_2O

Q. 9. An oxide is known to have the formula X_2O_7 and to contain 76.8% X by mass. Which of the following would you use to determine the atomic mass of X?

(a) $\dfrac{76.8}{\left(\dfrac{23.2}{16.0}\right)\times\left(\dfrac{7}{2}\right)}$ (b) $\dfrac{76.8}{\left(\dfrac{16.0}{23.2}\right)\times\left(\dfrac{2}{7}\right)}$ (c) $\dfrac{76.8}{\left(\dfrac{23.2}{16.0}\right)\times\left(\dfrac{2}{7}\right)}$ (d) $\dfrac{76.8}{\left(\dfrac{16.0}{23.2}\right)\times\left(\dfrac{7}{2}\right)}$

Q. 10. A chemist analyzed a sample of copper (II) sulfate pentahydrate for water of hydration by weighting the hydrate, heating it to convert it to anhydrous copper 9ii0 sulfate, and then weighing the anhydride. The %H_2O was determine to be 30%. The theoretical value is 33%. Which of the following choices is definitely NOT the cause of the error?

(a) After the chemist weighed the hydrate, a place of rust fell from the tongs into the crucible.
(b) Moisture driven from the hydrate condensed on the inside of the crucible cover before this chemist weighed the anhydride.
(c) All the weighings were made on a balance that was high by 10%
(d) The original sample was wet.

Q. 11. When 100 grams of butane gas (C_4H_{10}, MW = 58.14) is burned in excess oxygen gas, the theoretical yield of H_2O is:

(a) $\dfrac{54.14 \times 18.02}{100 \times 5}$ (b) $\dfrac{100 \times 5 \times 18.02}{58.14}$

(c) $\dfrac{4 \times 18.02}{13/2 \times 100} \times 100\%$ (d) $\dfrac{5 \times 58.14 \times 18.02}{100}$

Q. 12. Element Q occurs in compounds X, Y, and Z. The mass of element Q in 1 mole of each compound is as follows:

Compound	*Grams of Q in Compound*
X	38.00
Y	95.00
Z	133.00

Element Q is most likely:
(a) N (b) O (c) F (d) Ir

Q. 13. The degree of dissociation of $Ca(NO_3)_2$ in a dilute aqueous solution, containing 7.0 g of the salt per 100 g of water at 373 K is 70%. If the vapour pressure of water at 373 K is 760 mm. The vapour pressure of the solutions.
(a) 794.2 mm Hg (b) 774.2 mm Hg
(c) 720 mm Hg (d) 746.3 mm Hg

Q. 14. A solution of a non volatile solute in water freezes at –272.7 K. The vapour pressure of pure water at 298 K is 23.51 mm Hg and K_f for water is 1.86 degree/molal. The vapour pressure of this solution at 298 K is:
(a) 23.44 mm Hg (b) 20.44 mm Hg (c) 23.58 mm Hg (d) 20.58 mm Hg

Q. 15. A very small amount of a nonvolatile solute (that does not dissociate) is dissolved in 56.8 cm^3 of benzene (density 0.889 g cm^{-3}). At room temperature, vapour pressure of this solution is 98.88 mm Hg while that of benzene is 100 mm Hg. Find the molality of this solution. If the freezing temperature of this solution is 0.73 degree lower than that of benzene, what is the value of metal freezing point depression constant of benzene.
(a) 4.07 (b) 6.07 (c) 3.07 (d) 5.07

Q. 16. An aqueous solution containing 12 g of a non-volatile compound is 100 g of water boils at 373.34 K at 1 atm. pressure. Determine the molecular weight of the compound (K_b for water = 0.51)
(a) 160 $g.mol^{-1}$ (b) 180 $g.mol^{-1}$ (c) 225 $g.mol^{-1}$ (d) 200 $g.mol^{-1}$

Q. 17. Addition of 0.643 g of a compound to 50 mL of benzene (density 0.879 g/mL) lowers the freezing point from 278.51 K to 278.03 K. If K_f for benzene is 5.12, the molecular mass of the compound is:
(a) 256 g.mol^{-1} **(b)** 356 g.mol^{-1} **(c)** 456 g.mol^{-1} **(d)** 156 g.mol^{-1}

Q. 18. The boiling point of a solution containing 0.61 g benzoic acid in 50 g of carbon disulphide assuming 84% dimerization of the acid when the boiling point and K_b of CS_2 are 319.2 K and 2.3 K kg mol^{-1} respectively is:
(a) 319.33 K **(b)** 419.33 K **(c)** 219.33 K **(d)** 319.0 K

Q. 19. In a cold climate water gets frozen causing damage to the radiator of a car. Ethylene glycol is used as an antifreezing agent. The amount of ethylene glycol to be added to 4 kg of water to prevent it from freezing at –267 K/ (K_f for water = 1.85 K mol^{-1}) will be
(a) 402 g **(b)** 1804 g **(c)** 804 g **(d)** 201 g

Q. 20. A 1.2% solution of NaCl is isotonic with 7.2% solution of glucose, the Van't Hoff's factor of NaCl solution is:
(a) 1.95 **(b)** 2.95 **(c)** 0.51 **(d)** 1.02

Q. 21. One litre of an aqueous solution contains 15 g of a solute of mol. wt. 150. If the density of the solution is 1.005 g cm^{-3}, the mole fraction of the solute is:
(a) 0.0038 **(b)** 0.0018 **(c)** 0.038 **(d)** 0.018

Q. 22. The molality of 1 litre solution of 93% H_2SO_4 when the density of the solution is 1.84 g/mL will be:
(a) 132.56 m **(b)** 138.56 m **(c)** 135.56 m **(d)** 142.56 mm

Q. 23. The osmotic pressure of a 0.2% solution of an organic compound in water is 570 mm of Hg at 285 K the molecular mass of the compounds.
(a) 62.4 g.mol^{-1} **(b)** 124.8 g.mol^{-1} **(c)** 187.28 g.mol^{-1} **(d)** 311.0 g.mol^{-1}

Q. 24. 0.85% aqueous solution of $NaNO_3$ is apparently 90% dissociated at 300 K its osmatic pressure is

(R = 0.082 L atm K^{-1} mol $^{-}$1)

(a) 4.674 atm **(b)** 6.474 atm **(c)** 4.476 atm **(d)** 6.744 atm

Q. 25. Methanol (methyl alcohol) is a potential future automotive fuel, either pure or mixed with gasoline. Pure methanol has an octane rating of 106, and a 85% methanol–15% gasoline mixture is about ten percent more efficient a motor fuel than straight gasoline. **A pycnometer** is a simple device used for precise density determinations.

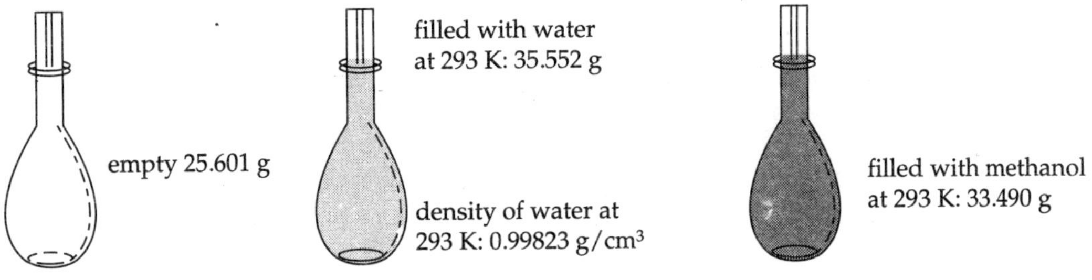

(i) The volume capacity of the pycnometer is
(a) 0.997 cm^3 **(b)** 9.97 cm^3 **(c)** 9.95 cm^3 **(d)** 99.7 cm^3

(ii) The density of methanol at 293 K is:
(a) 0.78 $g.cm^{-3}$ **(b)** 0.87 $g.cm^{-3}$ **(c)** 7.88 $g.cm^{-3}$ **(d)** none of these

Q. 26. Steel is a metallic mixture having iron as its principal element. A key minor element in steel is carbon. A stainless steel ball bearing, has a radius of 6.35 mm and is made of steel with 0.25% carbon. The steel has a density of 7.75 $g.cm^{-3}$.

(i) The volume of the ball bearing is
(a) 1.07 cm^3 **(b)** 2.07 cm^3 **(c)** 4.14 cm^3 **(d)** 0.75 cm^3

(ii) The mass of carbon in the ball bearing is:
(a) 21 g **(b)** 2.1 g **(c)** 0.21 g **(d)** 0.021 g

(iii) The total number of carbon atoms in the ball bearing is
(a) 0.11×10^{21} C atoms **(b)** 1.1×10^{21} C atoms
(c) 0.011×10^{21} C atoms **(d)** 11×10^{21} C atoms

(iv) Given that the natural abundance of C-13 is 1.108% thus, C-13 atoms that are present in the ball bearing are:
(a) 0.12×10^{19} C-13 atoms **(b)** 0.012×10^{19} C-13 atoms
(c) 1.2×10^{19} C-13 atoms **(d)** 12.0×10^{19} C-13 atoms

(v) If the mass of the isotope C-13 is 13.00335 u, then the mass of C-13 present in the ball bearing is:
(a) 0.26×10^{-4} g C-13 **(b)** 2.6×10^{-4} g C-13
(c) 0.026×10^{-4} g C-13 **(d)** 1.2 NO^{-4} g C-13

Q. 27. Copper sulfate is the most widely used copper compound, finding applications in electroplating, in pesticide mixtures, as a wood preservative, and as an algicide (for killing algae in lakes). When heated, the hydrated solid goes through the succession of changes.

A 2.574-g sample of hydrate copper (II) sulfate, $CuSO_4 \cdot xH_2O$, was heated to 413 K, cooled to room temperature, and reweighed. The resulting solid was reheated to 673 K, cooled to room temperature and reweighed. Finally, this solid was heated to 1273 K; the resulting black residue was cooled to room temperature and weighed for a final time. The measured masses were

original sample;	2.574 g
after heating to 413 K	1.833 g
after reheating to 673 K	1.647 g
after reheating to 1273 K	0.812 g

Now answer the following:

1. Assuming that all the water of hydration is driven off after heating to 673 K, what must be the formula of the hydrate $CuSO_4 \cdot xH_2O$?
(a) $CuSO_4 \cdot 3H_2O$ **(b)** $CuSO_4 \cdot 4H_2O$
(c) $CuSO_4 \cdot 5H_2O$ **(d)** $CuSO_4 \cdot 6H_2O$

2. What is the formula of the hydrate obtained after the original solid is heated
(a) $CuSO_4 \cdot H_2O$ **(b)** $CuSO_4 \cdot 3H_2O$
(c) $CuSO_4 \cdot 5H_2O$ **(d)** $CuSO_4 \cdot 7H_2O$

3. The black residue obtained at 1293 K is an oxide of copper so its percentage composition and its empirical formula is:

(a) Cu_2O; Cu = 90%, O = 10% **(b)** CuO; Cu = 81%, O – 19%
(c) CuO; Cu = 19%, O = 81% **(d)** Cu_2O; Cu = 10%, O – 90%

Q. 28. Sodium nitrite, $NaNO_2$, is used in the production of dyes for coloring fabrics; as a pre servative in meat processing (to prevent botulism) in low concentrations and in higher concentrations sodium nitrite is a carcinogen (cancer-producing agent).

Its also used, as a bleach for fibers, and in photography. It is prepared by passing nitrogen monoxide and oxygen gases into an aqueous solution of sodium carbonate. Carbon dioxide gas is another product of the reaction.

(i) The mass of $NaNO_2$ that should result from the reaction of 748 g Na_2CO_3 with an excess of NO(g) and O_2(g) is:

(a) 0.974 g $NaNO_2$ **(b)** 97.4 g $NaNO_2$
(c) 974 g $NaNO_2$ **(d)** 9740 g $NaNO_2$

(ii) In another reaction the reactants are 225 mL of 1.50 M Na_2CO_3(aq), 22.1 g NO, and a large excess of O_2. The mass of $NaNO_2$ produced, if the reaction has a 95% yield will be:

(a) 44.3 g $NaNO_2$ **(b)** 67.5 g $NaNO_2$
(c) 73.6 g $NaNO_2$ **(d)** 46.6 g $NaNO_2$

Q. 29. A solution of a moss-killer contained aqueous ion (II) ions. 25.0 cm^3 of this solution was acidified with dilute sulphuric acid and titrated with 0.0200 mol dm^{-3} potassium manganate (VII). It was found that 21.0 cm^3 of the $KMnO_4$ (aq) was needed to fully react with the Fe(aq) in the solution of the moss-killer. The concentration of Fe(aq) in the solution of the moss-killer is:

(a) 0.042 mol dm^{-3} **(b)** 0.021 mol dm^{-3}
(c) 0.084 mol dm^{-3} **(d)** 0.168 mol dm^{-3}

Q. 30. Sodium dithionite, $Na_2S_2O_4$, is an important reducing agent used in the textile, pulp and paper, and ceramics industries. One interesting use is the reduction of chromate ion to insoluble chromium (III) hydroxide by dithionite ion, $S_2O_4^{2-}$, in basic solution. An ionic expression representing the reaction of chromate and dithionite ions is:

$CrO_4^{2-}(aq) + S_2O_4^{2-}(aq) + OH^- \rightarrow Cr(OH)_3(s) + SO^{2-}(aq)$

The mass of $Na_2S_2O_4$ is consumed in reacting with 100.0 L of wastewater that is 0.0148 M in CrO_4^{2-} is:

(a) 148 g of $NO_2S_2O_4$ **(b)** 387 g of $NO_2S_2O_4$
(c) 193.5 g of NO_2SO_4 **(d)** 774 g of $NO_2S_2O_4$

Q. 31. Gasoline, used as a motor fuel, is mixture of a large number of hydrocarbons. The principal ones are various octanes, which have the formula C_8H_{18}. When gasoline is burnt in an automobile engine, to get complete conversion of the to hydrocarbons in CO_2 and H_2O it requires use the correct mass ratio of air to gasoline (called the ari/fuel ratio). If the mixture is fuel-rich there is insufficient air to complete the combustion of the gasoline and unburned hydrocarbons and carbon monoxide are produced. These are unwanted air pollutants. If the mixture is fuel-lean, there is an excess of air. This completes the combustion process, reducing the emissions of hydrocarbons and carbon monoxide.

The chemical equation for the complete combustion of octane (C_8H_{18}), is:

$$2C_8H_{18}(l) + 25\ O_2(g) \downarrow 16\ CO_2(g) + 18\ H_2O(l)$$

(i) The volume of $CO_2(g)$, measured at 295 K and 745/760 atm. pressure produced in the complete combussion of 100 g. C_8H_{18} is:
(a) 173 L **(b)** 273 L **(c)** 373 L **(d)** 73 L

(ii) The mass of $O_2(g)$ consumed in the complete combustion of 100 g C_8H_{18} is
(a) 250 g **(b)** 350 g **(c)** 450 g **(c)** 700 g

(iii) The percent, by mass, of oxygen in air is:
(a) 26.15% **(b)** 24.15% **(c)** 25.15% **(d)** 23.15%

(iv) The mass ratio of air to octane for a mixture in which the octane burns completely and there is no air in excess will be:
(a) 1.51×10 **(b)** 15.1 **(c)** 1.51×10^2 **(d)** 1.51

Q. 32. The number of moles in 56 g silicon are:
(a) 4 moles **(b)** 2 moles **(c)** 0.5 moles **(d)** 1 mole

Q. 33. The mass of 0.25 mol of iron is:
(a) 14 g **(b)** 24 g **(c)** 224 g **(d)** 28 g

Q. 34. Which of the following samplės contains the largest number of atoms?
(a) 1 g of Ni(s) **(b)** 1 g of Ca(s) **(c)** 1 g of $N_2(g)$ **(d)** 1 g of B(s)

Q. 35. Which of the following samples contains the smallest number of atoms?
(a) 1 g of $CO_2(g)$ **(b)** 1 g of $C_8H_{18}(\ell)$
(c) 1 go of $C_2H_6(g)$ **(d)** 1 go of LiF(s)

Q. 36. The ratio of oxygen atoms to hydrogen atoms in $(NH_4)_2SO_4$ is
(a) 1 : 1 **(b)** 2:1 **(c)** 1:2 **(d)** 1:4

Q. 37. A sample of ammonium phosphate, $(NH_4)_3PO_4$, contains 3.18 mol of hydrogen atoms. The number of moles of oxygen atoms in the sample is:
(a) 0.265 **(b)** 0.795 **(c)** 1.06 **(d)** 3.18

Q. 38. A sample of copper sulfate pentahydrate, $CuSO_4 \cdot 5H_2O$, contains 3.782 g of Cu. How many grams of oxygen are in this sample?
(a) 0.952 g **(b)** 3.809 g **(c)** 4.761 g **(d)** 8.570 g

Q. 39. Element X reacts with oxygen to produce pure sample of X_2O_3. In an experiment it is found that 1.0000 g of X produces 1.1596 g of X_2O_3. Using the known atomic weight of oxygen, 15.9994 $g \cdot mol^{-1}$, the atomic weight of X will be:
(a) 20.70 **(b)** 66.85 **(c)** 100.2 **(d)** 150.4

Q. 40. An ore contains 1.34% of the mineral argentite, Ag_2S, by weight. How many grams of this ore would have to be processed in order to obtain 1.00 g of pure solid silver, Ag?
(a) 74.6 g **(b)** 85.7 g **(c)** 107.9 g **(d)** 134.0 g

Q. 41. When pentane, C_5H_{12}, is burned in excess oxygen, the products of the reaction are $CO_2(g)$ and $H_2O(l)$. In the balanced equation for this combustion

$$C_5H_{12}(g) + \text{—}O(g) \rightarrow SCO_2(g) + 6H_2O(\ell)$$

the coefficient of oxygen should be
(a) 16 **(b)** 12 **(c)** 11 **(d)** 8

Q. 42. The complete of ethane gas yields $CO_2(g)$ and $H_2O(l)$ as indicated by the following equation:

$$C_2H_6(g) + \text{—}O_2(g) \rightarrow \text{—}CO_2(g) + \text{—}H_2O(l)$$

In this equation the ratio of the coefficients of CO_2 to H_2O is:

(a) 1:1 (b) 2:3 (c) 3:2 (d) 1:3

Q. 43. Ethanol, C_2H_5OH, is the substance commonly called "alcohol". The density of liquid ethanol is 0.7893 $g \cdot mL^{-1}$ at 293 K. If 1.2 mol of ethanol are needed for a particular experiment, what volume of ethanol should be measured out?

(a) 55 mL (b) 58 mL (c) 70 mL (d) 79 mL

Q. 44. What is the total number of atoms present in 25.0 mg of camphor, $C_{10}H_{16}O$?

(a) 9.89×10^{19} (b) 6.02×10^{20} (c) 9.89×10^{20} (d) 2.67×10^{21}

Q. 45. A compound having the empirical formula $(C_3H_4O)_n$ has a molecular weight of 170 ± 5. The molecular formula of this compound is:

(a) C_3H_4O (b) $C_6H_8O_2$ (c) $C_6H_{12}O_3$ (d) $C_9H_{12}O_3$

Q. 46. Analysis of a compound yields the following percentage composition by weight: 65.03% Ag, 15.68% cr, 19.29% O. The simplest formula of this compound is:

(a) Ag_2CrO_4 (b) $Ag_2Cr_2O_7$ (c) $AgCrO_2$ (d) $AgCr_2O_3$

Q. 47. Cortisone is a molecular substance containing 21 atoms of carbon per molecule. The weight percentage of carbon in cortisone is 69.98%. The molecular weight of cortisone is:

(a) 360.4 (b) 452.2 (c) 287.6 (d) 312.8

Q. 48. Methyl benzoate is prepared by the reaction between benzoic acid and methanol according to the equation

$$\underset{\text{Benzoic acid}}{C_6H_5COOH} + \underset{\text{Methanol}}{CH_3OH} \rightarrow \underset{\text{Methyl benzoate}}{C_6H_3COOCH_3} + H_2O$$

In an experiment 24.4 g of benzoic acid were reacted with 70.0 mL of CH_3OH. The density of CH_3OH is 0.791 $g \cdot mL^{-1}$. The methyl benzoate produced had a mass of 21.6 g. The percentage yield of the product is:

(a) 91.7% (b) 79.3% (c) 71.5% (d) 21.7%

Q. 49. An atom that has the same number of neutons as ^{85}Rb is"

(a) ^{85}Kr (b) ^{87}Rb (c) ^{85}Sr (d) ^{86}Sr

Q. 50. Which of the following samples contains 2.0×10^{23} atoms?

(a) 8.0 g O_2 (b) 3.0 g Be (c) 8.0 g C (d) 12.0 g He

Q. 51. The number of electrons in the telluride ion, Te^{2-}, is:

(a) 52 (b) 51 (c) 54 (d) 53

Q. 52. Which of the following gases is red brown in color?

(a) N_2O (b) NO_2 (c) NO (d) O

Q. 53. Which of the following substances is commonly called laughing gas?

(a) N_2O (b) NO_2 (c) NO (d) N_2

Q. 54. The atomic weight of Cu is 63.546. There are only two naturally occurring isotopes of copper, ^{63}Cu and ^{65}Cu. the natural abundance of the ^{63}Cu isotope must be approximately.

(a) 10% (b) 30% (c) 50% (d) 70%

Q. 55. The reaction between yttrium metal and dilute hydrochloric acid produces $H_2(g)$ and Ag_2S Y^{3+} ions. The molar ratio of yttrium used to hydrogen produced is

(a) 2:3 (b) 1:3 (c) 2:1 (d) 3:2

Q. 56. The mass of glycine (NH_2CH_2COOH) used to make 100 mL of its solution of molar concentration 0.01 M is:

(a) 0.075 grams (b) 0.04 grams
(c) 0.75 grams (d) 0.4 grams

Q. 57. A180 mL of conc. H_2SO_4 (sp. gr. = 1.84) was carefully added to water and the cold solution was made up to 1 litre exactly. The normality of the solution was found to be 6.61 and sp. gr. was found to be 1.198. The,

(i) Normality of the conc. acid is:

(a) 36.0 (b) 36.36 (c) 36.72 (d) 72.36

(ii) Molarity of the diluted solution:

(a) 3.305 m (b) 6.61 m (c) 13.22 m (d) 26.44 mm

(iii) % of H_2SO_4 by weight in conc. H_2SO_4:

(a) 97.8 (b) 105 (c) 48.9 (d) 76.3

(iv) % of H_2SO_4 by weight in diluted solution:

(a) 72 (b) 27 (c) 94 (d) 49

(v) Molarity of the diluted solution is:

(a) 3.78 (b) 7.38 (c) 8.37 (d) 3.87

Q. 58. The volume of oxygen required to burn 600 cm^3 of natural gas under room conditions, assuming that natural gas is pure methane is:

(a) 300 cm^3 (b) 600 cm^3 (c) 1800 cm^3 (d) 1200 cm^3

Q. 59. For preparing 2.0 *M* solution of glucose, the number of grams of glucose required is:

(a) 180 grams (b) 360 grams (c) 540 grams (d) 90 grams

Q. 60. The oxidation number of chromium in the dichromate ion $Cr_2O_7^{2-}$ is:

(a) +2 (b) +4 (c) +6 (d) +3

Q. 61. A coal-fired power station uses fuel that contains 0.5 percent sulphur by mass. The volume of sulphur dioxide at 298 K and 101 kPa is released into the atmosphere by burning 100 tonnes of coal is:

(a) 375×10^3 dm^3 (b) 275×10^3 dm^3
(c) 475×10^3 dm^3 (d) 175×10^3 dm^3

Q. 62. Magnesium oxide is the most extensively used magnesium compound, and its principal use, because of its high melting point, is as a refractory material—one able to withstand very high temperatures. It is used to line furnaces in steel-making and in electrical insulators and ceramic materials.

One method of preparing magnesium oxide is to burn pure magnesium in air. In one laboratory experiment 0.200 g Mg yielded 0.315 g of product Assuming that there was no loss material, could this product have been pure MgO?

(a) YES (b) NO (c) DON'T KNOW (d) Both YES and NO

Q. 63. You heat 2.53 grams of metallic mercury in air which produces 2.73 grams of a red-orange residue. Assume that the chemical change is the reaction of the metal with oxygen in air.

Mercury + oxygen → red-orange residue

The mass of oxygen that reacts when you strongly heat the red-orange residue and it decomposes to give back the mercury and release the oxygen, which you collect, will be:

(a) 5.36 gms (b) 0.20 gms (c) 2.73 gms (d) 2.53 gms

Q. 64. Nitrogen gas is the major component of air. A sample of nitrogen gas in a glass bulb weighs 243 mg. This mass in SI base units of mass (kilograms) is:

(a) 2.43×10^{-4} (b) 24.3×10^{-4} (c) 2.43×10^{-1} (d) All of these.

Q. 65. The world's oceans contain approximately 1.35×109 m^3 of water. The volume in liters is:

(a) 13.5×10^{9} dm^3 (b) 0.135×10^{21} dm^3
(c) 1.35×10^{18} dm^3 (d) 1.35×10^{21} dm^3

Q. 66. Chromium Cr, has the following isotopic masses and fractional abundances:

Mass Number	Isotopic Mass (amu)	Fractional Abundance
50	49.4961	0.0435
52	51.9405	0.8379
53	52.9407	0.0950
54	53.9389	0.0236

The atomic weight of chromium is:

(a) 519.9 amu (b) 5.199 amu (c) 51.99 amu (d) 5199 amu

Q. 67. (a) Chromium (III) oxide is used as a green paint pigment. It is a compound composed of Cr^{3+} and O^{2-} ions. The formula of chromium (III) oxide is:

(a) CrO_3 (b) Cr_2O_2 (c) CrO (d) Cr_2O_3

Q. 68. The mass in gram of a chlorine atom, Cl and the mass in grams of a hydrogen chloride molecule, is:

(a) 5.9×10^{-23} g; 6.06×10^{-23} g
(b) 6.06×10^{-23} g; 5.9×10^{-23} g
(c) 210×10^{23} g; 219×10^{23} g
(d) 219×10^{23} g; 210×10^{23} g

Q. 69. Zinc Iodide, ZnI_2, can be prepared by the direct combination of elements Zinc and Iodine.

A chemist determines from the amounts of elements that 0.0654 mol ZnI_2 can form. The grams of zinc iodide is:

(a) 319 g ZnI_2 (b) 20.9 g ZnI_2
(c) 0.021×10^{-2} ZnI_2 (d) None of above

Q. 70. Lead (II) chromate, $PbCrO_4$, is yellow paint pigment (called chrone yellow) prepared by a precipitation. In a preparation, 45.6 g of lead (II) chromate is obtained as a precipitate. The number of moles of $PbrO_4$ is:

(a) 45.6 g $PbCrO_4$ (b) 1.41 mol $PbCrO_4$
(c) 0.141 mol $PbrO_4$ (d) None of above

Q. 71. The no. of molecules in a 3.46-g sample of hydrogen chloride, HCl are:
(a) 5.71×10^{22} HCl molecules (b) 57.1×10^{22} HCl molecules
(c) 5.71×10^{22} HCl molecules (d) 5710×10^{22} molecules

Q. 72. Formaldehyde, CH_2O, is a toxic gas with a pungent odor. Large quantities are consumed in the manufacture of plastics and a water solution of the compound is used to preserve biological specimens. The mass percentages of the elements in formaldehyde are:
(a) C = 40%, H = 6.73%, O = 53.3% (b) C = 67.3%, H = 50%, O = 53.3%
(c) C = 53.3%, H = 40%, O = 67.3% (d) C = 40%, H = 67.3%, O = 53.3%

Q. 73. Acetic acid contains only C, H, and O A 4.24-mg sample of acetic acid is completely burned. It gives 6.21 mg of carbon dioxide and 2.54 mg of water. The mass percentage of each element in acetic acid is:
(a) C = 67.2% H = 53.4% O = 39.9% (b) C = 39.9% H = 67.2% O = 53.4%
(c) C = 53.4% H = 6.72% O = 39.9% (d) C = 39.9% H = 67.2% O = 53.4%

Q. 74. Zinc metal reacts with hydrochloric acid by the following reaction:

$$Zn(s) + 2HCl(aq) \rightarrow ZnCl_2(aq) + H_2(g)$$

If 0.30 mol Zn is added to hydrochloric acid containing 0.52 mol HCl, the number moles of H_2 are produced are:
(a) 0.26 mol H_2 (b) 0.30 mol H_2 (c) 0.52 mol H_2 (d) 0.60 mol H_2

Q. 75. In a process for producing acetic acid, oxygen gas is bubbled into acetaldehyde, CH_3CHO, containing manganese (II) acetate (catalyst) under pressure at 333 K

$$2CH_3CHO(l) + O_2(g) \rightarrow 2HC_2H_3O_2(l)$$

(i) In a laboratory test of this reaction, 20.0 g CH_3CHO and 10.0 g O_2 were put into a reaction vessel. The grams of acetic acid that can be produced by this reaction from these amounts of reactants is:
(a) 0.635 g $HC_2H_3O_2$ (b) 27.3 g $HC_2H_3O_2$
(c) 0.454 g $HC_2H_3O_2$ (d) 54.6 g $HC_2H_3O_2$

(ii) The number of grams of excess reactant which remains after the reaction is complete will be:
(a) 2.7 g O_2 (b) 27.0 g O_2 (c) 0.27 g O_2 (d) 7.26 g O_2

Q. 76. A sample of $NaNO_3$ weighing 0.38 g is placed in a 50.0-mL volumetric flask. The flask is then filled with water to the mark on the neck. The molarity of the resulting solution is:
(a) 0.89 M $NaNO_3$ (b) 8.9 M $NaNO_3$
(c) 0.089 M $NaNO_3$ (d) None of these

Q. 77. An experiment calls for the addition to a reaction vessel of 0.184 g of sodium hydroxide, NaOH, in aqueous solution. The number of milliliters of 0.150 *M* NaOH that should be added will be:
(a) 3.07 ml (b) 46.05 ml (c) 15.35 ml (d) 30.7 ml

Q. 78. A 1.000-L sample of polluted water was analyzed for lead(II) ion, Pb^{2+}, by adding and excess of sodium sulfate to it. The mass of lead(II) sulfate that precipitated was 229.8 mg. The mass of lead in a liter of the water is:
(a) 157.0 mg Pb $litre^{-1}$ (b) 15.7 mg Pb $litre^{-1}$
(c) 157.0 mg Pb $litre^{-1}$ (d) None of these

Q. 79. Consider the reaction of sulfuric acid, H_2SO_4, with sodium hydroxide, NaOH.

$$H_2SO_4(aq) + 2NaOH(aq) \rightarrow 2H_2O(l) + Na_2SO_4(aq)$$

Suppose a beaker contains 35.0 mL of 0.175 M H_2SO_4. How many millimeters of 0.250 *M* NaOH must be added to react completely with the sulfuric acid?

(a) 49.0 ml of 0.250 *M* NaOH **(b)** 49.0 ml of 0.250 *M* NaOH
(c) 35.0 ml of 0.250 *M* NaOH **(d)** None of these

Q. 80. Glucose, $C_6H_{12}O_6$, is the principal energy substance of the body and is transported to various cells by the blood.

Glucose, $C_6H_{12}O_6$, is a sugar that occurs in fruits. It is also known as "blood sugar" because it is found and is the body's main source of energy.

(i) The molarity of a solution containing 5.67 g of glucose dissolved in 25.2 g of water is:

(a) 12.5 *m* $C_6H_{12}O_6$ **(b)** 1.25 *m* $C_6H_{12}O_6$
(c) 0.125 *m* $C_6H_{12}O_6$ **(d)** 2.50 *m* $C_6H_{12}O_6$.

(ii) What are the mole fractions of glucose and water in a solution containing 5.67 g of glucose, $C_6H_{12}O_6$, dissolved in 25.2 g of water respectively.

(a) 0.978, 013 **(b)** 0.0220, 0.978 **(c)** 0.0220, 0.978 **(d)** 0.013, 0.978

(iii) An aqueous solution is 0.120 *m* glucose $C_6H_{12}O_6$. The mole fractions of each component in the solution are:

(a) 0.998, 0.02 **(b)** 0.002, 0.998 **(c)** 0.899, 0.101 **(d)** 0.101, 0.899

(iv) A solution is 0.150 mole fraction glucose, $C_6H_{12}O_6$, and 0.850 mole fraction water. The molality of glucose in the solution will be:

(a) 0.0098 m $C_6H_{12}O_6$ **(b)** 0.098 m $C_6H_{12}O_6$
(c) 0.9890 m $C_6H_{12}O_6$ **(d)** 9.80 m $C_6H_{12}O_6$

Q. 81. An aqueous solution is 0.273 *m* KCl. What will be the molar concentration of potassium chloride, KCl? The density of the solution 1.011×10^3 g/L.

(a) 0.0271 *M* KCl **(b)** 2.71 *M* KCl **(c)** 0.271 *M* KCl **(d)** 5.42 *M* KCl

Q. 82. An aqueous solution is 0.907 *M* $Pb(NO_3)_2$. What will be the molarity of lead nitrate, $Pb(NO_3)_2$, in this solution? The density of the solution is 1.252 g/mL.

(a) 3.00 m $Pb(NO_3)_2$ **(b)** 0.953 m $Pb(NO_3)_2$
(c) 0.853 m $Pb(NO_3)_2$ **(d)** None of these

Q. 83. A 50 ml H_2SO_4 when titrated with potassium hydroxide required 42.4 mL of 0.1354 MKOH the molarity of given sulphuric acid solution is:

(a) 0.047 *M* **(b)** 0.027 *M* **(c)** 0.037 *M* **(d)** 0.057 *M*

Chapter **5**

SOLUTIONS

KEY CONCEPT

Equations and relationships that you need to know:

- $$\text{mass percent solute} = \frac{\text{mass solute}}{\text{total mass solution}} \times 100\%$$

  $$\text{molality} = \frac{\text{moles solute}}{\text{kg solvent}}$$

  $$\text{molarity} = \frac{\text{moles solute}}{\text{liter solution}}$$

  In dilute aqueous solutions, molarity = molality

  $$M_1V_1 = M_2V_2$$

- extent of solubility
  1. nature of solute-solvent interactions
  2. temperature
  3. pressure of gaseous solute
- Henry's law: $C = kP$

  where

  P = partial pressure of gas solute over the solution (atm)

  C = concentration of dissolved gas (mole liter^{-1})

  k = Henry's constant (dependent on temperature, mole L^{-1} atm^{-1})
- $\Delta V_{aporization} = V_{vapor} - H_{liquid}$

  $\Delta H_{fusion} = H_{liquid} - H_{solid}$

  $\Delta H_{sublimation} = H_{fusion} + H_{vaporization}$

- Raolut's law: $P_1 = X_1 P_1^\circ$
  where
  P_1 = vapor pressure of solvent over the solution
  X_1 = mole fraction of solvent
  P_1° = vapor pressure of pure solvent
- Clausius – Clapeyron equation:

$$\text{In}(P_{vap}) = \frac{\Delta H_{vap}}{R}\left(\frac{1}{T}\right) + C$$ where C is a constant that is characteristic of a liquid

$$\log\left(\frac{P_2}{P_1}\right) = \frac{\Delta H_{vap}}{2.303\,R}\left(\frac{T_2 - T_1}{T_2 T_1}\right)$$

$$\ln\frac{P_1}{P_2} = \frac{\Delta H_{vap}}{R}\left(\frac{1}{T_2} - \frac{1}{T_1}\right)$$

- Intermolecular forces
  I. van der Waals
    (a) dipole-dipole $HBr—H_2S$
    (b) dipole-induced dipole $NH_3—C_6H_6$
    (c) dispersion He—He
  II. Ion-induced dipole $NO_3^-—I_2$
  III. Hydrogen bond $H_2O—H_2O$

- $\pi = \frac{\pi RT}{V} = iMRT$

  where
  π = osmotic pressure (atm)
  I = van't Hoff factor, should be 1 for all nonelectrolytes

$$= \frac{\text{actual number of particles in solution after dissociation}}{\text{number of formula units initially disolved in solution}}$$

$$\text{MW} = \frac{gRT}{\pi V}$$

- $\Delta T_b = i \cdot k_b \cdot m$ for water $k_b = 0.52°\text{C/m}$
  $\Delta T_f = i \cdot k_f \cdot m$ for water $k_f = 1.86°\text{C/m}$

PROBLEMS

Q. 1. A 10.0% sucrose solution has a density of 2.00 g mL^{-1}. What is the mass of sucrose dissolved in 1.00 liter of this solution?

(a) 1.00×10^2 g **(b)** 2.00×10^2 g **(c)** 5.00×10^2 g **(d)** 1.00×10^3 g

Q. 2. How many milliliters of a 50.0% (by mass) HNO_3 solution, with density of 2.00 grams per milliliter, are required to make 500 mL of a 2.00 M HNO_3 solution?

(a) 50.0 mL **(b)** 63.0 mL **(c)** 100. mL **(d)** 200. mL

Q. 3. What is the normality of a solution that contains 9.80 g of H_2SO_4 in 200. mL of solution, assuming 100% ionization $H_2SO_4(aq) \rightarrow 2H^+ (aq) + SO_4^{2-} (aq)$?

(a) 0.500 N **(b)** 1.00 N **(c)** 1.50 N **(d)** 2.00 N

Q. 4. Calculate the number of gram-equivalents of solute in 0.500 liter of a 3.00 N solution.

(a) 1.00 g-eq **(b)** 1.50 g-eq **(c)** 2.00 g-eq **(d)** 3.00 g-eq

Q. 5. What is the percentage (by mass) of NaCl (FW = 58.50) in a 10.0-molar solution?

(a) $\frac{10.0 \times 58.50}{1585}$ **(b)** $\frac{10.0 \times 58.50}{1000.00}$ **(c)** $\frac{2 \times 58.50 \times 10.0}{1000.00}$ **(d)** $\frac{10.0 \times 58.50}{100.00}$

Q. 6. When 5.92 grams of a nonvolatile, non-ionizing compound is dissolved in 186 grams of water, the freezing point (at normal pressure) of the resulting solution is –0.592°C. What is the molecular weight of the compound?

(a) 10.0 g mol^{-1} **(b)** 100. g mol^{-1} **(c)** 110. g mol^{-1} **(d)** 200. g mol^{-1}

Q. 7. Calculate the number of grams of glycerol, $C_3H_5(OH)_3$(MW = 92.1 g/mol), that must be dissolved in 520. grams of water to raise the boiling point to 102.00°C.

(a) 5.65 g **(b)** 92.0 g **(c)** 184 g **(d)** 194 h

Q. 8. In order to determine the molecular weight of a particular protein, 0.010 g of the protein was dissolved in water to make 2.93 mL of solution. The osmotic pressure was determined to be 0.821 torr at 293 K. What is the molecular weight of the protein?

(a) 3.8×10^3 g $mole^{-1}$ **(b)** 7.6×10^3 g $mole^{-1}$

(c) 3.8×10^4 g $mole^{-1}$ **(d)** 7.6×10^4 g $mole^{-1}$

Q. 9. A solution of NH_3 dissolved in water is 10.0 *m*. What is the mole fraction of water in the solution?

(a) 1.00/1.18 **(b)** 1.00/2.18 **(c)** 0.18/1.00 **(d)** 0.18/10.0

Q. 10. At 310 K and 1.00 atm of pressure, nitrogen dissolves in the blood at a solubility of 6.0×10^{-4} M. If a diver breathes compressed air where nitrogen gas constitutes 80, mole % of the gas mixture, and the total pressure at this depth is 3.0 atm, what is the concentration of nitrogen in the diver's blood.

(a) 1.4×10^{-4} M **(b)** 6.0×10^{-4} M **(c)** 1.0×10^{-3} M **(d)** 1.4×10^{-3} M

Q. 11. The vapor pressure of an ideal solution is 450. mm Hg. If the vapor pressure of the pure solvent is 1000. mm Hg. what is the mole fraction of the nonvolatile solute?

(a) 0.450 **(b)** 0.500 **(c)** 0.550 **(d)** 0.950

Q. 12. In an aqueous solution of strontium nitrate, the $[NO_3^-]$ is 0.080 M. This solution is correctly labeled as:

(a) 0.080 F $Sr(NO_3)_2$ **(b)** 0.160 F $Sr(NO_3)_2$
(c) 0.080 M Sr^{2+} **(d)** 0.040 F $Sr(NO_3)_2$

Q. 13. Which of the following solutions has the lowest freezing point?

(a) 0.10 m sucrose **(b)** 0.10 m $NiCl_2$
(c) 0.10 m $CuSO_4$ **(d)** 0.10 m NH_4NO_3

Q. 14. An aqueous solution of acetone, CH_3COCH_3, is 10.00% acetone by weight. What is the mole percentage of acetone in this solution?

(a) 3.332% **(b)** 5.000% **(c)** 10.00% **(d)** 11.11%

Q. 15. The density of an aqueous solution of acetone, CH_3COCH_3, that is 10.00% acetone by weight is 0.9867 $g \cdot mL^{-1}$ at 293 K. What is the molarity of acetone in this solution at 293 K.

(a) 0.1722 M **(b)** 0.9867 M **(c)** 1.699 M **(d)** 3.332 M

Q. 16. A 10.00-g sample of $CuSO_4 \cdot 5H_2O$ was dissolved in sufficient water to make 500.0 mL of solution. What is the molarity of Cu^{2+} in this solution?

(a) 0.04005 M **(b)** 0.06266 M **(c)** 0.8010 M **(d)** 0.1253 M

Q. 17. The freezing point of an aqueous solution of a nonelectrolyte is –0.14°C. The molality of this solution is

(a) 1.86 m **(b)** 1.00 m **(c)** 0.15 m **(d)** 0.075 m

Q. 18. 1,2-Benzanthracene is a yellow-brown nonvolatile solid with molecular weight 228.29 $g \cdot mol^{-1}$. An 18.2632-g sample of 1,2-bezanthracene is dissolved in 250.0 g of benzene, C_6H_6. What is the vapor pressure of this solution at 298 K, in millimeters of mercury, if the vapor pressure of pure benzene is 93.4 mm Hg at this temperature?

(a) 60.8 **(b)** 91.1 **(c)** 93.4 **(d)** 95.7

Q. 19. Assuming that all volumes are additive, how much water should be added to 25.00 mL of 6.00 F HNO_3 to prepare 0.500 F HNO_3?

(a) 350 mL **(b)** 325 mL **(c)** 300 mL **(d)** 275 mL

Q. 20. What is the $[NH_3^+]$ in 0.0520 F $(NH_4)_2SO_4$?

(a) 0.0260 M **(b)** 0.0520 M **(c)** 0.104 M **(d)** 0.520 M

Q. 21. An aqueous solution of ethanol, CH_3CH_2OH, that is 12.00% ethanol by weight, has a density of 0.9808 $g \cdot mL^{-1}$ at 293 K. What is the molality of ethanol in this solution?

(a) 0.05063 **(b)** 0.1200 **(c)** 2.555 **(d)** 2.960

Q. 22. The freezing point of a solution prepared by dissolving 20.5461 g of a nonvolatile nonelectrolyte with empirical formula $(C_3H_2)_n$ in 400.0 g of benzene is 4.33°C. The benzene used to prepare the solution froze at 5.48°C using the same thermometer. The correct molecular formula of this compound is

(a) C_3H_2 **(b)** $C_{18}H_{12}$ **(c)** C_9H_6 **(d)** $C_{15}H_{10}$

Q. 23. The boiling point of a 1.00 m solution of $CaCl_2$ should be elevated by

(a) exactly 0.51° **(b)** somewhat less than 1.02°
(c) exactly 1.02° **(d)** somewhat less than 1.53°

Q. 24. What volume of a 0.0500 *F* $K_2Cr_2O_7$ solution should you pipet into a 500.0 mL volumetric flask in order to dilute the more concentrated solution to 0.0200 F?

(a) 250 mL **(b)** 200 mL **(c)** 150 mL **(d)** 100 mL

Q. 25. The liquids benzene (molecular weight = 78.11) and toluene (molecular weight = 92.14) form ideal solutions. At 308 K the vapor pressure of benzene is 160.0 mm Hg, while that of toluene is 50.0 mmHg. If 64.05 g of benzene and 106.26 g of toluene are poured into a large container which is then covered and maintained at 308 K, what is the mole fraction of toluene *in vapor phase* when the system comes to equilibrium?

(a) 0.305 **(b)** 0.584 **(c)** 0.624 **(d)** 0.695

Q. 26. How many grams of potassium permanganate, $KMnO_4$, are needed to prepare 250.00 mL of a 0.1000 *F* solution?

(a) 3.951 g **(b)** 9.877 g **(c)** 15.80 g **(d)** 39.51 g

Q. 27. The concentration of a saturated solution of a certain polypeptide is 1.0×10^{-3} M at 298 K. The osmotic pressure of this solution, in millimeters of mercury, is

(a) 0.0245 **(b)** 0.760 **(c)** 18.6 **(d)** 24.5

Q. 28. A 20.00-mL portion of 0.100 *F* $Ba(NO_3)_2$ is mixed with 30.00 mL of 0.400 *F* NH_4NO_3. The $[NO_3^-]$ in the resulting solution is

(a) 0.250 *M* **(b)** 0.280 *M* **(c)** 0.320 *M* **(d)** 0.400 *M*

Q. 29. Of the following measurements, the one most suitable for the determination of the molecular weight of oxyhemoglobin, a molecule with a molecular weight of many thousands is

(a) the osmotic pressure

(b) the elevation of the boiling point

(c) the depression of the freezing point

(d) any of the previous four, us they are all equally good

Q. 30. A solution is prepared by dissolving 1.864 g of KCl and 8.293 g of K_2CO_3 in enough water to make the final volume 500.00 mL. What is the $[K^+]$ in the solution?

(a) 0.2900 *M* **(b)** 0.1200 *M* **(c)** 0.1450 *M* **(d)** 0.1700 *M*

Q. 31. At 308 K the vapor pressure of CS_2 is 512 mmHg. and of acetone, CH_3COCH_3, is 344 mmHg. A solution of CS_2 and acetone in which the mole fraction of CS_2 is 0.25 has a total vapor pressure of 600 mmHg. Which of the following statements about solutions of acetone and CS_2 is TRUE?

(a) A mixture of 100.00 mL of acetone and 100.00 mL of CS_2 has a volume of 200.00 mL.

(b) When acetone and CS_2 are mixed at 308 K, heat must be absorbed in order to produce a solution at 308 K.

(c) When acetone and CS_2 are mixed at 308 K, heat is released.

(d) Raoult's law is obeyed by both CS_2 and acetone for the solution in which the mole fraction of CS_2 is 0.25.

Q. 32. What is the maximum mass of $PbSO_4$ that can be precipitated by mixing 20.00 mL of 0.100 *F* $Pb(NO_3)_2$ with 30.00 mL of 0.1000 *F* Na_2SO_4?

(a) 0.1516 g **(b)** 0.3033 g **(c)** 0.6065 g **(d)** 0.7591 g

Q. 33. What is the $[Na^+]$ in a solution prepared by mixing 30.00 mL of 0.120 *F* NaCl with 70.00 mL of 0.150 *F* Na_2SO_4?

(a) 0.135 *M** **(b)** 0.141 *M* **(c)** 0.210 *M* **(d)** 0.246 *M*

Q. 34. Which of the following is weak electrolyte?

(a) CH_3COOH **(b)** CsOH **(c)** CH_3OH **(d)** $Ca(OH)_2$

Q. 35. At 273 K, 100.0 mL of a saturated solution of silver sulfate contains 0.57 g of dissolved silver sulfate. What is the silver ion concentration in this solution?

(a) 0.0057 *N* **(b)** 0.018 *M* **(c)** 0.028 *M* **(d)** 0.037 *M*

Q. 36. Which of the following substances is insoluble in water?

(a) $(NH_4)_2Cr_2O_7$ **(b)** C_6H_6 **(c)** CH_3CH_2OH **(d)** CH_3NH_2

Q. 37. What volume of 0.1250 *F* NHO_3 is required to completely neutralize 25.00 mL of 0.108 *F* $Ba(OH)_2$?

(a) 14.47 mL **(b)** 21.60 mL **(c)** 28.94 mL **(d)** 43.20 mL

Q. 38. In which of the following liquids would you expect the solubility of NaCl to be the smallest?

(a) HF **(b)** CH_3OH

(c) CCl_4 **(d)** H_2O

Q. 39. When 80.00 mL of 0.200 *F* HCl is mixed with 120.00 mL of 0.150 *F* KOH, the resultant solution is the same as solution of

(a) 0.160 *F* KCl and 0.0200 *F* HCl

(b) 0.0800 *F* KCl

(c) 0.0800 *F* KCl and 0.0100 *F* KOH

(d) 0.160 *F* KCl and 0.0100 *F* KOH

Q. 40. Equal volumes of 0.200 *F* $Ba(NO_3)_2$ and 0.120 *F* K_2CrO_4 are mixed. If the precipitation of yellow barium chromate is quantitative, what is the $[Ba^{2+}]$ in the final solution?

(a) 0.00 *M* **(b)** 0.0400 *M* **(c)** 0.0800 *M* **(d)** 0.100 *M*

Q. 41. In an experiment, 40.00 mL of 0.100 *F* $Pb(NO_3)_2$ was mixed with 60.00 mL of 0.300 *F* HCl, and a white precipitate of $PbCl_2$ was formed. The precipitate was collected by filtration, dried, and found to weigh 1.068 g. What was percentage yield of $PbCl_2$?

(a) 42.7% **(b)** 73.2% **(c)** 84.7% **(d)** 96.0%

Q. 42. A precipitate will be formed when an aqueous solution of hydrochloric acid is added to an aqueous solution of

(a) sodium nitrite **(b)** barium nitrate

(c) mercury(I) nitrate **(d)** zinc sulfate

Q. 43. Which of the following statements is correct?

(a) A 6.0 *M* solution of a weak electrolyte cannot be prepared.

(b) One can prepare a 1.0 *M* solution of any strong electrolyte.

(c) Proton-transfer reactions between acids and bases always go to completion.

(d) A significant number of strong electrolytes have very low solubilities in water.

Q. 44. Equal volumes of 0.200 *F* HCl and 0.400 *F* KOH are mixed. The concentrations of the principal ions in the resulting solution are.

(a) $[K^+] = 0.400\ M$ $[Cl^-] = 0.200\ M$ $[H^+(aq)] = 0.200\ M$
(b) $[K^+] = 0.200\ M$ $[Cl^-] = 0.100\ M$ $[OH^-] = 0.100\ M$
(c) $[K^+] = 0.100\ M$ $[Cl^-] = 0.100\ M$ $[OH^-] = 0.100\ M$
(d) $[K^+] = 0.200\ M$ $[Cl^-] = 0.100\ M$ $[OH^-] = 0.100\ M$

Q. 45. When solutions of iron (III) nitrate and sodium hydroxide are mixed, a red brown gelatinous precipitate of iron (III) hydroxide is formed. Which of the following mixtures will produce the largest precipitate of iron (III) hydroxide?

(a) 10.00 mL of 0.100 *F* $Fe(NO)_3$ and 90.00 mL of 0-.100 *F* NaOH
(b) 20.00 mL of 0.100 *F* $Fe(NO_3)_3$ and 80.00 mL of 0.100 *F* NaOH
(c) 25.00 mL of 0.100 *F* $Fe(NO_3)_3$ and 75.00 mL of 0.100 *F* NaOH
(d) 46.00 mL of 0.100 *F* $Fe(NO_3)_3$ and 54.00 mL of 0.100 *F* NaOH

Q. 46. In conductance titration of a 50.00-mL sample of a strontium hydroxide solution with 0.0452 FH_2SO_4, the minimum conductance was attained when 34.57 mL of acid had been added. The concentration of the original strontium hydroxide solution was

(a) 0.0313 *F* (b) 0.0625 *F* (c) 0.0654 *F* (d) 0.0904 *F*

Q. 47. What is the maximum number of milimoles of lanthanum oxalate. $La_2(C_2O_4)_3$, that can be precipitated by mixing 30.00 mL of 0.0860 *F* $La(NO_3)$ with 20.00 mL of 0.114 *F* $Na_2C_2O_4$?

(a) 0.760 (b) 1.29 (c) 1.52 (d) 2.28

Q. 48. An aqueous solution of barium chloride is added to an aqueous solution of each of the following reagents: sodium carbonate, silver nitrate, ammonium sulfate, and lead nitrate. In which cases will a precipitate from?

(a) Silver nitrate and lead nitrate.
(b) Silver nitrate and ammonium sulfate.
(c) A precipitate forms with each of the four reagents.
(d) Silver nitrate, lead nitrate, and ammonium sulfate.

Q. 49. Insoluble hydroxides can generally be dissolved in strong acids. Suppose you have solid samples of $Pb(OH)_2$, $Al(OH)_3$, $Sr(OH)_2$ and $Zn(OH)_2$. In which of the following strong acids could you dissolve all four solids to produce solely a clear solution?

(a) HCl (b) HNO_3 (c) HBr (d) H_2SO_4

Q. 50. Cinnamaldehyde C_9H_8O is a yellowish oily liquid with a strong odor of cinnamon. It is the chief constituent of cinnamon oil, which is obtained from the twigs and leaves of cinnamon trees grown in tropical areas. Cinnamon oil is used in the manufacture of food flavorings, perfumes, and cosmetics.

(i) Which would you choose as a solvent dissolve cinnamaldehyde—water or benzene?

(a) water (b) benzene (c) Both (d) None

(ii) A solution is prepared by dissolving 1.00 g of connamaldehyde in 50.00 g of benzene (C_6H_6). If the vapor pressure of pure benzene at 298 k is 95.1 mmHg, the partial pressure of benzene above this solution will be:

(a) 76.0 mmHg (b) 38.0 mmHg (c) 47.0 mmHg (d) 94.0 mmHg

(iii) The freezing point of benzene is 278.533 K and freezing point depression contort is 278.12 K. The freezing point of the solution prepared by dissolving 1.00 g of cinnamaldehyde in 50.00 g of benzene will be:

(a) 282.52 K (b) 279.306 K (c) 277.76 K (d) 273.773 K

(iv) If Cinnamaldehyde dissolves in water, to the extent of 1 part to about 700 parts of water, by mass. The osmotic pressure of a saturated solution of cinnamaldehyde in water at 298 K will be:

(a) 0.1 atm (b) 0.2 atm (c) 0.8 atm (d) 0.4 atm

Q. 51. The radiator of an automobile holds 10 litres of water. It is desired to lower the freezing point of water by 10 K. The quantity of ethylene glycol required is:
Given $K_f(H_2O) = 1.86$ K kg mol^{-1}

(a) 4.334 kg (b) 7.668 kg (c) 3.334 kg (d) 1.668 kg

Q. 52. The addition of 28.0 gm of a compound to 750 gm of tetrachloromethane, CCl_4, lowered the freezing point of the solvent by 5.4 K. The molar mass of the compound is: (Given: $K_f(CCl_4) = 30$ K kg mol^{-1})

(a) 20.74 gm mol^{-1} (b) 207.4 gm mol^{-1}
(c) 2074 gm mol^{-1} (d) 0.2074 gm mol^{-1}

Q. 53. The freezing point of an aqueous solution whose boiling point is 373 kg in terms of the ΔH_{vap} and ΔH_{fus} will be:

(a) $0.54 \dfrac{\Delta H_{fus}}{\Delta H_{vap}}$ (b) $0.45 \dfrac{\Delta H_{vap}}{\Delta H_{fus}}$ (c) $0.45 \dfrac{\Delta H_{fus}}{\Delta H_{vap}}$ (d) $0.54 \dfrac{\Delta H_{vap}}{\Delta H_{fus}}$

Q. 54. One litre aqueous solution of sucrose (molar mass = 342 g/mol) weighing 1015 gm is found to record an osmotic pressure of 4.82 atm at 298 K. The molality of the sucrose solution is:

Given $R = 0.082$ atm mol^{-1} K^{-1}.

(a) 0.28 m (b) 0.80 m (c) 0.208 (d) .82 m

Q. 55. The osmotic pressure of a urea solution is 500 mm of Hg at 283 K. The solution is diluted x times and its temperature raised to 298 K. It is now found that the osmotic pressure of the solution is reduced to 105.3 mm of Hg. The extent of dilution of the solution i.e. x is :

(a) 3 (b) 7 (c) 5 (d) 9

Q. 56. A 10% solution of sucrose ($C_{12}H_{22}O_{11}$) is isotomic with 1.754% solution of urea, the molecular mass of urea is:

(a) 30g mol^{-1} (b) 120g mol^{-1} (c) 90g mol^{-1} (d) 60g mol^{-1}

Q. 57. The osmotic pressure of a solution at 300 K is 3 atm. What will be the osmotic pressure, if the conc. is reduced to half and its temperature is raised to 400 K?

(a) 2 atm (b) 9 atm (c) 6 atm (d) 8 atm

Q. 58. The osmotic pressure at 300 K of solution containing 0.17 g of sucrose (MW = 342s), 0.18 gm glucose (MW = 180) and 0.06 of urea (MW = 60) in 50 mL of the solution is:

(a) 1.223 (b) 1.322 (c) 1.320 (d) 1.232

Q. 59. 0.25 gm of a polymer dissolved in 100 mL of benzene solution showed a rise of 2.4 mm when osmotic equilibrium to established at 310 K if the density of the solution is 0.88 g/cm^3, The molecular weigh of the polymer will be:

(a) 6.2×10^2 **(b)** 3.1×10^2 **(c)** 3.1×10^5 **(d)** 6.2×10^5

Q. 60. A solution is prepared by dissolving 0.131 of a substance in 25.4 g of water. The molality of the solution determined by freezing point depression found to be 0.056 m. The molecular weight of the substance is:

(a) 91 amu **(b)** 90 amu **(c)** 93 amu **(d)** 92 amu

Q. 61. Camphor is a white solid that melts at 452.5 K. It has been used to determine the molecular weights of organic compounds, because of its unusually large freezing point-depression constant (40 K/m), which allows ordinary thermometer to be used. The organic substance is dissolved in method, camphor, and then the milting point of the solution is determined. (a) A 1.07 mg sample of a compound was dissolved in 78.1 mg of camphor. The solution melted at 449 K. the molecular weight of the compound is:

(a) 164 amu **(b)** 160 mau **(c)** 168 amu **(d)** 162 amu

Q. 62. The formula for low-molecular weight starch is $(C_6H_{10}O_5)_n$, where n averages 2.00×10^2. When 0.798 g of starch is dissolved in 100.0 mL of water solution, the osmotic pressure at 298 K will be:

(a) 458 mmHg **(b)** 45.8 mmHg **(c)** 4.58 mmHg **(d)** 760 mmHg

Chapter **6**

THERMODYNAMICS

KEY CONCEPT

Equations and relationships that you need to know:

- **Endothermic reaction has + ΔH; $H_{products} > H_{reactants}$** (Heat absorbed)
  Exothermic has $-\Delta H$; $H_{products} < H_{reactants}$ (Heat released)

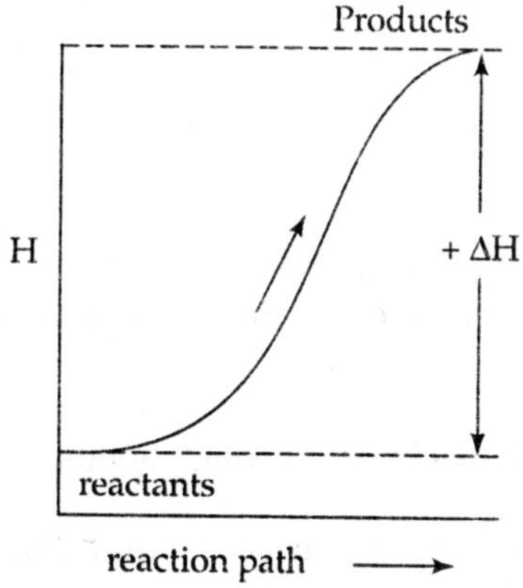

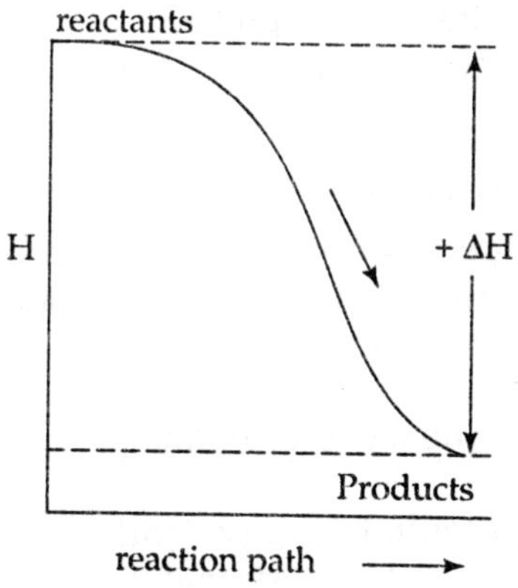

$\Delta E = \Delta H - RT\Delta n = \Delta H - P\Delta V = q + w$
$\Delta H = \Delta E + \Delta(PV) = q - w + P\Omega V$

$$w = -P\Delta V = -nRT\ln\frac{V_f}{V}$$

	Done on a system	Done by a system
Work, w	$+w$	$-w$
Heat, q	$+q$ (endothermic)	$-q$ (exothermic)

- "bomb" calorimeter
  $\Delta E = q_v$ (valid with constant volume: "bomb" calorimeter)
  $q_{reaction} = -(q_{water} + q_{bomb})$
  $q_{bomb} = C \cdot \Delta t$
  where C = calorimeter constant (heat capacity) in JK^{-1}.
- "coffee cup" calorimeter
  $\Delta H = q_p$ (valid with constant pressure: "coffee cup" calorimeter)
  $q = m \cdot c \cdot \Delta t$
  $$C_p = \frac{\Delta H}{\Delta t}$$
  $\Delta H_{reaction} = -q_{water}$
  $\Delta H° = \Sigma H^{\circ}_{f\,products} - \Sigma H^{\circ}_{f\,reactants} = q_p$
  where q_p = heat flow, constant p
- First law of thermodynamics
  $\Delta E = q + w = q_p - P\Delta V = \Delta H - P\Delta V$
  In any process, the total change in energy of the system. ΔE, is equal o the sum of the heat absorbed, q, and the work, w, done on the system.
- Second law of thermodynamics
  $\Delta S_{univ} = \Delta S_{sys} + \Delta S_{surr} > 0$ spontaneous
  $\Delta S_{univ} = \Delta S_{sys} + \Delta S_{surr} < 0$ nonspontaneous
  $\Delta S_{univ} = \Delta S_{sys} + \Delta S_{surr} = 0$ equilibrium
  The entropy of the universe increases in a spontaneous and process and remains unchanged in an equilibrium process.
- Third law of thermodynamics
  $S° = q_p/T$
  $\Delta S° = \Sigma S^{\circ}_{products} - \Sigma S^{\circ}_{reactants}$
  The entropy of a perfect crystalline substance is zero at absolute zero.
- Hess's law: If reaction (1) has ΔH_1 and reaction (2) ΔH_2 and reaction (1) + reaction (2) = reaction (3), then
  $\Delta H_3 = \Delta H_1 + \Delta H_2$
- Bond breaking: potential energy (enthalpy) of bond is increased; "strong" bonds → "weak" bonds: $\Delta H > 0$
  Bond forming: potential energy (enthalpy) of bond is decreased; bond distance is decreased; "weak" bonds → "strong" bonds; $\Delta H < 0$.
  ΔH + Σ bond energy (reactants) –Σ bonds energy (products) = total energy input – total energy released
- $\Delta H° = \Sigma \Delta H^{\circ}_{f\,products} - \Sigma \Delta H^{\circ}_{f\,reactants}$

$\Delta S^\circ = \Sigma\Delta S^\circ_{f\,\text{products}} - \Sigma\Delta S^\circ_{f\,\text{reactants}}$

$\Delta S_{\text{univ}} = \Delta S_{\text{sys}} + \Delta S_{\text{surr}}$

$\Delta S_{\text{surr}} = \dfrac{-\Delta H_{\text{sys}}}{T}$ if system and surroundings are at same T which derived to $\Delta G = \Delta H - T\Delta S$

$\Delta G^\circ = \Sigma\Delta G^\circ_{\text{products}} - \Sigma\Delta G^\circ_{\text{reactants}}$

$\Delta G^\circ < 0$: spontaneous in forward direction
$\Delta G^\circ > 0$: nonspontaneous in forward direction
$\Delta G^\circ = 0$: equilibrium
$\Delta G^\circ = \Delta H^\circ - T\Delta S^\circ$
$\Delta E = q + w$
$\Delta G^\circ = -RT \ln$ where $R = 8.314$ J/K ($K \cdot$ mole) $= -2.303\ T \log K = -nE^\circ$
if $K > 1$, then ΔG° is negative; products favored at equilibrium
if $K = 1$; then $\Delta G^\circ = 0$; reactants and products equally favored at equilibrium
if $K < 1$, then ΔG° is positive; reactants are favored at equilibrium
$\Delta G = \Delta G^\circ + RT. \ln Q$ where Q = reaction quotient
$\Delta G = \Delta G^\circ + 2.303\ RT \log Q$
$\Delta G^\circ = -n \mathscr{F} E^\circ$, where $\mathscr{F}$ = Faradays
$1\ \mathscr{F} = 96{,}500\ \text{J} \cdot \text{mole}^{-1} \cdot V^{-1}$
E° = standard cell potential, volts (V)
n = number of electrons in the half-reaction

$\Delta S = 2.303\ C_p \log \dfrac{T_2}{T_1}$

$\dfrac{\Delta H^\circ (T_2 - T_1)}{2.303\ R \cdot T_1 \cdot T_2} = \log \dfrac{K_n}{K_n}$, where $R = 8.314\text{J} \cdot \text{K}^{-1}$

- Temperature's Effects on Spontaneity: $\Delta G^\circ = \Delta H^\circ - T\Delta S^\circ$

Case	ΔH°	ΔS°	ΔG°
I	–	+	–spontaneous at all temperatures ex: $2H_2O_2(l) \rightarrow 2H_2O(l) + O_2(g)$
II	+	–	+ nonspontaneous at all temperatures ex: $3O_2(g) \rightarrow 2O_3(g)$
III	+	+	+ nonspontaneous at low temperatures – spontaneous at high temperatures ex: $H_2(g)$ + 12(g) $\rightarrow$ 2HI(g)
IV	–	–	– spontaneous at low temperatures + nonspontaneous at high temperatures ex: $NH_3(g) + HCl(g) + NH_4Cl(s)$

PROBLEMS

Q. 1. Given the following information:

Reaction (1): $H_2(g) + \frac{1}{2}(g) \rightarrow H_2O(l)$ $\Delta H° = -286$ kJ

Reaction (2): $CO_2(g) \rightarrow C(s)\ O_2(g)$ $\Delta H° = 394$ kJ

Reaction (3): $2CO_2(g) + H_2O(l) \rightarrow C_2H_2(g) + \frac{5}{2}O_2(g)$ $\Delta H° = 1300$ kJ

$\Delta H°$ for the reaction $C_2H_2(g) \rightarrow 2\,C(s) + H_2(g)$ will be:

(a) –226 kJ **(b)** –113 kJ **(c)** 113 grams **(d)** 226 kJ

Q. 2. Given the following heat of reaction and the bond energies listed in the table below: All numerical values are in kilo Joules per mole, and all substances are in the gas phase.

$CH_3CHO + H_2 = CH_3CH_2OH$ $\Delta H° = -71.06$ kJ mol^{-1}

The energy of the C=O bond will be:

Bond	*O—H*	*C—H*	*C—C*	*C—O*	*H—H*
Bond Energy (kJ mole^{-1})	464	414	347	351	435

(a) 330 kJ mol^{-1} **(b)** 656 kJ mol^{-1} **(c)** 723 kJ mol^{-1} **(d)** 794 kJ mol^{-1}

Q. 3. Given the following heats of formation:

Substance	H°
acetic acid	–501.6 kJ mol^{-1}
carbon dioxide	–397.1 kJ mol^{-1}
water	–250.8 kJ mol^{-1}

$\Delta H°$ of combustion for acetic acid (CH_3COOH) will be:

(a) +794.2 kJ mol^{-1} **(b)** –794.2 kJ mol^{-1}

(c) –188.0 kJ mol^{-1} **(d)** +188.0 kJ mol^{-1}

Q. 4. For $H_2C = CH_2(g) + H_2(g) \rightarrow H_3C - CH_3(g)$, predict the enthalpy given the following bond dissociation energies:

H—C, 413 kJ mol^{-1} H—H, 436 kJ mol^{-1}

C=C, 714 kJ mol^{-1} C—C, 348 kJ mol^{-1}

(a) –656 kJ mol^{-1} **(b)** –343 kJ mol^{-1}

(c) –289 kJ mol^{-1} **(d)** –124 kJ mol^{-1}

Q. 5. Given these two standard enthalpies of formation:

Reaction (1) $S(s) + O_2(g) \rightleftharpoons SO_2(g)$ $\Delta H° = -295$ kJ mol^{-1}

Reaction (2) $S(s) + 3/2O_2(g) \rightleftharpoons SO_3(g)$ $\Delta H° = -395$ kJ mol^{-1}

What is the reaction heat for $2SO_2(g) + O_2(g) \rightleftharpoons 2SO_3(g)$ under the same conditions?

(a) –1380 kJ mol^{-1} **(b)** –690 kJ mol^{-1}

(c) –295 kJ mol^{-1} **(d)** –200. kJ mol^{-1}

Q. 6. In expanding from 3.00 to 6.00 liters at a constant pressure of 2.00 atmospheres, a gas absorbs 418 J calories (101 Joules = 1 liter · atm). The change in energy, ΔE, for the gas is:

(a) +836 J **(b)** –418 J **(c)** –188 J **(d)** +188 J

Q. 7. A gas which initially occupies a volume of 6.00 liters at 4.00 atm is allowed to expand to a volume of 14.00 liters at a pressure of 1.00 atm. Calculate the value of work, w, done by the gas on the surroundings

(a) –8.0 L · atm **(b)** –7.0 L · atm

(c) 6.0 L · atm **(d)** 7.0 L · atm

Q. 8. The molar heat of sublimation for molecular iodine is 62.30 kJ/mol at 298 K and 1.00 atm. Calculate the ΔH in ($J \cdot mol^{-1}$) for the reaction

$$I_2\ (s) \rightarrow I_2\ (g) \qquad R = 8.314\ \text{kJ mole}^{-1} \cdot \text{K}^{-1}$$

(a) $\dfrac{62.30}{8.314 \cdot 298}$ **(b)** 62.30 – (8.314)(298)

(c) 62.30 + (8.314)(298) **(d)** 62.30(1000) + 1 · (8.314)(298)

Q. 9. Given the following standard molar entropies measured at 298 K and 1 atm pressure, calculate $\Delta S°$ in (J/K) for the reaction

$$2\ Al(s) + 3\ MgO(s) \rightarrow 3\ Mg(s) \rightarrow Al_2O_3(s)$$

$$Al(s) = 28.0\ J.K^{-1}$$
$$MgO(s) = 27.0\ J.K^{-1}$$
$$Mg(s) = 33.0\ J.K^{-1}$$
$$Al_2O_3(s) = 51.0\ J.K^{-1}$$

(a) –29.0 $J.K^{-1}$ **(b)** –13.0 $J.K^{-1}$ **(c)** 13.0 $J.K^{-1}$ **(d)** 69.0 $J.K^{-1}$

Q. 10. For the given reaction and the following information, calculate $\Delta G°$ at 298 K

$$2\ PbO(s) + 2\ SO_2(g) \rightarrow 2\ PbS(s) + 3\ O_2(g)$$

Species	*ΔH° (kJ mol⁻¹) at 298 K and 1 atm*	*ΔS° (kJ mol⁻¹) at 298 K and 1 atm*
PbO(s)	–218.0	70.0
$SO_2(g)$	–297.0	248.0
PbS(s)	–100.0	91.0
$O_2(g)$	—	205.0

(a) 273.0 kJ **(b)** 438.0 kJ **(c)** 634.0 kJ **(d)** 782.0 kJ

Q. 11. Given the information that follows, calculate the standard free energy change, $\Delta G°$, for the reaction

$$CH_4(g) + 2\ O_2(g) \rightarrow 2\ H_2O(l) + CO_2(g)$$

Species	*ΔH° (kJ mol⁻¹) at 298 K and 1 atm*	*ΔS° (kJ mol⁻¹) at 298 and 1 atm*
$CH_4(g)$	–75.00	–51.00
$O_2(g)$	0	0
$H_2O(l)$	–286.00	–237.00
$CO_2(g)$	–394.00	–394.00

(a) –919.00 kJ mol^{-1} **(b)** –817.00 kJ mol^{-1}

(c) –408.50 kJ mol^{-1} **(d)** 459.50 kJ mol^{-1}

Q. 12. Calculate the approximate standard free energy change for the ionization of hydrofluoric acid, HF (K_a = 1.0×10^{-3}), at 298 K

(a) –9.0 kJ (b) 17 kJ (c) 0.050 kJ (d) 4.0 kJ

Q. 13. Arrange the following reactions according to increasing ΔS°_{rxa} values.

1. $H_2O(g) \rightarrow H_2O(l)$
2. $2\ HCl(g) \rightarrow H_2(g) + Cl_2(g)$
3. $SiO_2(s) \rightarrow Si(s) + O_2(g)$

lowest → highest

(a) $\Delta S^\circ(1) < \Delta S^\circ(2) < \Delta S^\circ(3)$ (b) $\Delta S^\circ(2) < \Delta S^\circ(3) < \Delta S^\circ(1)$

(c) $\Delta S^\circ(3) < \Delta S^\circ(1) < \Delta S^\circ(2)$ (d) $\Delta S^\circ(1) < \Delta S^\circ(3) < \Delta S^\circ(2)$

Q. 14. Given for reaction $Hg(l) \rightarrow Hg(g)$ that $\Delta H^\circ = 63.0\ kJ \cdot mole^{-1}$ and $\Delta S^\circ = 100.\ J \cdot K^{-1}\ mole^{-1}$, calculate the normal boiling point of Hg.

(a) 6.30 K (b) 63.0 K (c) 630 K (d) 6.30×10^3 K

Q. 15. Given the following data:

$Fe_2O_3(s) + 3\ CO(g) \rightarrow 2\ Fe(s) + 3\ CO_2(g)$ $\Delta H^\circ = -27\ kJ\ mol^{-1}$

$3\ Fe_2O_3(s) + CO(g) \rightarrow 2\ Fe_3O_4(s) + CO_2(g)$ $\Delta H^\circ = -61\ kJ\ mol^{-1}$

$Fe_3O_4(s) + CO(g) \rightarrow 3\ FeO(s) + CO_2(g)$ $\Delta H^\circ = 38\ kJ\ mol^{-1}$

Species	ΔS° (J K^{-1} *mole*$^{-1}$)
$Fe_2O_3(s)$	87.0
$CO(g)$	190.0
$Fe(s)$	27.0
$CO_2(g)$	214.0
$Fe_3O_4(s)$	146.0
$FeO(s)$	61.0

Calculate the approximate ΔG° (at 298 K for the reaction)

$$FeO(s) + CO(g) \rightarrow Fe(s) + CO_2(g)$$

(a) –26 kJ mol^{-1} (b) –13 kJ mol^{-1} (c) 13 kJ mol^{-1} (d) 26 kJ mol^{-1}

Q. 16. Given the balanced equation

$$H_2(g) + F_2(g) \leftrightarrow 2\ HF(g) \Delta G^\circ = -546\ kJ\ mol^{-1}$$

Calculate ΔG if the pressures were changed from he standard 1 atm to the following and the temperature was changed to 773 K

$H_2(g)$ = 0.50 atm $F_2(g)$ = 2.00 atm $HF(g)$ = 1.00 atm

(a) –1090 kJ mol^{-1} (b) –546 kJ mol^{-1}

(c) –273 kJ mol^{-1} (d) 546 kJ mol^{-1}

Q. 17. If ΔH° and ΔS° are both negative, then ΔG° is

(a) always negative

(b) always positive

(c) positive at low temperatures and negative at high temperatures

(d) negative at low temperatures and positive at high temperatures

Q. 18. Determine the entropy change that takes place when 50.0 grams of compound x are heated from 323 K to 3230 K. It is found that 290.7 kilojoules of heat are absorbed.

(a) –461 JK^{-1} (b) 0.00 JK^{-1} (c) 230 JK^{-1} (d) 461 JK^{-1}

Q. 19. Ten moles of an ideal gas expand isothermally and reversibly from a volume of dm^3 to a volume of 50 dm^3 at 298 K. The maximum work done is:
(a) +57 kJ (b) –57 kJ (c) +75 kJ (d) –75 kJ

Q. 20. One mole of an ideal gas expands reversibly and isothermally from a pressure of 5 atm to 1 atm at 298 K. The largest mass which can be lifted through a height of 10 meters in the expansion is:
(a) 65.04 kg (b) 4.06 kg (c) 40.65 kg (d) 6.5 kg

Q. 21. Ten moles of an ideal gas initially at 298 K and 1 atm pressure absorbs 41.84 J of heat during a reversible isothermal expansion.
Its find volume will be:
(a) 141.5 dm^3 (b) 14.15 dm^3 (c) 1415 dm^3 (d) 1.42 dm^3

Q. 22. A gas expands isothermally against a constant external pressure of 4 atm from a volume of 5 dm^3 to a volume of 25 dm^3. In this process it absorbs 500 J of thermal energy from its surroundings.
(a) +760.53 J (b) –760.53 J (c) +7605.3 J (d) –7605.3 J

Q. 23. A given mass of a gas at 298 K is compressed reversibly and adiabatically to a pressure 5 times the initial value. The final temperature of the gas will be:
(Specific heat ratio of the gas (γ) = 1.66).
(a) 564.5 K (b) 291.5 K (c) 1128.0 K (d) 583.0 K

Q. 24. The entropy change ΔS when 5 moles of an ideal gas are heated from temperature of 500 K to a temperature of 1000 K at constant pressure is: (Assume that $C_p = 5/2$ R)
(a) 72.04 JK^{-1} (b) 144.08 JK^{-1} (c) 720.4 JK^{-1} (d) 216.12 JK^{-1}

Q. 25. Ten moles of an ideal gas expands reversibly from a volume of 1 dm^3 and temperature 298 K to a volume of 2 dm^3 and temperature 323 K. Assuming $C_V = \frac{3}{2}R$, then the entropy change for the process will be:
(a) 67.68 JK^{-1} (b) 135.36 JK^{-1} (c) 203.04 JK^{-1} (d) 33.84 JK^{-1}

Q. 26. Given bond energy for C—H = 418 kJ mol^{-1}; bonding = 498 kJ mol^{-1}; C=O = 741 kJ mol^{-1} and —O = 464 kJ mol^{-1} & C—C = 343 kJ mol^{-1}
Estimate whether CH_4 or C_8H_{18} would produce more heat *per gram* upon complete combustion.
(Assume all reactants and products to be in the gaseous state).
(a) Both CH_4 and C_8H_{18} (b) C_8H_{18}
(c) CH_4 (d) NONE

Q. 27. Given that ΔH_f(H) = 218 kJ mol^{-1}, the H—H bond energy in kJ mol^{-1} is:
(a) 218 kJ mol^{-1} (b) 109 kJ mol^{-1} (c) 872 kJ mol^{-1} (d) 436 kJ mol^{-1}

Q. 28. For which of the following equations is the enthalpy change at 298 K and 1 atm equal to ΔH_f° (CH_3OH), (l)?
(a) $CO(g) + 2H_2(g) \rightarrow CH_3OH\ (l)$
(b) $C(g) + 4H(g) + O(g) \rightarrow CH_3OH(l)$
(c) $C(gr) + H_2O(g) + H_2(g) \rightarrow CH_3OH\ (l)$
(d) $C(gr) + 2H_2(g) + \frac{1}{2}O_2(g) \rightarrow CH_3OH\ (l)$

Q. 29. For which of the following substances does $\Delta H_f^\circ = 0$?
(a) $Br_2(g)$ (b) $N(g)$ (c) $C(g)$ (d) $Ne(g)$

Q. 30. The molar heat capacity at 298 K should be close to 25 J · mol^{-1} for all of the following elements EXCEPT
(a) Pt (b) Kr (c) W (d) K

Q. 31. At 298 K, ΔH° for the reaction $Ca(s) + \frac{1}{2}O_2(g) \rightleftharpoons CaO(s)$ is –635.1 kJ · mol^{-1}. How many grams of Ca must combine with oxygen to liberate 100 kJ of heat?
(a) 88.30 (b) 63.11 (c) 40.08 (d) 25.45

Q. 32. The standard heat of combustion of propane is –2220. 1 kJ · mol^{-1}. The standard heat of vaporization of liquid water is 44.0 kJ · mol^{-1}. What is ΔH°, in kilojoules per mole, for the reaction.

$$C_3H_8(g) + 5\ O_2(g) \rightarrow 3CO_2(g) + 4H_2O(g)$$

(a) –2220.1 (b) –2176.1 (c) –2044.1 (d) –2088.1

Q. 33. The standard enthalpy change in kilojoules per mole 298 K for the reaction

$$CH_4(g) + 2\ O_2(g) \rightleftharpoons CO_2(g) + 2H_2O(g)$$

It the value of:
(a) –393.51 – 2(241.82) + 74.81 (b) –393.51 – 2(285.83) – 74.81
(c) –393.51 – 2(285.83) + 74.81 (d) –393.51 + 2(241.82) – 74.81

Q. 34. The heat of formation C_2H_6 is by definition the enthalpy change for the reaction
(a) $C_2H_4(g) + H_2(g) \rightarrow C_2H_6(g)$ (b) $2C(s) + 3H_2(g) \rightarrow C_2H_6(g)$
(c) $2C(s) + 6H(g) \rightarrow C_2H_6(g)$ (d) $2CH_4(g) \rightarrow C_2H_6(g) + H_2(g)$

Q. 35. In the balanced equation for the combustion of 1 mol of butane, $C_4H_{10}(g)$, the coefficient of oxygen is
(a) $\frac{5}{2}$ (b) $\frac{9}{2}$ (c) 5 (d) $\frac{13}{2}$

Q. 36. If 100.0 J of heat are added to 1.00 mol of Ne(g) at 303 K and constant pressure, how much will its temperature rise?
(a) 3.3° (b) 4.8° (c) 8.0° (d) 30.0°

Q. 37. Given the standard enthalpies at 298 K, in kilojoules per mole, for the following two reactions

$$Fe_2O_3(s) + \frac{3}{2}C(s) \rightleftharpoons \frac{3}{2}CO_2(g) + 2Fe(s) \qquad \Delta H^\circ = +234.1$$

$$C(s) + O_2(g) \rightleftharpoons CO_2(g) \qquad \Delta H^\circ = -393.5$$

The ΔH° value for $4Fe(s) + 3\ O_2(g) \rightleftharpoons 2Fe_2O_3(s)$ is calculated as

(a) $\frac{3}{2}(-393.5) - 234.1$ (b) $\frac{3}{2}(-393.5) + 234.1$
(c) –393.5 – 234.1 (d) 3(–393.5) – 2(234.1)

Q. 38. A gas absorbs 100 J of heat and is simultaneously compressed by a constant external pressure of 1.50 atm from 8.00 to 2.00 L in volume. What is ΔE in joules for the gas?
(a) –812 **(b)** 812 **(c)** –1012 **(d)** 1012

Q. 39. Given the following data:

$$N_2O(g) + 3H_2(g) \rightarrow N_2H_4(l) + H_2O(l) \qquad \Delta H^\circ = -317.25 \text{ kJ}$$

the standard heat of formation of liquid hydazine, $\Delta H_f^\circ(N_2H_4(l))$, can be calculated as:
(a) –317.25 + 82.05 + 285.83 **(b)** +317.25 + 82.05 – 285.83
(c) –317.25 – 82.05 + 241.82 **(d)** –317.25 + 82.05 + 241.82

Q. 40. How much heat, in joules must be added to 0.250 mol of Ar(g) to raise its temperature from 293 to 309 K at constant pressure?
(a) 50.0 **(b)** 83.2 **(c)** 187 **(d)** 220

Q. 41. The standard heat of combustion of solid boron is equal to
(a) $\Delta H_f^\circ(B_2O_3)$ **(b)** $-\Delta H_f^\circ(B_2O_3)$ **(c)** $\frac{1}{2}\Delta H_f^\circ(B_2O_3)$ **(d)** $-\frac{1}{2}\Delta H_f^\circ(B_2O_3)$

Q. 42. An adiabatic process is one in which there is no transfer of heat across the boundary between system and surroundings. For such a process
(a) $P_{ext}\Delta V = 0$ **(b)** $q = w$ **(c)** $\Delta E = w$ **(d)** $\Delta H = 0$

Q. 43. A bomb calorimeter has a heat capacity of 2.47 kJ · K^{-1}. The combustion of a 0.105-g sample of ethylene, $C_2H_4(g)$, in this bomb causes a temperature rise of 2.14°. ΔE, in kilojoules per mole of ethylene, for the combustion reaction is:
(a) -1.41×10^3 **(b)** –50.3 **(c)** –572 **(d)** –660

Q. 44. What is ΔH in kilojoules for the vaporization of 10.00 g of liquid bromine at 298 K and 1 atm?
(a) 1.934 **(b)** 6.996 **(c)** 13.99 **(d)** 30.91
(e) 223.6

Q. 45. A gas expands against a constant external pressure of 2.00 atm, increasing its volume by 3.40 L. Simultaneously, the system absorbs 400 J of heat from its surroundings. What is ΔE, in joules, for this gas?
(a) –689 **(b)** –289 **(c)** +400 **(d)** +289

Q. 46. The standard heat of combustion of Al(s) is –837.8 kJ · mol^{-1} at 298 K. If Al reacts with O_2 at 298 K, which of the following releases 1046 kJ of heat?
(a) The reaction of 0.624 mol of Al.
(b) The formation of 0.624 mol of Al_2O_3.
(c) The reaction of 0.312 mol of Al.
(d) The formation of 0.150 mol of Al_2O_3.

Q. 47. The C—Cl bond energy can be calculated from
(a) $\Delta H_f^\circ(CCl_4, l)$ only
(b) $\Delta H_f^\circ(CCl_4, l)$ and $D(Cl_2)$
(c) $\Delta H_f^\circ(CCl_4, l)$, $D(Cl_2)$, and $\Delta H_f^\circ(C, g)$
(d) $\Delta H_f^\circ(CCl_4, l)$, $D(Cl_2)$, $\Delta H_f^\circ(C, g)$, and $\Delta H_f^\circ(CCl_4)$

Q. 48. Using bond enthalpies (symbolized by ∈) and estimated value of $\Delta H°$ for the reaction $H_2C{=}CH_2(g) + H_2(g) \rightarrow C_3{-}H_3(g)$ would be

(a) $\in_{C=C} + \in_{H-H} - 2\in_{C-H} - \in_{C-C}$
(b) $\in_{C=C} + \in_{H-H} - 6\in_{C-H} - \in_{C-C}$
(c) $\in_{C=C} - \in_{H-H} + 4\in_{C-H} - \in_{C-C}$
(d) $\in_{C=C} + \in_{H-H} - 4\in_{C-H} + \in_{C-C}$

Q. 49. Using only the following data:

$$Fe_2O_3(s) + 3CO(g) \rightleftharpoons 2Fe(s) + 3CO_2(g) \qquad \Delta H° = -26.8 \text{ kJ}$$

$$Fe(s) + CO_2(g) \rightleftharpoons FeO(s) + CO(g) \qquad \Delta H° = +16.5 \text{ kJ}$$

The $\Delta H°$ value, in kilojoules, for the reaction

$$Fe_2O_3(s) + CO(g) \rightarrow 2FeO(s) + CO_2(g)$$

(a) –43.2 (b) –10.3 (c) +6.2 (d) +10.3

(*Note:* The compound FeO is nonstoichiometric, so that its composition varies with preparation, and thermochemical data involving FeO have large uncertainities.)

Q. 50. Use Tables for any data needed **for problems 50 to 67.**
For the reaction $2C(gr) + 3H_2(g) \rightarrow C_2H_6(g)$ the entropy change at 298 K in joules per mole per kelvin ($J \cdot mol^{-1} K^{-1}$) is

(a) 229.5 (b) 93.2 (c) –173.7 (d) –162.3

Q. 51. For the reaction $2SO_2(g) + O_2(g) \rightleftharpoons 2SO_3(g)$ at 298 K, $\Delta S°$ is –188.0 $J \cdot mol^{-1} K^{-1}$, and $\Delta H°$ is –197.7 $kJ \cdot mol^{-1}$. $\Delta G°$ for this reaction in kilojoules per mole is:

(a) –253.8 (b) –193.0 (c) –141.6 (d) 5.586×10^4

Q. 52. If an ideal gas is expanded at constant temperature,

(a) $\Delta E > 0$ and $\Delta S > 0$ (b) $\Delta E = 0$ and $\Delta S > 0$
(c) $\Delta E = 0$ and $\Delta S < 0$ (d) $\Delta E < 0$ and $\Delta S > 0$

Q. 53. If ΔG_f° (HI, g) = +1.7 kJ, what is the equilibrium constant at 298 K for $2HI(g) \rightleftharpoons H_2(g) + I_2(s)$?

(a) 3.9 (b) 2.0 (c) 0.69 (d) 0.50

Q. 54. For the reaction $H_2O(l) \rightleftharpoons H_2O(g)$ at 373 K and 1-atm pressure, which of the following is true?

(a) $\Delta H = 0$ (b) $\Delta S = 0$ (c) $\Delta H = \Delta E$ (d) $\Delta H = T\Delta S$

Q. 55. Calculate the equilibrium constant at 298 K for the reaction

$$3C(gr) + 4H_2(g) \rightleftharpoons C_3H_8(g)$$

(a) 3.0×10^9 (b) 1.3×10^4 (c) 1.0 (d) 7.6×10^{-5}

Q. 56. For the reaction $A(g) \rightleftharpoons B(g) + C(g)$, the equilibrium constant at a certain temperature is 2.0×10^{-4} atm. A mixture of the three gases is placed in a flask and the initial partial pressures are $P_A = 20$ atm, $P_B = 0.50$ atm, and $P_C = 1.0$ atm. Which of the following is true at the instant of mixing?

(a) $\Delta G° = 0$ (b) $\Delta G° < 0$ (c) $\Delta G > 0$ (d) $\Delta G < 0$

Q. 57. For which of the following reactions is $\Delta S° < 0$?

(a) $NH_4Cl(s) \rightleftharpoons NH_3(g) + HCl(g)$

(b) $2NO_2(g) \rightleftharpoons N_2(g) + 2\ O_2(g)$

(c) $2\ Ibr(g) \rightleftharpoons I_2(s) + Br_2(l)$

(d) $(NH_4)_2CO_3(s) \rightleftharpoons 2\ NH_3(g) + H_2O(g) + CO_2(g)$

Q. 58. Which of the following statements is true if the reaction, Q, is equal to 1?

(a) $\Delta G = 0$ (b) $\Delta G° = 0$ (c) $\Delta S° = 0$ (d) $\Delta G = \Delta G°$

Q. 59. For the reaction $I_2(s) + Cl_2(g) \rightleftharpoons 2\ ICl(g)$, $\Delta G°_{298} = -10.9$ kJ. If both $Cl_2(g)$ and IC1(g) are put in a flask at 298 K with some solid I_2, and the initial partial pressures are $P_{IC1} = 0.800$ atm and $P_{Cl_2} = 2.1 \times 10^{-3}$ atm, what is ΔG in kilojoules at the instant of

(a) 14.2×10^3 (b) 1.18×10^3 (c) 3.3 (d) –4.7

Q. 60. For the reaction $\frac{1}{2}N_2(g) + O_2(g) \rightleftharpoons NO_2(g)$, $\Delta H°_{298} = +33.2$ kJ and $\Delta G°_{298} = -60.9\ J \cdot mol^{-1}\ K^{-1}$. Using these data the equilibrium constant for the reaction at 298 K is

(a) 14.2×10^3 (b) 1.18×10^{-3} (c) 8×10^{-1} (d) 4×10^2

Q. 61. If a process is both endothermic and spontaneous then

(a) $\Delta S > 0$ (b) $\Delta S < 0$ (c) $\Delta H < 0$ (d) $\Delta G > 0$

Q. 62. Assuming HCl obeys Trouton's rule, if $\Delta H°_{vap} = 16.2\ kJ \cdot mol^{-1}$ at the normal boiling point, a good approximation to the normal boiling temperature is

(a) 159 K (b) 184 K (c) 273.18 K (d) 366 K

(e) 457 K

Q. 63. For the reaction $HBr(g) = \frac{1}{2}H_2(g) + \frac{1}{2}Br_2(g)$, $\Delta G°$ in kilojoules is:

(a) –51.86 (b) –50.29 (c) +53.43 (d) +55.00

Q. 64. Assuming $\Delta H°$ and $\Delta S°$ are independent of temperature between 298 K and 373 K for the reaction $N_2(g) + 3H_2(g) \rightleftharpoons 2NH_3(g)$, what is $\Delta G°$ in kilojoules at 373 K

(a) 27.97 (b) –18.14 (c) –26.24 (d) –32.98

Q. 65. Assuming ΔH_{vap} and ΔS_{vap} are approximately constant between 298 K and the normal boiling point of CCl_4, estimate the temperature of the normal boiling point.

(a) 273.35 K (b) 348 K (c) 373 K (d) 447 K

Q. 66. For a certain reaction $\Delta G°_{800} = -40.00$ kJ. K_{eq} at 800 K for this reaction is:

(a) 1.02×10^6 (b) 409 (c) 6.01 (d) 2.45×10^{-3}

Q. 67. For the gas-phase decomposition $PCl_5(g) \rightleftharpoons PCl_3(g) + Cl_2(g)$

(a) $\Delta H < 0$ and $\Delta S < 0$ (b) $\Delta H > 0$ and $\Delta S > 0$

(c) $\Delta H > 0$ and $\Delta S < 0$ (d) $\Delta H < 0$ and $\Delta S > 0$

Q. 68. Predict whether the reaction between hydrogen and chlorine is spontaneous at 298 K.

$H_2(g) + Cl(g) \rightarrow 2HCl(g)$ $\Delta H° = -185\ kJ\ mol^{-1}$

$\Delta S° = 141\ JK^{-1}\ mol^{-1}$

(a) Yes (b) No (c) Can't say

Q. 69. Predict whether the thermal dissociation of ammonium chloride will proceed spontaneously at 298 K.

$$NH_4Cl(s) \rightarrow NH_3(g) + HCl(g) \qquad \Delta H^o = +176 \text{ kJ mol}^{-1}$$
$$\Delta S^o = +284 \text{ JK}^{-1} \text{ mol}^{-1}$$

(a) Yes (b) No (c) Can't say

Q. 70. The temperature at which the reaction between nitrogen and oxygen becomes spontaneous is:

Given for, $N_2(g) + O_2(g) \rightarrow 2NO(g)$ $\Delta H = +180$ kJ mol^{-1}; $\Delta S = +25$ J K^{-1} mol^{-1}

(a) 7360 K (b) 7260 K (c) 740 K (d) 760 K

Q. 71. The enthalpy of formation of Sucrose (Given that the enthalpies of combustion of carbon, hydrogen and sucrose are –393.5, –286.2 and –5644.2 kJ mol^{-1}) is:

(a) +2226 KJ (b) +2118.7 (c) –2226 KJ (d) –2118.7

Q. 72. The enthalpy change for the reaction

$$CH_2{=}CH_2(g) + H_2(g) \rightarrow CH_3 - CH_3(g) \text{ is:}$$

Given the bond energies as:

(C=C) = 606 kJ, (C—C) = 346.9 kJ, (C—H) = 413.8 kJ, H – H = 434.7 kJ

(a) –134.5 kJ (b) –1040 kJ (c) +134.5 kJ (d) 1175 kJ

Q. 73. The space Shuttle Orbiter uses metylhydrazine as fuel, and this is oxidised by dinitrogen tetroxide to provide the energy for propulsion.

The standard enthalpy change $\Delta H_r^\ominus$ for the reaction:

$$4CH_3NHNH_2(l) + 5N_2O_4(l) \rightarrow 4CO_2(g) + 12H_2O(l) + 9N_2(g) \text{ is:}$$

(Given the standard enthalpy changes of formation as:

$\Delta H_f^\ominus(CH_3NHNH_2(l)) = +54$ kJ mol^{-1} $\quad$ $\Delta H_f^\ominus(N_2O_4(l)) = -20$ kJ mol^{-1}

$\Delta H_f^\ominus(CO_2(g)) = -393$ kJ mol^{-1} $\quad$ $\Delta H_f^\ominus(H_2O(l)) = -286$ kJ mol^{-1}

(a) –4882 kJ mo1^{-1} (b) +4882 kJ mo1^{-1}

(c) –5120 kJ mol^{-1} (d) +5120 kJ mol^{-1}

Q. 74. The standard enthalpy change of formation of methanol,

given that: $\Delta H_c^\ominus(CH_3OH(1)) = -726$ kJ mol^{-1}

$\Delta H_c^\ominus(C(s)) = -393$ kJ mol^{-1}, $\Delta H_c^\ominus(H_2(g))$ –286 kJ mol^{-1} will be:

(a) +239 kJ mo1^{-1} (b) +1691 kJ mo1^{-1}

(c) –239 kJ mol^{-1} (d) –1691 kJ mol^{-1}

Q. 75. The partial burning of coal in the presence of O_2 and H_2O produces a mixture of CO(g) and $H_2(g)$ called synthesis gas. Synthesis gas can be used to produce a number of important organic compounds (e.g., methyl alcohol), burned as a fuel, or converted to a gaseous mixture, called substitute natural gas (SNG), that is predominantly methane (CH_4).

A particular synthesis gas is found to consist of 55% CO(g), 33% $H_2(g)$, and 11% noncombustible gases (mostly CO_2), by volume.

If 1.00 L sample of the gaseous mixture is measured at STP, then

(i) The moles of $CO(g)$ and $H_2(g)$, present in the 1.00 L sample are:
 (a) 0.0245 mol $CO(g)$; 0.0147 H_2 **(b)** 0.0446 mol CO and H_2 each
 (c) 0.0147 mol C(g); 0.0245 $CO(g)$ **(d)** None of these

(ii) the enthalpy (heat) of combustion of 0.0147 mol H_2 is:

$$(\text{given } H_2(g) + \frac{1}{2}O_2(g) \rightarrow H_2O(l) \qquad \Delta H° = -258.8 \text{ kJ})$$

 (a) 4.2 kJ **(b)** –2.1 kJ **(c)** 2.1 kJ **(d)** –4.2 kJ

(iii) the enthalpy (heat) of combustion of 0.0245 mol $CO(g)$.
 (a) 6.93 kJ **(b)** –1.1 kJ **(c)** 1.1 kJ **(d)** –6.93 kJ

(iv) the heat of combustion per liter (STP) of the synthesis gas is:
 (a) –2.73 kJ **(b)** +11.3 kJ **(c)** –11.3 kJ **(d)** +2.73 kJ

(v) The temperature at which 1.00 kG of water at 298 K be heated with the heat given off in the combustion of 1.00 L (STP) of the synthesis gas will be:
 Given q_{water} = 11.13 kJ.
 (a) 300.7 K **(b)** 295.3 K **(c)** 275.6 K **(d)** 270.3 K

Q. 76. Hydrogen bonding is an important intermolecular force of attraction in hydrazine. In the Space Shuttle Columbia, a derivative of hydrazine (methyldrazine, CH_3NHNH_2) was used as a fuel.

(i) The heat of vaporization of hydrazine is:
 (given vap. Pressure at 298 K is 14.4 mm Hg)
 (a) 42.9 kJ mol^{-1} **(b)** 4.29 Kg mol^{-1}
 (c) 0.429 kJ mol^{-1} **(d)** 429 Kg mol^{-1}

(ii) The heat of subliation fo hydrazine at its melting point is:
 (given ΔH_{fus} Hydrazine = 12.66 KJ mol^{-1})
 (a) 55.56 **(b)** 16.95K **(c)** 13.089 **(d)** 441.66

Q. 77. Synthesis of methanol from CO and H_2 at 500, K is of great important because methanol can be used directly as a motor fuel, Converted to gasoline for fuel use, or converted to other organic compounds. The synthesis reaction is

$CO(g) + 2\,H_2(g) \rightarrow CH_3OH(g)$ ....(A)

(i) Would you expect the reaction (A) to take place to any significant extent at 298 K
 (a) NO **(b)** YES **(c)** DON'T KNOW.

(ii) The value of *K* for the methanol synthesis reaction 298 Kg is:
 (a) 2.2×10^4 **(b)** 0.22×10^4 **(c)** 4.4×10^4 **(d)** 8.8×10^4

(iii) The value of the thermodynamic equilibrium constant, *K*, at 500.*K* is:
 (given that K_c = 14.5)
 (a) 2.15×10^{-3} **(b)** 4.3×10^{-3} **(c)** 12.9×10^{-3} **(d)** 8.61×10^{-3}

(iv) If the values of *K* at 298.15 K and 500 K are 2×10^4 & 8.61×10^{-3} respectively, the standard enthalpy change in this reaction, $\Delta H°$ will be:
 (a) –14.8 kJ mol^{-1} **(b)** –91.1 kJ mol^{-1}
 (c) +91.1 kJ mol^{-1} **(d)** +14.8 kJ mol^{-1}

Q. 78. A system consists of one mole of an ideal gas. Below are given three steps each of the three different reversible cycles I, II, III that the system undergoes. The initial and final stages of each step are marked as *A*, *B* and *C*.

Cycle	Stage A	Stage B	Stage C
I	300 K, 24.6 dm^3	600 K, 24.6 dm^3	300 K, 12.3 dm^3
I	1.013 bar, 24.6 dm^3	1.013 bar, 49.2 dm^3	0.5065 bar, 49.2 dm^3
III	1.013 bar, 300 K	2.026 bar, 300 K	2.026 bar, 600 K

Indicate whether the steps *A* to *B*; *B* to *C*; *C* to *A* are isothermal/isobaric is isochoric.

Cycle	*A* to *B*	*B* to *C*	*C* to *A*
I	Isothermal	Isobaric	isochoric.
II	Isothermal	Isobaric	isochoric.
III	Isothermal	Isobaric	isochoric.

Q. 79. Two moles of hydrogen at standard conditions are compressed adiabatically to a volume of 10 litres. Value of γ for hydrogen is 1.41.

Given: $P_1 = 1$ atm $\quad P_2 = x$

$V_1 = 2\ (22.4) = 44.8$ Litres $\quad V_2 = 10$ litres

$T_1 = 273.2$ K $\quad T_2 = y$

The final pressure and temperature of the gas will be:

(a) 1010 K (b) 1515 K (c) 505 K (d) 2020 K

Q. 80. The work done when one mole of ammonia gas expands (i) Isothermally irreversibly at 1 atm and (ii) isothermally reversibly assuming an ideal gas behaviour. (Given: the temperature 300 K, initial volume 246×10^{-3} m^3 to 24.6×10^{-3} m^3. The value of a = 0.6799 Nm^4 mol^{-2} and $b = 0.0564 \times 10^{-3}$ m^3 mol^{-1}) will be:

(i) (a) 2642.8 J mol^{-1} (b) 2492.0 J mol^{-1}
(c) 2742.8 J mol^{-1} (d) 2242.8 J mol^{-1}

(ii) (a) 5744 J mol^{-1} (b) 4475 J mol^{-1}
(c) 4574 J mol^{-1} (d) 7454 J mol^{-1}

Q. 81. One mole of an ideal gas is expanded isothermally and reversibly at 300 K from a volume of 2.28 m^3 to 4.56 m^3. *q*, *w*, ΔE and ΔH are:

	q	*w*	ΔE	ΔH
(a)	−1.728 kJ	+1.728 kJ	0	0
(b)	+1.728 kJ	−1.728 kJ	0	0
(c)	0	+1.728 kJ	0	0
(d)	−1.728 kJ	0	0	0

Q. 82. The minimum work which must be done to compress 16 g of oxygen at 300 K from a pressure 1.01325×10^3 Nm^{-2} to a pressure of 1.01325×10^5 Nm^{-2} will be:

(a) 7445 Joules (b) 7544 Joules (c) 4754 Joules (d) 5744 Joules

Q. 83. The amount of work done by 2 mole of an ideal gas at 298 K in reversible isothermal expansion from 10 L to 20 L will be:

(a) 3435 Joules (b) 3345 Joules (c) 3543 Joules (d) 3354 Joules

Q. 84. About 3 dm^3 gastric juice are produced from body fluids such as blood in one day in the human body. The pH of blood is 7.4 at 310 K. The minimum work required to produce 3 dm^3 gastric Juice at the body temperature 310 K will be: (The pH of gastric Juice is taken as unity.)

(a) 1.14 kJ (b) 0.114 kJ (c) 11.4 kJ (d) 114 kJ

Q. 85. One mole of nitrogen, initially at 300 K, is expanded from 10 dm^3 to 20 dm^3. The maximum work done on the system:

(i) by isothermal expansion is

(a) –1.729 kJ **(b)** –0.58 kJ **(c)** +1.729 kJ **(d)** +0.58 kJ

(ii) by the adiabatic expansion, assuming ideal behaviour (Assume $C_V = 5/2\ R$). will be:

(a) –1.509 kJ **(b)** –1.095 kJ **(c)** +1.509 kJ **(d)** +1.095 kJ

Q. 86. In a adiabatic process 5000 J of work are performed on a system. In the non-adiabatic process by which the system returns to its original state 1000 J of heat are added to the system. The workdone during the non-adiabatic process is

(a) +6000 J **(b)** –6000 J **(c)** +4000 J **(d)** –4000 J

Q. 87. For a system undergoing as irreversible process:

(a) $\int Tds = Q$ **(b)** $\int Tds = Q$ **(c)** $\int Tds > Q$ **(d)** $\int Tds < Q$

Q. 88. Air is flowing in a main at constant pressure and temperature of 5 atm and 300 K respectively. A value connecting the main an insulated. A value connecting the main to an insulated evacuated vessel is slowly opened and air flows from the main into the vessel. The final temperature of air in the vessel is:

(a) 393 K **(b)** 120 K **(c)** 273 K **(d)** 200 K

(Assume air to behave as an ideal gas with the ratio of specific heats equal to 1.4. Neglect the heat transfer between the air in the main and the air in the vessel as also the changes in kinetic and potential energies.

Q. 89. In a reversible isothermal process undergone by an ideal gas:

(a) Heat transfer is zero **(b)** The change in internal energy is zero

(c) Work transfer is zero **(d)** Heat transfer is equal to work transfer.

Q. 90. In a steady flow reversible isobaric heating of an ideal gas from state 1 to state 2, the equations for heat transfer and work are:

(a) $Q = C_p\ (T_2 - T_1);\ W = P(V_2 - V_1)$

(b) $Q = C_p\ (T_2 - T_1);\ W = 0$

(c) $Q = \int_1^2 CpdT;\ W = 0$

(d) None of the above.

Q. 91. The equation $\Delta s = R \ln (p_1/p_2)$ determine the changes in entropy of an ideal gas for a reversible process which is:

(a) Adiabatic **(b)** Isothermal **(c)** Isobaric **(d)** All of the above.

Q. 92. The loss of available energy associated with the transfer of 1000 kJ of heat from a constant temperature system at 600°K to another at 400°K when the environment temperature is 300°K, is

(a) 500 kJ **(b)** 250 kJ **(c)** 166.67 kJ **(d)** 250 kJ

Q. 93. The enthalpy of a pure substance is defined as

(a) $H = U + mRT$ **(b)** $H = U - mRT$ **(c)** $H = U + pV$ **(d)** $H = U - pV$.

Q. 94. Or an ideal gas, internal energy is a function of

(a) temperature and pressure only **(b)** temperature and volume only

(c) pressure only **(d)** temperature only.

Q. 95. A given mass of gas at 273 K is compressed suddenly to a pressure 20 times the initial value. The final temperature of the gas will be:
(Take specific heat ratio for the as 1.42).
(a) 662.2 K (b) 62.2 K (c) 862.2 K (d) 262.2 K

Q. 96. Ether boils at 306.5 K at one atmospheric pressure. At what temperature will it boil at a pressure of 750 mm, given that the latent heat of vaporisation of ether is 369.86 Joules per gram (R = 8.314 J°K^{-1} mole^{-1}, mol. wt. of ether is 74).
(a) 296 K (b) 326 K (c) 306 K (d) 316 K

Q. 97. A reversible Carnot engine takes in heat from there reservoir at 1000°K, 800°K and 600°K. The work done by the engine per minute is 142 kJoule. The engine rejects 89 kJoule of heat per minute. If the heat applied to the reservoir at 1000°K is 60% of the heat supplied by a reservoir at 600°K, find the quantities of heat supplied by each reservoir respectively.
(a) 80, 60, 100 kJ (b) 100, 80, 60 kJ
(c) 60, 80, 100 kJ (d) 100, 60, 80 kJ.

Q. 98. Five moles of an Ideal gas expand reversibly from a temperature of 323 find pressure of 760 mm Hg to a temperature of 298 K and pressure of 380mm Hg. The change in entropy for the process, will be: $\left(\text{Given } C_V = \frac{3}{2}R\right)$
(a) 40.88 JK^{-1} (b) 81.76 JK^{-1} (c) 20.44 JK^{-1} (d) 163.52 JK^{-1}

Q. 99. One mole of an ideal gas expands reversibly from a volume of 1 dm^3 to 10 dm^3 at 298 K, the change in entropy is:
(a) 57.42 JK^{-1} (b) 19.14 JK^{-1} (c) 38.28 JK^{-1} (d) 76.56 JK^{-1}

Q. 100. The maximum efficiency of an engine operating between 373 K and 293 K is
(a) 43% (b) 21.5% (c) 10.75% (d) 86%

Q. 101. If the temperature of a cold reservoir of a reversible engine is 350 K, then the temperature of its hot reservoir working with an efficiency of 80% will be:
(a) 306 K (b) 412 K (c) 825 K (d) 1750 K

Q. 102. Five moles of an ideal gas expand against a constant external pressure of 5 atm from a volume of 1 dm^3 to a volume of 5 dm^3. The work done by the gas is:
(a) –20.26 J (b) –2026 J (c) –2.026 J (d) –202.6 J

Q. 103. The enthalpy of combustion of ethane at 298 K from the following reactions is:

$$C(s) + O_2(g) \rightarrow CO_2(g);\ \Delta H^\circ = -393.5 \text{ kJ mol}^{-1}$$

$$H_2(g) + \frac{7}{2}O_2(g) \rightarrow H_2O(l);\ \Delta H^\circ = -285.8$$

$$2\ C(s) + 3\ H_2(g) \rightarrow C_2H_6(g);\ \Delta H^\circ = -84.6 \text{ kJ mol}^{-1}$$

(a) +1739 kJ mol^{-1} (b) –1559.8 kJ mol^{-1}
(c) –1739 kJ mol^{-1} (d) +1559.8 kJ mol^{-1}

Q. 104. The enthalpy of neutralization of acetic acid by sodium hydroxide, from the following data will be:

(i) $H^+ (aq) + OH^- (aq) \rightarrow N_2O(l);\ \Delta H^\circ = -57.32 \text{ kJ mol}^{-1}$

(ii) $CH_3COOH(aq) \rightarrow CH_3COO–(aq) + H+ (aq)$;

$\Delta H^{\circ}_{\text{ionization}} = +189 \text{ kJ mol}^{-1}$

(a) +55.43 kJ mol^{-1} (b) –59.21 kJ mol^{-1}
(c) +59.21 kJ mol^{-1} (d) –55.43 kJ mol^{-1}

Q. 105. The heat of formation of methane at constant pressure at –74.85 kJ/mol. the heat of formation at constant volume will be.

(a) 72.37 kJ mol^{-1} (b) 77.85 kJ mol^{-1}
(c) –77.85 kJ mol^{-1} (d) –72.37 kJ mol^{-1}

Q. 106. Given the enthalpy of formation of
Methane –74.8 kJ mol^{-1}; the bond energy of hydrogen, 435.5 kJ mol^{-1}; and the enthalpy of sublimation of carbon, 717.7 kJ mol^{-1}, the C—H bond energy in methane, will be?

(a) 1248.9 kJ mol^{-1} (b) 790.0 kJ mol^{-1}
(c) 1663.5 kJ mol^{-1} (d) 416.0 kJ mol^{-1}

Q. 107. The enthalpy of formation of NH_3 from bond energy data given below is:

$$N—H = 389 \text{ kJ mol}^{-1}; N \equiv N = 946 \text{ kJ mol}^{-1}; H—H = 436 \text{ kJ mol}^{-1}$$

(a) +40 kJ mol^{-1} (b) –80 kJ mol^{-1} (c) –40 kJ mol^{-1} (d) +80 kJ mol^{-1}

Q. 108. How much heat is evolved when 907 kg of ammonia is produced according to the following equation? (Assume that the reaction occurs at constant pressure.)

$$N_2(g) + 3H_2(g) \rightarrow 2NH_3(g); \Delta H = -91.8 \text{ kJ}$$

(a) 2.45×10^6 kJ (b) 24.5×10^6 kJ (c) 0.245×10^6 kJ (d) 254×10^6 kJ

Q. 109. ΔG° for the combustion of 1 mol of ethanol, C_2H_5OH, at 298 K for

$$C_2H_5OH(l) + 3O_2(g) \rightarrow 2CO_2(g) + 3H_2(O) + 3H_2O(g) \text{ is}$$

Given: The standard free energies of formation as:

$$\Delta G^{\circ}_f\, C_2H_5OH(l) = -174.6 \text{ kJ}$$
$$\Delta G^{\circ}_f\, CO_2(g) = -394.4 \text{ kJ}$$
$$\Delta G^{\circ}_f\, H_2O(g) = -228.6 \text{ kJ}$$

(a) –1649.4 kJ (b) 1649.4 kJ (c) +1299.8 kJ (d) –1299.8 kJ

Q. 110. ΔH° and ΔG° for the reaction

$$2KClO_3(s) \rightarrow 2KCl(s) + 3O_2(g) + 3O_2(g) \text{ respectively are:}$$

(a) +91.0 kJ, +237.8 kJ (b) 91.0 kJ, –237.8 kJ
(c) –91.0 kJ, –237.8 kJ (d) –9.0 kJ, +237.8 kJ

(Values of ΔH°_f (in kJ mol^{-1}) are as follows: $KClO_3(s)$, –391.2; KCl(s), –436.7 and values of ΔG°_f (in kJ mol^{-1}) are as follows: $KClO_3(s)$, –289.9; KCl(s), –408.8

Q. 111. In a bomb calorimeter, The heat of a reaction involving a gas (here the heat of combustion of graphite) is conveniently determined in a sealed vessel called a bomb. The reaction is started by an ignition coil running through the graphic sample.

Suppose 0.562 g of graphic is placed in a calorimeter with an excess of oxygen in 298 K and 1 atm pressure. The graphic is ignited and it burns according to the equation.

$$C_{(graphite)} + O_2(g) \rightarrow CO_2(g)$$

On reaction, the calorimeter temperature rises from 298 k to 298.89 k. The heat capacity of the calorimeter and its contents, determined in a separate experiment 20.7 kJ K^{-1}; the heat of reaction at 298 k and one atm pressure will be:

(a) -39×10^2 kJ **(b)** -0.39×10^2 kJ

(c) -0.039×10^2 kJ **(d)** -3.9×10^2 kJ

Q. 112. Tungsten carbide is very hard and is used to make current tools and rock drills.)

$$W(s) + C(graphic) \rightarrow WC(s)$$

The enthalpy change for this reaction is difficult to measure directly, because the reaction occurs at 1573 K. However, the heats of combustion of the elements and of tungsten carbide can be measured easily.

$$2W(s) + 3O_2(g) \rightarrow 2WO_3(s); \Delta H = -1680.6 \text{ kJ} \quad (1)$$

$$C(graphic) + O_2(g) \rightarrow CO_2(g); \Delta H = -393.5 \text{ kJ} \quad (2)$$

$$2WC(s) + 5O_2(g) \rightarrow 2WO_3(s) + 2CO_2(g); \Delta H = -2391.6 \text{ kJ} \quad (3)$$

The enthalpy of reaction, ΔH, for the formation of tungsten carbide, WC, from the element is:

(a) –19 kJ **(b)** –38.0 kJ **(c)** +38.0 kJ **(d)** –2429.06 kJ

Q. 113. Large quantities of ammonia are used to prepare nitric acid. The first step consists of the catalytic oxidation of ammonia to nitric oxide, NO.

$$NH_3(g) + 5O_2(g) \xrightarrow{pt} 4NO(g) + 6H_2O(g)$$

The standard enthalpy change for this reaction is:

Given: ΔH_f° (NO) = 90.3 kJ, ΔH_f° (NH_3) = –45.9 kJ, ΔH_f°(H_2O) = –241.8 kJ

(a) –1995.6 kJ **(b)** –906 kJ **(c)** +906 kJ **(d)** +1995.6 kJ

Q. 114. The heat of vaporization, ΔH_{vap}, of carbon tetrachloride, CCl_4, at 298 K is 43.0 kJ mol^{-1}.

$$CCl_4(l) \rightarrow CCl_4(g); \Delta H_{vap} = 43.0 \text{ kJ mol}^{-1}$$

If 1 mol of liquid tetrachloride at 298 K has an entropy of 214 JK^{-1}, the entropy of 1 mol of the vapour in equilibrium wit the liquid at this temperature will be

(a) 144 J. mol^{-1}. K^{-1} **(b)** 288 J. mol^{-1}. K^{-1}

(c) 358 J. mol^{-1}. K^{-1} **(d)** 716 J. mol^{-1}. K^{-1}

Q. 115. The following equation represents the essential change that takes place during the fermentation of glucose (grape sugar) to ethanol (ethyl alcohol).

$$\underset{\text{Glucose}}{C_6H_{12}O_6(s)} \rightarrow \underset{\text{ethanol}}{2C_2H_5OH(l)} + 2CO_2(g)$$

Is S° positive or negative.

(a) Positive **(b)** Negative **(c)** No change

Q. 116. The change of entropy, ΔS°, at 298 K for the reaction in which urea is formed from NH_3 and CO_2, is:

$$2NH_3(g) + CO_2(g) \rightarrow NH_2CONH_2(aq) + H_2O(l)$$

The standard entropy of NH_2CONH_2 (aq) is 174 J. mol^{-1}. K^{-1}

$\Delta S°$ $NH_3(g)$ = 193 J. mol^{-1}

$CO_2(g)$ = 214 J. mol^{-1}. K^{-1} $H_2O(l)$ = 70 J. mol^{-1}. K^{-1}

(a) +356 J.K^{-1} **(b)** –356 J.K^{-1} **(c)** +844 J.K^{-1} **(d)** –844 J.K^{-1}

Q. 117. The standard free-energy change, $\Delta G°$, for the following reaction at 298 K.

Given, $N_2(g) + 3H_2(g) \rightarrow 2NH_3(g)$ is:

ΔH_f° $NH_3(g)$ = –45.9 KJ

$\Delta S°$ $NH_3(g)$ = 193 JK^{-1} $\Delta S°$ $H_2(g)$ = 120.6 JK^{-1}

$\Delta S°$ $H_2(g)$ = 130.6 JK^{-1}

(a) –91.8 KJ **(b)** –33.1 KJ **(c)** –197 KJ **(d)** –104 KJ

Chapter 7

CHEMICAL KINETICS

KEY CONCEPT

Equations and relationships this you need to know:

• *Reaction Order*	*0*	*1*	*2*
Rate Law	k	$k[A]$	$k[A]^2$
Integrated rate Law	$[A] = -kt + [A]_0$	$\ln[A] = -kt + \ln[A]_0$	$\frac{1}{[A]} - \frac{1}{[A]_0} = kt$
Relationship between Concentration and Time	$[A]_0 - [A] = kt$	$\log_{10} \frac{[A]_0}{[A]} = \frac{kt}{2.30_3}$	$\frac{1}{[A]} - \frac{1}{[A]_0} = kt$
Half-life	$\frac{[A]_0}{2k}$	0.693/k	$1/k[A]_0$
Linear Plot Slope	$[A]$ vs. t $-k$	log $[A]$ vs. t $-k$	a/$[A]$ vs. t k

- Zero order: $m = 0$ rate is independent of the concentration of the reactant. Doubling the concentration of the reactant does not affect the rate.
  First order: $m = 1$ rate is directly proportional to the concentration of the reactant. Doubling the concentration of the reactant doubles the rate.
  Second order: $m = 2$, rate is proportional to the square of the concentration of the reactant. Doubling the concentration of the reactant increases the rate by a factor of 4.

- rate = $\dfrac{\Delta \text{ concentration}}{\Delta \text{ time}}$
- Arhenius equation

  $k = Ae^{-E_a/RT}$

  where

  k = rate constant
  e = base of natural logarithm
  R = universal gas constant
  A = Arhenius constant
  E_a = activation energy
  T = temperature (K)

  $\ln k = \dfrac{-E_a}{RT} + \ln A$

  or $\log k = \log A - \dfrac{E_a}{2.30\,RT}$; Here, slope = $\dfrac{-E_a}{2.303\,R}$

  $\ln\dfrac{k_1}{k_2} = \dfrac{E_a}{R}\left(\dfrac{T_1 - T_2}{T_1 T_2}\right)$;

  or $\log\dfrac{k_2}{k_1} = \dfrac{E_a}{2.303\,R} \cdot \dfrac{(T_2 - T_1)}{(T_1 T_2)}$
- $\Delta E = \Sigma E_{\text{products}} - \Sigma E_{\text{reactant}}$
- collision theory: rate = $f \cdot Z$

  where

  Z = total number of collisions
  f = fraction of total number of collisions that occur at sufficiently high energy for reaction
  $Z = Z_0[A]^n[B]^m$

  where

  Z_0 = collision frequency where all reactants are at unit concentration
- $\Delta H = E_a - E'_a$

  where

  E_a = forward reaction activation energy
  E'_a = reverse reaction activation energy

PROBLEMS

Q. 1. Acetaldehyde, CH_3CHO, decomposes into machine gas and carbon monoxide gas. This is a second-order reaction. The rate of decomposition at 313 K is 0.10 mole liter^{-1} · sec^{-1} when the concentration of acetaldehyde is 0.010 mol · liter^{-1}. The rate of reaction when the concentration of acetaldehyde is 0.50 mole.liter^{-1} is:

(a) $2.3 \cdot 10^2$ mol · liter^{-1} · sec^{-1} **(b)** 0.50 mol · liter^{-1} · sec^{-1}
(c) 1.0×10^1 mol · liter^{-1} · sec^{-1} **(d)** $1.5 \cdot 10^2$ mol · liter^{-1} · sec^{-1}

Q. 2. The rate of the chemical reaction between substances A and B is found to follow the rate law

$$\text{rate} = k[A]^2[B]$$

where k is the rate constant. The concentration of A is reduced to half its original value. To make the reaction proceed at 50% of its original rate, the concentration of B should be

(a) decreased by ¼ **(b)** halved **(c)** kept constant **(d)** doubled

Q. 3. Which of the following charges will decrease the rate of collisions between gaseous molecules of type A and B in a closed container?

(a) Decrease the volume of the container.
(b) Increase the temperature of the system.
(c) Add A molecules.
(d) Take away B molecules.

Q. 4. For a certain decomposition reaction, the rate is 0.50 mol · liter^{-1} · sec when the concentration of the reactant is 0.10 M. If the reaction is second order, what will the new rate be when the concentration of the reactant is increased to 0.40 M?

(a) 0.50 mol · liter^{-1} · sec^{-1} **(b)** 1.0 mol · liter^{-1} · sec^{-1}
(c) 8.0 mol · liter^{-1} · sec^{-1} **(d)** 16 mol · liter^{-1} · sec^{-1}

Q. 5. The rate-determining step of a several-step reaction mechanism has been determined to be

$$3X(g) + 2Y(g) \rightarrow 4Z(g)$$

When 3.0 moles of gas X and 2.0 moles of gas Y are placed in a 5.0-liter vessel, the initial rate of the reaction is found to be 0.45 mol · liter^{-1} · min^{-1}. What is the rate constant for the reaction?

(a) $\dfrac{0.45}{\left(\frac{3.0}{5.0}\right)^3\left(\frac{2.0}{5.0}\right)^2}$ **(b)** $\dfrac{0.45}{(3.0)(2.0)}$ **(c)** $\dfrac{0.45}{\left(\frac{3.0}{5.0}\right)^2\left(\frac{2.0}{5.0}\right)^3}$ **(d)** $\dfrac{0.45}{\left(\frac{3.0}{5.0}\right)\left(\frac{2.0}{5.0}\right)}$

Q. 6. The reaction $2X + Y \rightarrow 3Z$ was studied and the following data were obtained:

Experiment	X	Y	*Rate (mol · liter^{-1} · sec^{-1})*
1	3.0	1.5	1.8
2	1.5	3.0	0.45
3	1.5	1.5	0.45

What is the proper rate expression?
(a) rate = $k[X][Y]$ **(b)** rate = $k[Y]^2$ **(c)** rate = $k[X]^2$ **(d)** rate = $k[X]^2[Y]$

Q. 7. The activation energy is represented by
(a) A **(b)** B **(c)** C **(d)** $B - A$

Q. 8. The enthalpy of the reaction is represented by
(a) $B - (C - A)$ **(b)** B **(c)** $C - A$ **(d)** $B - C$

Q. 9. Given the following reaction mechanism, express the rate of the overall reaction.

$$2A + 3C \rightarrow 4D + 5E \text{ slow}$$
$$A + 2D \rightleftharpoons 2A + 2E \text{ fast}$$
$$2E + C \rightarrow 3D + 2C \text{ fast}$$

(a) $k \cdot 2[A]^2 3[C]^3$ **(b)** $k \cdot [D]^4[E]^5/[A]^2[C]^3$
(c) $k \cdot [A]^2[C]^3$ **(d)** $k \cdot [A]^3[C]^2$

Q. 10. Referring to problem 9, express the overall reaction.
(a) $3C + 2D \rightarrow 2A + 4E$ **(b)** $3A + 3C \rightarrow 3D + 4E$
(c) $2A + 3C \rightarrow A + 4E$ **(d)** $A + 2C \rightarrow 5D + 5E$

Q. 11. If for first order reaction, the time required for 99.9% reaction to take place is x times the time required for the completion of half of the reaction. Then x is equal to:
(a) 8 **(b)** 10 **(c)** 6 **(c)** 12

Q. 12. A substance decomposes according to the first order rate law with a half-life period of 50 minute. The fraction of the sample which, remains after 100 minutes of the reaction will be:
(a) 0.50 **(b)** 0.30 **(c)** 0.75 **(d)** 0.25

Q. 13. If the decomposition of hydrogen peroxide in aqueous solution is an x order reaction, then from the following data x is
(a) Zero **(b)** Second **(c)** First **(d)** Third

Volume V in (ml)	25	20	15.7	12.5	9.6
Time in minutes	0	10	20	30	40

where V is number of mL of $KMnO_4$ required to decompose a definite volume of H_2O_2 solution.

Q. 14. The reaction $SO_2Cl_2 \rightarrow SO_2 + Cl_2$ is a first order gas reaction with $K = 2.2 \times 10^{-5}$ sec^{-1} at 593 K. The percentage of SO_2Cl_2 that decomposes on heating this gas for 90 minutes will be
(a) 11.74 **(b)** 11.80 **(c)** 11.26 **(d)** 11.20

Q. 15. A first order reaction is 40% complete in 50 minutes, the value of the rate constant and time taken to complete 80% of reaction will be respectively:
(a) 0.0201 min^{-1}, 157.8 min **(b)** 0.0201 min^{-1}, 187.5 min
(c) 0.0102 min^{-1}, 157.8 min **(d)** 0.0102 min^{-1}, 187.5 min

Q. 16. If the order of a reaction is n, and the time of half-completion for the reaction is found to charge from 50 sec to 25 sec, when the initial concentration is changed from 0.5 to 1 M, then n is:
(a) 2 **(b)** 3 **(c)** 1 **(d)** 0

Q. 17. The reaction 2 NO + O_2 follows the rate law $r = k[NO]^2 [O_2]$. If K = 2.0×10^{-6} s^{-1} mol^{-1} L^2, The rate of reaction when [NO] = 0.04 mol L^{-1} and $[O_2]$ = 0.2 mol^{-1} L^{-1} is:
(a) 1.6×10^{-8} mol^{-1} sec^{-1} **(b)** 3.2×10^{-9} mol^{-1} sec^{-1}
(c) 6.4×10^{-10} mol^{-} sec^{-1} **(d)** 4.8×10^{-10} mol^{-1} sec^{-1}

Q. 18. The effect of adding more $H_2(g)$ to a constant-volume equilibrium mixture of N_2, H_2, and NH_3 for the reaction; $N_2(g) + 3H_2(g) \rightleftharpoons 2NH_3(g)$ will be:
(a) Shifting equilibrium to left **(b)** Shifting equilibrium to right
(c) Increasing the amount of NH_3 **(d)** Decreasing the amount of NH_3

Q. 19. The effect on equilibrium in the calcination (decomposition) of limestone, $CaCO_3(s) \rightleftharpoons CaO(s)$, produced by adding a small quantity of $CaCO_3(s)$ is:
(a) Increase in $[CaCO_3]$ **(b)** Increase in [CaO(s)]
(c) Increase in $[CO_2(g)]$ **(d)** No effect

Q. 20. The reaction,

$$2SO_2(g) + O_2(g) \rightleftharpoons 2SO_3(g) \ \Delta H^\circ = -180 \text{ kJ is favoured at:}$$

(a) High temperature **(b)** Low temperature
(c) Zero°C **(d)** Zero Kelvin

Q. 21. For the reaction $N_2 + 3H_2 \rightleftharpoons 2NH_3$ + 22 kcal
the effect of this equilibrium of halving the volume by doubling the pressure is
(a) Shifting of equilibrium to right **(b)** Shifting of equilibrium of left
(c) No change **(d)** None of these

Q. 22. The effect of having the pressure by doubling the volume on the following system at 773 K is:

$$H_2(g) + I_2(g) \rightleftharpoons 2HI(g)$$

(a) Shifting the equilibrium to right once **(b)** Shifting the equilibrium to left once
(c) No change **(d)** Shifting of equilibrium to left twice

Q. 23. The effect on the following equilibrium by the addition of a Calalyst is:

$$\frac{1}{2}N_2 + O_2 \rightleftharpoons NO_2 + \text{heat}$$

(a) Shifting equilibrium to left once **(b)** Shifting equilibrium of right, once
(c) No change **(d)** Shifting of equilibrium to right, twice

Q. 24. Chlorine is added to equilibrium mixture of the system

Heat $+SO_2Cl_2 \rightleftharpoons SO_2 + Cl_2$ at constant volume. The temperature of the system will:
(a) increase **(b)** decrease **(c)** no change **(d)** none of the above

Q. 25. The value of K when the reaction goes farthest to completion is:
(a) k = 10^{-10} **(b)** K = 10^{-100} **(c)** K = 10^{10} **(d)** K = 1.

Q. 26. The equilibrium constant K_C for $A(g) \rightleftharpoons B(g)$ is 1.1. Which gas has a molar concentration greater than 1?
(a) *A* **(b)** *B* **(c)** Both *A* & *B* **(d)** None of these

Q. 27. Read the paragraph and answer the question (i) to (iii)

Peroxyacetyl nitrate (PAN) is an air pollutant produced in photochemical smog by the reaction of hydrocarbons, oxides of nitrogen, and sunlight.

It produces eye irritations and causes plant damage. PAN is unstable and dissociates into peroxyacetyl radicals and $NO_2(g)$. Its presence in polluted air, then is a kind of reservoir for NO_2 storage.

$$\underset{\text{PAN}}{CH_3\overset{\overset{O}{||}}{C}OONO_2} \longrightarrow \underset{\text{peroxyacetyl radical}}{CH_3\overset{\overset{O}{||}}{C}OO} + NO_2 \qquad \ldots (A)$$

(i) A sample of polluted air is analyzed for its PAN content, which is reported as molecules PAN per liter of air. From the following data at 298 K, the order of the PAN decomposition reaction (A) will be:

Time:	0.0	10.0	20.0	30.0	40.0	50.0	60.0	min
Molecules of PAN/L air:	5.0	4.0	3.2	2.6	2.1	1.7	1.3×10^{11}	

(a) Zero order **(b)** Third order
(c) Second order **(d)** First order

(ii) What is the value of the rate constant k for reaction (A) at 298 K
(a) 0.045 per min **(b)** 0.023 min.
(c) 0.069 per min **(d)** 0.092 per min

(iii) The activation energy (E_n) for the decomposition of PAN, if the half-life at 273 K is found to be 35 hours will be:
(a) 3.6×10^2 kJ **(b)** 1.2×10^2 kJ
(c) 1.2×10^5 kJ **(d)** 3.6×10^5 kJ

Q. 28. If concentration are measured in moles per liter and time in minutes the units for the rate of a third-order reaction are
(a) $mol \cdot L^{-1}\ min^{-1}$ **(b)** $L^2\ mol^{-2}\ min^{-1}$
(c) $L \cdot mol^{-1}\ min^{-1}$ **(d)** min^{-1}

Q. 29. If concentrations are measured in moles per liter and time in minutes, the units for the specific rate constant of a third-order reaction are
(a) $mol \cdot L^{-1}\ min^{-1}$ **(b)** $L^2\ mol^{-2}\ min^{-1}$
(c) $L \cdot mol^{-1}\ min^{-1}$ **(d)** $mol^2\ L^{-2}\ min^{-1}$

Q. 30. Which of the following statements about the order of a reaction is TRUE?
(a) The order of a reaction must be positive integer.
(b) A second-order reaction is also bimolecular.
(c) The order of a reaction can only be determined by experiment.
(d) The order of a reaction increased with increasing temperature.

Q. 31. In an experiment to study the reaction $A + 2B \rightarrow C + 2D$, the initial rate $-D[A\}dt$ at $t = 0$, was found to be $2.6 \times 10^{-2} \times 10^{-2}\ Ms^{-1}$. The value of $-d[B]/d$ at $t = 0$ in Ms^{-1} will be:
(a) 2.6×10^{-2} **(b)** 5.2×10^{-2} **(c)** 1.3×10^{-2} **(d)** 1.0×10^{-1}

Q. 32. The reaction $A(g) + 2B(g) \rightarrow C(g) + D(g)$ is an elementary process. In an experiment, the initial partial pressure of A and B are $P_A = 0.60$ atm and $P_a = 0.80$ atm. When $P_C = 0.20$ atm, the rate of the reaction, relative to the initial rate, is

(a) $\frac{1}{48}$ **(b)** $\frac{1}{24}$ **(c)** $\frac{1}{6}$ **(d)** $\frac{3}{4}$

Q. 33. For the reaction between chlorine and nitric oxide,

$$Cl_2(g) + 2NO(g) \rightarrow 2NOCl(g)$$

it is fond the doubling the concentration of both reactants increases the rate by a factor of 8. If only the concentration of Cl_2 is doubled, the rate increases by a factor of 2. The order of this reaction with respect to NO is

(a) 0 (b) 1 (c) 2 (d) 3

Following Information is for Use in Question 34 through 36

In an "iodine-clock" experiment, the time *t*, it takes for the blue color of the starch-iodine complex to appear is a measure of the initial rate of formation of I_3. The rate of the reaction.

$$3I^- + S_2O_4^{2-} \rightarrow I_3^- + 2SO_4^{2-}$$

was studied using the iodine-clock technique, and the following data were obtained:

Run Number	*$[I]_o$ (M)*	*$[S_2O_8^{2-}]_o$ (M)*	*r(s)*
1	0.0800	0.0400	44.0
2	0.0800	0.0800	22.1
3	0.1600	0.0200	43.9
4	0.0400	0.0400	88.0

Q. 34. The order of this reaction with respect to $S_2O_8^{2-}$ is

(a) 0 (b) 3 (c) 1 (d) 2

Q. 35. The overall order of this reaction is

(a) 0 (b) 3 (c) 1 (d) 2

Q. 36. If a run is made with $[I]_o$ – 0.120 *M* and $[S_2O_8^{2-}]_o$ = 0.0400 *M*, you would expect to see the blue color appear in approximately

(a) 29 s (b) 33 s (c) 58 s (d) 66 s

Q. 37. For a hypotetical reaction A + 2B → 3C + D. $d[C]/dt$ is equal to

(a) $-d[A]/dt$ (b) $-d[B]/dt$ (c) $+3d[A]/dt$ (d) $\frac{-3}{2}d[B]/dt$

Q. 38. Which of the following statement TRUE?

(a) Endothermic reactions have higher activation energies than exothermic reactions.

(b) The rate law for a reaction depends on the concentration of all reactants that appear in the stoichometric equation.

(c) The rate of a catalyzed reaction is independent of the concentration of the catalyst.

(d) The specific rate constant for a reaction is independent of the concentrations of the reacting species.

Q. 39. The first-order rate constat for the decomposition of N_2O_3 in CCl_4 solution is 6.2×10^{-4} at 318 K and 2.1×10^{-3} at 328 K. What is the value of the activation energy for this reaction in kilojoules per mole?

(a) 46 (b) 1.1 10^2 (c) 2.5×10^3 (d) 2.5 10^4

Q. 40. For the reaction A + 2 B → 2C, the rate law for formation of C is

(a) rate = $k[A][B]^2$ (b) impossible to state from the data given

(c) rate = $[C]^2/[A][B]^2$ (d) rate = $k[A]^2[B]$

Q. 41. The rate law for the dimerization of NO_2 is

$-d[NO_2]/dt = k[NO_2]^2$

Which of the following changes will change the value of the specific rate constant, k?

(a) Running the reaction in CCl_4 solution rather than in the gas phase.

(b) Doubling the volume of the container in which the reaction is occurring.

(c) Adding more O_2 to the reaction mixture.

(d) Adding more NO_2 to the reaction mixture.

Q. 42. For a reaction for which the activation energies of the forward and reverse directions are equal in value,

(a) the stoichiometry is the mechanism

(b) $\Delta H = 0$

(c) $\Delta S = 0$

(d) the order is zero

Following Information is for Use in Questions 43 through 45

The rate law for the hydrolysis of thioacetamide, CH_3CSNH_2

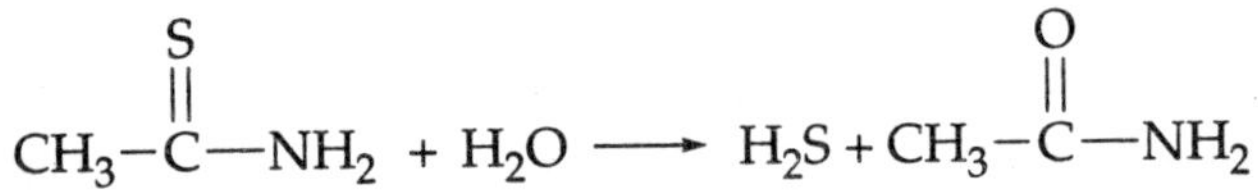
S
CH3—C—NH2 + H2O
H2S + CH3—C—NH2
O

is are = $k[H^+][TA]$, where TA = thioacetamide.

Q. 43. In which of the following solutions all at 298 K, will the rate of hydrolysis of thioacetamide (TA) be LEAST?

(a) 0.10 M in TA, 0.20 F in HNO_3

(b) 0.15 Min TA, 0.15 F in HCl

(c) 0.15 M in TA, 0.10 Fin CH_3COOH

(d) 0.060 M in TA, 0.10 F in HNO_3

Q. 44. If some sodium acetate is added to a solution that is 0.10 M in both TA and H^+(aq) at 298 K.

(a) The reaction rate decreases but k remains the same.

(b) Both the reaction rate and k decrease.

(c) The reaction rate increases but k remains the same.

(d) Both the reaction rate and k increase

Q. 45. If the temperature rate decreases but k remains the same.

(a) The reaction rate decreases but k remains the same.

(b) Both the reaction rate and k decrease.

(c) The reaction rate increases but k remains the same.

(d) Both the reaction rate and k increase.

Q. 46. An experiment is carried out to determine the reaction kinetics of this reaction:

$$BrO_3^-(aq) + 6I^-(aq) + 6H^+(aq) \rightarrow 3I_2(s) + 3H_2O(l) + Br^-(aq)$$

The results obtained are given below:

Mixture	*[BrO_3^-(aq)]/mol dm^{-3}*	*[I^-(aq)]/mol dm^{-3}*	*[H^+(aq)]/mol dm^{-3}*	*Relative initial rate*
A	0.10	0.60	0.60	1
B	0.20	0.60	0.60	2
C	0.20	1.20	0.60	4
D	0.20	0.60	1.20	8

the orders of reaction with respect to each reactant, respectively are:
(a) 1, 1, 1, 3 **(b)** 2, 1, 1, 4 **(c)** 1, 2, 1, 4 **(d)** 1, 1, 2, 4

Q. 47. The reaction between nitrogen monoxide and oxygen was investigated using the initial rate method. The stoichometric equation is

$$2NO(g) + O_2(g) \rightarrow 2NO_2(g)$$

The result obtained at a particular temperature are given

Experiment	*[NO(g)]/mol dm^{-3}*	*(O_2(g)]/mol dm^{-3}*	*Initial rate of production of NO_2(g)/mol dm^{-2} s^{-1}*
1	1.00×10^{-3}	3.00×10^{-3}	4.00 ´ 10^{-4}
2	1.00×10^{-3}	6.00×10^{-3}	8.00 ´ 10^{-4}
3	2.00×10^{-3}	3.00×10^{-3}	1.60 ´ 10^{-3}

the orders of the reaction with respect to NO(g) and O_2(g) and the value of the rate consistent respectively are:
(a) 1,2; 0.33×10^5 dm^5 mol^{-2} s^{-1} **(b)** 1,1; 0.33×10^{-5} dm^5 mol^{-2} s^{-1}
(c) 2,1; 1.33×10^5 dm^5 mol^{-2} s^{-1} **(d)** 2,2; 1.33×10^5 dm^5 mol^{-2} s^{-1}

Q. 48. Consider the reaction of nitrogen dioxide with fluorine to give nitryl fluoride, NO_2F.

$$2NO_2(g) + F_2(g) \rightarrow 2NO_2F(g)$$

The rate of formation of NO_2F related to the rate of reaction of flourine is:

(a) $\frac{\Delta[NO_2F]}{\Delta t} = \frac{-\Delta[F_2]}{\Delta t}$ **(b)** $\frac{1}{2}\frac{\Delta[NO_2F]}{\Delta t} = -\frac{[\Delta F_2]}{\Delta t}$

(c) None of these

Q. 49. The average rate of decomposition of N_2O_5, $-\Delta[N_2O_5]/\Delta t$, by the reaction

$$2N_2O_5(g) \rightarrow 4NO_2(g) + O_2(g)$$

during the time interval from t = 600 s to t = 1200 s (regard all time figures as significant), using the following data is:

Time	**[N_2O_5]**
600 s	1.24×10^{-2} *M*
1200 s	0.93×10^{-2} *M*

(a) 5.2×10^{-6} Ms^{-1} **(b)** 6.2×10^{-6} · Ms^{-1}
(c) 7.2×10^{-6} Ms^{-1} **(d)** 4.2×10^{-6} · Ms^{-1}

Q. 50. Iodide ion is oxidized in acidic solution to triodide ion, I_3^-, by hydrogen peroxide.

$$H_2O(aq) + 3I^-(aq) + 2H^+(aq) \rightarrow I_3^-(aq) + 2H_2O(l)$$

A series of four experiments was run at different concentrations, and the initial rates of I_3^- formation were determined (i) From these data, the reaction orders with respect to H_2O_2, I^-, and H^+ and, (ii) the rate constant is":

	Initial Concentrations (mol/L)			*Initial Rate*
	H_2O_2	I^-	H^+	[mol $L^{-1} \cdot s^{-1}$
Exp. 1	0.010	0.010	0.00050	1.15×10^{-6}
Exp. 2	0.020	0.010	0.00050	2.30×10^{-6}
Exp 3	0.010	0.020	0.00050	2.30×10^{-6}
Exp. 4	0.010	0.010	0.00100	1.15×10^{-6}

(a) 1, 1, 2; 2.4×10^{-2} L mol^{-1} s^{-1} **(b)** 1, O, 2; 4.8×10^{-2} L mol^{-1} s^{-1}
(c) 1, 2, O; 3.6×10^{-2} L mol^{-1} s^{-1} **(d)** 1, O, 2; 1.2×10^{-2}, L mol^{-1}, s^{-1}

Q. 51. Sulfuryl chloride SO_2Cl_2, is a colorless, corrosive liquid whose vapor decomposes in a first-order reaction to sulfur dioxide and chlorine

$$SO_2Cl_2(g) \rightarrow SO_2(g) + Cl_2(g)$$

At 593 K the rate constant is 2.20×10^{-5}/s. What is the half-life of SO_2Cl_2 vapor at this temperature and how long (in hours) would it take for 75% of the SO_2Cl_2 to decompress.

(a) 17.5 Hrs, 8.75 Hrs **(b)** 8.75 Hrs, 4.75 Hrs
(c) 8.75 Hrs; 17.5 Hrs **(d)** 4.75 Hrs, 8.75 Hrs

Q. 52. The rate constant for the formation of hydrogen iodide from the elements

$$H_2(g) + I_2(g) \rightarrow 2HI(g)$$

is 2.7×10^{-4} L · mol^{-1} · s^{-1} at 600 K and 3.5×10^{-3} L mol^{-1} · s^{-1} at 650 K. The activation energy E_o, and the rate constant is 700 K, will be:

(a) 3.2×10^5 J · mol^{-1}, 1.66×10^{-2} L · mol^{-1} · s^{-1}
(b) 1.66×10^5 J · mol^{-1}, 1.2×10^2 L · mol^{-1} · s^{-1}
(c) 3.2×10^5 J · mol^{-1}, 1.66×10^{-2} L · mol^{-1} · s^{-1}
(d) 1.66×10^5 J · mol^{-1}, 3.2×10^{-2} L · mol^{-1} · s^{-1}

Chapter **8**

EQUILIBRIUM

KEY CONCEPT

Equation and relationships that you need to know:

- $\Delta G° = -RT \ln K = -0191\ T \log\ K_p$
  $\Delta G° \ll 0$, K is large; $\Delta G° \gg 0$, K is very small
- $pK_a = \log K_a$
  $HA_{(aq)} + H_2O_{(l)} \rightleftharpoons H_3O^+_{(aq)} + A^-_{(aq)}$ (acid, base, conjugate acid, conjugate base); $K_a = \dfrac{[H_2O^+][A^-]}{[HA]} \rightleftharpoons \dfrac{[H^+][A^-]}{[HA]}$

  $B_{(aq)} + H_2O_{(l)} \rightleftharpoons BH^+_{(aq)} + OH^-_{(aq)}$ (acid, base, conjugate acid, conjugate base); $K_b = \dfrac{[BH^+][OH]}{[B]}$
- $K_a \times K_b = 1.0 \times 10^{-14} = K_w = [H^+][OH^-]$
  $K = K_1 \times K_2 \times \ldots$
- Q = ion product (or reaction quotient) = calculated as K_{ap} except using initial concentrations
  $Q > K_{sp}$, supersaturated solution, precipitate forms until such point that $Q = K_{sp}$
  $Q < K_{sp}$, unsaturated solution, no precipitate
  $Q = K_{sp}$, saturated solution, no precipitate forms
- Henderson-Hasselbalch: $\text{pH} = pK_a + \log \dfrac{[A^-]}{[HA]} = pK_a + \log \dfrac{[\text{base}]}{[\text{acid}]}$
- van't Hoff equation; $\ln \dfrac{K_2}{K_1} = -\dfrac{\Delta H°}{R}\left[\dfrac{1}{T_2} = \dfrac{1}{T_1}\right]$
- Clausius-Clapeyron equation: $\ln \dfrac{P_2^°}{P_1^°} = \dfrac{\Delta H^°_{vap}}{R}\left[\dfrac{1}{T_2} - \dfrac{1}{T_1}\right]$
- $K_p = K_e(RT)^{\Delta n}$
  where Δn = moles gaseous products—moles gaseous reactants

PROBLEMS

Q. 1. 6.0 moles of chlorine gas are placed in a 3.0-liter flask at 1250 K. At this temperature the chlorine molecules begin to dissociate into chlorine atoms. What is the value for K_c if 50% of the chlorine molecules dissociate when equilibrium has been achieved?

(a) 1.0 (b) 3.0 (c) 4.0 (d) 6.0

Q. 2. 6.00 moles of nitrogen gas and 6.00 moles of oxygen gas are placed in a 2.00-liter flask at 773 K and the mixture is allowed to reach equilibrium. What is the concentration, in moles per liter, of nitrogen monoxide at equilibrium if the equilibrium constant is found to be 4.00?

(a) 3.00 M (b) 6.00 M (c) 8.00 M (d) 10.0 M

Q. 3. Solid carbon reacts with carbon dioxide gas to produce carbon monoxide. At 1773 K the reaction is found to be at equilibrium with a K_p value of 0.50 and a total pressure of 3.5 atm. What is the proper expression for the partial pressure (in atmospheres) of the carbon dioxide.

(a) $\dfrac{-0.5 \pm \sqrt{(0.50)^2 - 4(1)(-3.5)]}}{2(1)}$ (b) $\dfrac{-0.5 \pm \sqrt{(0.50)^2 - 4(1)(-1.75)]}}{2(1)}$

(c) $\dfrac{-0.5 \pm \sqrt{(0.50) - 4(1)(-7.5)]}}{2(1)}$ (d) $\dfrac{-0.5 \pm \sqrt{(0.50)^2 - 2(1)(3.5)]}}{2(1)}$

For Problems 4 and 5, use the following information:

A student prepared a 1.00 M acetic acid solution ($HC_3H_3O_2$). The student found the pH of the solution to be 2.00.

Q. 4. What is the K_a value for the solution?

(a) 3.00×10^{-7} (b) 2.00×10^{-6} (c) 2.00×10^{-5} (d) 1.00×10^{-4}

Q. 5. What is the % dissociation of the acetic acid? (Use the 5% rule.)

(a) 0.05% (b) 1.00% (c) 1.50% (d) 2.00%

Q. 6. Given that the first, second, and third dissociation constants for H_2PO_4 are 7.0×10^{-3}, 6.0×10^{-8}, and 5.0×10^{-13}, respectively, calculate K for the overall reaction.

(a) 2.10×10^{-32} (b) 2.10×10^{-28} (c) 2.10×10^{-22} (d) 2.10×10^{-11}

Q. 7. A buffer is found to contain 0.35 M NH_3 ($K_b = 1.8 \cdot 10^{-5}$) and 0.20 M NH_4Cl. What would be the mathematical expression for K_b in terms of $[NH_4^+]$, $[OH^-]$, and $[NH_3]$?

(a) $1.8 \times 10^{-5} = \dfrac{(0.35 + x)(x)}{0.20}$ (b) $1.8 \times 10^{-5} = \dfrac{(0.35 - x)(0.20x)}{0.20}$

(c) $1.8 \times 10^{-5} = \dfrac{(0.20 + x)(x)}{(0.35 - x)}$ (d) $1.8 \times 10^{-5} = \dfrac{(0.20)(x)}{0.35}$

Q. 8. Copper(II) iodate has a solubility of 3.3×10^{-3} M at 298 K. Calculate its K_{sp} value.

(a) 1.4×10^{-7} (b) 1.1×10^{-5} (c) 3.3×10^{-3} (d) 5.1×10^{-1}

Q. 9. Lead iodide has a K_{sp} value of 1.08×10^{-7} at 293 K. Calculate its solubility at 293 K will be:

(a) 5.00×10^{-8} **(b)** 3.00×10^{-6} **(c)** 1.00×10^{-4} **(d)** 3.00×10^{-3}

Q. 10. Will a precipitate from when one mixes 75.0 mL of 0.050 M K_2CO_4 solution with 75.0 mL of 0.10 M $Sr(NO_3)_2$? K_{sp} for $SrCrO_4 = 3.6 \times 10^{-5}$.

(a) Yes, a precipitate will form, $Q > K_{sp}$
(b) Yes, a precipitate will form, $Q < K_{sp}$
(c) Yes, a precipitate will from, $Q = K_{sp}$
(d) No, a precipitate will not form, $Q > K_{sp}$

Q. 11. Some solid NH_2HS is placed in a flask containing 0.5 atm of NH_3. What would be the pressure of NH_3 and H_2S when equilibrium is reached?

$NH_4HS(s) \rightleftharpoons NH_3(g) + H_2S(g)$, $K_p = 0.11$

(a) 0.598 atm; 0.098 atm **(b)** 0.598; 0.089 atm
(c) 0.985 atm; 0.089 atm **(d)** 0.098; 0.589 atm

Q. 12. An equilibrium mixture at 300 K contains N_2O_4 and NO_2 at 0.28 and 1.1 atm pressure, respectively. If the volume of the container is doubled, the new equilibrium pressure of the two gases will be:

(a) 0.95; 0.064 atm **(b)** 0.64; 0.095 atm
(c) 0.064; 0.95 atm **(d)** 0.095; 0.64 atm.

Q. 13. At 319 K, K_p of the reaction,

$$N_2O_4(g) \rightleftharpoons 2NO_2(g)$$

is 0.66 Compute the percent dissociation of N_2O_4 at 319 K and a total pressure of 0.5 atm. The partial pressure of N_2O_4 and NO_2 at equilibrium will be:

(a) 0.661; 0.33 **(b)** 0.616; 0.33 **(c)** 0.166; 0.33 **(d)** 0.33; 0.616

Q. 14. An equilibrium mixture for the reaction,

$$2H_2S(g) \rightleftharpoons 2H_2(g) + S_2(g)$$

has 1 mole H_2S, 0.2 mole of H_2 and 0.8 mole of S_2 in a 2 litre vessel. K_c for the reactions:

(a) 0.0016 **(b)** 0.16 **(c)** 0.016 **(d)** 1.6

Q. 15. The percent dissociation of $H_2S(g)$ if 0.1 mole of H_2S is kept in 0.4 litre vessel at 1,000 K, for the reaction,

$$2HS(g) \rightleftharpoons 2H_2(g) + S_2(g) \text{ will be:}$$

(the value of K_c is 1.0×10^{-6})

(a) 2% **(b)** 1% **(c)** 1.5% **(d)** 2.5%

Q. 16. When sulphur in the form of S_8 is heated at 900 K the initial pressure of 1 atm falls by 29% of equilibrium. This is because of conversion of some S_8 to S_2. The value of equilibrium constant for this reaction is:

(a) 166.8 atm^3 **(b)** 266.8 atm^3 **(c)** 366.8 atm^3 **(d)** 66.8 atm^3

Q. 17. The composition of the equilibrium mixture ($Cl_2 \rightleftharpoons 2Cl$), which is attained at 1473 K, is determined by measuring the rate of effusion through a pin-hole. It is observed that at 1.80 mm Hg pressure, the mixture effuses 1.16 times as fast as Krypton effuses under the some conditions. Calculate the fraction of chlorine molecules dissociated in to atoms (At. wt. of *Kr* = 84).

(a) 0.713 **(b)** 0.317 **(c)** 0.173 **(d)** 0.137

Q. 18. One mole of H_2 two moles of I_2 and three moles of HI are injected in a 1 litre flask. What will be the concentration of H_2, I_2 and *HI* at equilibrium at 763 K. The equilibrium constant for the reaction at 763 K is 45.9.

(a) $[H_2] = 1.316$ mol · liter^{-1}
$[I_2] = 0.316$ mol · liter^{-1}
$[HI] = 4.368$ mol · liter^{-1}

(b) $[H_2] = 0.316$ mol · liter^{-1}
$[I_2] = 1.316$ mol · liter^{-1}
$[HI] = 4.368$ mol · liter^{-1}

(c) $[H_2] = 0.316$ mol · liter^{-1}
$[I_2] = 4.368$ mol · liter^{-1}
$[HI] = 1.316$ mol · liter^{-1}

(d) $[H_2] = 1.316$ mol · liter^{-1}
$[I_2] = 4.368$ mol · liter^{-1}
$[HI] = 0.316$ mol · liter^{-1}

Q. 19. A sample of air consisting of N_2 and O_2 was heated to 2500 K_5 until the equilibrium

$$N_2(g) + O_2(g) \rightleftharpoons 2NO(g)$$

ws established with an equilibrium constant $K_c = 2.1 \times 10^{-3}$. At equilibrium the mole % of NO was 1.8 the initial composition of air in mole fraction of N_2 and O_2. will be:

(a) $N_2 = 0.205$ mol
$O_2 = 0.795$ mol

(b) $N_2 = 0.795$ mol
$O_2 = 0.205$ mol

(c) $N_2 = 0.196$ mol
$O_2 = 0.786$ mol

(d) $N_2 = 0.786$ mol
$O_2 = 0.196$ mol

Q. 20. The concentration equilibrium constant, K_c, for the reaction;

$$N_2 + 3H_2 \rightleftharpoons 2NH_3$$

at 400 K is 0.5. K_P ($R = 0.082$ L atm K^{-1} mol^{-1}) will be:

(a) 2.65×10^{-14} · atm^{-2} **(b)** 3.65×10^{-4} · atm^{-2}

(c) 5.65×10^{-4} · atm^{-2} **(d)** 4.65×10^{-4} · atm^{-2}

Q. 21. At 700 K, hydrogen and bromine react to form hydrogen bromide. The value of equilibrium constant for this reaction is 5×10^8. The amount of H_2, Br_2 and HBr at equilibrium if a mixture of 0.6 mol of H_2 and 0.2 mol of Br_2 is heated to 700 K will be:

(a) $[H_2] = 0.6$ mol · L^{-1}
$[Br_2] = 0.4$ mol · L^{-1}
$[HBr] = 0.6$ mol · L^{-1}

(b) $[H_2] = 0.5$ mol · L^{-1}
$[Br_2] = 0.0$ mol · L^{-1}
$[HBr] = 0.5$ mol · L^{-1}

(c) $[H_2] = 0.4$ mol · L^{-1}
$[Br_2] = 0$ mol · L^{-1}
$[HBr] = 0.4$ mol · L^{-1}

(d) $[H_2] = 7.0$ mol · L^{-1}
$[Br_2] = 0$ mol · L^{-1}
$[HBr] = 0.7$ mol · L^{-1}

Q. 22. At 540 K, 0.10 mol of PCl_5 is heated in a 8.0 L flask. The pressure of the equilibrium mixture is found to be 1.0 atm. K_P and K_c for the reaction, respectively is:

(a) 2.78, 0.04 **(b)** 4.78; 0.06 **(c)** 3.78, 0.06 **(d)** 1.78; 0.04

Q. 23. At temperature *T*, a compound $AB_2(g)$ dissociates according to the reaction:

$$2AB_2(g) \rightleftharpoons 2AB(g) + B_2(g)$$

with a degree of dissociation s, which is small compared with unity. The expression for *x* in terms of the equilibrium constant, K_P and the total pressure *P* is:

(a) $\frac{2k_p}{p}$ (b) $\sqrt{\frac{2k_p}{p}}$ (c) $3\sqrt{2\frac{k_p}{p}}$ (d) $\sqrt{\frac{p}{2k_p}}$

Q. 24. At some temperature and under a pressure of 4 atm, PCl_5 is 10% dissociated. The pressure at which PCl_5 will be 20% dissociated, (temperature remaining same) will be

(a) 0.96 atm (b) 0.69 atm (c) 0.72 atm (d) 0.27 atm

Q. 25. The equilibrium constant K_c for the reaction

$$N_2(g) + 3H_2(g) \rightleftharpoons 2NH_3(g)$$

is 0.495 at 673 K and 2 atmospheric pressure. The K_P for the reaction is:

(a) 1.62×10^{-4} (b) 1.62×10^{-2} (c) 1.62×10^{-5} (d) 1.62×10^{-1}

Q. 26. A mixture of hydrogen and iodine are heated in 1 dm^3 vessel at 723 K in the mole ratio of 2:4 until equilibrium is attained, the mole of HI formed at equilibrium, given $K_c = 50$ at 723 K will be:

(a) 0.93 (b) 1.86 (c) 7.44 (d) 3.72

Q. 27. How much ester would be formed when 0.5 mole of ethyl alcohol is mixed with 1.0 mole of acetic acid?

Given $K_c = 4$ at 298 K

(a) 0.85 (b) 1.7 (c) 0.425 (d) 3.4

Q. 28. In the manufacture of ammonia the chief source of hydrogen gas for the reaction $N_2(g) + 3H_2(g) \rightleftharpoons 2NH_3(g)$ is from

the reforming of methane at elevated temperature i.e.

$$CH_4(g) + 2H_2)g) \rightleftharpoons CO_2(g) + 4H_2(g) \qquad \ldots\ldots (A)$$

If the following data are available

(a) $CO(g) + 2H_2O(g) \rightleftharpoons CO_2(g) + H_2(g)$

$\Delta H = -40.$ kJ $K_c = 1.4$ at 1000 K

(b) $CO(g) + 3H_2(g) \rightleftharpoons H_2O(g) + CH_4(g)$

$\Delta H = -230.$ kJ $K_C = 190.$ at 1000 K

Then answer the following (i) for (iv)

(i) The value of K_c for reaction (A) at 1000 K is:

(a) 5.5×10^{-3} (b) 4.7×10^{-3} (c) 5.5×10^{-4} (d) 7.4×10^{-3}

(ii) The value of K_P for reaction (A) 1000 K is

(a) 74 (b) 470 (c) 50 (d) 350

(iii) 1.00 mol each of $CH_4(g)$ and $H_2(g)$ are allowed to come to equilibrium in a 10.0 L vessel at 1000 K. The number of moles of $H_2(g)$ present at equilibrium will be

(a) 0.46 (b) 0.92 (c) 0.23 (d) 0.69

(iv) Match the qualitative effect that each of the following actions would have on the amount of $H_2(g)$ in the 10.0 L equilibrium mixture of part (iii) (a) add a catalyst; (b) add 24.0 g CH_3: (c) add 10.0 g CO_2; (d) remove 6.00 g H_2O; (e) raise the temperature to 1200 K; (f) transfer the mixture to a 15-L vessel; (g) add 10.0 g He(g) at constant volume.

(a) increase (b) increase (c) decrease (d) increase

(e) not change (f) not change (g) decrease

Q. 29. For which of the following reactions is the equilibrium constant called a solubility product?

(a) $CaC_2O_4(s) + 2H^+(aq) \rightleftharpoons Ca^{2+}(aq) + H_2C_2O_4$

(b) $La(OH)_3(s) \rightleftharpoons La^{3+}(aq) + 3OH^-(aq)$

(c) $Ag^+(aq) + Cl^-(aq) \rightleftharpoons AgCl(s)$

(d) $Cu(OH)_2(s) + 4NH_3(aq) \rightleftharpoons Cu(NH_3)_4^{2+}(aq) + 2OH^-(aq)$

Q. 30. For which of the following reactions are the numerical values of K_p and K_c the same?

(a) $2NOCl(g) \rightleftharpoons 2NO(g) + Cl_2(g)$

(b) $N_2 + 3H_2(g) \rightleftharpoons 2NH_3(g)$

(c) $H_2(g) + Cl_2(g) \rightleftharpoons 2HCl(g)$

(d) $H_2(g) + I_2(s) \rightleftharpoons 2HI(g)$

Q. 31. If the system $CaCO_3(s) \rightleftharpoons CaO(s) + CO_2(g)$ is at equilibrium at constant temperature, and the number of moles of CaO in the vessel is doubled.

(a) The reaction quotient Q, is doubled.

(b) The reaction quotient, Q, is halved

(c) The partial pressure of CO_2 in the vessel remains unchanged.

(d) The number of moles of $CaCO_3$ in the vessel remains unchanged.

Q. 32. At a given temperature the equilibrium constant for the reaction

$$PCl_5(g) \rightleftharpoons PCl_3(g) + Cl_2(g)$$

is 2.4×10^{-3}

What is the equilibrium constant for the reaction

$$PCl_3(g) + Cl_2(g) \rightleftharpoons PCl_5(g)$$

at the same temperature?

(a) 2.4×10^{-3} **(b)** -2.4×10^{-3} **(c)** 4.8×10^{-2} **(d)** 4.8×10^{2}

Q. 33. Which of the following statements about the reaction quotient, Q, is FALSE?

(a) The reaction quotient, Q, and the equilibrium constant K_{eq}, always have the same numerical value.

(b) The reaction quotient may sometimes be zero.

(c) The reaction quotient may be smaller than the equilibrium constant.

(d) The numerical value of the reaction quotient charges with time as a reaction proceeds.

Q. 34. At constant temperature, for which of the following system at equilibrium will, doubling the volume cause a shift to the right?

(a) $H_2(g) + Cl_2(g) \rightleftharpoons 2HCl(g)$

(b) $2CO(g) + O_2(g) \rightleftharpoons 2CO_2(g)$

(c) $N_2(g) + 2H_2(g) \rightleftharpoons 2NH_3(g)$

(d) $PCl_5(g) \rightleftharpoons PCl_3(g) + Cl_2(g)$

Q. 35. For the system $NH_4Cl(s) \rightleftharpoons NH_3(g) + HCl(g)$, if the concentration of NH_3 is doubled, the equilibrium constant will
(a) double
(b) increase, but by less than a factor of 2
(c) be halved
(d) remain the same

Q. 36. For which of the following reactions is the equilibrium constant called an acidity constant, K_a?
(a) $HCOOH(aq) + NH_3(aq) \rightleftharpoons HCOO^-(aq) + NH_4^+(aq)$
(b) $H_3O^+(aq) + OH^-(aq) \rightleftharpoons 2H_2O$
(c) $HONO(aq) + H_2O \rightleftharpoons ONO^-(aq) + H_3O^+(aq)$
(d) $CH_3COOH(aq) + OH^-(aq) \rightleftharpoons CH_3COO^-(aq) + H_2O$

Q. 37. If the pressure on an equilibrium mixture of the three gases NO, Cl_2, and NOCl,

$$2NO(g) + Cl_2(g) \rightleftharpoons 2NOCl(g)$$

is suddenly decreased by doubling the volume of the container at constant temperature, when the system returns to equilibrium
(a) The concentration of NOCl will have increased
(b) The value of the equilibrium constant K_c will have increased.
(c) The number of moles of Cl_2 will have increased.
(d) The number of moles of NOCl will have increased

Q. 38. For which of the following system at equilibrium at constant temperature will decreasing the volume cause no shift?
(a) $H_2(g) + CO_2(g) \rightleftharpoons H_2O(g) + CO(g)$
(b) $2NO(g) + O_2(g) \rightleftharpoons 2NO_2(g)$
(c) $2NO_2(g) \rightleftharpoons N_2O_4(g)$
(d) $SO_3(g) \rightleftharpoons SO_2(g) + \frac{1}{2}O_2(g)$

Q. 39. The decomposition of phosgene, $COCl_2(g) \rightleftharpoons CO(g) + Cl_2(g)$, is an endothermic process. Which of the following factors will cause the value of the equilibrium constant to increase?
(a) adding Cl_2
(b) adding He(g)
(c) decreasing the temperature
(d) none of the above

Q. 40. At a given temperature K, for the reaction

$$H_2(g) + CO_2(g) \rightleftharpoons H_2O(g) + CO(g)$$

is 3.24. If 0.800 mol of both $H_2(g)$ and $CO_2(g)$ are placed in a 1.00-1 container at this temperature, when the system comes to equilibrium the concentration of CO(g) will be
(a) 1.60 *M* (b) 0.800 *M* (c) 0.611 *M* (d) 0.514 *M*

Q. 41. For a reaction of the type $A(s) + 2B(g) \rightleftharpoons 2C(g)$ an equilibrium mixture consists of 3.0 mol of A, 0.80 mol of B and 0.40 mol of C in a 2.00-L flask. What is the value of K_c for this reaction?

(a) 5.0×10^{-1} **(b)** 3.3×10^{-1} **(c)** 2.5×10^{-1} **(d)** 1.7×10^{-1}

Q. 42. An equilibrium mixture of the reaction

$$2NO(g) + O_2(g) \rightleftharpoons 2NO_2(g)$$

contains 0.120 mol of NO_2. 0.080 mol of NO_2 and 0.640 mol of O_2 in a 4.00-L bulb at a temperature *T*. What is the value of K_c for this reaction at this temperature?

(a) 88 **(b)** 14 **(c)** 9.4 **(d)** 3.5

Q. 43. At a certain temperature the equilibrium constant, K_p, for the reaction

$$H_2(g) + I_2(g) \rightleftharpoons 2HI(g)$$

is 9.0. At this temperature, an equilibrium mixture of these three gases contained 0.60 mol of HI and 0.40 mol of H_2 in a 2.00-L flask. How many moles of I_2 were in this equilibrium mixture?

(a) 0.40 **(b)** 0.17 **(c)** 0.10 **(d)** 0.085

Q. 44. The K_c for the esterification reaction between ethanol and ethanoic acid is 4.0. The yield of the ester when 1.0 mole of ethanol and 1.0 mole of ethanoic acid are left to reach equilibrium will be:

(a) 0.33 moles **(b)** 0.18 moles **(c)** 0.67 moles **(d)** 0.82 moles

Q. 45. Carbon monoxide and hydrogen react according to the following equation:

$$CO(g) + 3H_2(g) \rightleftharpoons CH_4(g) + H_2(g)$$

When 1.0 mol CO and 3.0 mol H_2 are placed in a 10.00-L vessel at 1200 K and allowed to come to equilibrium, the mixture is found to contain 0.387 and H_2O. The molar composition of the equilibrium mixture will be:

(a) 0.613 mol CO, 1.839 mol H_2, 0.387 mol CH_4, 0.387 mol H_2O
(b) 0.387 mole CO, 0.387 mol H_2, 0.613 mol CH_4, 1.839 mol H_2O
(c) 0.387 mole CO, 0.387 mol H_2, 1.839 mol CH_4, 0.613 mol H_2O
(d) 1.839 mol CO, 0.613 mol H_3, 0.387.9 mol CH_4, 0.387 mol H_2O

Q. 46. Hydrogen HI, decomposes at moderate temperature according to the equation

$$2HI(g) \rightleftharpoons H_2(g) + I_2(g)$$

The amount of I_2 in the reaction mixture can be determined from the intensity of the violet color of I_2: the more intense the color, the more I_2 in the reaction vessel. When 4.00 mol HI was placed in a 5.00-L vessel at 731 K, the equilibrium mixture was found to contain 0.442 mol I_2. The value of K_c for the decomposition of HI at this temperature is:

(a) 0.623 **(b)** 0.0884 **(c)** 0.0402 **(d)** 0.0201

Q. 47. The reaction

$$CO(g) + H_2O(g) \rightleftharpoons CO_2(g) + H_2(g)$$

is used to increase the ratio of hydrogen in synthesis gas (mixture of CO and H_2). Suppose you start with 1.00 mol each of carbon monoxide and water in a 50.0-L vessel. The number of moles of each substance are in the equilibrium mixture at 1273 K are: (The equilibrium constant K_c are this temperature is 0.58.)

(a) 0.57 mol CO, 0.43 mol H_2O, 0.43 mol CO_2, 0.57 mol H_2

(b) 0.43 mol CO, 0.43 mol H_2O, 0.57 mol CO_2, 0.57 mol H_2
(c) 0.57 mol CO, 0.57 mol H_2O, 0.43 mol CO_2, 0.43 mol H_2
(d) 0.43 mol CO, 0.57 mol CO, 0.57 mol CO_2, 0.43 mol H_2

Q. 48. Hydrogen and iodine react according to the equation

$$H_2(g) + I_2(g) \rightleftharpoons 2HI(g)$$

Suppose 1.00 mol H_2 mol 2.00 mol I_2 are placed in a 1.00 L vessel. The no. of moles of substances are in the gaseous mixture when it comes to equilibrium at 731 K will be: (The equilibrium constant K_c at this temperature to 49.7).
(a) 0.7 mol H_2, 0.107 mol I_2, 0.186 mol HI
(b) 0.07 mol H_2, 1.07 mol I_2, 1.86 mol HI
(c) 1.07 mol I_2, 0.7 mol H_2, 0.07 mol HI
(d) 0.107 mol I_2, 0.186 mol H_2, 0.7 mol HI

Q. 49. Find the value of the equilibrium constant K at 298 K for the reaction

$$2NH_3(g) + CO_2(g) \rightleftharpoons NH_2CONH_2(aq) = H_2O(l)$$

The standard free energy change, $\Delta G°$, at 298 K equals –13.6 kJ
(a) 2.4×10^2 **(b)** 2.4×10^{-2} **(c)** 4.8×10^2 **(d)** 4.8×10^{-2}

Q. 50. The equilibrium constant K_{sp} = 298 K for the reaction

$$AgCl(s) \rightleftharpoons Ag^+(aq) + Cl^-(aq) \text{ is:}$$

(The standard free energies of formation are:

$$\Delta G_f^\circ \; AgCl(s) = -109.7, \; Ag^+(aq) = 77.1 \text{ kJ}, \; Cl^-(aq) = -132.2 \text{ kJ}$$

(a) 1.8×10^{-8} **(b)** 1.8×10^{-10} **(c)** 1.8×10^{-12} **(d)** 1.8×10^{-14}

Q. 51. (i) Given:

	$CaCO_3(s)$	$CaO(s)$	$CO_2(g)$
ΔH_f°	–1206.9	–365.1	–393.5 kJ
$S°$	92.9	38.2	213.7 J/K

$\Delta G°$ at 1273 K for the following reaction

$$CaCo_3(s) \rightleftharpoons C_2O(s) + CO_2(g) \text{ will be:}$$

Is this reaction spontaneous at 1273 K and 1 atm?
(a) +24.1 kJ, spontaneous **(b)** –24.1 kJ, non spontaneous
(c) –24.1 kJ spontaneous **(d)** +24.1 kJ, Non spontaneous

(ii) The value of K_p at 1273 K for this reaction and the partial pressure of CO_2 are:
(a) 9.75; 9.75 atm **(b)** 3.6; 3.6 atm
(c) 7.95; 7.95 atm **(d)** 6.7; 6.7 atm

Q. 52. For the reaction of XO with O_2 to form XO_2 the equilibrium constant at 398 K is 1.0×10^{-4} L mol^{-1}. If 1.0 mol XO and 2.0 mol of O_2 are placed in a 1.0-L vessel and allowed to come to equilibrium the equilibrium concentration of each of the species will be:
(a) $[XO_2] = 1.4 \times 10^{-2}$ mol · L^{-1}; [XO] = 1.0 mol · L^{-1}; $[O_2]$ = 2.0 mol · L^{-1}
(b) $[XO_2]$ = 2.0 mol · L^{-1}; [XO] = 2.0 mol · L^{-1}; $[O_2]$ = 1.0 mol · L^{-1}
(c) $[XO_2]$ = 1.0 mol · L^{-1}; [XO] = 1.4×10^{-2} mol · L^{-1}; $[O_2]$ = 2.0 mol · L^{-1}
(d) $[XO_2]$ = 2.0 mol · L^{-1}; [XO] = 1 mol · L^{-1}; $[O_2] = 1.4 \times 10^{-2}$ mol · L^{-1}

Q. 53. A theoretically computed equilibrium constant for the polymerization of formaladehyde, HCHO, to glucose $C_6H_{12}O_6$, in aqueous solution is 6×10^{22}.

$$6HCHO \rightarrow C_6H_{12}O_6$$

If a 1.0 M solution of glucose were to reach dissociation equilibrium with respect is the above equation the concentration of formaldehyde in the solution is:

(a) 1.6×10^{22} M
(b) 1.6×10^{-4} M
(c) 16×10^{-8} M
(d) 0.16×10^{-14} M

Q. 54. Which of the following statements is correct for a reversible process in a state of equilibrium?

(a) $\Delta G = -2.30\ RT \log K$
(b) $\Delta G = 2.30\ RT \log K$
(c) $\Delta G° = -2.30\ RT \log K$
(d) $\Delta G° = 2.30\ RT \log K$

Q. 55. The value of the equilibrium constant for the following reaction at 300 K and constant pressure

$$A(g) + B(g) \rightleftharpoons C(g) + D(g) + E(g)$$

using the ideal gas law and the following data is:

$$\Delta E° = -376 \text{ KJ } \Delta S° = 408 \text{ KJ}^{-1}$$

(a) $K = 0.3 \times 10^{86}$
(b) $K = 0.3 \times 10^{85}$
(c) $K = 3 \times 10^{86}$
(d) $K = 3 \times 10^{85}$

Q. 56. The following solutions are acidic, basic, or pH neutral: (i) NaCN(aq); (ii) KCl(aq); (iii) NH_4CN(aq). All three together have one correct property in column:

		(a)	(b)	(c)	(d)
NaCN(aq)	—	basic	neutral	acidic	basic
Kcl (aq)	—	acidic	basic	basic	neutral
NH_4CN(aq)	—	basic	acidic	neutral	basic

Q. 57. For the reaction $CaCO_3(g) \rightleftharpoons CaO(s) + CO_2(g)$, the value pressure of CO_2 in equilibrium with a mixture of $CaCO_3$ and CaO at 298 K is:

Given that

$$\Delta G_f^\circ (CO_2) = -394 \text{ KJ}; \Delta G_f^\circ\ CaCO_3 = -1127 \text{ KJ and } \Delta G_f^\circ(EaO) = -603 \text{ KJ}$$

(a) 1.6×10^{-23} atm
(b) 16×10^{-23} atm
(c) 0.16×10^{-25} atm
(d) -22.8 atm

Chapter 9

IONIC EQUILIBRIA

KEY CONCEPT

Equations and relationships that you need to know:

- $pH = -\log[H^+]$

$$pOH = -\log[OH^-]$$
$$K_w = [H^+][OH^-] = 10^{-14} = K_a \cdot K_b$$
$$pK_w = pH + pOH = 14(\text{exact})$$

$$K_a = \frac{[H^+][A^-]}{[HA]} \text{ for the equation } HA(aq) = H(aq)^+ + A(aq)^-$$

$$pK_a = -\log K_a$$

$$PH = pK_a + \left[\frac{\text{conjugate base}}{\text{conjugate acid}}\right] = pK_a + \log\frac{[A^-]}{[B]}$$

$$K_b = \frac{[BH^+][OH^-]}{[B]} \text{ for the equation } B(aq) = H_2O(l) = BH^+(aq) + OH^-(aq)$$

$$pK_b = -\log K_b$$

$$pOH = pK_b + \log\left[\frac{\text{conjugate acid}}{\text{conjugate base}}\right] = pK_b + \log\frac{[BH^+]}{[B]}$$

$$\%\text{ ionization} = \frac{\text{amount ionized (M)}}{\text{initial concentration (M)}} \cdot 100\% = \frac{[H^+]}{[HA]_0} \cdot 100\%$$

$$\text{conjugate acid} \underset{\text{gain of } H^+}{\overset{\text{loss of } H^+}{\rightleftharpoons}} \text{conjugate base}$$

- Acid-Base Theory

	Arrhenius	*Brønsted-Lowry*	*Lewis*
acid	H^+ supplied to water	donates H^+	accepts electron pair
base	OH^- supplied to water	accepts H^+	donates electron pair

- Acid-Base Properties of Common Ions in Aqueous Solution

	Anions (–)	*Cations (+)*
Acidic	HSO_4^-, H_2PO^{4-}	NH_4^+, Mg^{2+}, Al^{3+} transition metal ions
Basic	$C_2H_3O_2^-$, CN^-, CO_3^{2-}, F^-, HCO_3^-, HPO_4^{2-}, HS^-, NO_2, PO, S^{2-}	NONE
Neutral	Cl^-, Br^-, I^-, ClO_4^-, NO_2^-, SO_4^{-2}	Li^+, Na^+, K^+, Ca^{2+}, Ba^{2+}

PROBLEMS

Q. 1. What is the OH^- concentration (M) of a solution that contains 5.00×10^{-3} mole of H^+ per liter? $K_w = 1.00 \times 10^{-14}$.

(a) 7.00×10^{-14} M **(b)** 1.00×10^{-12} M
(c) 2.00×10^{-12} M **(d)** 1.00×10^{-11} M

Q. 2. From the following equation, identify the conjugate acid found in the products.

$$NH_3\ (g) + H_2O(l) \rightarrow NH_4^+\ (aq) + OH^-\ (aq)$$

(a) NH_3 **(b)** H_2O **(c)** NH_4^+ **(d)** OH^-

Q. 3. The oxyacids HClO, HIO, HBrO, $HClO_3$, $HClO_2$ in order of decreasing acid strength will be:

(a) HClO > HIO > HBrO > $HClO_3$ > $HClO_2$
(b) HClO > $HClO_2$ > $HClO_3$ > HBrO > HIO
(c) $HClO_3$ > $HClO_2$ > HClO > HBrO > HIO
(d) HBrO > HClO > $HClO_3$ > $HClO_2$ > HIO

Q. 4. From the following reversible equation, determine which species is or are Brønsted acids.

$$CO_3^{2-}(aq) + H_2O(l) \rightleftharpoons HCO_3^-(aq) + OH^-_{(aq)}$$

(a) CO_3^{2-} (aq) **(b)** $H_2O(l)$ and OH^-(aq)
(c) $H_2O(l)$ and HCO_3^-(aq) **(d)** $H_2O(l)$

Q. 5. Which of the following salts contains a basic anion?

(a) NaCl **(b)** $Ba(HSO_4)_2$ **(c)** KI **(d)** Li_2CO_3

Q. 6. The net ionic products(s) produced when solutions of potassium bicarbonate ($KHCO_3$) and hydrobromic acid (HBr) are mixed are:

(a) KBr and H_2CO_3 **(b)** H_2CO_3
(c) KBr, H_2O and CO_2 **(d)** K^+, Br^-, H_2O and CO_2

Q. 7. All of the following choices are strong base EXCEPT:

(a) CsOH **(b)** $Mg(OH)_2$ **(c)** $Ca(OH)_2$ **(d)** $Ba(OH)_2$

Q. 8. A solution is prepared by adding 0.600 liter of 1.0×10^{-3} M HCl to 0.400 liter of 1.0×10^{-3} M HNO_3. The pH of the final solution will be:

(a) 1.00 **(b)** 2.00 **(c)** 3.00 **(d)** 4.00

Q. 9. Suppose that 0.500 liter of 0.0200 HCl is mixed with 0.100 liter of 0.100 M $Ba(OH)_2$. The pH in the final solution after neutralization has occurred is:

(a) 3.00 **(b)** 5.00 **(c)** 7.00 **(d)** 12.00

Q. 10. A student wants to make up 250 mL of an HNO_3 solution that has a pH of 2.00. How many milliliters of the 2.00 M HNO_3 should the student use? (The remainder of the solution is pure water.)

(a) 0.50 mL **(b)** 0.75 mL **(c)** 1.0 mL **(d)** 1.3 mL

Q. 11. The mass of 12 equivalent of $Sr(OH)_2$, (assuming complete ionization) is:

$$Sr(OH)_{2(aq)} \rightarrow Sr^{2+}_{(aq)} + 2OH^-_{(aq)}$$

(a) 15.21 g **(b)** 30.41 g **(c)** 60.82 g **(d)** 121.64. g

Q. 12. Which of the following species is a stronger acid than formic acid in aqueous solution?
(a) CH_3COOH (b) NH_4^+ (c) H_2SO_3 (d) HPO_4^{2-}

Q. 13. All of the following species can exist in aqueous solution EXCEPT.
(a) NH_4^+ (b) NO_3^- (c) NO_2^- (d) NH_2^-

Q. 14. The pH of a solution of 0.10 F CH_3COOH increases when which of the following substances is added?
(a) $NaHSO_4$ (b) $HClO_4$ (c) NH_4NO_3 (d) K_2CO_3

Q. 15. A solution has pH = 10.82. The $[OH^-]$ in this solution is
(a) 6.6×10^{-4} (b) 6.6×10^{-10} (c) 1.0×10^{-7} (d) 1.9×10^{-5}

Q. 16. All of the following are acid-base conjugate pairs EXCEPT
(a) $HONO, NO_2^-$ (b) H_3O^+, OH^-
(c) $CH_3NH_3^+, CH_3NH_2$ (d) HS^-, S^{2-}

Q. 17. Which of the following species is an ampholyte?
(a) CH_3COO^- (b) H_2O (c) NH_4^+ (d) C_6H_5COOH
(e) S^{2-}

Q. 18. The pH of a 0.050 *F* HA solution is 5.35. What is K_a for HA?
(a) 2.0×10^{-11} (b) 4.0×10^{-10} (c) 4.5×10^{-6} (d) 8.9×10^{-5}

Q. 19. The correctly balanced net ionic equation for the reaction that occurs when an aqueous solution of formic acid is mixed with an aqueous solution of potassium hydroxide is
(a) $HCCOH(aq) + H_2O \rightleftharpoons HCOO^-(aq) + H_3O^+(aq)$
(b) $H_3O^+(aq) + OH^-(aq) \rightleftharpoons 2H_2O$
(c) $HCOOH(aq) + KOH \rightleftharpoons H_2O + HCOOK$
(d) $HCOOH(aq) + OH^-(aq) \rightleftharpoons H_2O + HCOO^-(aq)$

Q. 20. The symbol $K_b(HS^-)$ is the equilibrium constant for the reaction:
(a) $HS^-(aq) + OH^-(aq) \rightleftharpoons S^{2-}(aq) + H_2O$
(b) $HS^-(aq) + H_2O \rightleftharpoons H_2S(aq) + OH^-(aq)$
(c) $HS^-(aq) + H_2O \rightleftharpoons H_3O^+(aq) + S_2^-(aq)$
(d) $HS^-(aq) + CH_3COOH(aq) \rightleftharpoons H_2S(aq) + CH_3COO^-(aq)$

Q. 21. The symbol $K_a(HS^-)$ is the equilibrium constant for the reaction:
(a) $HS^-(aq) + OH^-(aq) \rightleftharpoons S^{2-}(aq) + H_2O$
(b) $HS^-(aq) + H_2O \rightleftharpoons H_2S(aq) + OH^-(aq)$
(c) $HS^-(aq) + H_2O \rightleftharpoons H_3O^+(aq) + S^{2-}(aq)$
(d) $HS^-(aq) + H_3O^+(aq) \rightleftharpoons H_2S(aq) + H_2O$

Q. 22. Which of the following solutions has the highest pH?
(a) 2.0 *F* $HClO_4$ (b) 0.20 *F* CH_3COOH
(c) 0.020 *F* HCl (d) 2.0 *F* NaCl

Q. 23. Which of the following solutions has the lowest pH?
(a) 2.0 *F* $HClO_4$ (b) 0.20 *F* CH_3COOH
(c) 0.020 *F* HCl (d) 2.0 *F* NaCl

Q. 24. When K_2O is added to water, the solution is basic because it contains a significant concentration of

(a) K^+ **(b)** K_2O **(c)** OH^- **(d)** O_2^{2-}

Q. 25. The correctly balanced net ionic equation for the reaction that occurs when a solution of acetic acid is mixed with a solution of sodium carbonate is

(a) $CH_3COOH(aq) + CO_3^{2-}(aq) \rightleftharpoons CH_3COO^-(aq) + HCO_3^-(aq)$

(b) $H_3O+(aq) + CO_2^{2-}(aq) \rightleftharpoons HCO_3^-(aq) + H_2O$

(c) $CH_3COOH(aq) + Na_2CO_3(s) \rightleftharpoons CH_3COO^-(aq) + NaHCO_3(s) + Na^+(aq)$

(d) $CO_3^{2-}(aq) + H_2O \rightleftharpoons HCO_3^-(aq) + OH^-(aq)$

Q. 26. A solution has $[OH^-] = 4.8 \times 10^{-3}$. Its pH is

(a) 11.7 **(b)** 8.4 **(c)** 4.8 **(d)** 3.7

Q. 27. The pH of 0.20 *F* HCOOH is between

(a) 1.5 and 2.0 **(b)** 2.0 and 2.5 **(c)** 2.5 and 3.0 **(d)** 3.0 and 3.5

Q. 28. A solution that 0.500 *F* HCl is saturated with H_2S at 298 K. Its pH is

(a) 0.30 **(b)** 0.50 **(c)** 1.5 **(d)** 2.3

Q. 29. Which of the following solutions has the highest pH?

(a) 0.10 *F* CH_3NH_2 **(b)** 0.10 *F* NaCN

(c) 0.10 *F* NH_4Cl **(d)** 0.10 *F* NaOAc

Q. 30. What is the degree of dissociation of 0.048 *F* C_6H_5COOH?

(a) 6.3% **(b)** 4.8% **(c)** 3.6% **(d)** 2.5%

Q. 31. What is the pH of a 0.080 F CH_3COOH solution?

(a) 0.80 **(b)** 1.10 **(c)** 2.92 **(d)** 4.76

Q. 32. Which of the following 0.20 F solutions is the most basic?

(a) Nal **(b)** NaF **(c)** NaCN **(d)** $NaNO_3$

Q. 33. The pH of a solution prepared by dissolving 0.100 mol of NaOH and 0.100 mol of NH_3 in enough water to make 1 L is:

(a) 14.0 **(b)** 13.0 **(c)** 11.1 **(d)** 7.0

Q. 34. The pK_a of nitrous acid is:

(a) 4.65 **(b)** 3.35 **(c)** 4.00 **(d)** 3.65

Q. 35. The conjugate base of CH3OH is:

(a) OH^- **(b)** CH_3O^- **(c)** O^{2-} **(d)** CH_4

Q. 36. The equilibrium constant for the reaction

$$HONO(aq) + CN\text{–}(aq) \rightleftharpoons HCN(aq) + ONO^-(aq)$$

is $1.1 \times 10^{+6}$. From the magnitude of this K_{eq} one can conclude that:

(a) CN^- is a stronger base than ONO^-.

(b) HCN is a stronger acid that HONO.

(c) The conjugate base of HONO is ONO^-.

(d) The conjugate acid of CN^- is HCN.

Q. 37. To prepare a buffer with pH to 3.4, you could use a mixture of

(a) NH_4NO_3 and NH_3 **(b)** HOCl and NaOCl

(c) HOAc and NaOAc **(d)** HNO_2 and $NaNO_2$

Q. 38. To prepare a buffer with pH to 9.0, you could use a mixture of
(a) NH_4NO_3 and NH_3 (b) HOCl and NaOCl
(c) HOAc and NaOAc (d) HNO_2 and $NaNO_2$

Q. 39. In titrating 0.100 *F* HCOOH (formic acid) with 0.100 F NaOH, the solution at the equivalence point is
(a) 0.100 *F* HCOONa (b) 0.0500 *F* HCOOH
(c) 0.0500 *F* HCOONa (d) 0.0500 *F* NaOH

Q. 40. A buffer that is mixture of acetic acid and potassium acetate has pH = 5.24. The $[Oac^-]/[HOAc]$ ratio in this buffer is
(a) 1 : 1 (b) 3 : 1 (c) 4 : 1 (d) 1 : 3

Q. 41. A 40.00-mL sample of benzoic acid of unknown concentration was titrated with 0.1025 *F* KOH. The equivalence point was reached when 29.38 mL of base has been added. The concentration of benzote ion at the equivalence point is
(a) 0.1025 *M* (b) 0.0753 *M* (c) 0.0591 *M* (d) 0.0434 *M*

Q. 42. The net ionic equation for he reaction that occurs when a solution of nitrous acid is titrated with potassium hydroxide is
(a) $HNO_2(aq) + OH^-(aq) \rightleftharpoons NO_2^-(aq) + H_2O$
(b) $H_3O^+(aq) + OH^-(aq) \rightleftharpoons 2H_2O$
(c) $NO_2^-(aq) + H_2O \rightleftharpoons HNO_2(aq) + OH^-(aq)$
(d) $HNO_2(aq) + H_2O \rightleftharpoons NO_2^-(aq) + H_3O^+(aq)$

Q. 43. A 50.00-mL sample of 0.0100 *F* $Ba(OH)_2$ is titrated with 0.0100 *F* HCl. The solution at the equivalence point is
(a) 3.33×10^{-3} *F* $BaCl_2$ (b) 5.00×10^{-3} *F* $BaCl_2$
(c) 1.00×10^{-2} *F* $BaCl_2$ (d) 2.50×10^{-3} *F* $BaCl_2$

Q. 44. A flask contains 100.0 mL of 0.100 *F* HOAc. To prepare a buffer with pH = 4.75, which of the following sampled of barium acetate solution should be added to the flask?
(a) 50.00 mL of 0.400 *F* $Ba(OAc)_2$
(b) 25.00 mL of 0.200 *F* $Ba(OAc)_2$
(c) 50.00 mL of 0.200 *F* $Ba(OAc)_2$
(d) 100.00 mL of 0.100 *F* $Ba(OAc)_2$

Q. 45. A 50.00-mL sample of acetic acid was titrated with 0.1200 *F* KOH, and 38.62 mL of base were required to reach the equivalence point? The pH of the titration mixture when 19.31 mL of base is added will be:
(a) 2.9 (b) 3.5 (c) 4.7 (d) 5.7

Q. 46. Which of the following mixtures is a buffer with pH close to 9?
(a) 50.00 mL of 0.100 *F* NH_4Cl and 50.00 mL of 0.100 *F* NaOH
(b) 50.00 mL of 0.100 *F* NH_4Cl and 25.00 mL of 0.100 *F* NaOH
(c) 50.00 mL of 0.100 *F* NH_4Cl and 50.00 mL of 0.100 *F* HCl
(d) 50.00 mL of 0.100 *F* NH_4Cl and 25.00 mL of 0.100 *F* HCl

Q. 47. A solution prepared by dissolving 0.0100 mol of benzoic acid and 0.0200 mol of sodium benzoate in 500.0 mL of water has a pH between
(a) 2 and 3 (b) 3 and 4 (c) 4 and 5 (d) 5 and 7

Q. 48. If 50.00 mL of 0.200 F KOH is added to 40.00 mL of 0.500 F HCOOH, the pH of the resulting solution is

(a) 7.00 (b) 5.50 (c) 4.00 (d) 3.75

Q. 49. A 40.00-mL sample of an aqueous solution of the weak base methylamine, CH_3NH_2, at 298 K is titrated with 0.150 *F* HCl, and the equivalence point is reached when 39.26 mL of the acid have been added. What is the pH at the equivalence point of this titration?

(a) 8.35 (b) 7.00 (c) 5.87 (d) 6.37

Q. 50. When 65.00 mL of 0.100 *F* HNO_3 have been added to 45.00 mL of 0.100 *F* NaOH, the pH of the resulting solution is

(a) 1.0 (b) 1.2 (c) 1.4 (d) 1.6

Q. 51. Carbon dioxide is dissolved in water under pressure until $[CO_2] = 0.050\ M$. The pH of this carbonated water is:

(a) 2.8 (b) 3.8 (c) 4.8 (d) 5.8

Q. 52. Which of the following mixtures will be a buffer solution when dissolved in 500.00 mL of water

(a) 0.200 mol of aniline and 0.200 mol of HCl
(b) 0.200 mol of aniline and 0.400 mol of NaOH
(c) 0.200 mol of NaCl and 0.100 mol of HCl
(d) 0.200 mol of aniline and 0.100 mol HCl

Q. 53. A 25.00-mL sample of a monoprotic weak acid is titrated with 0.1016 *F* NaOH. The equivalence point is reached when 28.54 mL of base have been added. If K_a of this acid is 4.0×10^{-6}, a suitable indicator to use to detect the equivalence point is

(a) phenolphthalein (b) thymolphthalein
(c) methyl red (d) bromthymol blue

Q. 54. A weak base, B has basicity constant $K_b = 2 \times 10^{-5}$. The pH of any solution in which $[B] = [BH^+]$ is

(a) 4.7 (b) 7.0 (c) 9.3 (d) 9.7

Q. 55. The net ionic equation for the reaction that occurs when barium hydroxide is titrated with sulfuric is

(a) $Ba^{2+}(aq) + 2\ OH^-(aq) + 2H^+(aq) + SO_4^{2-}(aq) \rightleftharpoons BaSO_4\downarrow + 2H_2O$

(b) $Ba(OH)_2(s) + 2H^+(aq) \rightleftharpoons 2H_2O + Ba^{2+}(aq)$

(c) $Ba^{2+}(aq) + SO_4^{2-}(aq) \rightleftharpoons BaSO_4\downarrow$

(d) $H_3O^+(aq) + OH^-(aq) \rightleftharpoons 2H_2O$

Q. 56. A 20.00-mL sample of a 0.200 *F* solution of the weak diprotic acid H_2X titrated with 0.250 *F* KOH. The solution at the *second* equivalence point is

(a) 0.100 *F* KHX (b) 0.111 *F* KHX
(c) 0.0769 *F* K_2X (d) 0.0667 *F* K_2X

Q. 57. A solution saturated with H_2S at 298 K and 1-atm pressure is 0.10 *M* in H_2S. What is the $[HS^-]$ in this solution?

(a) $1.0 \times 10^{-1}\ M$ (b) $1.0 \times 10^{-2}\ M$ (c) $1.0 \times 10^{-4}\ M$ (d) $1.0 \times 10^{-7}\ M$

Q. 58. The pH of 0.10 *M* H_2S at 298 K is:

(a) 1.0 (b) 2.0 (c) 4.0 (d) 6.0

Q. 59. The $[S^{2-}]$ in 0.10 M H_2S at 298 K is:
(a) 1×10^{-20} M (b) 1×10^{-13} M (c) 1×10^{-8} M (d) 1×10^{-7} M

Q. 60. The $[H_3O^+]$ in a solution made by mixing 60.0 mL of 1.00 F sodium acetate with 40.0 L of 0.500 F HCl is:
(a) 5.4×10^{-5} M (b) 3.6×10^{-5} M (c) 1.8×10^{-5} M (d) 9.0×10^{-6} M

Q. 61. It is necessary to maintain a solution at pH = 3.3. Which of the following pairs could be used to make a buffer with this pH?
(a) NH_4^+/NH_3 (b) $HCOOH/HCOO^-$
(c) HCO_3^-/CO_3^{2-} (d) CH_3COOH/CH_3COO^-

The following options are to be used for answers to **Questions 62-67**. All solutions are at 298 K.

Q. 62. A mixture of 1 mol of KCl and 1 mol of HCl in enough water to make a liter of solution.
(a) weak acid (b) weak base (c) acidic (d) basic

Q. 63. A mixture of 1 mol of Na_2CO_3 and 1 mol of CH_3COONa in enough water to make a liter of solution.
(a) weak acid (b) weak base (c) acidic (d) basic

Q. 64. A mixture of 0.50 mol of HOAc and 1 mol of NaOAc in enough water to make a liter of solution.
(a) acidic (b) basic (c) buffer (d) No buffer

Q. 65. A mixture of 0.50 ml of NaOH and 0.10 mol NH_3 in enough water to make a liter of solution.
(a) weak acid (b) weak base (c) buffer (d) No buffer

Q. 66. A mixture of 40.00 mL of 1.00 F HCl with 40.00 mL of 1.00 F KOH. The p^H of soln. is:
(a) 3 (b) 8 (c) 7 (d) 11

Q. 67. A mixture of 50.00 mL of 1.00 F HCl with 25.00 mL of 1.00 F NaOAc. The p^H of soln. is:
(a) <7 (b) >7 (c) 7 (d) None

Q. 68. The solubility product expression for Hg_2Cl_2 is
(a) $[Hg^+]^2\,[Cl^-]^2$ (b) $[Hg_2^{2+}]\,[Cl^-]^2$
(c) $[Hg^+]\,[Cl^-]$ (d) $[Hg_2^{2+}] + [2Cl^-]$

Q. 69. The anion of the insoluble salt MX is the conjugate base of the weak acid HX. The relationship between the equilibrium constant for the reaction

$$MX(s) + H^+(aq) \rightleftharpoons M^+(aq) + HX(aq)$$

The solubility product of MX, and the acidity constant of HX, is that K_{eq} equals
(a) $K_{sp}(MX)/K_a(HX)$ (b) $K_a(HX)/K_{sp}(MX)$
(c) $K_a(HX) + K_{sp}(MX)$ (d) $K_{sp}(MX) - K_a(HX)$

Q. 70. The molar solubility, s, of $Mn(OH)_2$, in water in terms of its K_{sp} is
(a) $s = (K_{sp})^{1/2}$ (b) $s = (K_{sp})^{1/3}$
(c) $s = (K_{sp}/4)^{1/3}$ (d) $s = (K_{sp}/6)^{1/3}$

Q. 71. If 50.0 L of 0.050 *F* $BaCl_2$ is mixed with 50.0 mL of 0.10 *F* K_2SO_4.

(a) Potassium chloride will precipitate

(b) The $[Ba^{2+}]$ in the resulting solution will be 0.0125 *M*.

(c) No reaction will occur.

(d) The $[K^+]$ in the resulting solution will be 0.050 *M*.

Q. 72. Which of the following insoluble electrolytes has the largest molar solubility in water?

(a) $SrCO_3$ (b) MnS (c) $PbSO_4$ (d) FeS

Q. 73. The maximum concentrating Ba^{2+} that can exist in a solution in which the sulfate ions concentration is 5.5×10^{-4} *M* is

(a) 2.0×10^{-7} *M* (b) 5.5×10^{-4} *M*

(c) 2.0×10^{-5} *M* (d) 6.0×10^{-14} *M*

Q. 74. The relationship between the K_{sp} of AgBr and the molar solubility, *z*, of AgBr in 0.20 *F* KBr is that K_{sp} equals

(a) z^2 (b) $z/0.20$ (c) $z^{1/2}$ (d) $0.20\ z$

Q. 75. Which of the following insoluble substances has the largest molar solubility in water?

(a) AgI (b) PbI_2 (c) $SrSO_4$ (d) $Zn(OH)_2$

Q. 76. When 80.0 mL of 0.120 *F* $Ca(NO_3)_2$ are mixed with 20.0 mL of 0.100 *F* K_2CO_3, the resulting solution has:

(a) $[Ca^{2+}] = 0.096$ *M* (b) $[K^+] = 0.040$ *M*

(c) $[NO_3^-] = -0.096$ *M* (d) $[CO_3^{2-}] = 0.100$ *M*

Q. 77. At the temperature at which the K_{sp} of $PbSO_4$ is 1.7×10^{-8}, the molar solubility of $PbSO_4$ in 0.10 *F* K_2SO_4 is

(a) 1.7×10^{-7} *M* (b) 1.7×10^{-8} *M*

(c) 1.3×10^{-4} *M* (d) 1.7×10^{-9} *M*

Q. 78. At the temperature at which the K_{sp} of $PbSO_4$ is 1.7×10^{-8}, the molar solubility of $PbSO_4$ in pure water is

(a) 1.7×10^{-7} *M* (b) 1.7×10^{-8} *M* (c) 1.3×10^{-4} *M* (d) 1.7×10^{-9} *M*

Q. 79. In which of the solutions below is $CaCO_3$ the most soluble?

(a) water (b) 0.20 *F* $CaCl_2$

(c) 0.20 *F* HCl (d) 0.20 *F* Na_2CO_3

Q. 80. When 60.00 mL of .010 *F* $Ca(MO_3)_2$ are mixed with 40.00 mL of 0.025 *F* K_2CrO_4.

(a) $CaCrO_4$ will precipitate.

(b) The $[NO_3^-]$ in the resulting solution will be 6.0×10^{-3} *M*.

(c) No reaction will occur.

(d) The $[CrO_4^{2-}]$ in the resulting solution will be 0.025 *M*.

Q. 81. How many moles of SrF_2 will dissolve in 2 L of water if $K_{sp}(SrF_2) = 7.9 \times 10^{-10}$?

(a) 2.8×10^{-5} (b) 7.9×10^{-10} (c) 5.8×10^{-4} (d) 5.8×10^{-5}

Q. 82. How many moles of SrF_2 will dissolve in 1 L of 0.10 *F* NaF if $K_{sp}(SrF_2) = 7.9 \times 10^{-10}$?

(a) 2.8×10^{-5} (b) 7.9×10^{-8} (c) 7.9×10^{-9} (d) 2.0×10^{-8}

Q. 83. How many moles of SrF_2 will dissolve in 1 L of 0.10 *F* $Sr(NO_3)_2$ if $K_{sp}(SrF_2) = 7.9 \times 10^{-10}$?

(a) 4.4×10^{-5} (b) 7.9×10^{-8} (c) 7.9×10^{-9} (d) 4.0×10^{-9}

Q. 84. If $K_{sp}(PbSO_4) = 1.8 \times 10^{-8}$ and $K_a(HSO_4^-) = 1.0 \times 10^{-2}$, the equilibrium constant for the reaction

$$PbSO_4(s) + H^+(aq) \rightleftharpoons HSO_4^-(aq) + Pb^{2+}(aq);\text{ is:}$$

(a) 1.8×10^{-6} (b) 1.8×10^{-10} (c) 2.8×10^{-10} (d) 3.2×10^{-14}

Q. 85. Which of the following in soluble electrolytes will NOT be significantly more soluble in 1.0 *F* HNO_3 than it is in pure water?

(a) FeS (b) $SrCO_3$ (c) AgCN (d) AgI

Q. 86. At the temperature at which the molar solubility of $PbBr_2$ in water is 2.3×10^{-2} *M*, the K_{sp} of $PbBr_2$ is:

(a) 5.3×10^{-4} (b) 4.9×10^{-5} (c) 2.4×10^{-5} (d) 2.3×10^{-2}

Q. 87. The relationship between the molar solubility, *y*, of CaF_2 in 0.080 *F* NaF and the K_{sp} of CaF_2 is that *y* equals

(a) $K_{sp}/0.080$ (b) $K_{sp}/(0.080)^2$ (c) $(K_{sp}/4)^{1/3}$ (d) $K_{sp}^{1/2}$

Q. 88. At the temperature at which a saturated solution $Zn(OH)_2$ has a pH of 8.30, the value of K_{sp} for $Zn(OH)_2$ is

(a) 8.0×10^{-18} (b) 4.0×10^{-18} (c) 1.6×10^{-17} (d) 4.0×10^{-12}

Q. 89. A solution is 0.120 *M* in Pb^{2+}. If $K_{sp}(PbCrO_4) = 1.8 \times 10^{-14}$, in order to precipitate 99.9% of all the Pb^{2+} present the $[CrO_4^{2-}]$ must be maintained at

(a) 1.8×10^{-17} *M* (b) 1.8×10^{-14} *M*

(c) 1.5×10^{-13} *M* (d) 1.5×10^{-10} *M*

Q. 90. The relationship between the molar solubility, *z*, of MgF_2 in 0.10 *F* $Mg(NO_3)_2$ and the K_{sp} of MgF_2 is that *z* equals

(a) $(K_{sp}/0.10)^{1/2}$ (b) $(K_{sp}/0.40)^{1/2}$ (c) $(K_{sp}/4)^{1/2}$ (d) $(K_{sp})^{1/2}$

Q. 91. When 20.00 mL of 0.100 *F* $Pb(O_3)_2$ and 30.00 mL of 0.150 *F* $Na_2C_2O_4$ are mixed, PbC_2O_4 precipitates. The $[C_2O_4^{2-}]$ in the resulting solution is

(a) 0.0450 *M* (b) 0.0500 *M* (c) 0.0900 *M* (d) 0.100 *M*

Q. 92. When 60.00 mL of 0.100 F $Ca(NO_3)_2$ and 40.00 mL of 0.125 F Na_2CO_3 are mixed, $CaCO_3$ precipitates. If $K_{sp}(CaCO_3) = 5 \times 10^{-9}$, the $[CO_3^{2-}]$ in the resulting solution is

(a) 0.10 *M* (b) 0.050 *M* (c) 5×10^{-7} *M* (d) 5×10^{-8} *M*

Q. 93. If Ksp of $Cu(OH)_2$ is 2×10^{-19} and the Ksp of CuS is 8.7×10^{-36}, the equilibrium constant for the reaction

$$Cu(OH)_2(s) + S^{-2}(aq) \rightleftharpoons CuS(s) + 2\,(OH)^-(aq)\text{ is:}$$

(a) 2×10^{-55} (b) 2×10^{16} (c) 4×10^{17} (d) 2×10^{18}

94. Which of the following insoluble solids is appreciably more soluble in 0.50 *M* HCl than it is in water?

(a) $ZnCO_3$ **(b)** AgI **(c)** $Ba(IO_3)_2$ **(d)** Hg_2Cl_2

Q. 95. Oxalic acid, $H_2C_2O_4$, is present in many plants, usually as the potassium or calcium salt. It is poisonous to humans because it forms calcium oxalate which can cause kidneys failure. Oxalic acid is a weak *diprotic* acid that is moderately soluble in water.

$$H_2C_2O_4 + H_2O \rightleftharpoons H_3O^+ + HC_2O_4^- \qquad K_{a_1} = ?$$

$$HC_2O_4^- + H_2O \rightleftharpoons H_3O^+ + HC_2O_4^{2-} \qquad K_{a_2} = ?$$

A saturated aqueous solution of $H_2C_2O_4$ at 298 K is found to have pH = 0.67 and $[C_2O_4^{2-}] = 5.4 \times 10^{-5}$ M.

A 5.00-mL sample of this saturated $H_2C_2O_4$(aq) is added to 100. mL of water and then titrated, 28.77 mL of 0.3650 M NaOH is required in the following titration

$$H_2C_2O_4(aq) + 2\ OH^-(aq) \rightarrow 2\ H_2O(l) + C_2O_4^{2-}(aq)$$

(i) The molarity of saturated $H_2C_2O_4$(aq) at 298 k is

(a) 1.05 m $H_2C_2O_4$ **(b)** 0.105 *M* $H_2C_2O_4$

(c) 10.5 M $H_2C_2O_4$ **(d)** 0.0105 M $H_2C_2O_4$

(ii) The value of K_{a_1} for oxalic acid is

(a) 5.2×10^{-4} **(b)** 5.2×10^{-2} **(c)** 4.37×10^{-2} **(d)** 4.37×10^{-4}

(iii) The concentration of $C_2O_4^{2-}$ in the neutralized solution produced in the titration in part (i) will be

(a) 0.00525 *M* **(b)** 0.0525 *M* **(c)** 0.00392 *M* **(d)** 0.0392 *M*

(iv) Is the sodium oxalate solution described in part (iii) acidic, basic, or pH neutral?

(a) BASIC **(b)** ACIDIC **(c)** BOTH **(d)** NEUTRAL

Q. 96. *p*-hydroxybenzoic acid, HOC_6H_4COOH, is a weak *diprotic* acid. It is used as a food preservative. Other compounds derived from the acid are used as preservatives in cosmetics and pharmaceuticals. Its structure is

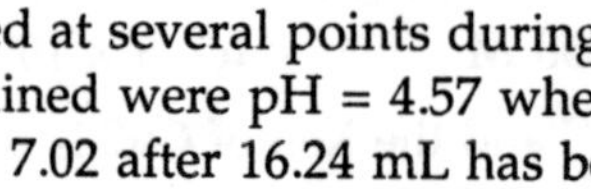

(i) 25.00 mL of a dilute aqueous solution *p*-hydroxybenzoic acid is titrated with NaOH(aq); 16.24 mL of 0.0200 M NaOH is required to reach the first equivalence point. The molarity of the *p*-hydroxybenzoic acid solution is:

(a) 0.00130 M **(b)** 0.130 M **(c)** 0.0130 M **(d)** 1.30 M

(ii) The pH was determined at several points during the titration referred to in part (i) Two of the values obtained were pH = 4.57 when 8.12 mL of 0.0200 M NaOH had been added, and pH = 7.02 after 16.24 mL has been added (the equivalence point). Use these data to determine K_{a_1} and K_{a_2} for p-hydroxybenzoic respectively; which is:

(a) 2.7×10^{-5}, 3.4×10^{-5} **(b)** 2.7×10^{-5}, 3.4×10^{-10}

(c) 2.7×10^{-10}, 3.4×10^{-5} **(d)** 2.7×10^{-10}, 3.4×10^{-10}

(iii) The pH at the point in the titration where 10.00 mL 0.0200 M has been added to the 25.00 mL of *p*-hydroxybenzoic acid solution is

(a) 9.43 **(b)** 4.57 **(c)** 4.77 **(d)** 9.23

Q. 97. Lime (quicklime), CaO; is the cheapest basic substance available. It is obtained by heating limestone ($CaCO_3$). When treated with water, CaO produces $Ca(OH)_2$ (slaked lime). Important environment uses of lime are in sewage treatment, neutralizing acid mine drainage, and removing oxides of sulfur from stack gases the principal method of producing sodium hydroxide, NaOH involved conversion of $Ca(OH)_2$ to $CaCO_3$ using Na_2CO_3.

$$Ca(OH)_2(s) + 2\,Na^+(aq) + CO_3^-(aq) \rightleftharpoons CaCO_3(s) + 2\,Na^+(aq) + 2\,OH^-(aq)$$

(i) The maximum pH that can be established in $Ca(OH)_2(aq)$ is:

(Given: $K_{sp} = 5.5 \times 10^{-6}$)

(a) 1.66 **(b)** 12.34 **(c)** 10.68 **(d)** 3.32

(ii) Given $Ca(OH)_2(s) \rightleftharpoons 2a^{2+}(aq) + 2\,OH^-(aq)\; K_{sp} = 5.5 \times 10^{-6}$

$2\,NH_4^+(aq) + 2\,OH^-(aq) \rightleftharpoons 2\,NH_3(aq) + 2\,H_2O \quad K = 1/(k_b)^2 = 1/(1.74 \times 10^{-5})^2$

$Ca(OH)_2(s) + 2\,NH_4^+(aq) \rightleftharpoons Ca^{2+}(aq) + 2\,NH_3(aq) + 2\,H_2O,\; K = 1.8 \times 10^4$

Is $Ca(OH)_2(s)$ more soluble in $NH_4Cl(aq)$ than it is in pure water?

(a) NO **(b)** YES **(c)** DON'T KNOW

(iii) Suppose that excess $Ca(OH)_2(s)$ is stirred into a 1.0 M Na_2CO_3 solution. The pH of the NaOH solution produced as a result of reaction is:

(a) –0.3 **(b)** 0.3 **(c)** 14.3 **(d)** 13.7

Q. 98. The pH of 0.1 *M* CH_3COOH. (*Ka* is 1.8×10^{-5}) is:

(a) 1.87 **(b)** 3.87 **(c)** 4.87 **(d)** 2.87

Q. 99. The pH of 0.10 mol dm^{-3} ethanoic acid. When 1.3 per cent of the ethanoic acid molecules have dissociated will be:

(a) 1.9 **(b)** 2.9 **(c)** 3.9 **(d)** 5.9

Q. 100. The pH of 0.10 mol dm^{-3} ethanoic aid when the acid dissociation constant K_a for ethnoic acid 1.8×10^{-5} mol dm^{-3} will be:

(a) 1.9 **(b)** 3.9 **(c)** 2.9 **(d)** 5.9

Q. 101. The solubility product for calcium carbonate given that the solubility of calcium carbonate at 298 K is 6.9×10^{-5} mol dm^{-3} will be.

(a) 6.9×10^{-5} mol^2 dm^{-2} **(b)** 2.4×10^{-9} mol^2 dm^{-6}

(c) 3.5×10^{-5} mol^2 dm^{-4} **(d)** 4.8×10^{-9} mol^2 dm^{-6}

Q. 102. The solubility of barium chromate (VI) in water and in 1.0 mol dm^{-3} potassium chromate (VI) respectively will be: (given that the solubility product of barium chromate is 2×10^{-10} mol^2 dm^{-6} at 298 K.

(a) 1.4×10^{-5}; 2×10^{-10} mol dm^{-3} **(b)** 2×10^{-5}; 1.4×10^{-10} mol dm^{-3}

(c) 2.8×10^{-5}; 4×10^{-10} mol dm^{-3} **(d)** 4×10^{-5}; 2.8×10^{-10} mol dm^{-3}

Q. 103. The concentration of hydrogen ions in a sample of tap water at a temperature of 298 K is 2.5×10^{-8} mol dm^{-3}. The concentration of aqueous hydroxide ions in the water sample is:

(a) 2.5×10^{7} mol dm^{3} **(b)** 4.0×10^{7} mol dm^{3}
(c) 2.5×10^{-7} mol dm^{-3} **(d)** 4.0×10^{-7} mol dm^{-3}

Q. 104. The pH of 0.150 m dm^{-3} aqueous potassium hydroxide is:

(a) 0.8 **(b)** 13.2 **(c)** 4.7 **(d)** 7.3

Q. 105. The pH of 0.200 mol dm^{-3} aqueous barium hydroxide is

(a) 7.3 **(b)** 0.4 **(c)** 13.6 **(d)** 3.7

Q. 106. The pH of 0.10 mol dm^{-3} aqueous ammonia ($K_b = 1.8 \times 10^{-5}$ mol dm^{-3}) is:

(a) 11.1 **(b)** 7.8 **(c)** 2.9 **(d)** 6.2

Q. 107. The solubility of silver chloride in water at 298 K the solubility of silver chloride in the presence of 0.1 M NaCl solution, respectively is:
(given K_{sp} (AgCl) $= 1.6 \times 10^{-10}$)

(a) 1.3×10^{-5} mol · L^{-1}, 4×10^{-9} mol · L^{-1}
(b) 2.6×10^{-5} mol · L^{-1}, 4×10^{-9} mol · L^{-1}
(c) 1.3×10^{-5} mol · L^{-1}, 2×10^{-9} mol · L^{-1}
(d) 2.6×10^{-5} mol · L^{-1}, 2×10^{-9} mol · L^{-1}

Q. 108. The solubility product of Ag_2 Cr O_4 is 9×10^{-12} at 98 K solubility of Ag_2CrO_4 is

(a) 1.3×10^{-6} mol · dm^{-3} **(b)** 1.3×10^{-4} mol · dm^{-3}
(c) 2.6×10^{-6} mol · dm^{-3} **(d)** 3.9×10^{-4} mol · dm^{-3}

Q. 109. The solubility of Mg $(OH)_2$ in pure water when K_{sp} for Mg $(OH)_2 = 1.2 \times 10^{-11}$ is:

(a) 2.884×10^{-4} dm^{-3} **(b)** 0.721×10^{-4} dm^{-3}
(c) 1.442×10^{-4} dm^{-3} **(d)** 4.326×10^{-4} dm^{-3}

Q. 110. Suppose a solution contains I^- and Cl^- ions and Ag^+ ions are added as $AgNO_3$. At what concentration of Ag^+ ions, precipitation of ions (I^- and Cl^-) will occur as Agl and AgCl)?

Given K_{sp} (Agl) $= 1.5 \times 10^{-16}$
and K_{sp} (AgCl) $= 1.56 \times 10^{-10}$

(a) 10.26×10^{7} **(b)** 9.6×10^{-7} **(c)** 9.6×10^{7} **(d)** 10.26×10^{-7}

Q. 111. The pH value of a solution made by dissolving 0.049 gm of H_2SO_4 in water to prepare 200 ml solution is:

(a) 3.7 **(b)** 11.7 **(c)** 10.3 **(d)** 2.3

Q. 112. The pH of a solution obtained by mixing 150 ml of 0.1 *N* NaOH and 150 ml of 0.2 *N* HCl. is:

(a) 1.3 **(b)** 12.7 **(c)** 0.5 **(d)** 13.5

Q. 113. The pH of the solution produced by mixing equal volumes of solutions of pH = 5 and pH = 3. is:

(a) 4.3 **(b)** 5.3 **(c)** 2.3 **(d)** 3.3

Q. 114. At 298 *K*, the pH of 0.23 *M* weak acid having ionization constant = 7.3×10^{-6} is:

(a) 5.78 (b) 2.89 (c) 8.22 (d) 11.11

Q. 115. The pH of 10^{-8} *M* HCl solution is:

(a) 5.98 (b) 8.02 (c) 6.98 (d) 7.02

Q. 116. The pH of a solution obtained by mixing one gram of acetic acid and 1.5 gm. of sodium acetate and making the volume equal to 100 ml is: (Given Dissociation constant of acetic acid = 1.75×10^{-5} at 258 K)

(a) 2.55 (b) 11.45 (c) 4.78 (d) 9.22

Q. 117. The pH and pOH of a buffer solution that contains 0.10 mol dm^{-3} ammonia and 0.20 dm mol^{-3} ammonium chloride respectively are:

(a) 9, 5 (b) 5, 9 (c) 3, 11 (d) 11, 3

Q. 118. A buffer solution is 0.10 mol dm^{-3} ethanoic acid and 0.20 mol dm^{-3} sodium ethanoate. Its pH is:

(a) 5.0 (b) 8.0 (c) 6.0 (d) 9.0

Q. 119. The number of grams of NH_4Cl that must be added to 1 litre of 0.05 *M* NH_3 to prepare a solution buffered at pH 11.45 at 298 K is:

(a) 1.77×10^{-2} gms (b) 3.77×10^{-2} gms
(c) 2.77×10^{-2} gms (d) 4.77×10^{-2} gms

Q. 120. A buffer solution contains 0.1 M acetic acid and 0.01 M sodium acetate. The dissociation constant of acetic acid at 298 K is 1.75×10^{-5}.

(i) the pH of the buffer solution. is:

(a) 3.76 (b) 4.23 (c) 3.67 (d) 4.32

(ii) the pH of the solution if 0.2 ml of 1M HCl is added to it, neglect the volume change on addition of HCl, is

(a) 3.57 (b) 3.75 (c) 4.86 (d) 4.68

(iii) the pH after the addition of 0.002 moles of NaOH to one litre of the buffer solution is:

(a) 3.647 (b) 3.757 (c) 3.845 (d) 3.947

Q. 121. The maximum pH of a solution 0.10 M in Mg^{2+} from which $Mg(OH)_2$ will not precipitate is:

(a) 10.04 (b) 4.96 (c) 9.04 (d) 5.76

Q. 122. K_{sp} of $PbBr_2$ is 8×10^{-5}. If the salt is 80% dissociated in solution, the solubility of salt in g per litre will be:

(a) 124.8 g $litre^{-1}$ (b) 12.48 g $litre^{-1}$
(c) 1.248 g $litre^{-1}$ (d) 15.0 g $litre^{-1}$

Q. 123. A sample of hard water contains 0.005 of $CaCl_2$ per litre. The minimum concentration of Na_2SO_4 which must be added for removing Ca^{2+} ions from this water sample is:

(K_{sp} for $CaSO_4$ is 24×10^{-5} at 298 K)

(a) 4.8×10^{-3} mol $litre^{-1}$ (b) 8.4×10^{-3} mol $litre^{-1}$
(c) 0.48×10^{-3} mol $litre^{-1}$ (d) 0.84×10^{-3} mol $litre^{-1}$

Q. 124. The $[OH^-]$ of a solution after 100 mL of 0.1 M $MgCl_2$ is added to 100 mL of 0.2 M NaOH is: (K_{sp} of $Mg(OH)_2$ is $1.\ 2 \times 10^{-11}$)

(a) 1.4×10^{-4} **(b)** 2.8×10^{-4} **(c)** 4.2×10^{-4} **(d)** 5.6×10^{-4}

Q. 125. The silver ion concentration in a solution prepared y shaking solid Ag_2S with saturated H_2S (0.10 M) in 0.15 m H_3O^+ until equilibrium is established will be:

Given K_{sp} for $Ag_2S = 1.6 \times 10^{-49}$

(a) 6×10^{-15} M **(b)** 1.6×10^{-49} M **(c)** 4×10^{-29} M **(d)** 4×10^{-21} M

Q. 126. A solution is prepared by dissolving 0.10 mole of NH_3 gas in enough water to make 1.0 litre of solution. The $[OH^-]$ in the resulting solution will be: (Given *K* for the solution $NH_4OH \rightleftharpoons NH_4^+ + OH^-$ is 1.8×10^{-5} mole L^{-1})

(a) 1.3×10^{-3} M **(b)** 1.3×10^{-5} M **(c)** 1.3×10^{-8} M **(d)** 1.3×10^{-11} M

Q. 127. To 35 cm^3 of 0.02 mole dm^{-3} acetic acid were added 10 cm^3 of 0.05 mole dm^{-3} sodium hydroxide. The pH of the solution; when pKa of acetic acid = 4.74 will be:

(a) 2.14 **(b)** 7.14 **(c)** 5.14 **(d)** 11.14

Q. 128. How many moles of sodium formate ($NaHCO_2$) must be added to 250 ml of a 1.0 M $HCHO_2$ solution to give a pH of exactly 4.00?

(a) 0.53 **(b)** 0.93 **(c)** 0.73 **(d)** 0.33

Q. 129. A 1.0 M Al $(OH)_3$ solution is found to have a pH 2.70. K_a for the reaction will be: (Given, $Al^{3+} + HOH \rightleftharpoons Al(OH)^{2+} + H^+$)

(a) 4.0×10^{-6} **(b)** 11.0×10^{-6} **(c)** 7.0×10^{-6} **(d)** 9.0×10^{-6}

Q. 130. (i) $AgNO_3$ is slowly added to a 0.020 m solution of $NaC_2H_2O_2$, (i) Will a precipitate of $AgC_2H_2O_2$ form, when conc. $[Ag^+] = 1 \times 10^{-2}$ M; $K_{sp} = (2 \times 10^{-3})$

(a) Yes **(b)** No **(c)** Can't say

(ii) At approximately that $[Ag^+]$ will a precipitate start forming?

(a) 0.2 M **(b)** 0.1 M **(c)** 0.3 M **(d)** 0.4 M

Q. 131. Assuming that the addition of a small quantity of solid does not cause the volume of the solution to change, than the concentration of hydroxide ions in a solution made by dissolving 0.020 mole of ammonium chloride in 100 ml of 0.15 M ammonia will be: (Given, K_{eq} for the reaction is 1.8×10^{-5}.)

(a) 1.4×10^{-6} M **(b)** 1.4×10^{-4} M **(c)** 1.4×10^{-5} M **(d)** 1.4×10^{-8} M

Q. 132. Forty milliliters of 0.10 molar NaOH was added to ten milliliter of 0.45 molar HCl. The pH of the resulting solution is: (Both HCl and NaOH are string electrolytes ($K_w = 1 \times 10^{-14}$ mole^2 litre^{-2})

(a) 3 **(b)** 2 **(c)** 4 **(d)**6

Q. 133. The $[H_3O^-]_{eq}$ in a 0.20 M formic acid, $HO\overset{O}{\overset{||}{C}}H$, ($K_a$ for formic acid is 1.76×10^{-4}) is:

(a) 5.9×10^{-3} M **(b)** 8.9×10^{-3} M **(c)** 7.9×10^{-3} M **(d)** 11.9×10^{-3} M

Q. 134. The pH of a 0.100 M NH_3 solution at 298 K will be: (K_b is 1.76×10^{-5} for NH_3 at 298 K)
(a) 13.124 (b) 11.124 (c) 12.124 (d) 10.124

Q. 135. For the reaction

$$H_2(g) + S(s) \rightleftharpoons H_2S(g)$$

K_p at 925 K was found to be 18.5. K_p at 1000 *K* is:
(Given, Enthalpy change in this range of temperature is –71.09 kJ mol^{-1})
(a) 9.25 (b) 9.25×10^{-2} (c) 9.25×10^{-4} (d) 9.25×10^{-3}

Q. 136. The pH of 0.56 mole dm aqueous solution of KOH assuming ideal behaviour will be:
(a) 0.25 (b) 13.75 (c) 1.75 (d) 12.25

Q. 137. The pH of a 1.00×10^{-7} molar solution of HCl when the ionic product of water is 10^{-14} will be:
(a) 10.28 (b) 3.72 (c) 6.79 (d) 7.21

Q. 138. The pH value of a solution in which $[H_3O^+] = 3 \times 10^{-3}$ g ion L^{-1} is:
(a) 9.66 (b) 4.34 (c) 2.52 (d) 11.48

Q. 139. Calculate (i) the $[H_3O^+]$ and (ii) $[OH^-]$ of a solution that pass a pH value of 6.5?
(i) (a) 3×10^{-7} (b) 2×10^{-7} (c) 4×10^{-7} (d) 1×10^{-7}
(ii) (a) 2×10^{-8} (b) 4×10^{-8} (c) 3×10^{-8} (d) 7×10^{-8}

Q. 140. The pH of a solution for which $[OH^-]$ = 0.15 M is:
(a) 11.4 (b) 2.6 (c) 13.2 (d) 0.8

Q. 141. The $[H_3O^+]$, and percentage ionization of 0.2 *M* acetic acid solution respectively, when the ionization constant of acetic acid is 1.8 ´ 10^{-5} will be :
(a) 3.9×10^{-3}, 95% (b) 1.9×10^{-3}, 0.95%
(c) 2.9×10^{-3}, 9.5% (d) 4.9×10^{-3}, 9.0%

Q. 142. The $[H^+]$ of a solution with a pH of 10.6 is:
(a) 2.5×10^{-11} (b) 2.5×10^{-12} (c) 2.5×10^{-10} (d) 2.5×10^{-14}

Q. 143. To 10 cm^3 of 0.1 mole dm^{-3} HCl was added 9.6 cm^3 of 0.1 mole dm^{-3} NaOH. Thus the pH of the final solution will be:
(a) 1.69 (b) 3.69 (c) 2.69 (d) 5.69

Q. 144. The average of several determinations of the solubility of silver chloride in water at 298 K was 0.00194 g/$liter^{-1}$. The pK_{sp} of AgCI by using this value will be:
(a) 1.73 (b) 7.73 (c) 9.73 (d) 13.73

Q. 145. How many moles of sodium hydroxide can be added to one litre of a solution of 0.001 *M* in NH_3 and 0.100 *M* in NH_4Cl without changing pOH by more than one unit? (Assume no change in Volume $K_b = 1.8 \times 10^{-5}$ and $pK_1 = 4.75$)
(a) 0.182 (b) 0.082 (c) 0.282 (d) 0.482

Q. 146. (i) If K_a for an acid is 2×10^{-5}, what is its pK_a.
(a) 3.7 (b) 4.7 (c) 5.7 (d) 6.7

(ii) The pH of 0.0001 N HCI solution will be:
(a) 4.0 (b) 3.0 (c) 2.0 (d) 1.0

(iii) The pOH of a solution containing 0.4 g of NaOH per litre of solution will be
(a) 1.0 (b) 3.0 (c) 2.0 (d) 4.0

Q. 147. (i) The pH of a solution is 5.2. The concentration of hydrogen ions in mol dm^{-3} will be
(a) 5.3×10^{-6} (b) 4.31×10^{-6} (c) 6.31×10^{-6} (d) 7.31×10^{-6}

(ii) If $pKw = 13.02$ at 333 K. Then K_a will be:
(a) 8.55×10^{-13} (b) 7.55×10^{-12} (c) 9.55×10^{-14} (d) 6.55×10^{-11}

Q. 148. (i) pK_a of acetic acid = 4.75. What is the pH of a solution containing 0.5 M acetic acid and 0.5 M sodium acetate.
(a) 3.75 (b) 4.75 (c) 2.75 (d) 5.75

(ii) When more of salt is added to solution to make it 0.6M with respect to salt them pH will be:
(a) 4.83 (b) 6.83 (c) 5.83 (d) 7.83

Q. 149. A solution composed of 0.5 M of an organic acid and 0.5M of its sodium salt, gives a pH of 5.5 at 298 K. The dissociation constants of the acid will be:
(a) 5.162×10^{-5} (b) 8.162×10^{-5} (c) 7.162×10^{-5} (d) 3.162×10^{-5}

Q. 150. $[H_3O^+]$ in pure water at 298 K will be:
(a) 2.0×10^{-7} M (b) 4.0×10^{-7} M (c) 1.0×10^{-7} M (d) 8.0×10^{-7} M

Q. 151. The pH of a solution in which $[H_3O^+] = 6.38 \times 10^{-6}$ M will be:
(a) 5.195 (b) 9.195 (c) 7.195 (d) 11.195

Q. 152. A solution is made 1.00×10^{-3} M in Ag^+ by adding enough water to 1.00×10^{-3} moles of $AgNO_3$ to make 1 litre of solution. Will any AgCI precipitate if 10.0×10^{-4} moles of NaCI is added (Ksp for AgCI is 2.8×10^{-10} $mole^2$ L^{-2}).
(a) No (b) Yes (c) can't say

Q. 153. A solution is 0.010 M in cuprous ion (Cu^+) and 0.0010 M in lead ion (Pb^{2+}). Chloride ion is slowly added. Will $PbCI_2$ (K_{sp} = 1.6×10^{-5} $mole^3L^{-3}$) or CuCl (Ks_p = 3.2×10^{-7} $mole^2$ L^{-2}) precipitate first?
(a) Yes (b) No (c) Can't say

Q. 154. The solubility of thallous chloride at 293 K will be ($K_{sp} = 1.7 \times 10^{-4}$)
(a) 1.3×10^{-3} mol. $litre^{-1}$ (b) 1.3×10^{-4} mol. $litre^{-1}$
(c) 1.3×10^{-5} mol. $litre^{-1}$ (d) $1.3 \times 10^{-}2$ mol. $litre^{-1}$

Q. 155. The solubility of silver ethanoate in moles per dm^3 will be: (the solubility product at 303 K is 4.9×10^{-3} mol^2 dm^{-6}
(a) 4×10^{-3} (b) 4×10^{-2} (c) 7×10^{-2} (d) 7×10^{-3}

Q. 156. One litre of buffer (pH = 5.20) is to be prepared from NH_4CI and NH_4OH. If 10 mole of NH_4Cl is used, how many moles of NH_3 should be added to the solution? (Assume an activity coefficient of one)
(a) 8.8016×10^{-6} moles (b) 8.9016×10^{-6} moles
(c) 8.9160×10^{-6} moles (d) 8.9610×10^{-6} moles

Q. 157. The pH of a solution prepared by mixing 0.10 mole of sodium acetate with enough water to make 1.0 litre of solution, will be: (Assume the activity coefficient of unity.)

(a) 7.86 **(b)** 9.86 **(c)** 8.86 **(d)** 10.86

Q. 158. The degree of hydrolysis of sodium acetate in its 0.01 *M* solution is 0.023%. The hydrolysis constant, K_h and concentration of OH^- ions are:

(a) 5.29×10^{-10} *M*; 2.3×10^{-6} *M* **(b)** 2.3×10^{-4} *M*; 5.29×10^{-6} *M*
(c) 5.28×10^{-4} *M*; 2.3×10^{-4} *M* **(d)** 2.3×10^{-10} *M*; 5.29×10^{-6} *M*

Q. 159. For titration cm^3 of 0.1 N acetic acid with 0.1 N NaOH, the pH when the degree of neutralisation $f = 0.$ $f = 0.5$ and $f = 1.0$ respectively will be: (Given Dissociation constant of acetic acid = 1.85×10^{-5} *M* and f = number of equivalents of alkali added to the number of acid present.)

(a) 8.72; 2.87; 4.73 **(b)** 4.73; 2.87; 8.72 **(c)** 2.87; 4.73; 8.72 **(d)** 8.72; 4.73; 2.87

Q. 160. An indicator is a weak acid and the pH range of its colour change is 3.1 to 4.5 of the neutral point of the indicator is in the centre of this pH range, calculate the ionization constant of the indicator.

(a) 4.1×10^{-4} **(b)** 4.1×10^{-8} **(c)** 4.1×10^{-6} **(b)** 4.1×10^{-10}

Q. 161. The pH of a buffer made by adding 10.3 g of sodium acetate to 500. mL of 0.150 M aceric acid will be:

(a) 5.97 **(b)** 4.97 **(c)** 6.97 **(d)** 8.97

Q. 162. A solution contains 0.15 M HCOOH (formic acid) and M 0.12 M HCOONa (sodium formate). The pH of the solution and the pH following addition of 0.02 m HCl respectively is:

(a) 3.5, 3.6 **(b)** 3.4, 3.5 **(c)** 3.6, 3.5 **(d)** 3.5, 3.4

Q. 163. The pH of the buffer containing (equimolar) 1 M acid and 1 M CH_3 COONa is equal to pK_a of the acid, i.e. 4.74 so the change in pH when 0.1 mole of HCl and 0.01 mole of NaOH is added to a litre of the buffer solution will be:

(a) +0.01; –0.01 **(b)** +0.02; –0.02 **(c)** –0.01; +0.01 **(d)** –0.02; +0.02

Q. 164. $CaSO_4$ at water has a solubility of 4.9×10^{-5} M, i.e., enough $CaSO_4$ dissolves in water to make a solution which has a normal concentration of 4.9×10^{-3} M. Nominal meaning we period that the $CaSO_4$ remains as a molecular unit in water when we quote this number, its equilibrium constant K_{sp} will be:

(a) 2.4×10^{-6} **(b)** 2.3×10^{-7} **(c)** 2.4×10^{-5} **(d)** 2.4×10^{-8}

Q. 165. Lead II Iodide, PbI_2, has a $K_{sp} = 1.4 \times 10^{-8}$. Its molar solubility is:

(a) 1.55×10^{-3} *M* **(b)** 1.4×10^{-8} *M* **(c)** 3×10^{-9} *M* **(d)** 2.56×10^{-2} *M*

Q. 166. Given the K_{sp} for silver chromate = 1.1×10^{-12}. Then, the solubility of Ag_2CrO_4 in pure water will be:

(a) 6.5×10^{-12} *M* **(b)** 6.5×10^{-10} *M* **(c)** 6.5×10^{-5} *M* **(d)** 1.4×10^{-7} *M*

Q. 167. The solubility product constant of $PbCl_2$ is 1.7×10^{-5} at 298 K. The solubility of lead (II) chloride in water at 298 K is:

(a) 4.3×10^{-6} *M* **(b)** 1.6×10^{-2} *M* **(c)** 1.6×10^{-4} *M* **(d)** 1.7×10^{-5} *M*

Q. 168. The solubility of $PbCl_2$ in 1.00 M HCl will be: (Given K_{sp} of $\times 10^{-5}$)

(a) 1.6×10^{-2} M (b) 4.3×10^{-6} M (c) 1.6×10^{-4} M (d) 4.3×10^{-2} M

Q. 169. Estimate the solubility if barium sulfate in a 0.020 M sodium sulfate solution. Given, the solubility product constant for barium sulfate is 1.1.

(a) 5.5×10^{-8} M (b) 5.5×10^{-10} M (c) 5.5×10^{-9} M (d) 5.5×10^{-11} M

Q. 170. The concentration of hydrogen ion and hydroxide ion at 298 K in: (i) 0.15 M HNO_3 is:

(i) (a) 0.15 M; 6.7×10^{-14} M (b) 0.25 M; 6.9×10^{-14} M

(c) 0.35 M; 6.5×10^{-14} M (d) 0.50 M; 6.3×10^{-14} M

(ii) In 0.010 M $Ca(OH)_2$ it is:

(a) 5.2×10^{-13} M; 0.04 M (b) 5.5×10^{-13} M; 0.01 M

(c) 5.0×10^{-13} M; 0.02 M (d) 5.7×10^{-13} M; 0.005 M

Q. 171. A sample of orange juice has a hydrogen-ion concentration of 2.9×10^{-4} M. The pH of solution is:

(a) 6.46 (b) 10.54 (c) 10.46 (d) 3.54

Q. 172. The pH of human arterial blood is 7.40. The hydrogen –ion concentration is:

(a) 10×10^{-8} M (b) 7.40×10^{-8} M (c) 4.0×10^{-8} M (d) 2.0×10^{-8} M

Q. 173. Nicotinic acid (niacin) is a monoprotic acid with the formula $HC_6H_4NO_2$. A solution that is 0.012 M in the nicotinic acid has a pH of 3.39 at 298 K. The acid-ionization constant, K_a, for this acid at 298 K and the degree of ionization of nicotinic acid in this solution respectively is:

(a) 1.4×10^{-5}; 3.4% (b) 2.6×10^{-55}; 4.3%

(c) 4.1×10^{-45}; 7.2% (d) 7.8×10^{-45}; 2.7%

Q. 174. (i) The concentrations of nicotinic acid, hydrogen ion, and nicotinate ion in a solution of 0.10 M nicotinic acid, $HC_6H_4NO_2$, at 298 K is:

(a) 0.10 M, 0.012 M, 0.012 M (b) 0.10 M, 0.0012 M, 0.0012 M

(c) 0.10 M, 0.12 M, 0.12 M (d) 0.10 M, 1.2 M, 1.2 M

(ii) The pH of the solution and the degree of ionization of nicotinic acid respectively is: The acid-ionization constant, $K_a = 1.4 \times 10^{-5}$.

(a) 6.92, 0.12 (b) 2.92, 0.012

(c) 7.96, 1.2 (d) 1.08, 0.0012

Q. 175. Aspirin (acetylsalicyclic acid) is a well-known headache remedy. The pH at 298 K of the solution obtained by dissolving a 5.00-grain tablet of aspirin (acetylsalicyclic acid) in 0.500 L of water will be : (the tablet contains 0.3256 g, of acetysalicyclic acid, $HC_9H_7O_4$. The acid is monoprotic, and K_a equals 3.3×10^{-4} at 298 K).

(a) 4.04 (b) 9.96 (c) 12.97 (d) 3.03

Q. 176. Morphine, $C_{17}H_{19}NO_3$, is administered medically to relieve pain. It is a naturally occurring base, or alkaloid. Some other well-known alkaloids and their plant sources are nicotine (tobacco), cocaine (coco bush), mescaline (species of cactus), and caffeine (coffee and tea).

The pH of a 0.0075 M solution of morphine at 298 K will be:- (The base-ionization constant, K_b, is 1.6×10^{-6} at 298 K.)

(a) 1.47 (b) 12.63 (c) 10.04 (d) 3.96

Q. 177. The degree of ionization of acetic acid, $HC_2H_3O_2$, in a 0.10 M aqueous solution at 298 K is 0.013. K_a at this temperature is 1.7×10^{-5}. The degree of ionization of $HC_2H_3O_2$ in a 0.10 M solution at 298 K to which sufficient HCl is added to make it 0.010 M HCl is:

(a) 0.0017 (b) 0.17 (c) 0.017 (d) 0.00017

Q. 178. Students were told to make up a buffer by mixing 60. ml (0.0060 L) of 0.100 M NH_3 with 40. ml (0.040 L) of 0.100 M NH_4Cl. The pH of this buffer is:

(a) 10.6 (b) 9.43 (c) 3.96 (d) 5.28

Q. 179. The pH of 75 mL of the buffer solution of 0.10 M $HC_2H_3O_2$ and 0.20 M $NaC_2H_3O_2$ to which 9.5 mL of 0.10 M hydrochloric of acid is added will be:

(a) 2.96 (b) 3.96 (c) 1.96 (d) 4.96

Q. 180. A liter of solution saturated at 298 K with calcium oxalate, CaC_2O_4, is evaporated to dryness, giving a 0.0061-g residue of $CaCO_2O_4$. The solubility product constant for this salt at 298 is:

(a) 4.8×10^{-5} (b) 4.8×10^{-9} (c) 2.3×10^{-5} (d) 2.3×10^{-9}

Q. 181. **(i)** By experiment, it is found that 1.2×10^{-3} mol of lead(II), iodide, PbI_2, dissolves in 1 L of aqueous solution at 298 K. The solubility product constant at this temperature is:

(a) 1.2×10^{-3} (b) 1.2×10^{-9} (c) 6.9×10^{-3} (d) 6.9×10^{-9}

(ii) The mineral fluorine is calcium fluoride, CaF_2. The solubility (in grams per liter) of calcium fluoride in water from the solubility product constant 3.4×10^{-11} is:

(a) 1.6×10^{-2} g CaF_2 L^{-1} (b) 1.6×10^{-4} CaF_2 L^{-1}

(c) 2.0×10^{-2} g CaF L^{-1} (d) 1.6×10^{-4} CaF_2 L^{-1}

Q. 182. What is the molar solubility of calcium oxalate in 0.15 M calcium chloride. Given the solubility product constant for calcium oxalate is 2.3×10^{-9}.

(a) 2.3×10^{-8} M (b) 2.3×10^{-9} M (c) 1.5×10^{-8} M (d) 2.3×10^{-9} M

Q. 183. Sulfate ion, SO_4^{2-} in solution is often determined quantitatively by precipitating it as barium sulfate, $BaSO_4$. The sulfate ion may have been formed from a sulfur compound. Analysis for the amount of sulfate ion then indicates the percentage of sulfur in the compound. Is a precipitate expected to form at equilibrium when 50.0 mL of 0.0010 M $BaCl_2$ is added to 50.0 mL of 0.00010 M Na_2SO_4. The solubility product constant for barium sulfate is 1.1×10^{-10}. Assume that the total volume of solution, after mixing, equals the sum of the volumes of the separate solutions.

(a) Yes (b) No (c) Don't Know

Q. 184. **(i)** When $BaCl_2$ and Na_2SO_4 are mixed, a precipitate of $BaSO_4$ is formed. Before precipitation, the solution contained 0.00050 mol Ba^{2+} per liter and 0.000050 mol SO_4^{2-} per liter. Thus (i) the molar concentration of SO_4^{2-} ion left after $BaSO_4$ precipitate is:

(Given K_{sp} for $BaSO_4$ = 1.1×10^{-10}).

(a) 2.4×10^{-7} *M* (b) 5.5×10^{-5} *M*
(c) 1.1×10^{-10} *M* (d) 4.8×10^{-7} *M*

(ii) What percentage of SO_4^{2-} is not precipitated?
(a) 0.76% (b) 0.48% (c) 0.67% (d) 0.84%

Q. 185. Consider two slightly soluble salts, calcium carbonate and calcium sulfate. Which of these would have its solubility more affected by the addition HCl, a strong acid?
(a) $CaSO_4$ is more soluble in acidic solution
(b) $CaCO_3$ is more soluble in acidic solution
(c) Both are more soluble in acidic solution
(d) Both are less soluble in acidic solution

Q. 186. Consider a solution 0.10 *M* Zn^{2+} and 0.10 *M* Pb^{2+} saturated with H_2S. What range of H^+ concentration will give a precipitate of one of the metal sulfides, leaving the other metal ion in solution? The concentration of H_2S in a saturated solution is 0.10 *M*.
(a) H^+ conc. is less then 0.32 *M*
(b) H^+ conc. is greater than 0.32 *M*
(c) H^+ conc. is equal to 0.32 *M*
(d) None of these

Q. 187. The concentration of Ag^+(aq) in 0.010 *M* $AgNO_3$ that is also 1.00 *M* NH_3 will be :
(K_f for $Ag(NH_3)_2^+$ ion is 1.7×10^7).
(a) 5.9×10^{-8} *M* (b) 6.1×10^{-8} *M* (c) 6.1×10^{-10} *M* (d) 5.9×10^{-10} *M*

Q. 188. If K_{sp} for AgCl is 1.8×10^{-10}, then
(i) Will silver chloride precipitate from a solution that is 0.010 *M* $AgNO_3$ and 0.010 *M* NaCl?
(a) Yes (b) No
(ii) Will silver chloride precipitate from this solution if it is also
(a) Yes (b) No

Q. 189. The molar solubility of AgCl in 1.0 *M* NH_3 will be: (Given K_{sp} (AgCl) = 1.8×10^{-10} and $K_f(AgNH_3)^{2+} = 1.7 \times 10^7$.)
(a) 0.50 *M* (b) 0.050 *M* (c) 0.0050 M (d) None of these

Q. 190. For each of the following identity the acids and bases and their conjugates.
(a) $HClO_2 + H_2O \rightleftharpoons H_3O^+ + ClO_2^-$
(b) $OCl^- + H_2O \rightleftharpoons HOCl + OH^-$
(c) $NH_3 + H_2PO_4 \rightleftharpoons NH_4^+ + HPO_4^{2-}$
(d) $HCl + H_2PO_4^- \rightleftharpoons H_3PO_4 + Cl^-$

Q. 191. The percentage ionization of $HC_2H_3O_2$ of molarity *M* is given by

(a) $\left(\frac{K_a}{M}\right) \times 100$ (b) $\left(\frac{M}{K_a}\right) \times 100$

(c) $\left(\sqrt{\frac{K_a}{M}}\right) \times 100$ **(d)** $\left(\sqrt{\frac{M}{K_a}}\right) \times 100$

Q. 192. The value of the equilibrium constant for the reaction $N_2 + 2O_2 \rightleftharpoons 2NO_2$ + heat, when there are present at equilibrium 5.0 mol of N_2, 7.0 mol of O_2, and 0.10 mol of NO_2 in a 1.5-L vessel at a certain temperature is:

(a) 6.1×10^{-5} **(b)** 16.1×10^{-5} **(c)** 0.61×10^{-5} **(d)** 1.61×10^{-5}

Q. 193. 20 mL of 0.2 *M* NaOH are added to 50 mL of 0.2 *M* acetic acid ($K_a = 1.8 \times 10^{-5}$). The pH of solution is

(a) 4.8665 **(b)** 45.686 **(c)** 9.4314 **(d)** 4.5686

Q. 194. The K_w for $2H_2O \rightleftharpoons H_3O^+ + OH^-$ changes from 10^{-14} at 298 *K* to 9.62×10^{-14} at 333 *K*. The pH of water at 333 *K* will be:

(a) 3.10×10^{-7} **(b)** 7.49 **(c)** –6.51 **(d)** 6.51

Q. 195. The $[H^+]$ in a 0.20 *M* solution of dichloroacetic acid ($K_a = 5 \times 10^{-2}$) that also contains 0.1 *M* sodium dichloroacetate will be. (Neglect hydrolysis of sodium salt.)

(a) 0.005 **(b)** 5.0 **(c)** 0.05 **(d)** 5×10^{-4}

Q. 196. Is the NH_3–NH_4Cl solution; a bufer solution

(a) No **(b)** Yes **(c)** neutral **(d)** Can't say

Q. 197. In the preparation of quicklime from limestone, the reaction is

$CaCO_3(s) \rightleftharpoons CaO(s) + CO_2(g)$

Experiments carried out between 1123 *K* and 1223 *K* led to a set of K_p values fitting an empirical equation

$$\log K_p = 7.282 - \frac{8500}{T}$$

Where *T* is the absolute temperature. If the reaction is carried out in quiet air, the temperature from this equation for the complete decomposition of the limestone is:

(a) 1176 K **(b)** 1165 K **(c)** 1167 K **(d)** 1156 K

Q. 198. A quantity of PCl_5 was heated in a 12-L vessel at 523 K

$PCl_5(g) \rightleftharpoons PCl_3(g) + Cl_2(g)$

At equilibrium the vessel contains 0.21 mol PCl_5, 0.32 mol PCl_3, and 0.32 mol Cl_2, so the equilibrium constant *K* for the dissociation of PCl_5 at 523 K when concentrations are referred to the standard state of 1 mol L^{-1} is.

(a) 0.410 **(b)** 0.014 **(c)** 0.040 **(d)** 0.041

Q. 199. A sample of slid AgCl is shaken with water at 298 K a solution containing 1.0×10^{-5} *M* silver ions is produced. The K_{sp} will be:

(a) 1.0×10^{-10} **(b)** 1.0×10^{-5} **(c)** 1.0×10^{-8} **(d)** 1.0×10^{-2}

Q. 200. The solubility of A_2X_3 in pure water, assuming that neither kind of ion reacts with water when for A_2X_3, $K_{sp} = 1.1 \times 10^{-23}$, will be:

(a) 0.5×10^{-5} mol L^{-1} **(b)** 2.0×10^{-5} mol L^{-1}

(c) 1.5×10^{-5} mol L^{-1} **(b)** 1.0×10^{-5} mol L^{-1}

Q. 201. Given, K_{sp} for $PbI_2 = 4 \times 10^{-9}$ and that 25 mL of a sample of clear saturated solution of PbI_2 required 10mL of a certain $AgNO_3(aq.)$ For its titration. The molarity of this $AgNO_3(aq.)$ is

(a) 0.5×10^{-2} **(b)** 5×10^{-3} **(c)** 0.005 **(b)** All of them

Q. 202. The $[Ag^+]$ of a solution is 4×10^{-3}. The $[Cl^-]$ that must be exceeded before AgCl can precipitate is:

(a) 1×10^{-10} **(b)** 0.2×10^{-8} **(c)** 2×10^{-8} **(d)** 2×10^{-10}

Q. 203. The mass of $BaSO_4$ that will dissolve in 450 mL of aqueous solution is

(a) 0.01 *g* **(b)** 0.001 *g* **(c)** 0.1 *g* **(d)** 0.0001 *g*

Q. 204. The solubility of AgCl in 0.10 *M* $BaCl_2$ is:

(a) 5×10^{-10} *M* **(b)** 0.5×10^{-8} *M* **(c)** 2×10^{-10} *M* **(d)** 0.2×10^{-8} *M*

Q. 205. If equal volumes of 0.02 *M* $CaCl_2$ and 0.0004 *M* Na_2SO_4 are mixed. Will a precipitate form?

K_{sp} for $CaSO_4 = 2.4 \times 10^{-5}$

(a) No **(b)** Yes **(c)** Don't Know

Chapter **10**

ELECTRO-CHEMISTRY

KEY CONCEPT

Equations and relationships that you need to know:

- $E° = E°_{ox} + E°_{red}$
- For the reaction: $aA + bB \rightarrow cC + dD$
  Nernst equation:

  $$E = E° - \frac{0.0592}{n}\log\frac{(C)^c\,(D)^d}{(A)^a\,(B)^b} = E° - \frac{0.0592}{n}\ \log Q \text{ at 298 K}$$

  $E° = (RT/n\ F)\ \ln Q = E° - (2.303\ RT/n\ F)\ \log Q$
  Total charge = $n\ F$
  1 F = 96,487 coulombs mol^{-1} = 96,487 $JV^{-1} \cdot mol^{-1}$
- potential(V) = $\frac{\text{work}}{\text{charge}} J.C^{-1}$

  I ampere = rate of flow of electrons, measured in coulombs sec^{-1} or amps
  electrical energy (work) = volts × coulombs = joules
- $E = \frac{-w}{q}$ where w = work, q = joules
- $\Delta G = w_{max} = -n\ FE$
- At equilibrium: $Q = K$; $E_{cell} = 0$

  $$\ln K = \frac{n \cdot E°_{cell}}{0.0257}$$

- OIL RIG: Oxidations is losing; Reduction is Gaining (electron)
- AN OX (A node is where OXidation occurs)
  RED CAT (REDuction occurs at the CAThode)
- Electrolytic cell—electrical energy is used to bring about a nonspontaneous electrical change, Anode is (+) electrode; cathode is (–) electrode.
- Voltaic (chemical) cell—electrical energy is produced by spontaneous redox reaction Anode is (–) terminal; cathode is (+) terminal.
- Reduction Between ΔG, K, and E°_{cell}

ΔG	K	E°_{cell}	*Reaction Under Standard State Conditions*
–	>1	+	Spontaneous
0	=1	0	Equilibrium
+	<1	–	Nonspontaneous, spontaneous in reverse direction

PROBLEMS

Q. 1. The mass of copper that would be produced by the reduction of the $Cu^{2+}(aq)$ ion by passing 96.487 amperes of current through a solution of copper (II) chloride for 100.00 minutes is: (1 Faraday = 96.487 coulombs)

(a) 95.325 g **(b)** 190.65 g **(c)** 285.975 g **(d)** 381.30 g

Q. 2. $E°$ for a cell composed of silver and gold electrodes in 1 molar solutions of their respective ions is; E°_{red} AG = +0.7991 volts, E°_{red} Au = +1.68 volts.

(a) –0.44 volt **(b)** 0 volt **(c)** 0.44 volt **(d)** 0.88 volt

Q. 3. $...FeCl_2 + ... KMnO_4 + ... HCl \rightarrow ... FeCl_3 + ... KCl + ... MnCl_2 + ?\ H_2O$

When the equation for this reaction is balanced with the lowest whole-number coefficients, the coefficient for H_2O is:

(a) 1 **(b)** 2 **(c)** 3 **(d)** 4

Q. 4. Given the following notation for an electrochemical cell;

$$Pt(s)\ |H_2(g)\ |H^+(aq)\ ||\ Ag^+(aq)|\ Ag(s)$$

Which of the following represents the overall balanced (net) cell reaction?

(a) $H_2(g) + Ag^+(aq) \rightarrow 2H^+(aq) + Ag(s)$

(b) $H_2(g) + Ag(s) \rightarrow H^+(aq) + Ag^+(aq)$

(c) $Ag(s) + H^+(aq) \rightarrow Ag^+(aq) + H_2(g)$

(d) none of the above

Q. 5. Given the following information, which of the statements is true?

$Cu^+(aq) + e^- \rightarrow Cu^+(aq)$	$E^{\circ}_{red} = 0.34$ V
$2H^+(aq) + 2e^- \rightarrow H_2(g)$	$E^{\circ}_{red} = 0.00$ V
$Fe^{2+}(aq) + 2e^- \rightarrow Fe(s)$	$E^{\circ}_{red} = -0.44$ V
$Ni(s) \rightarrow Ni^{2+}(aq) + 2e^-$	$E^{\circ}_{ox} = 0.25$ V

(a) $Cu^{2+}(aq)$ is the strongest oxidizing agent.

(b) $Cu^{2+}(aq)$ is the weakest oxidizing agent

(c) Ni(s) is the strongest oxidizing agent.

(d) Fe(s) would be the weakest reducing agent.

Q. 6. A cell has been set up as shown in the following diagram, and $E°$ has been measured as 1.00 V at 298 K. $\Delta G°$ for the reaction is:

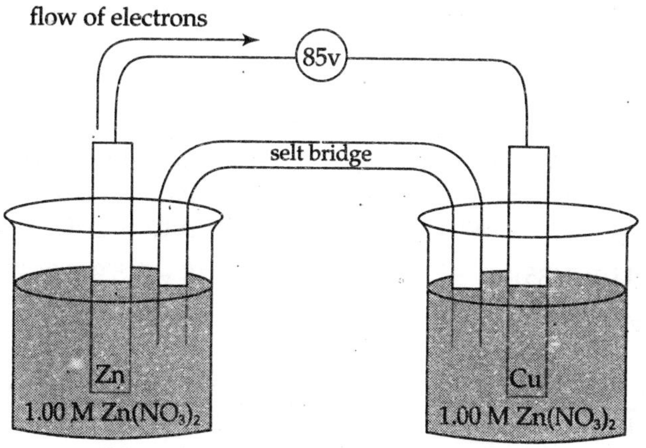

(a) –386 kJ **(b)** –193 kJ **(c)** 1.00 kJ **(d)** 386 kJ

Q. 7. Given the following diagram, determine which of the following statements is FALSE.

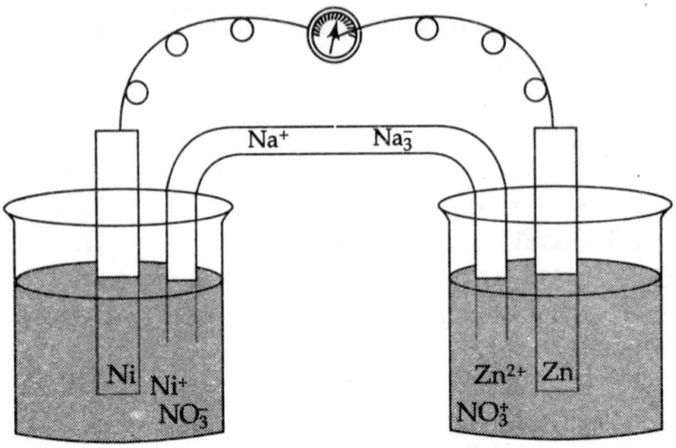

$$N_1^{2+}(aq) + Zn(s) \longrightarrow Ni(s) + Zn^{2+}(aq)$$

(a) The solid zinc electrode is being reduced to $Zn^{2+}(aq)$
(b) The nickel electrode is the cathode.
(c) The zinc electrons is the anode.
(d) The nickel(II) ions are being reduced.

Q. 8. Given that

$$Zn^{2+}(aq) + 2e^- \rightarrow Zn(s) \quad E^\circ_{red} = -0.76 \text{ V}$$
$$Cr^{3+}(aq) + 3e^- \rightarrow Cr(s) \quad E^\circ_{red} = -0.74 \text{ V}$$

the equilibrium constant K for the following balanced reaction is:

$$3Zn(s) + 2Cr^{3+}(aq) \rightarrow 3Zn^{2+}(aq) + 2Cr(s)$$

(a) $K = e^{-0.02}$ **(b)** $K = e^{0.02}$ **(c)** $K = e^{4.7}$ **(d)** $K = e^{8.0}$

Q. 9. An electric current is applied to an aqueous solution of $FeCl_2$ and $ZnCl_2$. Which of the following reactions occurs at the cathode?

(a) $Fe^{2+}(aq) + 2e^-(aq) \rightarrow Fe(s)$ $E^\circ_{red} = -0.44$ V
(b) $Fe(s) \rightarrow Fe^{2+}(aq) + 2e^-$ $E^\circ_{red} = 0.44$ V
(c) $Zn^{2+}(aq) + 2e^-(aq) \rightarrow Zn(s)$ $E^\circ_{red} = -0.76$ V
(d) $Zn(s) \rightarrow Zn^{2+}(aq) + 2e^-$ $E^\circ_{ox} = 0.76$ V

Q. 10. For the reaction

$$Pb(s) + PbO_2(s) + 4H^+(aq) + 2SO_4^{2-}(aq) \rightarrow 2PbSO_4(s) + 2H_2O(l)$$

which is the overall reaction in a lead storage battery, $\Delta H^\circ = -315.9$ kJ mol^{-1} and $\Delta S^\circ = 263.5$ J · K^{-1} mol^{-1}. What is the proper setup to find E° at 348 K.

(a) $\dfrac{-315.9 - 348(0.2635)}{-2(96.487)}$ **(b)** $\dfrac{-348 + 315.9(0.2635)}{2(96.487)}$

(c) $\dfrac{-348 + 315.9(0.2635)}{96.487}$ **(d)** $\dfrac{-2(-348) + 263.5}{96.487 + 315.9}$

Q. 11. The cell potential for the following cell is:

$$Co(s) \mid Co^{2+}(aq) \parallel Cr^{3+}(aq) \mid Cr(s)$$

(given : E_{red} $CO^{2+}(aq) = -0.28$ V

E_{red} $Cr^{3+}(aq) = -0.74$ V)

(a) −1.06 V (b) +0.5 V (c) 1.06 V (d) −0.5 V

Q. 12. The cell potential of the cell that has the following overall reaction is:

$$Zn + I_2 \rightarrow Zn^{2+} + 2I^- \text{ is:}$$

given $E_{oxid} Zn^{2+} = 0.76$ V

E_{red} $I^- = 0.54$ V

(a) −1.30 V (b) +0.22 V (c) + 1.30 V (d) −0.22 V

Q. 13. The free energy charge for an electrochemical cell based on the following cell reaction is:

$$Mn(s) + Cd^{2}(aq) \rightarrow Mn^{2+}(aq) + Cd(s),$$

Given, Oxidation half reaction, $Mn(s) \rightarrow Mn^{2+}(aq) + 2e^-$ $E^{\circ}_{oxid} = +1.18$ V

Reduction half reaction $Cd^{2+}(aq) + 2e^- \rightarrow Cd(s)$ $E^{\circ}_{red} = -0.40$ V

(a) +75.25 kJ mol^{-1} (b) −150.5 kJ mol^{-1}
(c) +150.5 kJ mol^{-1} (d) −75.25 kJ mol^{-1}

Q. 14. Predict whether the following cell reaction should proceed spontaneously:

$$Cr^{3+}(aq) + Fe^{2+}(aq) \rightarrow Cr^{2+}(aq) + Fe^{3+}(aq)$$

(a) Yes (b) No

Q. 15. A Zinc rod is placed in a 0.01 M solution of zinc sulphate at 298 K. The potential of the electrode at this temperature, (assuming 96% dissociation of $ZnSO_4$ and E° $(Zn^{2+}/Zn) = 0.76$ V) will be:

(a) −0.79 V (b) −0.97 V (c) +0.79 V (d) +0.97 V

Q. 16. The potential of the following electrochemical cell at 298 K
$Cu(s)$ $Cu^{2+}(aq)$ (0.50 M) $\parallel$ H^+ (0.01) $\mid$ H_2(0.95 atm); is:

(Given: $E^{\circ}_{cathode} = 0.00$ V and $E^{\circ}_{anode} = 0.34$ V)

(a) +0.944 V (b) −0.944 V (c) − 0.449 V (d) +0.449 V

Q. 17. The concentration x of Cd^{2+} ions in the following electrochemical cell,

$$Fe, Fe^{2+}(0.1M) \mid Cd^{2+} (xm), Cd \text{ is:}$$

(assuming that are of equal concentrations and given the EMF of the cell $E = -0.02$ V and $E^{\circ} = 0.04$ V at 298 K

(a) 0.01 M (b) 0.001 M (c) 0.1 M (d) 0.0001 M

Q. 18. The equilibrium constant for the reduction of riboflavin with rubredoxin in the reaction riboflavin (ox) + rubredoxin (red) $\rightleftharpoons$ riboflavin (red) + rubredoxin (ox) is: (given that at pH = 7, the standard potential for riboflavin and rubredoxin are −0.21 V and −0.06 V respectively)

(a) 8.24×10^{-6} (b) 2.84×10^{-6} (c) 4.28×10^{-6} (d) 8.42×10^{-6}

Q. 19. Arrange the following metals in the decreasing order in which they displace each other, Al, Cu, Fe, Mg, Zn. The standard reduction potentials of these metals are −1.66 V, +0.34 V, −0.44 V, −2.37 V and −0.76 V respectively. The arrangement will be:

(a) Mg > Zn > Al > Cu > Fe
(b) Al > Mg > Fe > Zn > Cu
(c) Mg > Al > Zn > Fe > Cu
(d) Cu > Zn > Fe > Mg > Al

Q. 20. K_{eq} for $Cu(s) + 2Ag^+(aq) \rightleftharpoons Cu^{2+}(aq) + 2Ag(s)$ at 298 K will be:
(Given $E°Cu^{2+}/Cu = 0.34$ V and $E°Ag^+/Ag = 0.80$ V)
(a) 7.8×10^{17} (b) 8.8×10^{18} (c) 6.8×10^{16} (d) 9.8×10^{19}

Q. 21. Given $E°(Cu^+, Cu) = +0.52$ V and $E°(Cu^{2+}, Cu^+) = +0.16$ V
The equilibrium constant for the reaction

$$2Cu^+(aq) \rightleftharpoons Cu(s) + Cu^2(aq) \text{ at 298 K is:}$$

(a) 1.2×10^6 (b) 2.4×10^6 (c) 3.6×10^6 (d) 9.6×10^6

Q. 22. Given $E°(Sn^{2+}/Sn) = -0.14$ V and $E°(Pb^{2+}/Pb) = -0.13$ V
The equilibrium constant for the reaction

$$Sn^{2+}(aq) + Pb(s) \rightleftharpoons Sn(s) + Pb^{2+}(aq) \text{ at 298 K is:}$$

(a) 0.046 (b) 0.46 (c) 0.0046 (d) 4.6

Q. 23. Given $Cd^2 + 2e^- \rightarrow Cd(s)$; $E° = -0.40$ V

$$Ag^+ + e^- \rightarrow Ag(s);\ E° = +0.78 \text{ V}$$

The overall cell reaction is: $Cd(s) + 2Ag^+(aq) \rightarrow Cd^2(aq) + 2Ag(s)$
The emf of the cells is:
(a) 0.83 V (b) 1.81 V (c) 0.38 V (d) 1.18 V

Q. 24. For the cell:

$$Zn \mid Zn^{2+} (1\ M) \parallel Fe^{2+} (1\ M), Fe^{3+} (1\ M) \mid Pt,$$

Given $E°$ $(Fe^{3+} \mid Fe^{2+}) = +0.77$ V and $E°$ $(Zn \mid Zn^{2+}) = +0.76$ V, the value of $E°_{cell}$ for the above cells is:
(a) 2.3 V (b) 0.01 V (c) −0.01V (d) 1.53 V

Q. 25. A hydrogen electrode with hydrogen gas at 1 atm and hydrogen ion at unit activity is suitably connected to a half-cell consisting of copper metal immersed in 1.00 M copper (II) sulphate solution. For the cell $Pt \mid H_2$ (1 atm) $\mid H^+$ (1 M) $\parallel$ (1 M) $\parallel Cu^{2+}$ (1 M) $\mid$ Cu, the measured potential to 0.34 V. the standard electrode potential for the Cu $\mid Cu^{2+}$ half-cell is:
(a) 0 (b) 0.34 V (c) +0.68 V (d) 1 V

Q. 26. The potential of a silver/silver chloride electrode measured with respect to a saturated calomel electrode is −0.022 V. The value of the standard reduction potential for the silver/silver chloride electrode is:
(a) 0.222 V (b) 22.2 V (c) 2.22 V (d) 0.246 V

Q. 27. Two copper rods are placed in copper sulphate solution of concentration 0.1 M and 0.01 M separately in the form of a cell. The emf of the cell at 298 K is:
(a) 0.0296 V (b) 0.0692 V (c) 0.0962 V (d) 0.0926 V

Q. 28. The valency of mercurous ions for the following cell when the emf is measured at 298 K is 0.029 volt (F = 96,500) is:

Hg	0.001 N mercurous nitrate in 0.1 N HNO_3	0.01 N mercurous nitrate in 0.1 N HNO_3
	Anode	*Cathode*

(a) 1 (b) 2 (c) 3 (d) 5

Q. 29. Time taken by 5 amperes of current to deposit 2 gm of copper from a solution of copper sulphate is: (Given chemical equivalent of copper = 32 and F = 96,500 coulombs)
(a) 100 min, 31 sec **(b)** 10 min, 3 sec **(c)** 20 min, 6 sec **(d)** 50 min, 15 sec

Q. 30. The EMF of the following cell at 298 K is:
$Ag(s)|AgNO_3$ (0.018 m) || Ag No_3 (1.2m)|Ag(s)
(a) 0.108 V **(b)** 0.045 V **(c)** 0.054 V **(d)** 0.09 V

Q. 31. A battery system that may be used to power automobiles in the future is the aluminum-air battery. This is a *flow battery* in which oxidation occurs at an aluminum anode and reduction at a carbon—air cathode. The electrolyte circulated through the battery is NaOH(aq); the ultimate reaction product is $Al(OH)_3(s)$, which is removed from the battery as it is formed. In operation the battery can be kept charged by feeding AI anode stags and water into the battery; oxygen is drawn from the air (see Figure). The battery can power an automobile several hundred miles between charges. The $Al(OH)_3(s)$ removed from the battery can be converted back to aluminum in an aluminum manufacturing facility.

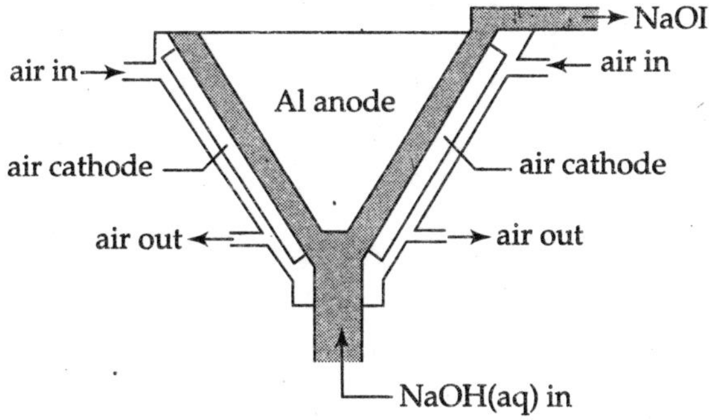

In actual practice Al^{3+} produced at the anode does not precipitate as $Al(OH)_3(s)$ but is obtained as the complex ion $[Al(OH)_4]^-$ in the presence of NaOH(aq). $[Al(OH)_3$ is precipitated from the circulating NaOH(aq) electrolyte *outside* the battery.] The equations for the oxidation and reduction half-reactions and the net reaction that occur in the battery are:

Oxidation: $4\{Al(s) + 4\ OH^-(aq) \rightarrow [Al(OH)_4]^- + 3e^-\}$

Reduction: $3\{O_2(g) + 2H_2O + 4e^- \rightarrow 4OH^-(aq)\}$

Net: $4Al(s) + 3O_2(g) + 6H_2O + 4OH^- \rightarrow 4[Al(OH)_4]^-$ . . . . (A)

(i) If the theoretical voltage of the aluminum—air cell is +2.73 V, and $E°$ for the reduction is +0.401 V then for the reaction

$[Al(OH)_4]^- + 3e^- \rightarrow Al(s) + 4OH^-$ $E° = ?$

(a) +3.173 V **(b)** −3.173 V **(c)** −2.33 V **(d)** +2.33 V

(ii) Given that E°_{cell} for reaction (A) is +2.73 V, that $\Delta G_f^\circ[OH^-] = -157$ kJ mol^{-1}, and that $\Delta G_f^\circ[H_2O(l)] = -237.2$ kJ mol^{-1}, the free energy of formation, ΔG_f°, of the aluminate ion, $[Al(OH)_4]^-$ will be:

(a) $+1.3 \times 10^3$ kJ mol^{-1} **(b)** -5.2×10^3 kJ mol
(c) -1.3×10^3 kJ mol^{-1} **(d)** $+5.2 \times 10^3$ kJ mol^{-1}

(iii) Mass of aluminum is consumed if 10.0 A of electric current is drawn from the battery for 4.00 h is:
(a) 40.2 g Al (b) 53.6g Al (c) 26.8 g Al (d) 13.4g Al
See tables in textbook for $\mathscr{E}^\circ$ values

Q. 32. A substance that will reduce Ag^+ to Ag but will not reduce Ni^{2+} to Ni is:
(a) Zn (b) Pb (c) Mg (d) Cd

Q. 33. The EMF of the cell Zn(s) | Zn^{2+}(1.0 *M*) || Pb^{2+}(1.0 *M*) | Pb(s) is
(a) 0.889 V (b) 0.637 V (c) 0.511 V (d) –0.637 V

Q. 34. The value of the reaction quotient, *Q*, for the cell

Zn(s) Zn^{2+}(0.0100 *M*) || Ag^+(1.25 *M*) | Ag(s)

is
(a) 156 (b) 125 (c) 1.25×10^{-2} (d) 6.40×10^{-3}

Q. 35. During the electrolysis of aqueous zinc nitrate.
(a) Zinc plates out at the cathode.
(b) Zinc plates out at the anode.
(c) Nitrogen gas, N_2, is evolved at the cathode.
(d) Oxygen gas, O_2, is evolved at the anode.

Q. 36. How many faradays are required to reduce a mole of MnO_4^- to Mn^{2+}?
(a) 5 (b) 2 (c) 3 (d) 4

Q. 37. If all species are in their standard states, which of the following is the strongest oxidizing agent?
(a) Br^- (b) Zn^{2+} (c) Fe^{2+} (d) Co^{3+}

Q. 38. Which of the following statements about a reaction occurring in a galvanic cell is true?
(a) If $\Delta\mathscr{E}_{cell} > 0, \Delta G < 0$
(b) If $\Delta\mathscr{E}^\circ_{cell} < 0, \Delta G < 0$
(c) If $\Delta\mathscr{E}^\circ_{cell} < 0, K > 1$
(d) If $\Delta\mathscr{E}_{cell} < 0, K > 1$

Q. 39. For the net cell reaction of the cell Zn(s) | Zn^{2+} || Cd^{2+} | Cd(s) ΔG° in kilojoules at 298 K is:
(a) 112.5 (b) 69.47 (c) –34.73 (d) –69.47

Q. 40. Consider the cell Cd(s) | Cd^{2+} (1.0 *M*) || Cu^{2+}(1.0 *M*) | Cu(s). If we wanted to make a cell with a more positive voltage using the same substances, we should:
(a) Increase both the [Cd^{2+}] and [Cu^{2+}] to 2.00 *M*.
(b) Increase only the [Cd^{2+}] to 2.00 *M*
(c) Decrease both the [Cd^{2+}] and [Cu^{2+}] to 0.100 *M*
(d) Decrease only the [Cd^{2+}] to 0.100 *M*

Q. 41. Of the following metals

Ag Cd Cu Hg Cr

those that plate out at the cathode during an electrolysis of an aqueous solution of one of their salts are:
(a) Ag, Cu and Hg
(b) Ag, Cd and Cu
(c) Cd, Cu and Cr
(d) Ag, Hg, and Cr

Q. 42. When a dilute aqueous solution of Li_2SO_4 is electrolyzed, the products formed at the anode and cathode, respectively, are:
(a) S and Li (b) O_2 and Li (c) SO_2 and H_2 (d) O_2 and H_2

Q. 43. Which of the following statements about the spontaneous reaction occurring in galvanic cell is always true?

(a) $\Delta\mathscr{E}^{\circ}_{cell} > 0. \Delta G^{\circ} < 0$, and $Q < K$ (b) $\Delta\mathscr{E}^{\circ}_{cell} > 0$, $\Delta G^{\circ} > 0$, and $Q < K$

(c) $\Delta\mathscr{E}^{\circ}_{cell} > 0$, $\Delta G < 0$, and $Q < K$ (d) $\Delta\mathscr{E}^{\circ}_{cell} > 0$, $\Delta G < 0$, and $Q > K$

Q. 44. For the cell Zn(s) | Zn^{2+} || Cu^{2+} | Cu(s), the standard cell voltage, $\Delta\mathscr{E}^{\circ}_{cell}$, is 1.10V. When a cell using these reagents was prepared in the lab the measured cell voltage was 0.98 V. One possible explanation for the observed voltage is

(a) There were 2.00 mol of Zn^{2+} but only 1.00 mol of Cu^{2+}

(b) The Zn electrode had twice the surface of the Cu electrode.

(c) The $[Zn^{2+}]$ was larger than the $[Cu^{2+}]$

(d) The volume of the Zn^{2+} solution was larger than the volume of the Cu^{2+} solution

Q. 45. If a steady current of 15.0 A is passed through an aqueous solution of $CuSO_4$, how many minutes will it take to deposit 0.250 mol of Cu at the cathode, assuming 10.0 efficiency?

(a) 3.22×10^3 (b) 1.61×10^3 (c) 53.6 (d) 26.8

Q. 46. The base 10 logarithm of the equilibrium constant, log *K*, of the net cell reaction the cell Cd(s) Cd^{2+} || Ag^+ | Ag(s) is

(a) 67.6 (b) 40.6 (c) 20.3 (d) 13.4

Q. 47. A steady current of 30.0 A for 70.2 min corresponds to a passage of

(a) 1.26×10^5 electrons (b) 1.31 C

(c) 1.26×10^5 $\mathscr{F}$ (d) 1.31 $\mathscr{F}$

Q. 48. Which of the following changes will increase the EMF of the cell

$$Co(s) \mid CoCl_2(M_1) \parallel HCl(M_2) \mid H_2(g)\ Pt(s)$$

(a) Increase the volume of the $CoCl_2$ solution from 100 to 200 mL.

(b) Increase M_2 from 0.010 to 0.500 *M*.

(c) Increase the pressure of the H_2(g) from 1.00 to 2.00 atm.

(d) Increase the mass of the Co electrode from 10.0 to 20.0 g

Q. 49. A 0.200 *F* KOH solution was electrolyzed for 1.50 h using a current of 8.00 A. How many moles of O_2 was produced at the anode?

(a) 0.448 (b) 0.224 (c) 0.112 (d) 2.24×10^{-2}

Q. 50. What is the value of the reaction quotient *Q*, for the cell

Ni(s) | $Ni(NO_3)_2$(0.10 *M*) || KCl(0.40 *M*) | Cl_2(g, 0.50 atm) | Pt(s)

(a) 3.1×10^{-1} (b) 1.3×10^{-1} (c) 8.0×10^{-2} (d) 3.2×10^{-2}

Q. 51. Warm dilute nitric acid will oxidize

(a) Pt but not Au (b) Cu but not Ag

(c) Au but not Ag (d) Ag but not Pt

Q. 52. Silver is electrodeposited on a metallic vessel of surface area 800 cm^2 by passing a current of 0.2 A for 3 hours. The thickness of silver deposited is:

[Given, Atomic mass of Ag = 107.92 amu and Density of Ag = 10.5 g/cm^3]

(a) 1.44×10^{-3} cm (b) 0.72×10^{-3} cm

(c) 5.76×10^{-3} cm (d) 2.88×10^{-3} cm.

Q. 53. A cell uses Zn^{2+} | Zn and Ag^+ | Ag electrodes.

Given $E^{\circ}Zn^2$ | Zn = 0.76 V and $E^{\circ}Ag^+$ | Ag = 0.8 V) The EMF of the cell is:

(a) 1.56 V (b) 5.56 V (c) 2.56 V (d) 3.56 V

Q. 54. **(i)** The cell reaction for following voltaic cell

$Tl(s) \mid Tl \mid^{+} (aq) \parallel Sn^{2+} (aq) \mid Sn(s)$--- is:

(a) $Tl(s) + Sn^{+}(aq) \rightarrow Tl^{+}(aq) + Sn((s)$

(b) $2Tl(s) + Sn^{2+}(aq) \rightarrow 2Tl^{+}(aq) + Sn(s)$

(c) $Tl(s) + Sn^{3+}(aq) \rightarrow Tl^{3+}(aq) + Sn(s)$

(d) $Tl(s) + 3Sn^{+}(aq) \rightarrow 2Tl^{+}(aq) + 2Sn(s)$

(ii) The cell reaction for the following voltaic cell

$Zn(s) \mid Zn^{2}(aq) \parallel Fe^{3+}(aq), Fe^{2+}(aq) \mid Pt$ . . . is:

(a) $2Zn(s) + 2Fe^{3+}(aq) \rightarrow 2Zn^{+}(aq) + 2Fe^{2+}(aq)$

(b) $Zn(s) + 3Fe^{3+}(aq) \rightarrow 2Zn^{+}(aq) + 2Fe^{2+}(aq)$

(c) $Zn(s) + 2Fe^{3+}(aq) \rightarrow Zn^{+}(aq) + 2Fe^{2+}(aq)$

(d) $2Zn(s) + 3Fe^{2+}(aq) \rightarrow 3Zn^{+}(aq) + 2Fe^{3+}(aq)$

Q. 55. The EMF of a particular voltaic cell with cell reaction

$Hg_2^{2+}(a) + H_2(g) \rightleftharpoons 2Hg(l) + 2H^{+}(aq)$

is 0.650V. The maximum electrical work of this cell when 0.5 g H_2 is consumed will be:

(a) -6.02×10^5 J **(b)** -4.01×10^3 J **(c)** -5.8×10^6 J **(d)** -3.09×10^4 J

Q. 56. **(i)** Order the following oxidizing agents by increasing strength under standard state condition: $Cl_2(g)$, $H_2O_2(aq)$, $Fe^{3+}(aq)$

(a) $H_2O_2(aq), Fe^{3+}(aq), Cl_2(g)$ **(b)** $Fe^{3+}(Aq), Cl_2(g), H_2O_2(aq)$

(c) $Fe^{3+}(aq), H_2O_2(aq), Cl_2(g)$ **(d)** $H_2O_2(aq), Cl_2(g), Fe^{3+}(aq)$

(ii) Order the following reducing agents by increasing strength

(a) $Al(s), H_2(g), Cu(s)$ **(b)** $Al(s), Cu(s), H_2(g)$

(c) $Cu(s), Al(s), H_2(g)$ **(d)** $Cu(s), H_2 (g), Al(s)$

Q. 57. The standard emf of the following voltaic cell at 298 K using standard electrode potential (given as $E^{\circ}_{Al} = -1.66$ V, $E^{\circ}_{Fe} = -0.412$ V) is:

(a) 2.07 V **(b)** 1.25 V **(c)** – 1.25 V **(d)** –2.07 V

Q. 58. The standard the energy charge at 25°C for the reaction

$Zn(s) + 2Ag^{+}(aq) \rightarrow Zn^{2+}(aq) + 2Ag(s)$is:

(a) -6.02×10^5 J **(b)** $+3{,}01 \times 10^5$ J **(c)** -3.01×10^5 J **(d)** $+6.02 \times 10^5$ J

Q. 59. Suppose the reaction of the zinc metal and chlorine gas is utilized in a voltaic cell in which zinc ions and chloride ions are formed in aqueous solution.

$$Zn(s) + Cl_2(g) \xrightarrow{H_2O} Zn^{2+}(aq) + 2Cl^{-}(aq)$$

To standard emf for this cell at 298 K from standard free energies of formation is:
(Given, ΔG°_f: Zn^{2+}(mg) = –147 kJ; ΔG°_f: Cl^{-}(aq) = –131 kJ)

(a) 3.18 V **(b)** 1.06 V **(c)** 4.24 V **(d)** 2.12 V

Q. 60. The standard emf for the following voltaic cell is 1.10 V.

$Zn(s) \mid Zn^{2+}(aq) \parallel Cu^{2+}(aq) \mid Cu(s)$

The equilibrium constant K_c for the reaction

$Zn(s) + Cu^{2+}(aq) \rightleftharpoons Zn^{2+}(aq) + Cu(s)$ is:

(a) 4.8×10^{37} **(b)** 3.2×10^{37} **(c)** 1.6×10^{37} **(d)** 6.4×10^{37}

Q. 61. The emf of the following voltaic cell at 298 K is:

$$Zn(s) \mid Zn^{2+} (1.00 \times 10^{-5}\ M) \parallel Cu^{2+} (0.100\ M) \mid Cu(s)$$

(The standard emf of the cell is 1.10 V)

(a) 1.22 V (b) 2.44 V (c) 0.98 V (d) 1.34 V

Q. 62. When an aqueous solution of copper(II) sulfate, $CuSO_4$, is electrolyzed, copper metal is deposited.

$$Cu^{2+}(aq) + 2e^- \rightarrow Cu(g)$$

(The other electrode reaction gives oxygen $2H_2O \rightarrow O_2 + 4H^+ + 4e^-$). If a constant current was passed for 5.00 h and 404 mg of copper meld was deposited what was the current?

(a) 3.4×10^{-2} A (b) 2.45×10^{-2} A (c) 6.81×10^{-2} A (d) 20.43×10^{-2} A

Q. 63. When an aqueous solution of potassium iodide is electrolyzed using platinum electrodes the half-reactions are

$$2I^-(aq) \rightarrow I_2(aq) + 2e^-$$
$$2H_2O(l) + 2e^- \rightarrow H_2(g) + 2HO^-(aq)$$

The number of gms of iodine that are produced when a current of 8.52 mA flows through the cell for 10 min. are:

(a) 13.46×10^{-3} g I_2 (b) 6.73 NO^{-3} g I_2
(c) 0.73×10^{-3} g I_2 (d) 3.36×10^{-3} g I_2

Q. 64. Zinc granules are needed in excess to a 500 mL of 1.0 *M* nickel nitrite solution at 298 K until the equilibrium is reached. If the standard deduction potentials of Zn^{2+}/Zn and Ni^2/Ni are –0.75 V and –0.24 V respectively; the concentration of Ni^{2+} in solution at equilibrium will be:

(a) 5.62×10^{-14} M (b) 5.62×10^{-16} M
(c) 5.62×10^{-18} M (d) 5.62×10^{-20} M

Q. 65. The standard reduction potential of Cu^{+2+}/Cu and Ag^+/Ag electrodes are 0.337 and 0.799 volt respectively. Construct a galvanic cell using these electrodes so that its standard emf is positive. For what concentration of Ag^+ will the emf. of the cell at 298 K be zero if the concentration of Cu^{++} is 0.01 *M*?

(a) 5.7×10^{-19} M (b) 1.52×10^{-19} M
(c) 6.08×10^{-19} M (d) 8.8×10^{-19} M

Q. 66. The equilibrium constant for the reaction

$$Fe^{2+} + Ce^{4+} \rightleftharpoons Fe^{3+} + Ce^{3+}$$

(given $E^\circ_{Ce^{+4}/Ce^{3+}} = 1.44$ V; $E^\circ_{Fe^{3+}/Fe^{2+}} = 0.68$ V) is:

(a) 7.61×10^{12} (b) 6.71×10^{10} (c) 6.71×10^{12} (d) 7.61×10^{10}

Q. 67. A current of 1 amp is passed in series through a solution of copper sulphate and another solution of a salt of metal, M. After certain time, 1.228 g of Cu and 4.176 g of M are deposited. Calculate the equivalent weight of M. What is the time taken for the deposition? (At, wt, Cu = 63.54; 1*F*= 965000 Coulomb mol^{-1})

(a) 3370 sec. (b) 3730 sec. (c) 7330 sec. (d) 4200 sec

Q. 68. If a constant current was passed for 50 hr and 404 mg of Cu was deposited. The amount of current passed is:

(a) 0.0682 amp (b) 0.0826 amp (c) 0.2086 amp (d) 0.0286 amp

Q. 69. A cell contains two hydrogen electrodes. The negative electrode is in contact with a solution of 10^{-6} M hydrogen ions. The e.m.f. of the cell is 0.118 V at 298 K. The concentration of hydrogen ions of the positive electrode is:

(a) 10^{-2} M (b) 10^{-6} M (c) 10^{-4}M (d) 10^{-3} M

Q. 70. The standard reduction potential for Cu^{+3} | Cu is +0.34 V. Calculate the reduction potential at pH = 14 for the above couple K_{sp} of $Cu(OH)_2$ is 10×10^{-19}.

(a) −0.11 V (b) +0.22 V (c) +0.11 V (d) −0.22 V

Q. 71. The standard redox potential of the following half cell is 0.78 V

$$NO_3 + 2H^+ + e^- \rightarrow NO_2(g) + H_2(C)$$

the redox potential of 8 M H^+ ions, and the redox potential of neutral solution will be: (Assume the concentration of all other species as unity)

(a) 0.886; −0.047 (b) 0.047, − 0.886 V
(c) 0.886 + 0.047 V (d) −0.047 + 0.886 V

Q. 72. Calculate the equilibrium constant for the reaction:

$$2Fe^{3+} + 3I^- \rightleftharpoons 2Fe^{2+} + I_3^-$$

The standard reduction potentials in acidic condition are 0.77 and 0.54 V respectively for Fe^{3+}/Fe^{2+} and I_3^-/I^- couples.

(a) 4.26×10^7 (b) 5.26×10^7 (c) 3.26×10^7 (d) 6.26×10^7

Q. 73. The pH of the following half cell Pt, H_2/H_2SO_4 when the oxidation electrode potential is +0.3 V is:

(a) 2.08 (b) 4.08 (c) 3.08 (d) 5.08

Chapter 11

NUCLEAR CHEMISTRY

KEY CONCEPT

- $t_{1/2} = \frac{0.693}{k}$
- $\ln\left(\frac{x_o}{x}\right) = Kt$

  where

  x_o = original number of nuclides at time zero
  x = number of nuclides at time t

- $\Delta E = \Delta mc^2$

  where

  $c = 3.00 \times 10^8 \text{ m} \cdot \text{sec}^{-1}$

  or

  $\Delta E = 9.00 \times 10^{10} \text{ kJ} \cdot g^{-1} \cdot \Delta\text{m}$

- Symbols

proton	${}_1^1p$ *or* ${}_1^1H$
neutron	${}_0^1n$
electron (β particle)	${}_{-1}^{0}e$ or ${}_1^0\beta$ or e^-
positron	${}_1^0e$ or ${}_1^0\beta$
(α particle)	${}_2^4\text{He}$ or ${}_2^4\alpha$
gamma rays	γ
deuteron	${}_1^2H$
triton	${}_1^3H$

PROBLEMS

For Questions 1–3, Choose answer from the following:

(a) γ

(b) $2{}^{1}_{0}n$

(c) $.{}^{1}_{0}n$

(d) ${}^{4}_{2}\text{He}$

Q. 1. ${}^{54}_{26}\text{Fe} + ? \rightarrow {}^{56}_{26}\text{Fe} + 2{}^{1}_{1}\text{H}$

Q. 2. ${}^{65}_{29}\text{Cu} + {}^{1}_{0}\text{n} \rightarrow {}^{64}_{29}\text{Cu} + ?$

Q. 3. ${}^{14}_{7}\text{N} + {}^{1}_{1}\text{H} \rightarrow {}^{15}_{8}\text{O} + ?$

Q. 4. Choose the one FALSE statement.

(a) Nuclei with an even number of protons and an even number of neutrons tend to be stable.

(b) "γ"-rays are high-energy photons.

(c) Nuclei with too few neutrons per proton tend to undergo positron (${}^{0}_{1}\beta$) emission

(d) Nuclei with too many neutrons per proton tend to undergo β- particle (${}^{0}_{-1}e$) emission.

For Questions 5–8 choose answer from the following:

(a) ${}^{201}_{80}\text{Hg} + {}^{0}_{-1}\text{e} \rightarrow {}^{201}_{79}\text{Au} + {}^{0}_{0}\gamma$

(b) ${}^{11}_{6}\text{C} \rightarrow {}^{0}_{1}\text{e} + {}^{11}_{5}\text{B}$

(c) ${}^{237}_{99}\text{Np} \rightarrow {}^{4}_{2}\text{He} + {}^{233}_{91}\text{Pa}$

(d) ${}^{235}_{92}\text{U} + {}^{1}_{0}\text{n} \rightarrow {}^{141}_{56}\text{Ba} + {}^{92}_{36}\text{Kr} + 3{}^{1}_{0}\text{n}$

Q. 5. Alpha (α)- particle production.

Q. 6. Electron capture.

Q. 7. Fission.

Q. 8. Positron production.

Q. 9. The half-life of C is 5770 years. What percent of the original radioactivity would be present after 28,850 years?

(a) 1.56% (b) 3.12% (c) 6.26% (d) 12.5%

Q. 10. Which of the following choices correctly describes the decreasing ability of the radiation to penetrate a sheet of lead that is 3 inches thick?

(a) alpha particles > beta particles > gamma rays

(b) gamma rays > alpha particles > beta particles

(c) alpha particles > gamma rays > beta particles

(d) gamma rays > beta particles > alpha particles

Q. 11. A radioactive sample has an initial of 28 dis/min; $\frac{1}{2}$h later the activity is 14 dis/min. How many atoms of the radioactive nuclide were there originally? ($\lambda_{t1/2}$ = 0.693)
(a) 1000 (b) 1400 (c) 1200 (d) 800

Q. 12. ^{18}F is found to undergo 90% radioactive decay in 366 min. Its half life is
(a) 11 min. (b) 55 min. (c) 22 min. (d) 110 min

Q. 13. A piece of (old) wood, was burned, and 7.32g CO_2 was collected. The total radioactivity in the CO_2 was 10.8 dis/min How old was the wood sample?
(a) 8610 years (b) 6180 years (c) 6801 years (d) 8160 years

Q. 14. Given 2.00kg of $^{238}_{92}U$ (half life 4.50×10^9 y). The ultimate decay product in this series is $^{206}_{82}Pt$. The weight of $^{238}_{92}U$ left after 3.00×10^9 y, and the weight of $^{206}_{82}Pt$ produced in this time respectively will be:
(a) 0.64kg; 1.26kg. (b) 1.26kg; 1.26kg. (c) 1.26kg; 0.64kg. (d) 0.64kg; 0.64kg.

Q. 15. The bones of a prehistoric bison were found to have a ^{14}C activity of 2.80 dis/min.g carbon Approximately how long ago did the animal live?
(a) 140 years (b) 14000 years (c) 14 years (d) 1400 years

Q. 16. A pure radiochemical preparation was observed to disintegrate at the rate of 4280 counts/min at 1 : 35 p.m. At 4 : 55 p.m. of the same day, the disintegration rate of the sample was only 1070 counts/min. The disintegration rate is proportional to the number of radioactive atoms in the sample. What is the half-life of the material?
(a) 200 min (b) 50 min (c) 100 min (d) 400 min

Q. 17. An atomic battery for pocket watches has been developed which uses the beta particles from ^{147}Pm as the primary energy source,. The half-life of ^{147}Pm is 2.62 y. How long would it take for the rate of beta emission in the battery to be reduced to 10% of its initial value?
(a) 4.1 years (b) 17.4 years (c) 2.6 years (d) 8.7 years

Q. 18. If the limit of a particular detection system is 0.002 dis/s for a 1 g sample, what would be the maximum half-life that this system could detect on a 1 g sample of nuclide whose mass number is around 200?
(a) 3×10^{16} y (b) 2×10^{16} y (c) 4×10^{16} y (d) 5×10^{16} y

Q. 19. ^{227}Ac has a half life of 22.0 yrs. with respect to radioactive decay. The decay follows two parallel paths, one leading to ^{227}Th and the other to ^{223}Fr. The percentage yields of these two daughter nuclides are 2.0 and 98.0 respectively. The decay constants (λ) for each of the separate paths respectively are:
(a) 1.84×10^{-4}/year; 3.13×10^{-2}/year
(b) 3.13×10^{-4}/year; 1.84×10^{-2}/year
(c) 3.68×10^{-4}/year; 6.26×10^{-2}/year
(d) 6.26×10^{-4}/year; 3.68×10^{-2}/year

Q. 20. A sample of ^{238}U (half-life 4.5×10^9 yrs.) ore is found to contain 23.8 g of ^{238}U and ^{206}Pb the age of the ore is:
(a) 4.49×10^7 years (b) 4.49×10^{11} years
(c) 4.49×10^9 years (a) 4.49×10^5 years

Q. 21. The half-life of 99^m Tc is 6.0 h. The delivery of a sample of 99^m Tc from the reactor to the nuclear medicine lab of a certain hospital takes 3.0 h. What is the minimum amount of 99^m Tc that must be shipped in order for the lab to receive 10.0 mg?
(a) 20.0 mg (b) 18.6 mg (c) 15.0mg (d) 14.1 mg

Q. 22. Which of the following pairs are isobars?
(a) $^{138}_{58}Ce$, $^{140}_{58}Ce$ (b) $^{106}_{48}Cd$, $^{106}_{46}Pd$ (c) $^{83}_{36}Kr$, $^{84}_{36}Kr$ (d) $^{74}_{32}Ge$, $^{75}_{33}As$

Q. 23. Of the following particles, the one with the greatest mass is the
(a) triton (b) proton (c) deutron (d) α-particles

Q. 24. Aluminium-25 decays by emitting a positron. The species immediately produced has
(a) 12 p, 13 n, 13 e^- (b) 13 p, 12 n, 13 e^-
(c) 12 p, 13 n, 12 e^- (d) 14 p, 11 n, 14 e^-

Q. 25. The most likely mode of decay $^{13}_{7}N$ is
(a) α emission (b) β^- emission (c) β^+ emission (d) γ emission

Q. 26. The most likely mode of decay for $^{25}_{11}Na$ is
(a) α emission (b) β^- emission (c) β^+ emission (d) γ emission

Q. 27. Of the following nuclei, the one that has the greatest binding energy per nucleon is
(a) $^{56}_{26}Fe$ (b) $^{2}_{1}H$ (c) $^{238}_{92}U$ (d) $^{239}_{94}Pu$

Q. 28. The half-life of 99^m Tc is 6.0 h. If the total residual activity in a patient 36.0 h after receiving an injection containing 99^m Tc must be no more than 0.01 μCi, what is the maximum activity (in μCi) that the sample injected can have?
(a) 0.06 (b) 0.16 (c) 0.32 (d) 0.64

Q. 29. The name thermonuclear reaction is applied to fusion reactions because
(a) They are initiated by thermal neutrons.
(b) They are exothermic.
(c) They produce more neutrons than the number used to initiate the reaction.
(d) They occur only at very high temperatures.

Q. 30. Of the following particles, the one with the smallest mass is the
(a) neutrino (b) proton (c) positron (d) H atom

Q. 31. One atomic mass unit kilograms is
(a) $1/N_A$ (b) $12/N_A$ (c) $1 \times 10^{-3}/N_A$ (d) $12 \times 10^{-3}/N_A$
(e) $1/12N_A$

Q. 32. The missing fission product in the reaction

$$^{235}_{92}U + ^{1}_{0}n \rightarrow ^{146}_{57}La + \_\_\_\_ + 3(^{1}_{0}n)$$

(a) $^{86}_{32}Br$ (b) $^{87}_{35}Br$ (c) $^{87}_{32}Ge$ (d) $^{89}_{35}Br$

Q. 33. The transuration element nobelium was made by bombarding another trans-uranium element with ^{12}C. The reaction is

$$\_\_\_\_ + ^{12}_{6}C \rightarrow ^{254}_{102}No + 4(^{1}_{0}n)$$

The target atom of the bombardment is

(a) $^{246}_{96}Cm$ (b) $^{246}_{100}Fm$ (c) $^{243}_{96}Cm$ (d) $^{243}_{100}Fm$

Q. 34. Which of the following atoms has only a single stale nuclide?

(a) $_{20}Ca$ (b) $_{28}Ni$ (c) $_{36}Kr$ (d) $_{45}Rh$

Q. 35. When $^{118}_{51}Sb$ emits a positron, the product (daughter) nuclide is

(a) $^{117}_{50}Sn$ (b) $^{118}_{50}Sn$ (c) $^{119}_{50}Sn$ (d) $^{118}_{52}Te$

Q. 36. Titrium, $^{3}_{1}H$, has a half-life of 12.26 yr. A 5.00-mL sample of tritiated water has an activity of 2.40×10^9 cpm. How many years will it take for the activity to fall to 3.00×10^8 cpm?

(a) 6.13 (b) 12.26 (c) 24.52 (d) 36.78

Q. 37. The activity of ^{14}C in living bones is 15.3 dpm $\cdot$ g^{-1} of carbon. The half-life of ^{14}C is 5730 yr. A fossil animal bone has an activity of 3.83 dpm $\cdot$ g^{-1} of carbon. How many years ago did the animal die?

(a) 5730 (b) 8600 (c) 11,400 (d) 14,300

Q. 38. When $^{226}_{88}Ra$ emits on α-particle, the nuclide formed is

(a) $^{222}_{87}Fr$ (b) $^{222}_{86}Rn$ (c) $^{224}_{86}Rn$ (d) $^{224}_{84}Po$

Q. 39. A radioisotope decays at such a rate that after 96.0 min only $\frac{1}{8}$ of the original amount remains. The half-life of this nuclide, in minutes, is.

(a) 12.0 (b) 24.0 (c) 32.0 (d) 48.0

Q. 40. In a series of three steps in a radioactive disintegration sequence starting with $^{228}_{88}Ra$, the particles emitted are, successively, a β^- particle, a β^- particle and an α-particle. The resulting product is an isoptope of

(a) $_{92}U$ (b) $_{90}Th$ (c) $_{88}Ra$ (d) $_{86}Rn$

Q. 41. One of the nuclides in each of the following three pairs is radioactive; the other is stable.

(i) $^{208}_{84}Po$, $^{209}_{83}Bi$ (ii) $^{39}_{19}K$, $^{40}_{19}K$ (iii) $^{71}_{31}Ga$, $^{76}_{31}Ga$

Now, choose the correct statement from the following:

(a) $^{209}_{83}Bi$ is radioactive, $^{71}_{31}Ga$ is stable, $^{40}_{19}K$ is stable

(b) $^{39}_{19}K$ is radioactive, $^{209}_{83}Bi$ is stable, $^{71}_{31}Ga$ is stable

(c) $^{208}_{84}Po$ is radioactive, $^{39}_{19}K$ is stable, $^{76}_{31}Ga$ is radioactive

(d) $^{40}_{19}K$ is radioactive, $^{208}_{84}Po$ is stable, $^{71}_{31}Ga$ is radioactive

Q. 42. The type of radioactive decay for the following radioactive nuclides: (i) ${}^{47}_{20}Ca$; (ii) ${}^{25}_{12}Al$, will be:

(a) α-emission (b) β-emission (c) γ-emission (d) Positron emission

Q. 43. A 1.0-mg sample of technetium-99 has an activity of 1.7×10^{-5} Ci (Ci = curies), decaying by beta emission. The decay constant for ${}^{99}_{43}Tc$ Will be

(a) $3.0 \times 10^{-13}\ s^{-1}$ (b) $2.0 \times 10^{-13}\ s^{-1}$ (c) $0.5 \times 10^{-13}\ s^{-1}$ (d) $1.0 \times 10^{-13}\ s^{-1}$

Q. 44. Tritium, ${}^{3}_{1}H$ is a radioactive nucleus of hydrogen. It is used in luminous watch dials. Tritium decays by beta emission with a half-life of 12.3 years. The decay constant (in/s) and the activity (in Ci) of a sample containing 2.5 μg of tritium is: (The atomic mass of ${}^{3}_{1}H$ is 3.02 amu.

(a) 3.88×10^{8} s, 5.0×10^{17} Ci
(b) $1.79 \times 10^{-9}\ s^{-1}$, 9.0×10^{8} Ci
(c) $1.79 \times 10^{-9}\ s^{-1}$, 0.024 Ci
(d) 3.88×10^{8} s, 1.79 Ci

Q. 45. The half life period of 1st order reaction is 15 minutes. The rate constant and the time taken to complete 80% of the reaction respectively will be:

(a) 0.462 min^{-1}, 3.48 min
(b) 0.00462 min^{-1}, 0.348 min
(c) 0.0462 min^{-1}, 34.8 min
(b) 4.62 min^{-1}, 348 min

Q. 46. The amount of heat evolved by a curie of Rn (an alpha emitter) in one hour and after its mean life respectively is [Decay constant for Rn = $2 \times 10^{-6}\ s^{-1}$ and Kinetic energy of 1 α-particle = 5.5 MeV]

(a) 114.5 J, 1.03×10^{4} J
(b) 1.03 J, 114.5×10^{4} J
(c) 11.45 J, 10.3×10^{4} J
(d) 10.3 J, 11.45×10^{4} J

Q. 47. The radioactivity of an element decreases to half its initial value in 5 hours; still how long it will take for the activity to become 1/8th of the initial activity?

(a) 5 Hours (b) 15 Hours (c) 1.66 Hours (d) 10 Hours

Q. 48. One of the hazards of nuclear explosion is the generation of ${}^{90}Sr$ and its subsequent incorporation in bones. This nuclide has a half life of 28.1 years. Suppose one microgram was absorbed by a new-born child; how much ${}^{90}Sr$ will remain in his bones after 20 years?

(a) 0.08 μg (b) 0.05 μg (c) 0.07 μg (d) 0.06 μg

Q. 49. The half-life of radioactive element is 6930 days. The average life of the element is:

(a) 3465 days (b) 9979.2 days (c) 5000 days (d) 13860.0 days

Q. 50. A radioactive isotope has a half-time of 25 days starting with 4 g of isotope what will be the mass remaining after 75 days.

(a) 2.0 g (b) 1.0 g (c) 0.5 g (d) 1.5 g

Q. 51. In a sample of a radioactive element radium disintegrated at an average rate if 2.24×10^{13} α-particles per minute. Each α-paerticle takes up 2 electrons from the air and becomes a neutral from the air and becomes a neutral helium atom. After 420 days the helium gas collected was 0.5 mL measured at 300 K and 750 mm of mercury pressure. From the above data, the value of Avogadro's constant will be:

(a) 8.28×10^{23} (b) 5.25×10^{23} (c) 9.95×10^{23} (d) 6.76×10^{23}

Q. 52. The decay constant for a radioactive sample whose activity reduces to 1/64th of its original value in 2 hours, will be:
(a) 3.465×10^{-2} min^{-1} (b) 34.65×10^{-2} min^{-1}
(c) 6.93×10^{-2} min^{-1} (d) 69.3×10^{-2} min^{-1}

Q. 53. The velocity with which an α-particle travels towards the nucleus of a copper atom so as to arrive at a distance 10^{-13} metre from the nucleus of the copper atom is:
(a) 6.318×10^{7} m s^{-1} (b) 6.318×10^{8} m s^{-1}
(c) 6.318×10^{5} m s^{-1} (d) 6.318×10^{6} m s^{-1}

Q. 54. If a piece of copper weights 0.635 g, how many atoms does it contain.
(a) 6.023×10^{20} (b) 6.023×10^{22} (c) 6.023×10^{21} (d) 6.023×10^{23}

Q. 55. The weight of one litre of a sample of ozonide oxygen at N.T.P. was found to be 1.5 g. When 100 mL of this mixture at N.T.P. was treated with turpentine, the volume was reduced to 90 mL; the molecular weight of ozone is:
(a) 32 g mol^{-1} (b) 64 g mol^{-1} (c) 16 g mol^{-1} (d) 48 g mol^{-1}

Q. 56. The weight of iron which will be converted into its oxide by the action of 18 g of steam is:
(a) 42g Fe (b) 21g Fe (c) 63g Fe (d) 84g Fe

Q. 57. $^{234}_{90}Th$ disintegrates to give $^{206}_{82}Pb$ as the final product. How many alpha and beta particles are emitted during this process respectively?
(a) 76, 6 (b) 28, 54 (c) 6, 76 (d) 54, 28

Q. 58. One gram of $^{198}_{79}Au$ ($t_{1/2}$ = 65 hrs) decays by β-emission to produce stable mercury: How much mercury will be present after 260 hrs?
(a) 0.54 g (b) 0.93 g (c) 0.76 g (d) 0.82 g

Q. 59. Radioactive is a first order process. Radioactive carbon in wood sample decays with a half life of 5700 years. What is the rate constant (in $year^{-1}$) for the decay? What fraction would remain after 11540 years?
(a) 37% (b) 50% (c) 25% (d) 75%

Q. 60. You have 0.1 g atom of a radioactive isotope $^{A}_{Z}X$ (half life = 5 days). How many number of atoms will decay during the eleventh day?
(a) 0.195×10^{19} atoms (b) 0.195×10^{23} atoms
(c) 0.195×10^{18} atoms (d) 0.195×10^{22} atoms

Q. 61. A piece of charcoal from a tree killed by the volcanic eruption gave 7.0 disintegrations of carbon-14 nuclei per minute per gram of total carbon. Present-day carbon (in living matter) gives 15.3 disintegrations per minute per gram of total carbon. The date of the volcanic eruption; when the half-life of carbon-14 is 5730 years, is:
(a) 6500 B.C. (b) 450 B.C. (c) 650 B.C. (d) 4500 B.C.

Q. 62. Deuterium is an isotope of hydrogen ($^{2}_{1}H$) which is made up of one proton and one neutron. What is the missing mass of deuterium, and what energy is released in its formation? Given the actual mass of a deuterium nucleus = $3.343\ 586 \times 10^{-27}$ kg;

Mass of one proton= 1.672623×10^{-27} kg;

Mass of one neutron= 1.674928×10^{-27} kg;

and Speed of light = 3.00×10^{8} ms^{-1}

(a) 0.3965×10^{-27} kg, 0.357×10^{-13} J
(b) 0.3965×10^{-27} kg, 0.0357×10^{-13} J
(c) 0.003965×10^{-27} kg, 3.57×10^{-13} J
(d) None of these

Q. 63. In experiment, cobalt-60 has been used to irradiate strawberries to prevent them developing mould. The half-life of cobalt-60 is 5.3 years. If a food company were to start with 2.00 g of the isotope, how many would be left after 21.2 years?
(a) 0.125 g (b) 0.15 g (c) 0.75 g (d) 1.25 g

Q. 64. All the isotopes of technetium are radioactive, they do not occur naturally on earth and all are artificially made. Thchnetium-99 has a half-life of 6 hours. It is used in medicine to assess the damage to heart muscles after a heart attack. A sample has an initial count rate of 3000 counts per minute (c.p.m.). How long will it take for the count rate to fall to 94 c.p.m?
(a) 3 Hours (b) 0.3 Hours (c) 30 Hours (d) 300 Hours

Chapter 12

CRYSTALOGRAPHY

PROBLEMS

Q. 1. Potassium metal crystallizes in the body-centered cubic structure. The number of nearest neighbour atoms for each potassium atom in the solid is

(a) 4 (b) 6 (c) 8 (d) 10

Q. 2. Which of the following crystals has the largest lattice energy?

(a) KCl (b) RbI (c) LiBr (d) MgO

Q. 3. The number of molecules per unit cell for a compound that crystallizes in an orthorhombic end-centered lattice with a molecule at each lattice site is

(a) 1 (b) 2 (c) 4 (d) 6

Q. 4. A compound contains two types of atoms, X and Y. Its crystal structure is a cubic lattice with X atoms at the corners of the unit cells and Y atoms at the body centers. The simplest formula of this compound is

(a) X_8Y (b) X_2Y (c) XY (d) XY_8

Q. 5. Which of the following crystals has the largest lattice energy?

(a) CaI_2 (b) NiS (c) BaS (d) SrO

Q. 6. The melting point RbBr is 955 K while that of NaF is 1261 K. The principal reason that the melting point of NaF is much higher than that of RbBr is that

(a) The two crystals are not isomorphous

(b) The molar mass of NF is smaller than that of RbBr.

(c) The internuclear distance, $r_c + r_a$, is greater for RbBr than for NaF.

(d) The difference in electronegativity between Rb and Br is smaller than the difference between Na and F.

Q. 7. If the radius of metal atom is 2.00 Å and its crystal structure is cubic close packed (face-centered cubic lattice), what is the volume (in cm^3) of one unit cell?
(a) 1.8×10^{-22} (b) 1.60×10^{-23} (c) 2.26×10^{-23} (d) 3.20×10^{-23}

Q. 8. For which of the following ionic crystalline solids does the cation-anion bond have the largest amount of covalent character?
(a) CdS (b) NaBr (c) SrS (d) BaO

Q. 9. For a certain crystal the unit cell axial lengths are found to be a = 5.62 Å, b = 7.41 Å, and c = 10.13 Å. The three coordinates axes are mutually perpendicular. The crystal system to which this crystal belongs is the
(a) tetragonal (b) orthorhombic (c) monoclinic (d) triclinic

Q. 10. Which of the following ions has the smallest radius?
(a) Ti^{2+} (b) Pt^{2+} (c) Ni^{2+} (d) Zr^{2+}

Q. 11. When the following five anions are arranged in order of decreasing ionic radius, the correct sequence is
(a) Se^{2-}, I^-, Br^-, $O^{2-}F^-$ (b) I^-, Se^{2-}, O^{2-}, Br^-, F^-
(c) Se^{2-}, I^-, Br^-, F^-O^{2-} (d) I^-, Se^{2-}, Br^-, O^{2-}, F^-

Q. 12. Fow which of the following crystals would you expect the assumption of anion – anion contact to be valid?
(a) CsBe (b) NaF (c) KCl (d) NaI

Q. 13. Which of the following ions has the largest heat of hydration?
(a) Na^+ (b) Al^{3+} (c) F^- (d) Sr^{2+}

Q. 14. Which of the following anions is most easily polarized?
(a) Cl^- (b) Se^{2-} (c) Br^- (d) Te^{2-}

Q. 15. Which of the following substances has the highest melting point?
(a) Cs_2O (b) SeO_2 (c) NiO (d) CaF_2

Q. 16. Of the five ΔH values needed to calculate a lattice energy using the Born-Haber cycle, the one that is most difficult to measure is
(a) The heat of sublimation of the metal.
(b) The heat of formation of gaseous atoms of the non metal.
(c) The ionization energy of the metal.
(d) The electron affinity of the nonmetal.

Q. 17. An ionic crystalline solid, MX_3, has a cubic unit cell. Which of the following arrangements of the ions is consistent with a the stoichiometry of the compound?
(a) M^{3+} ions at the corners, X^- ions at the face centers.
(b) M^{3+} ions at the corners, X^- ions at the body centers.
(c) X^- ions at the corners, M^{3+} ions at the face centers.
(d) X^- ions at the corners, M^{3+} ions at the body centers.

Q. 18. If the three interaxial angles defined the unit cell are all equal in magnitude, the crystal cannot belong to the
(a) orthorhombic system (b) hexagonal system
(c) tetragonal system (d) cubic system

Q. 19. Which of the following statements about metallic elements is FALSE?
(a) These elements from ions larger in size than the neutral atoms.
(b) These elements have relatively low ionization energies.
(c) These elements have few outer shell electrons.
(d) These elements have relatively low electronegativities.

Q. 20. Of the following crystal lattices, the one that has the largest packing fraction is
(a) body-centered cubic **(b)** rhombohedral
(c) face-centered cubic **(d)** simple tetragonal

Q. 21. Nickel crystallizes in an *fcc* lattice with a unit cell length of 3.5238 Å. The density of Ni (in $g \cdot cm^{-3}$) is
(a) 17.82 **(b)** 9.92 **(c)** 8.91 **(d)** 4.45

Q. 22. A certain metal crystallizes in a face-centered cube measuring 4.00×10^2 picometers on each edge. What are the radius of the atom? (1 picometer (pm) = 1×10^{-12} meter)
(a) 141 pm **(b)** 173 pm **(c)** 200 pm **(d)** 282 pm

Q. 23. An imaginary metal crystallizes in a cubic lattices. The unit cell edge length is 100 picometers (1 picometer = 1×10^{-12} meter). The density of this metal is $200 \cdot g.cm^{-3}$. The atomic mass of the metal is $60.2\ g \cdot mol^{-1}$. How many of these metal atoms are there within a unit cell?
(a) 1.00 **(b)** 2.00 **(c)** 4.00 **(d)** 6.00

Q. 24. X-ray diffraction from crystals provides one of the most accurate ways of determining Avogadro's number. Silver crystallizes in a face-centered cubic lattice with all atoms at the lattice points. The length of an edge of the unit cell was determined by x-ray diffraction to be 408.6 pm (4.086 Å). The density of silver is $10.50\ g\ cm^{-3}$. The mass of a silver atoms is:
(a) 0.79×10^{-22} g **(b)** 1.79×10^{-22} g **(c)** 3.79×10^{-22} g **(d)** 2.79×10^{-22} m

Q. 25. Platinum crystallises in a face-centered cubic lattice with all atoms at the lattice points. It has a density of $21.45\ g.cm^{-3}$ and an atomic weight of 195.08 amu. The length of a unit-cell edge as obtained from x-ray diffraction is:
(a) 3.129×10^{-25} m **(b)** 6.024×10^{-25} m **(c)** 6.024×10^{-30} m **(d)** 3.924×10^{-10} g

Q. 26. An element (atomic mass = 60) having face-centered cubic crystal has a density of 6.23 $g\ cm^{-3}$. The edge length of the unit cell is: (Avogadro constant, $N_0 = 6.902 \times 10^{23}$ mol^{-1}).
(a) 3.0×10^{-10} m **(b)** 2.0×10^{-10} m **(c)** 4.0×10^{-10} m **(d)** 5.0×10^{-10} m

Q. 27. Chromium metal crystallizes with a body centered cubic lattice. The length of the unit cell edge is found to be 287 pm. The atomic radius and the density of chromium in $g.cm^{-3}$ will be:
(a) 1.24 Å, 7.3 $g.cm^{-3}$ **(b)** 2.14 Å, 3.7 $g.cm^{-3}$
(c) 2.48 Å, 7.3 $g.cm^{-3}$ **(d)** 2.48 Å, 14.6 $g.cm^{-3}$

Q. 28. An electron beam can undergo diffraction by crystals. Through what potential should a beam of electrons be accelerated so that its wavelength becomes equal to 1.54 Å.
(a) 9046 Volts **(b)** 7046 Volts **(c)** 8046 Volts **(d)** 6046 Volts

Q. 29. Soduim metal crystallises in body centered cubic lattice with the cell edge, a = 4.29 Å. What are the radius of sodium atom?

(a) 2.86 Å (b) 4.86 Å (c) 0.86 Å (d) 1.86 Å

Q. 30. An element crystallizes into a structure which may be described by a cubic type of unit cell having one atom on each corner of the cube and two atoms on one of its diagonals. If volume of this unit cell is 24×10^{-24} cm^3 and density of the element is 7.2 g cm^{-3}, calculate the number of atoms present in 200 g of the element.

(a) 3.47×10^{26} atoms (b) 3.47×10^{24} atoms
(c) 3.47×10^{25} atoms (d) 3.47×10^{22} atoms

Q. 31. Metallic gold crystallizes in the face-centered cubic lattice. The length of the cube unit cell (see Fig.) is a = 4070 Å. Now answer (i) to (iii) below:

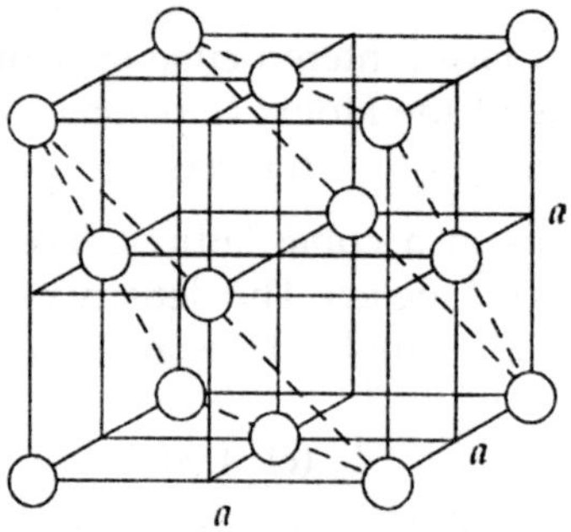

(i) the closest distance between gold atoms is:
(a) 4.070 Å (b) 1.0 Å (c) 0.5 Å (d) 2.878 Å

(ii) the density if gold is:
(a) 4.85 g.cm^{-3} (b) 9.70 g.cm^{-3} (c) 14.55 g.cm^{-3} (d) 19.4 g.cm^{-3}

(iii) The Packing factor for gold is:
(a) 0.3255 (b) 0.5205 (c) 0.2205 (d) 0.7405

Q. 32. Krypton crystallizes in a structure that has four Kr atoms in each unit cell, and the unit cell is a cube. The edge length of the unit cell is 0.559 m. the density of crystalline Kr, in kg.m^{-3} will be:

(a) 3190 kg.m^{-3} (b) 6380 kg.m^{-3} (c) 1595 kg.m^{-3} (d) 2000 kg.m^{-3}

Q. 33. The platinum-chlorine distance has been found to be 2.32 Å in several crystalline compounds. If this value applied to both of the compounds shown in Fig. below. The chlorine-chlorine distance in structure (i) and structure (ii) is respectively as:

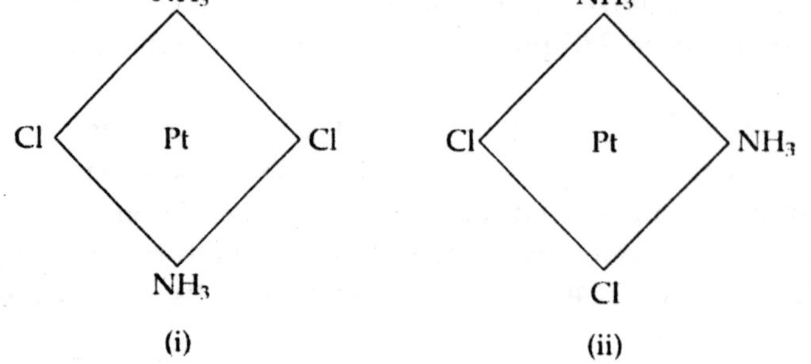

(i) (ii)

(a) 3.28 Å; 4.64 Å (b) 4.64 Å; 3.28 Å
(c) 3.28 Å; 3.28 Å (d) 4.64 Å; 4.64 Å

Q. 34. If the C—C single-bond distance is 1.54 Å then the distance between the terminal carbons in propane, C_3H_8 is: (Assume that the four bonds of any cabron atom are pointed toward the corners of a rectangular tetrahedron.)
(a) 5.02 Å (b) 1.255 Å (c) 2.51 Å (d) None of these

Q. 35. A plant virus was examined by the electron microscope and was found to consist of uniform cylindrical particles 150 Å in diameter and 5000 Å long. The virus has a specific volume of 0.75 cm^3g^{-1}. If the virus particle is considered to be one molecule, then its molecular weight is:
(a) 7.10×10^7 g · mol^{-1} (b) 1.18×10^{-16} g · $molecule^{-1}$
(c) 7.10×10^{-16} g · $molecule^{-1}$ (d) 1.18×10^7 g · mol^{-1}

Q. 36. A unit cell of sodium chloride has four formula units. The edge length of the unit cell is 0.564 nm. What is the density of sodium chloride?
(a) 4.165 g · cm^{-3} (b) 2.165 g · cm^{-3} (c) 1.165 g · cm^{-3} (d) 3.165 g · cm^{-3}

Q. 37. CsCl crystallizes in a cubic structure that has a Cl ionic radius = 1.81 Å at each corner and a Cs^+ ionic radius = 1.69 Å at the center of the unit cell. The value for a, the lattice constant is:
(a) 2.13 Å (b) 4.04 Å (c) 8.26 Å (d) 16.5 Å

Q. 38. The density of crystalline CsCl is 4.988 g · cm^{-3}, the volume occupied by a single CsCl pair in the crystal is:
(a) 7.014 cm^{-3} (b) 7.014×10^{-11} cm^3
(c) 7.014×10^{-23} cm^3 (b) 7.014×10^{-18} cm^3

Q. 39. The smallest Cs-to-Cs internuclear distance, equal to the length of the side of a cube corresponding to the volume of one CsCl ion pair will be
(a) 4.125 Å (b) 0.242 Å (c) 0.6125 Å (d) 0.4125 Å

Q. 40. The smallest Cs-to-Cs internucleas distance in the crystal, assuming each Cs^+ ion to be locate din the center of a cube with Cl^- ions at each corner of the cube will be:
(a) 2.537 Å (b) 5.372 Å (c) 3.725 Å (d) 3.752 Å

Q. 41. Iron crystallizes in several modifications. At about 1183 K, the body-centered cubic α from undergoes a transition to the face-centered cubic γ form. Assuming that the distance between nearest neighbours is the same in the two forms at the transition temperature, then, the ratio of the density of γ iron to that of α iron at the transition temperature will be.
(a) 1.09 (b) 2.90 (c) 3.09 (d) 0.97

Q. 42. The ZnS zinc blende structure is cubic. The unit cell may be described as a face-centered sulfide ion sublattice with zinc ions in the centers of alternating minicubes made by partitioning the main cube into 8 equal parts. The minimum r_+/r_- ratio needed to avoid anion-anion contact, if closest cation-anion pairs are assumed to tough will be:
(a) 0.775 (b) 0.365 (c) 0.225 (d) 0.542

Q. 43. The packing factor for spheres occupying a body-centered cubic and a simple cubic structure, where closest neighbours in both cases are in contact, respectively will be:
(a) 0.68; 0.524 (b) 0.068; 0.0524 (c) 0.524; 0.68 (d) 0.0524; 0.068

Q. 44. A compound formed by elements A and B crystallizes in the cubic structure, where A atoms are the corners of a cube and B atoms are at the face centres. The formula of the compound is?
(a) AB_3 (b) A_2B_3 (c) $A\ B$ (d) A_2B_2

Q. 45. A cubic solid is made of two elements *P* and *Q*. Atoms *Q* are at the corners of the cube and *P* at the body centre. The formula of the compound and the coordination numbers of *P* and *Q* respectively are:
(a) PQ, 4 (b) P_2Q_2, 4 (c) PQ, 8 (d) P_2Q_2 8

Q. 46. A compound formed by elements *X* and *Y* crystallizes in the cubic structure, where *Y* atoms are at the corners of a cube and *X* atoms are the alternate faces. The formula of the compound is?
(a) XY (b) X_2Y_2 (c) X_3Y (d) X_3Y_2

Q. 47. Zinc sulphide crystallizes with zinc ions occupying 1/2 of the tetrahedral holes in a closest packed array of sulphide ions. The formula of zinc sulphide is?
(a) ZnS_2 (b) ZnS (c) Zn_3S_2 (d) Zn_2S

Q. 48. Aluminium oxide crystallizes with aluminium ions in 2/3 of the octahedral holes in closest packed array of oxide ions. The formula of aluminium will be?
(a) AlO (b) Al_2O (c) Al_2O_3 (d) Al_2O_2

Q. 49. The ionic radii of Na^+ and Cl^- are 95 and 181 pm, respectively. The edge length of the unit cell of NaCl is?
(a) 138 pm (b) 276 pm (c) 1104 pm (d) 552 pm

Q. 50. One form of tungsten crystallizes in a body-centered cubic unit cell with tungsten atom on each lattice point. If the edge length of the unit cell is 3.165Å, the atomic radius of tungsten in this structure is:
(a) 3.16 Å (b) 1.37 Å (c) 12.66 Å (d) 2.74 Å

Q. 51. An element occurs in the *BCC* structure with a cell-edge of 288 pm. The density of the element is 7.2 g cm^{-3}. How many atoms are present in 208 g of the element?
(a) 2.42×10^{24} atoms (b) 1.21×10^{24} atoms
(c) 3.63×10^{24} atoms (d) 4.84×10^{24} atoms

Q. 52. A face-centered cubic element (atomic mass 60 amu) has a cell-edge of 400 pm. Its density is?
(a) 1.56 g · cm^{-3} (b) 6.23 g · cm^{-3} (c) 3.12 g · cm^{-3} (d) 9.36 g · cm^{-3}

Q. 53. Find out the rank of the unit cell of sodium chloride lattice if the edge length of the unit cell is 562 pm and the density of NaCl is 2.165 g cm^{-3}.
(a) 8 (b) 2 (c) 4 (d) 0

Q. 54. An element crystallizes in a structure having *F.C.C.* unit cell of an edge 200 pm. Its density; if 200 g of this element contains 24×10^{22} atom will be:
(a) 10.42 g cm^{-3} (b) 41.66 g cm^{-3} (c) 20.84 g cm^{-3} (d) 83.32 g cm^{-3}

Q. 55. The number of atoms in a cube based will unit cell having one atom on each corner and two atoms on its each diagonal will be:
(a) 4 (b) 8 (c) 2 (d) 9

Q. 56. A metal crystallizes as body centered cubic lattice with the edge length of a unit cell 305 pm. If the density of the metal is 5.955 g cm^{-3}, its molar mass will be:
(a) 101.8 g mol^{-1} **(b)** 25.5 g mol^{-1} **(c)** 50.9 g mol^{-1} **(d)** 203.6 g mol^{-1}

Q. 57. If NaCl is doped with 10^{-3} mol% of $SrCl_2$, the concentration of cation vacancies will be:
(a) 6.023×10^{18} vac. mol^{-1} **(b)** 6.023×10^{20} vac. mol^{-1}
(c) 6.023×10^{21} vac. mol^{-1} **(d)** 6.023×10^{23} vac. mol^{-1}

Q. 58. An element '*A*' crystallizes in FCC structure, 208 g of it has 4.2832×10^{24} atoms. The edge of the unit cell if density of '*A*' is 7.2 g cm^{-3} will be?
(a) 270 pm **(b)** 290 pm **(c)** 280 pm **(d)** 240 pm

PART II

Chapter 13

ORGANIC CHEMISTRY

OBJECTIVE

KEY POINTS

TO THE STUDENT

Always read the problem carefully. In trying to solve a problem quickly, many students read the question fast. It is important to read the question carefully.

"OH I KNOW THIS"; when you see a question, you may recall a similar question solved before instead of solving it you assume a lot of stuff. One small change will change the question completely. So concentrate yourself on the question. Lack of patience and concentration is like rushing to arrive at a conclusion. It should be avoided. While solving a problem do not think about the previous question but focus at the current problem.

You should chose the question to be attempted and try to solve it right instead of attempting a lot of questions. Accuracy is important. 50% accuracy means you have wasted 50% time.

Always understand the facts of a problem. Get to the bottom of the facts, follow the given instructions.

1. Alkanes

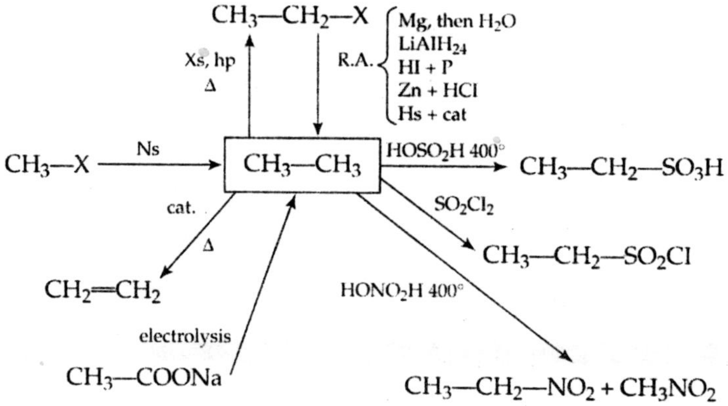

2. Alkens

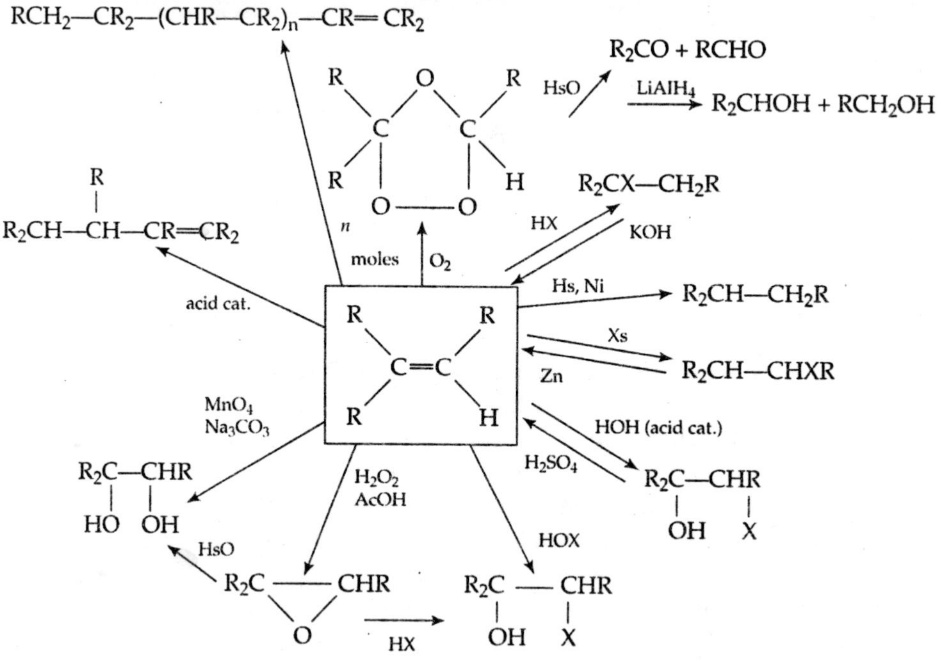

Alkyns

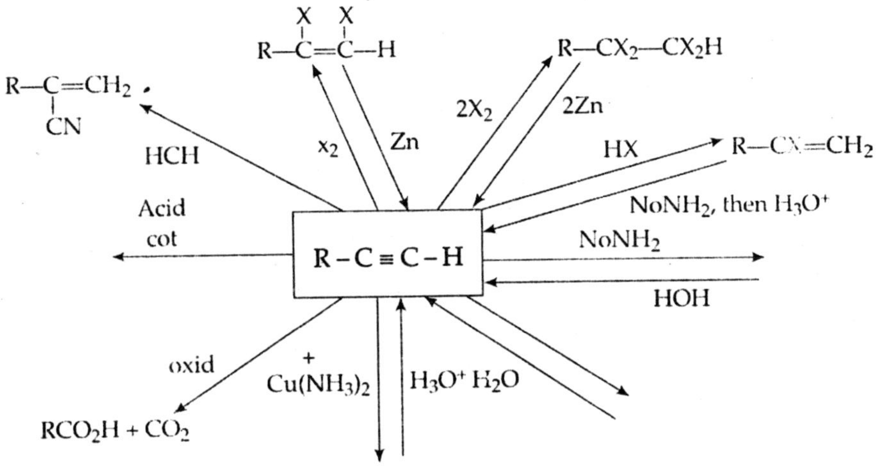

Alcohals

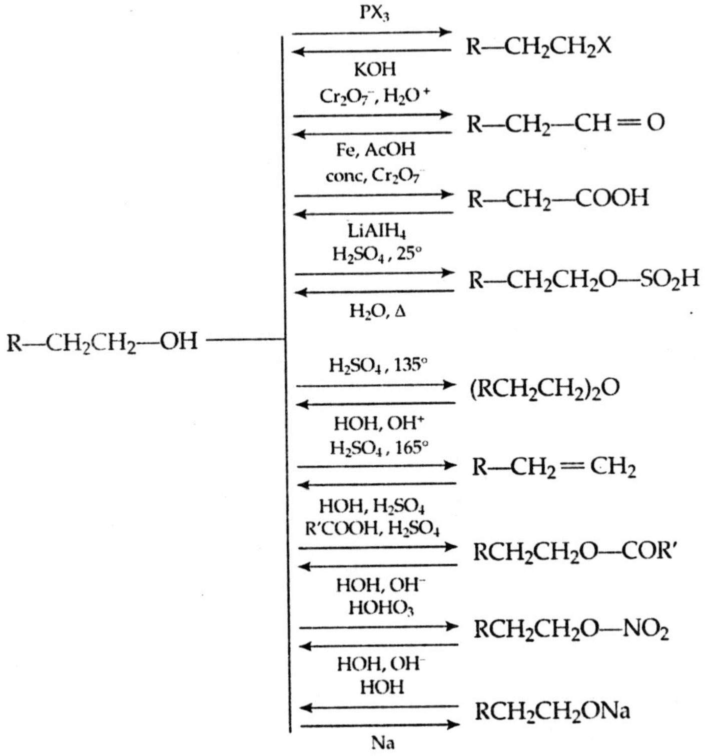

Alkylhalides

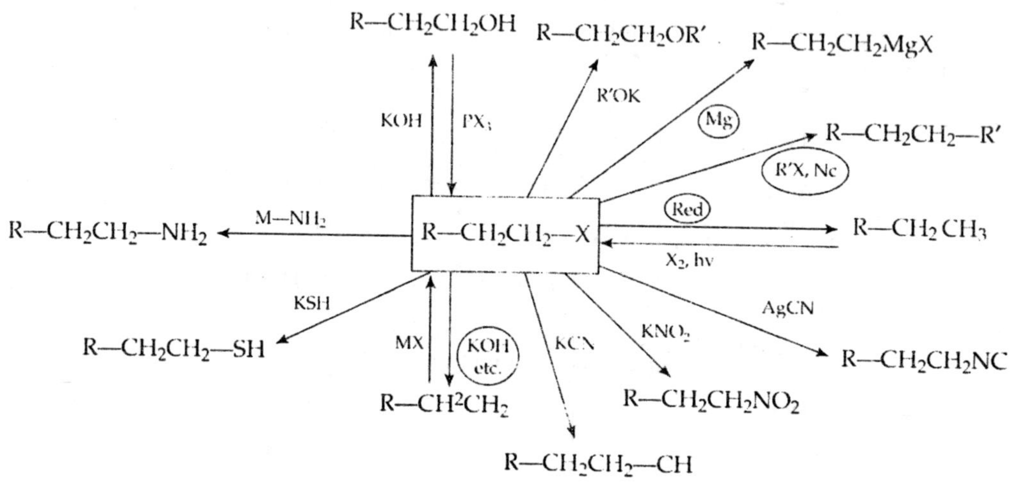

Grignard Reagent

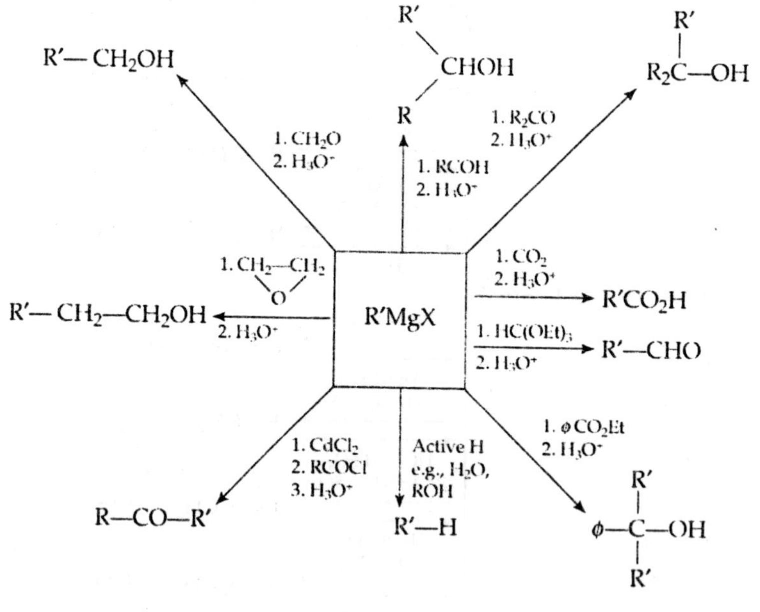

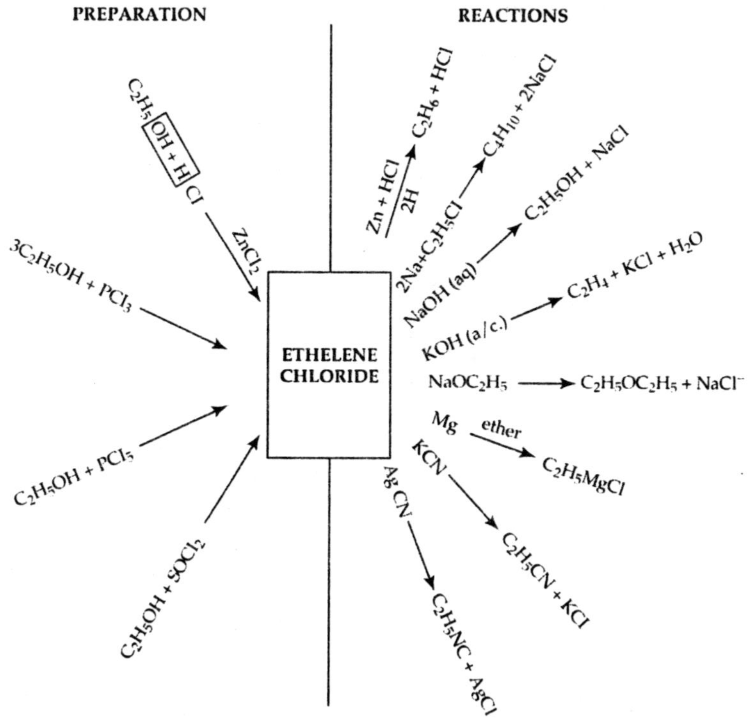
PREPARATION
REACTIONS
ETHELENE CHLORIDE
C_2H_5 OH + H Cl
$ZnCl_2$
$3C_2H_5OH + PCl_3$
$C_2H_5OH + PCl_5$
$C_2H_5OH + SOCl_2$
Zn + HCl
2H
$C_2H_6 + HCl$
$2Na+C_2H_5Cl$
$C_4H_{10} + 2NaCl$
NaOH (aq)
$C_2H_5OH + NaCl$
KOH (a/c.)
$C_2H_4 + KCl + H_2O$
$NaOC_2H_5$
$C_2H_5OC_2H_5 + NaCl^-$
Mg
ether
C_2H_5MgCl
KCN
$C_2H_5CN + KCl$
Ag CN
$C_2H_5NC + AgCl$

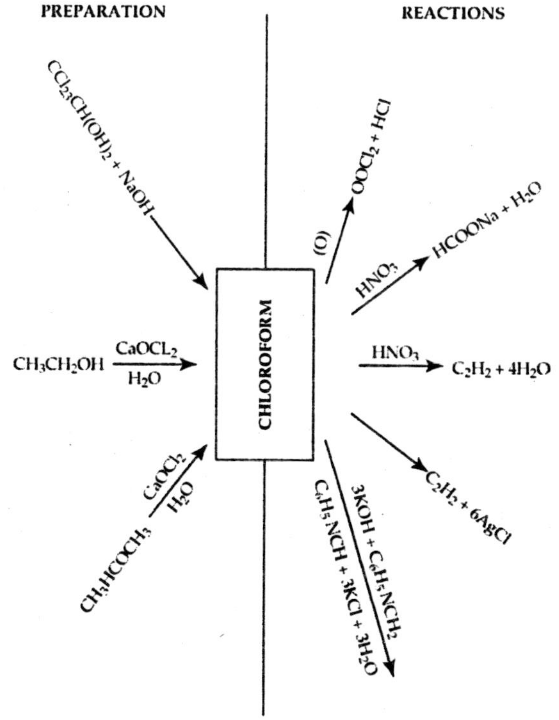
PREPARATION
REACTIONS
CHLOROFORM
$CCl_3CH(OH)_2 + NaOH$
CH_3CH_2OH
$CaOCL_2$
H_2O
CH_3HCOCH_3
$CaOCl_2$
H_2O
(O)
$OOCl_2 + HCl$
HNO_3
$HCOONa + H_2O$
HNO_3
$C_2H_2 + 4H_2O$
$C_2H_2 + 6AgCl$
$3KOH + C_6H_5NCH_2$
$C_6H_5NCH + 3KCl + 3H_2O$

PREPARATION **REACTIONS**

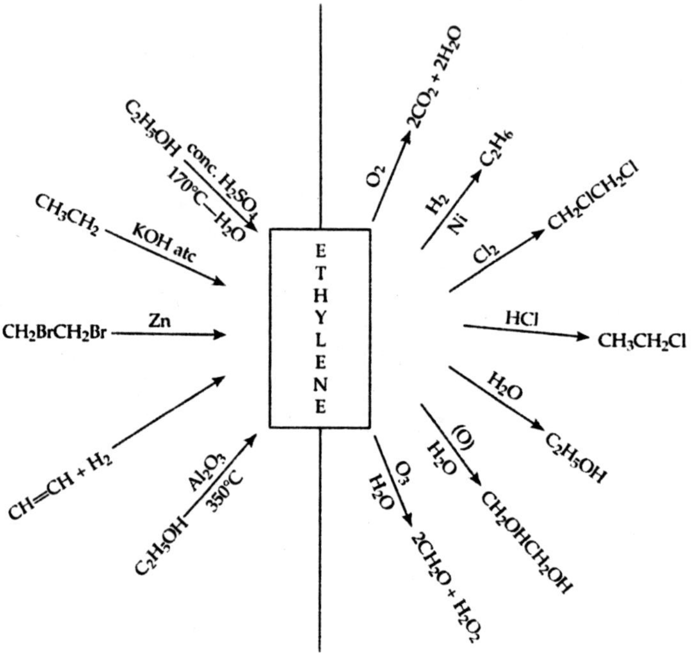
ETHYLENE
C2H5OH conc. H2SO4 170°C—H2O
CH3CH2 KOH alc
CH2BrCH2Br Zn
CH≡CH + H2
C2H5OH Al2O3 350°C
O2 2CO2 + 2H2O
H2 Ni C2H6
Cl2 CH2ClCH2Cl
HCl CH3CH2Cl
H2O C2H5OH
(O) H2O CH2OHCH2OH
O3 H2O 2CH2O + H2O2

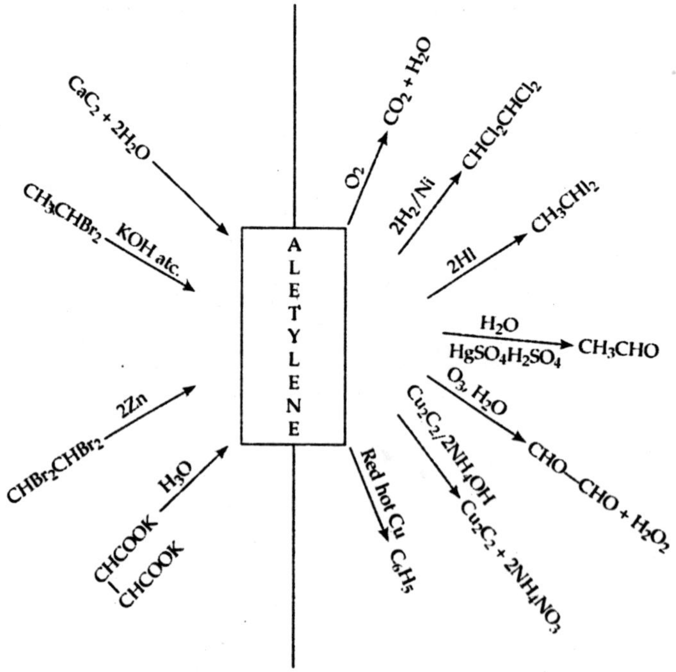
ALETYLENE
CaC2 + 2H2O
CH3CHBr2 KOH alc.
CHBr2CHBr2 2Zn
CHCOOK | CHCOOK H2O
O2 CO2 + H2O
2H2/Ni CHCl2CHCl2
2HI CH3CHI2
H2O HgSO4H2SO4 CH3CHO
O3, H2O CHO—CHO + H2O2
Cu2C2/2NH4OH Cu2C2 + 2NH4NO3
Red hot Cu C6H5

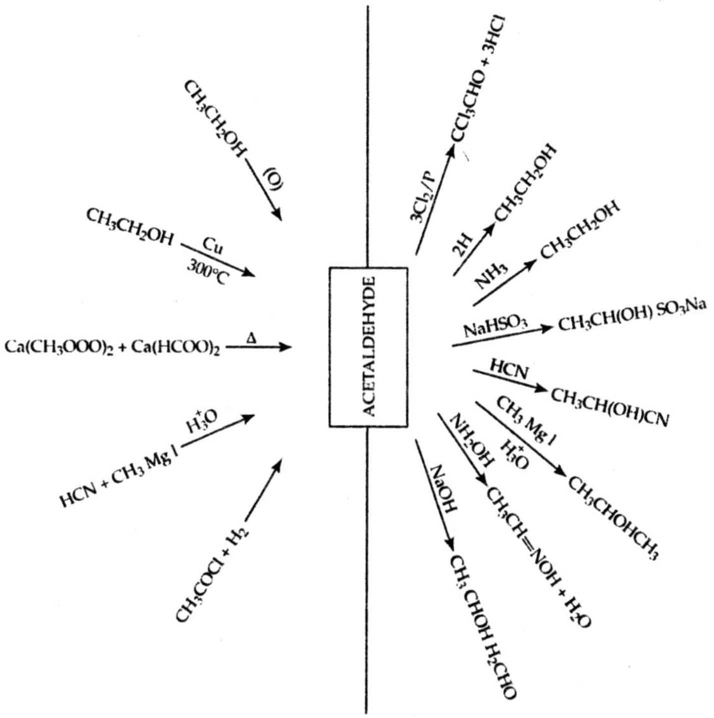
ACETALDEHYDE
CH_3CH_2OH (O)
CH_3CH_2OH Cu 300°C
$Ca(CH_3OOO)_2 + Ca(HCOO)_2$ Δ
$HCN + CH_3MgI$ $H_3\overset{+}{O}$
$CH_3COCl + H_2$
$3Cl_2/P$ $CCl_3CHO + 3HCl$
2H CH_3CH_2OH
NH_3 CH_3CH_2OH
$NaHSO_3$ $CH_3CH(OH)SO_3Na$
HCN $CH_3CH(OH)CN$
CH_3MgI $H_3\overset{+}{O}$ $CH_3CHOHCH_3$
NH_2OH $CH_3CH{=}NOH + H_2O$
NaOH $CH_3CHOH\,H_2CHO$

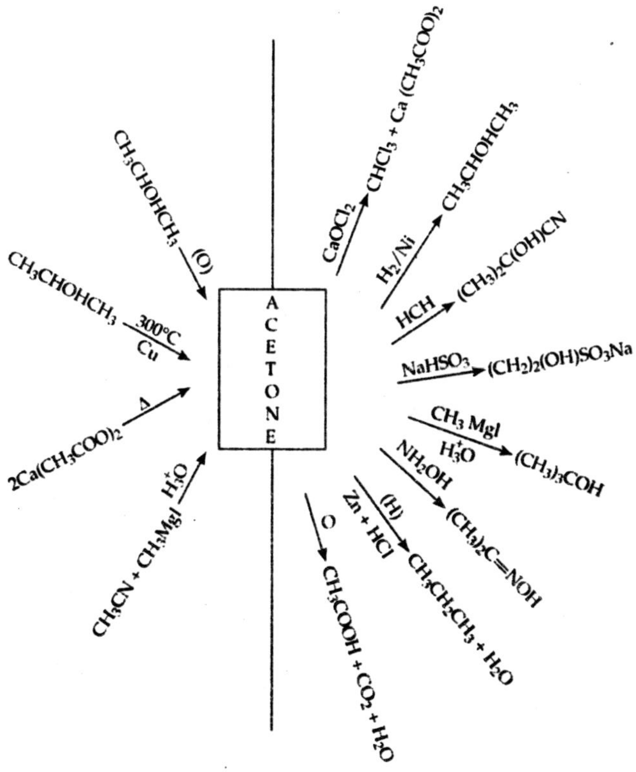
ACETONE
$CH_3CHOHCH_3$ (O)
$CH_3CHOHCH_3$ 300°C Cu
$2Ca(CH_3COO)_2$ Δ
$CH_3CN + CH_3MgI$ $H_3\overset{+}{O}$
$CaOCl_2$ $CHCl_3 + Ca(CH_3COO)_2$
H_2/Ni $CH_3CHOHCH_3$
HCH $(CH_3)_2C(OH)CN$
$NaHSO_3$ $(CH_2)_2(OH)SO_3Na$
CH_3MgI $H_3\overset{+}{O}$ $(CH_3)_3COH$
NH_2OH $(CH_3)_2C{=}NOH$
(H) Zn + HCl $CH_3CH_2CH_3 + H_2O$
O $CH_3COOH + CO_2 + H_2O$

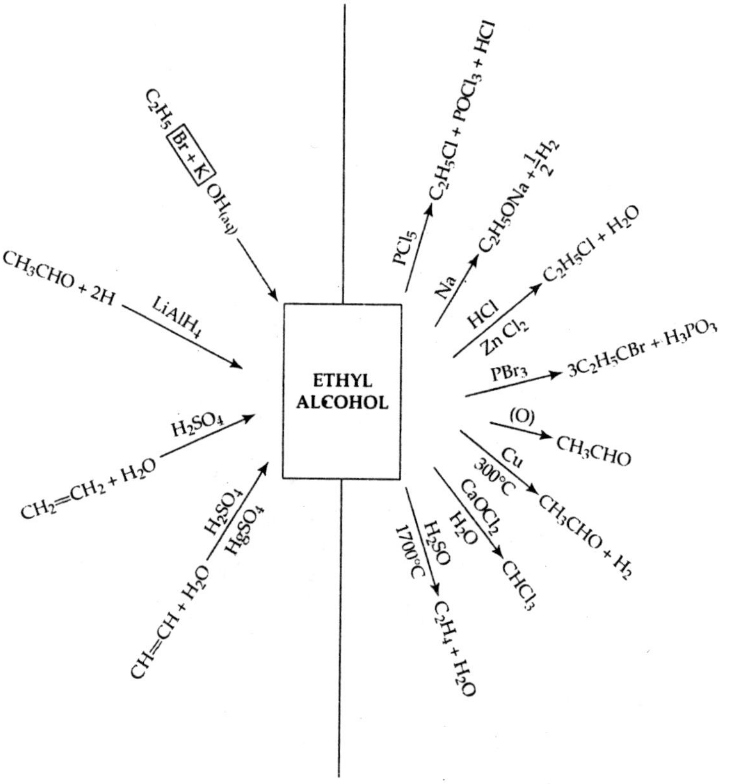
ETHYL ALCOHOL
C_2H_5 Br + K $OH_{(aq)}$
$CH_3CHO + 2H$ $LiAlH_4$
H_2SO_4
$CH_2{=}CH_2 + H_2O$
H_2SO_4 $HgSO_4$
$CH{\equiv}CH + H_2O$
PCl_5 $C_2H_5Cl + POCl_3 + HCl$
Na $C_2H_5ONa + \frac{1}{2}H_2$
HCl Zn Cl_2 $C_2H_5Cl + H_2O$
PBr_3 $3C_2H_5CBr + H_3PO_3$
(O) CH_3CHO
Cu 300°C $CH_3CHO + H_2$
$CaOCl_2$ H_2O $CHCl_3$
H_2SO 1700°C $C_2H_4 + H_2O$

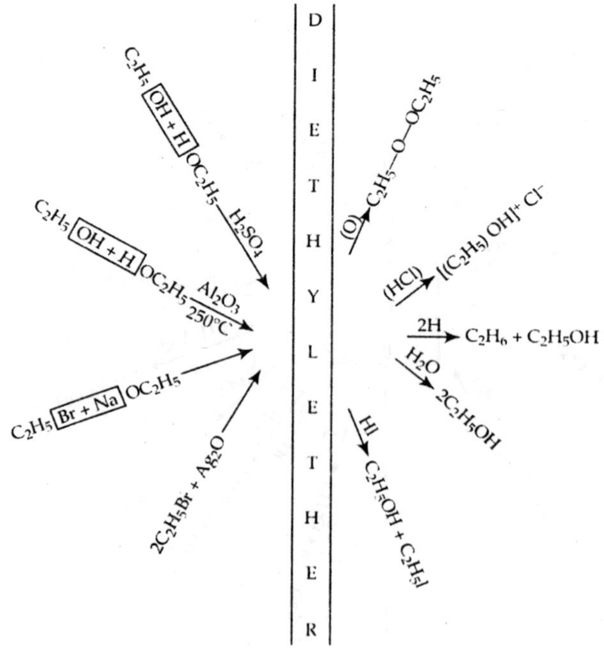
DIETHYL ETHER
C_2H_5 OH + H OC_2H_5 H_2SO_4
C_2H_5 OH + H OC_2H_5 Al_2O_3 250°C
C_2H_5 Br + Na OC_2H_5
$2C_2H_5Br + Ag_2O$
(O) $C_2H_5{-}O{-}OC_2H_5$
(HCl) $[(C_2H_5)\,OH]^+\,Cl^-$
2H $C_2H_6 + C_2H_5OH$
H_2O $2C_2H_5OH$
HI $C_2H_5OH + C_2H_5I$

AROMATIC
BENZENE/NITROBENZENE/ANILINE

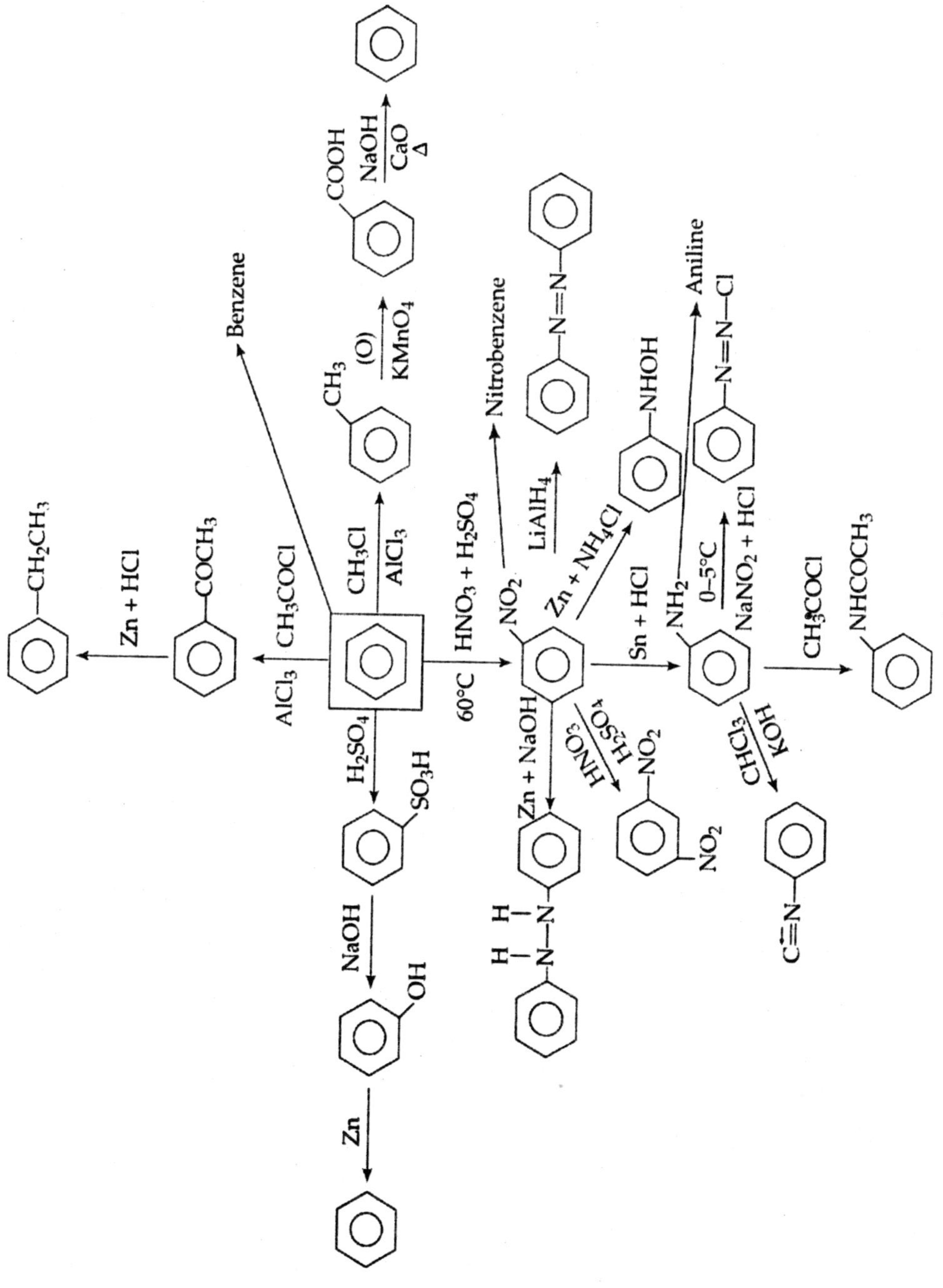

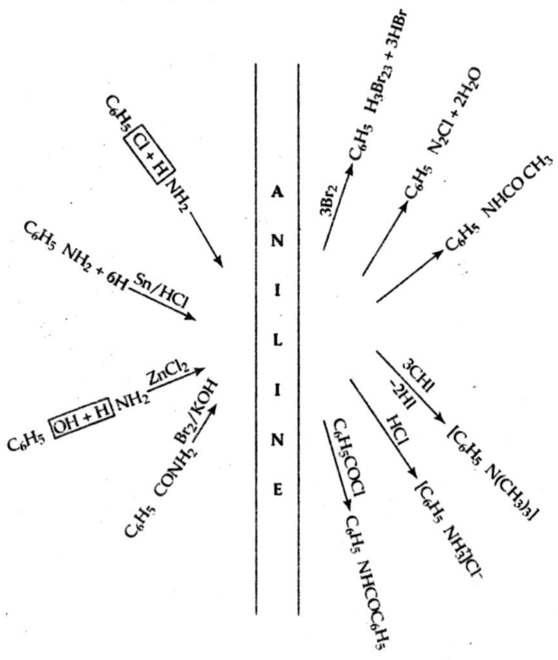
C6H5 Cl + H NH2
C6H5 NH2 + 6H Sn/HCl
C6H5 OH + H NH2 ZnCl2
C6H5 CONH2 Br2/KOH
A N I L I N E
3Br2 C6H5 H3Br23 + 3HBr
C6H5 N2Cl + 2H2O
C6H5 NHCO CH3
3CHI -2HI [C6H5 N(CH3)3]
HCl [C6H5 NH3]Cl-
C6H5COCl C6H5 NHCOC6H5

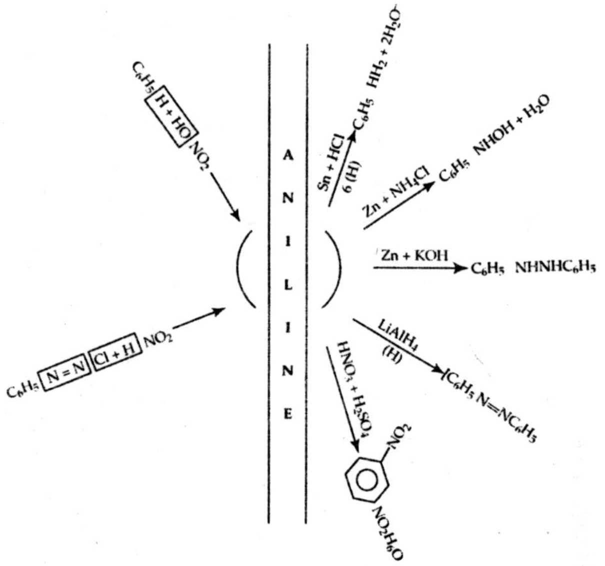
C6H5 H + HO NO2
C6H5 N = N Cl + H NO2
A N I L I N E
Sn + HCl 6 (H) C6H5 HH2 + 2H2O
Zn + NH4Cl C6H5 NHOH + H2O
Zn + KOH C6H5 NHNHC6H5
LiAlH4 (H) C6H5 N=NC6H5
HNO3 + H2SO4
NO2
NO2H6O

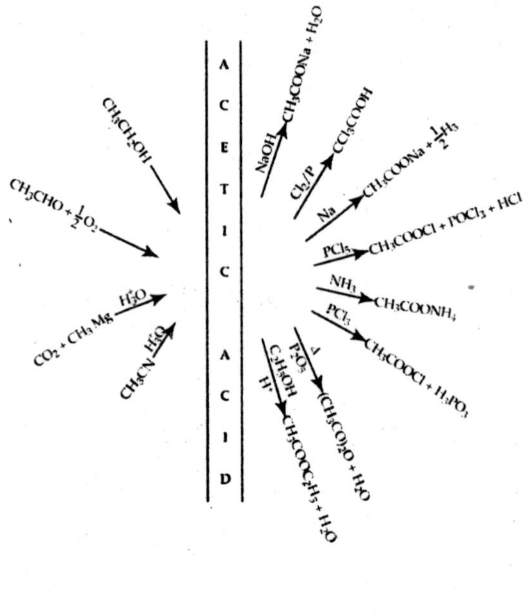
CH3CH2OH
CH3CHO + 1/2 O2
CO2 + CH3 Mg H2O
CH3CN H2O
A C E T I C A C I D
NaOH CH3COONa + H2O
Cl2/P CCl3COOH
Na CH3COONa + 1/2 H2
PCl5 CH3COOCl + POCl3 + HCl
NH3 CH3COONH4
PCl3 CH3COOCl + H3PO3
Δ P2O5 (CH3CO)2O + H2O
C2H5OH H+ CH3COOC2H5 + H2O

REDUCTION REACTIONS

Metal-hydride Reductions

(A) (lithium aluminium hydride and sodium borohydride)

The lithium aluminium hydride reductions listed as follows:

$$\left.\begin{matrix} RCOOR' \\ RCOOH \\ RCOCl \\ RCHO \end{matrix}\right\} \xrightarrow{LiAlH_4} RCH_2OH$$

$$R_2C{=}O \xrightarrow{LiAlH_4} R_4CHOH$$

$$\left.\begin{matrix} RCN \\ RCH_2NO_3 \\ RCONH_2 \end{matrix}\right\} \xrightarrow{LiAlH_4} RCH_4NH_2$$

$$RCONR_4 \xrightarrow{LiAlH_4} RCH_2NR_4$$

$$R{-}X \xrightarrow{LiAlH_4} R{-}H$$

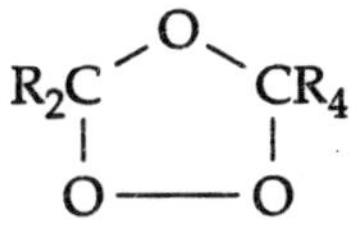

$$\xrightarrow{LiAlH_4} 2R_4CHOH$$

$$2\phi NO_2 \xrightarrow{LiAlH_4} \phi{-}N{=}{=}N{-}\phi$$

(b) Other type of reductions are

(i) lithium gallium hydride reduces aliphatic, but not aromatic aldehydes, ketones, and nitrites

(ii) sodium borohydride reduces acid chlorides, but not carboxylic acid and anhydrides

(iii) $LiALH_4$ reduces acid chlorides to the corresponding alcohols;

(iv) lithium tri-4-butoxyaluminohydride ($Li(t\text{-}BuO)_4AIH$) reduces acid ethorides to aldehydes; (v) by combinding the metal hydrides with Lewis acids such as aluminum chloride.

DEGRADATION METHODS

$$RCOOH \xrightarrow[\Delta]{NaOH,\ CaO} R{-}H \text{ (for electronegative R groups)} \quad (1)$$

$$\left.\begin{matrix} RCOOH \\ RCOCl \\ RCOOR' \end{matrix}\right\} \xrightarrow{HN_2} R{-}NH_4 \quad (2)$$

$$RCONH_4 \xrightarrow[\Delta]{Br_2,\ CCl_4} R{-}NH_4 \quad (3)$$

$$RCOOAg \xrightarrow[\Delta]{X_5\ ,\ CCl_4} R{-}X \quad (4)$$

$$R{-}COCl \xrightarrow[\Delta]{AgO,\ X_2} R{-}X \quad (5)$$

$$RCOCH_3 \xrightarrow[\Delta]{X_2,\ NaOH} RCOONa \quad (6)$$

CHAIN-LENGTHENING METHODS

(i) $R—X \xrightarrow{KCN} R—CN \xrightarrow{36\,H_2O} RCOOH$

$\quad R—CN \xrightarrow{Na\ in} RCH_2NH_2$

(ii) $RC≡CH \xrightarrow{1.\,NaNH_2\ \ 2.\,CHaI} RC≡CH—CH_4$

(iii) $R—MgX— \xrightarrow{1.\,H_2CO\ \ 2.\,H_2O^+} R—CH_2OH$

$\quad R—MgX— \xrightarrow{1.\,CO_2\ \ 2.\,H_2O^+} RCOOH$

$\quad R—MgX— \xrightarrow[\Delta]{1.\,HC(OEt)_2\ \ 2.\,H_2O^+} R—CHO$

(iv) $R—CHO \xrightarrow{HCN} R—CH(OH)—CN \xrightarrow[\Delta]{H_2O^+} R—CH(OH)—CO_2H$

$\quad R—CH(OH)—CN \xrightarrow[dry\ HCl]{ErOH} R—CH(OH)—CONH_2$

(v) $R—COCl \xrightarrow[2.\,AgO,\,H_2O]{1.\,CH_2N_2} R—CH_2—COOH$

REACTIONS PRODUCING CARBONIUM IONS

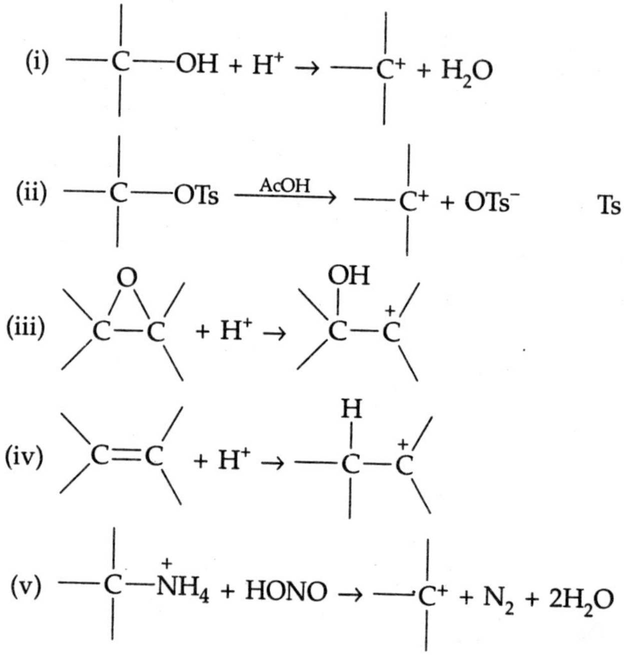

Ts = tosylate

ORGANIC REDOX REACTIONS

Organic Redox Reactions the student should indicate a least one regent which can be used for carrying out the following oxidations or reductions ($\phi = C_6H_5$)

Oxidations

1. ϕ—$CH_2OH \rightarrow \phi$—CHO
2. ϕ_2C—H $\rightarrow \phi_2C$—OH
3. ϕ—$CH_3 \rightarrow \phi$—CHO
4. ϕ—CH==$CHCH_2OH$ $\rightarrow \phi$—CH==CHCHO
5. ϕ—CH==CH—R $\rightarrow \phi$—CHO
6. ϕ—$NH_2 \rightarrow \phi$—NO_2
7. ϕ—CH==CH—ϕ $\rightarrow \phi$—CHOH—CHOH—ϕ
8. $\phi_2C(OH)$—$C(OH)\phi_2$—$2\phi_2CO$
9. CH_4—$C_6H_{11} \rightarrow CH_2$—$C_4H_4$
10. ϕ—$CH_2Cl \rightarrow \phi$—CHO
11. ϕ—N—NH—$\phi \rightarrow \phi$—N==N—ϕ
12. ϕ—NHOH $\rightarrow \phi$—NO
13. ϕ—$CH_2 \rightarrow \phi$—COOH
14. R—SH $\rightarrow$ R—S—S—R
15. Phenol $\rightarrow$ Quinone
16. $R_2CHOH \rightarrow R_2CO$
17. R—CHO $\rightarrow$ R—COOH
18. R—$COOCH_3 \rightarrow$ R—COOH
19. R—$CH_2CH_2COOH \rightarrow$ R—COOH
20. R—CO—CHOH—R $\rightarrow$ R—CO—CO—R

Reductions

1. ϕ—CHO $\rightarrow \phi$—CH_2OH
2. ϕ—$NO_2 \rightarrow \phi$—NH_3^+
3. $\phi N_2^+ \rightarrow \phi$—N==N—$\phi$
4. ϕ—$NO_2 \rightarrow \phi$—NH—NH—ϕ
5. ϕ—$NO_2 \rightarrow$ —NH—NH—ϕ
6. ϕ—$NO_2 \rightarrow \phi$—$NH^+(O^-)$==N—ϕ
7. ϕ—$NO_2 \rightarrow \phi$—NHOH
8. ϕ—COOR $\rightarrow \phi$—CH_2OH
9. ϕ—$CONH_2 \rightarrow \phi$—CH_2NH_2
10. ϕ—$CH_2CH_2NO_4 \rightarrow \phi$—$CH_2CH_4NH_4$
11. ϕ—$CH_2Cl \rightarrow \phi$—CH_2
12. ϕ—COCl $\rightarrow \phi$—CH_3
13. $\phi CN \rightarrow \phi$—CH_2NH_2
14. ϕ—COCl $\rightarrow \phi$—CH_2OH
15. Quinone $\rightarrow$ hydroquinone
16. $R_2CO \rightarrow R_2CH_2$
17. $R_2CO \rightarrow R_2CHOH$
18. $2R_4CO \rightarrow R_4C(OH)$—$C(OH)R_4$
19. 3-Hexene ozonide $\rightarrow 2CH_3CH_2CHO$
20. 3-Hexene ozonide $\rightarrow 2CH_4CH_2CH_2OH$
21. *p*-O_2N—C_4H_4—CH==CH—CHO $\rightarrow$ *p*-O_2N—C_6H_4—CH==CH—CH_2OH
22. Benzene $\rightarrow$ 1,4-cycholexadine

An example of the variation n products obtained with a variety of oxidizing agents is found in the oxidation of cyclopentene. If hydrogen peroxide in glacial acetic acid is used, the intrial product is cyclopentente oxide, which can be hydrolyzed to *trans*-1,2-cyclopentaneliol.

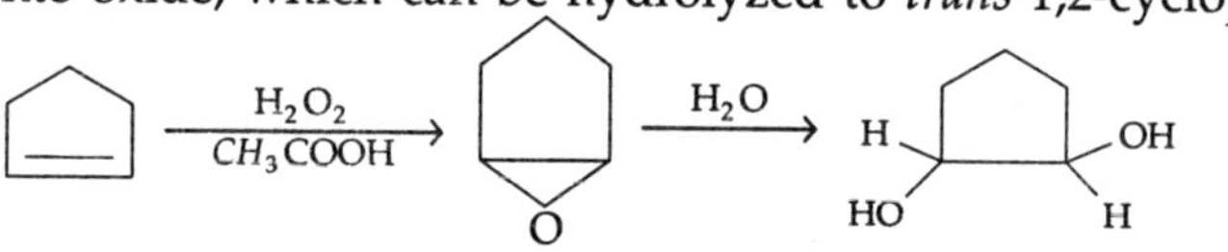

H_2O_2
CH_3COOH
O
H_2O
H
OH
HO
H

Oxidation of the olefin with hydrogen peroxide in the presence of osmium tetroxide yields *cis*-1,2-cyclopentanediol.

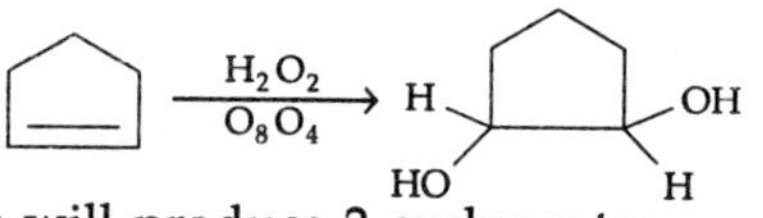

H_2O_2
O_8O_4
H
OH
HO
H

Oxidation with selenium dioxide will produce 2-cyclopentenone.

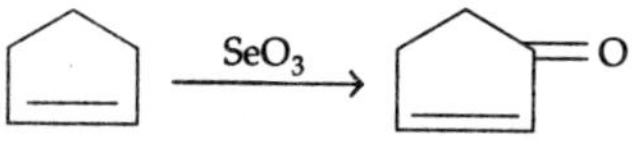

SeO_3
O

It lead tetracetate is used, among the products is 3-acetoxycyelopentene, which can e hydrolyzed to 3-cyclopentenol.

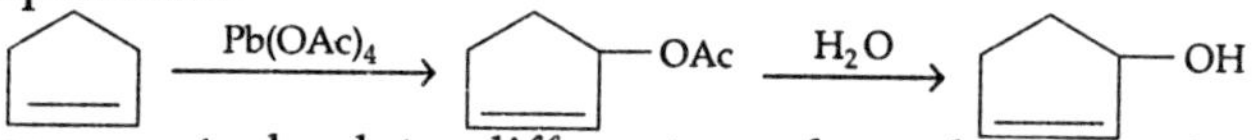

$Pb(OAc)_4$
OAc
H_2O
OH

The several oxidizing agents lead to different products because they react by different mechanisms.

PROBLEMS

Q. 1. How many isomers of C_6H_{14} are possible. (Draw & Name them)
(a) 1 **(b)** 3 **(c)** 5 **(d)** 7

Q. 2. Which of the following compounds exist as pairs of enantiomers? For the ones that do, draw the enantiomers. **(a)** (i), (ii), (v) **(b)** (i), (ii), (iii) **(c)** (ii), (iv), (v) **(d)** (ii), (iii), (v)

(i) $CH_3-CH(Cl)-CH_3$

(ii) $CH_3-CH(Cl)-CH_2Cl$

(iii) H_2N-CH_2-COOH

(iv) $H_2N-CH(CH_2OH)-COOH$

(v) $CH_3-CH_2-C(Br)(I)-CH_2Br$

Q. 3. The following five compounds, **(i)** dimethyl ether, **(ii)** formic acid **(iii)** propane, **(iv)** ethanol, **(v)** acetaldehyde have similar molecular weights, but a wide range of boiling points.
Their order of increasing boiling point is :
(a) (iii)<(i)<(v)<(iv)<(ii) **(b)** (iv)<(i)<(iii)<(v)<(ii)
(c) (iii)<(ii)<(i)<(iv)<(v) **(d)** (v)<(iv)<(iii)<(ii)<(i)

Q. 4. Each of the following reactions modifies the substituents on one or more carbon atoms. For the carbon atoms involved, indicate the oxidation state in the starting material and in the product. Has the carbon atom been oxidized or reduced?

(a) $CH_3-CH{=}CH-CH_3 \xrightarrow{KMnO_4} CH_3-CH(OH)-CH(OH)-CH_3$

(b) 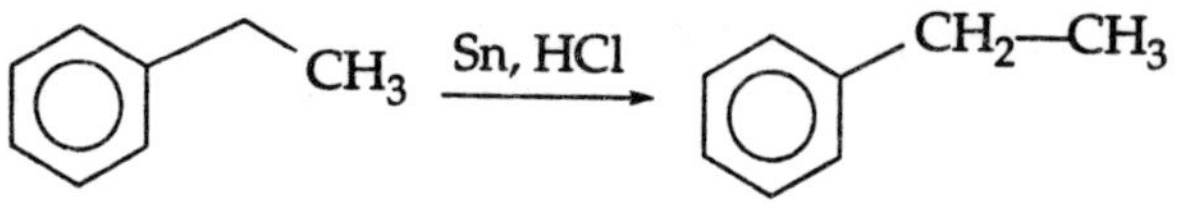

(c) $CH_3-CH{=}CH-CH_3 \xrightarrow{H_2,\,Pt} CH_3-CH_2-CH_2-CH_3$

(d) $CH_3CH_2CH_2OH \xrightarrow{CrO_3,\ pyridine} CH_3-CH_2-C(H){=}O$

(e) 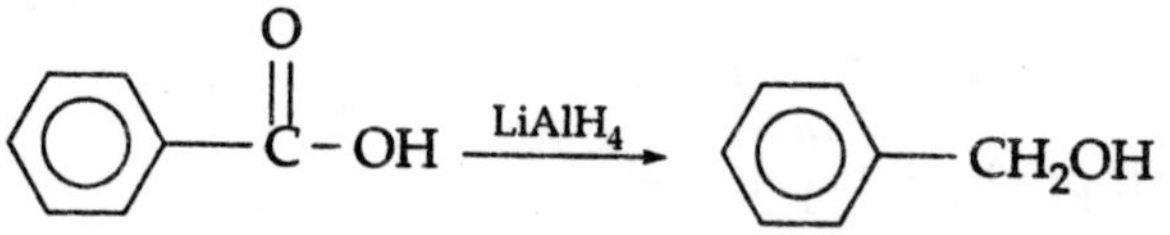

Q. 5. Identify the amino acids that have combined to make piece of a peptide is shown below. Classify each of the six side chains that extent from the peptide linkages as hydrophilic, phydrophobic, or in neither category.

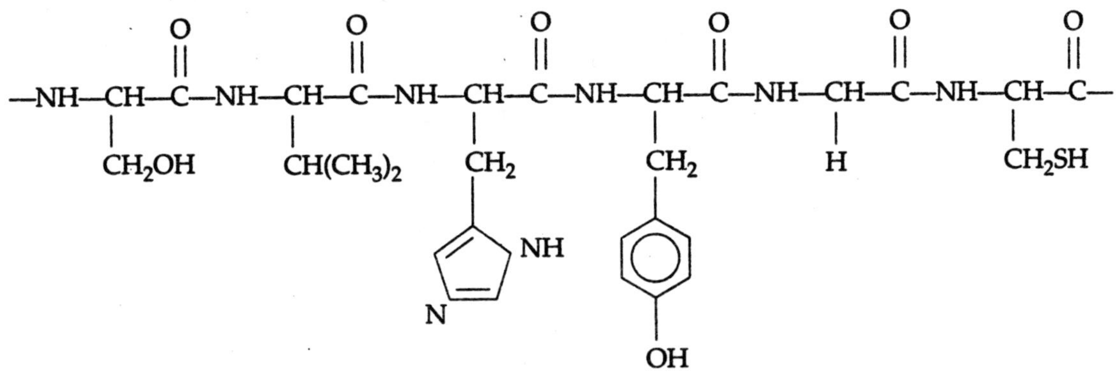

Q. 6. State whether each of the polymers shown is an addition (I) or a condensation (II) polymer.

(a)

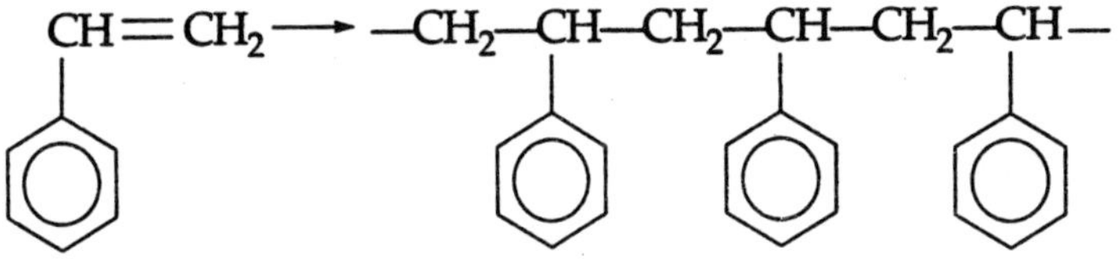

Styreme Polystyrene

(b)

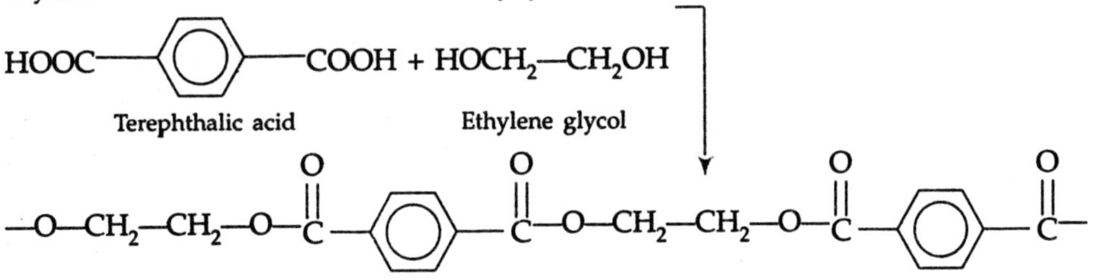
Terephthalic acid
Ethylene glycol

Dacron polyester

(c)

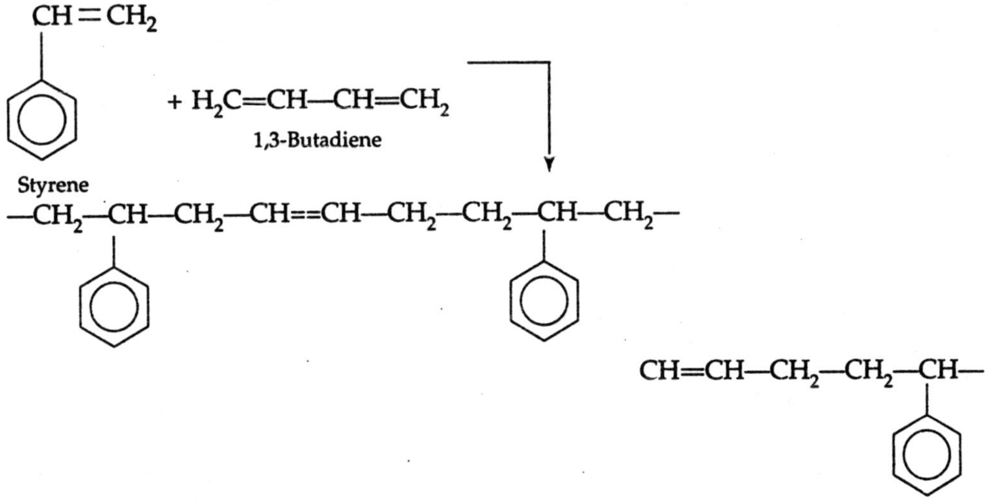
Styrene
1,3-Butadiene

Q. 7. There are six isomers with the molecular formula C_4H_8. Their structures are given below. Match the name of each compound.
The structure of six isomers of C_4H_8 are

Structures | **Names**

(a) 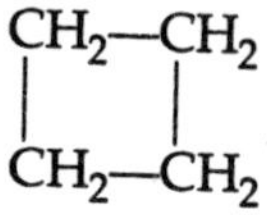 (i) 1-butene

(b) 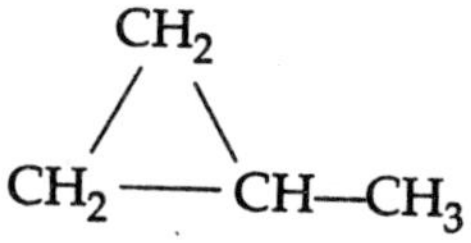 (ii) cis-2-butene

(c) $CH_2{=}CH{-}CH_2{-}CH_3$ (iii) 2-methyl propene

(d) 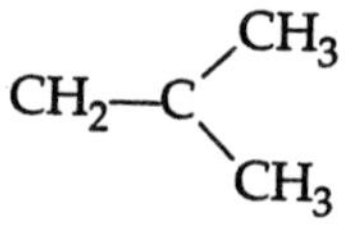 (iv) trans-2-butene

(e) 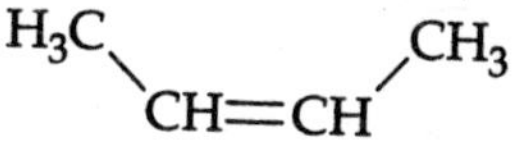 (v) cyclobutane

(f) 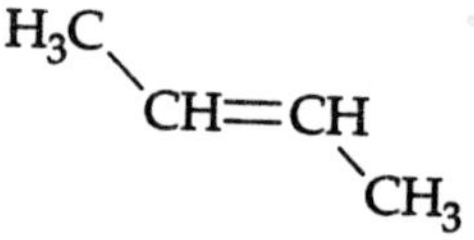 (vi) methyl cyclopropane

Q. 8. There are six *isomeric alkenes* with the molecular formula C_5H_{10}. Their structures are given below. Match the name of each compound.
The six *alkene isomers* with formula C_5H_{10} are

Structures | **Names**

(a) $H_2C{=}CH{-}CH_2{-}CH_2{-}CH_3$ (i) trans-2-pentene

(b) 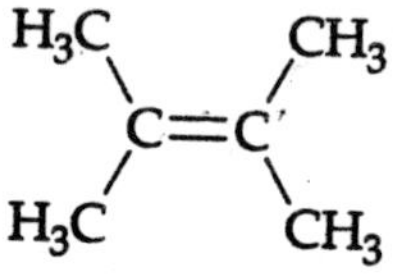 (ii) 1-pentene

(c) 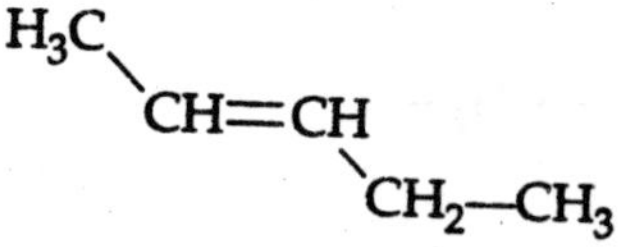 (iii) cis-2-pentene

(d) 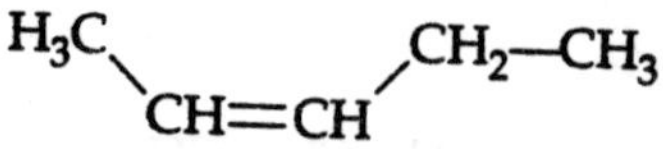 (iv) 3-methyl-1-butene

(e) 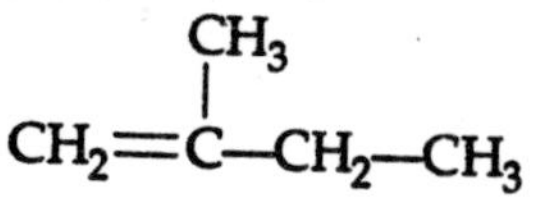 (v) 2-methyl-2-butene

(f) 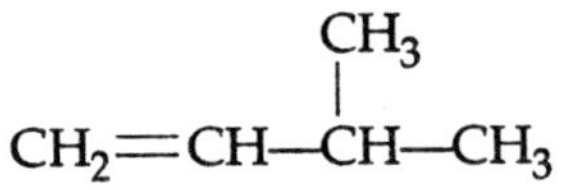 (vi) 2-methyl-1-butene

Q. 9. There are five cyclic alkane isomers with the molecular formula C_5H_{10}, including geometric isomers. Their structures are given below. Match the name of each compound.

Structures	Names
(a) 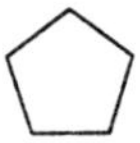	(i) 1, 1-dimethyl cyclopropane
(b) 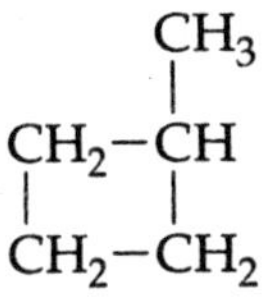	(ii) cis-1, 2, cyclopropane
(c) 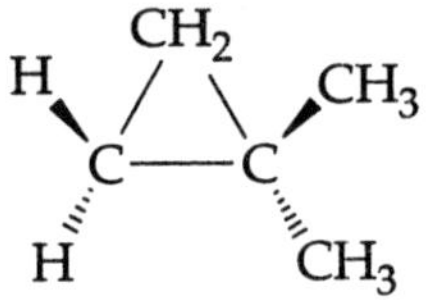	(iii) trans-1, 2, cyclopropane
(d) 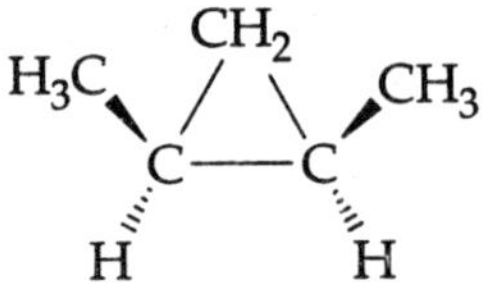	(iv) methyl cyclobutane
(e) 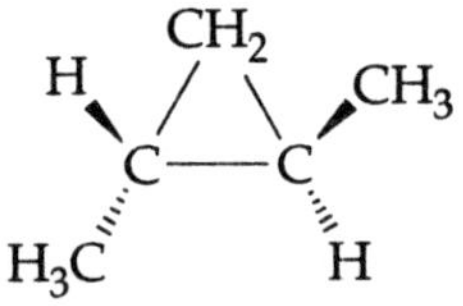	(v) cyclopentane

Q. 10. Which compound in each of the following pairs would have the higher boiling point?

(a) (i) Ethylalcohal (CH_3CH_2OH) or (ii) methyl ether (CH_3OCH_3)

(b) (i) Pentane 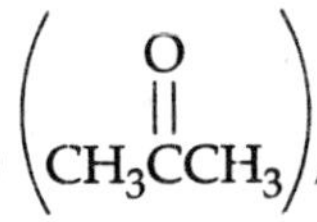, or (ii) Heptane (C_7H_{16})

(c) (i) Acetone 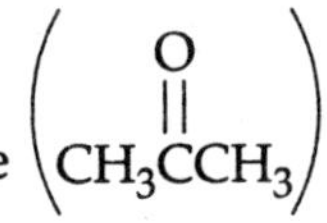 or (ii) 1-propanol ($CH_3CH_2CH_2OH$)

(d) (i) *cis*-1,2-Dichloroethene 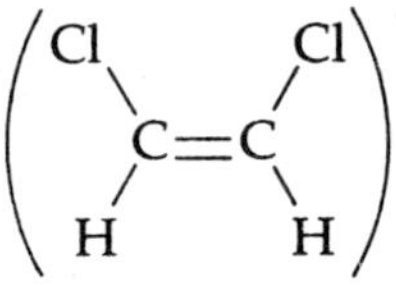 or

(ii) trans-1,2-dichloroethene 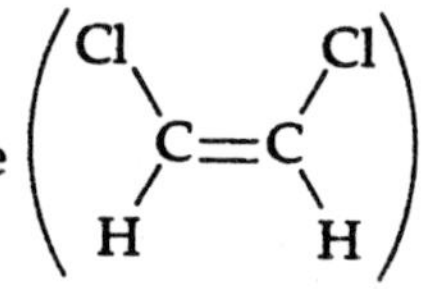
Cl
Cl
C=C
H
H

(e) (i) Propionic acid $\left(CH_3CH_2\overset{\overset{O}{\|}}{C}OH\right)$ or (ii) methyl acetate $\left(CH_3\overset{\overset{O}{\|}}{C}—OCH_3\right)$

Q. 11. Match the structural formulas for the products Y(☐ + ☐) of the following reactions Column (z). if more than one product is possible tell which one would be the major product.

Column X	Column Y	Column Z
(a) $CH_3\overset{\overset{CH_3}{\vert}}{\underset{\underset{Cl}{\vert}}{C}}CH_2CH_2CH_3 \xrightarrow[C_2H_5OH]{C_2H_5ONa}$	☐ + ☐	(i) $CH_3CCH_3{=}CHCH_2CH_3$
(b) $CH_3\overset{\overset{CH_3}{\vert}}{\underset{\underset{Cl}{\vert}}{C}}CH_2CH_2CH_3 \xrightarrow[(CH_3)_2\,COH]{(CH_3)_3\,COK}$	☐ + ☐	(ii) $CH_3C(CH_3){=}CHCH_2CH_3$ (major)
(c) $CH_3\overset{\overset{CH_3}{\vert}}{\underset{\underset{OH}{\vert}}{C}}CH_2CH_2CH_3 \xrightarrow[Heat]{H^+}$	☐ + ☐	(iii) $CH_2{=}C(CH_3)\,CH_2CH_2CH_3$
(d) 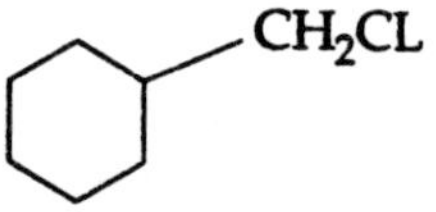 $\xrightarrow[(CH_3)_3\,COH]{(CH_3)_3\,COK}$	☐ + ☐	(iv) $CH_2{=}C(CH_3)CH_2CH_2CH_3$ (major)
(e) Product of (d) $\xrightarrow{Br_2}$	☐ + ☐	(v) 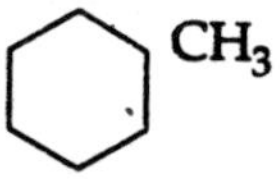
(f) Product of (e) $\xrightarrow[Acetone]{NaI}$	☐ + ☐	(vi) 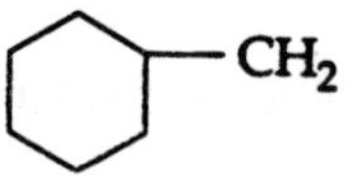
(g) Product of (f) $\xrightarrow[Pt]{H_2}$	☐	(vii) 
(h)  $\xrightarrow[Heat]{Pt}$	☐	(viii) 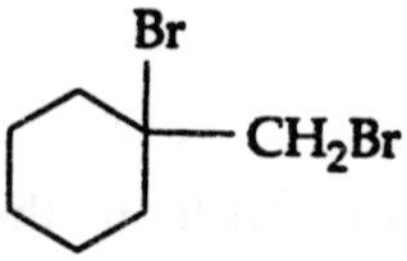

CH₂CL
CH₃
CH₂
Br
CH₂Br

Q. 12. Consider the following pairs of structures. Identity the relation between them by describing them as representing enantiomers, diastereomers, structural isomers, or two numbers, or two molecules of the same compound.

Column I | **Column II**

(a) 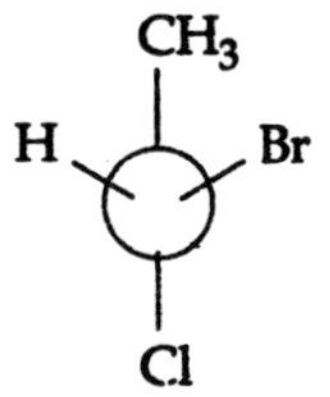
CH3
H
Br
Cl
and 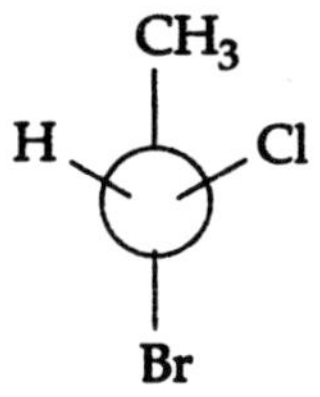
CH3
H
Cl
Br

(b) 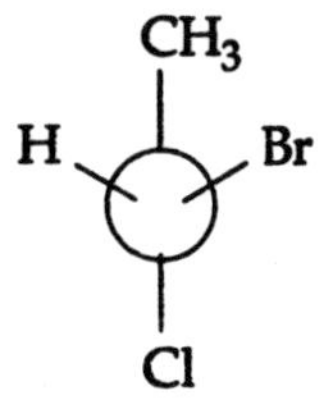
CH3
H
Br
Cl
and 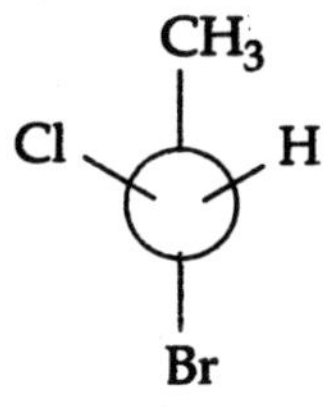
CH3
Cl
H
Br

(c) 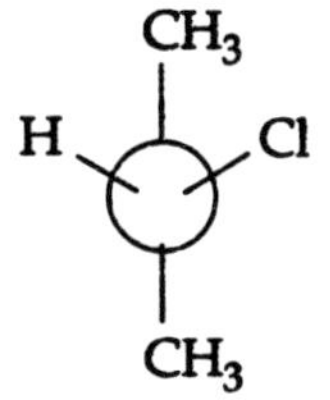
CH3
H
Cl
CH3
and 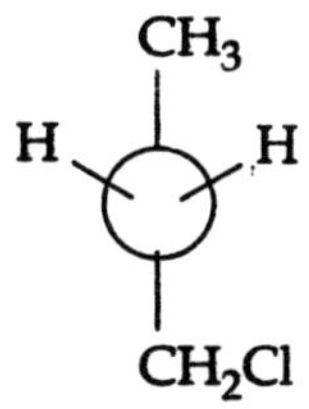
CH3
H
H
CH2Cl

(d) 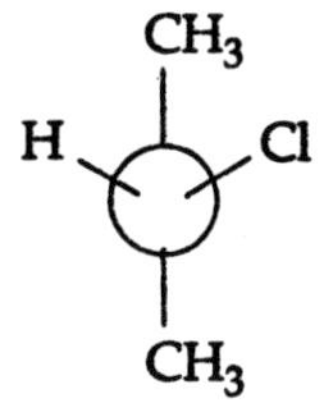
CH3
H
Cl
CH3
and 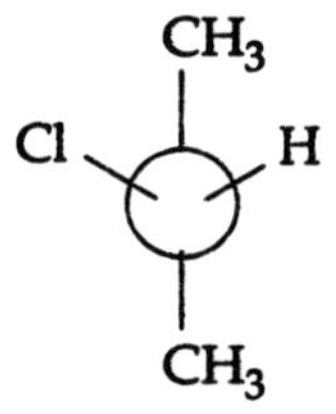
CH3
Cl
H
CH3

(e) 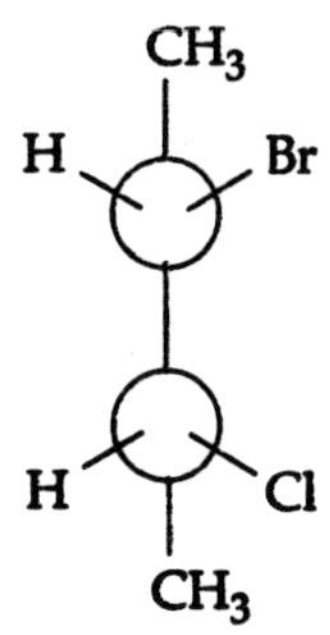
CH3
H
Br
H
Cl
CH3
and 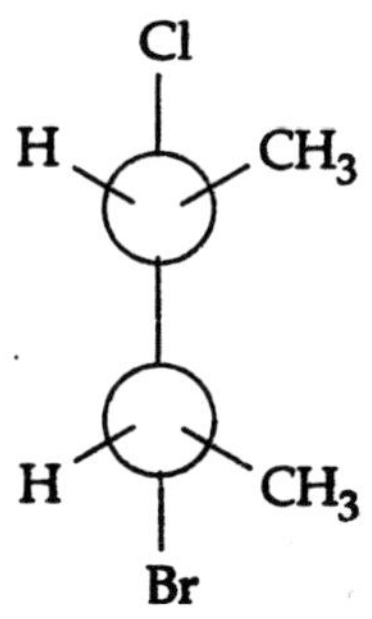
Cl
H
CH3
H
CH3
Br

(f) 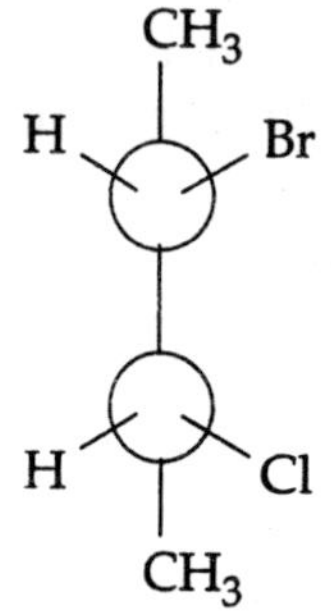
CH3
H
Br
H
Cl
CH3
and
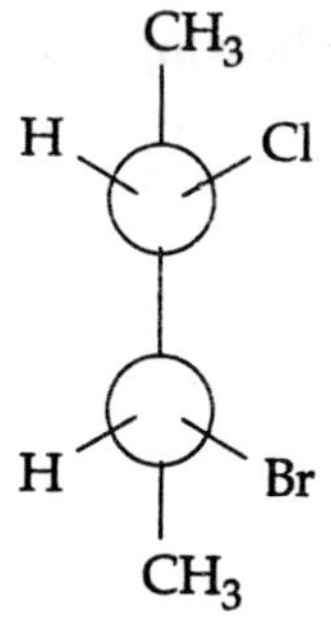
CH3
H
Cl
H
Br
CH3

(g) 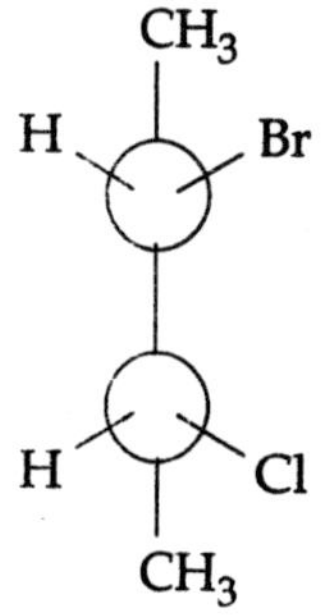
CH3
H
Br
H
Cl
CH3
and
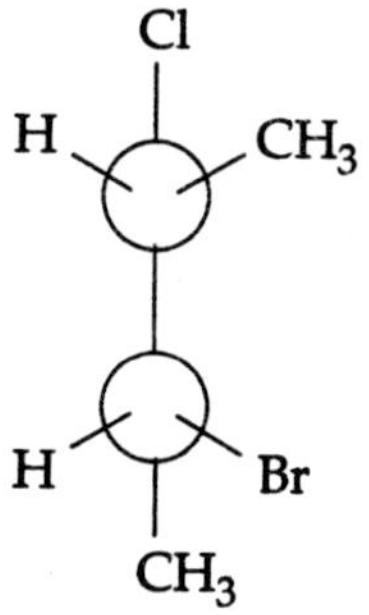
Cl
H
CH3
H
Br
CH3

(h) 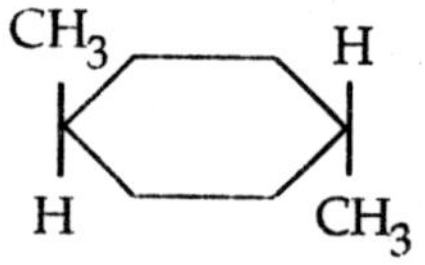
CH3
H
H
CH3
and
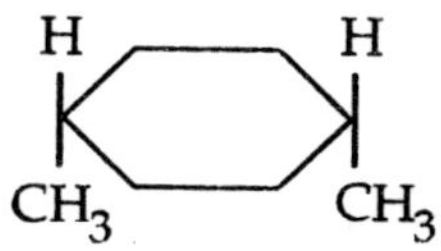
H
H
CH3
CH3

(i) 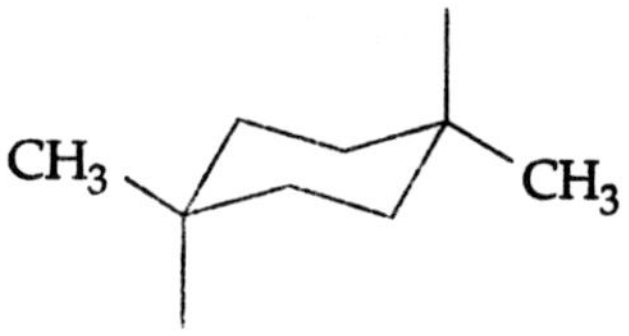
CH3
CH3
and
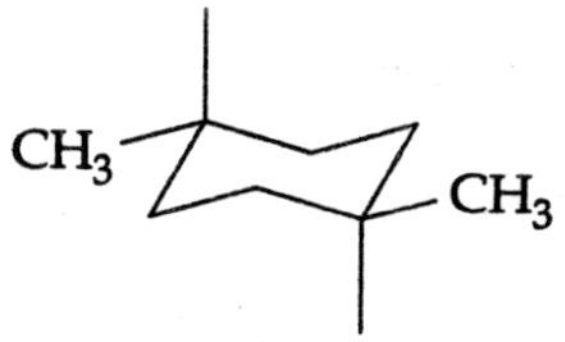
CH3
CH3

(j) 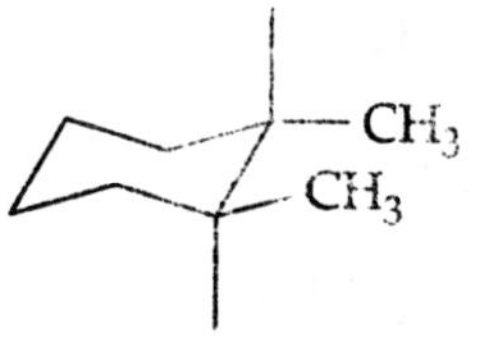
CH3
CH3
and
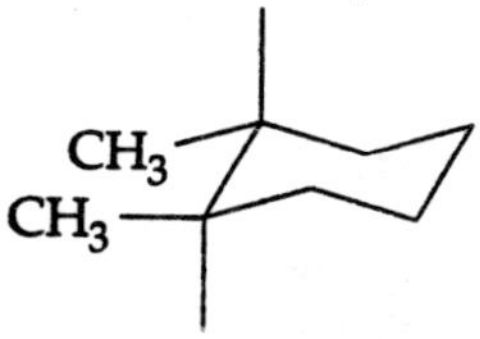
CH3
CH3

(k) 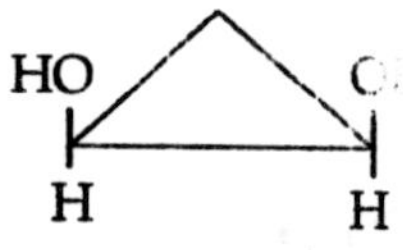
HO
H
H
and
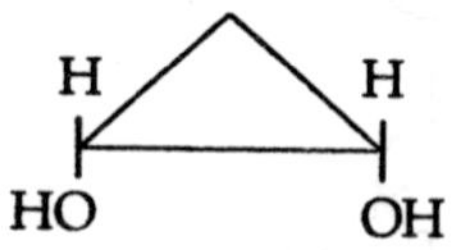
H
H
HO
OH

(l) 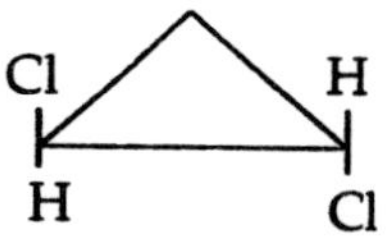
Cl
H
H
Cl
and

H
Cl
Cl
H

(m) 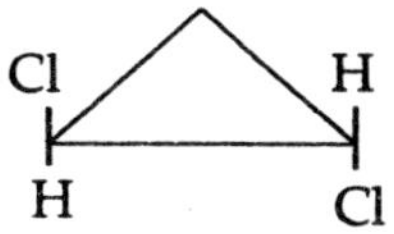 and 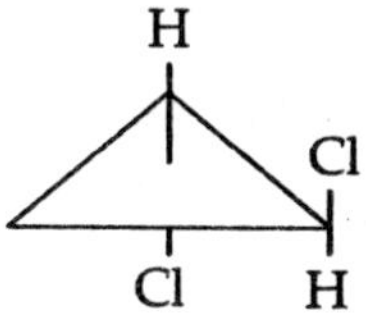

(n) 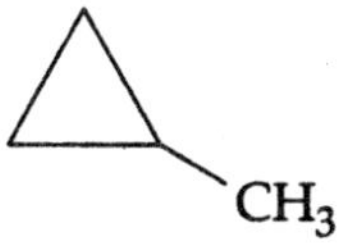 and 

(o) 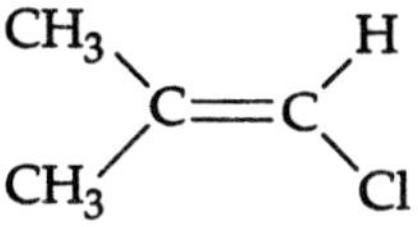 and 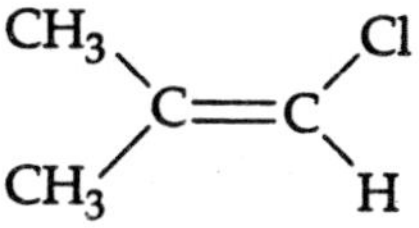

(p) 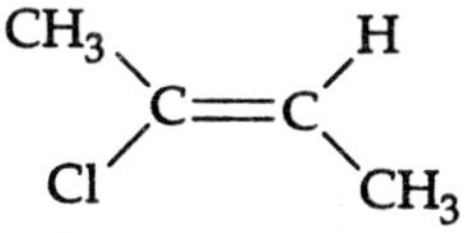 and 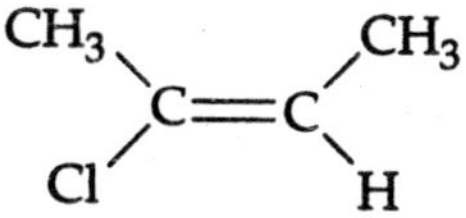

(q) 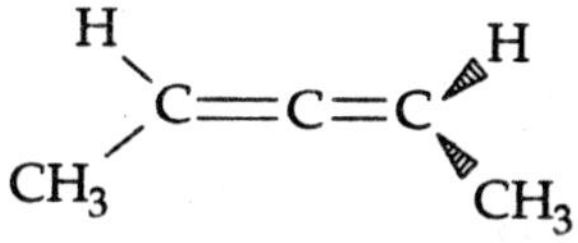 and 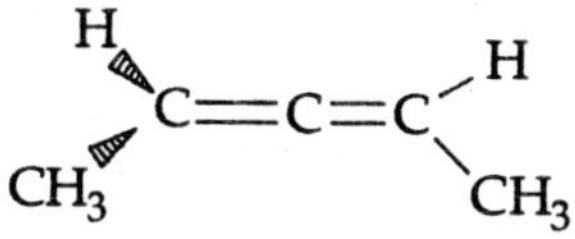

Q. 13. In each of the following sets of compounds, tick the one from (i), (ii) or (iii) you expect will have the lowest boiling point.

(a) (i) 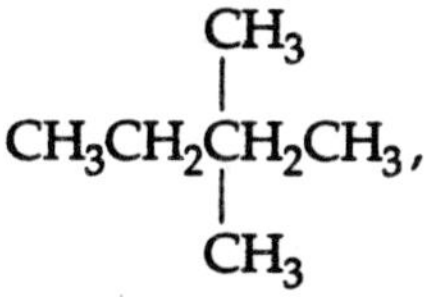, (ii) 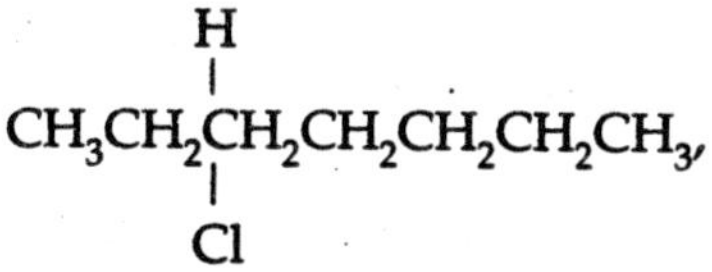,

(iii) 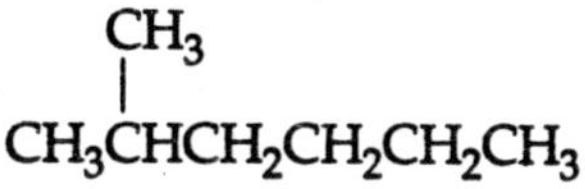

(b) (i) 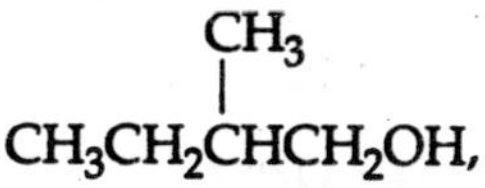, (ii) 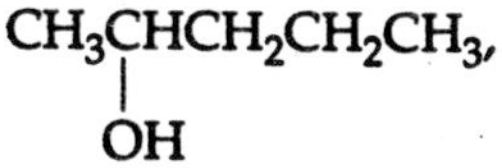,

(iii) 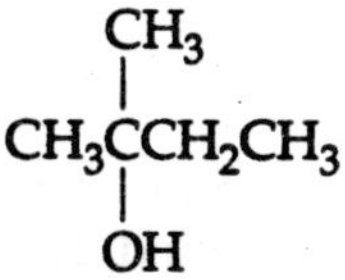

(c) (i) $CH_3CH_2CH_2CH_2CH_2OH$, (ii) 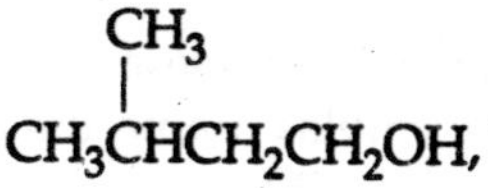,

CH_3
|
(iii) CH_3CCH_2OH
|
CH_3

CH_3
|
(d) (i) CH_3COCH_3,
|
CH_3

(ii) $CH_3CH_2CHCH_2CH_3$,
|
OH

(iii) $CH_3CH_2CH_2CH_2CH_2OH$

O
||
(e) $CH_3CH_2CH_2CH$,

(ii) 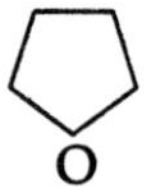

(iii) 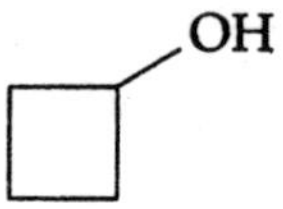

Q. 14. Adding hydrogen to the compound with the molecular formula C_7H_{12} gives a C_7H_{16} compound, which is identified as being 2-methylhexane. In light of this evidence how many structures are possible for the C_7H_{12} compound?
(a) 2 **(b)** 4 **(c)** 8 **(d)** 12

Q. 15. Determine whether each of the following compounds contains a stereocenter. If so, draw the two enantiomers of the compound. Give answer in Yes or No.

(a) $CH_3CH_2CH_2Br$

(b) CH_3CHCH_2Cl
|
OH

CH_3
|
(c) $CH_3CH_2CHCH_2Cl$

(d) 

(e) $CH_3CHCH_2CH_2CH_3$
|
Br

CH_3
|
(f) $CH_3CHCH_2CHCH_3$
|
OH

Q. 16. How many stereoisomers are possible for 2, 3, 4-tribromohexane. Draw all possible stereoisomers for the compound, and identify the enantiomers and diastereomers.
(a) 2 **(b)** 8 **(c)** 4 **(d)** 12

Q. 17. (i) How many stereoisomers are possible 1,3-dimethylcyclopethane? Draw the structural formulas, name the compounds.
(a) 2 **(b)** 5 **(c)** 4 **(d)** 3

(ii) How many stereoisomers are possible for 1-bromo-2-methylcyclopentane? Draw structural formulas, and give them names that designate configurations at the stereocenters.

(a) 2 **(b)** 5 **(c)** 4 **(d)** 3

(iii) How many different configurational isomers are possible for 1-bromo-2-methylcyclohexane? Draw structural formula and give their names:

(a) 2 **(b)** 5 **(c)** 4 **(d)** 3

Q. 18. Predict which compound in (a) (b) and (c) i.e. (i), (ii) or (iii) will have the highest solubility in water and which will have the lowest respectively.

(a) (i) $CH_3CH_2CH_2CH_2\overset{O}{\overset{||}{C}}OH$ (ii) $CH_3CH_2\overset{O}{\overset{||}{C}}OCH_2CH_3$

(iii) $CH_3CH_2CH_2CH_2\overset{O}{\overset{||}{C}}O^-Na^+$

(b) (i) $CH_3CH_2CH_2\overset{O}{\overset{||}{C}}OH$ (ii) $CH_3CH_2CH_2CH_2OH$

(iii) $CH_2CH\overset{O}{\overset{||}{C}}OCH_2CH_3$

(c) (i) $CH_3CH_2CH_2\overset{O}{\overset{||}{C}}OCH_2CH_3$ (ii) $CH_3CH_2CH_2\overset{O}{\overset{||}{C}}NH_2$

(iii) $CH_3CH_2CH_2\overset{O}{\overset{||}{C}}O(CH_2)_4CH_3$

Q. 19. The following amides (a) and (b) were needed for a study on the stereochemistry. They were each made from an acid chloride (i)-(iii) and an amine (iv)-(vi). Which acids chloride and which amines would be needed to prepare these amides respectively?

(a) 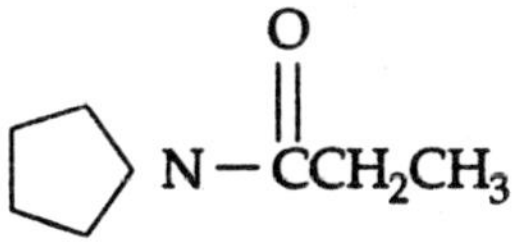

(b) 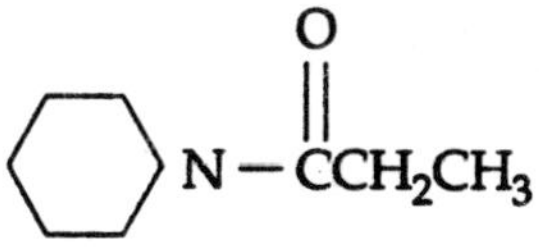

Acidchlorides	Amides
(i) CH_3COCl	(iv) C_6H_5NH = ?
(ii) CH_3CH_2COCl	(v) 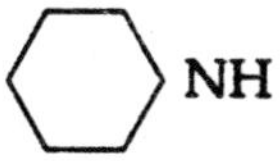
(iii) $CH_3CH_2CH_2COCl$	(vi) 

Q. 20. Match the reagents x given, for the following transformations. Some of them may require more than one reaction.

(a) $CH_3CH_2CH{=}CH_2 \xrightarrow{?} CH_3CH_2\underset{\displaystyle SH}{\underset{|}{C}}HCH_3$

(b) $CH_3CH{=}CHCH_3 \xrightarrow{?} CH_3CH_2\underset{\displaystyle Cl}{\underset{|}{C}}HCH_3$

(c) $CH_2{=}CH_3 \xrightarrow{?} CH_3CH_3\overset{+}{N}H_3I^-$

(d) $CH_3\underset{\displaystyle Br}{\underset{|}{C}}HCH_2CH_2CH_3 \xrightarrow{?} CH_3\underset{\displaystyle CN}{\underset{|}{C}}HCH_2CH_2CH_3$

Reagents (x)

(i) HCl
(ii) HBr
(iii) HI
(iv) NH_3
(v) NaCN
(vi) NaSH

Q. 21. Match the line structural formulas for the following compounds:

(a) $CH_3CH_2CH_2CH{=}CH_2$

(b) 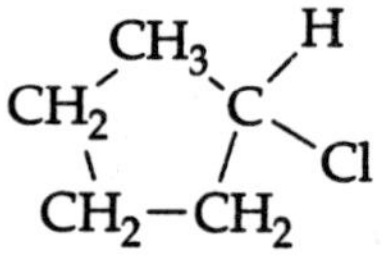

(c) 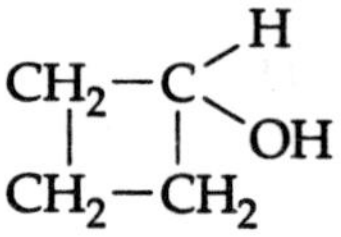

(d) $CH_3CH_2\overset{\displaystyle CH_3}{\overset{|}{\underset{\displaystyle Br}{\underset{|}{C}}}}HCHC{\equiv}CCH_2CH_3$

(e) $CH_3\overset{\displaystyle CH_3}{\overset{|}{C}}HCH_2CH_2\overset{\displaystyle O}{\overset{||}{C}}OH$

(i) 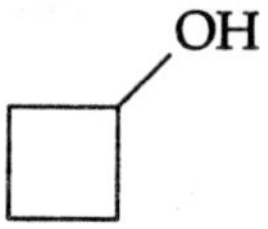

(ii) 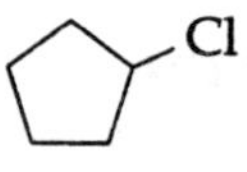

(iii) 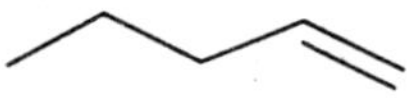

(iv) 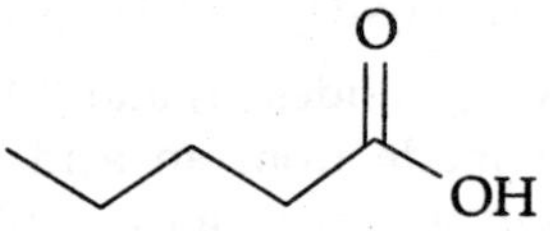

(v) 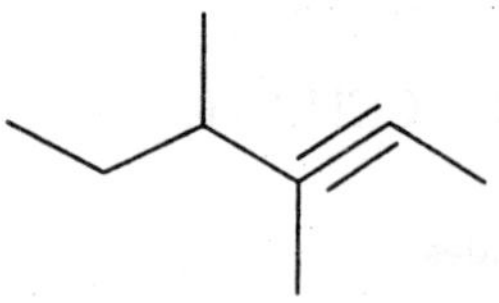

Q. 22. Match the condensed structural formulas for the following compounds.

(a) 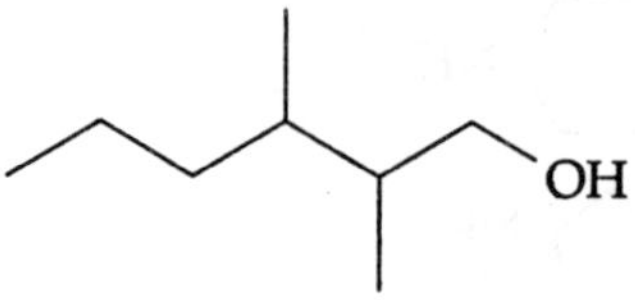

(i) $CH_3CH_2CH_2CH_2COCHBrCH_3$

(b) 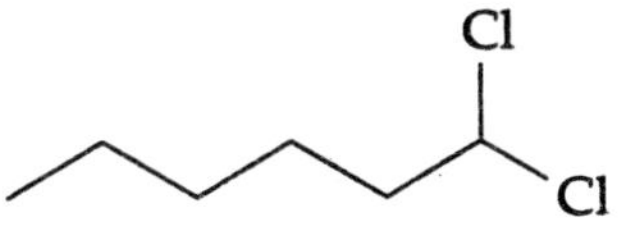 (ii) $CH_3CH_2CH(CH_3)CH(CH_3)CH_2OH$

(c) 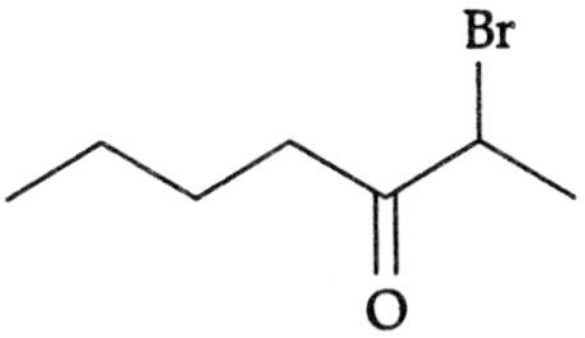 (iii) $CH_3CH_2CH_2CH_2CH(CH_3)COOH$

(d) 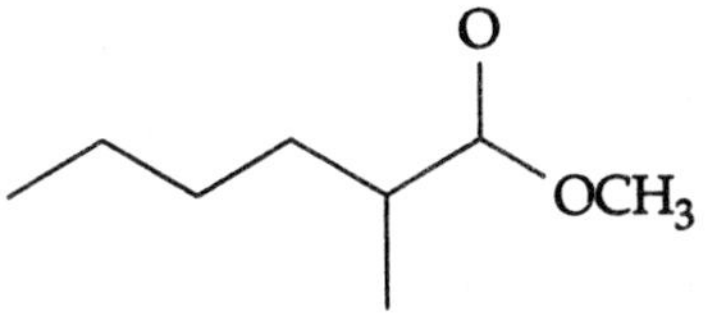 (iv) $CH_3CH_2CH_2CH_2CH_2CHCl_2$

Q. 23. Match the names of the following compounds with column ⊗

column ⊗

(a) 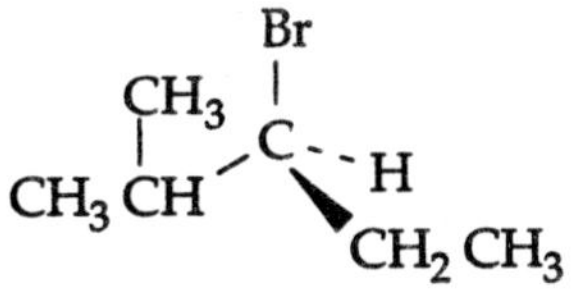 (i) (*S*)-3-methyl-3-hexanol.

(b) 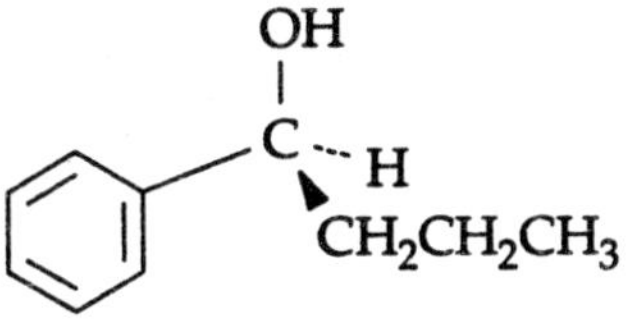 (ii) butane.

(c) 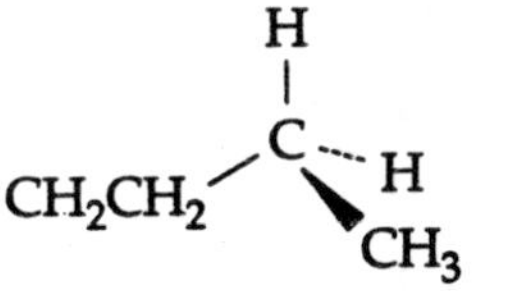 (iii) (2*R*, 3*S*)-2, 3-dibromobutane.

(d) 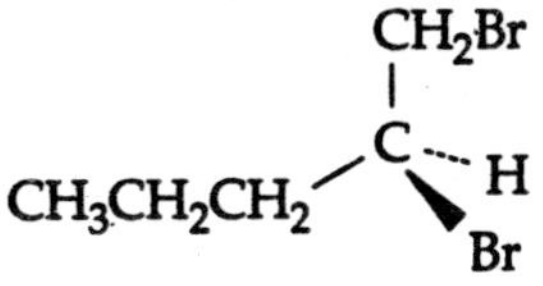 (iv) (*S*)-3 Bromo-2-methyl pentane.

(e) 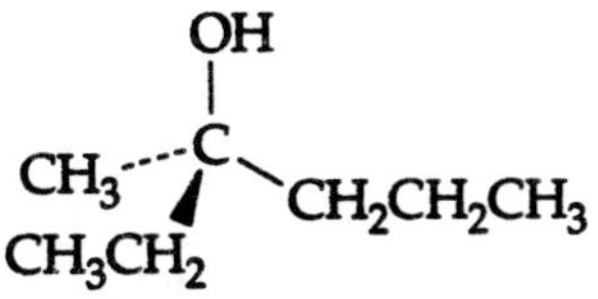 (v) (2*R*, 3*S*)-2, 3-dichloropentane.

(f) 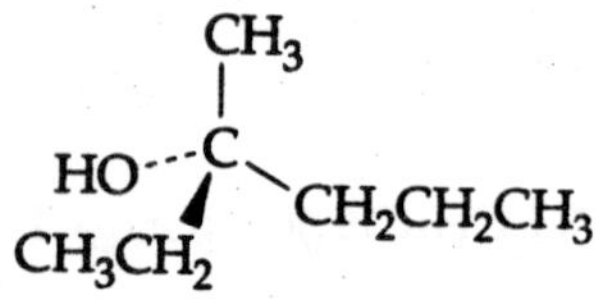

(vi) (1*S*, 2*S*, 4*S*)-1, 2-dibromo-4-methylcyclohexane.

(g) 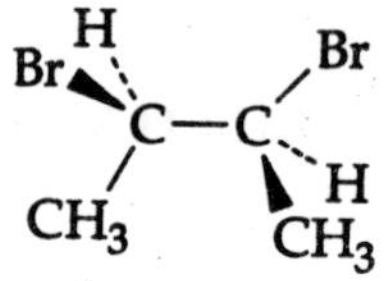

(vii) (1*R*, 2*R*)-2-bromocyclo pentanol.

(h) 

(viii) (*S*)-1-phenyl-1-butanol

(i) 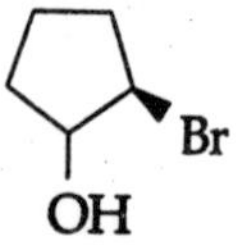

(ix) (*R*)-3-methyl-3-hexanol.

(j) 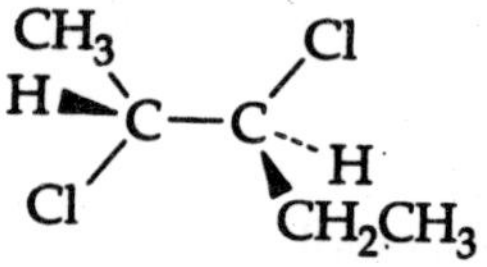

(x) (*S*)-1, 2-dibromopentane.

Q. 24. For each of the following pairs of structural formulas, tell whether the two represent identical molecular species (I); conformers of the same species (II); constitutional isomers (III); enantiomer, or diastereomers (IV). Tick in the chart. Check answer only after completing the chart.

(a) 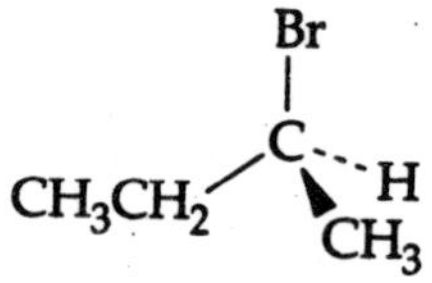 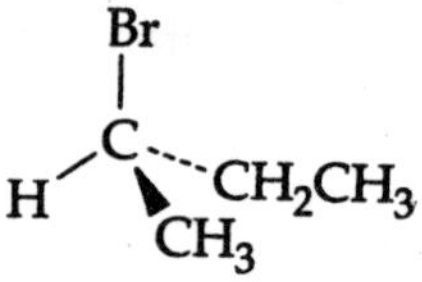

(b) 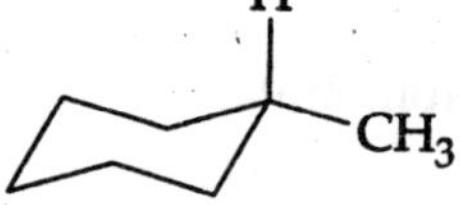 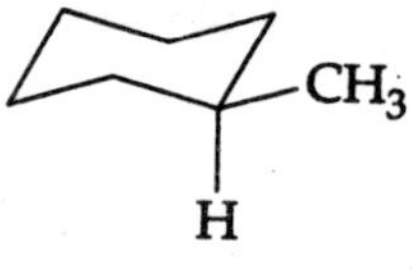

(c) 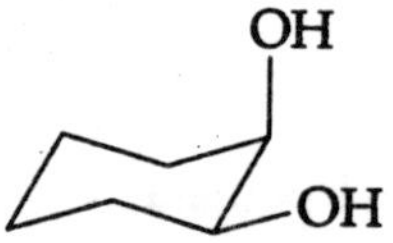 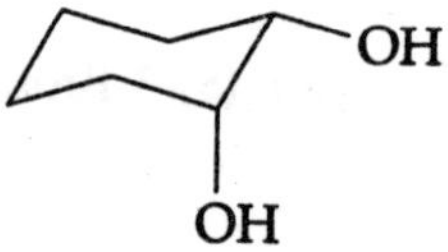

(d) 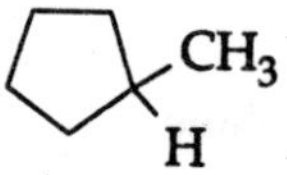 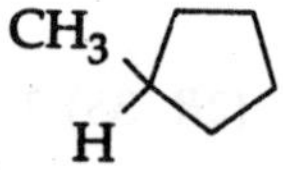 (d)

	I	II	III	IV
(a)				
(b)				
(c)				
(d)				
(e)				
(f)				
(g)				
(h)				
(i)				
(j)				
(k)				
(l)				

Chart

(e) 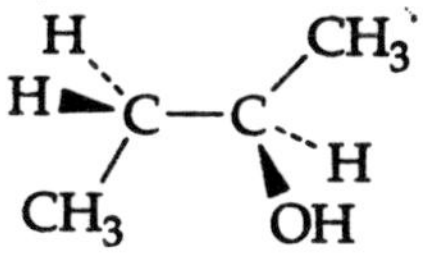 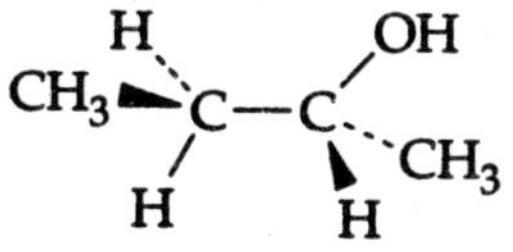

(f) 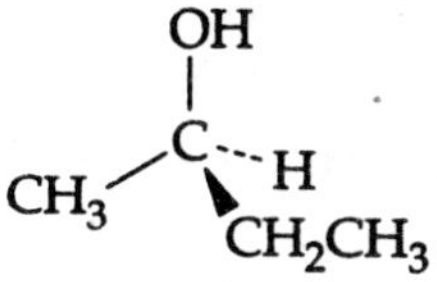 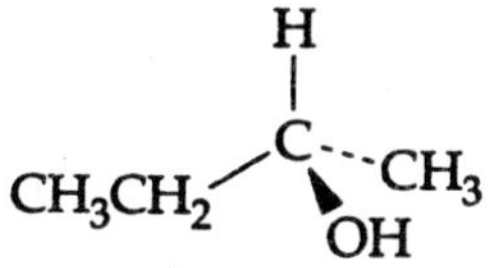

(g) 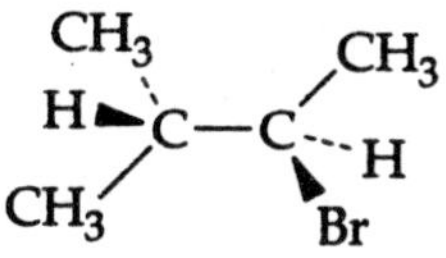 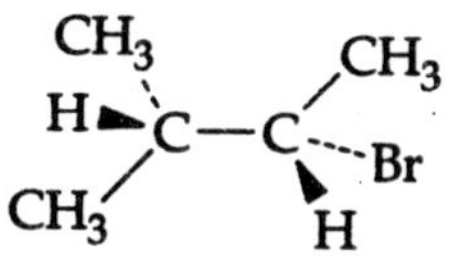

(h) 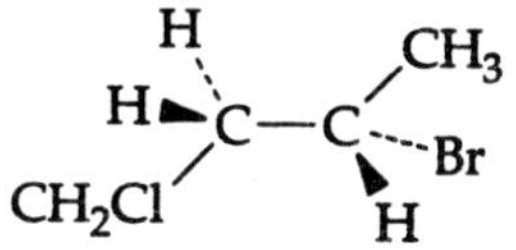 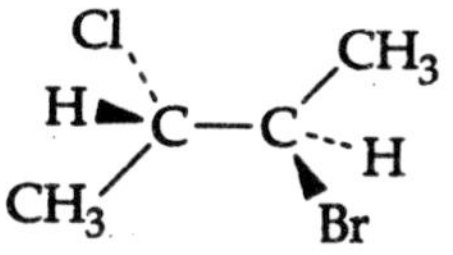

(i) 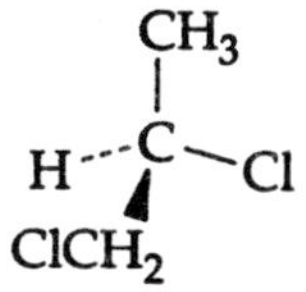 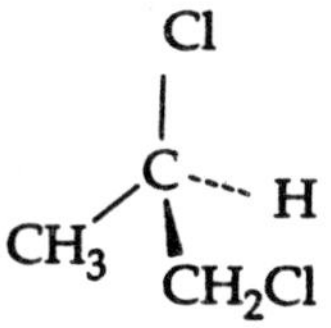

(j) 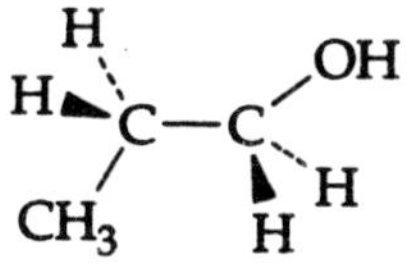 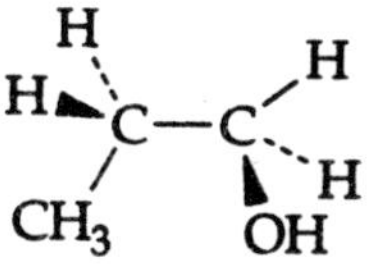

(k) 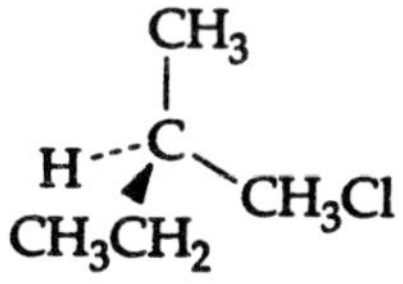 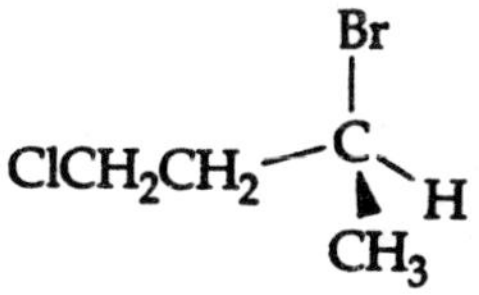

(l) 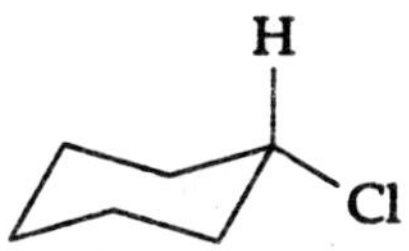 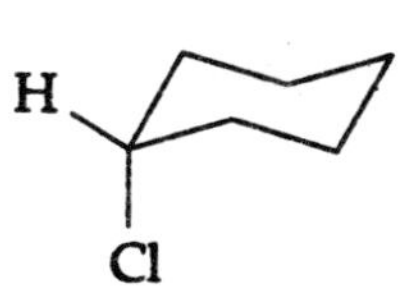

Q. 25. Match the following products A to I with Column ⊗

(a) 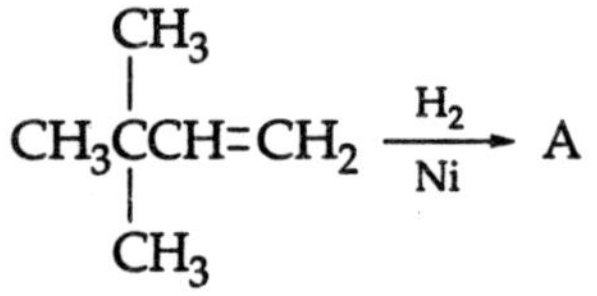

(b) 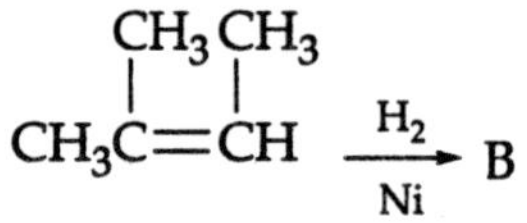

(c) 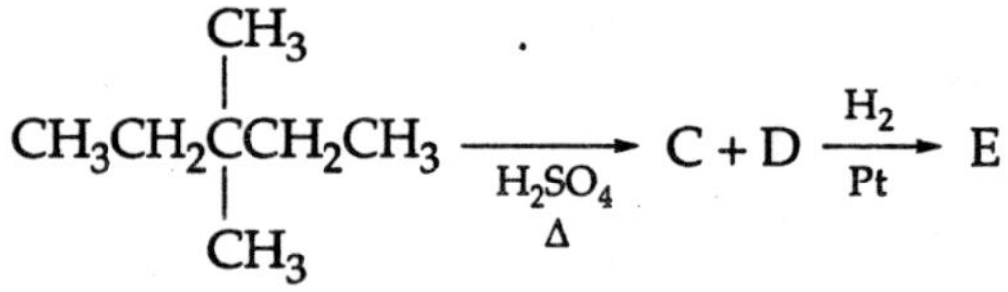

(d) 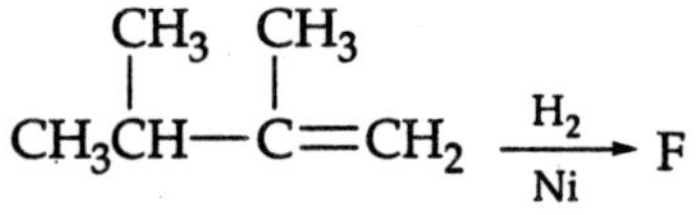

(e) 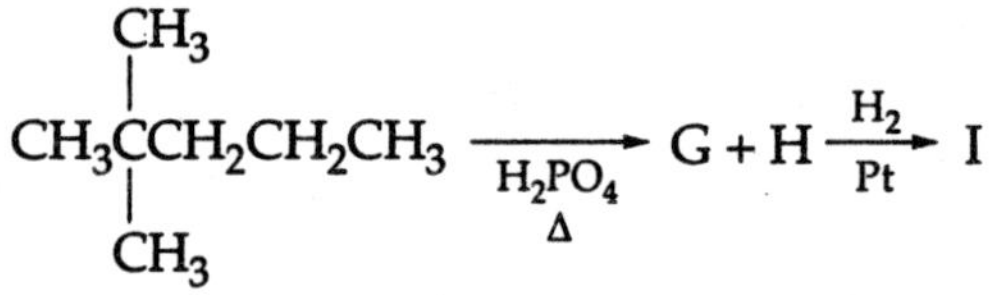

Column ⊗

(i) $CH_3CH(CH_3)CH(CH_3)CH_3$

(ii) $CH_3CH_2C{=}(CH_2)CH_2CH_3$

(iii) $CH_3CH(CH_3)CH_2CH_2CH_3$

(iv) $CH_3 \cdot (CH_3)\ C \cdot CH_2CH_3$

(v) $CH_3CH_2CH\ (CH_3)\ CH_2CH_3$

(vi) $CH_3(CH_3)CHCH(CH_3)CH_3$

(vii) $CH_2{=}\ C(CH_3)\ CH_2CH_2CH_3$

(viii) $CH_3C(CH_3){=}CHCH_2CH_3$

(ix) $CH_3CH{=}C(CH_3)CH_2CH_3$

Q. 26. Match the following symbols A to I with column ⊗

(a) $HC{\equiv}CH \xrightarrow{NaNH_2} A \xrightarrow{C_6H_5CH_2Br}$

$B \xrightarrow{CH_3CH_2MgBr} C$

(b) 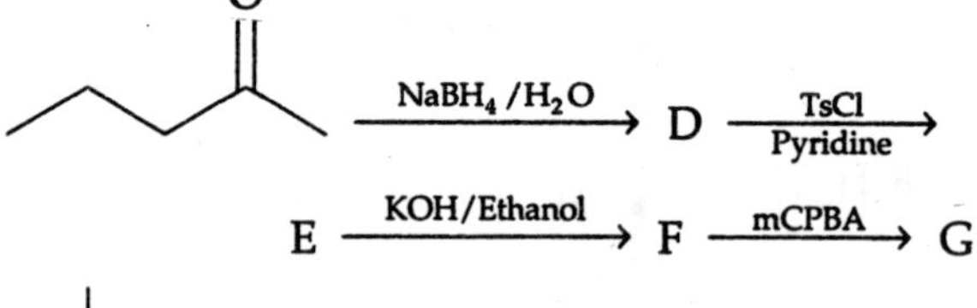

(c) 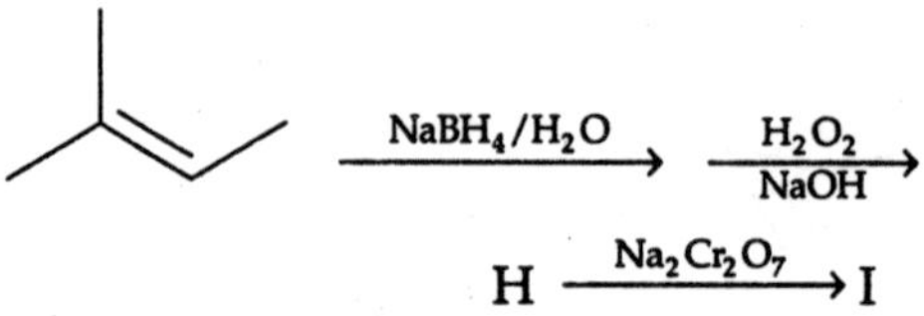

Column ⊗

(i) $H{-}C{\equiv}CNa$

(ii) 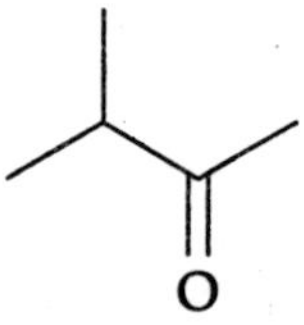

(iii) 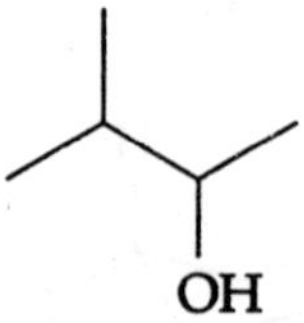

(iv) $C_2H_6 + C_6H_5CH_2C{\equiv}CMgBr$

(v) 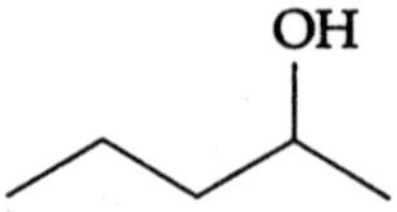

(vi) 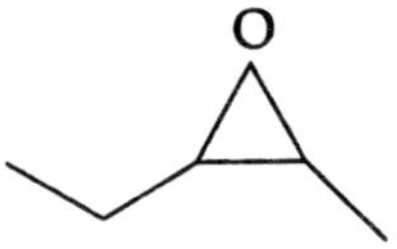

(vii) $C_6H_5CH_2C{\equiv}CH$

(viii) 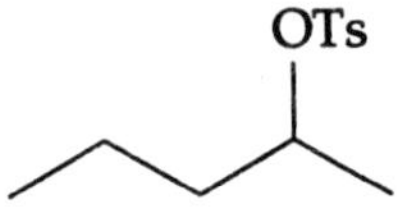

(ix) 

Q. 27. Match the following structures in column ⊗ with each of the indicated compound of the series of reaction (i.e. A to H)

(a) 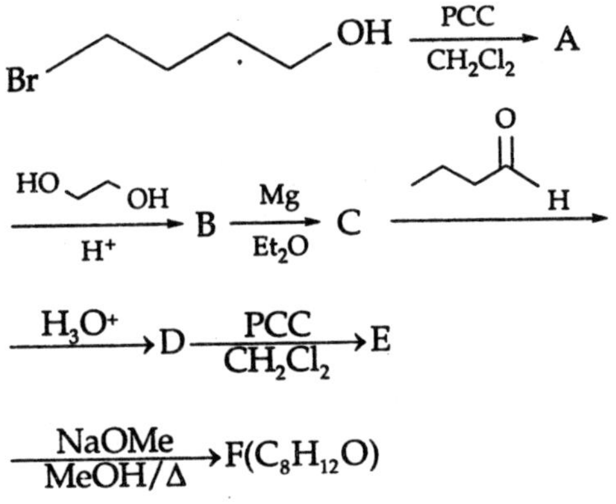

Column ⊗

(i) 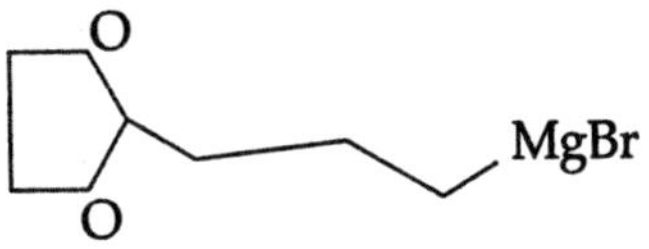

(ii) 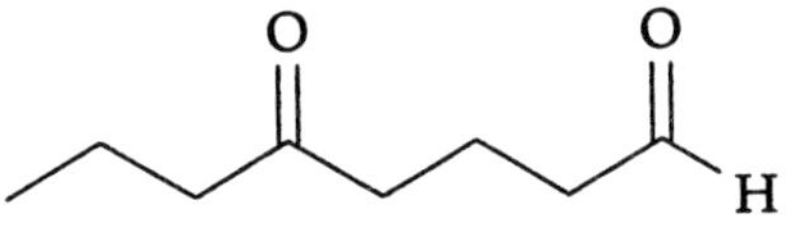

(iii) 

(iv) 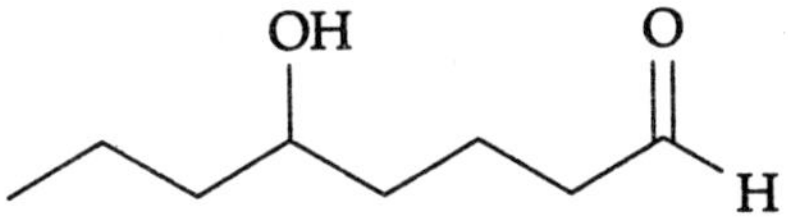

(v) 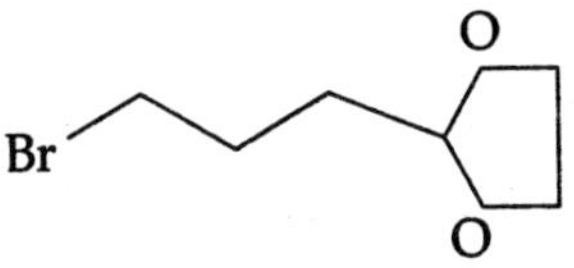

(vi) 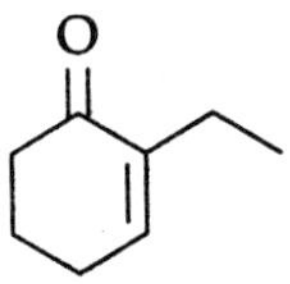

Column ⊗

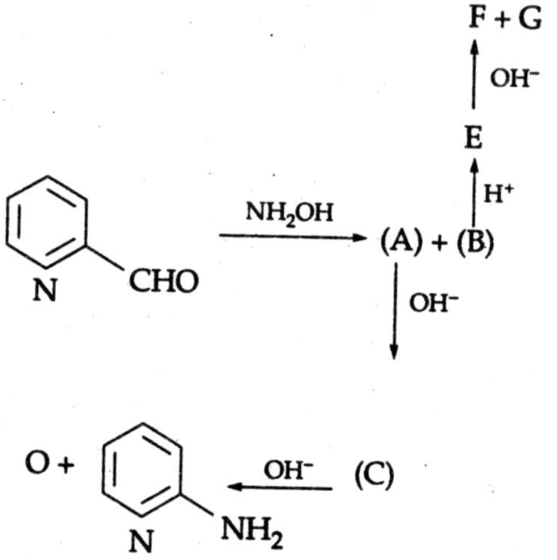

(D)

(i) 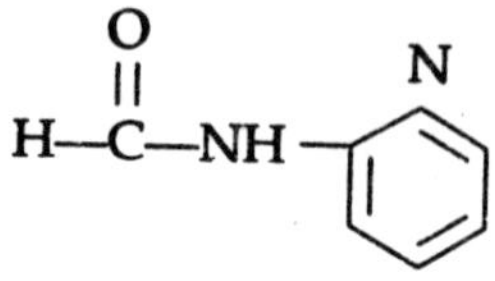

(ii) 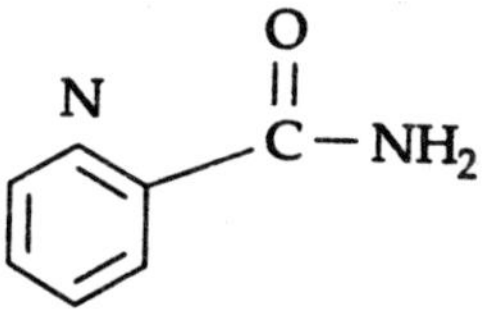

(iii) $+NH_3$

(iv) 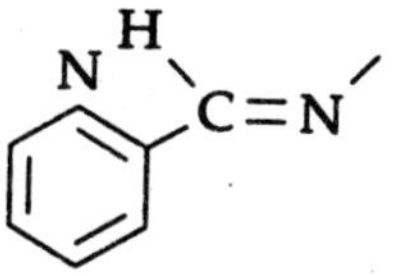

(v) 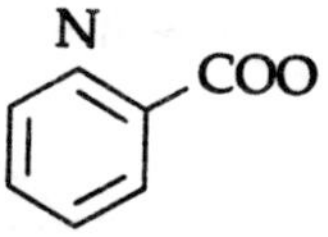

(vi) $HCOO^-$

(vii) 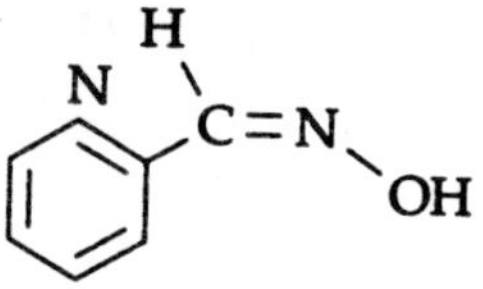

Q. 28. Which of the compounds below can exhibit geometric isomerism?

(a) $CH_3—CH=C(CH_3)_2$

(b) $CH_3—CH=CHCl$

(c) $CH_3—CH=CBrCl$

(d) 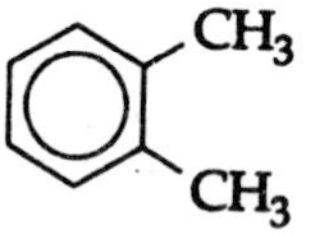

(e) 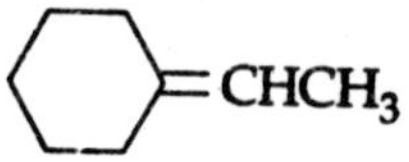

(f) 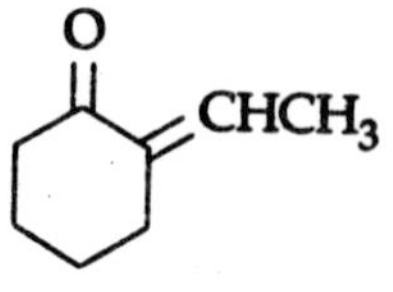

Q. 29. How many isomers of dibromobenzene ($C_6H_4Br_2$) are possible? Draw them.

(a) 2 **(b)** 3 **(c)** 4 **(d)** 6

Q. 30. How many isomers of dibromopyridine ($C_5H_3Br_2N$) are possible? Draw them.
(a) 2 **(b)** 3 **(c)** 4 **(d)** 6

Q. 31. Match the names of the following compounds with column ⊗

(a) 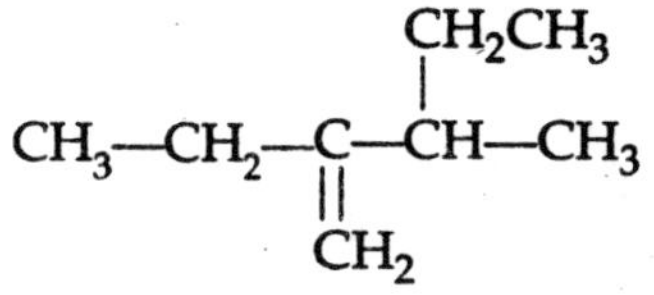

(b) 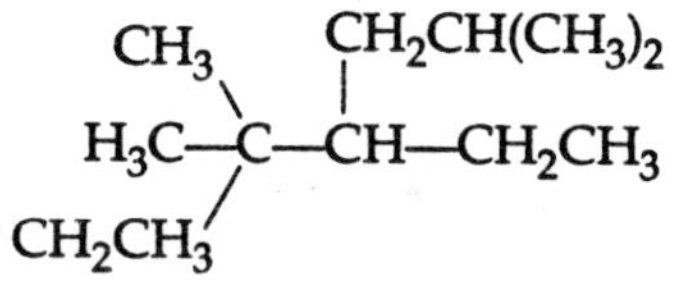

(c) 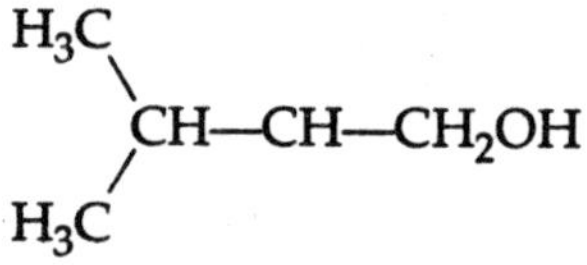

(d) 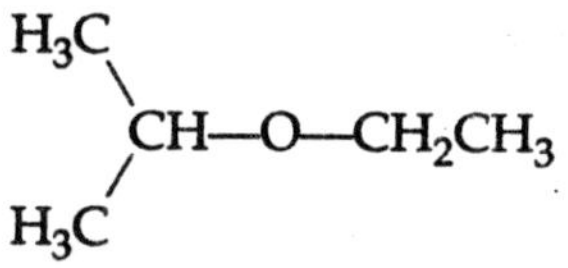

(e) 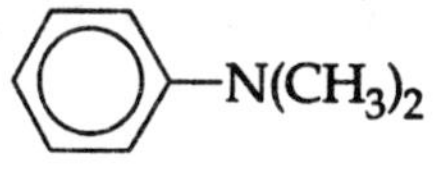

(f) 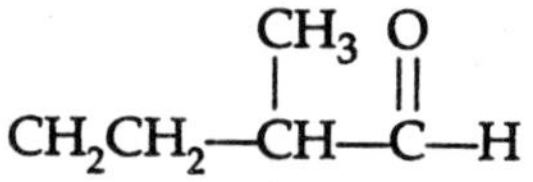

(g) 

(h) 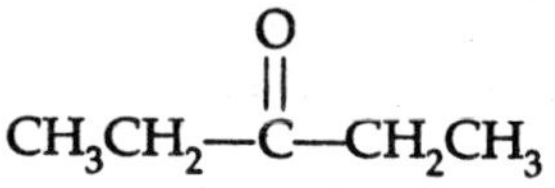

(i) 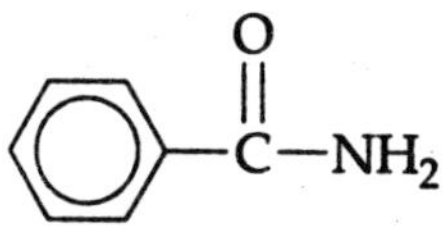

Column ⊗

(i) 3-methyl-1-butanol
(ii) 2-methyl butanal
(iii) 2-ethyl-3-methyl-1-pentene
(iv) benzamide
(v) 4-ethyl 1-2, 5, 5-trimethylheptane
(vi) 3-pentanone
(vii) ethyl isopropyl ether
(viii) pentanoic acid
(ix) N, N-dimethyl aniline

Q. 32. The following reaction were carried out during research into the stereochemistry of the hydroboration reaction.

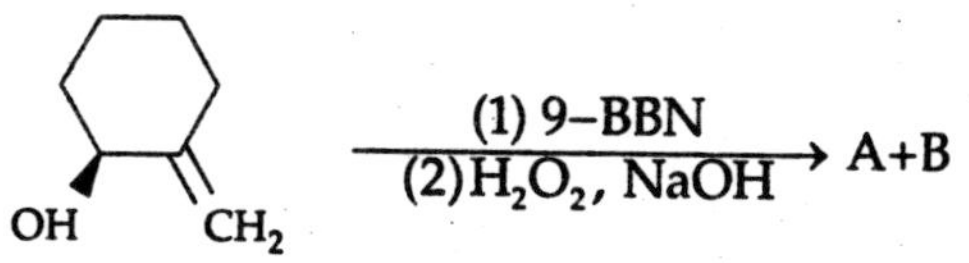
OH
CH₂
(1) 9–BBN
(2)H₂O₂, NaOH
A+B

Compounds *A* and *B* are stereoisomers of each other and are formed in roughly equal amounts. Compound *A* is thermodynamically more stable than compound B. Match the structural formulas for compounds *A* and *B* with following:

(a) 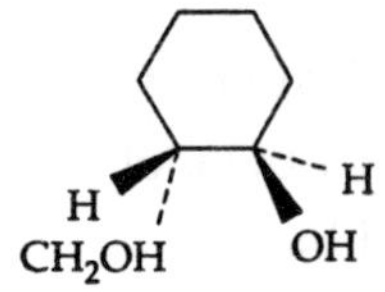

(b) 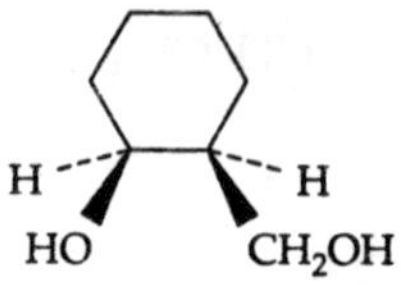

(c) 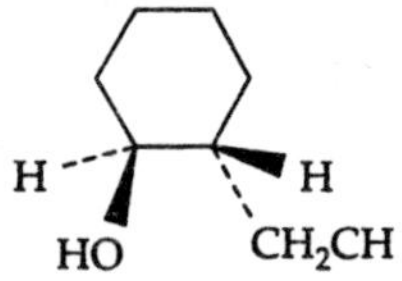

(d) 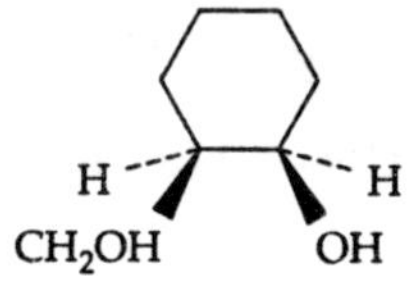

Q. 33. Compound A, $C_{13}H_{10}O$, named Carlina-oxidem is hydrogenated. The product is compound, B, $C_{13}H_{14}O$, which has the following structural formula:

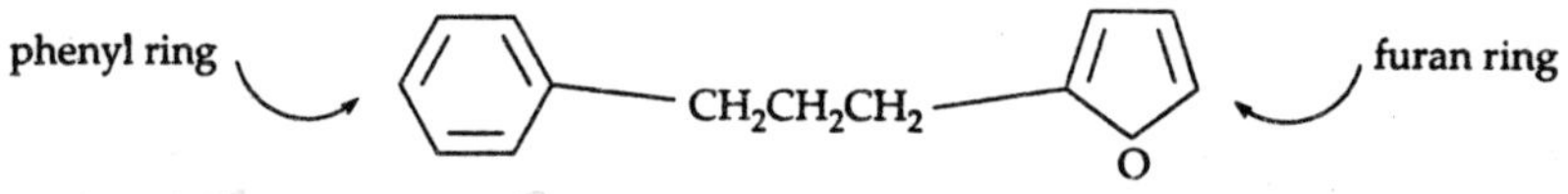
phenyl ring
CH2CH2CH2
furan ring
O

compound B

Compound *B* has a three-carbon chain with a phenyl ring at one end and a furan ring at the other. The furan ring is similar to the phenyl ring in resistance to hydrogenation at low pressures and temperatures.

How many units of unsaturation does Carlina-oxide contain and how many of them were saturated by reaction with hydrogen respectively. Give structures:

(a) 9, 7 **(b)** 10, 3 **(c)** 10, 8 **(d)** 9, 2

Q. 34. Chemists are always interested in new ways to construct five-membered ring compounds because such rings are found in many important natural products. The following reactions were carried out during research into such synthesis. Match the structural formula for any missing reagents, starting materials or products indicated by letters A–D with column ⊗.

(a) 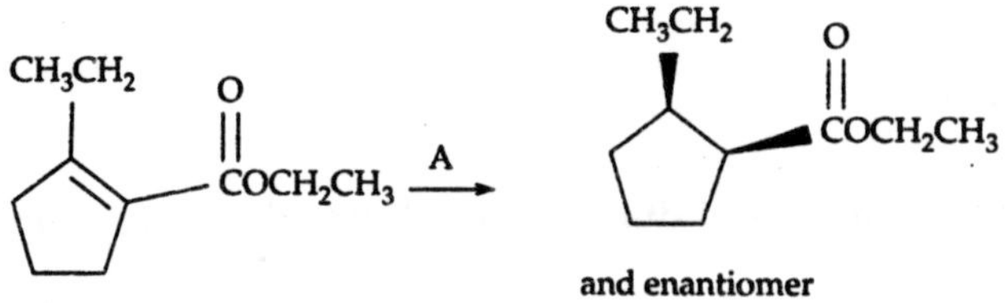
CH3CH2
O
COCH2CH3
A
CH3CH2
O
COCH2CH3

and enantiomer

(b) 

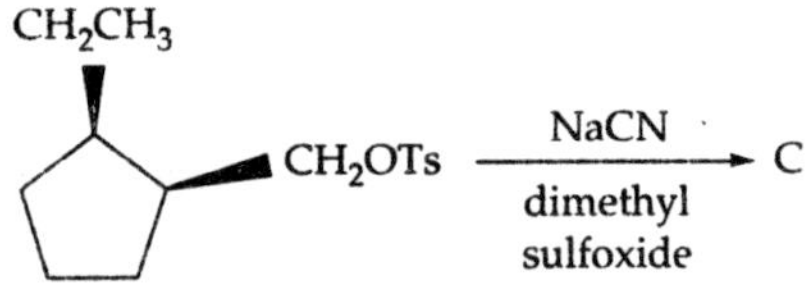

Column ⊗

Reagents/Starting Material/Products

(i) 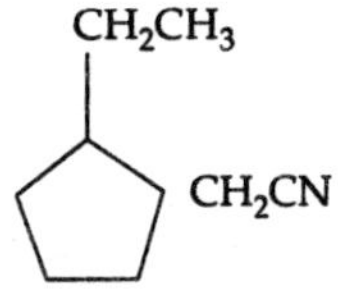

(ii) $ICH_2CH_2CH_2Cl$ or $BrCH_2CH_2CH_2Cl$

(iii) H_2/Catalyst

(iv) 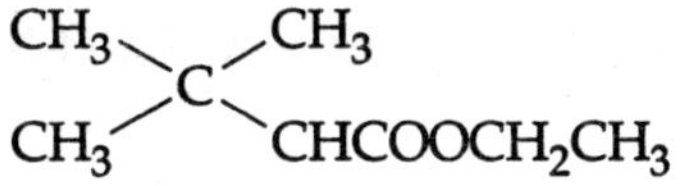

(v) TsCl/Pyridine.

(c) 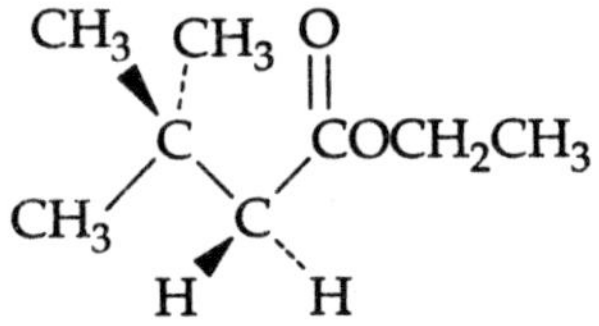

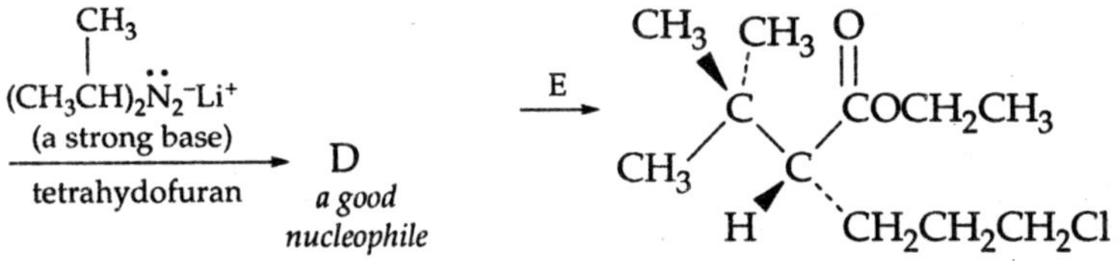

Q. 35. Tick the following groups as ortho (I) para (II) or meta (III) directing.

(a) CF_3—	(a) I/II/III
(b) $(CH_3)_3\overset{+}{P}$—	(b) I/II/III
(c) O_2NCH_2—	(c) I/II/III
(d) $(CH_3)_2S$—	(d) I/II/III
(e) 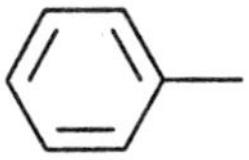	(e) I/II/III
(f) $CH_3CH{=}CH$—	(f) I/II/III

Q. 36. In the following transformations, decide whether the starting material has undergone oxidation or reduction. Match them with the serial numbers (i) to (v) of column ⊗. Specify whether an oxidising or reducing agent would be needed, respectively to carry out the change that is shown.

(a) 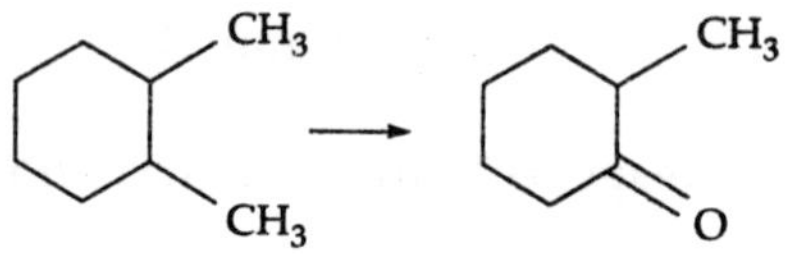

(b) 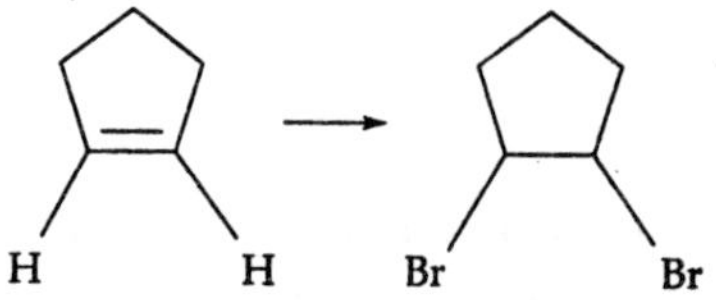

(c) $CH_3CH_2Br \rightarrow CH_3CH_2NH_2$

(d) 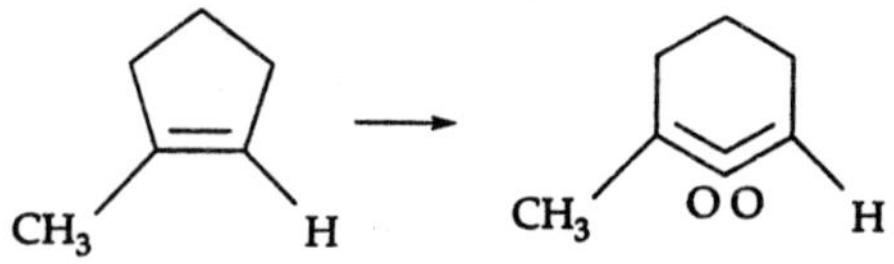

(e) 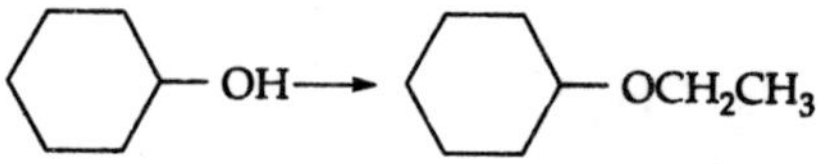

(f) $CH_3CH_2\overset{\overset{O}{||}}{C}H \rightarrow CH_3CH_2CH_2OH$

(g) $CH_3CH{=}CHCH_3 \rightarrow CH_3\underset{\underset{OH}{|}}{C}H{-}\underset{\underset{OH}{|}}{C}HCH_3$

(h) $CH_3C{\equiv}CCH_2CH_3 \rightarrow CH_3CH{=}CHCH_2CH_3$

(i) $CH_3C{\equiv}CCH_3 \rightarrow CH_3\overset{\overset{O}{||}}{C}{-}\overset{\overset{O}{||}}{C}CH_3$

(j) $CH_3CH_2CH_2OH \rightarrow CH_3CH_2\overset{\overset{O}{||}}{C}OH$

(k) $CH_3\overset{\overset{O}{||}}{C}H \rightarrow CH_3\overset{\overset{OCH_3}{|}}{C}HOCH_3$

Column ⊗

(i) Oxidation
(ii) Reduction
(iii) Nochange
(iv) An oxidising agent
(v) An reducing agent.

Q. 37. Match the reagents would you use to carry out each step of the following transformation; with column ⊗.

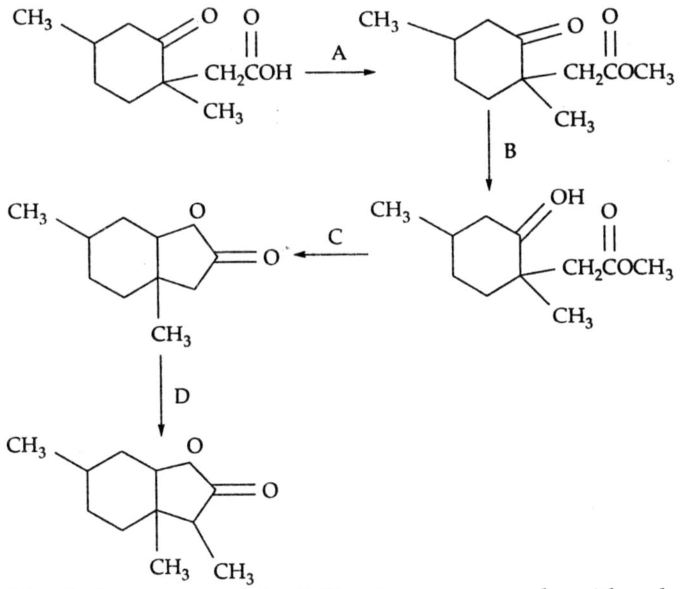
CH3
O
O
CH2COH
CH3
A
CH3
O
O
CH2COCH3
CH3
B
OH
O
CH2COCH3
CH3
C
CH3
O
O
CH3
D
CH3
O
O
CH3 CH3

column ⊗

Reagents

(i) NaH
(ii) $NaBH_4$
(iii) H_2SO_4
(iv) CH_3I
(v) CH_3OH
(vi) △
(vii) Tetrahydrofuran
(viii) $(CH_3CH_2)_2\overset{-}{N}\overset{+}{Li}$ with CH_3 below

Q. 38. Match the names of the following compounds with column ⊗

(i) 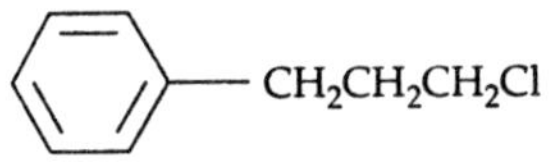
CH2CH2CH2Cl

(ii) 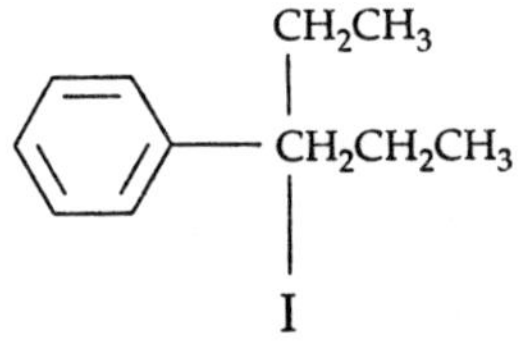
CH2CH3
CH2CH2CH3
I

(iii) 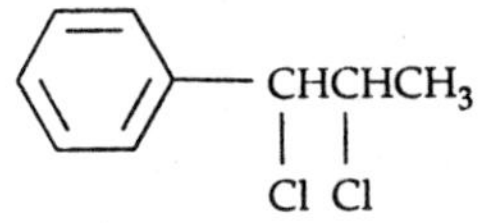
CHCHCH3
Cl Cl

(iv) 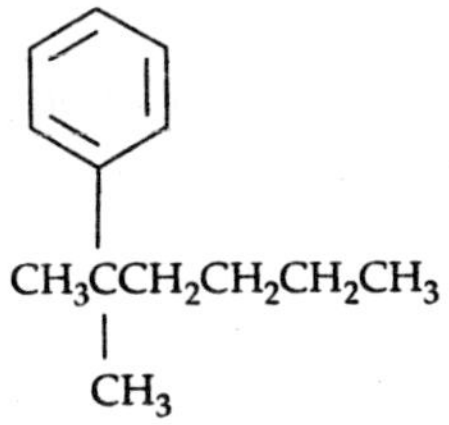
CH3CCH2CH2CH2CH3
CH3

(v) 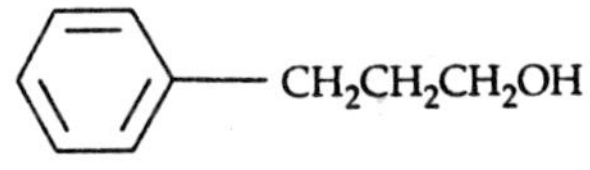
CH2CH2CH2OH

(vi) 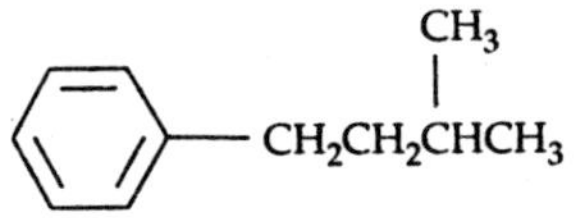
CH3
CH2CH2CHCH3

Column ⊗

(a) 1, 2-dichloro-1-phenylpropane
(b) 2-meythyl-2-phenyl hexane
(c) 3-methyl-1-phenyl butane
(d) 1-chloro-3-phenyl propane
(e) 3-iodo-3-phenyl hexane
(f) 3-phenyl-1-propanol..05

Q. 39. Match the names of the following compounds. With column ⊗

(a) 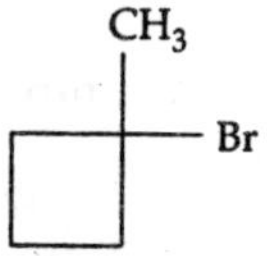

(b) 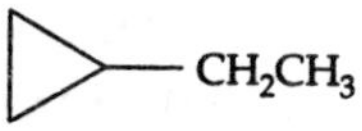

(c) 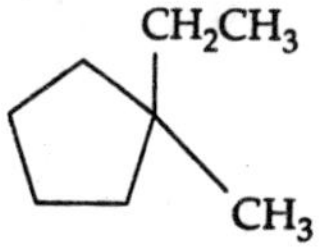

(d) 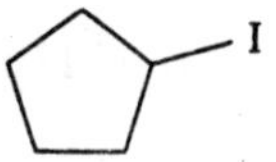

(e) 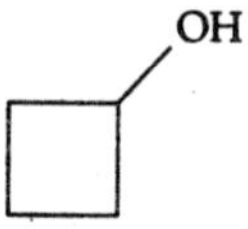

(f) 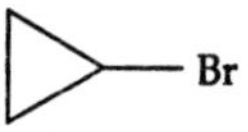

Column ⊗

(i) 1-ethyl-1-methyl cyclopentane
(ii) iodocyclo pentane
(iii) cyclobutanol
(iv) 1-bromo-1-methyl cyclobutane
(v) ethyl cyclopropane
(vi) bromocyclo propane

Q. 40. Match the names of the following compounds with column ⊗.

(a) 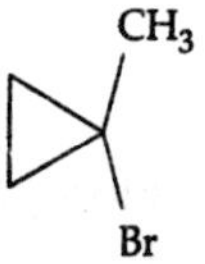

(b) 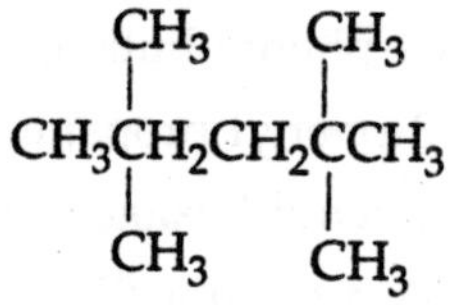

(c) 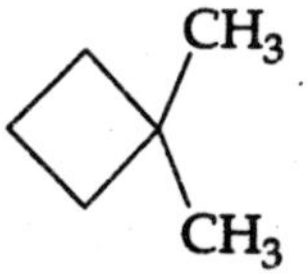

(d) 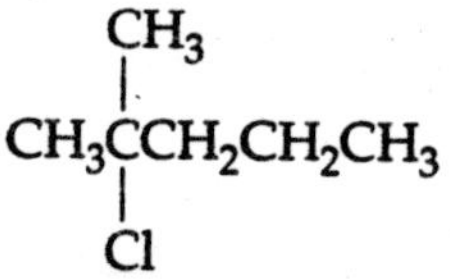

(e) 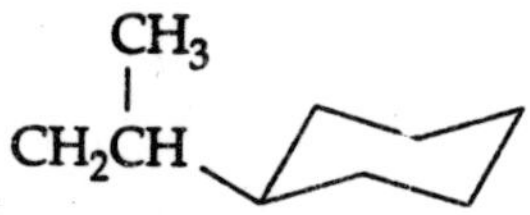

Column ⊗

(i) 1-phenyl-1-cyclohexane
(ii) 2-cyclohexyl propane
(iii) 3-bromo-3-ethyl-1-heptanol
(iv) 1-bromo-1-methyl cyclopropane
(v) 1-chloro-1-ethyl cyclohexane
(vi) 2-chloro-2-methyl pentane
(vii) 2,2,5,5-tetramethyl hexane
(viii) 1-ethyl-1-cyclopentanol
(ix) 6,7-dichloro-3-ethyl-2,2-dimethyl octane
(x) 1,1-dimethyl cyclobutane

(f) 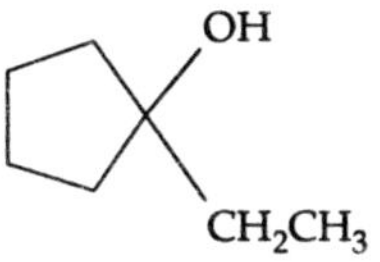

(g) 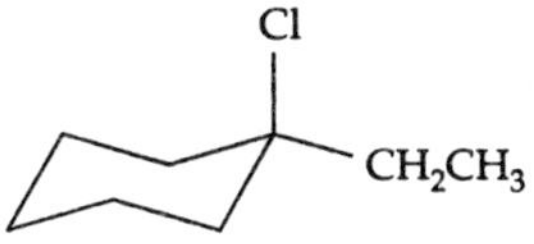

(h) 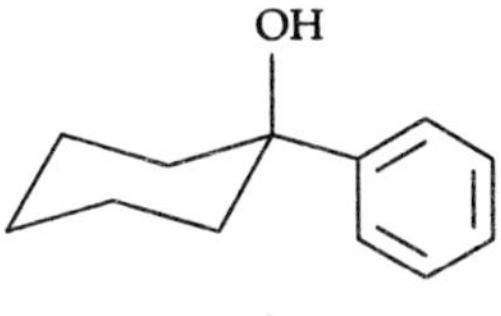

(i) $CH_3CH_2CH_2CH_2\overset{CH_2CH_3}{\underset{Br}{C}}CH_2CH_2OH$

(j) $CH_3CH_2\overset{\overset{CH_3}{CH_3CCH_3}}{C}HCH_2CH_2\underset{Cl}{C}H\underset{Cl}{C}HCH_3$

Q. 41. Match the following the names of the following compounds according to the IUPAC rules, including stereochemistry wherever it is shown with column ⊗.

(a) $CH_3CH_2\underset{OH}{C}H\overset{CH_3}{C}H—\overset{O}{\overset{\|}{C}}H$

(b) $CH_3\overset{O}{\overset{\|}{C}}CH_2CH_2CH_3$

(c) 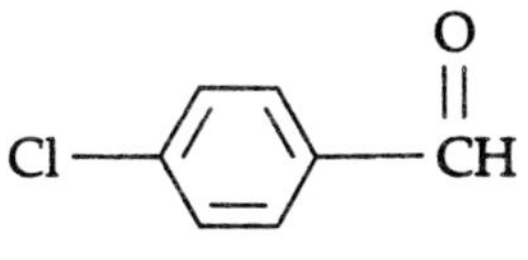

(d) $CH_2CH_2C\equiv C\overset{O}{\overset{\|}{C}}CH_2CH_3$

Column ⊗

(i) 4-heptyne-3-one propen-1-one
(ii) 4-pentanedione
(iii) 2-pentanone
(iv) (E)-1,3-diphenyl-2-nonen-2-one
(v) 2-ethyl-2-methyl cyclo butanone
(vi) 4-chlorobenzaldehyde
(vii) 2-2-dibromocyclohexanone
(viii) 3-hydroxy-2-methyl pentanol
(ix) trichloroethanol
(x) cis-2-propylcyclopentane carbaldehyde
(xi) 3, 3-dimethyl cyclohaxane carboldehyde

(e) 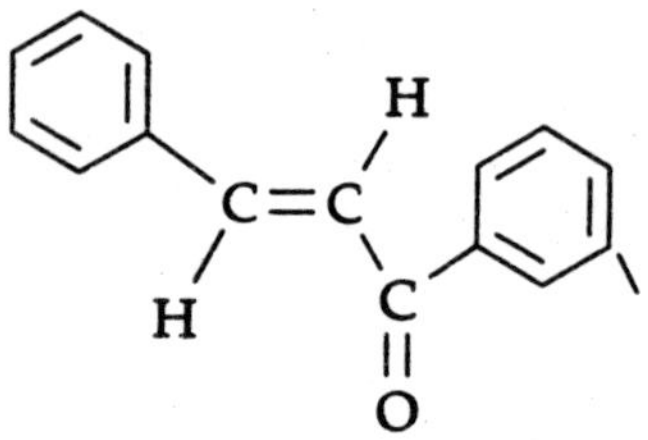

(f) 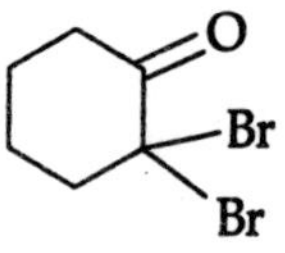

(g) $Cl_3\overset{O}{\overset{\|}{C}}H$

(h) 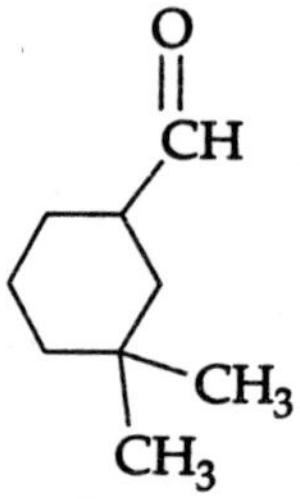

(i) $CH_3\overset{O}{\overset{\|}{C}}CH_2\overset{O}{\overset{\|}{C}}CH_3$

(j) 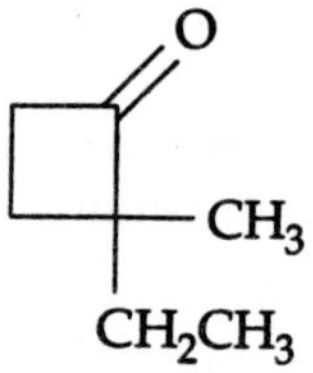

(k) 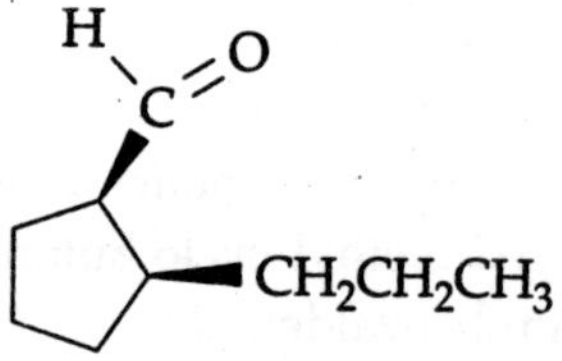

Q. 42. Match the names of the following compounds with column ⊗.

(a) 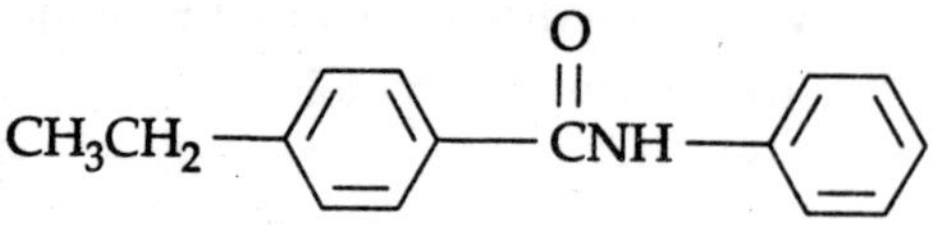

(b) 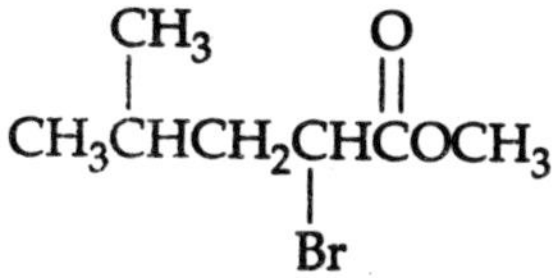

(c) 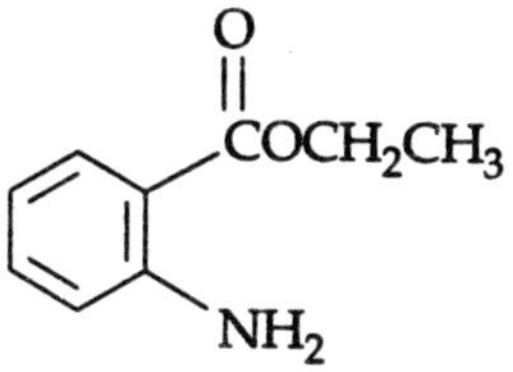

(d) 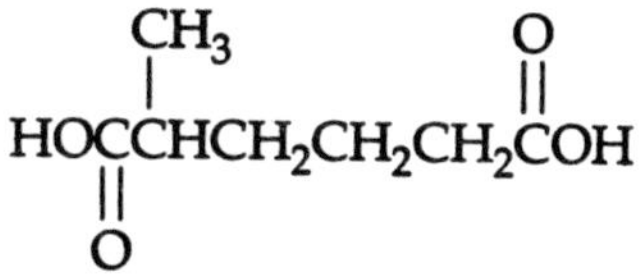

(e) 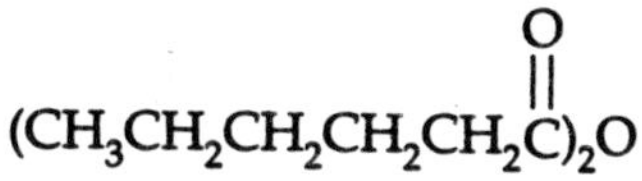

(f) 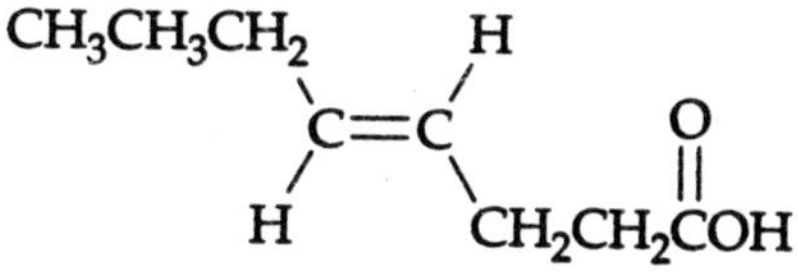

(g) 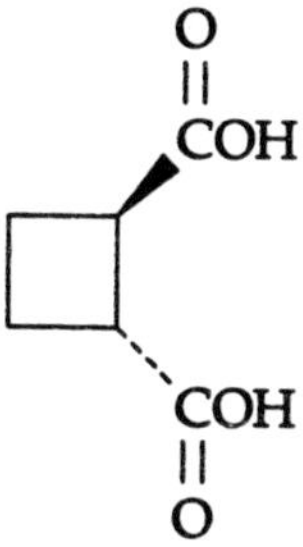

Column ⊗

(i) 4-ethyl benzanilide
(ii) (1R, 2R)-cyclo butane dicarboxlic acid
(iii) methyl-2-bromo-4-methyl pentanoate
(iv) (E)-4-octenoic acid
(v) ethyl 2-amino benzoate (or 2-methyl adipic acid)
(vi) hexanoic anhydride
(vii) 2-methyl hexanedioic acid

Q. 43. Match the names of the following compounds, specifying the stereochemistry where appropriate.

(a) 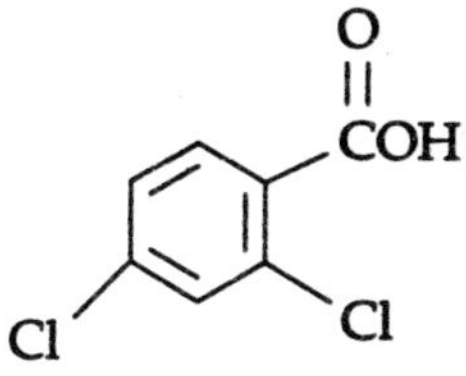

Column ⊗

(i) ethyl (1R,2R)-2 ethyl cyclo hexane-Carboxylate
(ii) 2-chloro pentanoyl chloride
(iii) 2-methyl butanedioic anlydride
(iv) 2,4-dichlorobenzoic acid
(v) N, N-dimethyl pentanamide
(vi) 4-bromohexonic acid
(vii) 2-methyl pentane nitrile
(viii) R-2-hydroxy pentanoic acid

(b) $CH_3CH_2\underset{\underset{Br}{|}}{C}HCH_2CH_2\overset{\overset{O}{||}}{C}OH$

(c) CH_3 (cyclic anhydride: five-membered ring with ring O and two C=O groups, CH_3 on ring carbon)

(d) $CH_3CH_2CH_2\overset{\overset{CH_3}{|}}{C}HC{\equiv}N$

(e) (cyclohexane ring bearing CH_2CH_3 (wedge) and $\overset{}{\underset{\underset{O}{||}}{C}}OCH_2CH_3$ (dashed) on adjacent carbons)

(f) $CH_3CH_2CH_2CH_2\overset{\overset{O}{||}}{C}\underset{\underset{CH_3}{|}}{N}CH_3$

(g) $\overset{\overset{O}{||}}{C}OH$ attached to C, with H (dashed), HO (wedge) and $CH_2CH_2CH_3$

(h) $CH_3CH_2CH_2CH_2\overset{\overset{O}{||}}{C}\underset{\underset{Cl}{|}}{N}Cl$

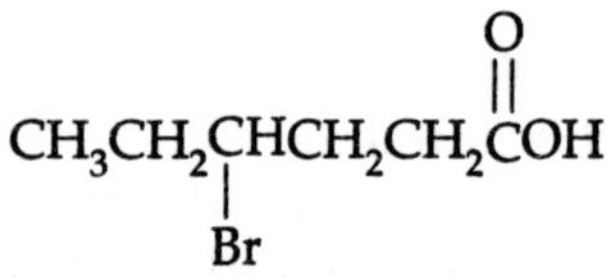

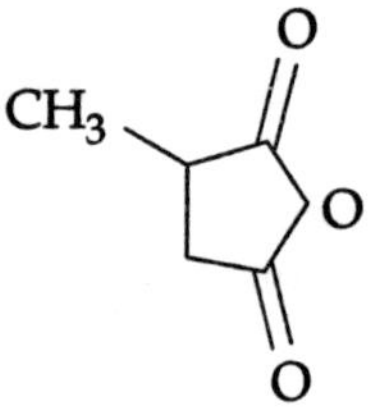

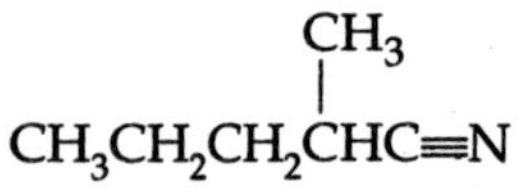

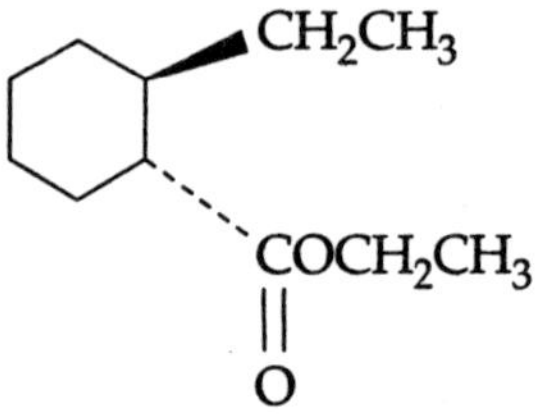

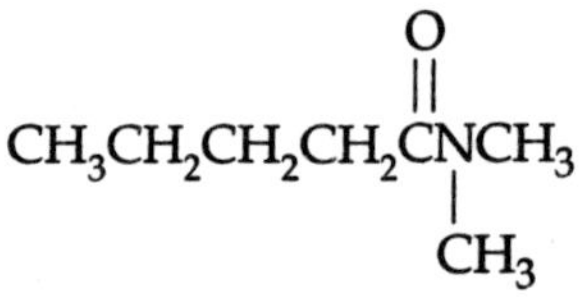

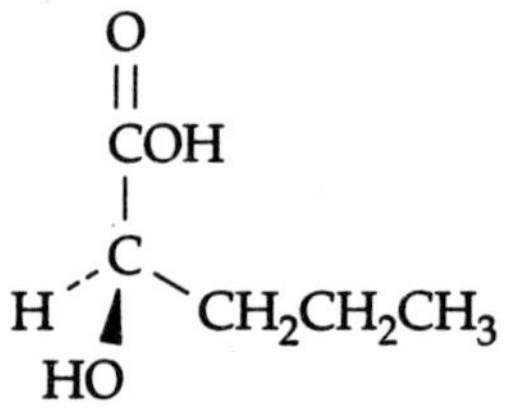

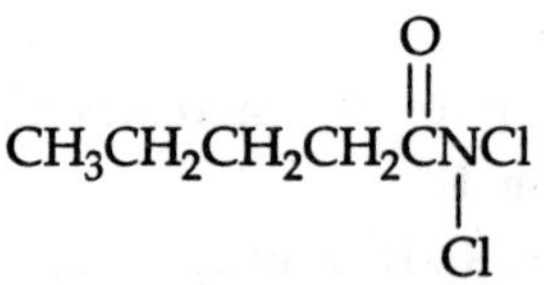

Q. 44. Name the following compounds according to the IUPAC rules, and match it with column ⊗.

(a) (cyclopentane ring)–$\overset{\overset{O}{||}}{C}H$

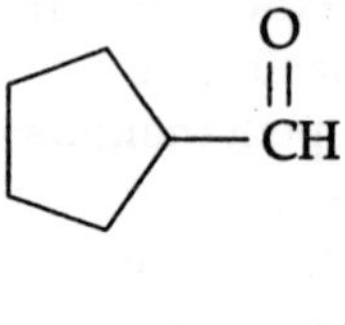

(b) 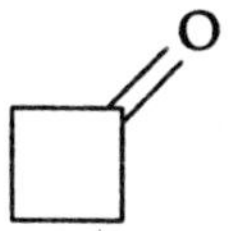

(c) 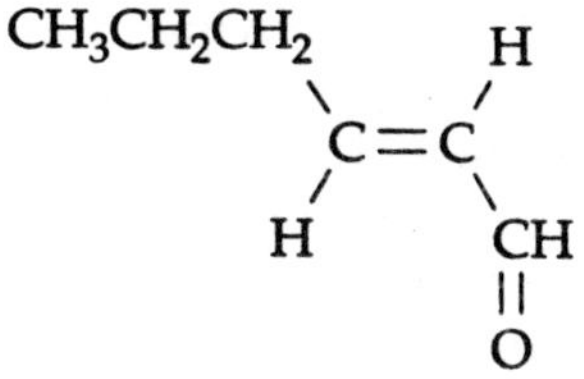

(d) 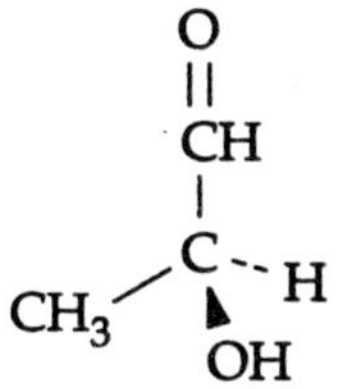

(e) 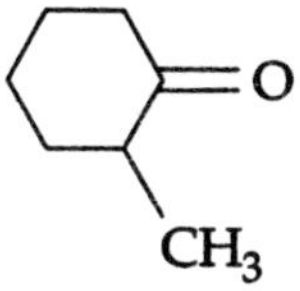

(f) 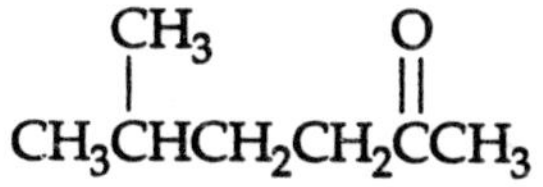

(g) 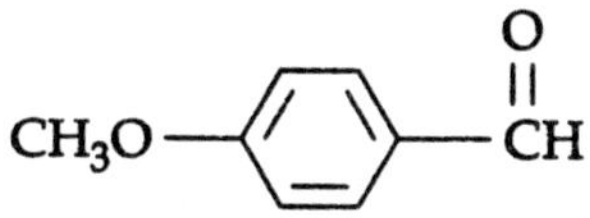

(h) 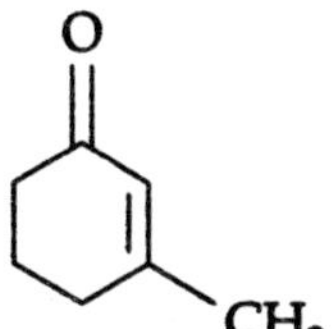

(i) 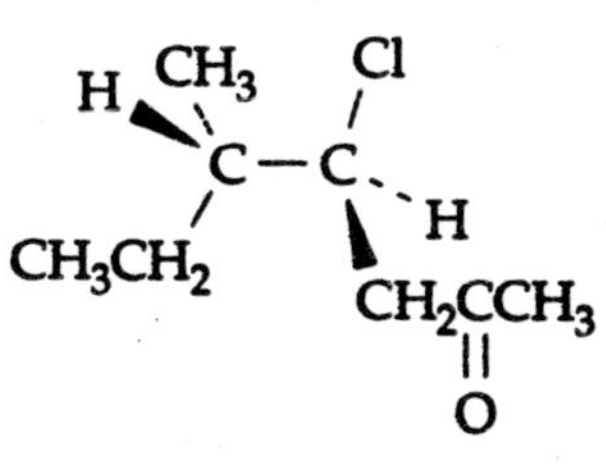

(j) 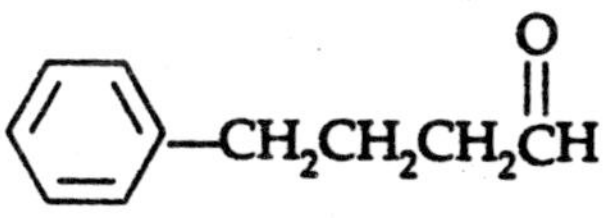

Column ⊗

(i) 5-methyl-2-hexanone
(ii) 4-penylbutanal
(iii) 4-methyoxy benzaldehyde
(iv) cyclobutanone
(v) 2-methyl cyclohexanone
(vi) 3-methyl-2-cyclohexanone
(vii) cyclopentane carbaldehyde
(viii) S-2-hydroxy propanal
(ix) (4S, 5S)-4-chloro-5-methyl-2-heptanone
(x) E-2-hexanol

Q. 45. Match the names of the following compounds with column ⊗.

(a)

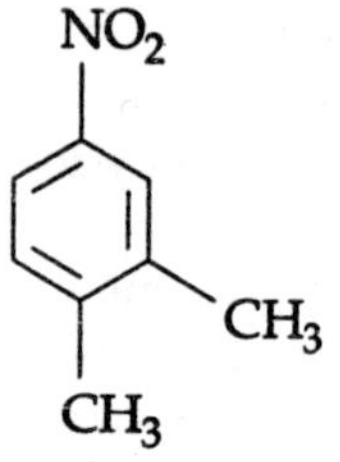

(b)

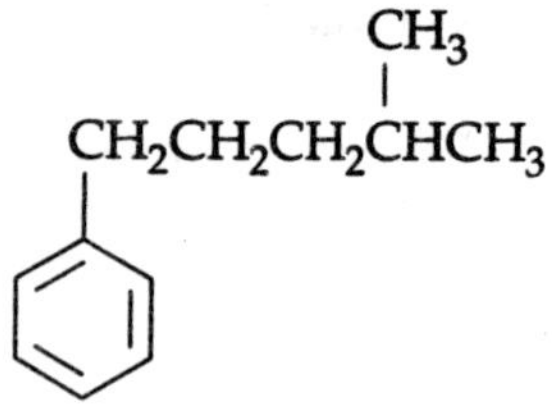

(c)

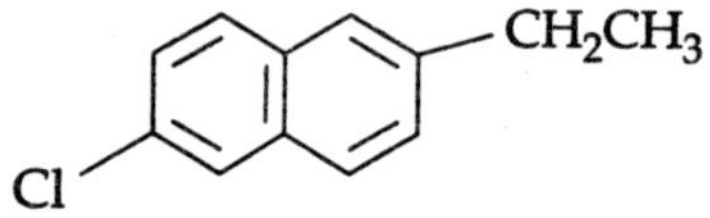

(d)

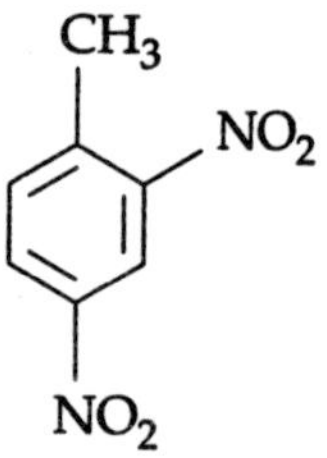

(e)

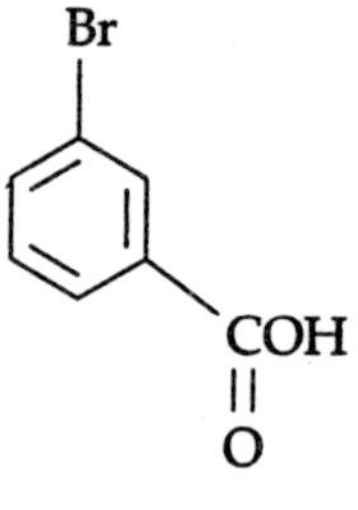

(f)

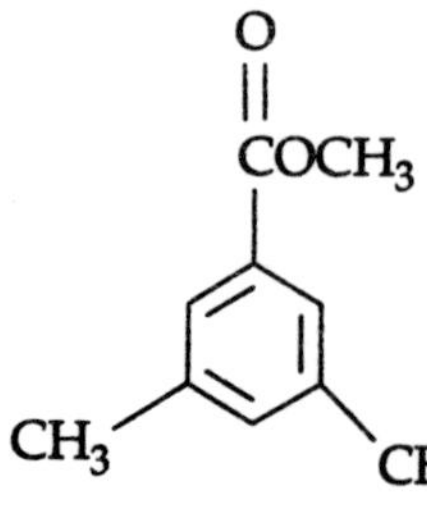

Column ⊗

(i) 2, 4-dinitrotoluene
(ii) methyl 3-5 dimethyl benzoate
(iii) 4-methyl-1-phenylpentane
(iv) 1-(4-methyl phenyl)-1-butanone
(v) 3,4-dimethyl nitrobenzene
(vi) 4-bromobenzaldehyde
(vii) 2-chloro-6-ethyl napthalene
(viii) m-bromobenzoic acid

(g) 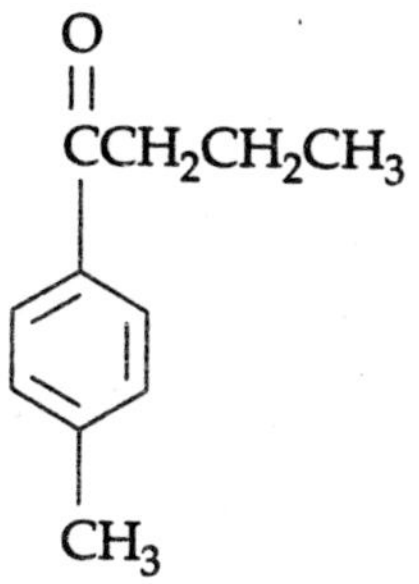
O
CCH2CH2CH3
CH3

(h) 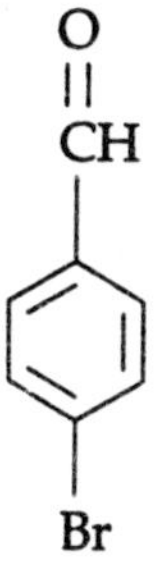
O
CH
Br

Q. 46. Match the names of the following compounds with column ⊗.

(a) 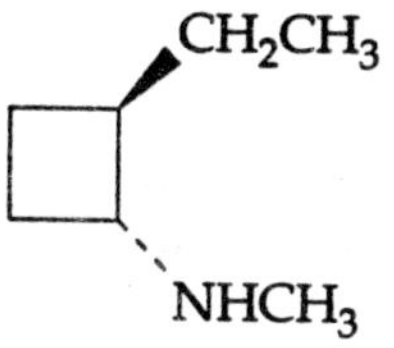
CH2CH3
NHCH3

(b) 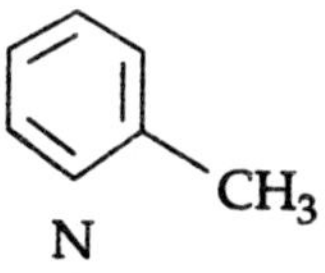
N
CH3

(c) $CH_3CH_2\underset{}{N}(CH_2CH_3)CH_2CH_3$ (i.e. $CH_3CH_2NCH_2CH_3$ with CH_2CH_3 on N)

(d) 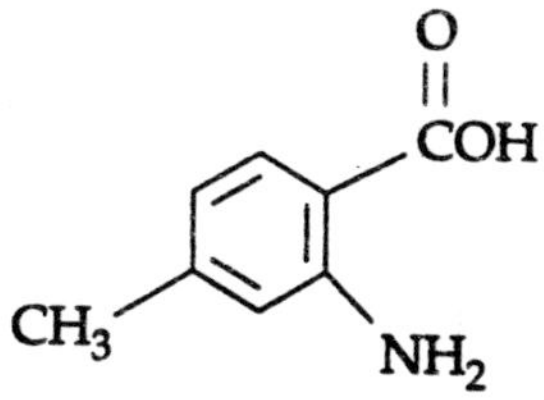
O
COH
CH3
NH2

(e) $H_3\overset{+}{N}CH_2CH_2CH_2\overset{O}{\overset{||}{C}}O^-$

Column ⊗

(i) 2, 4, 6-trichloroaniline
(ii) 4-amino butanoic acid
(iii) N-N-diethyl-4-nitro aniline
(iv) cis-2-amino cyclohexanol
(v) triethylamine
(vi) 2-amino-4-methyl benzoic acid
(vii) 2-amino-5-methyl haxane
(viii) 2-methyl pyridine
(ix) trans-2-ethyl-1 (N-methyl amino) cyclobutane

(f) 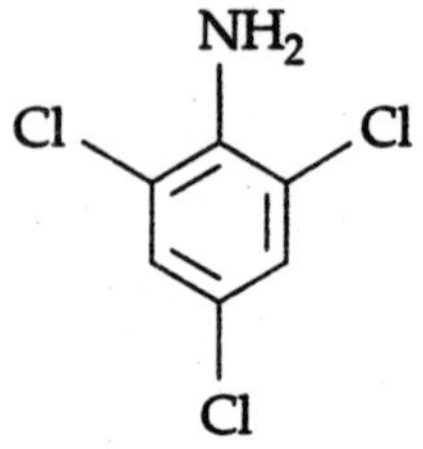

(g) 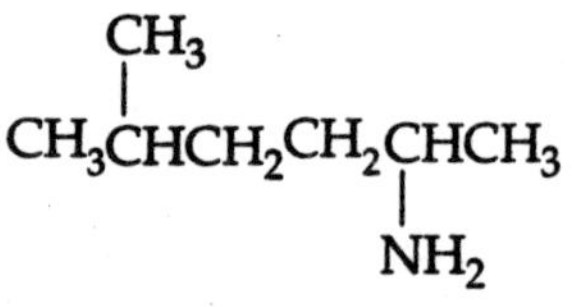

(h) 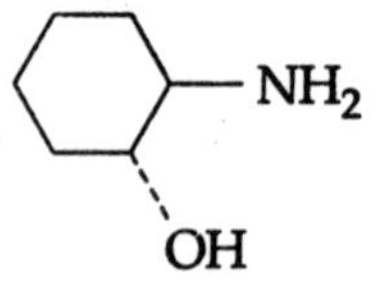

(i) 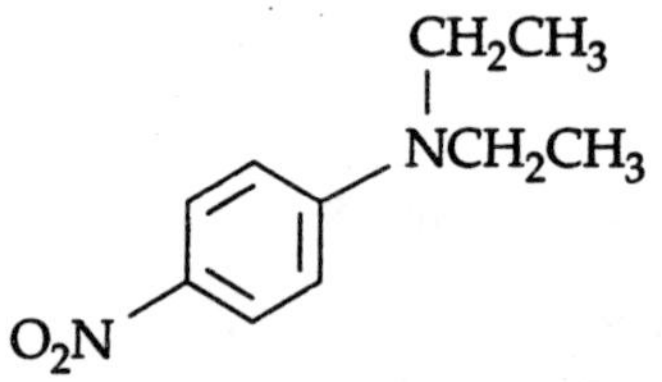

Q. 47. Compounds having the following general structure are known as aryloxypropanolamines and belong to a class of medications known as β-blockers, which are used to treat hypertension and heart disease.

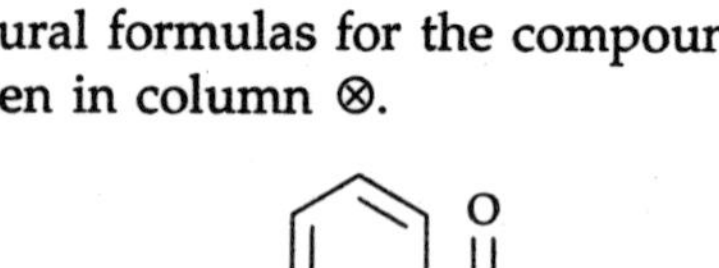

Using the following reactions, chemists synthesized model compounds in which the conformational relationships between the functional groups of the medications were known.

Thus match the structural formulas for the compounds indicated by letters, *A*, *B*, *C*, *D* and *E*; with those given in column ⊗.

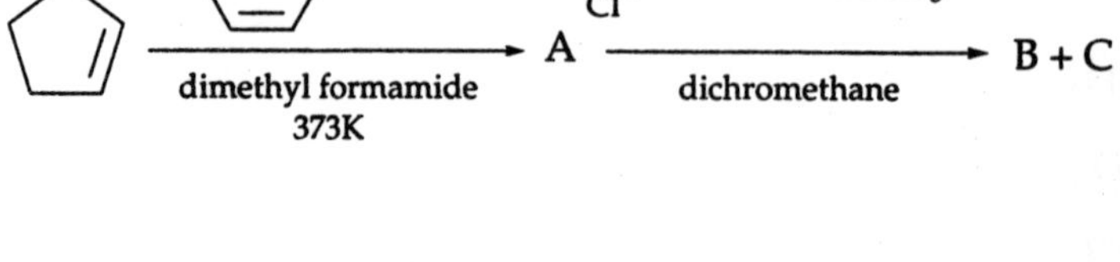

Column ⊗

(i) 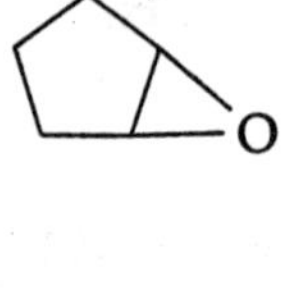

B $\xrightarrow[\text{ethanol}]{CH_3CH(CH_3)NH_2}$ D (major product)

C $\xrightarrow[\text{ethanol}]{CH_3CH(CH_3)NH_2}$ E (major product)

(ii) 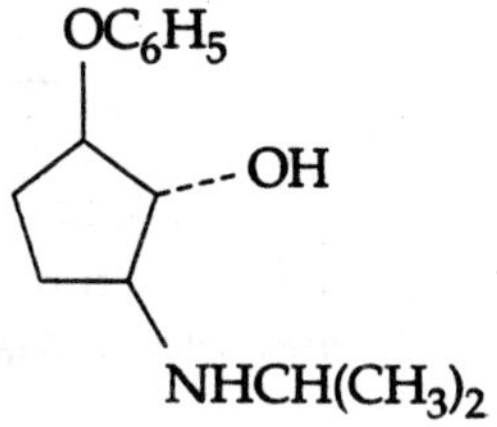

(iii) 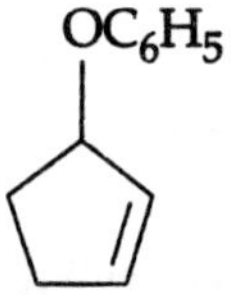

(iv) 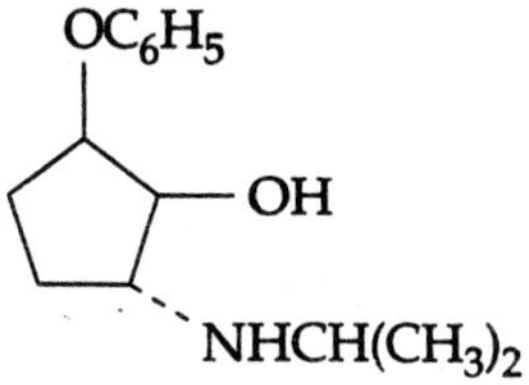

(v) 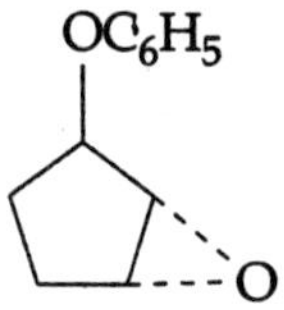

Q. 48. Match the names of the following compounds with those in column ⊗.

(a) $CH_3CH(CH_3)CH_2NH_2$

(b) 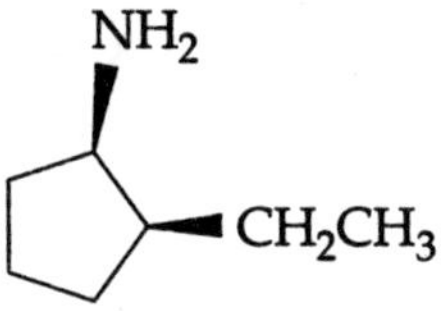

(c) $CH_3CH_2N(CH_2CH_3)CH_2CH_2CH_2CH_3$

(d) 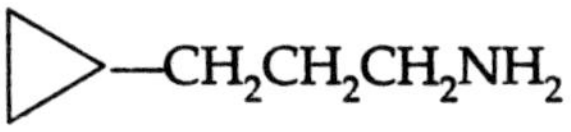

(e) 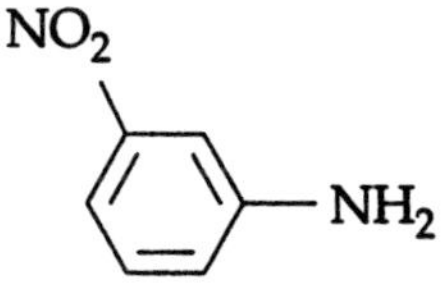

Column ⊗

(i) 3-nitroaniline
(ii) 2, 4-dibromoaline
(iii) 2 methyl-1-propanamine
(iv) N, N-dimethyl butan amine
(v) 3-cyclopropyl-1-propylamine
(vi) Cis-1-amino-2-ethyl cyclopentane

(f) 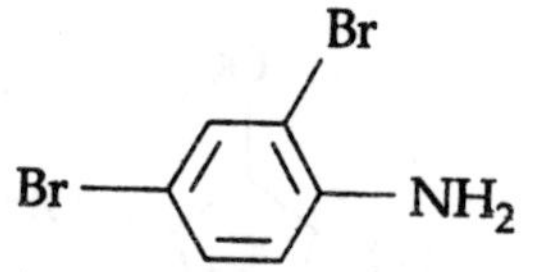

Q. 49. Supply structural formulas for the reagents, intermediates, and products designated by capital letters. Match them with those given in column ⊗.

(I) (a) 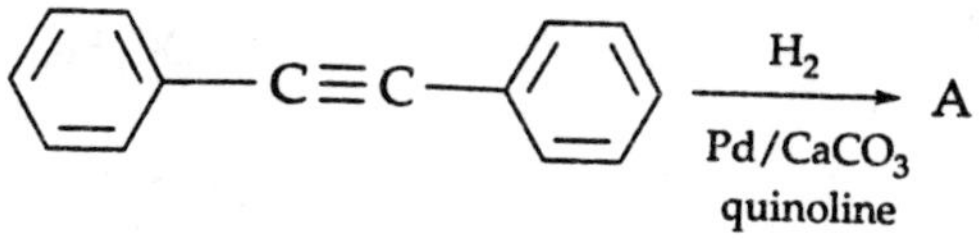

(b) $CH_2{=}CHBr \xrightarrow{HBr} B$

(c) $CH_2{=}CHCH_2Br \xrightarrow[\text{carbon tetrachloride, 273 K}]{Br_2} C$

(d) 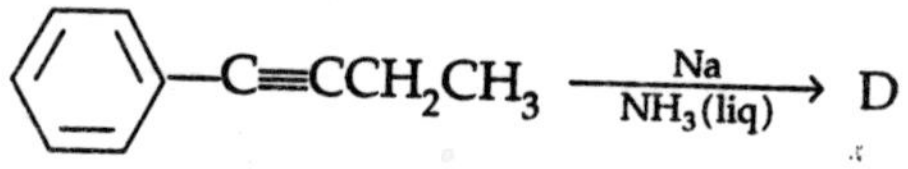

(e) 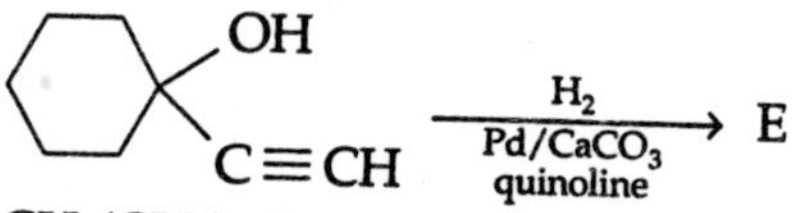

(f) $CH_3(CH_2)_7C{\equiv}C(CH_2)(CH_2)_7CO_2H \xrightarrow[\text{carbon tetrachloride}]{O_3} \xrightarrow{H_2O} F + G$

(g) 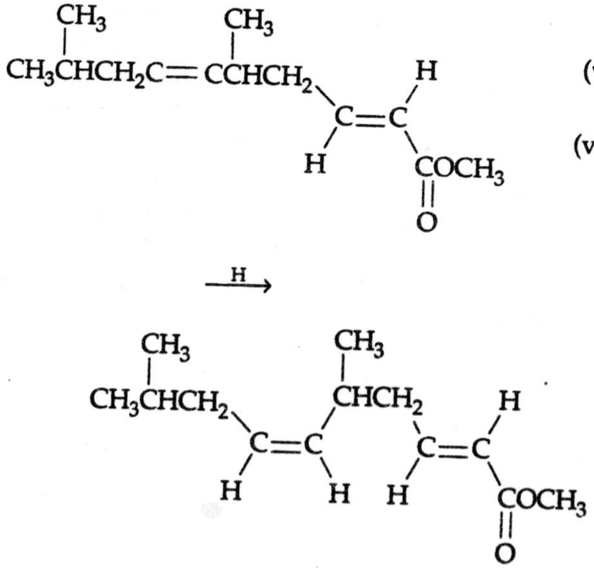

Column ⊗

(i) $CH_3(CH_2)_7COOH$

(ii) $H_2/PdCaCO_3$(quinoline)

(iii) CH_3CHBr_2

(iv) $COOH(CH_2)_7COOH$

(v) 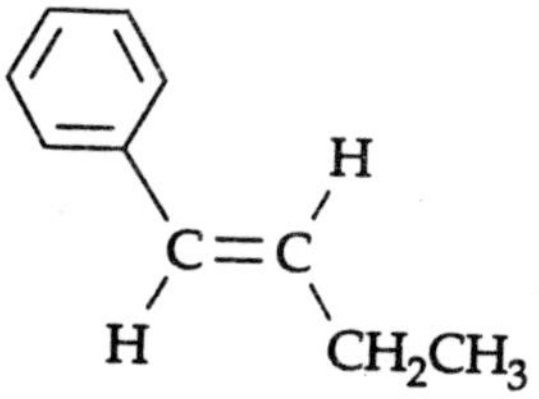

(vi) 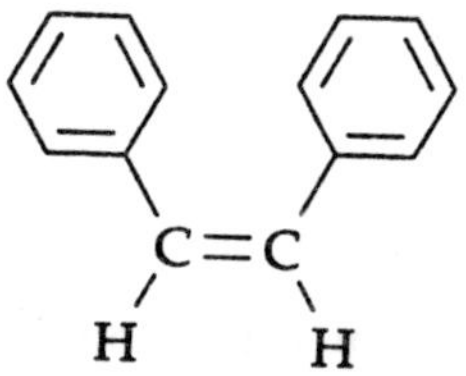

(vii) 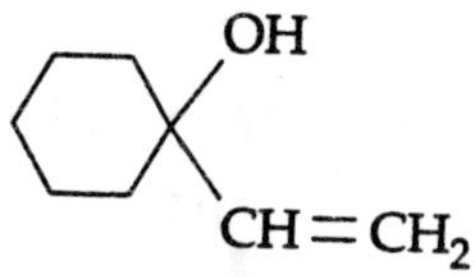

(viii) $BrCH_2CHBrCH_2Br$

II (a) $CH_3C{\equiv}CCH_3 \xrightarrow{A} CH_3\overset{O}{\overset{\|}{C}}CH_2CH_3$

(b) 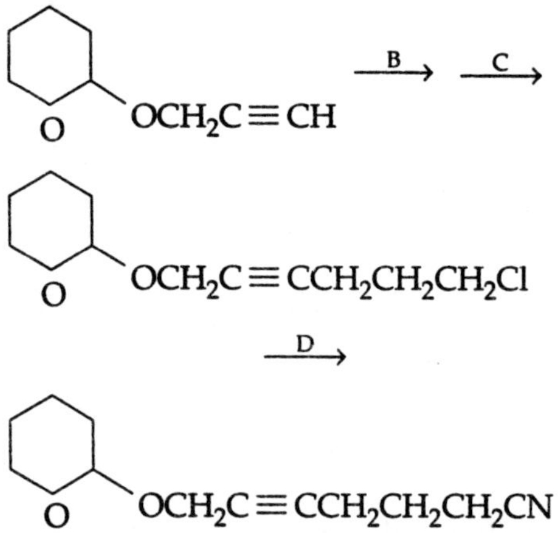

(c) $CH_2{=}CHCH_2CH_2Br \xrightarrow[NH_2(liq)]{HC{\equiv}C^-Na^+} E$

(d) 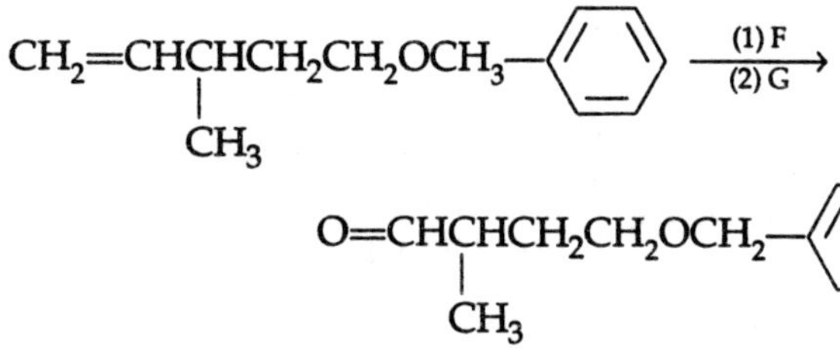

(e) 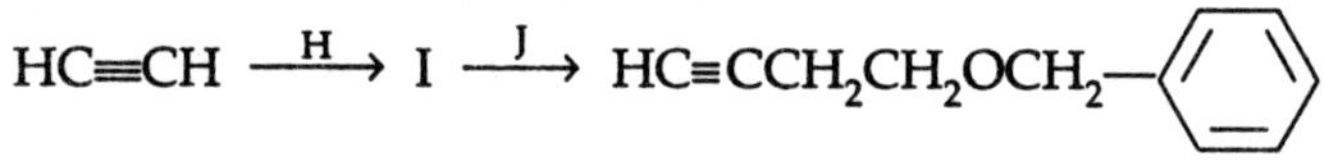

(f) 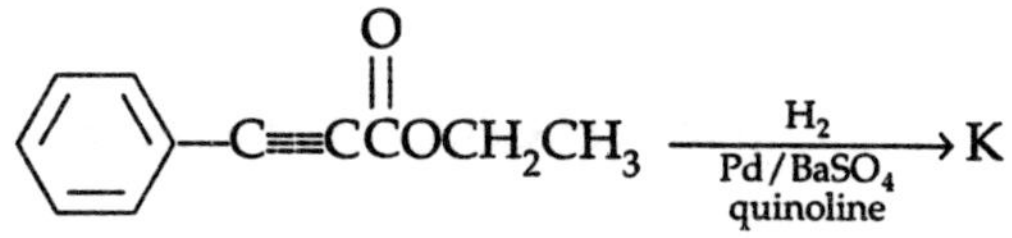

Column ⊗

(i) O_3
(ii) $(CH_3)_2S$
(iii) $BrCH_2CH_2CH_2Cl$
(iv) $CH_2{=}CHCH_2CH_2C{\equiv}CH$
(v) $HC{\equiv}C^-Na^+$
(vi) $NaNH_2/NH_3$(liq)
(vii) NaCN
(viii) H_2O/H_2SO_4
(ix) 
(x) 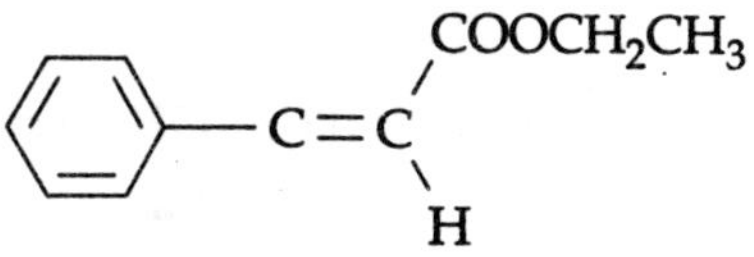
(xi) $Na^+NH_2^-/H$

Q. 50. Match the products obtained by Using **2-pentanone** as a typical *ketone* with the following reagents in column ⊗. Also write equations showing the reactions that you predicts will occur with these. Write "no reaction" if you expect none.

(a) $NaBH_4$, H_2O

(b) 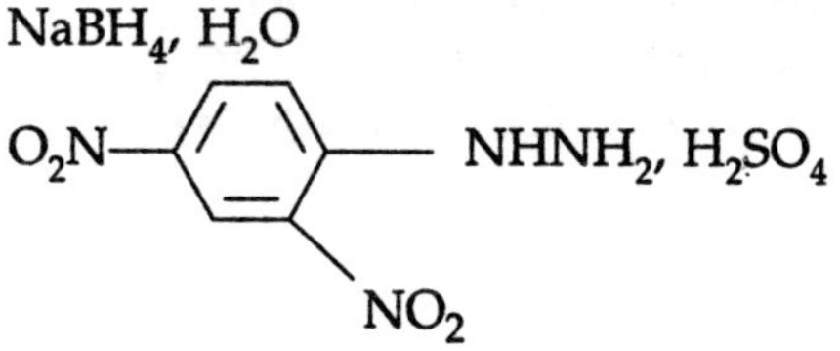

(c) 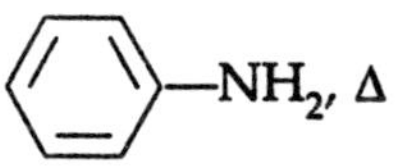

Column ⊗

(i) $CH_3CH_2CH_2C(CH_3)(OH)CN$
(ii) $CH_3CH_2CH_2C(CH_3)(OH)CH_2CH_3$
(iii) $CH_3CH_2CH_2CH(OH)CH_3$
(iv) $CH_3CH_2CH_2C(CH_3)NOH$
(v) $CH_3CH_2CH_2CH_2CH_3$
(vi) $CH_3CH_2CH_2CH_2CH_3$

(d) (1) —Li, tetrahydrofuran;

(2) NH_4Cl, H_2O

(e) NaCN, H_2SO_4(aq)

(f) $HONH_3^+Cl^-$, $CH_3\overset{O}{\overset{\|}{C}}O^-Na^+$

(g) (1) CH_3CH_2MgBr, diethyl ether;

(2) H_3O^+

(h) (1) $CH_3C{\equiv}CNa$;

(2) NH_4Cl, H_2O

(i) H_2NNH_2, KOH, diethylene glycol, Δ

(j) Zn(Hg), HCl, Δ

(vii) $CH_3CH_2CH_2C(CH_3){=}N$— 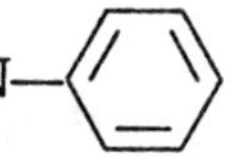

(viii) $CH_3CH_2CH_2C(CH_3){=}$

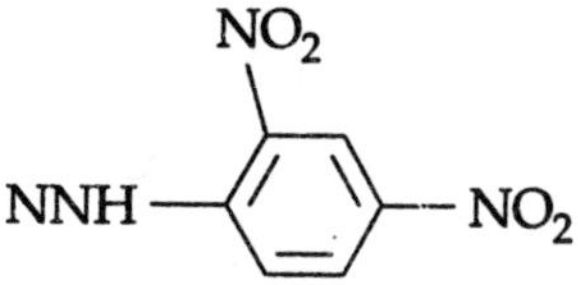

(ix) $CH_3CH_2CH_2C(CH_3)(OH)C{\equiv}CCH_3$

(x) $CH_3CH_3CH_2 \cdot C(CH_3)(OH)$— 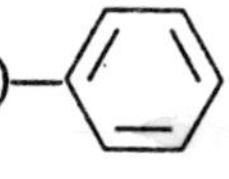

Q. 51. Match the products obtained by Using **butanol** as a typical **aldehyde** with the following reagents. In column ⊗.

Also write equations showing the reactions that you predict will occur with these. Write "no reaction" if you expect none.

(a) $NaBH_4$, H_2O

(b) 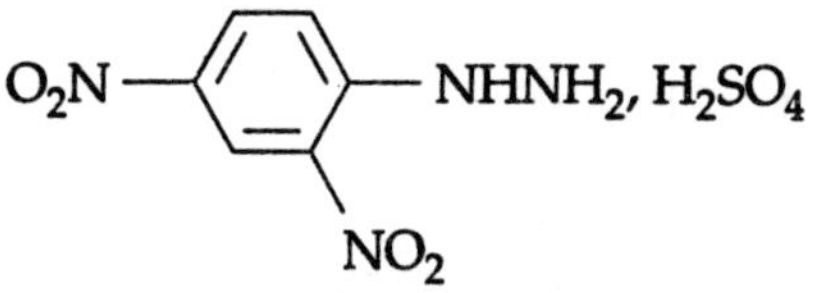

(c) (1) $CH_3CH_2CH_2CH_2Li$;

(2) H_3O^+

(d) 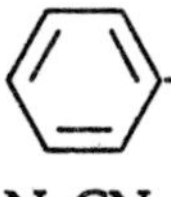—NH_2, Δ

(e) NaCN, H_2SO_4(aq)

(f) $Na_2Cr_2O_7$, H_2O, H_3O^+

(g) $HONH_3^+Cl^-$, $CH_3\overset{O}{\overset{\|}{C}}O^-Na^+$

(h) (1) CH_3MgI, diethyl ether,

(2) H_3O^+

(i) (1) $LiAlH_4$, diethyl ether,

(2) H_3O^+

(j) H_2NNH_2, KOH, diethylene glycol, Δ

(k) Zn(Hg), HCl, Δ

Column ⊗

(i) $CH_3CH_2CH_2CH_3$

(ii) $CH_3CH_2CH_2COOH$

(iii) $CH_3CH_2CH_2CH(OH)CH_2CH_2CH_2CH_3$

(iv) $CH_3CH_2CH_2CH_2OH$

(v) $CH_3CH_2CH_2CH_3$

(vi) $CH_3CH_2CH_2CH(OH)CH_3$

(vii) $CH_3CH_2CH_2CH{=}NNH$— 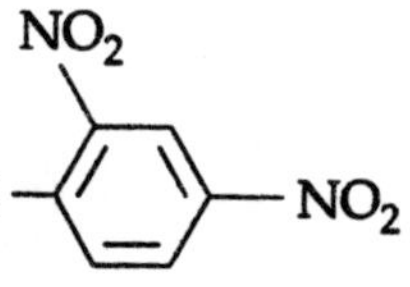

(viii) $CH_3CH_2CH_2CH{=}N$— 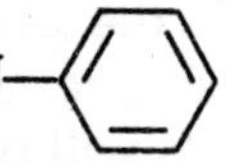

(ix) $CH_3CH_2CH_2CH_2OH$

(x) $CH_3CH_2CH_2CH(OH)CN$

(xi) $CH_3CH_2CH_2CH{=}NOH$

Q. 52. An intermediate in the synthesis of a natural product that has a antitumor activity is prepared by the following sequence of steps.

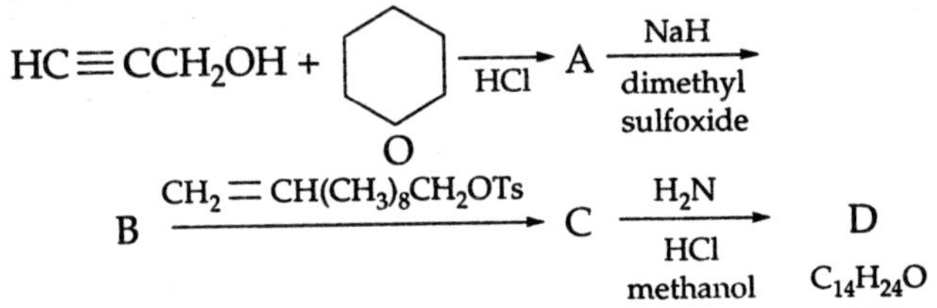

Match the structures of compounds *A, B, C* and *D* with column ⊗.

Column ⊗

(i) $CH_2{=}CH_2(CH_2)_8CH_2C{\equiv}CCH_2OH$

(ii) 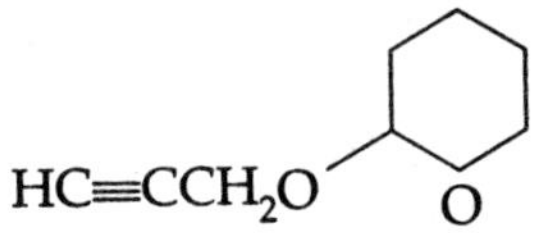

(iii) 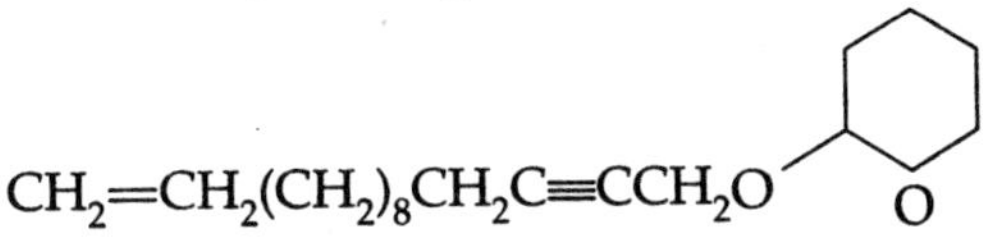

(iv) 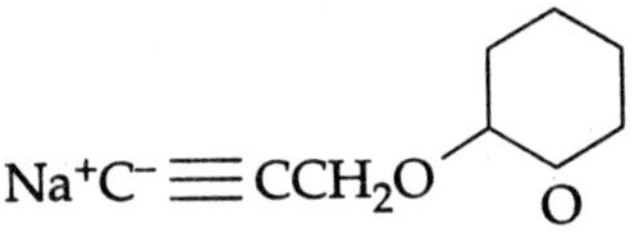

Q. 53. Match the structures of all compounds symbolized by a letter (A–M) in the following equations with those given in column ⊗.

(a) 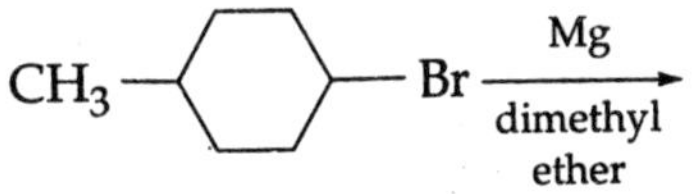

$$A \xrightarrow{CO_2} B \xrightarrow{H_3O^+} C$$

(b) $CH_3CH_2CH_2CH(CH_3)CH_2CH_2OH$

$$\xrightarrow[H_2O,\ \Delta]{NaOH,\ KMnO_4} D \xrightarrow{H_2O^+} E$$

(c) 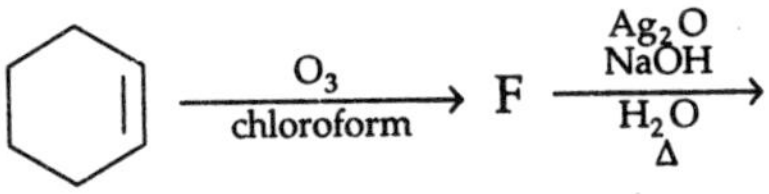

$$G \xrightarrow{H_2O^+} H$$

Column ⊗

(i) 

(ii) $CH_3CH_2CH_2CH(CH_3)CH_2COOH$

(iii) 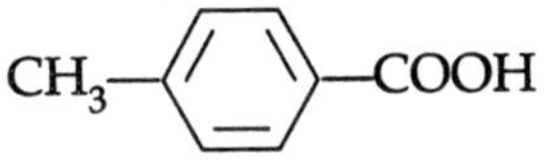

(iv) 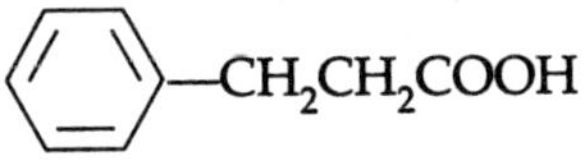

(v) $CH_3(CH_2)_5COOH$

(vi) 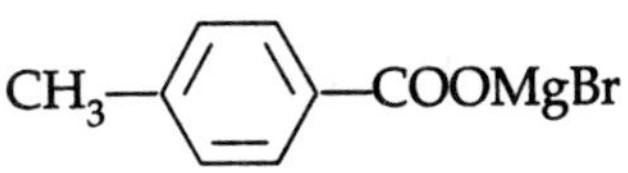

(vii) $NaOOC(CH_2)_4COONa$

(viii) $CH_3CH_2CH_2CH(CH_3)CH_2COONa$

(d) 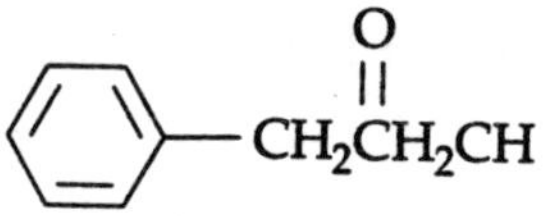

$\xrightarrow[H_2O,\ \Delta]{Ag_2O,\ NaOH}$ I $\xrightarrow{H_2O^+}$ J

(e) $CH_3(CH_2)_5\overset{O}{\overset{\|}{C}}H \xrightarrow[H_2O]{CrO_3,\ H_2SO_4}$ K

(f) $CH_3(CH_2)_9CH{=}CH_2$

$\xrightarrow[chloroform]{O_3} \xrightarrow{H_2O_2,\ NaOH} \xrightarrow{H_3O^+}$ L + M

(ix) HCOOH

(x) $CH_3(CH_2)_9COOH$

(xi) $HOOC{-}(CH_2)_4COOH$

(xii) 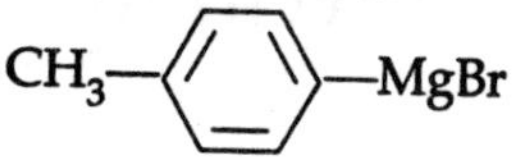

(xiii) 

Q. 54. Complete the following and match (A–P) with the products in column ⊗. There may be more than one product.

(a) 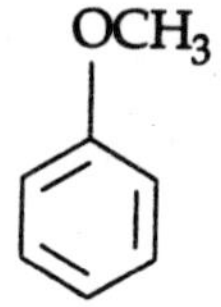 $\xrightarrow[Fe]{Br_2}$ A + B

(b) 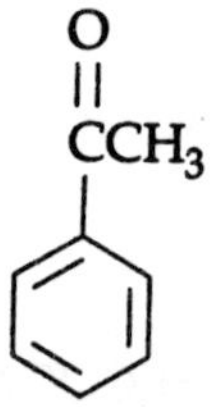 $\xrightarrow[H_2SO_4]{HNO_3}$ C

(c) 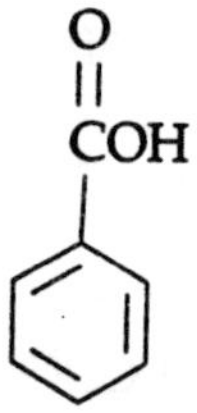 $\xrightarrow[Fe]{Br_2}$ D

(d) 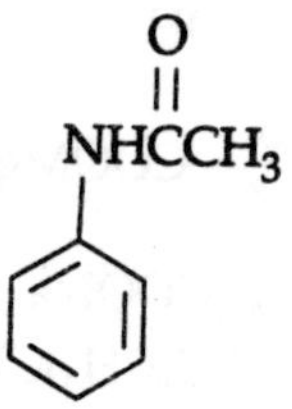 $\xrightarrow[H_2SO_4]{HNO_3}$ E + F

Column ⊗

(i) 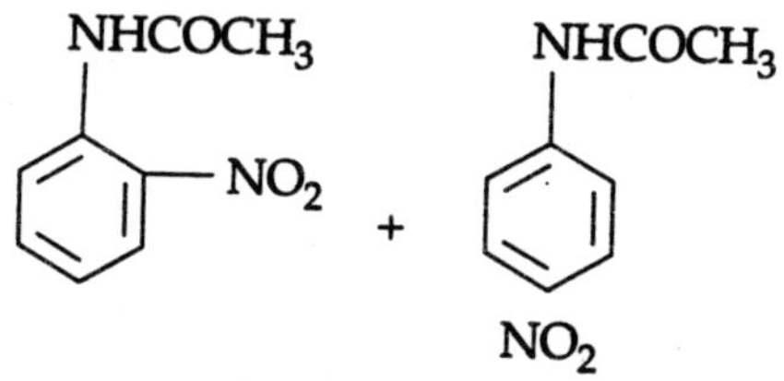

(ii) 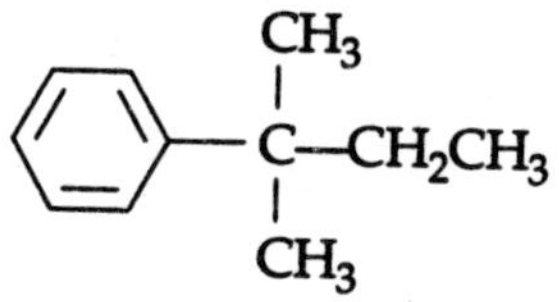

(iii) 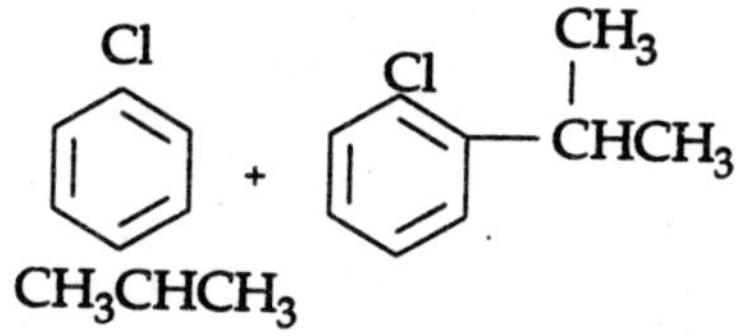

(iv) 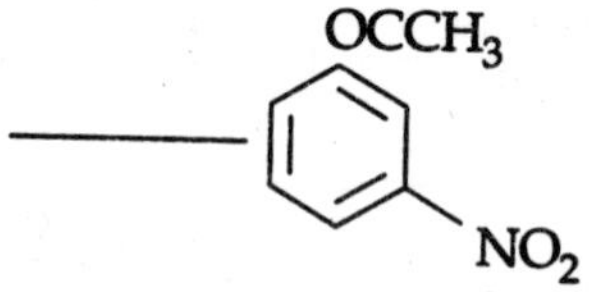

(v) 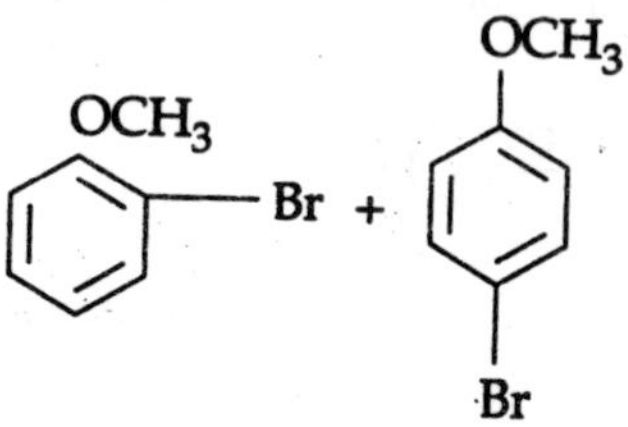

(e) 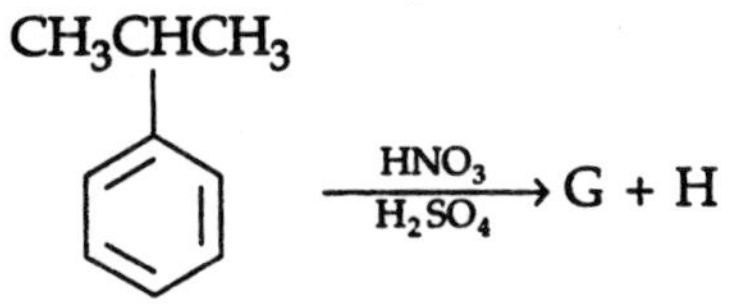

CH_3CHCH_3
HNO_3
H_2SO_4
G + H

(f) 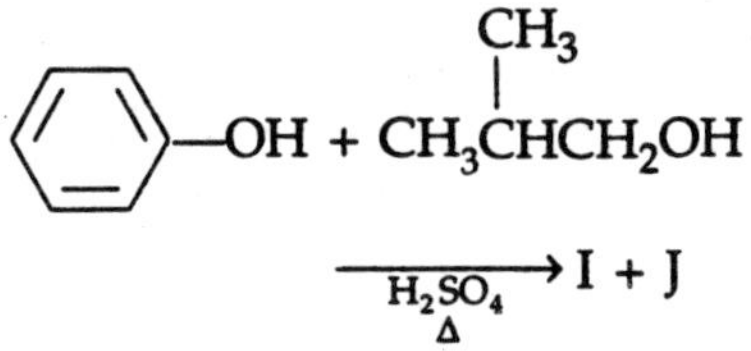

OH + CH_3CHCH_2OH
CH_3
H_2SO_4
Δ
I + J

(g) 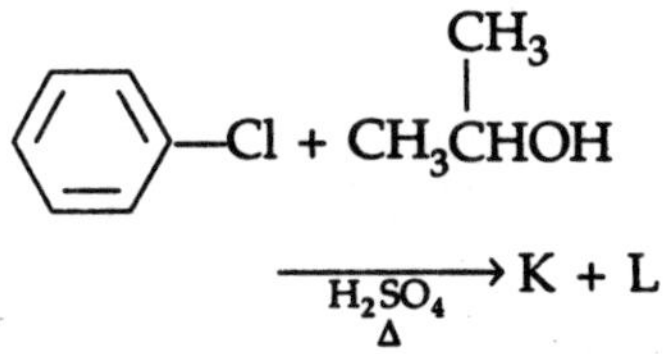

Cl + CH_3CHOH
CH_3
H_2SO_4
Δ
K + L

(h) 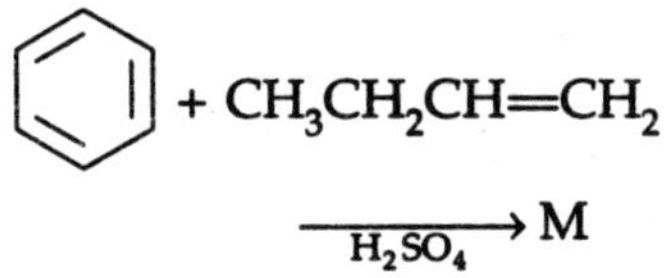

+ $CH_3CH_2CH{=}CH_2$
H_2SO_4
M

(i) 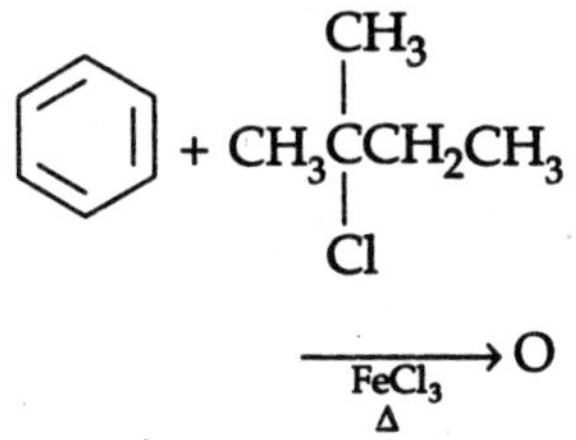

+ $CH_3CCH_2CH_3$
CH_3
Cl
$FeCl_3$
Δ
O

(j) 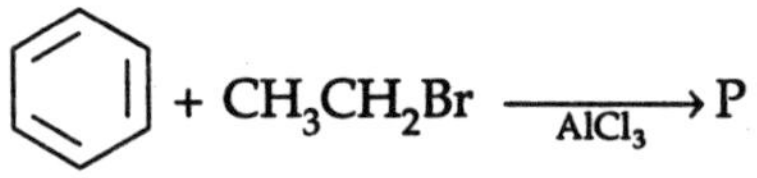

+ CH_3CH_2Br
$AlCl_3$
P

(vi) 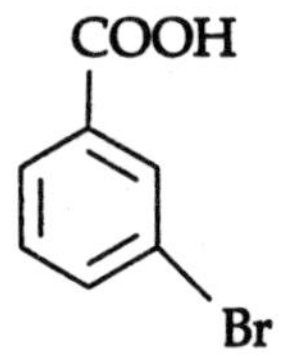

COOH
Br

(vii) 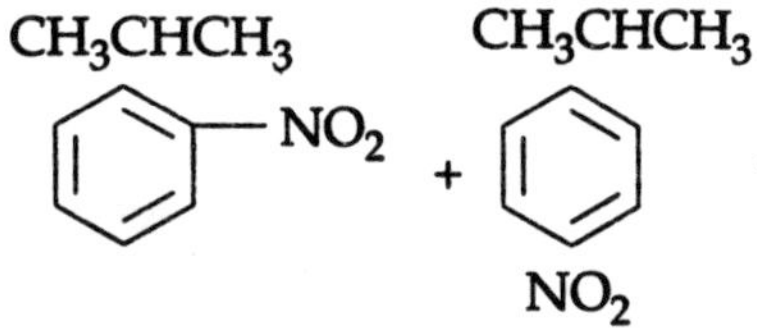

CH_3CHCH_3
NO_2
+
CH_3CHCH_3
NO_2

(viii) 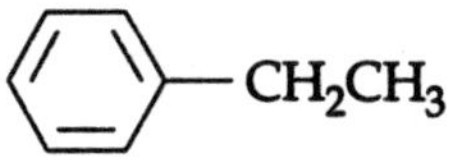

CH_2CH_3

(ix) 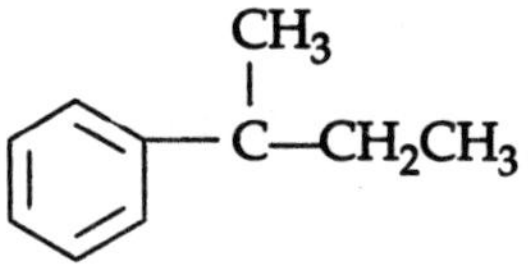

CH_3
C—CH_2CH_3

(x) 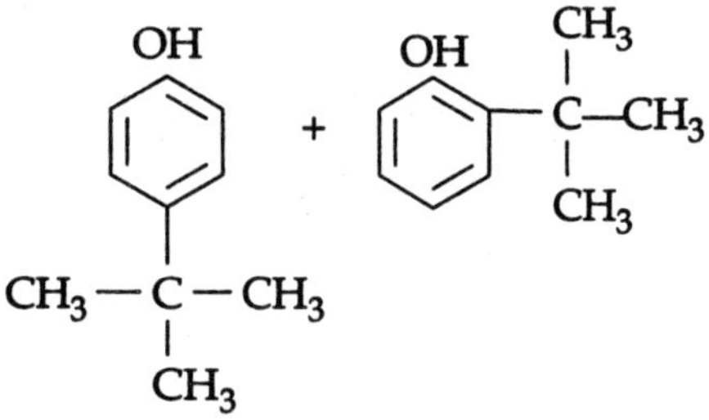

OH
CH_3—C—CH_3
CH_3
+
OH
CH_3
C—CH_3
CH_3

Q. 55. Match the structural formulas for all compounds designated by letters (A–M) in the following equation with those given in column ⊗

(a) 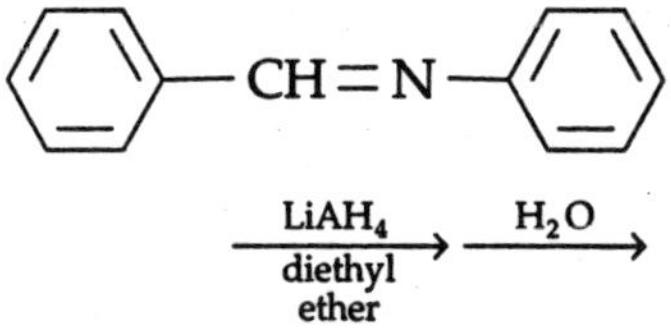

CH=N
$LiAH_4$
diethyl ether
H_2O
A

(b) 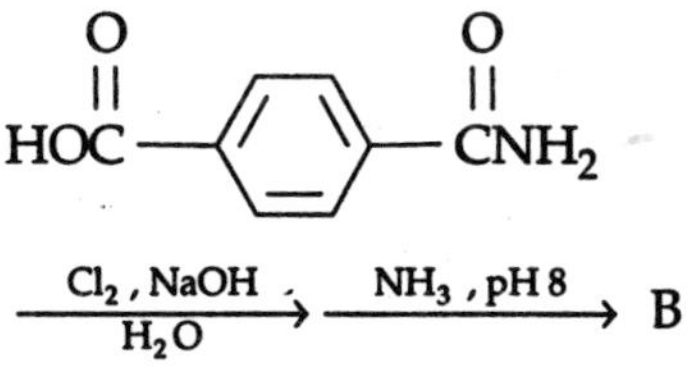

O
HOC
O
CNH_2
Cl_2, NaOH
H_2O
NH_3, pH 8
B

Column ⊗

(i) $CH_3COCH_2C(CH_3)(NH_2)CH_3$

(ii) $CH_3CH_2OCONHCH_2CH_2CH_3$

(iii) $CH_3CH_2CH_2CONHCH_2CH_3$

(iv) $CH_3CH_2—\overset{\displaystyle HON}{\overset{|}{C}}—CH_2CH_2CH_3$

(v) $CH_3CH_2—\overset{\displaystyle NOH}{\overset{|}{C}}—CH_2CH_2CH_3$

(c) $CH_3\overset{O}{\overset{\|}{C}}CH{=}\overset{CH_3}{\overset{|}{C}}CH_3 \xrightarrow[H_2O]{NH_3} C$

(d) $CH_3\underset{CH_3}{\overset{CH_3}{C}}CH_2CH_2\overset{O}{\overset{\|}{C}}OCH_3$

$\xrightarrow[\substack{Ni \\ \Delta}]{H_2} D\ (C_6H_{11}NO)$

(e) 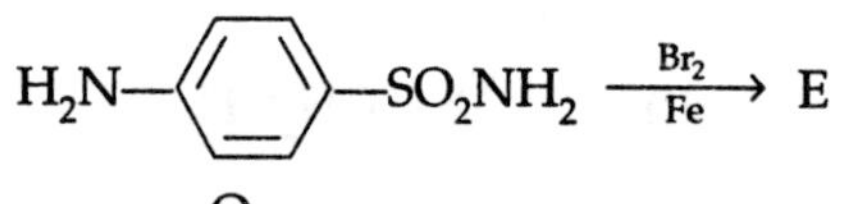

(f) $CH_3CH_2\overset{O}{\overset{\|}{C}}CH_2CH_2CH_3$

$\xrightarrow[Na_2CO_3]{HO\overset{+}{N}H_3\,HSO_4^-} F + G \xrightarrow{H_2SO_4} H + I$

(g) $CH_3\overset{O}{\overset{\|}{C}}CH_2CH_2CH_2CH_2CH_3$

$\xrightarrow[\substack{Ni \\ \Delta}]{NH_3,\ H_3} J$

(h) $CH_3CH_2CH_2NO_2 \xrightarrow[NaOH]{H\overset{O}{\overset{\|}{C}}H}$

$K \xrightarrow{H_2SO_4,\ Fe} L \xrightarrow{H_2SO_4} M$

(vi) ______ 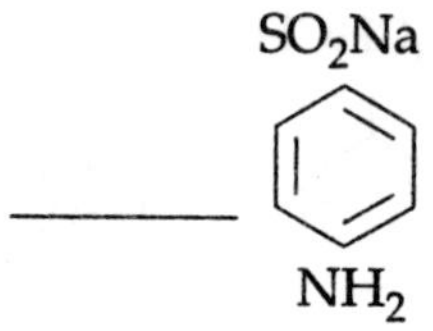

(vii) $CH_3CH_2CH(NO_2)CH_2OH$

(viii) $CH_3CH_2CH(NH_3^+HSO_4^-)CH_2OH$

(ix) 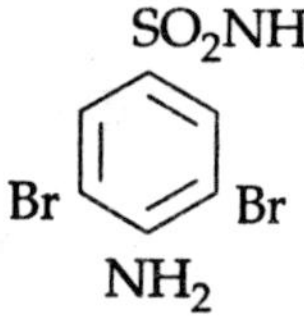

(x) ______ $CH_3CH_2CH(NH_2)CH_2OH$

(xi) $CH_3CH(CH_2)_4CH_3$

(xii) 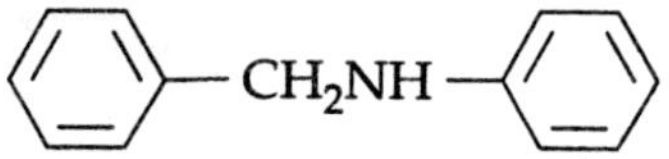

(xiii) 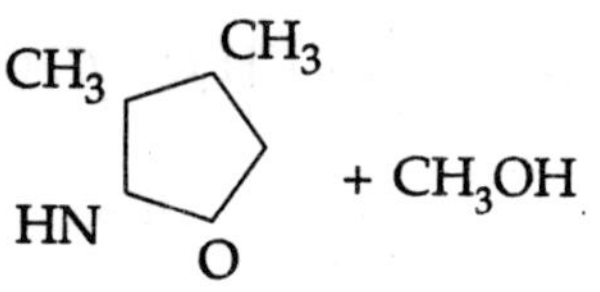

Q. 56. Match the formulas for all products and intermediates designed by letters in the following equation with those given in column ⊗

(a) 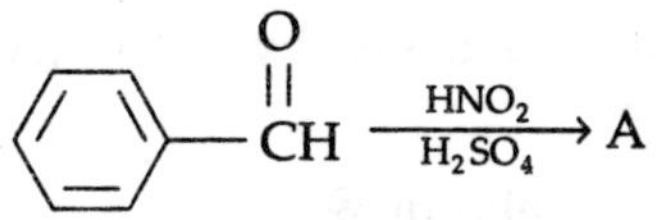

$\xrightarrow[HCl]{SnCl_2} B \xrightarrow[\substack{H_2O \\ 273\text{-}278\ K}]{NaNO_2,\ HCl} C \xrightarrow[HCl]{CuCl} D$

(b) 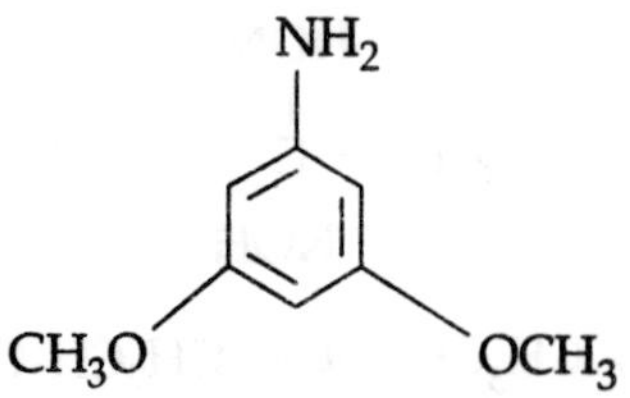

Column ⊗

1. 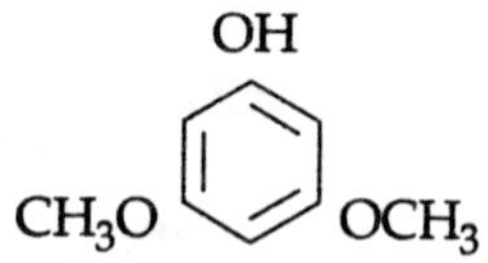

2. 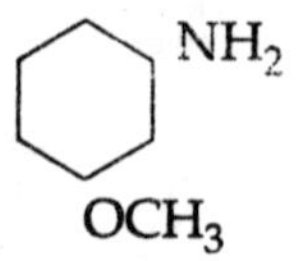

$$\xrightarrow[H_2O,\ 273\text{-}278K]{NaNO_2,\ H_2SO_4} E \xrightarrow[H_2SO_4]{H_2O} F$$

(c) 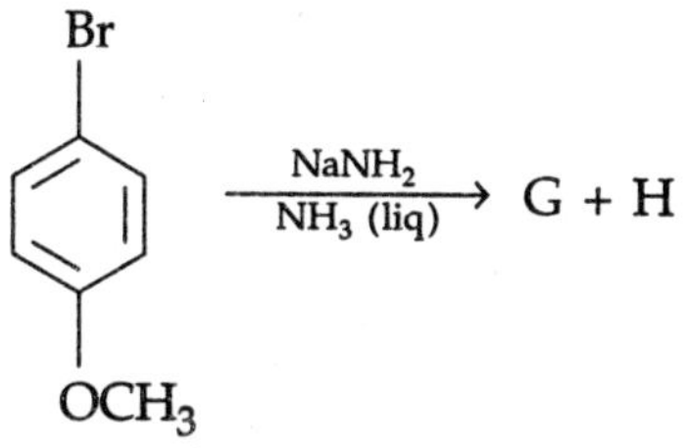
Br
OCH₃
NaNH₂
NH₃ (liq)
G + H

(d) 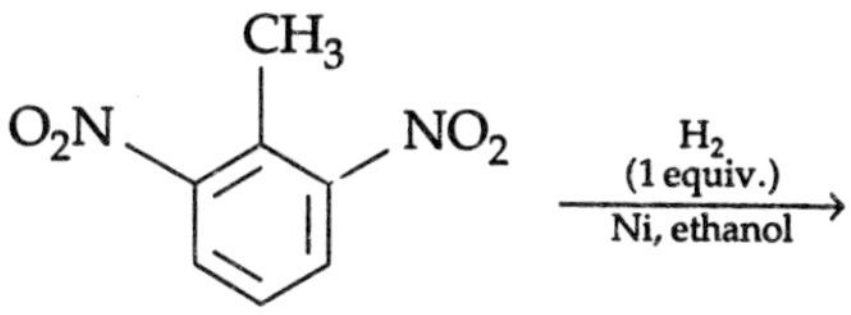
CH₃
O₂N
NO₂
H₂
(1 equiv.)
Ni, ethanol

$$I \xrightarrow[H_2O,\ 273\text{-}278\ K]{NaNO_2,\ H_2SO_4} J \xrightarrow[H_2SO_4]{H_2O} K$$

(e) 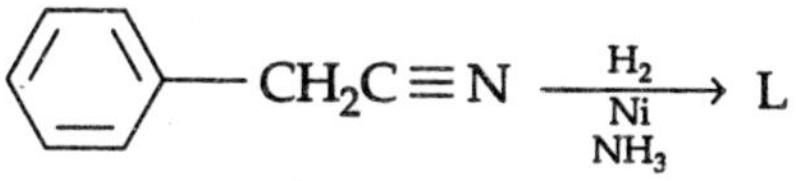
CH₂C≡N
H₂
Ni
NH₃
L

(f) 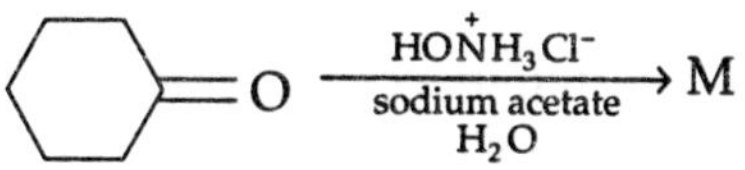
O
HONH₃Cl⁻
sodium acetate
H₂O
M

$$\xrightarrow[ethanol]{Na} N$$

(g) 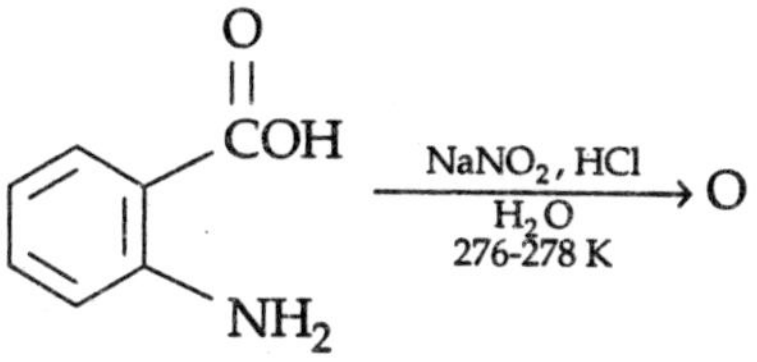
O
COH
NH₂
NaNO₂, HCl
H₂O
276-278 K
O

(h) 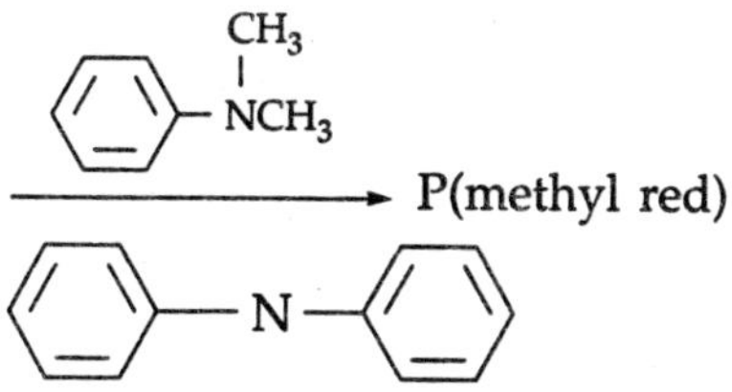
CH₃
NCH₃
P(methyl red)
N

$$\xrightarrow[H_2O,\ 276\text{-}278\ K]{NaNO_2,\ HCl} Q$$

3. 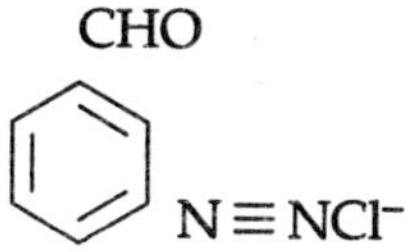
CHO
N≡NCl⁻

4. 
CH₂CH₂NH₂

5. 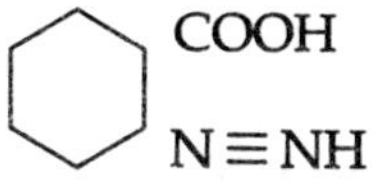
COOH
N≡NH

6. 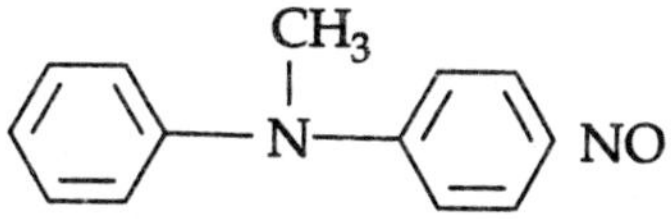
CH₃
N
NO

7. 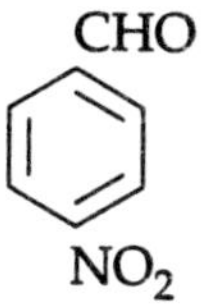
CHO
NO₂

8. 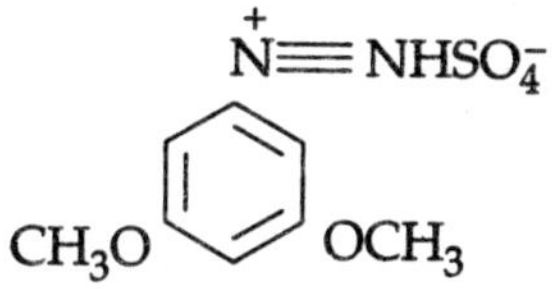
N≡NHSO₄⁻
CH₃O
OCH₃

9. 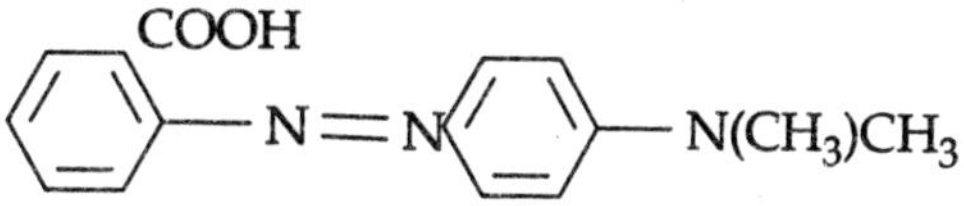
COOH
N=N
N(CH₃)CH₃

10. 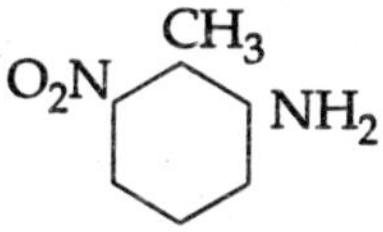
CH₃
O₂N
NH₂

11. 
CH₂NO₂

12. 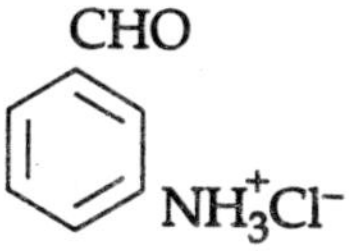
CHO
NH₃Cl⁻

13. 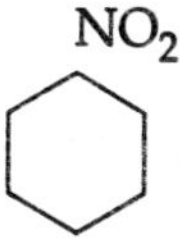
NO₂

(i) C_6H_5–CH_2Br $\xrightarrow[\text{dimethyl-formamide}]{NaNO_2}$ R

(j) O_2N–C_6H_3(OH)–$\overset{O}{\overset{||}{C}}OCH_2CH_3$ $\xrightarrow[\text{Ni, ethyl acetate}]{H_2}$ S

14. H_2N–C_6H_3(OH)–$COOCH_2CH_3$

15. O_2N, CH_3–C_6H_3–$\overset{+}{N}\equiv NHSO_4^-$

16. CH_3, O_2N–C_6H_3–OH

17. CHO, Cl–C_6H_4

18. NOH (cyclohexane)

19. NH_2, C_6H_4, OCH_3

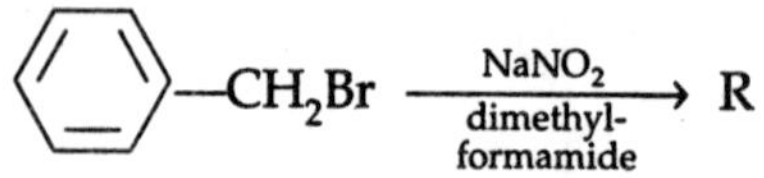

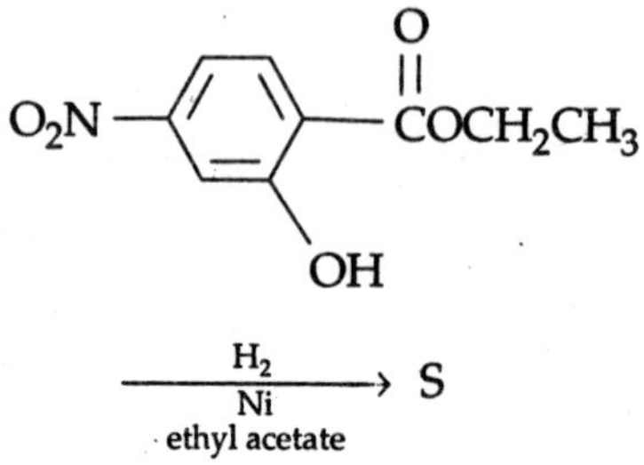

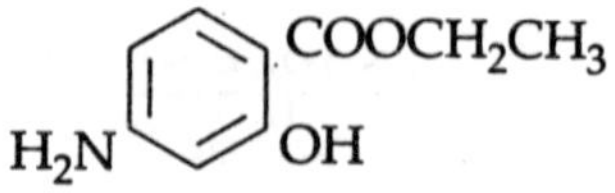

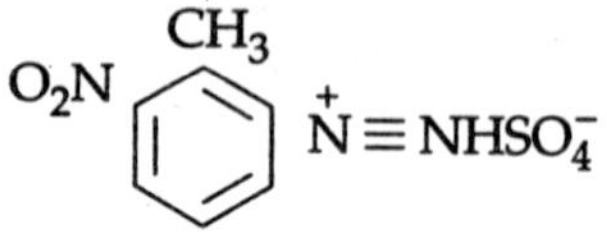

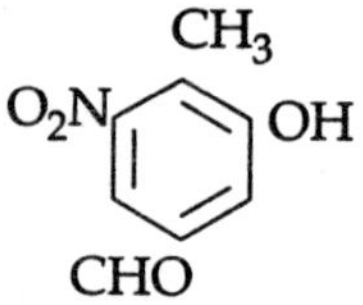

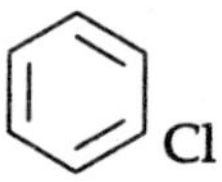

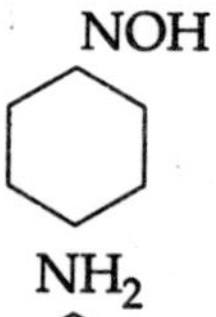

Q. 57. The following series of reaction was carried out to prepare compounds A–H, Match the structures labeled A to H with those given in column ⊗.

(a) C_6H_5–CH_2 $\xrightarrow{2CH_3Li}$

$\xrightarrow{H_2O}$ A $\xrightarrow[\text{diehyl ether}]{LiAH_2}$ $\xrightarrow{H_3O^+}$ B

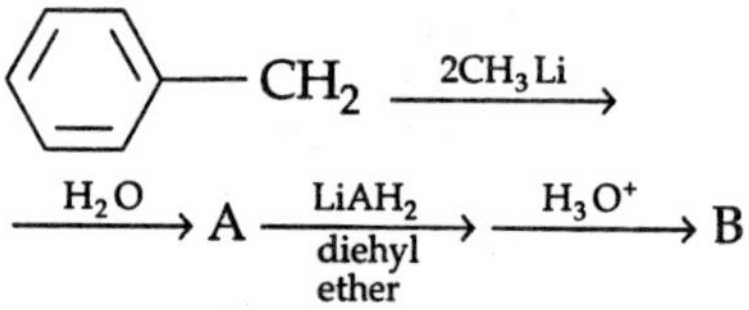

(b) B $\xrightarrow[\text{pyridine}]{TsCl}$ C $\xrightarrow{H\overset{O}{\overset{||}{C}}OH}$;

D $\xrightarrow[\text{diethyl ether}]{LiAlH_2}$ $\xrightarrow{H_3O^+}$ B

(c) B $\xrightarrow{HBr}$ E

(d) B $\xrightarrow{SOCl_2}$ F

(e) $\xrightarrow[\text{NaOH, } H_2O, \Delta]{KMnO_4}$ G + H

Column ⊗

(i) C_6H_5–CH_2–CH(COOH)CH_3

(ii) C_6H_5–$CH_2CH(OH)CH_3$

(iii) C_6H_5–$CO\bar{O}\overset{+}{N}a$

(iv) C_6H_5–CH_2CH (Cl) CH_3

(v) $\overset{+}{N}a\bar{O}$–$COCH_3$

(vi) C_6H_5–$CH_2CH(OTs)$ CH_3

(vii) C_6H_5–CH_2COCH_3

(viii) C_6H_5–$CH_2CH(Br)CH_3$

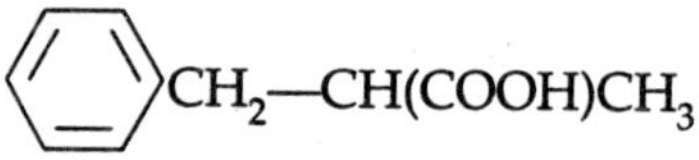

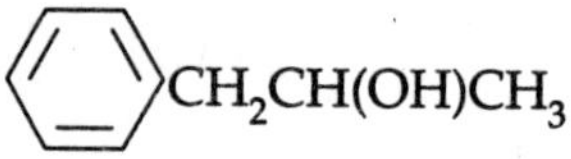

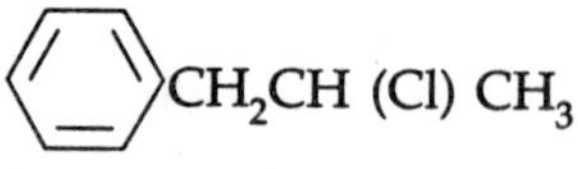

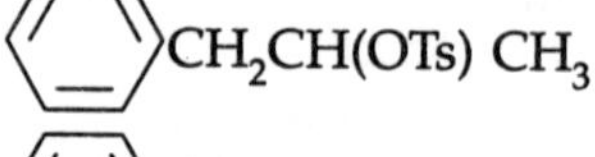

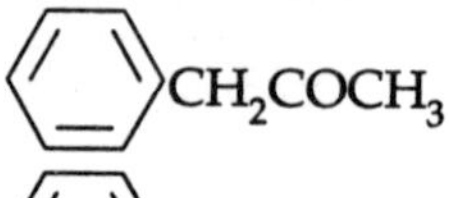

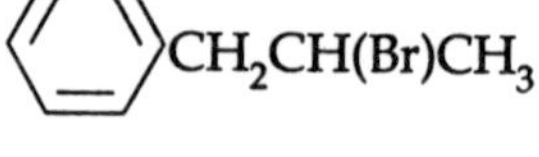

Q. 58. Arrange each of the following groups of compounds (I/II/III/IV) in order of decreasing acidity. Match your answer with (i), (ii), (iii) or (iv).

I II III IV

(a) 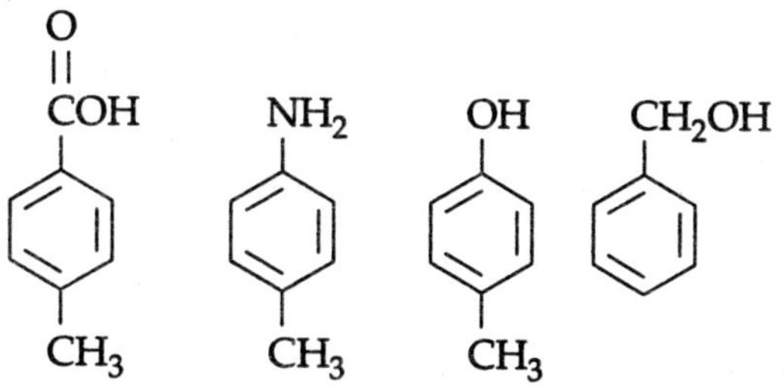

(i) IV > III > I > II
(ii) II > IV > III > I
(iii) I > III > IV > II
(iv) III > I > II > IV

(b) 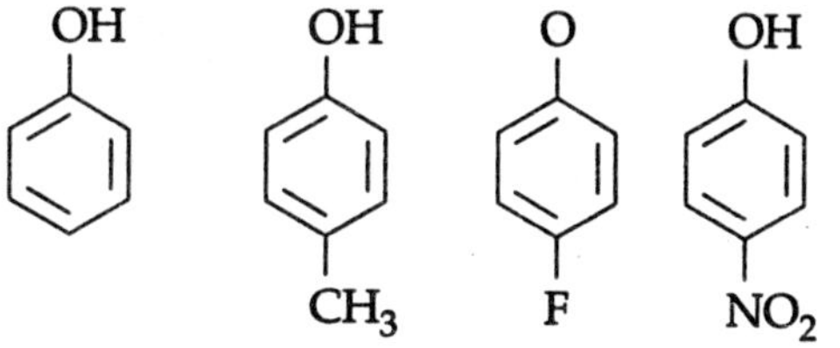

(c) 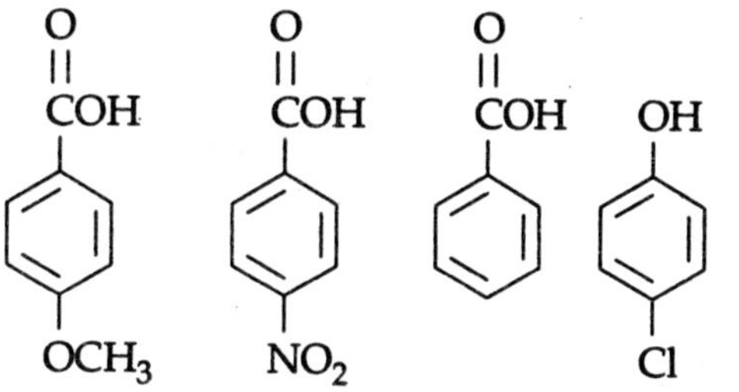

Q. 59. The $-\overset{O}{\overset{\|}{C}}-\overset{H}{\overset{|}{N}}-$ linkage occurs in all of the following substances EXCEPT

(a) nylon **(b)** glycogen **(c)** protein **(d)** valylglycine

Q. 60. Which of the following can exist as cis and trans isomers?

(a) $CH_2{=}CHCl$ **(b)** $CH_2{=}CH{-}CH{=}CH_2$
(c) $CH_3CH_2CH{=}CHCl$ **(d)** $(CH_3)_2C{=}O$

Q. 61. A racemic mixture contains

(a) Equal amounts of cis and trans isomers.
(b) Both straight-chain and branched-chain alkanes.
(c) Equal amounts of a pair of enantiomers.
(d) Equal amounts of a primary and a secondary amine.

Q. 62. All of the following substances are water soluble EXCEPT

(a) $CH_3CH_2OCH_3$ **(b)** $CH_3CH_2CH_2OH$
(c) $(CH_3)_2C{=}O$ **(d)** $CH_3CH_2CH{=}O$

Q. 63. The correct systematic name for

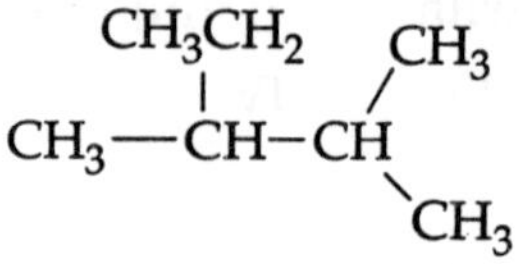

(a) 2-methyl-3-ethylbutane **(b)** 1,1-dimethyl-2-ethylpropane
(c) 2,3dimethylpentane **(d)** 1,1,2-trimethylbutane

Q. 64. A heterocyclic compound
(a) Contains two or more fused benzene rings.
(b) Contains both polar and nonpolar groups.
(c) Is a reactive compound due to ring strain.
(d) Has a ring of atoms, at least one of which is not carbon.

Q. 65. $CH_3CH{=}CHCH_3$ $(CH_3)_3CH$

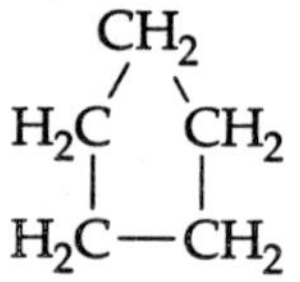

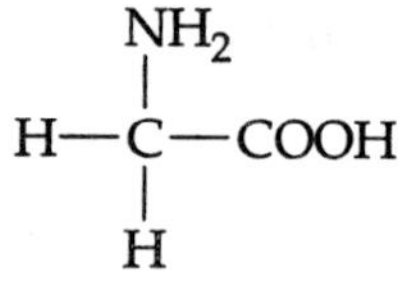

I II III IV

Which of the compounds listed can exist as optical isomers?
(a) None **(b)** I and II **(c)** III and IV **(d)** I, III and IV

Q. 66. The correct systematic name of

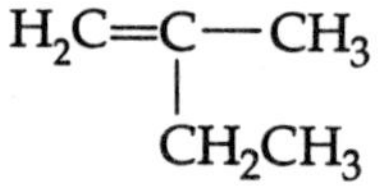

is

(a) 2-ethyl-1-propene **(b)** 1-methyl-1-ethyl-ethylene
(c) 2-ethyl-2-propene **(d)** 2-methyl-1-butene

Q. 67. The monomer from which the polymer $(–CH_2CCl_2CH_2CCl_2–)_n$ (Saran) is made is
(a) $H_2C{=}CCl_2$ **(b)** $CHCl{=}CHCl$ **(c)** $Cl_2C{=}CCl_2$ **(d)** $H_2C{=}CHCl$

Q. 68.

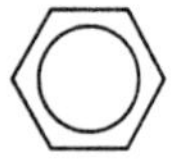

$CH_3CH{=}CH_2$

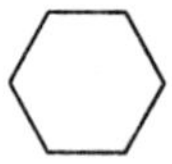

$CH_3CH_2CH_3$

I II III IV

Of the compounds listed, which would react readily with HBr?
(a) I **(b)** I and II **(c)** II **(d)** II and IV

Q. 69. The correct systematic name for

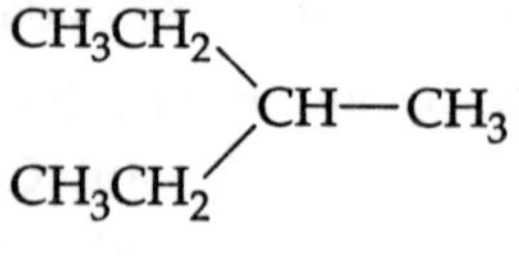

is

(a) 3-ethylbutane **(b)** 1, 1-diethylethane **(c)** 3-methylhexane **(d)** 3-methylpentane

Q. 70. $CH_3CH_2CH_2CH_2OH$ (I) $CH_3CH_2—O—CH_2CH_2CH_3$ (II)

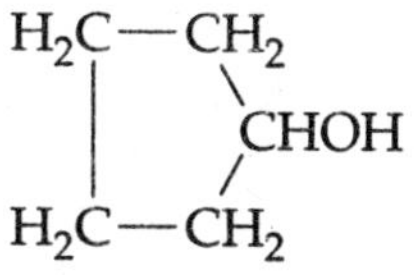

III

$CH_3CH{=}CHCH_2CH_2OH$

IV

Which of the compounds listed above are isomers?

(a) III and IV **(b)** I and IV **(c)** II and III **(d)** I and II

Q. 71. Which of the following compounds is a tertiary amine?

(a) $CH_3CH_2NH_2$ **(b)** $(CH_3)_2NH$ **(C)** $(CH_3)_3N$ **(d)** $(CH_3CH_2)_4NCI$

Q. 72. Which of the following compounds can exist as a pair of enantiomers?

(a) $CH_3—CHCI_2$

(b) 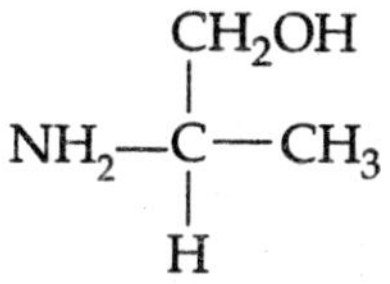

(c) $FCH_2—COOH$

(d) 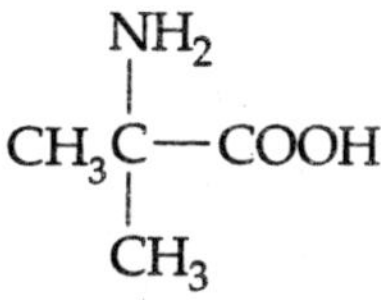

Q. 73. In which of the following compounds is there a carbon atom in the + 3 oxidation state?

(a) CH_3CH_2OH **(b)** $CH_3—O—CH_3$ **(c)** CH_3COOH **(d)** 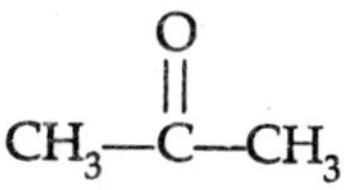

Q. 74. 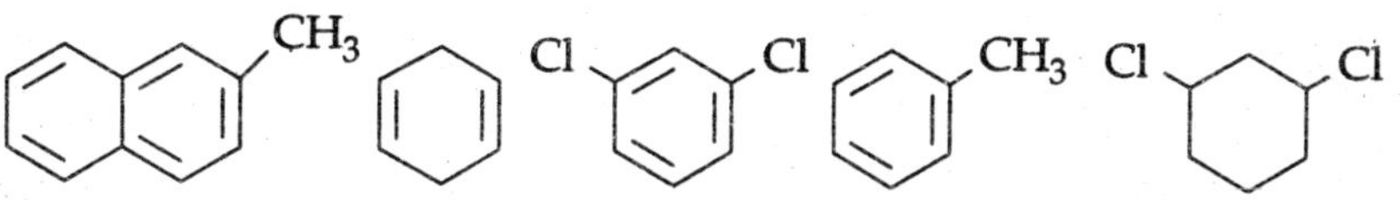

I II III IV V

Which of the compounds listed above are aromatic?

(a) III and IV **(b)** I, III and V **(c)** I, II and IV **(d)** I, III and IV

Q. 75. A secondary amine is

(a) An organic compounds with two—NH_2 group.

(b) A compound with two carbon atoms and an—NH_2 group.

(c) A compound with an—NH_2 group on the carbon atom in the number 2 position.

(d) A compound which two of the hydrogen atoms of NH_3 have been replaced by organic groups.

Q. 76. The product of the reaction of 2-butene with HBr is
(a) 1-bromobutane (b) 2-bromobutane
(c) 3-bromobutane (d) 1, 3-dibromobutane

Q. 77. Which of the following is a polysaccharide?
(a) nylon (b) amylose (c) polyvinyl chloride (d) ribose

Q. 78. (i) Compound (*A*) gives positive Lucas test in 5 minutes. When 6.0 g of (*A*) is treated with sodium metal, 1120 mL of hydrogen is evolved at STP. Assuming (*A*) to contain one atom of oxygen per molecule, write structure formula of (*A*) from following:

(a) CH_3OH
(b) $CH_3CH_2CH_2OH$
(c) CH_2CH_2OH
(d) $CH_3CHOHCH_3$

(ii) Compound (*A*) when treated with PBr_3 gives (*B*) which when treated with benzene in presence of anhydrous aluminum chloride gives (C). What are (*B*) and (*C*) respectively in the following:

(a) $CH_3CHBrCH_3$; $C_6H_5CH(CH_3)_2$
(b) $CH_3CH_2CH_2Br$; $C_6H_5CH(CH_3)_2$
(c) $CH_3CHBrCH_3$; $C_6H_5CH_2CH_2CH_3$
(d) $CH_3CH_2CH_2Br$; $C_6H_5CH_2CH_2CH_3$

Q. 79. An organic liquid A with sweet smell and b. pt. 351 k contais C, H and O. On heating with conc. H_2SO_4 gives a gaseous product (*B*) of empirical formula CH_2. Compound (*B*) decolourizes Br_2 of water and alk. $KMnO_4$. (*B*) also reacts with one mole of H_2 in presence of Ni. What are (*A*) and (*B*) respectively in the following:

(a) CH_3OH, CH_2
(b) CH_3OH, C_3H_6
(c) C_2H_5OH, C_2H_4
(d) C_2H_5OH, C_3H_6

Q. 80. Compound *A* contains, C, *H* and oxygen and it reacts with sodium with evolution of hydrogen. Controlled oxidation of (*A*) gives (*B*) which forms a precipitate with sodium bisulphite but does not give Ag mirror test. A solution of 7.41g of (*A*) in 100 g of water freezes at –1.86k. The formulae of (*A*) and (*B*) (*K* for H_2O = 18.58), respectively in the following:

(a) $CH_3CH_2CH_2CH_2OH$; $CH_3CH_2CH_2CHO$
(b) $CH_3CH_2CHOHCH_3$; $CH_3CH_2COCH_3$
(c) $CH_3CH_2CH_2CHO$; $CH_3CH_2CH_2CH_2OH$
(d) $CH_3CH_2COCH_3$; $CH_3CH_2CHOHCH_3$

Q. 81. (i) 0.037 g of an alcohol, ROH, was added to CH_3Mgl and the gas evolved measured 11.2 cm^3 at STP. The molecular weight of ROH is:

(a) 148 (b) 37
(c) 296 (d) 74

(ii) On dehydration ROH gives an alkene which an ozonolysis gives acetone as one of the product. ROH on oxidation easily gives an acid containing the same number of carbon atoms. The structures of ROH and the acid respectively from the following is:

(a) 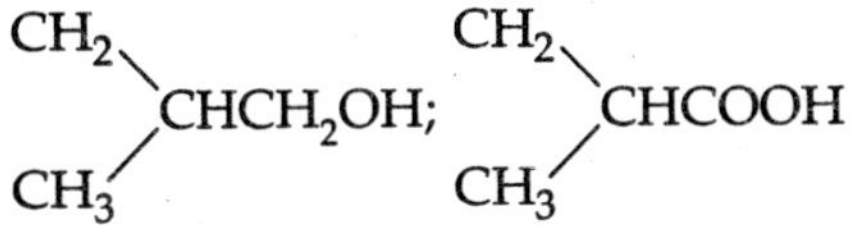

(b) $CH_3CH_2CH_2CH_2OH$; 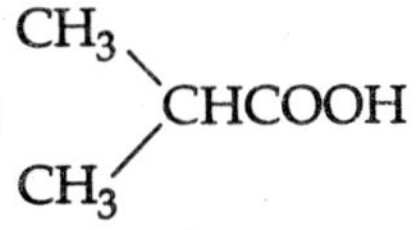

(c) $CH_3CH_2CH_2CH_2OH$; $CH_3CH_2CH_2COOH$

(d) 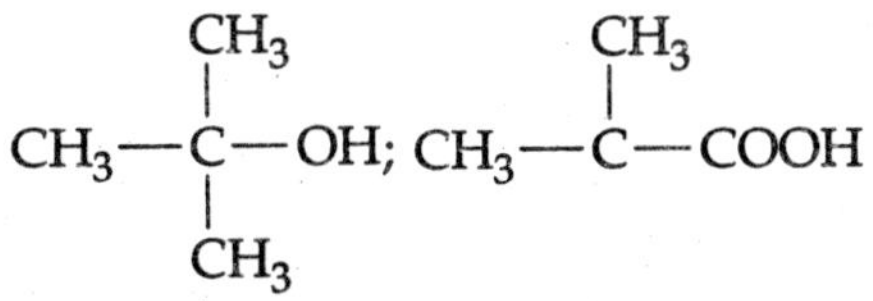

Q. 82. A chloro compound (*A*) showed the following properties:

- Decolourized bromine CCl_4;
- Absorbed hydrogen catalystically;
- Gave a precipitate with ammonical cuprous chloride and
- When vaporised 1.49 g of (*A*) gave 448 mL of vapours at STP. The compound *A* is:

(a) 1-chloropropyne (b) 2-chloropropyne
(c) 3-chloropropyne (d) 4-chloropropyne

Q. 83. A hydrocarbon (*A*) was found to contain 85.7% carbon and 14.3% H. This compound consumes 1 molar equivalent of hydrogen to give a saturated hydrocarbon. 1 g of (*A*) just decolourized 38.05 g of 5% solution by weight of Nr_2 in CCl_4. Compound (*A*) on oxidation with conc. $KMnO_4$ gave a compound (C). C_4H_8O and (*B*) acetic acid. Compound (C) can easily be prepared by the action of acidic aqueous mercuric sulphate on 2-butyne. (*A*) and (C), respectively is:

(a) Butanone, 3-methyl pent-2-ene (b) Butanone; Ethanoic acid
(c) 3-methylpent-2-ene; Ethanoic acid (d) 3-methyl pent-2-ene; Butanone

Q. 84. 0.535 g ethanol and acetaldehyde mixture when heated with Fehling solution gave 1.2 g of a red precipitate. The percentage of acetaldehyde in the mixture is: *(Given At. wt. of Cu = 63.8)*

(a) 31.27 (b) 65.635
(c) 68.73 (d) 34.365

Q. 85. Compound (x) (C_5H_8O) does not react appreciably with Lucas reagent at room temperature but gives a precipitate with ammonical silver nitrate. With excess of MgBr, 0.42 g of (x) gives 224 mL CH_4 at STP. Treatment of (x) with H_2 in presence of Pt. Catalyst followed by boiling with excess HI gives n-pentane. The structure of x is:

(a) Pent-2-yn-2-ol **(b)** Pent-4-yn-1-ol
(c) Pent-3-yn-1-ol **(b)** Pent-3-yn-2-ol

Q. 86. 10 g of a mixture of hexane and ethanol are reacted with sodium to give 200 mL hydrogen at 300 k and 760 mm pressure. The percentage of ethanol in the mixture is.

(a) 14.96 **(b)** 7.48 **(c)** 29.92 **(d)** 3.74

Q. 87. One mole of a hydrocarbon (A) reacts with 1 mole of bromine having a dibromo compound $C_6H_{10}Br_2$. (A) on treatment with cold dilute alkaline $KMnO_4$ solution forms a compound $C_5H_{12}O_2$. On ozonolysis (A) gives equimolar quantities of propanone and ethanol. The structure of (A) will be.

(a) n-butyne **(b)** 2-methylbut-1-ent
(c) 2-methylbut-2-ene **(b)** 2-pentene

Q. 88. When t-butanol and n-butanol are separately treated with a few drops of dilute $KMnO_4$, in one case only, the purple colour disappears and a brown precipitate is formed. Which of the two alcohols gives the above reaction?

(a) t-butanol **(b)** n-butanol **(c)** none

Q. 89. Compound (A) with molecular weight 108 contained 88.89% C and 11.11% H. It gave a white precipitate with ammonical silver nitrate. Complete hydrogenation of (A) gave another compound (B) with molecular weight 112. Oxidation of (B) gave an acid of molecular weight 128. Decarboxylation of this acid gave cyclohexane. The structures of (A) and (B) respectively are:

(a) 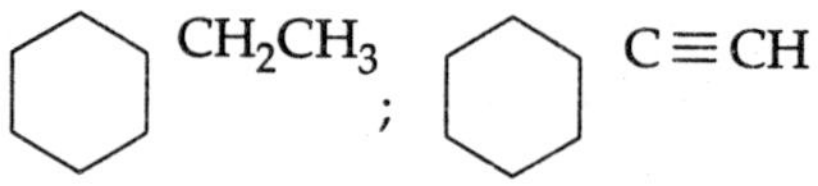

(b) 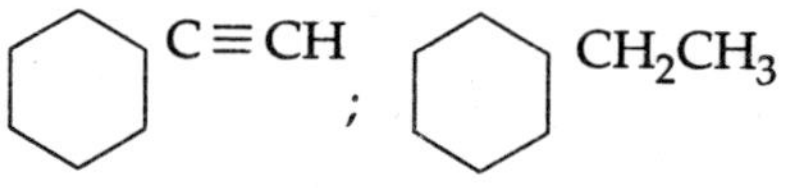

(c) HC≡C—$(CH_2)_4$; CH=CH_2; 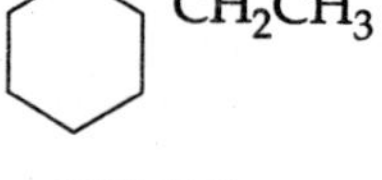

(d) CH_3—(CH_2)—CH=CH_3; 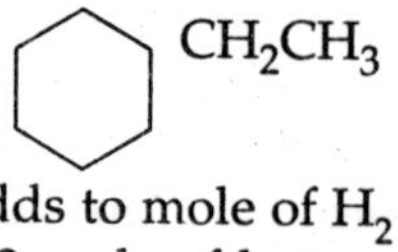

Q. 90. An organic compound C_8H_{12} adds to mole of H_2 per mole of hydrocarbon. It undergoes reductive ozonolysis to give a 2 mole of butan –1, 4-dial. The hydrocarbon is:

(a) Butane-1, 4 dial **(b)** Butan-1, 3 dial
(c) Butan-1, 2 dial **(d)** n-butane

Q. 91. 0.0852 g of an organic halide (A) when dissolved in 2.0 g of camphor, the melting point of the mixture was found to be 440 K. Halide (A) when heated with sodium gives a gas (B). 280 mL of gas (B) at STP weighs 0.375 g.

(i) The structural formula of (A) and (B) respectively will be: (Given K_f for camphor = 4; m. pt. of camphor = 452 k)

(a) CH_3I, C_3H_8 **(b)** CH_3I, C_2H_6
(c) CH_3Br, CH_4 **(d)** CH_3Br, C_2H_6

(ii) The mol. wt. of (A) is:

(a) 16 **(b)** 30 **(c)** 40 **(d)** 44

Q. 92. It required 0.7 g of hydrocarbon (A) to react completely with Br_2(2.0 g). On treatment of (A) with HBr it yielded monobromo alkane. This same compound was obtained when (A) was treated with HBr in presence of peroxide. The structure formulae of (A) is:

(a) n-Butane **(b)** But-2-ene **(c)** But-2-yne **(d)** But-3-yne

Q. 93. Compound X with molecular formula $C_9H_{10}O$ forms a semicarbazone and gives negative Tollen's and Iodoform tests. Upon reduction it gives n-propyl benzene. The structure of X is:

(a) $C_6H_5COCH_2CH_3$
(b) $C_6H_5COCH_3$
(c) $CH_3—CH{\equiv}C—CH{=}CH—CO—CH{=}CH—CH_3$
(d) $C_6H_5CH_2COCH_3$

Q. 94. An optically active alcohol (A) ($C_6H_{10}O$) absorbs two mole of hydrogen per mole of (A) upon catalytic hydrogenation and gives a product (B). The compound (B) is resistant to oxidation by CrO_3 and does not show any optical activity. The structure of (A) will be:

(a) 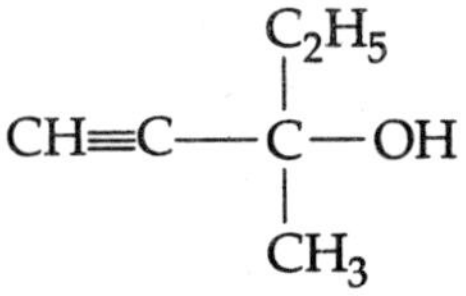

(b) 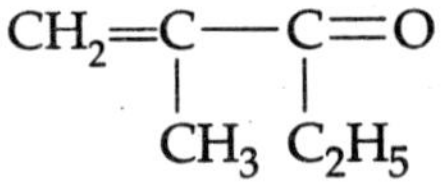

(c) 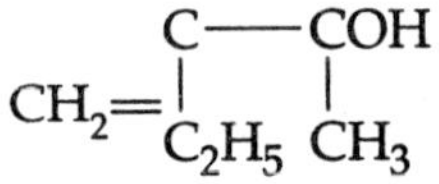

(d) 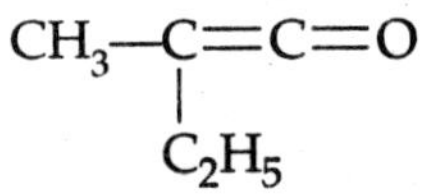

Q. 95. A compound (D), ($C_8H_{10}O$) upon treatment with alkaline solution of iodine gives a yellow precipitate. The filtrate on acidfication gives a white solid (E), ($C_7H_6O_2$). The structures of (D), will be:

(a) $CH_3—C{\equiv}C—CH_2—\overset{O}{\overset{\|}{C}}—CH{=}CH—CH_3$ **(b)** $C_6H_5CHOHCH_3$
(c) $C_6H_5COCH_3$ **(d)** C_6H_5COOH

Q. 96. An alkyl halide, (X) of formula $C_6H_{13}Cl$ on treatment with potassium tertiary butoxide gives two isomeric alkenes (C_6H_{12}). Both alkenes on hydrogenation give an hydrocarbon (Y) the structures of (X), and Y respectively from *Column A* and *B* are:

Column A	*Column B*
(a) 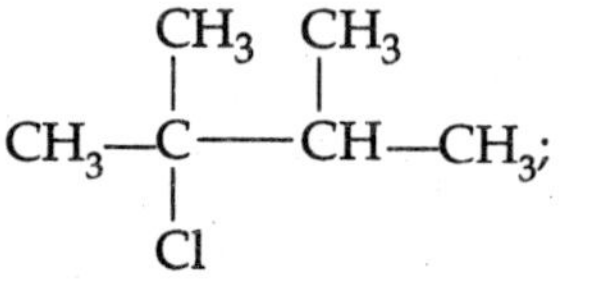	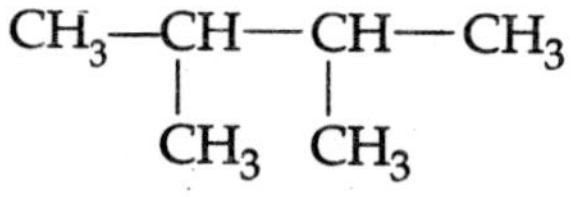
(b) 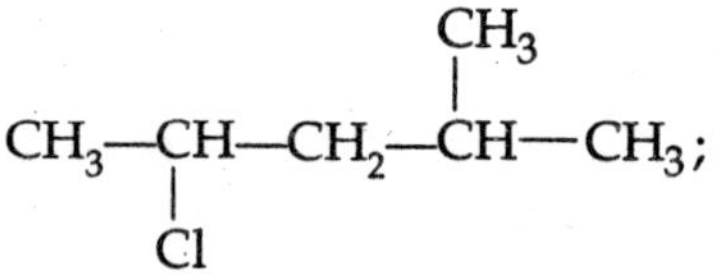	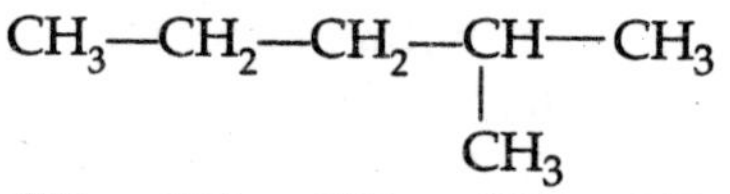
(c) 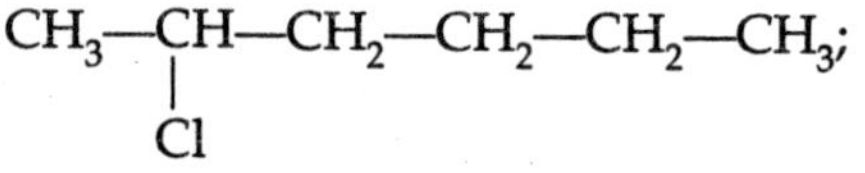	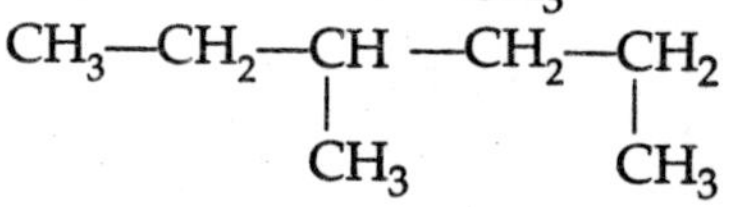
(d) 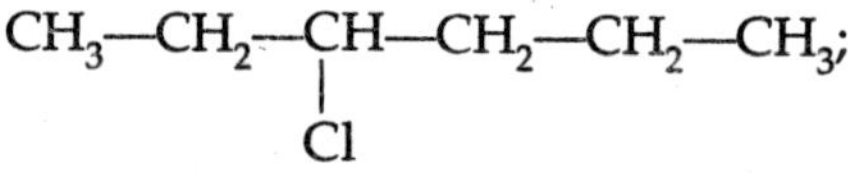	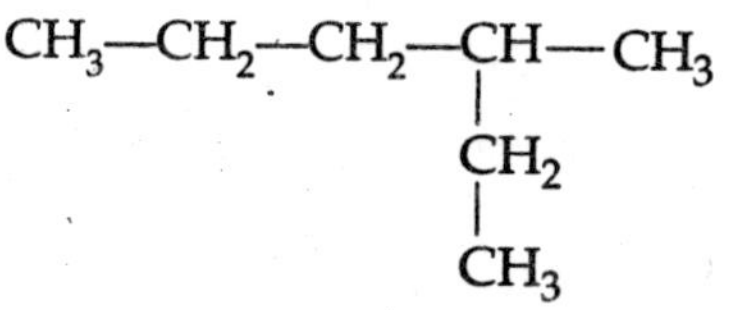

Q. 97. 448 mL of a hydrocarbon (*A*) having C = 87.80%, H = 12.19% weigh 1.64 g at STP. On hydrogenation, it gives 2-methylpentane. Treatment of (*A*) with acidic $HgSO_4$ gives a new compound (*B*) $C_6H_{12}O$. Compound (*A*) does not react with ammonical $AgNO_3$. So the compound *A* is.

(a) 3-methyl pent-1-ene (b) 4-methyl pent-2-yne

(c) 2-hexane (d) 3-hexene

Q. 98. Isopropyl alcohol on treatment with excess of conc. H_2SO_4 gives (X) which on treatment with HBr give (Y). When (Y) dissolved in ether and treated with sodium gives (Z) therefore (Z) is:

(a) *n*-hexyne (b) 3-ethyl, methyl propene

(c) 2, 2-dimethyl pentane (d) 2, 3, dimethyl butane

Q. 99. An organic compound (*A*) $C_7H_{15}Cl$ on treatment with alcoholic caustic gives a hydrocarbon (*B*) C_7H_{14}. (*B*) on treatment with ozone and subsequent hydrolysis gives acetone and butyraldehyde (*A*) and (*B*) respectively are as:

(a) 3-chloro-2-methylhexane; 2 methyl hex-3-ene

(b) 2-chloro-2 methylhexene; 2 methylhex-3-ane

(c) 2-chloro-2 methyl hexane; 2-methyl hex-2-ene

(d) 4-chloro-2 methylhexene; 2-methylhex-2-ane

Q. 100. An organic compound C_4H_6 on ozolysis gives formaldehyde and glyoxal. The compound C_4H_6 is:

(a) 2-Butyne (b) 2-methyl propene
(c) Buta-1, 2-diene (d) Buta-1, 3-diene

Q. 101. An organic compound C_4H_6 on ozolysis gives HCHO, CO_2 and CH_3CHO. The compound C_4H_6 is:

(a) Buta 1, 3-diene (b) Buta 1, 2-diene
(c) 2-methyl propene (d) 2-Butyne

Q. 102. An organic compound $A(C_5H_8)$ on hydrogenation gives compound $B(C_5H_{12})$. Compound A on ozonolysis gives formaldehyde and 2-ketopropanal. The structure of compound A is:

(a) $CH_3—C(=CH_2)—CH=CH_2$ (b) $CH_3—C(CH_3)=C=CH_2$
(c) $CH_3—CH_2—CH=C=CH_2$ (d) none of above

Q. 103. Compound (*A*) C_6H_{12} gives a positive test with bromine in carbon tetrachloride. Reaction of (*A*) with alkaline $KMnO_4$ yield only (*B*) which is the potassium salt of an acid. Thus, (*A*) and (*B*) are respectively as:

(a) hex-2-ene, pot. butanote (b) hex-2-yne, pot. propanoate
(c) hex-3-ene, pot. propanoate (b) hex-3-yne, pot. butanote

Q. 104. An halide $C_5H_{11}X$ on treating with alc. KOH gives pentene-1. The halide is:

(a) 1-Chloro hexane (b) 3-Iodo pentane
(c) 2-Bromo butane (d) 1-halo pentane

Q. 105. An acid *A* contains carbon = 40.7%; hydrogen = 5.1% and its silver salt contains 65.1% Silver. The ethylester of *A* has V.D. of 87(H_2 = 1). The structure *A* is:

(a) CH_3COOH (b) $—CH_2—COOH$ | $CH_2—COOH$
(c) $CH_3—CH_2—COOH$ (d) $CH_3—CH(COOH)(COOH)$

Q. 106. An organic compound (*A*) C_6H_{12} does not decolourizes Br_2 water and shows same addition product if HBr is added in presence or in absence of O_2. However (*A*) does not give a propanal on ozonolysis. The compound (*A*) is:

(a) 3-hexane (b) 2, 3-methyl but-2-ene
(c) 3-methyl pentene (d) 2-methyl pentane

Q. 107. A hydrocarbon (*A*) decolourizes Br_2 in CCl_4 and adds up a molecule of H_2 in presence of Pt catalyst to give hexane. When (*A*) is oxidized vigorously with $KMnO_4$, a single molecule of carboxylic acid salt with three carbon atoms is isolated. The name of (*A*) and the salt formed respectively is:

(a) hex-2-ene; pot. propanoate (b) hex-3-ene; pot. butanoate
(c) hex-4-ene; pot. propanoate (d) hex-5-ene; pot. hexaanote

Q. 108. A hydrocarbon *A* of the formula of C_7H_{12} on ozonolysis gives a compound *B* which undergoes aldol condensation giving 1-acetyl-cyclopentene. The compound *B* has

(a) $C_7H_{12}O$ **(b)** $C_7H_{12}O_2$ **(c)** $C_7H_{12}O_3$ **(d)** $C_7H_{12}O_4$

Q. 109. 36.4 g of 1, 2, 3, 2-tetrachloropropane was heated with zinc dust and the product was bubbled through ammonical $AgNO_3$. The weight of precipitate obtained is. Give equations for the reactions involved.

(a) 42.9 g **(b)** 39.2 g **(c)** 29.4 g **(d)** 22.9 g

Q. 110. He silver salt of an unknown alkyne contains 67.08% of Ag. Assuming no other functional group present, what is the structure of alkyne?

(a) Prop-1-yne **(b)** Prop-2-yne **(c)** But-1-yne **(d)** But-2-yne

Q. 111. A halide $C_5H_{11}X$, on treating with (alc.) KOH gives only pentene-2. The halide is:

(a) 1-halopentene **(b)** 2-halopentyne
(c) 3-halopentane **(d)** 2-halo, 1-methyl-butane

Q. 112. How many g of Br_2 will react with 5 g of
Pent-1-ene, Pent-1-yne, Pentene? respectively.

(a) 11.11; 11.43; 23.53 **(b)** 11.11; 23.53; 11.11
(c) 23.53; 11.11; 11.43 **(d)** 23.53; 11.43; 11.11

Q. 113. How many asymmetric carbon atoms are created during the complete reduction of benzil (PbCOCOPh) with $LiAlH_4$ and the number of possible stereoisomers in the product respectively.

(a) 1, 2 **(b)** 2, 2 **(c)** 2, 3 **(d)** 3, 2

Q. 114. A hydrocarbon (*A*) was found to have vapour density 36. It forms only single monochloro substitution product. *A* is thus:

(a) 1-methyl butene **(b)** 2, 2-dimethyl propane
(c) *n*-pentane **(d)** All of above

Q. 115. An alkane *A* gaving a molecular weight of 86; forms two monobromo alkanes? The alkane *A* is names as:

(a) 3-ethyl, 3-methyl propane **(b)** *n*-hexane
(c) 2, 2, dimethyl butane **(d)** 2, 2, dimethyl pentane

Q. 116. For the compounds (i) and (ii) below, the no. of asymmetric carbon in each optically active compound and the total no. of isomers respectively is:

(i) $CH_3(CHOH)_2COOH$

(a) 1; 2 **(b)** 2; 4 **(c)** 2; 3 **(d)** 1; 4

(ii) $COOH(CHOH)_2COOH$

(a) 2; 4 **(b)** 3; 2 **(c)** 2; 3 **(d)** 3; 4

Q. 117. The compound $C_4H_8C_{12}$(*A*) on hydrolysis gives a compound C_4H_8O, (*B*). The compound (*B*) reacts with hydroxylamine and gives a negative test with Tollen's reagent. Compounds (*A*) and (*B*)? respectively are:

(a) 1, 3-dichlorobutane; Butane-1-one
(b) 2, 3-dichlorobutane; Butane-2-one
(c) 2, 3-dichlorobutane; Butane-1-one
(d) 1, 3-dichlorobutane; Butane-2-one

Q. 118. Compound (*A*) $C_5H_{10}O$ forms phenyl hydrozone and gives negative Tollen's and Iodoform tests. Compound (*A*) on reduction gives n-pentane. The compound *A* is named as:

(a) Butan-3-one **(b)** Pentan-2-one **(c)** Pentan-3-one **(b)** Hexan-3-one

Q. 119. An unknown compound of carbon, hydrogen and oxygen contains 69.77% H, and has a molecular weight of 86. It does not reduce Fehling solution but forms a bisulphite addition compound and gives a positive iodoform test thus the names of the possible structures are

(a) 3-methyl Butan-3-one; Pentan-2-one

(b) Pentan-2-one; 3-methylbutan-2-one

(c) Pentan-3-one; 3-methylbutan-2-one

(d) 3-methyl Butan-3-one; Pentan-3-one

Q. 120. An Compound (*A*) $C_4H_8O_2$ It reacts with sodium bicarbonate to liberate CO_2; On fusion with alkali gives propane and with $Ca(OH)_2$ it gives $C_8H_{14}O_4Ca$ which on heating decomposes to di isopropyl ketone. The compound *A* is:

(a) Butanoic acid **(b)** isopropyl ketone

(c) 2-methylpropionic acid **(d)** Di-isopropyl ketone

Q. 121. A compound (*A*) does not react with sodium but reacts with PCl_5. (*A*) also undergoes haloform reaction and gives 3-methylbut-2-enoic acid on acidification of products. On hydrogentaion (*A*) gives 2-methylpentane. The compound *A* is:

(a) 2-methyl pentane **(b)** Hex-4-en-3-one

(c) 3, 3-dimethyl but-3-ene-2-one **(d)** 4-methylpent-3-en-2-one

Q. 122. Two organic compounds (*A*) and (*B*) have some empirical formula CH_2O. Vapour density of (*B*) is twice the vapour density of (*A*). (*A*) reduced Fehling solution but does not react with $NaHCO_3$. Compound (*B*) neither reacts with $NaHCO_3$ nor reduces Fehling solution. What are (*A*) and (*B*) respectively.

(a) $HCHO, CH_3COOH$ **(b)** $CH_3COOH, HCHO$

(c) CH_3COOH, CH_3CH_2COOH **(d)** CH_3CH_2COOH, CH_3COOH

Q. 123. A dihalogen derivative '*X*' of a hydrocarbon with '3' carbon atoms in a molecule reacts with alcoholic KOH giving another hydrocarbon which gives a red ppt. with ammonical cuprous chloride. '*X*' gives an aldehyde when treated with aqueous KOH. What is the original compound '*X*'?

Q. 124. A chloride derivative '*X*' on treatment with Zn-Cu couple gives a hydrocarbon with five carbon atoms in the molecule. When '*X*' is dissolved in ether and treated with sodium, 2, 2, 5, 5-Tetramethylhexane is obtained. Identity the compound '*X*'.

(a) Methyl chloride **(b)** Ethyl chloride

(c) Tetramethyl chloride **(d)** Isobutyl chloride

Q. 125. Match the product of the following reactions with column ⊗

(a) [cyclohexene]CH + HOCl → (A)

(b) H_3C–CH(CH_3)–CH(CH_3)–CH_3 + CH_2=C(CH_3)–CH_3 $\xrightarrow{\Delta}$ (B)

(c) [bi(cyclohexenyl)] + CH_2=CH–NO_2 → (C)

(d) [1-methylcyclopentene] CH_3 $\xrightarrow[\text{(ii) Oxidation}]{\text{(i) } BH_3}$ (D)

(e) [cyclohexane] Br–Br $\xrightarrow{[OH^-]}$ (E)

(f) [naphthalene] $\xrightarrow[333\text{ K}]{NaOCl}$ (F)

(g) [cyclopentene] + C_6H_6 $\xrightarrow{HF}$ (G)

(h) [cyclohexanone]=O + H_2 $\xrightarrow[\text{Pressure}]{Ni}$ (H)

column ⊗

(i) [benzene]

(ii) [naphthalene] COOH

(iii) [benzene] OH

(iv) H_3C, H_3C [benzene] CH_3, CH_3

(v) [tricyclic structure]

(vi) [cyclohexane] CH_3, OH, Cl, H

(vii) [phenylcyclopentane]

(viii) [cyclopentane] H_3C, H, H, OH

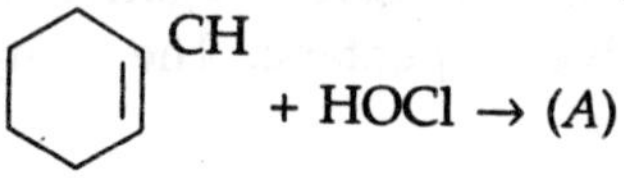

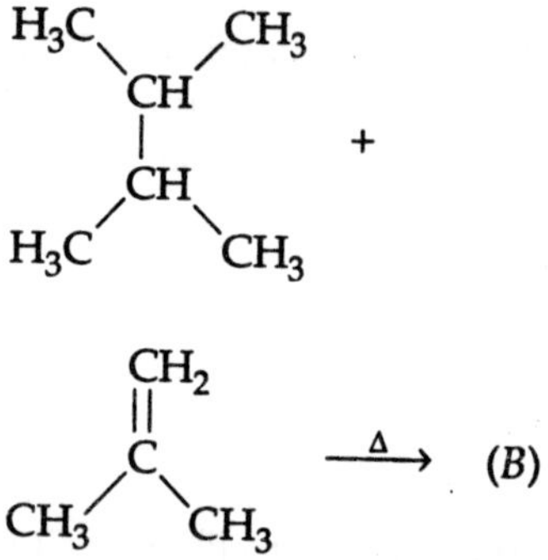

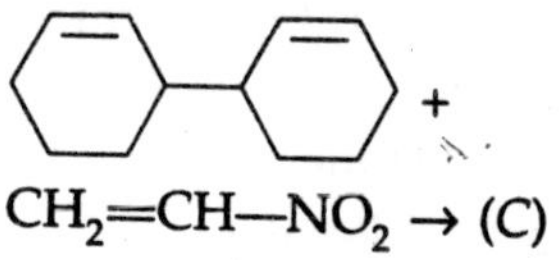

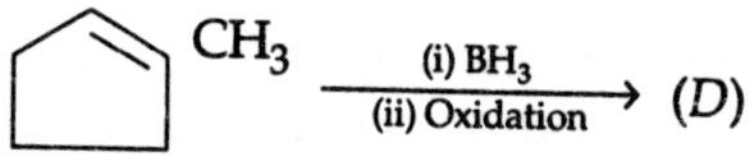

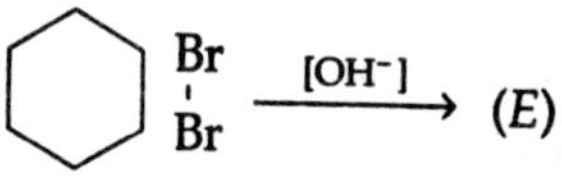

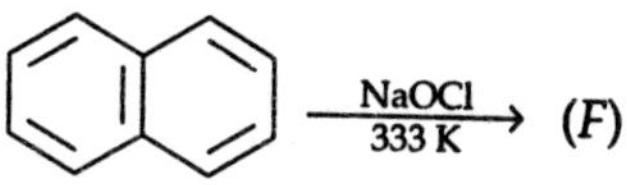

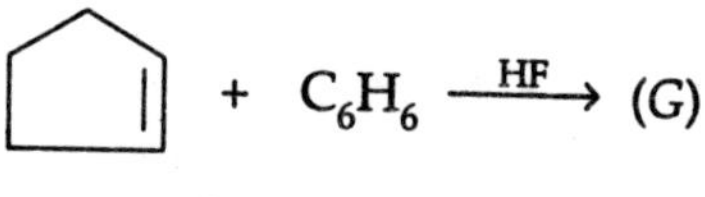

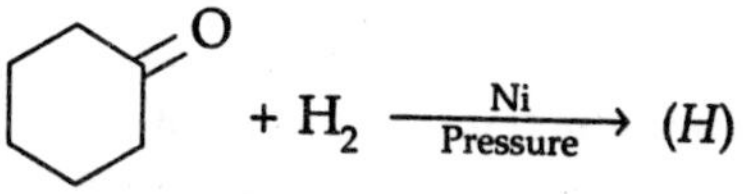

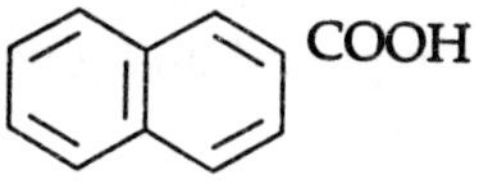

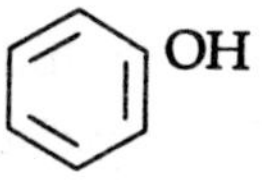

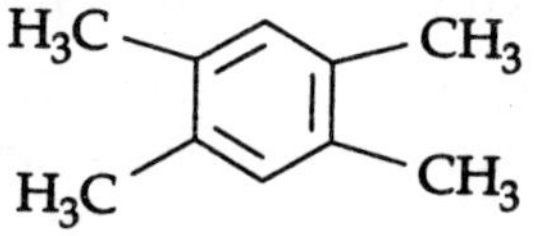

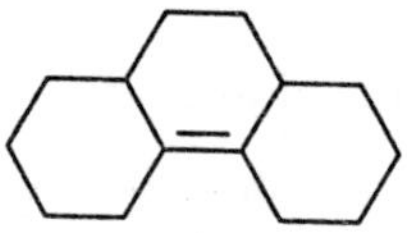

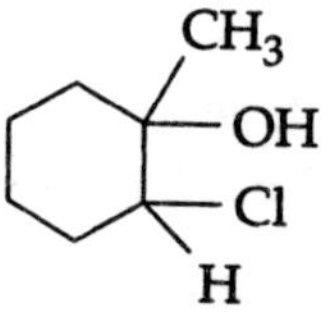

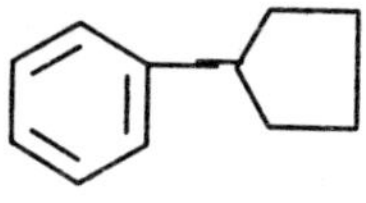

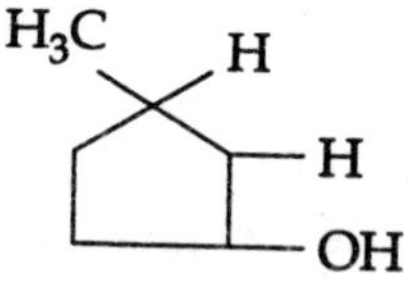

Q. 126. An aromatic compound (A) of m. wt. 106, contains C = 90.6% and H = 9.4%. On oxidation of (A) compound (B) is obtained containing C = 57.8%, H = 3.6%. (B) can form inner anhydride. What are (A) and (B)?

(i) [benzene] CO–O–CO

(ii) [benzene] COOH, COOH

(iii) [benzene] CH_3, CH_3

(iv) [benzene] C_2H_5, CH_3

Q. 127. How many optically active forms are possible for the following?

(i) 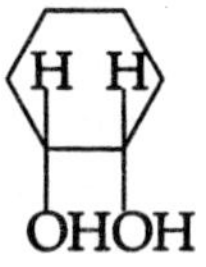

(ii)
```
                Cl
                |
CH3CH2CHCHCH3
              |
             CH3
```

(a) 2 (b) 3 (c) 4 (d) 6

Q. 128. A liquid (*A*) C_7H_6O forms an oxime, does not reduce. Fehling solution but undergoes Cannizzaro reaction. What is *A*?

(i) $C_6H_5CH_2OH$ (ii) C_6H_5CHO
(iii) $C_6H_5CH{=\!=}NOH$ (iv) $C_6H_5COCH_3$

Q. 129. An organic compound (*A*) contains 69.42% carbon 5.78% H and 11.57% nitrogen. Its vapour density is 60.5. It evolves NH_3 when boiled with caustic potash. On heating with P_2O_5 it give compound (*B*) containing 81.85% carbon, 4.85% hydrogen and 13.599 nitrogen. On reduction with sodium and alcohal, (*B*) forms a base which reacts with HNO_2 giving off nitrogen and yielding an alcohal (C). The alcohal can be oxidized to benzoic acid. A, B and C respectively is:

(i) $C_6H_5CH_2OH$; C_6H_5CN; C_6H_5COOH.
(ii) $C_6H_5CONH_2$; $C_6H_5CH_2OH$; C_6H_5CN.
(iii) C_6H_5COOH; C_6H_5OH; $C_6H_5CONH_2$
(iv) C_6H_5CN; C_6H_5COOH; $C_6H_5CONH_2$

Q. 130. An organic compound (A) $C_9H_{13}N$ dissolves in dil. HCl and releases N_2 with HNO_2 giving an optically active alcohol (B). (B) on oxidation gives dicarboxylic acid, which on heating form anhydride. What are (A) and (B)?

(i) $C_9H_{11}OH$ (ii) $C_9H_{11}COOH$ (iii) $C9H_{11}CN$ (iv) $C_9H_{11}NH_2$

Q. 131. Hydrocarbon (A) C_8H_{10} on refluxing with alkaline $KMnO_4$ forms a monobasic acid (B), which on heating with sodalime gives C_6H_6. what are (A) and (B)?

(i) C_6H_5CN (ii) $C_6H_5CH_2CH_3$ (iii) $C_6H_5NH_2$ (iv) C_6H_5COOH

Q. 132. Hydrocarbon (A) with molecular formula C_8H_8 gave the following reactions:

1. On shaking with bromine, a bromo derivative (B) $C_8H_{10}Br_2$ was formed.
2. Vigorous oxidation of the hydrocarbon with alkaline $KMnO_4$ gave a monobasic acid (C).
3. Acid (C) on distillation with sodalime gave C_6H_6. The structures of A, B and C are:

(i) $C_6H_5CH{=}CH_2$ (ii) C_6H_5COOH
(iii) $C_6H_5CHBr{-}CH_2Br$ (iv) C_6H_5OH

Q. 133. An organic compound $C_2H_7NO_2$, on slow distillation loses water and gives (B) which on further distillation with P_2O_5 forms (C). (C) on hydrolysis gives (D) which on treating with P + Cl_2 followed with hydrolysis forms glycollic acid. What are (A) , (B), (C) and (D)?

(i) CH_3CONH_2 (ii) $CH_2ClCOOH$
(iii) CH_3COONH_4 (iv) $CH_2OHCOOH$ (v) CH_3CN

Q. 134. There are three isomeric compounds (A), (B) and (C) having molecular formula $C_8H_8O_2$. When heated with sodalime (A) gives C_6H_6 while (B) and (C) gives toluene. (A) is neutral compound while (B) and (C) are monobasic acid (B) on chlorinationin sunlight gave a compound containing 3 chlorine atom per molecule which is easily hydrolysed to a dibasic acid (D). (D) with sodalime gave benzene the photo chlorination of (C) gave a product which contains two Cl atoms per molecule and is strong acid identify (A), (B), (C) and (D).

(i) $C_6H_5CH_2COOH$ **(ii)** $C_6H_5COOCH_3$

(iii) $C_6H_5CH_2CH_3$ **(iv)** $C_6H_4(COOH)_2$ (COOH, COOH)

(v) C_6H_4 (CH$_3$, COOH)

Q. 135. An organic compound (A) $C_2H_5NO_2$ gave on hydrolyses an alcohol (B) of molecular weight 46 and an acid. On reduction it gave the same alcohol (B) and NH_3 gas. Thus compounds (A) is

(i) Ethyi Nitrite **(ii)** Niboethane

(iii) Phenylalcohol **(iv)** ethyl Alcohol

Q. 136. Two isomeric compounds (A) and (B) have $C_4H_{11}N$ molecular formula. Both on separately treating with HNO_2 lose their N_2 producing two isomeric alcohols (C) and (D) respectively of molecular formula $C_4H_{10}O$. (C) reacts with Lucas reagent immediately and resists oxidation. (D) does not react with Lucas reagent in cold but can be easily oxidized. Complete methylation of either (A) or (B) is made which on decomposition does not produce butene-1. Match *A, B, C, D* with following:

(i) 2-methyl propanol **(ii)** 2-methyl propan-2-amine

(iii) 2-methyl propan-1-amine **(iv)** 3-methyl propanol

Q. 137. From analysis and molecular weight determination, the molecular formula of (A) is C_3H_7NO. The compound gave following reactions:

1. On hydrolysis it gives an amine (B) and a carboxylic acid (C).
2. Amine (B) reacts with benzene sulphonyl chloride and gives a product which is insoluble in aqueous sodium hydroxide solution.
3. Acid (C) on reaction with Tollen's reagent gives a silver mirror, Match (A), (B) and (C)?

(i) $(CH_3)_2NH$ **(ii)** Acetic acid

(iii) Formic acid **(iv)** N, N'-dimethyl form-amide

Q. 138. 0.450g of an aromatic compound (A) on ignition gave 0.905 CO_2 and 0.185g H_2O. 0.350g of (A) on boiling with HNO_3 and adding $AgNO_3$ solution gives 0.574g of AgCI. The vapour density of (A) is 87.5.(A) on hydrolysis with Ca $(OH)_2$ yields (B) which on mild reduction gives an optically active compound (C). On heating (C) with I_2 + NaOH iodoform is produced alongwith (D). With HCI (D) gives a solid which is markedly more soluble in hot eater than cold. Identify (A) to (D) and match them with following:

(i) $C_6H_5CHOHCH_3$
(ii) C_6H_5COONa
(iii) $C_6H_5C.CI_2CH_3$
(iv) $C_6H_5COCH_3$

Q. 139. An organic compound (A) containing C 81.6%, H = 4.8%, N =13.6% has vapour density 51.5. It gives of ammonia upon heating with KOH. Reaction of (A) with Na and ethanol gives a basic compound (B) which gives of N_2, and forms an alcohol (C) when reacted with HNO_2. The alcohol (C) can be oxidised to benzoic acid. Match (A) (B) and (C) with following:

(i) Benylalcohol
(ii) Phenyl cyamide
(iii) Benzyylamine
(iv) Benzoic acid

Q. 140. Complete the following by matching the structures of (A), (B), (C), (D) and (E) from column ⊗

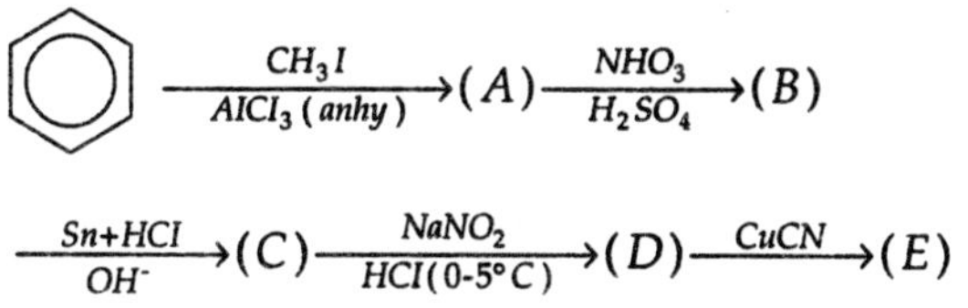

Coloumn ⊗

(i) 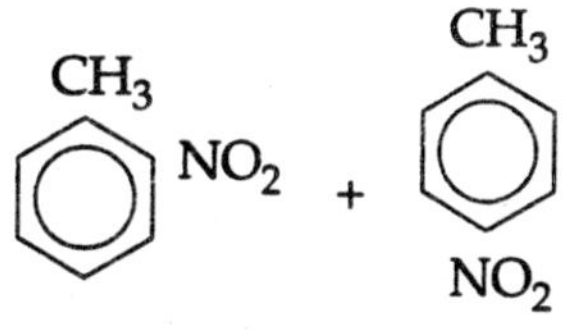

(ii) 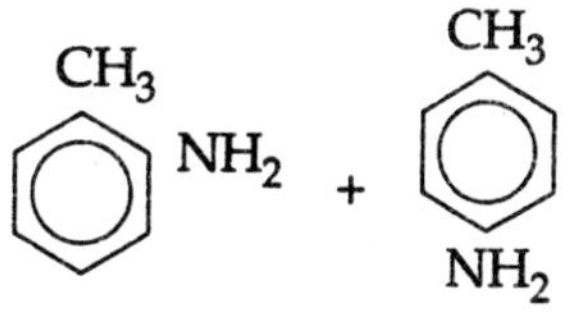

(iii) 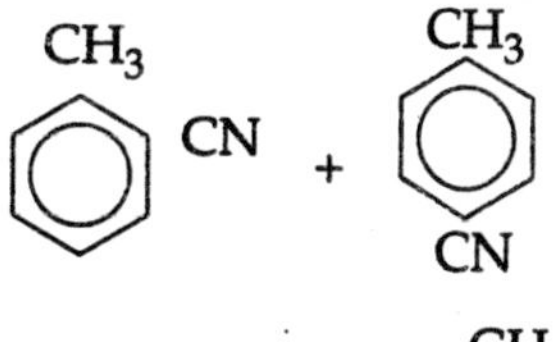

(iv) 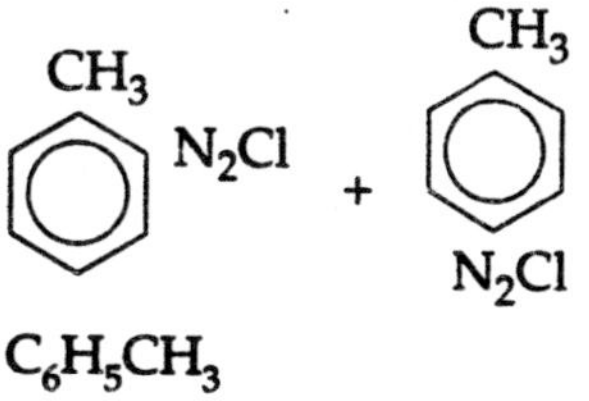

(v) $C_6H_5CH_3$

PART III

Chapter 14

PRACTICE PAPERS

PROBLEMS

Directions: Each group of lettered answer choices below refers to the numbered statements or questions that immediately follow. For each questions or statement, select the one lettered choice that is the best answer and fill in the corresponding circle on the answer sheet. An answer choice may be used once, more than once, or not at all in each set of questions.
(See solutions only after you have attempted the full paper)

Questions 1–3

(a) F
(b) Co
(c) Sr
(d) Be

Q. 1. Which has the lowest ionization energy?

Q. 2. Of those elements with negative oxidation states, which has the fewest such states?

Q. 3. Which has the smallest ionic radius?

Questions 4–8

(a) SO_2
(b) SiH_4
(c) CO_2
(d) Be_2
(e) NO

Q. 4. In which of the choices is there polar double bronding in a nonpolar molecule?

Q. 5. In which molecule (s) does resonance occur?

Q. 6. In which molecule is the bond order 2½?

Q. 7. Which of the molecules has four sp^3 hybrid bonds?

Q. 8. Which molecule would you expect to be unstable on the basis of molecular orbital theory?

Questions 9–11

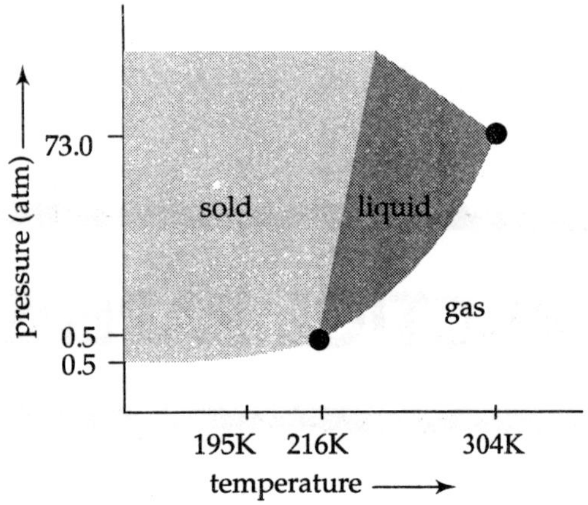

(a) 195K
(b) 216K, 5.1 atm
(c) 304K, 73 atm
(d) None of the above.

Q. 9. What does the phase diagram above show to be the normal boiling point of carbon dioxide?

Q. 10. Which point represents the critical point?

Q. 11. Which point represents the triple point?

Questions 12–16

Directions: Predict the change in entropy.

(a) The change in entropy will be positive.
(b) The change in entropy will be zero
(c) The change in entropy will be negative.
(d) The change in entropy cannot be determined from the information given.

Q. 12. $Cl_2 \rightarrow 2Cl(g)$

Q. 13. $H_2(g)$ at 5.0 atm $\rightarrow$ $H_2(g)$ at 1.0 atm

Q. 14. Sublimation of solid CO_2

Q. 15. $2H_2(g) + O_2(g) \rightarrow 2H_2O(g)$

Q. 16. $PCI_5(g) \leftrightarrow PCI_3(g) + CI_2(g)$

Q. 17. Unknown element X combines with oxygen to form the compound XO_2. If 44.0 grams of element X combine with 8.00 grams of oxygen, what is the atomic mass of element X?

(a) 16amu (b) 44amu (c) 88amu (d) 176amu

Q. 18. When water evaporates at constant pressure, the sign of the change in enthalpy
(a) Is negative
(b) Is negative
(c) Depends on the temperature
(d) Depends on the volume of the container

Q. 19. Which of the following statements is NOT correct?
(a) At constant temperature, the pressure of a certain amount of gas increases with increasing volume.
(b) At constant volume, the pressure of a certain amount of gas increases with increasing temperature.
(c) At constant pressure, the volume of a certain amount of gas increases with increasing temperature.
(d) In dealing with gas laws, the most convenient scale of temperature to use is the Kelvin temperature scale.

Q. 20. Within a period, an increase in atomic number is usually accompanied by
(a) A decrease in atomic radius and an increase in electronegativity
(b) An increase in atomic radius and an increase in electronegativity
(c) A decrease in atomic radius and a decrease in electronegativity
(d) An increase in atomic radius and a decrease in electronegativity

Q. 21. A molecule exhibits sp^3d^2 hybridization in its bonding structure. The most probable geometric shape of this molecule is
(a) Triangular bipyramidal
(b) T - shaped
(c) Octahedral
(d) Linear

Q. 22. What is the proper name of $[Co(NH_3)_5Br]\ Cl_2$
(a) Cobaltpentaamine bromodichloride
(b) Pentaamminecobalt (iii) bromodichloride
(c) Dichlorocobalt (v) bromodichloride
(d) Pentaaminebromocobalt (iii) chloride

Q. 23. Which one of the following does NOT show hydrogen bonding?
(a) Ammonia, NH_3
(b) Hydrazine, N_2H_4
(c) Hydrogen peroxide, H_2O_2
(d) Demethyl ether, CH_3OCH_3

Q. 24. How many moles of solid Ca $(NO_3)_2$ should be added to 450 milliliters of 0.35 M AL $(NO_3)_3$ to increase the concentration of the NO_3^- ion to 1.7 M? (Assume that the volume of the solution remains constant.)
(a) 0.07 mole (b) 0.15 mole (c) 0.29 mole (d) 0.45 mole

Questions 25–29

A(g) + 2B(g)3C(g) → 4D(g) + 5E(g)

$$\text{Rate of formation of E} = \frac{d[E]}{dt} = k[A]^2[B]$$

Q. 25. If one were to double the concentration of B, the rate of the reaction shown above would increase by a factor of

(a) ½ **(b)** 1 **(c)** 2 **(d)** 4

Q. 26. $\frac{-d[B]}{dt}$ is equal to

(a) $\frac{-d[A]}{dt}$ **(b)** $\frac{-d[C]}{dt}$ **(c)** $\frac{-d[D]}{dt}$ **(d)** $\frac{-d[E]}{dt}$

Q. 27. To decrease the rate constant k, one could

(a) Increase [E] **(b)** Decrease [B]

(c) Decrease the temperature **(d)** Increase the volume

Q. 28. If one were to reduce the volume of a container by $\frac{1}{3}$ the rate of the reaction would increase by a factor of

(a) 3 **(b)** 9 **(c)** 16 **(d)** 27

Q. 29. Sulfur trioxide gas dissociates into sulfur dioxide gas and oxygen gas at 1523K. In an experiment, 3.60 moles of sulfur trioxide were placed into an evacuated 3.0 – liter flask. The concentration of sulfur dioxide gas measured at equilibrium was found to be 0.20 M. What is the equilibrium constant, k, for the reaction?

(a) 1.6×10^{-4} **(b)** 1.0×10^{-3} **(c)** 2.0×10^{-3} **(d)** 4.0×10^{-3}

Q. 30. A solution has a pH of 11.0. What is the hydrogen ion concentration?

(a) 1.0×10^{-11} M **(b)** 1.0×10^{-3} M **(c)** 0.0 M **(d)** 1.0×10^{-3} M

Q. 31.

Species	DH_f° *(kj.mole^{-1}) at 298K and 1 atm*	*ΔS° (j.mole^{-1} k^{-1})*
Ba CO_3 (s)	–1170.	100.00
Ba O (s)	–600.	70.00
CO_2 *g*	–400.	200.00

At what temperature does ΔG° become zero for the following reaction:

$$\text{Ba CO}_3\,(s) \rightarrow \text{BaO}(s) + \text{CO}_2\ \gamma$$

(a) 0K **(b)** 1.0×10^1 K **(c)** 1.0×10^2 K **(d)** 1.0 × 103 K

Q. 32. Calculate the rate constant for the radioactive disintegration of an isotope that has a half - life of 6930 years.

(a) 1.00×10^{-5} yr^{-1} **(b)** 1.00×10^{-4} yr^{-1}

(c) 1.00×10^{-3} yr^{-1} **(d)** 1.00×10^{3} yr^{-1}

Questions 33–37

(a) Amide
(b) Amine
(c) Ketone
(d) Thiol
(e) Salt

Q. 33. 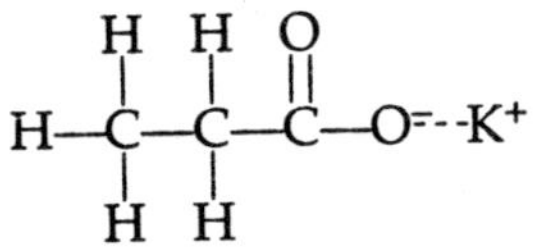

Q. 34. 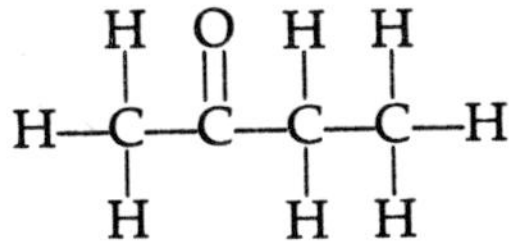

Q. 35. 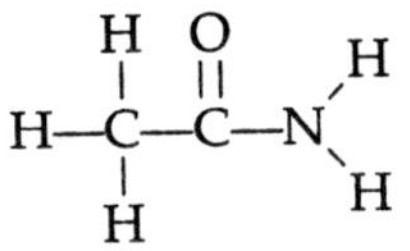

Q. 36. 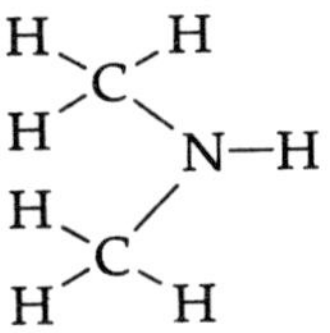

Q. 37. 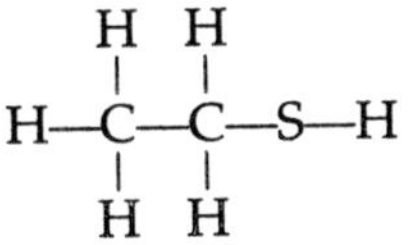

Q. 38. A mining company supplies an ore that is 15.915% chalcocite, Cu_2S, by weight. How many metric tons of ore should be purchased in order to produce 6.0×10^2 metric tons of an alloy containing 12.709% Cu?

(a) 5.0×10^{-1} metric tons **(b)** 6.0×10^{-2} metric tons
(c) 2.0×10^{-2} metric tons **(d)** 3.0×10^{-2} metric tons

Q. 39. At constant temperature and pressure, the heats of formation of H_2O(g), CO_2(g), and C_2H_6 (g) (in K.J.mol^{-1}) are as follows:

Species	<(D)H_f (K.J.mol^{-1})
H_2O (g)	– 250.8
CO_2 (g)	– 392.9
C_2H_6 (g)	– 83.6

If Δ H values are negative for exothermic reactions, what is Δ H (in K.J.mol^{-1}) for 1 mole of C_2H_6 gas to oxidize to carbon dioxide gas and water vapor (same temperature and pressure)?

(a) –5601.2 K.J.mol^{-1} **(b)** –2909.28 K.J.mol^{-1}
(c) –1454.6 K.J.mol^{-1} **(d)** –560.12 K.J.mol^{-1}

Q. 40. Which one of the following is NOT an assumption of the kinetic theory of gases?
(a) Gas particles are negligibly small.
(b) Gas particles undergo a decrease in kinetic energy when passed from a region of high pressure to a region of low pressure.
(c) Gas particles are in constant motion.
(d) Gas particles don't attract each other.

Q. 41. How many electrons can be accommodated in all the atomic orbitals that correspond to the principal quantum number 4?
(a) 2 (b) 8 (c) 18 (d) 32

Q. 42. A linear molecule can have the general formulas AA, AB, OR AB_2. Given a molecule with the general formula AB_2 which one of the following would be the most useful in determining whether the molecule was bent or linear?
(a) Ionization energies (b) Electron affinities
(c) Dipole moments (d) Electronegativities

Q. 43. What is the charge of Zn in Zn $(H_2O)_3(OH)^+$?
(a) 0 (b) + 1 (c)+ 2 (d) +3

Q. 44. A certain organic compound has a vapor pressure of 132 mm Hg at 327K. To determine the vapor pressure of 2.00 moles of the compound at 310K, taking the heat of vaporization for the compound to be 4.33×10^4 j $mole^{-1}$, you would use
(a) The Arrhenius equation
(b) The Clausiss – Clapeyron equation
(c) The combined gas lows
(d) The ideal gas low

Q. 45. What is the molality of a 10.% (by weight) C_6H_2O (MW = 90.) solution?
(a) 0.012 m (b) 0.12 m (c) 1.2 m (d) 12 m

Q. 46. If a 10. cm^3 sample of unknown contains 1 cm^3 of 0.1 M $AlCl_3$, then the concentration of Al^{3+} in the unknown is about?
(a) 0.001 m (b) 0.01 m (c) 0.1 m (d) 1 m

Q. 47. If a reactant concentration is doubled, and the reaction rate increases by a factor of 8, the exponent for that reactant in the rate law should be?
(a) ½ (b) ½ (c) 2 (d) 3

Q. 48. For the reaction

$$2NO(g) + O_2(g) \rightarrow 2NO_2\ (g)$$

which two of the following possible intermediate mechanisms would support this reaction?
(i) $2NO(g) \rightarrow N_2O_2(g)$
(ii) $NO(g) + O_2(g) \rightarrow NO_3(g)$
(iii) $2\ NO_2(g) \rightarrow N_2O_2(g) + O_2(g)$
(iv) $NO_3(g) + NO(g) \rightarrow 2NO_2(g)$
(v) $NO(g) \rightarrow NO(g) + O_2\ (g)$
(a) 1 and 2 (b) 2 and 3 (c) 3 and 4 (d) 2 and 4

Q. 49. The value of K_a for lactic acid, HLac, is 1.5×10^{-5}. What is the value of K_b for the lactate anion, Lac^-?
(a) 1.0 (180) 10^{-14} **(b)** 8.5 (180) 10^{-10} **(c)** 6.7 (180) 10^{-10} **(d)** 8.5 (180) 10^{-10}

Q. 50. How many grams of NaOH are required to neutralize 700 ml of 3.0 N HCI?
(a) 2.1 grams **(b)** 21 grams **(c)** 42 grams **(d)** 84 grams

Q. 51. Solid calcium carbonate decomposes to produce solid calcium oxide and carbon dioxide gas. The value of $\Delta G°$for this reaction is 130.24 $K.J.mole^{-1}$. Calculate ΔG at 373K for this reaction if the pressure of the carbon dioxide gas is 1.00 atm.
(a) – 998.56 $K.J.mol^{-1}$ **(b)** – 604.2 $K.J.mol^{-1}$
(c) – 56.31 $K.J.mol^{-1}$ **(d)** –130.24 $K.J.mol^{-1}$

Q. 52. Which of the following choices represents ${}^{239}_{94}Pu$ producing a positron?

(a) ${}^{239}_{94}Pu \rightarrow {}^{235}_{94}Pu + {}^{4}_{2}H$

(b) ${}^{239}_{94}Pu \rightarrow {}^{0}_{-1}e + {}^{239}_{93}Np$

(c) ${}^{239}_{94}Pu + {}^{0}_{-1}e \rightarrow {}^{239}_{99}Np$

(d) ${}^{239}_{94}Pu + {}^{0}_{1}e \rightarrow {}^{239}_{93}Np$

Q. 53. $Ni(s) | Ni^{2}(aq) || Ag^{+}(aq) | Ag(s)$

$Ni^{2+} + 2e^- \rightarrow Ni(s) E^{\circ}_{mol} = -0.25$ volts
$Ag^+ + e^- \rightarrow Ag(s) E^{\circ}_{mol} = 0.80$ volt
Which of the following statements is true of the above reaction?
(a) The reaction is spontaneous,
$E° = 1.05$ volts.
(b) The reaction is nonspontaneous,
$E° = -1.05$ volts.
(c) The reaction is spontaneous,
$E° = -1.05$ volts.
(d) The reaction is spontaneous,
$E° = 0.55$ volts.

Q. 54. An excess of $S_8(s)$ is heated with a metallic element until the metal reacts completely. All excess sulfur is combusted to a gaseous compound and escapes from the crucible. Given that information that follows, determine the most probable formula for the residue.
mass of crucible, lid and metal = 55.00 grams
mass of crucible and lid = 41.00 grams mass of crucible, lid, and residue = 62.00 grams
(a) CuS **(b)** Cu_2S **(c)** FeS **(d)** Fe_2S_3

Q. 55. When 3.00 grams of a certain metal are completely oxidized, 3.80 grams of its oxide are produced. The specific heat of the metal is 0.052 cal $g^{-1} \cdot K^{-1}$. What is the approximate atomic weight of this metal?
(a) 35 g $mole^{-1}$ **(b)** 65 g $mole^{-1}$ **(c)** 124 g $mole^{-1}$ **(d)** 150 g $mole^{-1}$

Q. 56. The density of gas is directly proportional to its
(a) pressure **(b)** volume **(c)** kinetic energy **(d)** temperature

Q. 57. The valence electron configuration of element A is $3s^2\ 3p^1$ and that of B is $3s^2 3p^4$. What is the probable empirical formula for a compound of thew two elements?
(a) A^2B **(b)** AB_2 **(c)** A_3B_2 **(d)** A_2B_3

Q. 58. Which formula correctly represents the diamminediaquadibromo chromium (III) ion?
(a) $[Cr(H_2O)_2(NH_3)_2Br_2]^+$ **(b)** $[(NH_3)_2(H_2O)Br_2Cr]^{3+}$
(c) $[Cr(H_2O)_2(NH_3)_2Br_2]^{3+}$ **(d)** $[(NH_3)_2(H_2O)_2Br_2 + Cr]^+$

Q. 59. According to Raoult's law, which statement is false?
(a) The vapor pressure of a solvent over a solution is less than that of the pure solvent.
(b) Ionic solids ionize in water, increasing the effect of all colligative properties.
(c) The vapor pressure of a pure solvent when measured over a solution decreases as the mole fraction increases.
(d) The solubility of a gas increases as the temperature decreases.

Q. 60. Calculate the volume a 36.45% solution of hydrochloric acid (density = 1.50 g/mL) required to prepare 9.0 liters of a 5.0-molar solution.
(a) 0.5 liter **(b)** 1.0 liter **(c)** 2.0 liter **(d)** 3.0 liter

Q. 61. Dinitrogen pentoxide decomposes according to the following balanced equation:

$$N_2O_5(g) \rightarrow 2\ NO_2(g) + \frac{1}{2}O_2(g)$$

The rate of decomposition was found to be 0.80 mole · liter^{-1} · sec^{-1} at a given concentration and temperature. What would the rate be for the formation of oxygen gas under the same conditions?
(a) 0.20 mole · liter^{-1} · sec^{-1} **(b)** 0.40 mole · liter^{-1} · sec^{-1}
(c) 0.80 mole · liter^{-1} · sec^{-1} **(d)** 1.60 mole · liter^{-1} · sec^{-1}

Q. 62. The K_{sp} of lead (II) chloride is 2.4×10^{-4}. What conclusion can be made about the concentration of [Cl$^-$] in a solution of lead chloride if [Pb^{2+}] = 1.0 M?
(a) [Cl$^-$] can have any value. **(b)** [Cl$^-$] cannot be greater than $\frac{1}{2}K_{sp}$.
(c) [Cl$^-$] cannot be less than $\frac{1}{2}K_{sp}$. **(d)** [Cl$^-$] cannot be equal to $\frac{1}{2}K_{sp}$.

Direction: For Question 63–65, use the following information:
25.0 mL of a sample of vinegar (a solution of $HC_2H_3O_2$, MW = 60.00 g · mol^{-1}) is neutralized by 50.00 mL of a 0.50 N NaOH solution.

Q. 63. What is the normality of the acid?
(a) 0.25 N **(b)** 0.50 N **(c)** 0.75 N **(d)** 1.0 N

Q. 64. Calculate the number of grams of acetic acid per of the vinegar.
(a) 5.0 grams **(b)** 25.0 grams **(c)** 30.0 grams **(d)** 60.0 grams

Q. 65. Calculate the weight percentage of acetic acid in the vinegar. The vinegar has a density of 1.0 g/mL.
(a) 1.0% **(b)** 1.5% **(c)** 2.5% **(d)** 6.0%

Q. 66. When a solid melts, which of the following is true?

(a) $\Delta H > 0, \Delta S > 0$ (b) $\Delta H > 0, \Delta S > 0$

(c) $\Delta H > 0, \Delta S > 0$ (d) $\Delta H > 0, \Delta S > 0$

Q. 67. For the following reaction

$$Zn(s) + 2\ Ag+(aq) \rightarrow Zn^{2+}(aq) + 2\ Ag(s)$$

the standard voltage E°_{cell} has bneen calculated to be 1.56 volts. To decrease the voltage from the cell to 1.00 volt, one could

(a) increase the siza of the zinc electrode

(b) reduce the coefficients of the reaction so that it reads

$$1/2\ Zn(s) + Ag^{+}(aq) \rightarrow 1/2\ Zn^{2+}(aq) + Ag(s)$$

(c) decrease the concentration of the silver ion in solution

(d) increase the concentration of the silver ion in solution

Q. 68. A radioactive isotope has a half-life of 6.93 years and decays by beta emission. Determine the approximate fraction of the sample that is left undecayed at the end of 11.5 years.

(a) 1% (b) 5% (c) 30% (d) 75%

Q. 69. The solubility product constant at 298K for AgCl is $1.6 \times 10^{-10}\ mol^2 \cdot L^{-2}$ and that for AgI is $8.0 \times 10^{-17}\ mol^2 \cdot L^{-2}$. Determine the equilibrium constant for the reaction of silver chloride with $I^-(aq)$.

(a) $1.3 \times 10^{-26}\ mol^2 \cdot L^{-2}$ (b) $5.0 \times 10^{-7}\ mol^2 \cdot L^{-2}$

(c) $1.0 \times 10^{3}\ mol^2 \cdot L^{-2}$ (d) $2.0 \times 10^{6}\ mol^2 \cdot L^{-2}$

Q. 70.

Species	*Bond Energy K.J.mol^{-1}*
F–F	137.9
H–H	430.5
H–F	564.3

Calculate the value of ΔH for the following reaction:

$$H_2(g) + F_2(g) \rightarrow 2\ HF(g)$$

(a) -1697.08 K.J.mol^{-1} (b) -1337.6 K.J.mol^{-1}

(c) -1132.78 K.J.mol^{-1} (d) -560.2 K.J.mol^{-1}

Q. 71. What happens to the velocities of different molecules as the temperature of a gas increases?

(a) The velocities of all component molecules increase equally.

(b) The velocity range among different molecules at higher temperature is smaller than that at lower temperatures.

(c) The effect on the velocities of the molecules depends on whether the pressure remains constant.

(d) The velocity range among different molecules at higher temperature is wider than the range at lower temperatures.

Q. 72. For the isoelectronic series S^{2-}, Cl^-, Ar, K^{0+}, Sc^{3+}, which species requires the least energy to remove an outer electron?

(a) S^{2-} (b) Cl^- (c) Ar (d) K^+

Q. 73. The silver ion in the complex $[Ag(CN_2)]^-$ has a coordination number of

(a) 2 (b) 3 (c) 4 (d) 5

Q. 74. Cesium (atomic radius = 0.255 nm) crystallizes with a body-centered cubic unit cell. What is the approximate length of a side of the cell? ($\sqrt{3}$ = 1.73)

(a) 0.4 nm (b) 0.5 nm (c) 0.6 nm (d) 0.8 nm

Q. 75. How many grams of H_3PO_4 are required to make 10.0 mL of a 0.100 N H_3PO_4 solution (Assume complete or 100% ionization?

(a) 0.0164 g (b) 0.164 g (c) 0.327 g (d) 0.654 g

Chapter 15

TEST PAPER 1

USEFUL DATA

mass of electron = **9.108** $\times 10^{-31}$ kg

Avogadro number = 6.023×10^{23}

Plank's constant, $h = 6.626 \times 10^{-34}$ Js

Gas constant R = 0.082 L atm K^{-1} mol^{-1} = 8.314 J K^{-1} mol^{-1}

Atomic Numbers: Cr = 24, Mn = 25, Fe = 26,

Co = 27, Pt = 78, Be 4

B = 5; C = 6 N = 7; O = 8; S = 16; Xe = 54

Li = 3 Be = 4, B = 5, C = 6

Atomic Mass: H = 1.008; O = 16.00; S = 32.07; Se = 78.96; Te = 1276

PROBLEMS

Q. 1. How many moles of electron weight one kilogram?

(a) 6.023×10^{23} (b) $\frac{1}{9.108} \times 10^{31}$

(c) $\frac{6.023}{9.108} \times 10^{54}$ (d) $\frac{1}{9.108 \times 6.023} \times 10^{8}$

Q. 2. Which of the following volume (V) temperature (T) plots represents the behaviour of one mole of an ideal gas at one atmospheric pressure?

(a)
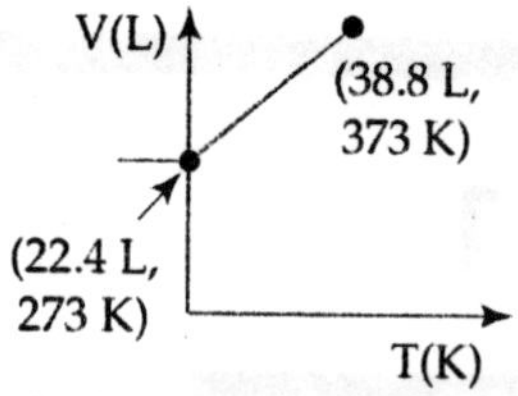

V(L)
(38.8 L,
373 K)
(22.4 L,
273 K)
T(K)

(b)
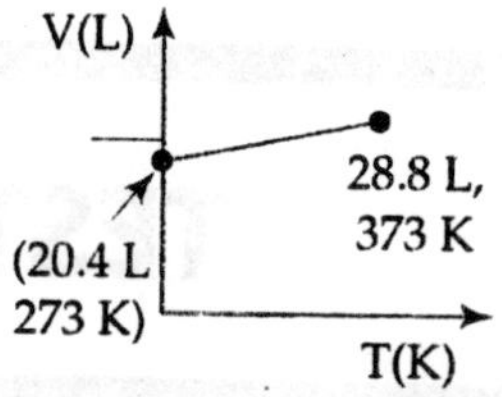

V(L)
28.8 L,
373 K
(20.4 L
273 K)
T(K)

(c)
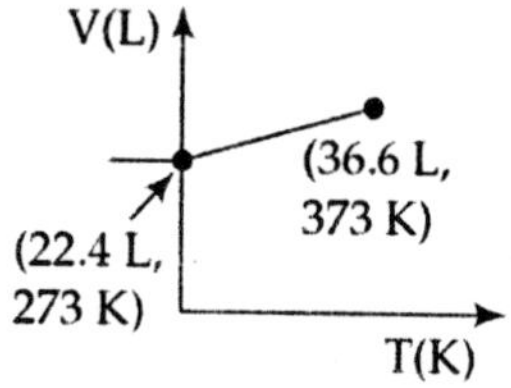

V(L)
(36.6 L,
373 K)
(22.4 L,
273 K)
T(K)

(d)
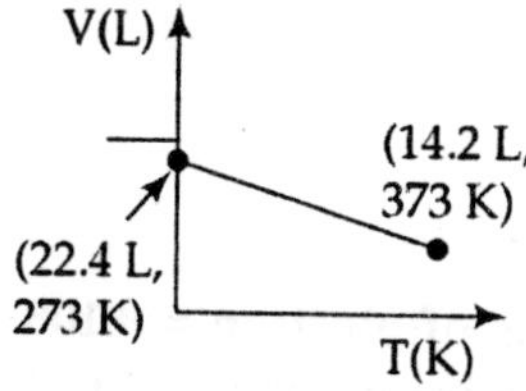

V(L)
(14.2 L,
373 K)
(22.4 L,
273 K)
T(K)

Q. 3. A substance $A_X B_Y$ crystallizers in a face central cube (F_{CC}) lattree in which atoms '*A*' occupy each corner of the cube and atoms '*B*' occupy the centres of each face of the cubic Identify the correct composition of the substance $A_X B_Y$

(a) AB_3 (b) A_4B_3

(c) A_3B (d) Composition cannot be specified

Q. 4. The order of reactivity of Phenyl Manesium Bromide with the following compounds is

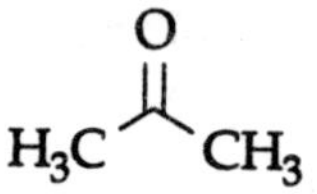

O
H3C
CH3

(I)

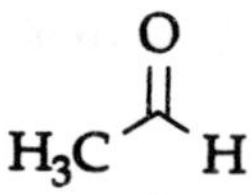

O
H3C
H

(II)

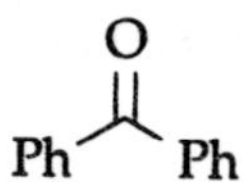

O
Ph
Ph

(III)

(a) (II) > (III) > (I) (b) (I) > (III)> (II)

(c) (II) > (I) > (III) (d) All react with the same rate

Q. 5. In the given conformation C_2 is rotated about C_2–C_3 bond anticlockwise by an angle of 120° then the conformation obtained is

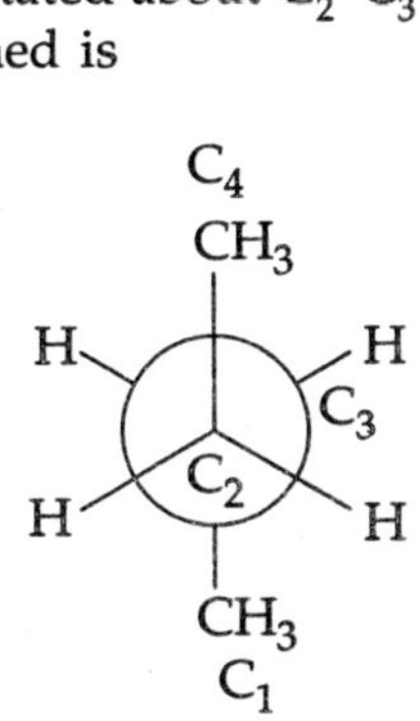
C4
CH3
H
H
C3
C2
H
H
CH3
C1

(a) Fully eclipsed conformation
(b) Partially eclipsed conformation
(c) Gauche conformation
(d) staggered conformation

Q. 6. One mole of a non-ideal gas undergoes a change of state (2.0 atm, 3.0 L, 95 K) → (4.0 atm, 5.0 L, 245 K) with a change in internal energy, ΔU = 3.0.0 L atm. The change in enthalpy (ΔH) of the process in L atm is:
(a) 40.0 **(b)** 42.3
(c) 44.0 **(d)** not defined, because pressure is not constant

Q. 7. Benzamide on treatment with $POCl_3$ gives
(a) Aniline **(b)** Banzonitrile
(c) Chlorobenene **(d)** Benzyl amine

Q. 8. 2-phenyl propene on acidic hydration gives
(a) 2-phenyl-2-propanol **(b)** 2-phenyl-1-propanol
(c) 3-phenyl-1-propanol **(d)** 1-phenyl-2-propanol

Q. 9. 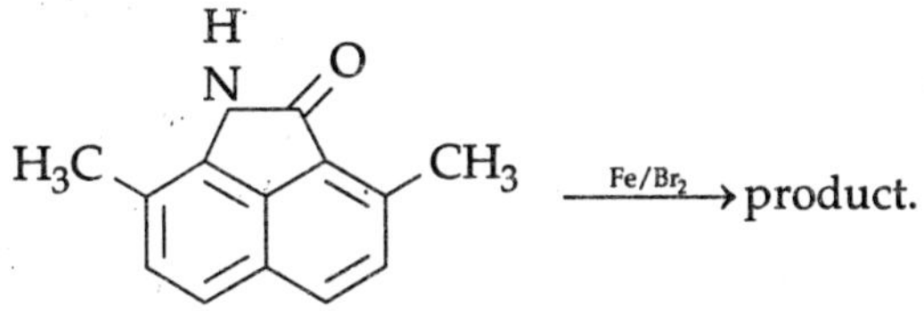
H
N
O
H3C
CH3
Fe/Br2
product.

On mono bromination of this the compound is

(a) 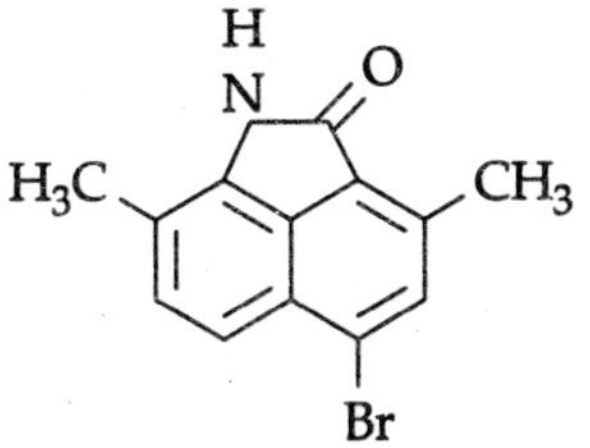
H
N
O
H3C
CH3
Br

(b) 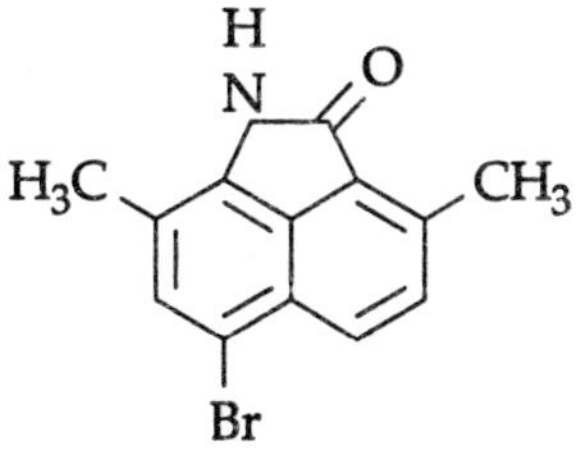
H
N
O
H3C
CH3
Br

(c) 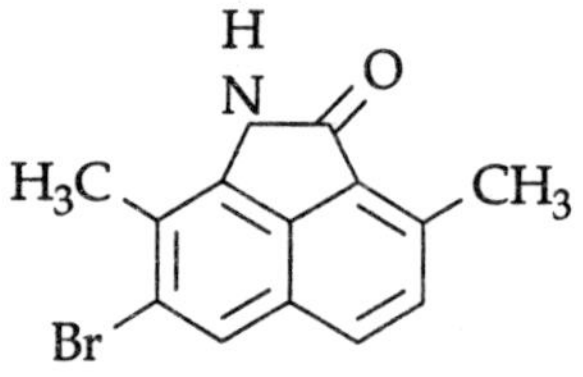

(d) 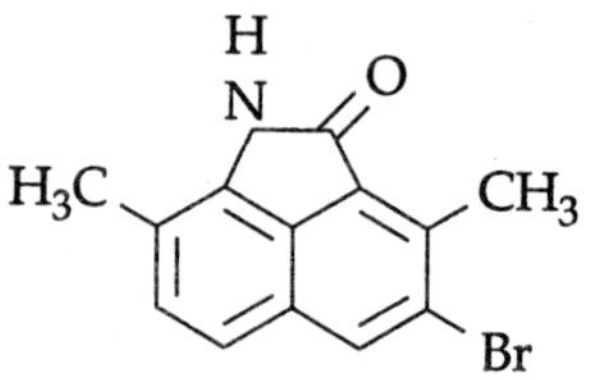

Q. 10. Standard electrode potential data are useful for understanding the suitability of an oxidant in a redox titration. Some half cell reactions and their standard potentials are given below:

$MnO_4^-(aq.) + 8H^+(aq.) + 5e^- \rightarrow (aq) + 4H_2O(l)$ E° = 1.51 V

$Cr_2O_7^{2-}(aq) + 14H^+(aq) + 6e^- \rightarrow Cr^{5+}(aq) + 7H_2O(l)$ E° = 1.38 V

$Fe^{3+}(aq) + e^- \rightarrow Fe^{2+}(aq)$ E° = 0.77 V

$Cl_2(g) + 2e^- \rightarrow 2Cl^-$ (aq) E° = 1.40 V

Identify the only Incorrect statement regarding the quantitative a estimation of aq. $Fe(NO_3)_2$

(a) MnO_4^- can be used in aqueous HCl

(b) $Cr_2O_7^{2-}$ can be used in aqueous HCl

(c) MnO_4^- can be used in aqueous H_2SO_4

(d) $Cr_2O_7^{2-}$ can be used in aqueous H_2SO_4

Q. 11. Zn | Zn^{2+} (a = 0.1 M) || Fe^{2+} (a = 0.01M) | Fe. The emf of the above cell is 0.2905 V. equilibrium constant for the cell reactions is

(a) $10^{0.32/0.0591}$ **(b)** $10^{0.32/0.0295}$ **(c)** $10^{0.26/0.0295}$ **(d)** $e^{0.3/0.295}$

Q. 12. $(NH_4)_2Cr_2O_7$ on heating gives a gas which is also given by

(a) Heating NH_4NO_2 **(b)** Heating NH_4NO_3

(c) $Mg_3N_2 + H_2O$ **(d)** Na (comp.) + H_2O_2

Q. 13. HX is a weak acid ($K_a = 10^{-5}$). It forms a salt NaX (0.1M) on reacting with caustic soda. The degree of hydrolysis of NaX is

(a) 0.01% **(b)** 0.0001% **(c)** 0.1% **(d)** 0.5%

Q. 14. For a monoatomic gas kinetic energy = E. The reaction with rms velocity is

(a) $u = \left(\frac{2E}{m}\right)^{1/2}$ **(b)** $u = \left(\frac{3E}{2m}\right)^{1/2}$ **(c)** $u = \left(\frac{E}{2m}\right)^{1/2}$ **(d)** $u = \left(\frac{E}{3m}\right)^{1/2}$

Q. 15. Spin only magnetic moment of the compound Hg $[Co(SCN)_4]$ is

(a) $\sqrt{3}$ **(b)** $\sqrt{15}$ **(c)** $\sqrt{24}$ **(d)** $\sqrt{8}$

Q. 16. Consider the following equilibrium in a closed container:

$$N_2O_4(g) \rightleftarrows 2NO_2(g)$$

As a fixed temperature, the volume of the reaction container is halved. For this change, which of the following statements holds true regarding the equilibrium constant (K_p) and degree of dissociation (α)?

(a) neither K_p nor α changes **(b)** both K_p and α change

(c) K_p changes, but α does not change **(d)** K_p does not change, but α changes

Q. 17. The compound having tetrahedral geometry is

(a) $[Ni(CN)_4]^{2-}$ (b) $[Pd(CN)_4]^{2-}$ (c) $[PdCl_4]^{2-}$ (d) $[NiCl_4]^{2-}$

Q. 18. Consider the chemical reaction, $N_2(g) + 3H_2(g) \rightarrow 2NH_3(g)$. The rate of this reaction can be expressed in terms of time derivative of concentration of $N_2(g)$. $H_2(g)$ or $NH_3(g)$. Identify the correct relationship amongst the rate expressions

(a) Rate $= -d[N_2]/dt = 1/3d[H_2]/dt = 1/2d\ [NH_3]/dt$

(b) Rate $= -d[N_2]/dt = -3d[H_2]/dt = 2d[NH_3]/dt$

(c) Rate $= d[N_2]/dt = 1/3d[H_2]/dt = 1/2d\ NH_3]/dt$

(d) Rate $= d[N_2]/dt = -d[H_2]/dt = d[NH_3]/dt$

Q. 19. A sodium salt of an unknown anion when treated with $MgCl_2$ gives white precipitate only on heating. The anion is

(a) SO_4^{2-} (b) HCO_3^- (c) CO_3^{2-} (d) NO_3^-

Q. 20. Which Hydrogen like species will have radius as that of Bohr orbit of Hydrogen atom?

(a) $n = 2$, Li^{+2} (b) $n = 2$, Be^{3+} (c) $n = 2$, He^+ (d) $n = 2$, Li^{2+}

Q. 21.

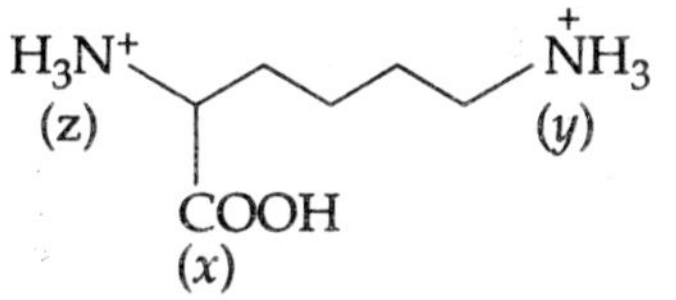

H_3N^+
(z)
$\overset{+}{N}H_3$
(y)
COOH
(x)

Arrange in order of increasing acidic strength

(a) $X > Z > Y$ (b) $Z < X > Y$ (c) $X > Y > Z$ (d) $Z > X > Y$

Q. 22. A 0.004 M solution of N_2SO_4 is isotonic with a 0.010 M solution of glucose at same temperature. The apparent degree of dissociation of N_2SO_4 is

(a) 25% (b) 50% (c) 75% (d) 85%

Q. 23. Identify the correct order of solubility of Na_2S, CuS and ZnS in aqueous medium

(a) CuS > ZnS > Na_2S (b) ZnS > Na_2S > CuS

(c) Na_2S > CuS > ZnS (d) Na_2S > ZnS > CuS

Q. 24. Which of the following progress is used in the extractive metallurgy of magnesium?

(a) fused salt electrolysis (b) self reduction

(c) aqueous solution electrolysis (d) thermite reduction

Q. 25. A gas '*X*' is passed through water to form a saturated solution. The aqueous solution on treatment with silver nitrate gives a while precipitate. The saturated aqueous solution also dissolves. The saturated aqueous solution also dissolves magnesium ribbon with evolution of a colourless gas '*Y*'. Identify '*X*' and '*Y*'

(a) $X = CO_2$, $Y = Cl_2$ (b) $X = Cl_2$, $Y = CO_2$

(c) $X = Cl_2$, $Y = H_2$ (d) $X = H_2$, $Y = Cl_2$

Q. 26. $\Delta H_{vap} = 30$ kJ/mole and $\Delta S_{vap} = 75$ J mol^{-1} K^{-1}. The temperature of Vapour at one atmosphere is?

(a) 400 K (b) 350 K (c) 298 K (d) 250 K

Q. 27. 2 mol of an ideal gas expanded isothermally & reversibly from 1 litre to 10 litres at 300 K. What is the enthalpy change?

(a) 4.98 kJ (b) 11.47 kJ (c) –11.47 kJ (d) 0 kJ

Q. 28. 2-hexyne gives trans-2-hyxene on treatment with

(a) $Li - NH_3/C_2H_5\,OH$ (b) $Pd/BaSO_4$

(c) $LiAl\,H_4$ (d) Pt/H_2

Q. 29. The reaction, $X \rightarrow$ Product follows first order kinetics. In 40 minutes the concentration of X changes from 0.1 M to 0.025 M. Then the rate of reaction when concentration of X is 0.01 M.

(a) 1.73×10^{-4} M min^{-1} (b) 3.47×10^{-5} M min^{-1}

(c) 3.47×10^{-4} M min^{-1} (d) 1.73×10^{-3} M min^{-1}

Q. 30. When I^- is oxidized by MnO_4^- in alkaline medium. I^- converts into

(a) IO_3^- (b) I_2 (c) IO_4^- (d) IO^-

Q. 31. How many chiral compounds are possible on mono chlorination of 2-methyl butane?

(a) 2 (b) 4 (c) 6 (d) 8

Q. 32. The chemical composition of 'slag' formed during the smelting process in the extraction of copper is

(a) $Cu_2O + FeS$ (b) $FeSiO_3$ (c) $CuFeS_2$ (d) $Cu_2S + FeO$

Q. 33. The common features among the species CN^-, CO and NO^+ are

(a) bond order three and isoelectronic

(b) bond order three and weak field ligands

(c) bond order two and π-acceptors

(d) isoelectronic and weak field ligands

Q. 34. Which of the following has O—O—linkage

(a) $H_2S_2O_6$ (b) $H_2S_2O_8$ (c) $H_2S_2O_3$ (d) $H_2S_4O_6$

TEST PAPER 2

Q. 1. The number of S—S bonds in sulphur trioxide trimer (S_3O_9) is

(a) three (b) two (c) one (d) zero

Q. 2. A aqueous solution of 6.3 g oxalic acid dihydrate is made up to 250 ml. The volume of 0.1 N NaOH required to completely neutralize 10 ml of this solution is

(a) 40 ml (b) 20 ml (c) 10 ml (d) 4 ml

Q. 3. In a solid 'AB' having the NaCl structure, '*A*' atoms occupy the corners of the cubic unit cell.

If all the face -centered atoms along one of the axes are removed, then the resultant stoichiometry of the solid is

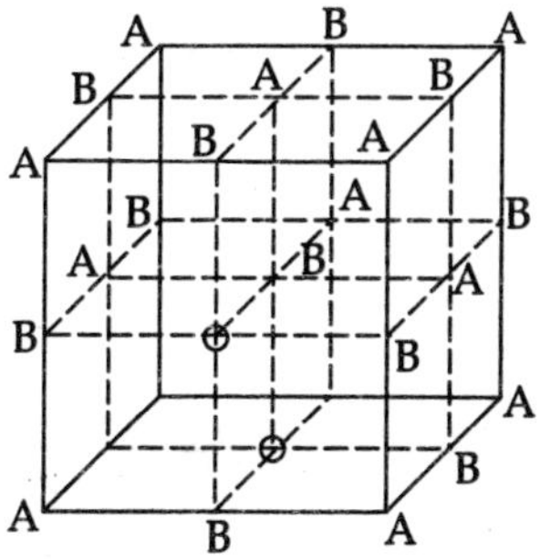

(a) AB_2 (b) A_2B (c) A_4B_3 (d) A_3B_4

Q. 4. The set with correct order of acidity is

(a) $HClO < HClO_2 < HClO_3 < HClO_4$
(b) $HClO_4 < HClO_3 < HClO_2 < HClO$
(c) $HClO < HClO_4 < HClO_3 < HClO_2$
(d) $HClO_4 < HClO_2 < HClO_3 < HClO$

Q. 5. 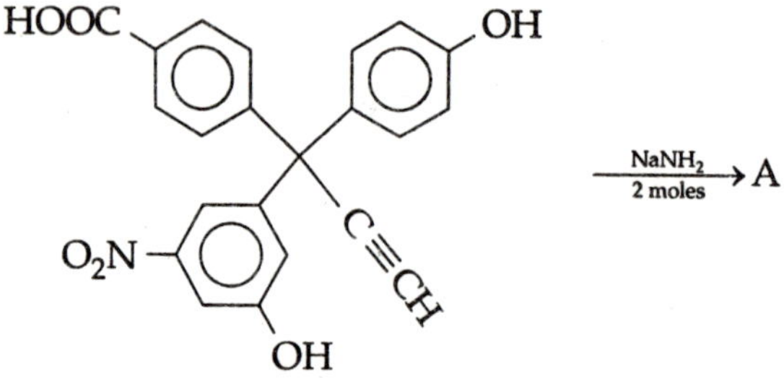
HOOC
OH
O_2N
OH
C≡CH
$NaNH_2$
2 moles
A

The product A will be:

(a) 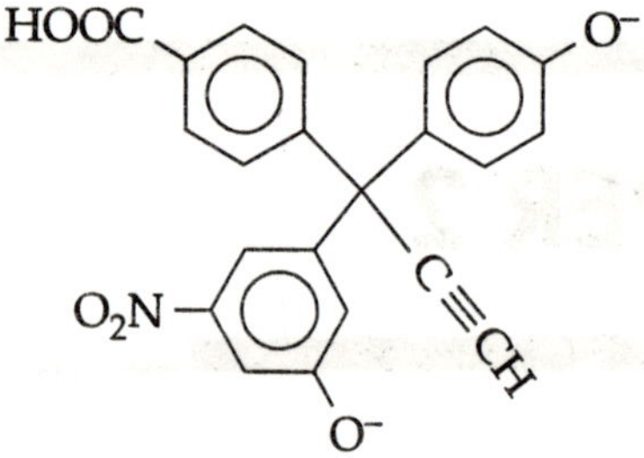
HOOC
O^-
O_2N
O^-
C≡CH

(b) 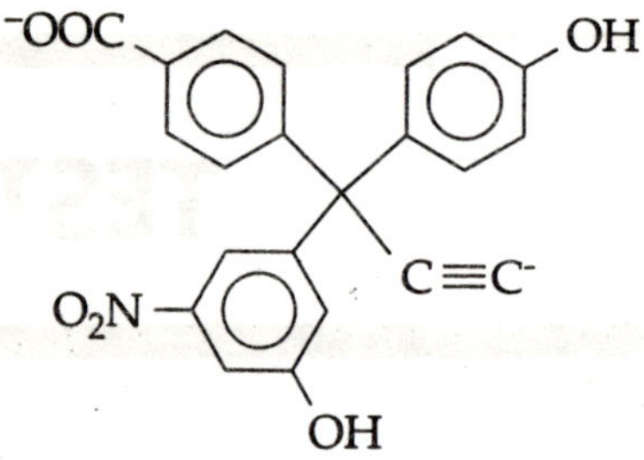
^-OOC
OH
C≡C^-
O_2N
OH

(c) 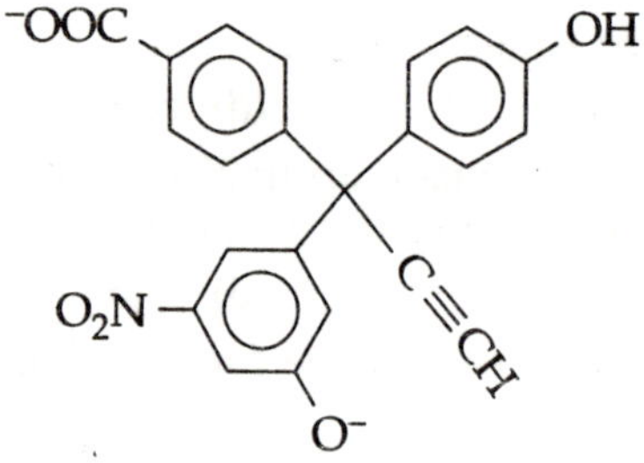
^-OOC
OH
O_2N
C≡CH
O^-

(d) 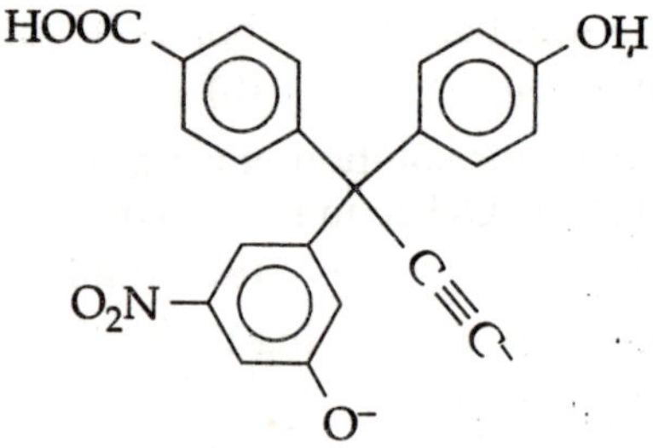
HOOC
OH
O_2N
C≡C^-
O^-

Q. 6. Given, [X] + H_2SO_4 → [Y] a colourless gas with irritating smell
[Y] + $K_2Cr_2O_2$ + H_2SO_4 → green solution. ∴ [X] and [Y] is:
(a) SO_3^{2-}, SO_2 **(b)** Cl^-, HCl **(c)** S^{2-}, H_2S **(d)** CO_3^{2-}, CO_2

Q. 7. An $S_N{}^2$ reaction at an asymmetric carbon of a compound always gives
(a) all enaniomer of the substrate
(b) a product with opposite optical rotation
(c) a mixture of diastereomers
(d) a single stereoisomer

Q. 8. A mixture of benzaldehyde and formaldehyde on heating with aqueous NaOH solution gives
(a) benzyl alcohol and sodium formate
(b) sodium benzoate and methyl alcohol
(c) sodium benzoate and sodium formate
(d) benzyl alcohol and methyl alcohol

Q. 9. The complex ion which has no *'d'* electrons in the central metal atoms is
(a) $[MnO_4]^-$ **(b)** $[Co(NH_3)_6]^{3-}$ **(c)** $[Fe(CN)_6]^3$ **(d)** $[Cr(H_2O)_6]^3$

Q. 10. The set representing the correct order of first ionization potential is

(a) $K > Na > Li$ **(b)** $Be > Mg > Ca$ **(c)** $B > C > N$ **(d)** $Ge > Si > C$

Q. 11. The correct order of hybridization of the central atom in the following species NH_3, $[PtCl_4]^{2-}$, PCl_5 and BCl_3 is

(a) dsp^2, dsp^3, sp^2 and sp^3 **(b)** sp^3, dsp^2, dsp^3, sp^2

(c) dsp^2, sp^2, sp^3, dsp^3 **(d)** dsp^2, sp^3, sp^2, dsp^3

Q. 12. Which of the following are isoelectronic and isostructural? NO_3^-, CO_3^{2-}, ClO_3^-, SO_3

(a) NO_3^-, CO_3^{2-} **(b)** SO_3, NO_3^- **(c)** ClO_3^-, CO_3^{2-} **(d)** CO_3^{2-}, SO_3

Q. 13. Which of the following represents the given mode of hybridization $sp^2 - sp^2 - sp - sp$ from left to right?

(a) $CH_2{=}CH{-}C{\equiv}N$ **(b)** $CH_2{=}C{=}C{=}CH_2$

(c) $CH_2{=}CH{-}CH{=}CH_2$ **(d)** $CH_2{=}C{=}CH{-}CH_3$

Q. 14. 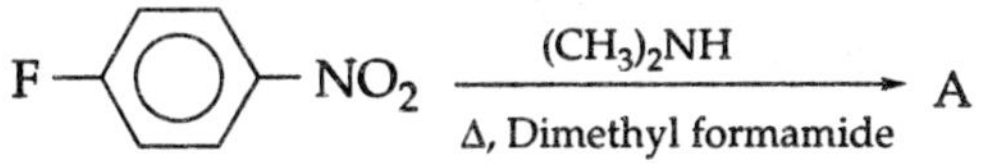

$$A \xrightarrow[\text{(ii) } H_2/Pt]{\text{(i) } NaNO_2/HCl \ (273\text{–}278 \ K)} B$$

Thus, B is

(a) 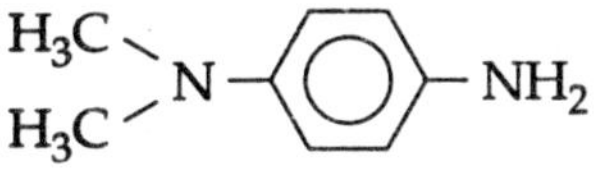 **(b)** 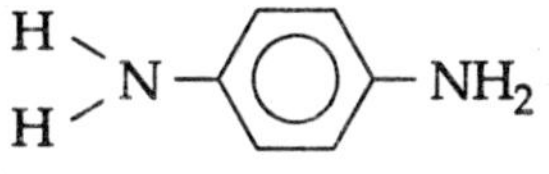

(c)  **(d)** 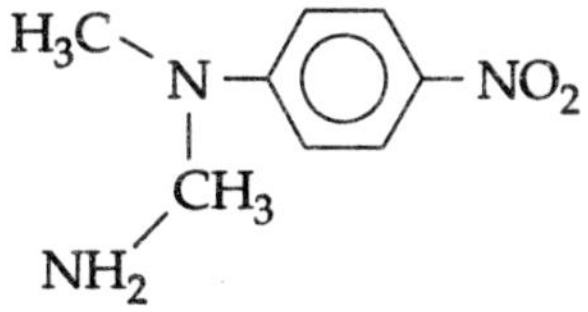

Q. 15. The product of acid hydrolysis of P and Q can be distinguished by

P = 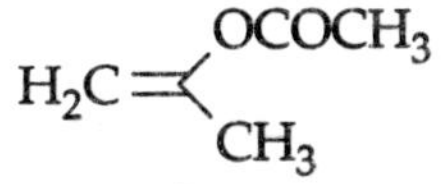 Q = 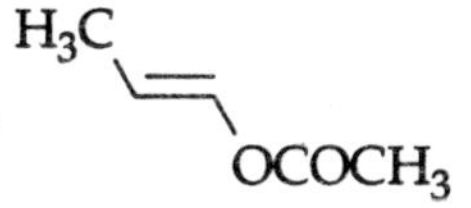

(a) Lucas Regent **(b)** 2, 4—DNP

(c) Fehling's Solution **(d)** $NaHSO_3$

Q. 16. 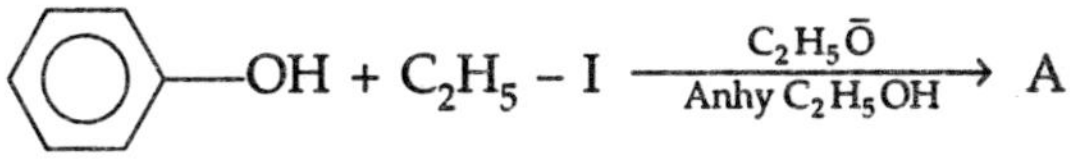

A, in the equation is :

(a) 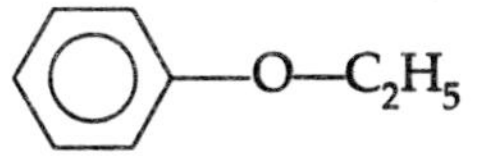 **(b)** $C_6H_5{-}O{-}C_6H_5$

(c) $C_2H_5{-}O{-}C_2H_5$ **(d)** C_6H_5I

Q. 17. H_3BO_3 is

(a) Monobasic and weak Lewis acid
(b) Monobasic and weak Bronsted acid
(c) Monobasic and strong Lewis acid
(d) Tribasic and weak Bronsted acid

Q. 18. In a first order reaction the concentration of reactant decreases from 800 mol/dm^3 to 50 mol/dm^3 is 2×10^4 sec. The rate constant of reaction in sec^{-1} is:

(a) 2×10^4 **(b)** 3.45×10^{-5} **(c)** 1.386×10^{-4} **(d)** 2×10^{-4}

Q. 19. 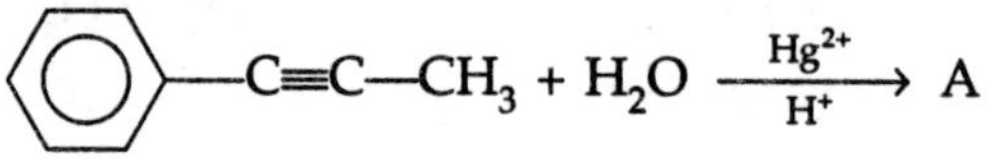
C≡C—CH3 + H2O
Hg2+
H+
A

Thus, A is:

(a) 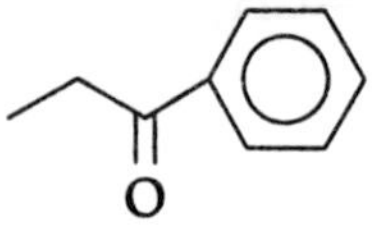
O

(b) 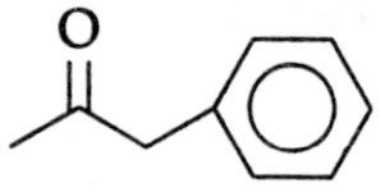
O

(c) 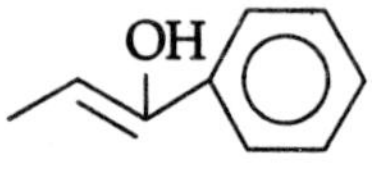
OH

(d) 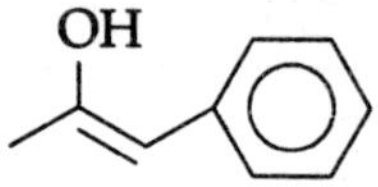
OH

Q. 20. Ethyl ester $\xrightarrow[\text{excess}]{CH_3MgBr}$ P. The product P will be

(a) 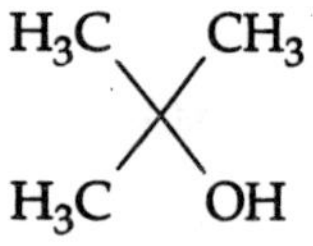
H3C CH3
H3C OH

(b) 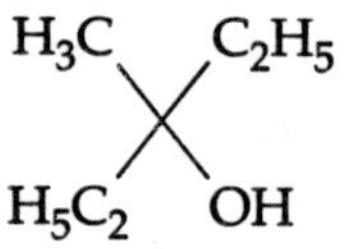
H3C C2H5
H5C2 OH

(c) 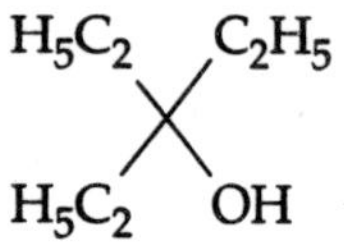
H5C2 C2H5
H5C2 OH

(d) 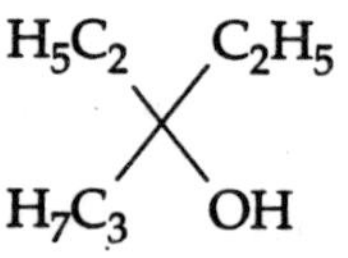
H5C2 C2H5
H7C3 OH

Q. 21. A solution which is 10^{-3} M each in Mn^{2+}, Fe^{2+}, Zn^{2+} and Hg^{2+} is treated, with 10^{-6} M sulphide ion. If K_{sp} of MnS, FeS, ZnS and HgS are 10^{-15}, 10^{-23}, 10^{-20} and 10^{-54} respectively, which one will precipitate first?

(a) FeS **(b)** MgS **(c)** HgS **(d)** ZnS

Q. 22. In the process of extraction of gold;

Roasted gold ore $CN^- + H_2O \xrightarrow{O_2} [X] + OH^-$

$[X] + Zn \rightarrow [Y] + Au$

Identify the complexes [X] and [Y]

(a) $X = [Au(CN)_2]^-$, $Y = [Zn(CN)_4]^{2-}$
(b) $X = [Au(CN)_4]^{3-}$, $Y = [Zn(CN)_4]^{2-}$

(c) $X = [Au(CN)_2]^- Y = Zn(CN)_6]^{2-}$
(d) $X= [Au(CN)_4]^-$, $Y = [Zn(CN)_4]^{2-}$

Q. 23. 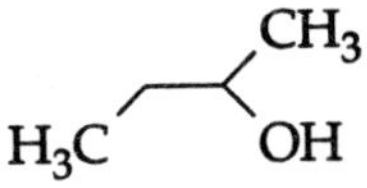

$\xrightarrow[-H_2O]{H^+} Z \xrightarrow[CCL_4]{Br_2} C_4H_8Br_2$

5 such products are possible

How many structures of Z is possible

(a) 2 (b) 3 (c) 5 (d) 6

Q. 24. Consider the following reaction:

$H_3C-CH(D)-CH(CH_3)-CH_3 + Br \rightarrow$ 'X' + HBr

Identify the structure of the major product 'X'

(a) $H_3C-CH(D)-CH(CH_3)-CH_2$

(b) $H_3C-CH(D)-\dot{C}H(CH_3)-CH_3$

(c) $H_2C-\dot{C}H(D)-CH(CH_3)-CH_3$

(d) $H_3C-\dot{C}-CH(CH_3)-CH_3$

Q. 25. Identify a reagant from the following list which can easily distinguish between 1-butyne and 2-butyne

(a) bromine, CCl_4
(b) H_2, lindlar catalyst
(c) dilute H_2SO_4, $HgSO_4$
(d) ammonical Cu_2Cl_2 solution

Q. 26. Identify the correct order of reactivity in electrophilic substitution reactions of the following compounds:

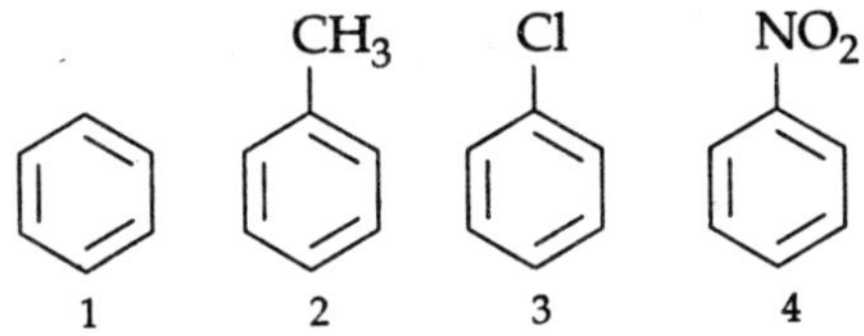

1 2 3 4

(a) 1 > 2 > 3 > 4 (b) 4 > 3 > 2 > 1 (c) 2 > 1 > 3 > 4 (d) 2 > 3 > 1 > 4

Q. 27. Compound 'A' (molecular formula C_3H_8O) is treated with acidified potassium dichromate to form a product 'B' (molecular formula C_3H_6O) 'B' forms a shining silver mirror on warming with ammonical silver nitrate. 'B' when treated with an aqueous solution of $H_2NCONHNH_2$ HCl and sodium acetate gives a product 'C'.

(a) $CH_3CH_2CH{=}NNHCONH_2$

(b) $CH_3C(CH_3){=}NNHCONH_2$

(c) $CH_3-C(CH_3){=}NCONHNH_2$

(d) $CH_3CH_2CH{=}NCONHNH_2$

Q. 28. Which of the following hydrocarbons has the lowest dipole moment?

(a) 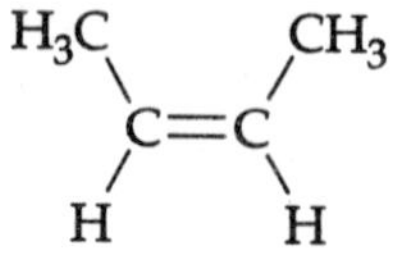

(b) $CH_3{=}CCH_3$

(c) $CH_3CH_2C{=}CH$

(d) $CH_2{=}CH{-}C{=}CH$

Q. 29. For the electrochemical Cell.

$M|M^+||X^-|X$, $E°(M^+/M) = 0.44$ and $E°(X/X^-) = 0.33V$. From this data one can deduce that

(a) $M + X \rightarrow M^+ + X^-$ is the spontaneous reaction

(b) $M^+ + X^- \rightarrow M + X^-$ is the spontaneous reaction

(c) $E_{cell} = 0.77$ V

(d) $E_{cell} = -0.77$ V

Q. 30. The correct order of radii is

(a) N < Be < B (b) $F^- < O^{2-} < N^{3-}$

(c) Na < Li < K (d) $Fe^{3+} < Fe^{2+} < Fe^{3+}$

Q. 31. Which one of the following will must readily be dehydrated in acidic condition?

(a) 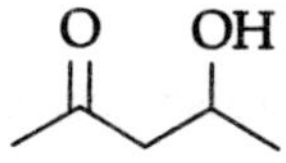

(b) 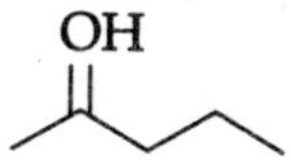

(c) 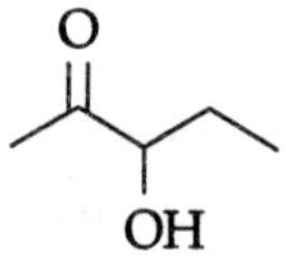

(d) 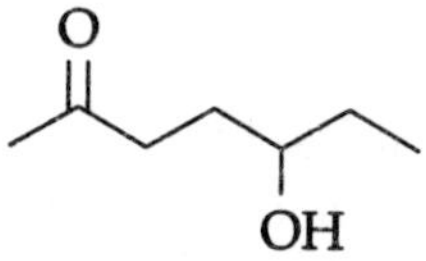

Q. 32. The ΔH_F° for $CO_2(g)$ and $H_2O(g)$ are –393.5, –110.5 and –241.8 kJ mol^{-1} respectively. The standard enthalpy change (in kJ) for the reaction;

$CO_2(g) + H_2(g) \rightarrow CO(g) + H_2O(g)$ is:

(a) –41.2 (b) –262.5 (c) 41.2 (d) 524.1

Q. 33. Which of the following compounds will exhibit geometrical isomerism?

(a) 1, 1-diphenyl-1-propene (b) 2-phenyl-1-butene

(c) 3-phenyl-1-butene (d) 1-phenyl-2-butene

Q. 34. For the revesible reaction

$N_2(g) + 3H_2(g) = 2NH_3(g)$ at 773 K, the value of K_p is 1.44×10^{-5} when partial pressure is measured in atmospheres. The corresponding value of K_c, with concentration in mole $litre^{-1}$, is

(a) $1.44 \times 10^{-5}/(0.082 \times 500)^{-2}$ (b) $1.44 \times 10^{-5}/(8.314 \times 773)^{-2}$

(c) $1.44 \times 10^{-5}/(0.082 \times 773)^{2}$ (d) $1.44 \times 10^{-5}/(0.082 \times 773)^{-2}$

TEST PAPER 3

SECTION A (SINGLE OPTION CORRECT)

Q. 1.. A monatomic ideal gas undergoes a process in which the ratio of P to V at any instant is constant and equals to 1. The molar heat capacity of the gas is:

(a) $\frac{4R}{2}$ (b) $\frac{3R}{2}$ (c) 5R/2 (d) 0

Q. 2. The reaction $B(OH)_3 + NaOH \rightleftharpoons NaBO_2 + Na[(B(OH)_4] + H_2O$
This reaction is made to proceed in forward direction by:
(a) addition of cis 1, 2, diol (b) addition of borax
(c) addition of trans 1, 2 diol (d) addition of Na_2HPO_4

Q. 3. A solution when diluted with H_2O and build, it gives a while precipitate. On addition of excess NH_4Cl/NH_4OH the volume of precipitate decreases leaving behind a while gelatinous precipitate. Identify the precipitate which dissolves in NH_4OH/NH_4Cl.
(a) $Zn(OH)_2$ (b) $Al(OH)_3$ (c) $Mg(OH)_2$ (d) $Ca(OH)_2$

Q. 4. When benzene sulfonic acid and *p*-nitrophenol are treated with $NaHCO_3$, the gases released respectively are
(a) SO_2, NO_2 (b) SO_2, NO (c) SO_2, CO_2 (d) CO_2, CO_2

Q. 5. (I) 1, 2-dihydroxy benzene (II) 1, 3-dihydroxy benzene
(III) 1, 4, dihydroxy benzene (IV) Hydroxy benzene
The increasing order of building points of above mentioned alcohols is
(a) I < II < III < IV (b) I < II < IV < III
(c) IV < I < II < III (d) IV < II < I < III

Q. 6. The IUPAC name of C_6H_5COCl is

(a) Benzoyl chloride (b) Benzene chloro ketone
(c) Benzene carbonyl chloride (d) Chloro phenyl ketone

Q. 7. $CH_3—CH{=}CH_2 + NOCl \rightarrow P$
Identify the adduct

(a) $CH_3-CH(Cl)-CH_2(NO)$ (b) $CH_3-CH(NO)-CH_2(Cl)$
(c) $CH_3-CH_2-CH(NO)_2$ (d) $CH_2(NO)-CH_2-CH_2(Cl)$

Q. 8. The direct conversion of *A* to *B* is difficult, hence it is carried out by the following shown path:

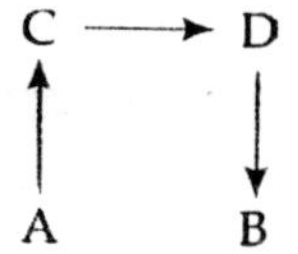

Given $\Delta S_{(A \rightarrow C)} = 50$ e.u.
$\Delta S_{(C \rightarrow D)} = 30$ e.u.
$\Delta S_{(B \rightarrow D)} = 20$ e.u.

where e.u. is entropy unit
then $\Delta S_{(A \rightarrow B)}$ is

(a) +100 e.u. (b) +60 e.u. (c) –100 e.u. (d) –60 e.u.

Q. 9. $Ag^+ + NH_3 \rightleftharpoons [Ag(NH_3)^+]; = 3.5 \times 10^{-3}$
$[Ag(NH_3)]^+ + NH_3 \rightleftharpoons [Ag(NH_3)_2]^+; k_2 = 1.7 \times 10^{-3}$
then the formation constant of $[Ag(NH_3)_2]^+$ is

(a) 6.08×10^{-6} (b) 6.08×10^{6} (c) 6.08×10^{-9} (d) None

Q. 10. $CH_3NH_2 + CHCl_3 + KOH \rightarrow$ Nitrogen containing compound + KCL + H_2O. Nitrogen containing compound is

(a) $CH_3—C{\equiv}N$ (b) $CH_3—NH—CH_3$
(c) $CH_3—N{\equiv}C$ (d) $CH_3N^+{\equiv}C^-$

Q. 11. $CuSO_4$ decolourises on addition of KCN, the product is

(a) $[Cu(CN)_4]^{2-}$ (b) Cu^{2+} get reduced to from $[Cu(CN)_4]^{3-}$
(c) $Cu(CN)_2$ (d) CuCN

Q. 12. $N_2 + 3H_2 \rightleftharpoons 2NH_3$
Which is correct statement if N_2 is added at equilibrium condition?

(a) The equilibrium will shift to forward direction because according to IInd law of thermodynamics the entropy must increases in the direction of spontaneous reaction.

(b) The condition for equilibrium is $G_{N_2} + 3G_{H_2} = 2G_{NH_3}$ where G is Gibbs free energy per mole of the gaseous species measured at the partial pressure. The condition of equilibrium is unaffected by the use of catalyst, when increases the rate of both the forward and backward reactions to the same extent.
(c) The catalyst will increase the rate of forward reaction by α and that of backward reaction by β.
(d) Catalyst will not after the rate of either of the reaction.

SECTION—B (MAY HAVE MORE THAN ONE OPTION CORRECT)

Q. 13. Which of the following reactants on reaction with conc. NaOH followed by acidification gives the following lactone as the only product?

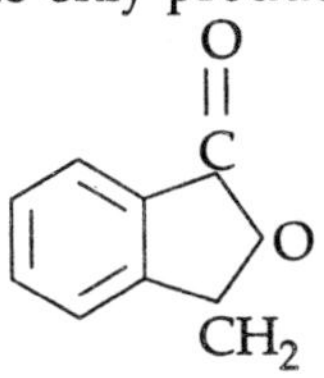

(a) 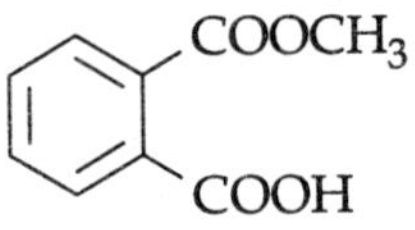
(b) 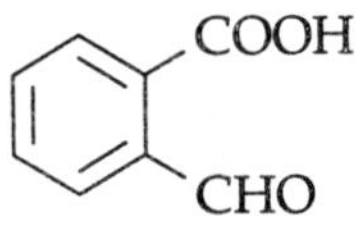
(c) 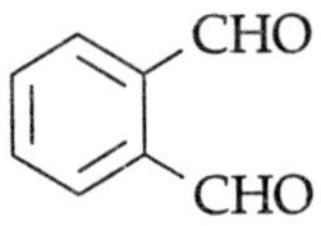
(d) 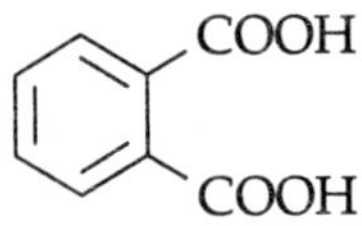

Q. 14. The species present in solution when CO_2 is dissolved in water are
(a) CO_2, H_2CO_3, HCO_3, CO_3^{2-}
(b) H_2CO_3, CO_3^{2-}
(c) CO_3^{2-}, HCO_3
(d) CO_2, H_2CO_3

Q. 15. If the bond length of CO bond in carbon monoxide is 1.128 Å, then what is the value of CO bond length in $Fe(CO)_5$?
(a) 1.15 Å (b) 1.128 Å (c) 1.72 Å (d) 1.118 Å

Q. 16.  + Cl—CH_2CH_2—CH_3 $\xrightarrow{AlCl_3}$ P $\xrightarrow[\text{(ii) } H_3O^-]{\text{(i) } O_2/\Delta}$ Q + Phenol

The major products P and Q are
(a) 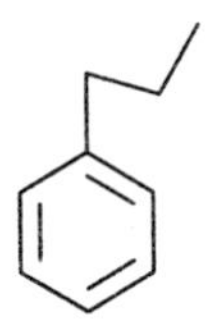 and CH_3CH_2CHO
(b) 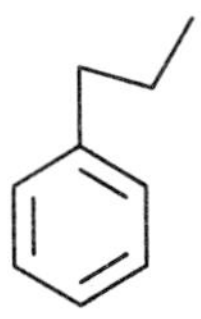 and CH_3COCH_3
(c) 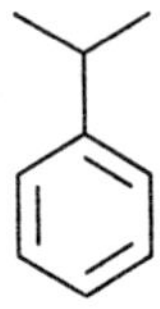 and CH_3COCH_3
(c) 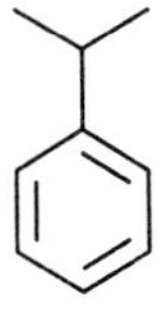 CH_3CH_2CHO

Q. 17. The smallest ketone and its next homologue are reacted with NH_2OH to form oxime.

(a) Two different oximes are formed
(b) Three different oximes are formed
(c) Two oximes are optically active
(d) All oxtimes are optically active

Q. 18. The given graph represents the variation of Z(compressibility factor = $\frac{PV}{nRT}$) versus P, for three real gases A, B, and C identify the only incorrect statement.

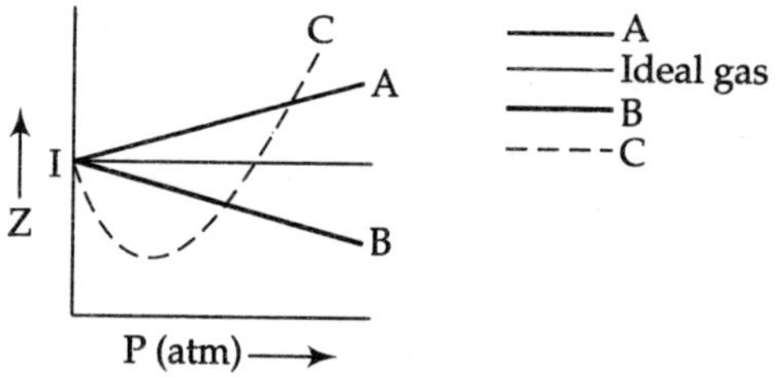

(a) For the gas A, $a = 0$ and its dependence on P is linear at all pressure.
(b) For the gas B, $b = 0$ and its dependence on P is linear at all pressure.
(c) For the gas C, which is typical real gas for which neither a nor $b = 0$. BY knowing the minima and the point of intersection, with Z = 1, a and b can be calculated.
(d) At high pressure, the slope is positive for all real gases.

Q. 19. 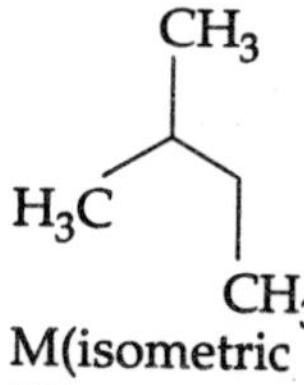 $\xrightarrow{Cl_2,\ h\upsilon}$ N(isometric products) $C_5H_{11}Cl$ $\xrightarrow{\text{fractional distillation}}$
M(isometric products)
What are N and M?

(a) 6, 6
(b) 6, 4
(c) 4, 4
(d) 3, 3

Q. 20. $MgSO_4$ on reaction with NH_4OH forms a white crystalline precipitate. What is its formula?

(a) $Mg(NH_4)PO_4$
(b) $Mg(PO_4)_2$
(c) $MgCl_2$, $MgSO_4$
(d) $MgSO_4$

SECTION C COMPREHENSION I

$RCONH_2$ is converted into RNH_2 by means of Hofmann bromamide degradation.

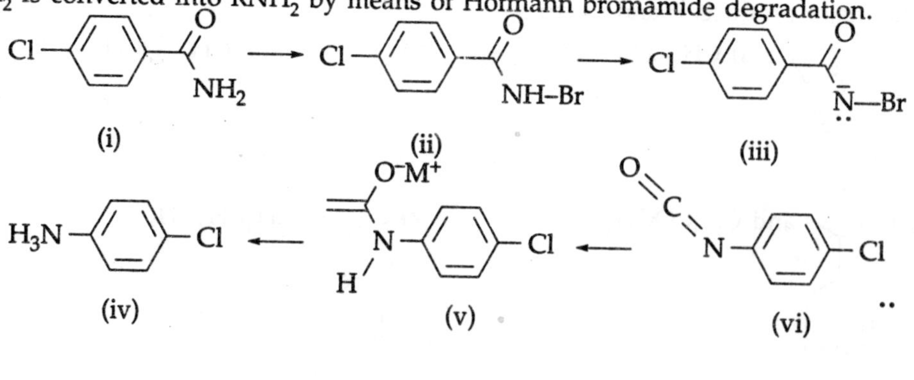

In this reaction RCONHBr is formed from which this reaction has derived its name. Electron donating group at phenyl activates the reaction. Hofmann degradation reaction is an intramolecular reaction.

Q. 21. How can the conversion of (i) to (ii) be brought about?

(a) KBr (b) KBr + CH_3ONa

(c) KBr + KOH (d) Br_2 + KOH

Q. 22. Which is the rate determining step in Hofmann bromamide degradation?

(a) Formation of (i) (b) Formation of (ii)

(c) Formation of (iii) (d) Formation of (iv)

Q. 23. What are the constituent amines formed when the mixture of (i) and (ii) undergoes Hofmann bromaide degradiation?

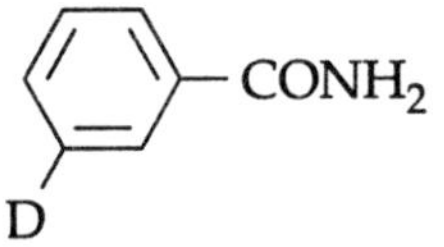
CONH2
D

(i)

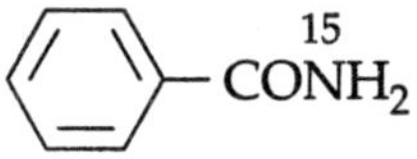
15
CONH2

(ii)

(a) 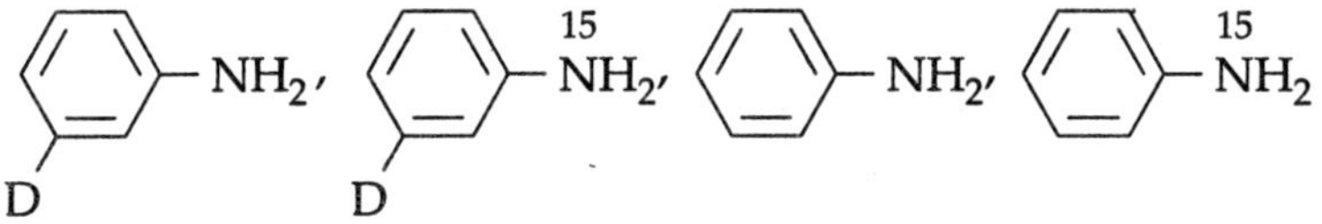
NH2,
15
NH2,
NH2,
15
NH2
D
D

(b) 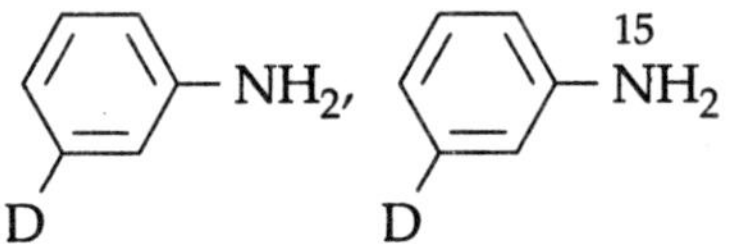
NH2,
15
NH2
D
D

(c) 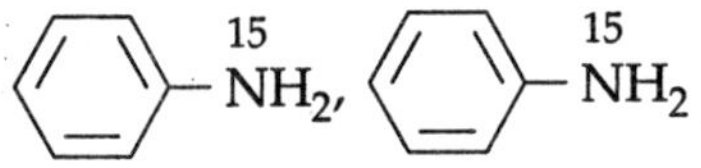
15
NH2,
15
NH2

(d) 
NHD,

COMPREHENSION II

The coordination number of Ni^{2+} is 4.

$NiCl_2$ + KCN (excess) → A (cyano complex)

$NiCl_2$ + Conc. HCl (excess) → (chloro complex)

Q. 24. The UPAC name of *A* and *B* are

(a) Potassium tetracyanonickelate (II), potassium tetrachloronickelate (II)

(b) Tetracyampotassiumnicketate (II), teterachlorpoatassiumnickelate (II)

(c) Foracyanornickel (II), tetrachloronickel (II)

(d) Potassium terracyamonickel (II), potassium teuachloronickel (II)

Q. 25. Predict the magnetic nature of *A* and *B*.

(a) Both are diamagnetic.
(b) *A* is diamagnetic and *B* is paramagnetic with one unpaired electron.
(c) *A* is diamagnetic and *B* is paramagnetic with two unpaired electrons.
(d) Both are paramagnetic.

Q. 26 The hybridization of *A* and *B* are

(a) dsp^2, sp^3 **(b)** sp^3, sp^3 **(c)** dsp^2, dsp^2 **(d)** sp^3d, d^3sp^3

From 27–29 read the following

Tollen's reagent is used for the detection of aldehyde when a solution of $AgNO_3$ is added to glucose with NH_4OH then gluconic acid is formed

$Ag^+ + e^- \rightarrow Ag, E^\circ_{red} = 0.8$ V

$C_2H_{12}O_6 + H_2O \rightarrow$ Glucomnic acid $(C_6H_{12}O_7) + 2H^+ + 2e^-$; $E^\circ_{oxd} = -0.05$ V

$Ag(NH_3)_2^+ + e \rightarrow Ag(s) + 2NH_3$; $E^\circ_{red} = 0.337$ V

[Use $2.303 \times \frac{RT}{F} = 0.0592$ and $\frac{F}{RT} = 38.92$ at 298 K]

Q. 27. $2Ag + C_6H_{12}O_6 + H_2O \rightarrow 2Ag(s) + C_6H_{12}O_7 + 2H^+$
Find in K of this reaction.

(a) 66.13 **(b)** 58.38 **(b)** 28.30 **(d)** 46.29

Q. 28. When ammonia is added to the solution, pH is raised, pH is raised to 11. Which half-cell reaction is affected by pH and by how much?

(a) E_{oxid} will increase by a factor of 0.65 from E°_{oxd}
(b) E_{oxd} will decreases by a factor of 0.65 from E°_{oxd}
(c) E_{red} will increases by a factor of 0.65 from E°_{red}
(d) E_{red} will decreases by a factor of 0.65 from E°_{red}

Q. 29. Ammonia is always is added in this reaction. Which of the following must be incorrect?

(a) NH_3 combines with Ag^+ to form a complex.
(b) $Ag(NH_3)_2^-$ is stronger oxidising regent than Ag^+
(c) In absence of NH_3 silver salt of gluconic acid is formed.
(d) NH_3 has affected the standard reduction potential of glucose/gluconic acid electrode.

For the problems 30-32 read the following

Carbon—14 is used to determine the age of organic material. The procedure is based on the formation of ^{14}C by neutron capture in the upper atmosphere.

$$^{14}N + {}_0n^1 \rightarrow {}^{14}_6C + {}_1n^1$$

^{14}C is absorbed by living organisms during photosynthesis. The ^{14}C content is constant in living organism once the plant or animal dies, the uptake of carbon dioxide by it cases and the level of ^{14}C in the dead being, falls due to the decay which C^{14} undergoes.

$$^{14}_6C \rightarrow {}^{14}_7N + \beta^-$$

The half life period of ^{14}C is 5770 years. The decay constant (l) can be calculated by using the following formula $\lambda = \frac{0.693}{t_{1/2}}$

The comparison of the β^- activity of the dead matter with that of the carbon still in circulation cubes measurement of the period of the isolation of the matcrial from the living cycle. The method however, ceases to be accurate over periods longer than 30,000 years. The proportion of ^{14}C to ^{12}C in living matter is $1 : 10^{12}$.

Q. 30. Which of the following option is correct?

(a) In living organisms, circulation of ^{14}C from atmosphere is high so the carbon content is constant is organism.

(b) Carbon during can be used to find out the age of earth crust and rocks

(c) Redioactive absorption due to cosmic radiation is equal to the rate of radioactive decay, hence the carbon content remains constant in living organism

(d) Carbon dating can not be used to determine concentration of ^{14}C in dead being.

Q. 31. What should be the age of fossil for meaningful determination of its age?

(a) 6 years **(b)** 6000 years

(c) 60,000 years **(d)** It can be used to calculate any age

Q. 32. a nuclear explosion has taken place leading to increase in concentration of C^{14} in nearby areas. C^{14} concentration is C_1 in nearby areas and C_2 areas for away. If the age of the fossil is determined to be T_1 and T_2 at the places, respectively then

(a) The age of the fossil will increase at the place where explosion has taken place and

$$T_1 - T_2 = \frac{1}{\lambda} \ln \frac{C_1}{C_2}$$

(b) The age of the fossil will decrease at the place where explosion has taken place and

$$T_1 - T_2 = \frac{1}{\lambda} \ln \frac{C_1}{C_2}$$

(c) The age of fossil will be determine to be same

(d) $\frac{T_1}{T_2} = \frac{C_1}{C_2}$

Q. 33. 75.2 g of C_6H_5OH (phenol) is dissolved in a solvent of $K_f = 14$. If the depression in freezing point is 7 K then the % of phenol the dimerises is

(a) 35% **(b)** 40% **(c)** 75% **(d)** 90%

Q. 34. For the reaction $2CO + O_2 \rightarrow 2CO_2$; $\Delta H = 560$ kJ. Two moles of CO and one mole of O_2 are taken in a container of volume 1 L. They completely from two moles of CO_2, the gases deviate appreciably from ideal behaviour. If the pressure in the vessel changes from 70 to 40 atm, find the magnitude (absolute value) of ΔU at 500K. (1 L atm = 01 kJ)

Q. 35. We have taken a saturated solution of AgBr, K_{ap} of AgBr is 12×10^{-44}. If 10^{-7} mole of $AgNO_3$ are added to 1 litre of this solution find conductivity (specific conductance) of this solution in terms of 10^{-7} S m^{-1} units.

Given $\lambda^\circ_{(Ag)} = 6 \times 10^{-3}$ S m^2 mol^{-6}. $\lambda^\circ_{(Br)} = 8 \times 10^{-3}$ S m^2 mol^{-3}, $\lambda^\circ_{NO_3} = 7 \times 10^{-3}$ S m^2 mol^{-1}.

Q. 36. The edge length of unit cell of a metal having molecular weight 75 g/mol is 5 Å which crystalizers in cube lattice. If the density is 2 g/cc then find the radius of metal atom. ($N_A = 6 \times 10^{23}$). Give the answer in pm.

Q. 37. Match the following

Column I	Column II
(a) CH_3—CHBr—CD_3 on treatement with al. KOH gives	(P) E1 reaction CH_2=CH—CD_3 as a major product
(b) Ph—CHBr—CH_3 reacts faster than Ph—CHBr—CD_3	(Q) E2 reaction
(c) Ph—CH_2—CH_2Br on treatment with C_2H_5OD/C_2H_5O gives Ph—CD—CH_2 as the major product	(R) Eleb reaction
(d) $PhCH_2CH_2Br$ and $PhCD_2CH_2Br$ react with same rate	(S) First order reaction

Q. 38. According to Bohr's theory,

E_o—Total energy
K_n—Kinetic energy
V_n—Potential energy
r_a—Radius of nth orbit

Match the following

Column I	Column II
(a) V_n/K_a = ?	(P) 0
(b) If radius of nth orbit $\propto E_n^x$, x = ?	(Q) −1
(c) Angular momentum in lowest orbital	(R) −2
(d) $\frac{1}{x^n} \propto Z^y$, y = ?	(S) 1

Q. 39. Match the extraction process listed in Column I with metals listed in Column II.

Column I	Column II
(a) Self reduction	(P) Lead
(b) Carbon reduction	(Q) Silver
(c) Complex formation and displacement by metal	(R) Copper
(d) Decomposition of iodide	(S) Boron

Q. 40. Match the following

Column I	Column II
(a) $Br^{-3} \rightarrow (BrO)^-$	(P) Heat
(b) $[AlO_2]^- \rightarrow Al(OH)_3$	(Q) Hydrolysis
(c) $SiO_4^{4-} \rightarrow Si_2O_7^{6-}$	(R) Acidification
(d) $(B_2O_7^{2-}) \rightarrow [B(OH)_3]$	(S) Dilution by water

ANSWERS

Chapter 1 : Atomic Structure

1. (b)	**2.** (d)	**3.** (c)	**4.** (d)	**5.** (b)	**6.** (b)	**7.** (d)	**8.** (c)
9. (b)	**10.** (a)	**11.** (d)	**12.** (b)	**13.** (d)	**14.** (b)	**15.** (c)	**16.** (d)
17. (d)	**18.** (a)	**19.** (a)	**20.** (c)	**21.** (d)	**22.** (d)	**23.** (c)	**24.** (c)
25. (a)	**26.** (c)						

Chapter 2 : Chemical Bonding

1. (c)	**2** (d)	**3.** (d)	**4.** (a)	**5.** (c)	**6.** (c)	**7.** (c)	**8.** (d)
9. (d)	**10.** (d)	**11.** (e)	**12.** (b)	**13.** (a)	**14.** (c)	**15.** (b)	**16.** (c)
17. (c)	**18.** (b)	**19.** (b)	**20.** (a)	**21.** (c)	**22.** (c)	**23.** (a)	**24.** (a)
25. (c)	**26.** (d)	**27.** (b)	**28.** (d)	**29.** (b)	**30.** (c)	**31.** (a)	**32.** (d)
33. (a)	**34.** (a)	**35.** (a)	**36.** (b)	**37.** (a)	**38.** (d)	**39.** (c)	**40.** (b)
41. (d)	**42.** (a)	**43.** (c)	**44.** (c)	**45.** (b)	**46.** (c)	**47.** (d)	**48.** (b)
49. (b)	**50.** (e)	**51.** (a)	**52.** (c)	**53.** (b)	**54.** (c)	**55.** (d)	**56.** (b)
57. (d)	**58.** (c)	**59.** (a)	**60. (i)** (d)	**60. (ii)** (b)	**60. (iii)** (a)	**61.** (a)	**62.** (c)
63. (c)	**64.** (b)	**65.** (d)	**66.** (c)	**67.** (b)	**68.** (b)	**69.** (a)	**70.** (c)
71. (c)	**72.** (a)	**73.** (d)	**74.** (b)	**75.** (a)	**76.** (c)	**77.** (c)	**78.** (b)
79. (d)	**80.** (c)	**81.** (a)	**82.** (b)	**83.** (d)	**84.** (b)	**85.** (a)	

Chapter 3 : Gas Laws

1. (b)	**2.** (b)	**3.** (c)	**4.** (c)	**5.** (c)	**6.** (d)	**7.** (a)	**8.** (a)
9. (d)	**10.** (a)	**11.** (d)	**12.** (b)	**13.** (d)	**14.** (b)	**15.** (b)	**16.** (a)
17. (b)	**18.** (c)	**19.** (d)	**20.** (d)	**21.** (d)	**22.** (b)	**23.** (d)	**24.** (b)
25. (d)	**26.** (c)	**27.** (a)	**28.** (c)	**29.** (d)	**30.** (a)	**31.** (b)	**32.** (c)
33. (b)	**34.** (b)	**35.** (c)	**36.** (b)	**37.** (d)	**38.** (c)	**39.** (c)	**40.** (d)
41. (a)	**42.** (d)	**43.** (c)	**44.** (d)	**45.** (b)	**46.** (d)	**47.** (b)	**48.** (c)
49. (c)	**50.** (a)	**51.** (a)	**52.** (b)	**53.** (b)	**54.** (b)	**55.** (b)	**56.** (a)
57. (c)	**58.** (d)	**59.** (a)	**60.** (c)	**61.** (a)	**63.** (d)	**64.** (c)	**65.** (b)
66. (c)	**67.** (b)	**68.** (c)	**69.** (a)	**70.** (b)	**71.** (d)	**72.** (b)	**73.** (a)
74. (c)	**75.** (d)	**76.** (a)	**77.** (c)	**78.** (c)	**79.** (d)	**80.** (c)	**81.** (d)
82. (b)	**83.** (d)	**84.** (c)	**85.** (c)	**86.** (b)	**87.** (a)		

Chapter 4 : Stoichiometry

1. (c)	**2.** (c)	**3.** (d)	**4.** (a)	**5.** (a)	**6.** (b)	**7.** (b)	**8.** (b)
9. (c)	**10.** (d)	**11.** (b)	**12.** (c)	**13.** (d)	**14.** (a)	**15.** (d)	**16.** (b)
17. (d)	**18.** (a)	**19.** (c)	**20.** (a)	**21.** (b)	**22.** (c)	**23.** (a)	**24.** (a)
25. (i) (d)	**(ii)** (a)	**26. (i)** (a)	**(ii)** (d)	**(iii)** (b)	**(iv)** (c)	**(v)** (b)	**27. (i)** (c)
27. (ii) (a)	**(iii)** (b)	**28. (i)** (c)	**(ii)** (a)	**29.** (c)	**30.** (b)	**31. (i)** (a)	**(ii)** (b)
31. (iii) (d)	**(iv)** (b)	**32.** (b)	**33.** (a)	**34.** (d)	**35.** (a)	**36.** (c)	**37.** (c)
38. (d)	**39.** (d)	**40.** (b)	**41.** (d)	**42.** (b)	**43.** (c)	**44.** (d)	**45.** (d)
46. (a)	**47.** (a)	**48.** (b)	**49.** (d)	**50.** (b)	**51.** (c)	**52.** (b)	**53.** (a)
54. (d)	**55.** (a)	**56.** (a)	**57. (i)** (c)	**(ii)** (a)	**(iii)** (a)	**(iv)** (b)	**(v)** (a)
58. (d)	**59.** (b)	**60.** (c)	**61.** (a)	**62.** (b)	**63.** (b)	**64.** (a)	**65.** (d)
66. (c)	**67.** (d)	**68.** (a)	**69.** (b)	**70.** (c)	**71.** (a)	**72.** (a)	**73.** (b)
74. (a)	**75.** (b)	**76.** (c)	**77.** (d)	**78.** (a)	**79.** (a)	**80. (i)** (b)	**(ii)** (c)
80. (iii) (b)	**(iv)** (d)	**81.** (c)	**82.** (b)	**83.** (d)			

Chapter 5 : Solutions

1. (b)	2. (b)	3. (b)	4. (b)	5. (a)	6. (b)	7. (c)	8. (d)
9. (a)	10. (d)	11. (c)	12. (d)	13. (b)	14. (a)	15. (c)	16. (c)
17. (d)	18. (b)	19. (d)	20. (c)	21. (d)	22. (b)	23. (d)	24. (b)
25. (a)	26. (a)	27. (c)	28. (c)	29. (d)	30. (a)	31. (b)	32. (c)
33. (d)	34. (a)	35. (d)	36. (b)	37. (d)	38. (c)	39. (c)	40. (b)
41. (d)	42. (c)	43. (d)	44. (b)	45. (c)	46. (a)	47. (a)	48. (c)
49. (b)	50. (i) (b)	(ii) (d)	(iii) (c)	(iv) (b)	51. (c)	52. (b)	53. (d)
54. (c)	55. (c)	56. (d)	57. (a)	58. (d)	59. (c)	60. (d)	61. (b)
62. (c)							

Chapter 6 : Thermodynamics

1. (a)	2. (c)	3. (b)	4. (d)	5. (d)	6. (c)	7. (a)	8. (d)
9. (c)	10. (d)	11. (b)	12. (b)	13. (a)	14. (c)	15. (b)	16. (b)
17. (d)	18. (c)	19. (b)	20. (c)	21. (a)	22. (d)	23. (a)	24. (a)
25. (a)	26. (c)	27. (d)	28. (d)	29. (d)	30. (b)	31. (b)	32. (c)
33. (a)	34. (b)	35. (d)	36. (b)	37. (d)	38. (d)	39. (a)	40. (b)
41. (c)	42. (c)	43. (a)	44. (a)	45. (b)	46. (b)	47. (d)	48. (a)
49. (c)	50. (c)	51. (c)	52. (b)	53. (a)	54. (d)	55. (b)	56. (c)
57. (c)	58. (d)	59. (c)	60. (a)	61. (a)	62. (b)	63. (d)	64. (b)
65. (b)	66. (b)	67. (b)	68. (a)	69. (b)	70. (a)	71. (c)	72. (a)
73. (c)	74. (c)	75. (i) (a)	(ii) (d)	(iii) (d)	(iv) (c)	(v) (a)	76. (i) (a)
76. (ii) (a)	77. (i) (b)	(ii) (a)	(iii) (d)	(iv) (b)	79. (c)	80. (i) (d)	(ii) (a)
81. (b)	82. (d)	83. (a)	84. (c)	85. (i) (a)	(ii) (a)	86. (a)	87. (c)
88. (b)	89. (d)	90. (a)	91. (b)	92. (a)	93. (c)	94. (d)	95. (a)
96. (c)	97. (b)	98. (c)	99. (b)	100. (b)	101. (d)	102. (b)	103. (b)
104. (d)	105. (d)	106. (d)	107. (c)	108. (a)	109. (d)	110. (c)	111. (d)
112. (b)	113. (b)	114. (c)	115. (a)	116. (b)	117. (b)		

Chapter 7 : Chemical Kinetics

1. (a)	2. (d)	3. (d)	4. (c)	5. (a)	6. (c)	7. (d)	8. (c)
9. (c)	10. (d)	11. (b)	12. (d)	13. (c)	14. (d)	15. (c)	16. (a)
17. (c)	18. (b) and (c)		19. (d)	20. (b)	21. (a)	22. (c)	23. (c)
24. (a)	25. (c)	26. (b)	27. (i) (d)	(ii) (b)	(iii) (b)	28. (a)	29. (b)
30. (c)	31. (b)	32. (c)	33. (c)	34. (c)	35. (d)	36. (a)	37. (d)
38. (d)	39. (b)	40. (b)	41. (a)	42. (b)	43. (c)	44. (a)	45. (d)
46. (d)	47. (c)	48. (c)	49. (a)	50. (d)	51. (c)	52. (d)	

Chapter 8 : Equilibrium

1. (c)	2. (a)	3. (b)	4. (d)	5. (b)	6. (c)	7. (c)	8. (a)
9. (d)	10. (a)	11. (a)	12. (d)	13. (c)	14. (c)	15. (a)	16. (b)
17. (d)	18. (b)	19. (a)	20. (d)	21. (c)	22. (d)	23. (c)	24. (a)
25. (a)	26. (d)	27. (c)	28. (i) (d)	(ii) (c)	(iii) (b)	29. (b)	30. (c)
31. (c)	32. (d)	33. (a)	34. (d)	35. (d)	36. (c)	37. (c)	38. (a)
39. (d)	40. (d)	41. (c)	42. (b)	43. (c)	44. (c)	45. (a)	46. (d)
47. (c)	48. (b)	49. (a)	50. (b)	51. (i) (c)	(ii) (a)	52. (a)	53. (b)
54. (c)	55. (c)	56. (d)	57. (a)				

Chapter 9 : Ionic Equilibria

1. (c)	2. (c)	3. (c)	4. (c)	5. (d)	6. (b)	7. (b)	8. (c)
9. (d)	10. (d)	11. (c)	12. (c)	13. (d)	14. (d)	15. (a)	16. (b)
17. (b)	18. (b)	19. (d)	20. (b)	21. (c)	22. (d)	23. (a)	24. (c)
25. (a)	26. (a)	27. (b)	28. (a)	29. (e)	30. (c)	31. (c)	32. (c)
33. (b)	34. (b)	35. (b)	36. (a)	37. (d)	38. (a)	39. (c)	40. (b)
41. (d)	42. (a)	43. (a)	44. (b)	45. (c)	46. (b)	47. (c)	48. (d)
49. (c)	50. (d)	51. (b)	52. (d)	53. (a)	54. (c)	55. (a)	56. (c)
57. (c)	58. (c)	59. (b)	60. (d)	61. (b)	62. (c)	63. (d)	64. (c)
65. (d)	66. (c)	67. (a)	68. (b)	69. (a)	70. (c)	71. (d)	72. (c)
73. (a)	74. (d)	75. (b)	76. (b)	77. (a)	78. (c)	79. (c)	80. (c)

81. (c)	82. (b)	83. (a)	84. (a)	85. (d)	86. (b)	87. (b)	88. (b)
89. (d)	90. (b)	91. (b)	92. (c)	93. (b)	94. (a)		
95. (i) (a)	(ii) (b)	(iii) (d)	(iv) (a)	96. (i) (c)	(ii) (b)	(iii) (c)	
97. (i) (b)	(ii) (b)	(iii) (c)	98. (d)	99. (b)	100. (c)	101. (d)	102. (a)
103. (d)	104. (b)	105. (c)	106. (a)	107. (c)	108. (b)	109. (c)	110. (b)
111. (d)	112. (a)	113. (d)	114. (b)	115. (c)	116. (c)	117. (a)	118. (a)
119. (a)	120. (i) (a)	(ii) (b)	(iii) (c)	121. (c)	122. (b)	123. (a)	
124. (b)	125. (a)	126. (a)	127. (c)	128. (a)	129. (a)	130. (i) (b)	(ii) (b)
131. (c)	132. (b)	133. (a)	134. (b)	135. (a)	136. (b)	137. (c)	138. (c)
139. (i) (a)	(ii) (c)	140. (c)	141. (b)	142. (a)	143. (c)	144. (c)	145. (b)
146. (i) (b)	(ii) (a)	(iii) (c)	147. (i) (c)	(ii) (c)	148. (i) (b)	(ii) (a)	149. (d)
150. (c)	151. (a)	152. (b)	153. (a)	154. (d)	155. (c)	156. (b)	157. (c)
158. (a)	159. (c)	160. (a)	161. (b)	162. (c)	163. (i) (c)	(ii) (a)	164. (c)
165. (a)	166. (c)	167. (c)	168. (a)	169. (c)	170. (i) (a)	(ii) (c)	171. (d)
172. (c)	173. (a)	174. (i) (b)	(ii) (b)	175. (d)	176. (c)	177. (a)	178. (b)
179. (c)	180. (d)	181. (i) (d)	(ii) (a)	182. (c)	183. (a)	184. (i) (a)	(ii) (b)
185. (b)	186. (b)	187. (c)	188. (i) (a)	(ii) (b)	189. (b)	190. (See solution)	
191. (c)	192. (a)	193. (d)	194. (d)	195. (c)	196. (b)	197. (c)	198. (d)
199. (a)	200. (d)	201. (d)	202. (c)	203. (b)	205. (a)	205. (a)	

Chapter 10 : Electro-Chemistry

1. (b)	2. (d)	3. (d)	4. (d)	5. (a)	6. (b)	7. (a)	8. (c)
9. (a)	10. (a)	11. (d)	12. (c)	13. (b)	14. (b)	15. (a)	16. (c)
17. (b)	18. (d)	19. (c)	20. (b)	21. (a)	22. (b)	23. (d)	24. (d)
25. (b)	26. (a)	27. (a)	28. (b)	29. (c)	30. (c)		
31. (i) (c)	(ii) (c)	(iii) (d)	32. (b)	33. (b)	34. (d)	35. (d)	36. (a)
37. (d)	38. (a)	39. (d)	40. (d)	41. (a)	42. (d)	43. (c)	44. (c)
45. (c)	46. (b)	47. (d)	48. (b)	49. (c)	50. (d)	51. (d)	52. (d)
53. (a)	54. (i) (b)	(ii) (c)	55. (d)	56. (i) (b)	(ii) (d)	57. (b)	58. (c)
59. (d)	60. (c)	61. (a)	62. (c)	63. (b)	64. (c)	65. (b)	66. (a)
67. (b)	68. (a)	69. (c)	70. (d)	71. (a)	72, (d)	73. (d)	

Chapter 11 : Nuclear Chemistry

1. (d)	**2.** (b)	**3.** (a)	**4.** (d)	**5.** (c)	**6.** (a)	**7.** (d)	**8.** (b)
9. (b)	**10.** (d)	**11.** (c)	**12.** (d)	**13.** (a)	**14.** (c)	**15.** (b)	**16.** (c)
17. (d)	**18.** (a)	**19.** (a)	**20.** (c)	**21.** (d)	**22.** (b)	**23.** (d)	**24.** (a)
25. (c)	**26.** (b)	**27.** (a)	**28.** (d)	**29.** (d)	**30.** (a)	**31.** (c)	**32.** (b)
33. (a)	**34.** (d)	**35.** (b)	**36.** (d)	**37.** (c)	**38.** (b)	**39.** (c)	**40.** (c)
41. (c)	**42.** (b)	**43.** (d)	**44.** (c)	**45.** (c)	**46.** (a)	**47.** (b)	**48.** (d)
49. (b)	**50.** (c)	**51.** (d)	**52.** (a)	**53.** (d)	**54.** (c)	**55.** (d)	**56.** (a)
57. (a)	**58.** (b)	**59.** (c)	**60.** (d)	**61.** (d)	**62.** (c)	**63.** (a)	**64.** (c)

Chapter 12 : Crystalography

1. (c)	**2.** (d)	**3.** (b)	**4.** (c)	**5.** (b)	**6.** (c)	**7.** (a)	**8.** (a)
9. (b)	**10.** (c)	**11.** (d)	**12.** (d)	**13.** (b)	**14.** (d)	**15.** (c)	**16.** (d)
17. (a)	**18.** (b)	**19.** (a)	**20.** (c)	**21.** (c)	**22.** (a)	**23.** (b)	**24.** (b)
25. (d)	**26.** (c)	**27.** (a)	**28.** (c)	**29.** (d)	**30.** (b)		
31. (i) (d)	**(ii)** (d)	**(iii)** (d)	**32.** (a)	**33.** (b)	**34.** (c)	**35.** (a)	**36.** (b)
37. (b)	**38.** (c)	**39.** (a)	**40.** (d)	**41.** (a)	**42.** (c)	**43.** (a)	**44.** (a)
45. (c)	**46.** (c)	**47.** (b)	**48.** (c)	**49.** (d)	**50.** (b)	**51.** (a)	**52.** (b)
53. (c)	**54.** (b)	**55.** (d)	**56.** (c)	**57.** (a)	**58.** (b)		

Organic Chemistry

1. (c)

2. (c)

3. (a)

4. **(a)** 1 to 0 state **(b)** is reduced
(c) is reduced **(d)** –1 to +1 state
(e) is reduced

6. **(a)** I **(b)** II **(c)** I

7. **(a)** (v) **(b)** (vi) **(c)** (i) **(d)** (iii)
(e) (ii) **(f)** (iv)

8. **(a)** (ii) **(b)** (v) **(c)** (i) **(d)** (iii)
(e) (vi) **(f)** (iv)

9.	(a) (v)	(b) (iv)	(c) (i)	(d) (ii)
	(e) (iii)			
10.	(a) (i)	(b) (ii)	(c) (ii)	(d) (i)
	(e) (i)			
11.	(a) (ii) and (iii)	(b) (iv) and (i)	(c) (i)	(d) (vi)
	(e) (viii)	(f) (vi)	(g) (v)	(h) (vii)
13.	(a) (i)	(b) (iii)	(c) (iii)	(d) (i)
	(e) (ii)			
14.	(c)			
15.	(a) No	(b) No	(c) No	(d) No
	(e) Yes	(f) Yes		
16.	(b)			
17.	(i) (d)	(ii) (c)	(iii) (c)	
18.	(a) (iii), (ii)	(b) (i), (iii)	(c) (ii), (iii)	
19.	(a) (ii), (vi)	(b) (ii), (v)		
20.	(a) (ii), (vi)	(b) (i)	(c) (iii), (iv)	(d) (v)
21.	(a) (iii)	(b) (ii)	(c) (i)	(d) (v)
	(e) (iv)			
22.	(a) (ii)	(b) (iv)	(c) (i)	(d) (iii)
23.	(a) (iv)	(b) (viii)	(c) (ii)	(d) (x)
	(e) (ix)	(f) (i)	(g) (iii)	(h) (vi)
	(i) (vii)	(j) (v)		
24.	(a) IV	(b) I	(c) I	(d) I
	(f) IV	(g) IV	(h) III	(i) I
	(j) II	(k) III	(l) II	
25.	(a) (iv)	(b) (vi)	(c) (ix)	(d) (ii)
	(e) (v)	(f) (i)	(g) (viii)	(h) (vii)
	(i) (iii)			
26.	(a) (i)	(b) (vii)	(c) (iv)	(d) (v)
	(e) (viii)	(f) (ix)	(g) (vi)	(h) (iii)
	(i) (ii)			
27.(a)	(a) (iii)	(b) (v)	(c) (i)	(d) (iv)
	(e) (ii)	(f) (vi)		
27.(b)	(a) (iv)	(b) (vii)	(c) (i)	(d) (vi)
	(e) (ii)	(f) (v)	(g) (iii)	

28.	**(b)** (c) and (f)				
29.	**(b)**				
30.	**(d)**				
31.	**(a)** (iii)	**(b)** (v)	**(c)** (i)	**(d)** (vii)	
	(e) (ix)	**(f)** (ii)	**(g)** (viii)	**(h)** (vi)	
	(i) (iv)				
32.	**A** (c)	**B** (b)			
33.	(c)				
34.	**A** (iii)	**B** (v)	**C** (i)	**D** (iv)	**E** (ii)
35.	**(a)** (III)	**(b)** (III)	**(c)** (III)	**(d)** (III)	
	(e) (I), (II)	**(f)** (I), (II)			
36.	**(a)** (i), (iv)	**(b)** (i), (iv)	**(c)** (iii)	**(d)** (i), (iv)	
	(e) (iii)	**(f)** (ii), (v)	**(g)** (i), (iv)	**(h)** (ii), (v)	
	(i) (i), (iv)	**(j)** (i), (iv)	**(k)** (iii)		
37.	**A** (iii) (v), (i)	**B** (ii), (v)	**C** (i)	**D** (vii), (viii), (iv)	
38.	**(i)** (d)	**(ii)** (e)	**(iii)** (a)	**(iv)** (b)	
	(v) (f)	**(vi)** (c)			
39.	**(a)** (iv)	**(b)** (v)	**(c)** (i)	**(d)** (ii)	
	(e) (iii)	**(f)** (vi)			
40.	**(a)** (iv)	**(b)** (vii)	**(c)** (x)	**(d)** (vi)	
	(e) (ii)	**(f)** (viii)	**(g)** (v)	**(h)** (i)	
	(i) (iii)	**(j)** (ix)			
41.	**(a)** (viii)	**(b)** (iii)	**(c)** (vi)	**(d)** (i)	
	(e) (iv)	**(f)** (vii)	**(g)** (ix)	**(h)** (xi)	
	(i) (ii)	**(j)** (v)	**(k)** (x)		
42.	**(a)** (i)	**(b)** (iii)	**(c)** (v)	**(d)** (vii)	
	(e) (vi)	**(f)** (iv)	**(g)** (ii)		
43.	**(a)** (iv)	**(b)** (vi)	**(c)** (iii)	**(d)** (vii)	
	(e) (i)	**(f)** (v)	**(g)** (viii)	**(h)** (ii)	
44.	**(a)** (vii)	**(b)** (iv)	**(c)** (x)	**(d)** (viii)	
	(e) (v)	**(f)** (i)	**(g)** (iii)	**(h)** (vi)	
	(i) (ix)	**(j)** (ii)			

45.	**(a)** (v)	**(b)** (iii)	**(c)** (vii)	**(d)** (i)	
	(e) (viii)	**(f)** (ii)	**(g)** (iv)	**(h)** (vi)	
46.	**(a)** (ix)	**(b)** (viii)	**(c)** (v)	**(d)** (vi)	
	(e) (ii)	**(f)** (i)	**(g)** (vii)	**(h)** (iv)	**(i)** (iii)
47.	**A** (iii)	**B** (v)	**C** (i)	**D** (ii)	**E** (iv)
48.	**(a)** (iii)	**(b)** (vi)	**(c)** (iv)	**(d)** (v)	
	(e) (i)	**(f)** (ii)			
49. I.	**A** (vi)	**B** (iii)	**C** (viii)	**D** (v)	
	E (vii)	**F** (i)	**G** (iv)	**H** (ii)	
	II. A (viii)	**B** (vi)	**C** (iii)	**D** (vii)	
	E (iv)	**F** (i)	**G** (ii)	**H** (xi)	
	I (v)	**J** (ix)	**K** (x)		
50.	**(a)** (iii)	**(b)** (viii)	**(c)** (vii)	**(d)** (x)	
	(e) (i)	**(f)** (iv)	**(g)** (ii)	**(h)** (ix)	
	(i) (vi)	**(j)** (v)			
51.	**(a)** (ix)	**(b)** (vii)	**(c)** (iii)	**(d)** (viii)	
	(e) (x)	**(f)** (ii)	**(g)** (xi)	**(h)** (vi)	
	(i) (iv)	**(j)** (v)	**(k)** (i)		
52.	**A** (ii)	**B** (iv)	**C** (iii)	**D** (i)	
53.	**A** (xii)	**B** (vi)	**C** (iii)	**D** (viii)	
	E (ii)	**F** (xiii)	**G** (vii)	**H** (xi)	
	I (i)	**J** (iv)	**K** (v)	**L** (x)	**M** (ix)
54.	**A** (v)	**B** (iv)	**C** (vi)	**D** (i)	
	E (vii)	**F** (x)	**G** (iii)	**H** (ix)	
	I (ii)	**J** (viii)			
55.	**A** (xii)	**B** (vi)	**C** (i)	**D** (xiii)	
	E (ix)	**F** (iv)	**G** (v)	**H** (ii)	
	I (iii)	**J** (xi)	**K** (vii)	**L** (viii)	**M** (x)
56.	**A** (vii)	**B** (xii)	**C** (iii)	**D** (xvii)	
	E (viii)	**F** (i)	**G** (xix)	**H** (ii)	
	I (x)	**J** (xv)	**K** (xvi)	**L** (iv)	

	M (xviii)	N (xiii)	O (xv)	P (ix)
	Q (vi)	R (xi)	S (xiv)	
57.	A (vii)	B (ii)	C (vi)	D (i)
	E (viii)	F (iv)	G (iii)	H (v)
58.	(a) (iii)	(b) (i)	(c) (ii)	
59.	(b)	60. (c)	61. (c)	62. (a)
63.	(c)	64. (d)	65. (a)	
66.	(d)	67. (a)	68. (c)	69. (d)
70.	(a)	71. (c)	72. (b)	73. (c)
74.	(d)	75. (d)	76. (b)	77. (b)
78.	(i) (d)	(ii) (a)	79. (c)	80. (b)
81.	(i) (d)	(ii) (a)	82. (c)	83. (d)
84.	(c)	85. (b)	86. (b)	87. (c)
88.	(b)	89. (b)	90. (a)	
91.	(i) (c)	(ii) (b)	(iii) (c)	
92.	(b)	93. (a)	94. (a)	95. (b)
96.	(a)	97. (b)	98. (d)	99. (c)
100.	(d)	101. (b)	102. (a)	103. (c)
104.	(d)	105. (d)	106. (b)	107. (b)
108.	(b)	109. (c)	110. (c)	111. (c)
112.	(b)	113. (c)	114. (b)	115. (c)
116.	(i) (b)	(ii) (c)	117. (b)	118. (c)
119.	(b)	120. (c)	121. (d)	122. (a)
123.	X is propanal	124. (c)		
125.	A (vi)	B (iv)	C (v)	D (viii)
	E (i)	F (ii)	G (vii)	H (iii)
126.	A (iii)	B (ii)		
127.	(i) (a)	(ii) (c)	128 (a) (ii)	
129.	A (ii)	B (iv)	C (i)	
130.	A (iv)	B (i)		
131.	A (ii)	B (iv)		

132.	**A** (i)	**B** (iii)	**C** (ii)	
133.	**A** (iii)	**B** (i)	**C** (v)	**D** (ii)
134.	**A** (ii)	**B** (v)	**C** (i)	**D** (iv)
135.	**A** (i)			
136.	**A** (ii)	**B** (iii)	**C** (i)	**D** (i)
137.	**A** (iv)	**B** (i)	**C** (iii)	
138.	**A** (iii)	**B** (iv)	**C** (i)	**D** (ii)
139.	**A** (ii)	**B** (iii)	**C** (i)	
140.	**A** (v)	**B** (i)	**C** (ii)	**D** (iv)
	E (iii)q			

Practice Papers

1. (c)	**2.** (a)	**3.** (d)	**4.** (c)	**5.** (a) or (c) or (e)		**6.** (e)	**7.** (b)
8. (d)	**9.** (e)	**10.** (c)	**11.** (b)	**12.** (a)	**13.** (a)	**14.** (a)	**15.** (c)
16. (b)	**17.** (d)	**18.** (b)	**19.** (a)	**20.** (a)	**21.** (c)	**22.** (e)	**23.** (d)
24. (b)	**25.** (c)	**26.** (c)	**27.** (c)	**28.** (d)	**29.** (d)	**30.** (a)	**31.** (d)
32. (b)	**33.** (e)	**34.** (c)	**35.** (a)	**36.** (b)	**37.** (d)	**38.** (e)	**39.** (c)
40. (b)	**41.** (d)	**42.** (c)	**43.** (c)	**44.** (b)	**45.** (c)	**46.** (b)	**47.** (d)
48. (d)	**49.** (c)	**50.** (d)	**51.** (d)	**52.** (d)	**53.** (a)	**54.** (a)	**55.** (c)
56. (a)	**57.** (d)	**58.** (a)	**59.** (c)	**60.** (e)	**61.** (b)	**62.** (b)	**63.** (d)
64. (e)	**65.** (d)	**66.** (a)	**67.** (c)	**68.** (c)	**69.** (d)	**70.** (d)	**71.** (d)
72. (a)	**73.** (a)	**74.** (c)	**75.** (c)				

Test Paper 1

1. (d)	2. (c)	3. (a)	4. (c)	5. (c)	6. (b)	7. (b)	8. (a)
9. (b)	10. (a)	11. (b)	12. (a)	13. (a)	14. (a)	15. (b)	16. (d)
17. (d)	18. (a)	19. (b)	20. (b)	21. (a)	22. (c)	23. (d)	24. (a)
25. (c)	26. (a)	27. (d)	28. (a)	29. (c)	30. (a)	31. (b)	32. (b)
33. (a)	34. (b)						

Test Paper 2

1. (d)	2. (a)	3. (d)	4. (a)	5. (c)	6. (a)	7. (d)	8. (a)
9. (a)	10. (b)	11. (b)	12. (a)	13. (a)	14. (a)	15. (c)	16. (a)
17. (a)	18. (c)	19. (a)	20. (a)	21. (c)	22. (a)	23. (b)	24. (b)
25. (d)	26. (c)	27. (b)	28. (b)	29. (b)	30. (b)	31. (a)	32. (c)
33. (d)	34. (d)						

Test Paper 3

1. (a)	2. (a)	3. (a)	4. (d)	5. (c)	6. (c)	7. (a)	8. (b)
9. (a)	10. (d)	11. (d)	12. (b)	13. (c)	14. (a)	15. (a)	16. (c)
17. (b)	18. (b)	19. (b)	20. (a)	21. (d)	22. (d)	23. (b)	24. (a)
25. (c)	26. (a)	27. (b)	28. (a)	29. (d)	30. (c)	31. (b)	32. (a)
33. (c)	34. 557KJ	35. 55 Sm^{-1}	36. 216.5 pm				

37. (a) Q: (b) Q: (c) R, S (d) P, S

38. (a) R: (b) Q: (c) P: d) S

39. (a) P, R: (b) P, R: (c) Q, (d) S

40. (a) Q: (b) R: (c) P: (d) Q, R

PART IV
SOLUTIONS

Chapter I : ATOMIC STRUCTURE

1. The following relationships are needed to solve this problem:

$$\Delta E = R_H\left(\frac{1}{n_1^2} - \frac{1}{n_2^2}\right) \text{and } \lambda = \frac{h \cdot c}{\Delta E}$$

Combining these equations to solve for λ gives the equation

$$\lambda = \frac{h \cdot c}{R_H\left(\frac{1}{n_1^2} - \frac{1}{n_2^2}\right)} = \frac{(6.63 \times 10^{-34} J \cdot sec)(3.00 \times 10^{10} cm \cdot sec^{-1})}{(2.18 \times 10^{-18} J)\left(\frac{1}{5^2} - \frac{1}{2^2}\right)}$$

2. The electron configuration for the Co^{3+} ion is $1s^2\, 2s^2\, 2p^6\, 3s^2\, 3p^6\, 3d^6$. The Co^{3+} ion would have a total of 24 electrons: 10 pairs of electrons and four unpaired electrons in the 3*d* orbitals. Atoms in which one or more electrons are unpaired are paramagnetic.
3. Atomic radius increases as one moves down a column (or group).
4. The word *unique* in this question means that *only* the alkali metals possess this particular characteristic. Of the choices listed, D is the only property that is unique to the alkali metals.
5. The photoelectric effect is the emission of electrons from the surface of a metal when light shines on it. Electrons are emitted, however, only when the frequency of that light is greater than a certain threshold value characteristic of the particular metal. The alkali metals, with only one electron in their valence shells, have the lowest threshold values.
6. Both the lithium ion, Li^+, and the hydride ion, H^-, have the configuration $1s^2$. Both species have 2 electrons, but the lithium ion has 3 protons, which cause a greater "pull" on the 2 electrons than the I proton found in the hydride ion.
7. The transition elements are filling the *d* orbitals. When completely filled, the *d* orbitals hold a maximum of 10 electrons.
8. The valence electron for the cesium atom is in the 6*s* orbital. In assigning quantum numbers, n = principal energy level = 6. The quantum number l represents the angular momentum (type of orbital) which *s* orbitals = 0, *p* orbitals = 1, *d* orbitals = 2, and so forth. In this case, $l = 0$. The quantum number m_l is known as the magnetic quantum number and describes the orientation of the orbital in space. For *s* orbitals (as in this case), m_l always equals 0. For *p* orbitals, m_l can take on the value of –1, 0, and +1. For *d* orbitals, m_l can take on the values –2, –1, 0, +1, and +2. The quantum number m_s is known as the electron spin quantum number and can take only two values, +1/2 and –1/2, depending on the spin of the electron.
9. Refer to Key Concepts for colours of flame tests in tables.
10. When atoms absorb energy, electrons move from ground states to excited levels. When the electrons move back to their ground state, they release energy (exothermic) at particular wave lengths (or frequencies).

11. Metals lose their electrons readily to become positively charged ions with charges of +1, +2, or +3.

12. E = hν and ν = $c\bar{\nu}$, so that E = hc$\bar{\nu}$. The energy is directly proportional to the wave number.

13. All other descriptions except D are incorrect for the spectrum of atomic H.

14. No information about the nuclear mass (number of neutrons) can be obtained from atomic emission spectra as they are determined by the electronic energy levels.

15.
$$E = \left(495.80 \frac{kJ}{mol}\right)\left(10^3 \frac{J}{kJ}\right)\left(\frac{1\,mol}{6.0221 \times 10^{23}\,atom}\right)$$

$$= 8.233\,5\ 10^{-19}\ J/atom$$

$$\nu = E/h = \frac{8.233 \times 10^{-19} J}{6.6261 \times 10^{-34} J \cdot s} = 1.243 \times 10^{15}\ s^{-1}$$

16. Thomson's experiment measured e/m for electrons. The charge to mass ratio for a hydrogen ion was already known.

17. The lower level quantum number is 4 for the Brackett series of lines in the spectrum of H atoms.

18. $\lambda = 1/\nu = 1/(1.028 \times 10^{-2}\ nm^{-1}) = 97.3$ nm.

19. Some oil drops are positively charged and some are negatively charged. The charge on some drops is as much as 5 to 9 times the magnitude of the electronic charge. In the absence of a field, the speed with which the drops fall is determined by both the viscosity of the air and the a^celeration of gravity. Oil drops are used because water evaporates so much more readily.

20. Options (a) and (d) are true statements, but were known before Rutherford's experiment. Option (b) is false. Option (c) is true, but was determined by combining Thomson's value of the e/m ratio with Millikan's value for the charge on an electron.

21. Convert from eV/atom to kJ/mol as follows :

$$\left(12.084 \frac{eV}{atom}\right)\left(1.602177 \times 10^{-19} \frac{J}{eV}\right)\left(6.0221 \times 10^{23} \frac{atom}{mol}\right)$$

$$= 1.166 \times 10^6\ J{\cdot}mol^{-1}) = 1166\ kJ{\cdot}mol^{-1}$$

22. In SI units, convert the wave number to m^{-1}

$(5331.6\ cm^{-1})\ (100\ cm.m^{-1}) = 5.3316 \times 10^5\ m^{-1}$

$\nu = c\bar{\nu} = (2.9979 \times 10^8\ m/s)\ (5.3316 \times 10^5\ m^{-1}) = 1.598 \times 10^{14}\ s^{-1}$

23. For an H-like ion

$$\Delta E = hcRz^2\left(\frac{1}{n_L^2} - \frac{1}{n_H^2}\right)$$

The ionization energy is the energy required to move the electron from $n_L = 1$ to $n_H = \infty$, or $hcRz^2$. Since Z = 1 for H and 3 for Li^{2+}, it requires $3^2 = 9$ times as much energy to ionize the electron in Li^{2+} as it does in H.

24. The de Broglie wavelength is $\lambda = h/mv$, where, must be in kg in SI units.

$$\lambda = \frac{6.6261 \times 10^{-34}\,\text{J}\cdot\text{s}}{(4.00 \times 10^{-2}\,\text{kg})(20.0\,\text{m/s})} = 8.28 \times 10^{-34}\ \text{m} = 8.28 \times 10^{-25}\ \text{nm}$$

25. $E = h\nu = hc/\lambda$

$$E = \frac{(6.626 \times 10^{-34}\ \text{J}\cdot\text{s})(2.9979 \times 10^{8}\,\text{m/s})}{(4.0 \times 10^{-6}\,\text{m})} = 5.0 \times 10^{-20}\ \text{J}$$

26. For the Brackett series, $n_L = 4$. The most intense line is the one with n_H just 1 more than n_L, i.e., $n_H = 5$. The state with n = 5 has a greater population than any state with n greater than 5 when H atoms are excited.

Chapter 2 : CHEMICAL BONDING

1. In forming ions, the transition metals lose their valence (outermost) shell electrons first, followed by their outer *d* electrons. *Note* : Inorder for transition metal ions to be colored, the *d* orbitals must be *partially* filled. In this case, the solution containing the Ni^{2+} ion would be colored (green).
2. Because the distance between the electrons and the nucleus is increasing, the electrons are becoming further away from the nucleus, making it easier to remove them by overcoming the electrostatic force attracting them to the nucleus. Also, there are more electrons in the way, increasing interference (the electron shielding effect).
3. Because chlorine is a halogen, it has an oridation number of –1. And because there are 6 chlorines, there is a total charge of –6 for the chlorines. The overall charge is –2, so algebraically, x + (–6) = –2. Solving this yields x = +4. Thus, the charge (or oxidation number) of platinum is +4.
4. Generally, expect an ionic bond whenever you have a metal (Cs) bonded to nonmetal (I). If you have a table of electronegativities to refer to, the electronegativity difference is greater than 1.7 for ionic compounds.
5. Note that all the ions have 10 electrons; that is, they are all isoelectronic with neon. Because they all have the same number of electrons, the only factor that will determine their size will be the nuclear charge—the greater the nuclear charge, the smaller the radius. Therefore, magnesium, with a nuclear charge of +2, has the smallest radius among these ions.
7. The compound is ionic—a metal (Al) bonded to a nonmetal (Cl). All ionic compounds are solids at room temperature and pressure. Aluminium has 13 electrons. As an ion, it will lose 3 electrons to become isoelectronic with neon. Thus the aluminium ion will have the electronic configuration $1s^2 2s^2 2p^6$.
8. Potassium is a metal, and the polyatomic anion, ClO_4 is a nonmetal; therefore, the compound is an ionic solid at room temperature. When the compound is dissolved in water, the ionic bond between the cation, K^+, and the polyatomic anion, ClO_4^-, is broken due to the polarity of the water molecule, resulting in the two aqueous ions, K^+ and ClO_4^-
9. In choice A, $KClO_3$ is potassium chlorate, not perchlorate. In choice B, CuO is copper (II) oxide, to distinguish it from copper (I) oxide, Cu_2O. In choice C, the formula for aluminium sulfate is $Al_2(SO_4)_4$. In D, the formula for magnesium phosphate is $Mg_3(PO_4)_2$.

10. 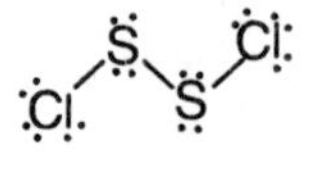

11. 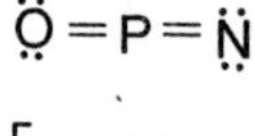

12. 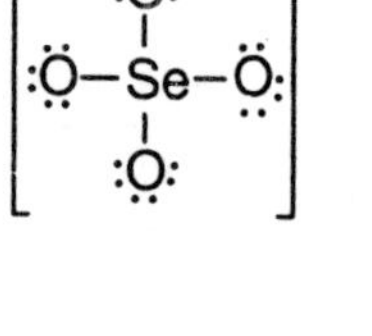

13. 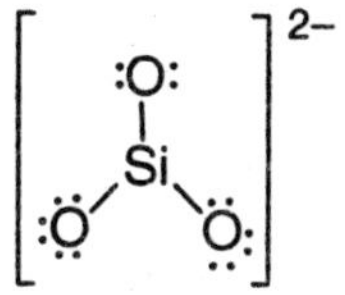

14. 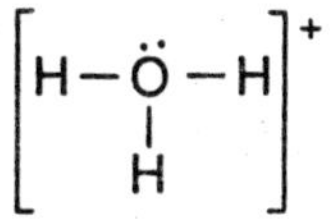

15. A. 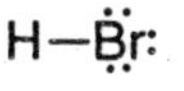 B. 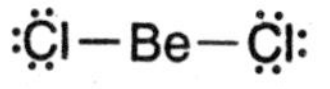 C. H—H D. 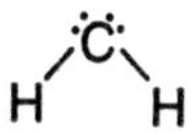

The Cl atom is more electronegative than the Be atom, resulting in a polar bond. However, because the molecule is linear and the two ends are identical, the overall molecule is nonpolar.

16. All single bonds are sigma (σ) in nature. Double bonds contain one sigma bond and one pi (π) bond. Triple bonds contain one sigma bond and two pi (π) bonds.

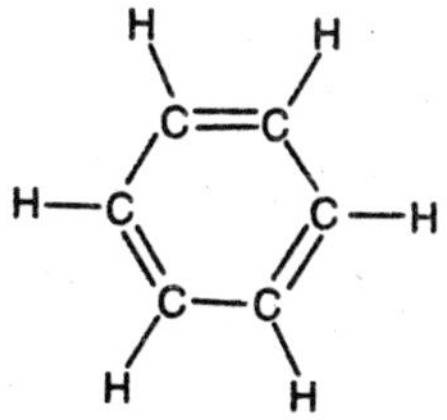

In the benzene molecule, there are 9 single bonds (9σ) and 3 double bonds (3σ and 3π) for a total of 12 sigma bonds.

17. There are no alternative ways of positioning electrons around the HI molecule.

18. 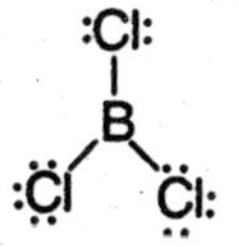

19. Electronegativity differences less than 1.7 are classified as covalent, *Unequal* differences in *sharing* electrons are know as polar covalent.

20. A coordinate covalent bond, also known as a dative bond, is a bond in which both electrons are furnished by one atom.

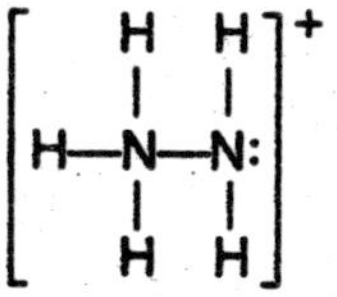

21. 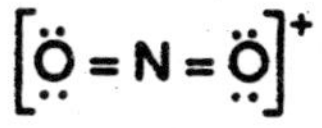

22. If the polarity of the bond A—B is about the same as those of the nonpolar bonds A—A and B—B then the bond energy of A—B can be taken as the average of the bond energies of A—A and B—B.

23. Since the ionisation energy of hydrogen atom is 1.312×10^3 kJ mol^{-1} hence the energy of first orbit of hydrogen atom is -1.312×10^6 J mol^{-1}

Energy of electron in second orbit,

$$E_2 = -\frac{1.312\times10^6 \text{ J mol}^{-1}}{2^2}$$

$$= -0.328\times10^6 \text{ J mol}^{-1}$$

Emission energy $= E_2 - E_1$

$= -0.328 \times 10^6 - (-1.312 \times 10^6)$

$= 0.984 \times 10^6$ J mol^{-1}

We know that, $E = N_A.h.v$

$$= N_A.h.\frac{c}{\lambda}$$

$$\text{or } \lambda = \frac{N_A.h.c.}{E}$$

$$= \frac{(6\times10^{23}\text{ mol}^{-1})(6.6\times10^{-34}\text{ kg m}^2\text{s}^{-1})(3\times10^8\text{ m s}^{-1})}{0.984\times10^6\text{ J mol}^{-1}}$$

$= 1207.3$Å

24. Bond energy of I_2 $= 240$ kJ mol^{-1}

$$= \frac{240\times10^3}{6.023\times10^{23}}$$

$= 3.98 \times 10^{-19}$ J molecules^{-1}

Energy absorbed $= hv = h.\frac{c}{\lambda}$

$$= \frac{6.626\times10^{-34}\text{ J s}\times10^8\text{ ms}^{-1}}{4500\times10^{-10}\text{ m}}$$

$= 4.417 \times 10^{-19}$J

Kinetic energy $= 4.417 \times 10^{-19}\text{J} - 3.98 \times 10^{-19}$ J

$= 0.437 \times 10^{-19}$J

$= 4.37 \times 10^{-20}$J

Kinetic energy of each atom

$$= \frac{4.37 \times 10^{-20}\,J}{2}$$

$$= 2.185 \times 10^{-20}J = 2.185 \times 10^{-23}KJ$$

25. $$\bar{v} = R_H\left(\frac{1}{n_1^2} - \frac{1}{n_2^2}\right)$$

or $$\frac{1}{\lambda} = R_H\left(\frac{1}{n_1^2} - \frac{1}{n_2^2}\right)$$

or $$\frac{1}{\lambda} = 1.1 \times 10^7\,m^{-1}\left(\frac{1}{1^2} - \frac{1}{4^2}\right)$$

$$= 1.1 \times 10^{7 m^{-1}}\left(1 - \frac{1}{16}\right)$$

or $$\lambda = \frac{16 \times 10^{-7}}{16.5} = 9.7 \times 10^{-8}\,m$$

26. Since,

$$\text{Density} = \frac{\text{mass}}{\text{volume}}$$

$$\therefore 13.6\,g\,cm^{-3} = \frac{200 \times 1.67 \times 10^{-24}\,g}{\alpha^3}$$

where α = edge length of the cube or diameter of an atom

$$\alpha = \left(\frac{200 \times 1.67 \times 10^{-24}\,g}{13.6\,g\,cm^{-3}}\right)^{1/3}$$

$$= 2.9 \times 10^{-8}\,cm = 2.9 \times 10^{-10}\,m$$

27. We know that,

$$\bar{v} = R_{Hv}Z^2\left[\frac{1}{n_1^2} - \frac{1}{n_2^2}\right]$$

$$= (10.9714 \times 10^6\,m^{-1})(1)^2\left(\frac{1}{1^2} - \frac{1}{3^2}\right)$$

$$= 9.7523 \times 10^6\,m^{-1}$$

$$\because v = \bar{v}c$$

$$= (9.7523 \times 10^6\,m^{-1})(3 \times 10^8\,m\,s^{-1})$$

$$= 2.9257 \times 10^{15}\,s^{-1}$$

$$\lambda = \frac{c}{v} = \frac{3 \times 10^8\,m\,s^{-1}}{2.9257 \times 10^{15}\,s^{-1}}$$

$$= 1025.4 \times 10^{-10}\,m = 1025.4\,\mathring{A}$$

28. We know that,

$$\frac{1}{\lambda}=R\left[\frac{1}{n_1^2}-\frac{1}{n_2^2}\right]$$

Substituting the value of $R = 1.09678 \times 10^7\ m^{-1}$, $n_1 = 1$, $n_2 = \infty$ in the above equation, we get

$$\frac{1}{\lambda}=1.09678\times10^7\ m^{-1}\left[\frac{1}{1^2}-\frac{1}{\infty^2}\right]$$

$$= 1.09678 \times 10^7\ m^{-1}$$

or $\lambda = 9.11 \times 10^{-6}\ m$

But $E = h\nu = \dfrac{hc}{\lambda}$

$$= (1.09678 \times 10^7\ m^{-1})\ (6.6256 \times 10^{-34}\ Js)\ (3.0 \times 10^8\ ms^{-1})$$

$$= 217.9 \times 10^{-20} J$$

$$= 217.9 \times 10^{-23}\ kJ$$

29. A molecule with an odd number of electrons is paramagnetic. Chlorine dioxide, CO_2, has 19 valence electrons. All the other molecules have an even number of C1 atoms and an even number of electrons.

30. Methane has 4 single C-H σ bonds, formed by the overlap of an sp^3 hybrid on C and the is atomic orbital of each H.

31. Benzene, C_6H_6, has planar hexagonal geometry and the vector sum of the six C-H bond dipoles is zero.

32. The Zr, Mo, and Sb atoms all have unpaired electrons in the ground state. See Table for the electronic configurations of all elements.

33. Phosphorus trifluoride has the same geometry as ammonia, NH_3, that is, trigonal pyramidal.

34. Phosphorus, in Group 15 (Group VA) has 5 valence electrons and each of the four H atoms has 1. One electron is lost on forming the PH_4+ cation, so there are a total of 8 valence electrons: 5 + 4 - 1 = 8.

35. There are 28 valence electrons for ICl_3, 7 from each of the four halogens. The structure is

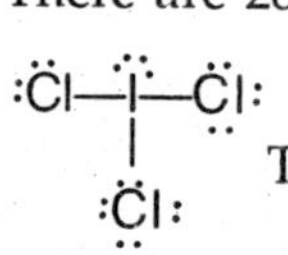

The formal charge on I is zero, as all 4 of the lone pair electrons plus 1 from each of the bonded pairs are assigned to the I atom. The molecule is T-shaped, an AX_3E_2 system.

36. In sulfur dioxide, SO_2, the S atom has 2 bonded pairs and one lone pair of electrons around it. Thus SO_2 is an AX_2E molecule, with a bond angle just slightly less than 120°. The S atom uses sp^2 hybrids in bonding to O and the lone pair is also in an sp^2 hybrid atomic orbital.

37. There are 26 valence electrons for the SO_3^{2-} ion : 6 from each O and from S, plus 2 from the -2 charge on the ion. In the Lewis structure there are 3 bonded pairs around S, and one lone pair. Thus SO_3^{2-}, is an AX_3E system, like NH_3, and is trigonal pyramidal.

38. In the best Lewis structure with only octets of electrons, there are four single bonds from Cl to each O. One electron from each bond is assigned to Cl, so that Cl has a formal charge of +3 (as there are 7 valence electrons around a gaseous Cl atom). In order to reduce the formal charge on Cl, one must include double-bonded structures, which have 10 or 12 electrons around Cl (for one or two double bonds. respectively).

39. Carbon tetrafluoride has tetrahedral geometry.

40. In a square planar AX_4 system, with 4 bonded pairs and no lone pairs on the central atom, the central atom uses dsp^2 hybrids. Count the number of atomic orbitals combined to obtain the number of hybrid orbitals. In an AX_4E_2 system, with two lone pairs on the central atom, octahedral d^2sp^3 hybrids are used, and the two lone pairs occupy the axial positions, so the moleculeor ion has square-planar geometry. In that sense, (d) is also a correct answer, although d^2sp^3 hybrids are octahedral.

41. Ozone has two bonded pairs and one lone pair around the central oxygen atom, so the bond angle in this AX_2E structure should be slightly less than 120°. Both CS_2 and N_2O are linear, and H_2O and OF_2 are AX_2E_2 systems, with bond angles slightly less than 109°28′, the tetrahedral angle.

42. The P atom has 5 valence electrons, each of the four O atoms has 6, and there are 3 added to make the -3 charge on the ion, for a total of 5 + 4(6) + 3 = 32.

43. There are 32 valence electrons in SIF_4, silicon tetrafluoride (or tetrafluoro silane), with 4 bonded pairs forming an octet around Si, and 1 bonded pair plus 3 lone pairs around each F. Hydrogen atoms never have 8 electrons in the valence shell. They have 2 electrons (He configuration) in bonded structures such as C_2H_4 and KH. There are an odd number of valence electrons, 17, in NO_2, and there are 12 electrons around I in IF_5, an AX_5E structure.

44. The electronegativity difference between Hand F is larger than that between any of the other pairs of atoms, so the HF bond has the greatest amount of ionic character of those listed.

45. The ions in strontium nitride, Sr_3N_2, are the Sr^{2+} and N^{3-} ions. The bonding is ionic as there is an electronegativity difference of 2.0 between N and Sr. The bonding is best described as ionic with some covalent character, however.

46. There are 26 valence electrons in SO_3^{2-}, with three bonded pairs and one lone pair around S. It is an AX_3E system, using tetrahedral sp^3 hybrid atomic orbitals.

47. Boron forms many electron deficient compounds. There are only 12 valence electrons for B_2H_6.

48. The bond order is 3 in CO and NO^+, 2.5 in O_2^+ and N_2^+, but only 1.5 in O_2^-, with 13 electrons to occupy molecular orbitals.

49. To describe the bonding in NO_3^-, three resonance structure are needed. For ClO_3^-, there is only one structure with an octet about each atom, but that does give the Cl atom a formal charge of +3. If one includes doublebonded structures with 10 or 12 electrons around Cl, then resonance structures are necessary.

50. Tetrachloroethylene, C_2Cl_4, has one π bond between the two C atoms. There are no π bonds in $RaCl_2$ or F_2, and 2 in CS_2 and HCN.

51. Carbon disulfide, like CO_2, has 2 double bonds, and is a linear molecule, $:\ddot{S} = C = \ddot{S}:$

52. The structure of hydrocyanic acid is H—C≡N:

53. Radium is an alkaline earth. The ions in $RaCl_2$ are Ra^{2+} and Cl^-.

54. At room temperature HCN is a gas, and as the molecule is asymmetric, it is polar. The CS_2, F_2, and C_2Cl_4 molecules are all nonpolar as they are symmetric. The ionic crystalline solid $RaCl_2$ does not consist of discrete molecules.

55. There is one σ bond in F_2. which is nonpolar.

56. Only $RaCl_2$, of those substances listed, is a solid. Both F_2 and HCN are gases, and CS_2 and $Cl_2C{=}CCl_2$ are both liquids.

57. There are 24 valence electrons for CO_3^{2-}. There are 3 bonded pairs around C, so the ion is trigonal planar.

58. Sulfur trioxide, SO_3, has polar bonds because S and O differ in electronegativity, but the sum of the bond-dipole vectors is zero, so this trigonal planar molecule is nonpolar. The NO_2, H_2O, and SO_2 molecules are bent and therefore they are polar. The HCl molecule is polar as it is asymmetric.

59. There are 20 valence electrons in OF_2. Each atom has a complete octet, with 2 lone pairs and 2 bonded pairs on the O atom. There are only 18 valence electrons for $SnCl_2$ with 1 lone paid and 2 bonded pairs on the Sn atom, so the Sn is electron deficient. Beryllium (Z=4) never has an octet of electrons. There are 19 valence electrons, and odd number, in ClO_2, and boron is electron deficient in BBr_3.

60. (i) Microwaves are a form of electromagnetic radiation and thus travel at the speed of light, 2.998×10^8 m s^{-1}.

$$\lambda = \frac{c}{v} = \frac{2.998 \times 10^8 \text{ m s}^{-1}}{2.45 \times 10^9 \text{ s}^{-1}} = 0.122 \text{ m}$$

(ii) Planck's equation, $E = hv$ yields the answer directly.

$E = hv = 6.626 \times 10^{-34}$ Js $\times 2.45 \times 10^9$ s^{-1} = 1.62×10^{-24} J

(iii) A mole of this radiation means one mole of photons. The energy content per mole of radiation is

$$\text{no. J/mol} = \frac{1.62 \times 10^{-24} \text{ J}}{\text{photon}} \times \frac{6.022 \times 10^{23} \text{ photons}}{1 \text{ mol photons}} = 0.976 \text{ J/mol}$$

Now, the quantity of heat energy required to raise the temperature of the water.

$$= 225 \text{ g water} \times \frac{4.18 \text{ J}}{\text{g water K}} \times (372.5 - 297.5)\text{K} = 7.05 \times 10^4 \text{ J}$$

Now we can determine the amount of radiation required. In doing so, we use a conversion factor based on the fact that 1 mol photons ⇔ 0.976 J.

$$\text{no. mol radiation} = 7.05 \times 10^4 \text{ J} \times \frac{1 \text{ mol photons}}{0.976 \text{ J}} = 7.22 \times 10^4 \text{ mol}$$

61. First you need to find out the atomic number of phosphorus. The atomic number, Z for phosphorus is 15, and its symbol is P.

$A = Z + N$, so $A = 15 + 16 = 31$

The full symbol notation is $^{31}_{15}P$

It specifies a particular atomic number and mass number, and hence represents an atom with a particular number of protons and number of neutrons. It is called a *nuclide*.

62. You start with the formula for the energy levels of the H atom, $E = \frac{-R_H}{n^2}$ and obtain the energy change for the transition. This energy difference will equal the energy of the photon, from which you can calculate frequency, then wavelength, of the emitted light.

From the formula for the energy levels, you know that

$$E_i = \frac{-R_H}{4^2} = \frac{-R_H}{16} \text{ and } E_f = \frac{-R_H}{2^2} = \frac{-R_H}{4}$$

You subtract the lower value from the higher value, to get a positive result. (The energy of the photon is positive. If your result is negative, it means you calculated $E_f - Ei$, rather than $E_i - E_f$. Simply reverse the substraction to obtain a positive result.) Because this result equals the energy of the photon, you equate it to hv.

$$\left(\frac{-R_H}{16}\right) - \left(\frac{-R_H}{4}\right) = \frac{-4R_H + 16R_H}{64} = \frac{-R_H + 4R_H}{16} = \frac{3R_H}{16} = hv$$

The frequency of the light emitted is:

$$v = \frac{3R_H}{16h} = \frac{3}{16} \times \frac{2.179 \times 10^{-18}\,\text{J}}{6.63 \times 10^{-34}\,\text{J.s}} = 6.17 \times 10^{14/s}$$

Since $\lambda = c/v$,

$$\lambda = \frac{3.00 \times 10^8\,\text{m}/\cancel{s}}{6.17 \times 10^{14/\cancel{s}}} = 4.86 \times 10^{-7}\,\text{m, or } 486\,\text{nm}$$

The colour is blue green.

63. First, you need to convert the wavelength to meters :

700 nm is 700×10^{-9} m, and the speed of light is 3.00×10^8 ms^{-1}.

Inserting these values into the equation :

$$300 \times 10^8 = \lambda \times 700 \times 10^{-9}$$

Rearranging : $\lambda = \frac{3.00 \times 10^8}{700 \times 10^{-9}} = 4.29 \times 10^{14}$

So the frequency of the red light is 4.29×10^{14} Hz.

64. Since,

$$\Delta x(M\Delta v) = \frac{\lambda}{4\pi}$$

$$(10^{-10}m)\ (9.1 \times 10^{-31}\ kg\ \Delta v) = \frac{6.6\times10^{-34}\ kg\,m^2s^{-1}}{4\times3.14}$$

$$\text{or } \Delta v = \frac{6.6\times10^{-34}\ kg\,m^2\,s^{-1}}{4\times3.14\times10^{-10}\ m\times9.1\times10^{-31}\ kg} = 0.577 \times 10^6 ms^{-1}.$$

65. Kinetic energy = $1/2\ mv^2 = 5 \times 10^{-25}$ Joules

$$\therefore v^2 = \frac{10\times10^{-25}\ J}{9.1\times10^{-31}\ kg}$$

or $$v = \sqrt{(1.0989 \times 10^6)}$$

$$= 1.04828 \times 10^3\ ms^{-1}$$

But

$$\lambda = \frac{h}{mv} = \frac{6.6\times 10^{-34}\ Js}{1.05\times10^3\ ms^{-1}\times9.1\times10^{-31}\ kg}$$

$$= 0.69\times10^{-6}\ m$$

66. Since K.E. = ½ mv^2

$$\therefore v = \sqrt{\frac{2.K.E.}{m}}$$

$$= \sqrt{\frac{2\times4.55\times10^{-25}\ J}{9.1\times10^{-31}\ kg}}$$

$$= 10^3\ m\,s^{-1} \quad (\because j = kg\,m^2\,s^{-2})$$

Now $$\lambda = \frac{h}{m\,v}$$

$$= \frac{6.62\times10^{-34}\ kg\,m^2\,s^{-1}}{(9.11\times10^{-31}\ kg)\ (10^3\ m\,s^{-1})}$$

$$= 7.2\times10^{-7}\ m = 720\ \text{Å}.$$

67. Since,

$$\Delta x(M\Delta v) = \frac{h}{2\pi}$$

$$\therefore (10^{-10} m)(M \times 5.27 \times 10^{-24} m s^{-1}) = \frac{6.625 \times 10^{-34} Js}{2(3.14)}$$

$$\therefore M = \frac{6.625 \times 10^{-34} Js}{(10^{-10} m)(5.27 \times 10^{-24} m s^{-1})(2)(3.14)} \quad (\because J = kg m^2 s^{-2})$$

$$= 0.2 kg = 200 mg.$$

68. Molecular weight of methane = 16

$\therefore$ 16 g methane contain 6.023×10^{23} molecules of methane

or 1.6 g methane contain 6.023×10^{22} molecules of methane.

Since 1 molecule of CH_4 contains 10 electrons

$\therefore$ 6.023×10^{22} of CH_4 contains

$= 6.023 \times 10^{23}$ electrons.

69. Since 14 g ^{14}C contain 6.023×10^{23} atoms

Hence 7 mg ^{14}C contain $= \frac{6.023 \times 10^{23} \times 7 \times 10^{-3} g}{14 g}$ atom

$= 3.012 \times 10^{20}$ atoms

Now, no. of neurons is one atom of $^{14}C = 14 - 6 = 8$

Hence, no. of neutron in 3.012×10^{20} atoms

$= 3.012 \times 10^{20} \times 8$

$= 24.096 \times 10^{20}$ neutrons

Now we have, the mass of one neutron = 1.675×10^{-27} kg

Hence, the mass of total neutrons

$= 1.675 \times 10^{-27}$ kg $\times 24.096 \times 10^{20}$

$= 4.036 \times 10^{-7}$ kg

70. Since

$$\frac{1}{\lambda} = R_H\left[\frac{1}{n_1^2} - \frac{1}{n_2^2}\right]$$

$$= R_H\left[\frac{1}{1^2} - \frac{1}{2^2}\right] = R_H\left[\frac{1}{1} - \frac{1}{4}\right] = \frac{3}{4}R_H$$

or $$\lambda = \frac{3}{3R_H} = \frac{4}{3} \times \frac{1}{1.09678 \times 10^7 m^{-1}}$$

$$\lambda = 1.216 \times 10^{-7} m$$

Now, $$v = \frac{c}{\lambda} = \frac{3 \times 10^8 m s^{-1}}{1.216 \times 10^7 m} = 2.47 \times 10^{15} s^{-1}$$

Since $E = hv$

Hence $E = 6.625 \times 10^{-34}\,Js \times 2.47 \times 10^{15}\,s^{-1}$

$= 16.36 \times 10^{-19}\,J$

71. $\lambda = 242\text{ nm} = 242 \times 10^{-9}\text{ m}$

$c = 3 \times 10^{8}\text{ m s}^{-1}$

Now $v = \dfrac{3\times10^{8}\,\text{m s}^{-1}}{242\times10^{-9}\,\text{m}} = 1.24\times10^{15}\,\text{s}^{-1}$

But $E = hv$

$= (6.626 \times 10^{-34}\text{ J s})\ (1.24 \times 10^{15}\text{ s}^{-1})$

$= 0.082 \times 10^{-17}\text{ J}$

$= 0.082 \times 10^{-17}\text{ J} \times 6.023 \times 10^{23}\text{ mol}^{-1}$

$= 494.5 \times 10^{3}\text{ J mol}^{-1}$

$= 494.5\text{ kJ mol}^{-1}$

72. In Balmer series of hydrogen atom electron comes from different energy levels to second orbit. Therefore, for wave number for shortest wavelength $n_1 = 2$ and $n_2 = \infty$.

We know that,

$$\bar{v} = R_H\left[\frac{1}{n_1^2}-\frac{1}{n_2^2}\right]$$

or

$$\bar{v} = 109677\,\text{cm}^{-1}\left[\frac{1}{2^2}-\frac{1}{\infty^2}\right] = 109677\times\frac{1}{4}\text{cm}^{-1}$$

$$= 27419.25\text{ cm}^{-1}$$

73. By Bohr's law

$$\frac{nh}{2\pi} = mvr$$

$$\therefore n = \frac{mvr}{h/2\pi} = \frac{2\pi r}{h/mv}$$

Now by de Broglie's equation

$\lambda = h/mv$

$2\pi r = \text{circumference}$

So $n = \dfrac{\text{circumference}}{\lambda}$

= No. of waves

Hence no. of waves in one complete revolution = 3.

74. The dipole moment of a 100% ionic "molecule" at the given internuclear distance would be

$$(1.60 \times 10^{-19}\ \text{C})\ (1.41 \times 10^{-10}\text{m}) = 2.26 \times 10^{-29}\ \text{C·m}$$

The actual dipole is less; the percent ionic character is given by

$$\frac{2.60 \times 10^{-30}\ \text{C·m}}{2.26 \times 10^{-29}\ \text{C·m}} \times 100\% = 11.5\%$$

76. O_2^-—1 unpaired electron, hence

↑↓ ↑ $\mu = \sqrt{n(n+1)} = \sqrt{1(1+2)} = 1.73\text{B.M.}$

↑↓ ↑↓

↑↓

77. O_2^{2-} has a bond order of 1; O_2 has a bond order of 2. Hence O_2^{2-} will have a longer and weaker bond.

78. Co has a bond order of 3. Others have, BN = 2, F_2 = 1, Ne_2 = 0.

79. Let us calculate the dipole moment of a hypothetical, fully ionized Li^+H^- ion pair with a separation of 1.596 Å.

μ(hypothetical) = (1 electronic charge) (separation) = $(1.602 \times 10^{-19}\ \text{C})\ (1.596 \times 10^{-10}\ \text{m}) = 2.557 \times 10^{-29}$ C·m. The fractional ionic character is the actual dipole moment divided by the hypothetical.

$$\text{Fractional ionic character} = \frac{1.964 \times 10^{-29}\ \text{C·m}}{2.557 \times 10^{-29}\ \text{C·m}} = 0.768$$

The bond is computed to be 76.8% ionic.

80. The distance from the center of the hexagon to any corner (apex) is 1.40 Å. Since the hexagon is regular, the lines from the center to two adjacent apices form an equilateral triangle.

(i) The distance from the enter of the hexagon to an iodine atom is (1.40 Å) + (0.77 Å) + (1.33 Å) = 3.50 Å. Since this distance is also the side of an equilateral triangle, the distance iodine to another is also 3.50 Å [Fig. (a)].

(ii) Half the I—I distance is $\sqrt{3}/2$ (3.50 Å); the entire distance is $\sqrt{3}$(3.50Å) = 6.06 Å [Fig. (b)].

(iii) 2 (3.50 Å) = 7.00 Å [Fig. (c)].

$$\frac{\sqrt{3}}{2}\ (3.50\ \text{Å})$$

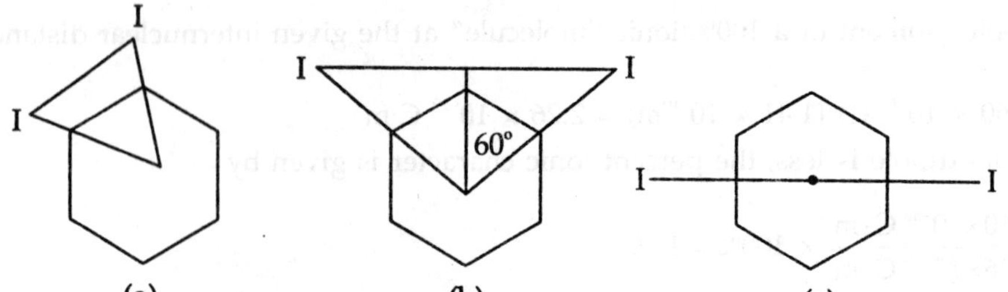

(a) (b) (c)

81. The width of the skeleton is merely the sum of two center-apex distances = 2.8Å. The length is equal to 8 times the length shown as x in Fig. below. That length is :

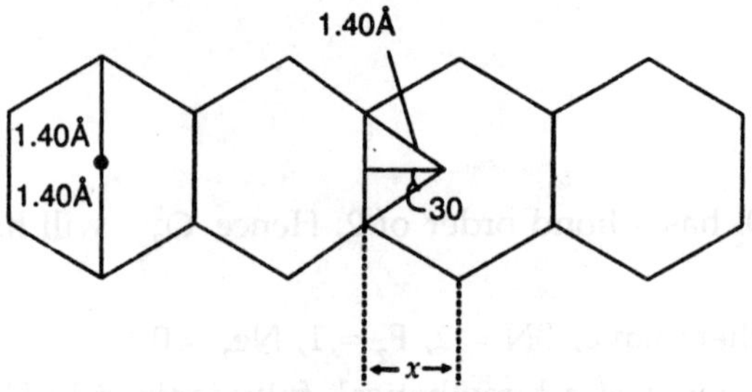

$$x = 1.40\ (\sqrt{3}/2) = 1.40\ (0.866) = 1.21\ \text{Å}$$
$$8x = 9.68\ \text{Å}$$

82. $\mu = 1.85\text{D} = 1.85 \times 10^{-18}$ esu·cm $= \delta d$

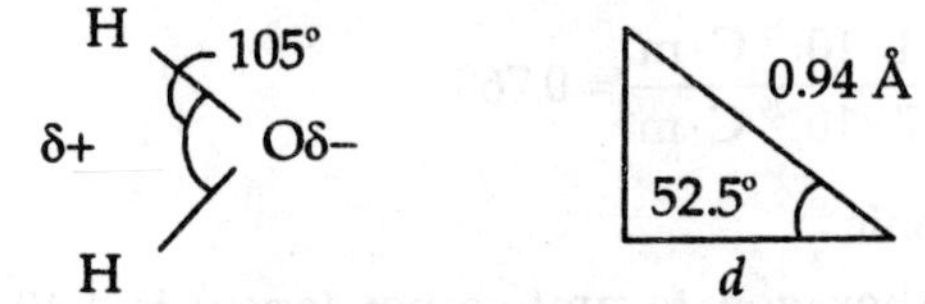

$$\cos 52.5^\circ = d/0.94\ \text{Å}$$
$$d = (0.609)\ (0.94\ \text{Å}) = 0.572\ \text{Å}$$

$$\delta = \frac{\mu}{d} = \frac{1.85\text{D}}{0.572\,\text{Å}} = \frac{1.85 \times 10^{-18}\ \text{esu}\cdot\text{sm}}{0.572 \times 10^{-8}\ \text{cm}} = 3.2 \times 10^{-10}\ \text{esu}$$

or 0.67 times the electronic charge (4.8×10^{-10} esu).

83. The sp^3d hybrids have 90° and 120° angles. Rearrangement of the bond angles could produce angles greater than 90° but at the expense of the highly symmetrical structure.

84. AX_5 is the only molecule with 5 atoms bonded to a central atom—it is the only possible molecule listed which could be trigonal bipyramidal, and in the absence of lone pairs, it must be dsp^3 or sp^3d. (An inner or a valence d orbital may be involved).

85. The electronic configuration of NO is shown in Fig. below. The 6 bonding and 1 antibonding electrons yield a net of 5 bonding electrons, for a bond order of $2\frac{1}{2}$.

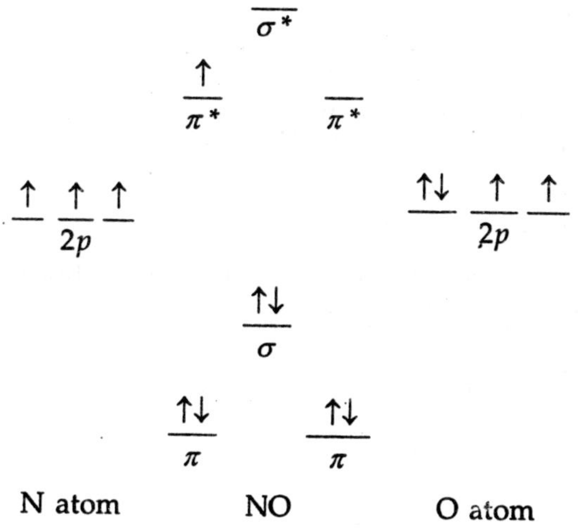
σ*
↑
π*
π*
↑ ↑ ↑
2p
↑↓ ↑ ↑
2p
↑↓
σ
↑↓
π
↑↓
π
N atom
NO
O atom

Chapter 3 : GAS LAWS

1. First, calculate the volume the CO_2 gas would occupy at 2.00 atm using the relationship

$$\frac{P_1 V_1}{T_1} = \frac{P_2 V_2}{T_2}$$

Since the temperature is remaining constant, we can use $P_1V_1 = P_2 V_2$, where initial conditions are at STP and final conditions are at 273 K and 2.0 atm.

(1.00 atm) (22.4 liters) = (2.00 atm) (V_2)

V_2 = 11.2 liters

Since the amount of gas has not changed from the initial STP conditions (1 mole or 44.01 grams), the density of the gas at 2.00 atm and 273 K would be

$$\frac{44.01 \text{ grams}}{11.2 \text{ liters}} = 4\,g/L$$

Another approach to this problem would be to use the ideal gas law, $PV = nRT$.

$$\text{density} = \frac{g}{V} = \frac{P \cdot MW}{RT}$$

$$= \frac{200\,\cancel{atm} \cdot 44.01\,g \cdot \cancel{mole}^{-1}}{0.08211 \cdot \cancel{atm} \cdot \cancel{mole}^{-1} \cdot \cancel{K}^{-1} \cdot 273\,\cancel{K}} \approx 4\,g/L$$

2. Begin by writing down a balanced equation.

$2CO(g) + O_2(g) \rightarrow 2CO_2(g)$

Next, use the factor-label method to solve the problem.

$$\frac{22\,\cancel{g\,CO_2}}{1} \times \frac{1\,\cancel{mole\,CO_2}}{44\,\cancel{g\,CO_2}} \times \frac{1\,\cancel{mole\,O_2}}{2\,\cancel{moles\,CO_2}} \times \frac{22.4\,L\,O_2}{1\,\cancel{mole\,O_2}} = 5.6\,L\ O_2$$

Cancel the 22 and the 44 first, leaving 1/4 times 22.4 = 5.61L.

3. Graham's law of effusion for two gases:

$$\frac{u_1}{u_2} = \frac{\sqrt{MW_2}}{\sqrt{MW_1}}$$

$$\frac{5.00 \times 10^4}{x} = \frac{\sqrt{4}}{\sqrt{16}} \qquad \therefore\ x = 1.00 \times 10^5 \text{ cm/sec}$$

4. Begin this problem by listing the known facts.

m = 2.00 g V = 0.821 liter T = 400.K P = 1.00 atm MW = ?

You will need to use the ideal gas law to solve the problem : $PV = nRT$. Because moles can be calculated by dividing the mass of the sample by its molecular weight, the ideal gas law becomes.

$$PV = \frac{m}{MW} R \cdot T$$

$$MW = \frac{m \cdot R \cdot T}{P \cdot V} = \frac{(2.00\,g)\,(\cancel{0.0821\,liter} \cdot atm)\,(400 \cdot K)}{(\cancel{1.00}\,atm)\,(\cancel{0.821}\,liter) \cdot mole \cdot K} = \frac{2 \times 400}{10} = 80.0\,g/mole$$

5. Begin by listing the information that is known.

   V = 8.00 liters H_2

   P = 720. mm Hg – 55 mm Hg = 665 mm Hg (corrected for vapor pressure)

   T = 313 K

   Using the ideal gas law, PV = nRT, and realizing that one can determine grams from moles, the equation becomes:

$$nH_2 = \frac{PV}{RT} = \frac{(665/760\ \cancel{atm}) \cdot 8.00\,L\,H_2}{(0.0821\ \cancel{L\,atm}/mole \cdot \cancel{K}) \cdot 313\ \cancel{K}}$$

   Since for every mole of hydrogen produced, one mole of zinc is consumed, the last step would be to convert these moles to grams by multiplying by the molar mass of zinc.

$$\frac{\cancel{moles\,H_2}}{1} \cdot \frac{1\ \cancel{mole\,Zn}}{1\ \cancel{mole\,H_2}} \cdot \frac{65.39\,g\,Zn}{1\ \cancel{mole\,Zn}} = \frac{(665/760) \times 8.00 \times 65.39}{(0.0821) \times 313}$$

6. Use the formula

$$P_1 = \frac{n_1}{n_{total}} P_{total}$$

   derived from Dalton's law of partial pressures. Find the number of moles of the two gases first.

$$\frac{8.0\ \cancel{g\,He}}{1} \times \frac{1\,mole\,He}{4.0\ \cancel{g\,He}} = 2.0\,moles\,He$$

$$\frac{16\ \cancel{g\,O_2}}{1} \times \frac{1\,mole\,O_2}{32\,g\,O_2} = 0.50\,mole\,O_2$$

$$n_{total} = 2.0\,moles + 0.50\,mole = 2.5\,moles$$

$$P_{He} = \frac{2.0\ \cancel{moles}}{2.5\ \cancel{moles}} \times 5.00\,atm = 4.00\,atm$$

7. The van der Waals constant *a* corrects for the attractive forces between gas molecules. The constant *b* corrects for particle volume. The attractive forces between gas molecules become pronounced when the molecules are closer together. Conditions which favor this are low temperatures (173K) and high pressures (5.0 atm).

8. Oxygen gas weighs 32 grams per mole, whereas helium gas weighs only 4 grams per mole. That there are roughly 3 moles of oxygen molecules and 25 moles of helium molecules. Gas

pressure is proportional to the number of molecules and temperature, and inversely proportional to the size of the container. Since there are more helium molecules, we expect a higher pressure in the helium container (with all other variables being held constant). As long as the temperatures of the two containers are the same, the average kinetic energies of the two gases are the same.

9. Manometer (D), the correct answer, shows the air pressure to be 10 mm Hg greater than that of the gas. Manometer (A) shows the gas pressure to be 10 mm Hg greater than air pressure. Manometer (B) shows the gas resource to be equal to that of the air pressure. Manometer (C) is a closed manometer showing that the pressure of the gas is 10 mm Hg greater than the unknown pressure of the left side of the manometer. Manometer (E) is also closed and shows that the gas on the right side (770 mm Hg) is exerting a pressure 10 mm Hg greater than the gas on the left side.

10. The mole fraction of CO_2 $= \dfrac{3.4}{3.4 + 4.5 + 1.1} = 0.38$

So: $pCO_2 = 0.38 \times 1.0 \times 1.0^5 = 3.8 \times 10^4 \text{Pa}$

The mole fraction of O_2 $= \dfrac{4.5}{3.4 + 4.5 + 1.1} = 0.50$

So: $pO_2 = 0.50 \times 1.0 \times 10^5 = 5.0 \times 10^4 \text{Pa}$

The mole fraction of N_2 $= \dfrac{1.1}{3.5 + 4.5 + 1.1} = 0.12$

So: $pN_s = 0.12 \times 1.0 \times 10^5 \text{Pa} = 1.2 \times 10^4 \text{Pa}$

11. As in other equilibrium questions, we need to start with a balanced equation. Under each term we write the partial pressure at equilibrium.

$$N_2(g) + 3H_2(g) \rightleftharpoons 2NH_3(g)$$

Partial pressures at equilibrium : x $3x$ 75 kPa

If the partial pressure of nitrogen is x kPa, then the partial pressure of hydrogen must be $3x$. This follows because the initial mole ratio was 1:3 and this remains unchanged through out the reaction, since the stoichiometry of the reaction indicates that for every mole of nitrogen that reacts, three moles of hydrogen react.

The total pressure is the sum of the partial pressures :

$$x + 3x + 75 = 800 \text{ kPa}$$

So: $x = 181$ kPa

Therefore: $pN_2 = 181$ kPa and $pH_2 = 544$ kPa.

Substituting these values into the expression for the equilibrium constant gives :

$$K_p = 1.93\ 5\ 10^{-7} \text{ kPa}^{-2}.$$

12. According to gas equation,

$$PV = nRT$$

or 6 atm × 3L = n × 0.0821 L atm K^{-1} mol^{-1} × 300 K or $$n = \frac{6 \times 3 \text{ mol}}{0.0821 \times 300} = 0.7308 \text{ mol}$$

Given that $n_{H_2} = 0.7$ mol

$$n_H = 0.7308 - 0.7 = 0.0308 \text{ mol}$$

Since $$\frac{r_{H_2}}{r_g} = \frac{0.7}{0.0308} = \sqrt{\frac{M_g}{M_{H_2}}}$$

$$\therefore \frac{0.7}{0.0308} = \sqrt{\frac{M_g}{2}}$$

or $$M_g = \frac{0.7 \times 0.7 \times 2}{0.0308 \times 0.0308} = 1033$$

13. The measurements are converted to SI units with the exception of the mass, which is left as grams. (The SI unit for mass is the kilogram).
$p = 100 \times 10^3$ kPa, $V = 100 \times 10^{-6}\text{m}^3$, $m = 0.178$g.
$M_r = ?$, $R = 8.314 \text{ JK}^{-1}\text{mol}^{-1}$ and $T = 298$ K.
Insert these values into the ideal gas equation :

$$pV = \frac{mRT}{M_r}$$

$$100 \times 10^3 \times 100 \times 10^{-6} = \frac{0.178}{M_r} \times 8.31 \times 298$$

Rearranging : $$M_r = \frac{0.178 \times 8.31 \times 298}{100 \times 10^3 \times 100 \times 10^{-6}} = 44.1$$

14. Mass of H_2 in bulb $= \frac{2.24 \text{ L}}{22.4 \text{ L mol}^{-1}} = 0.1 \text{ mol} = 0.2$ g

Mass of D_2 in bulb $= \frac{1.12 \text{ L}}{22.4 \text{ L mol}^{-1}} = 0.05 \text{ mol} = 0.2$g

Mass of D_2 left after opening the stop cock = 0.1 g

Mass of D_2 diffused = 0.2 – 0.1 = 0.1 g

Volume of D_2 diffused from the 1st bulb $= \frac{0.1 \text{ g} \times 22.4 \text{ L mol}^{-1}}{4\text{g mol}^{-1}} = 0.56$ L

Let the volume of H_2 diffused from the bulb = VL

$$V_{H_2} = \frac{V}{t}, V_{D_2} = \frac{0.56}{t}$$

According to Grahm's law of diffusion

$$\frac{r_1}{r_2} = \sqrt{\frac{D_2}{D_1}} = \sqrt{\frac{M_2}{M_1}}$$

$$\frac{V/t}{0.56/6} = \sqrt{\frac{4}{2}}$$

or $V = 0.56\sqrt{2} = 0.792$ L

Mass of D_2 in 2nd bulb $= 0.1$ g

Mass of H_2 in 2nd bulb $= \dfrac{0.792 \text{ L} \times 2\text{g mol}^{-1}}{22.4 \text{ L mol}^{-1}} = 0.0707$ g

Total mass of gases $= 0.1 + 0.0707 = 0.1707$ g

% of H_2 in gases $= \dfrac{0.707 \times 100}{0.1707} = 41.2\%$

% of D_2 in gases $= \dfrac{0.1 \times 100}{0.1707} = 58.8\%$

15. According to Grahm's law of diffusion

$$\frac{r_1}{r_2} = \sqrt{\frac{M_2}{M_1}}$$

or $$\frac{r_{SO_2}}{r_{O_2}} = \sqrt{\frac{M_{O_2}}{M_{SO_2}}}$$

Now $$r_{SO_2} = \frac{20\,dm^3}{60S}; r_{O_2} = \frac{V\,dm^3}{30\,S}; M_{O_2} = 32 \text{ and } M_{SO_2} = 64$$

$\therefore$ $$\frac{20/60}{V/30} = \sqrt{\frac{32}{64}}$$

or $$V = 10\sqrt{2} = 14.1\ dm^2$$

16. Let the air be taken 1 mole in the open vessel at 300 K and at T_2K 3/5 mole air in expelled. Thus the remaining air = (1 – 3/5) mole = 0.4 mole.

Since the vessel is open hence pressure is constant, volume is same as given. Therefore,

$$PV = n_1RT_1$$

and $$PV = n_2RT_2$$

$$\Rightarrow n_1RT_1 = n_2RT_2$$

or $$T_2 = \frac{n_1T_1}{n_2} = \frac{1 \times 300}{0.4} = 750 \text{ K}$$

17. We know that rate of effusion = $K \frac{P}{\sqrt{M}}$

where P is partial pressure of the gas

M = molar mass; M of He = 4, M of CH_4 = 16

$$\frac{\text{rate of effusion of He}}{\text{rate of effusion of } CH_4} = \frac{\frac{KP_{He}}{\sqrt{M_{He}}}}{\frac{KP_{CH_4}}{\sqrt{M_{CH_4}}}} = \frac{P_{He}}{P_{CH_4}} \times \sqrt{\frac{M_{CH_4}}{M_{He}}} = \frac{4}{1} \times \sqrt{\frac{16}{4}} = 8$$

18. Let V_2 be the volume of H_2 in cylinder at N.T.P. From gas equation,

$$\frac{P_1V_1}{T_1} = \frac{P_2V_2}{T_2}$$

or $$\frac{20 \text{ atm} \times 2.82 \text{ L}}{300 \text{ K}} = \frac{1 \text{ atm} \times V_2}{273 \text{ K}}$$

$$\therefore \quad V_2 = \frac{20 \text{ atm} \times 2.82 \text{ L} \times 273 \text{ K}}{300 \text{ K} \times 1 \text{ atm}} = 51324\text{L} = 51324 \text{ mL}$$

Actual volume transferred

$= 51324 - 2820 = 48504$ mL

Volume of each balloon

$= 4/3\ \pi r^3 = (4/3) \times 3.14 \times (10.5 \text{ cm})^3 = 4847$ mL

∴ No. of balloons which can be filled up

$$= \frac{48504 \text{ mL}}{4847 \text{ mL}} = 10$$

19. Volume of balloon,

$$V = \frac{4}{3}\pi r^3 = \frac{4}{3} \times 3.14 \times (10 \text{ m})^3$$

$= 4.187 \times 10^3 \text{ m}^3 = 4187 \text{ m}^3$

Mass of displaced air

$= 4187 \text{ m}^3 \times 1.2 \text{ Kg m}^{-3}$

$= 5024.4$ kg

$\therefore \quad PV = nRT$

$$\therefore \quad n = \frac{PV}{RT} = \frac{(1 \text{ atm}) \times 4187 \text{ m}^3}{(0.082 \text{ dm}^3 \text{ atm K}^{-1} \text{ mol}^{-1}) \times 300 \text{ K}}$$

$$= 170.2 \times 10^3 \text{ mol}$$

Mass of He present

$$= 170.2 \times 10^3 \times 4 \text{ g} = 680.8 \text{ kg}$$

Mass of empty balloon = 100 kg

Mass of the filled balloon

$$= 100 + 680.8 = 780.8 \text{ kg}$$

Pay-load = Mass of the air displaced – Total mass of balloon

$$= 5024.4 \text{ kg} - 780.8 \text{ kg} = 4243.6 \text{ kg}$$

20. According to gas equation,

$$P = \frac{nRT}{V} = \frac{wRT}{mV}$$

$$= \frac{(10 \text{ g})(0.082 \text{ L atm K}^{-1} \text{ mol}^{-1})(300 \text{ K})}{44 \text{ g mol}^{-1})(1 \text{ L})} = 5.59 \text{ atm}$$

Pressure inside bottle

$$= P + \text{atm, pressure}$$

$$= 5.59 + 1 = 6.59 \text{ atm}$$

21. According to gas equation,

$$PV = nRT$$

$$\therefore n = \frac{PV}{RT}$$

For NO_2

$$n = \frac{(1.053 \text{ atm})(0.25 \text{ L})}{(0.0821 \text{ L atm K}^{-1} \text{ mol}^{-1})(300 \text{ K})} = 0.0107 \text{ mole}$$

For O_2

$$n = \frac{(0.789 \text{ atm})(0.1 \text{ L})}{(0.0821 \text{ L atm K}^{-1} \text{ mol}^{-1})(300 \text{ K})} = 0.0032 \text{ mole}$$

From the reactions,

$$2NO + O_2 \rightarrow 2NO_2 \rightarrow N_2O_4$$

Since the reactions proceeds nearly to completion hence

Number of moles of O_2 = 0

Number of moles of NO = 0.0107 – 2 × 0.0032 = 0.0043

Number of moles of N_2O_4 = 0.0032

On cooling N_2O_4 condenses. Only NO will be present in 0.35 L. Therefore, Pressure of the gas

$$= \frac{nRT}{V}$$

$$= \frac{(0.0043 \text{ mole})\,(0.0821 \text{ L atm mol}^{-1}\text{ K}^{-1})\,(220 \text{ K})}{0.35 \text{ L}} = 0.22 \text{ atm}$$

22. For 1 mol of gas, Van der Waals' equation is $\left(P+\frac{a}{V^2}\right)(V-b) = RT$ . . . (1)

$$\Rightarrow PV + \frac{a}{V} - Pb - \frac{ab}{V^2} = RT$$

$$\Rightarrow V^3 - \frac{(Pb+RT)}{P}V^2 + \frac{a}{p}V - \frac{ab}{P} = 0 \quad \text{. . . (2)}$$

When $P = P_C$ and $T=T_C$ equation (2) must have the unique solution $V=V_C$. Hence the equation must be identical to

$$(V-V_C)^3 = V^3 - 3V^2V_C + 3VV_C^{\,2} - V_C^{\,3} = 0 \quad \text{. . . (3)}$$

Equating the coefficients of V^2 in equations (2) and (3), we get

$$\frac{P_Cb + RT_C}{P_C} = 3V_C$$

$$\Rightarrow b = \frac{3V_C P_C - RT_C}{P_C} = 3V_C - \frac{RT_C}{P_C} \quad \text{. . . (4)}$$

Equating the coefficients of V in equations (2) and (3), we get

$$\frac{a}{P_C} = 3V_C^2 \Rightarrow a = 3V_C^2 P_C \quad \text{. . . (5)}$$

Equating the coefficients of V in equations (2) and (3), we get

$$\frac{ab}{P_C} = V_C^3$$

$$\Rightarrow ab = V_C^3 P_C \Rightarrow b = \frac{V_C^3 P_C}{a} \quad \text{. . . (6)}$$

Substituting the value of a from equation (5) in equation (6), we get

$$b = \frac{V_C^3 P_C}{3V_C^2 P_C} \Rightarrow b = \frac{1}{3}V_C$$

23.
$$\frac{rN_2O}{rC_6H_{11}OBr} = \sqrt{\frac{M_{C_6H_{11}OBr}}{M_{N_2O}}} = \sqrt{\frac{178.9}{44}} = 2.016$$

Now $\frac{rN_2O}{rC_6H_{11}OBr} = \frac{x}{13-x}$

$\therefore \frac{x}{13-x} = 2.016 \Rightarrow \boxed{x = 8.69}$

24. The number of moles of gas and the volume are constant. Therefore, the pressure is directly proportional to the absolute temperature. Since the gas is cooled, the pressure must decrease. Only (a) and (b) are possible, and as the temperature has changed by only a relatively small amount, (b) is the most likely answer. To calculate the final pressure, use

$$P_2 = P_1\left(\frac{T_2}{T_1}\right) = (754.2 \text{ mmHg})\left(\frac{293.2 \text{ K}}{301.4 \text{ K}}\right) \text{ so } P_2 = 733.7 \text{ mmHg.}$$

25. Both n and V are constant, so P and T are directly proportional. Thus, $T_2 = T_1\,(P_2/T_1)$

$P_1 = (765.3/760)$ atm $= 1.007$ atm

$T_2 = (297.6 \text{ K})\ (2.50/1.007) = 738.8$ K

26. The molecular weight, M, of the gas is $M = \frac{\delta RT}{P}$.

$$M = \frac{(1.60\,g\cdot L^{-1})\,(62.364\,L\cdot mmHg\cdot mol^{-1}\,K^{-1})(299.7\,K)}{(680.2\,mmHg)}$$

$= 44.0$ g·mol^{-1}.

27. Since the gases are at the same temperature and pressure, the volume of each is directly proportional to the number of moles (Avogadro's Law). The molar ratio of CO_2 formed to C_4H_{10} burned is 4:1, by the conservation of mass. Thus the volume of CO_2 formed must be 4 times the volume of C_4H_{10} used, or 4(1.85 L) = 7.40 L.

28. $P_{CO_2} = n_{CO_2}(RT/V)$. From the equation, the number of moles of CO_2 formed is 12 times the number of moles of sucrose burned. The mol. wt. of $C_{12}H_{22}O_{11}$ is 342.3 g/mol. Thus the number of moles of sucrose is

$$\frac{0.8763}{342.3\,g\cdot mol^{-1}} = 2.560\times10^{-2}\,mol$$

and

$$\text{No. of moles of } CO_2 = 12(2.560\times10^{-3}) = 3.072\times10^{-2}$$

$$P_{CO_2} = \frac{(3.072\times10^{-2}\,mol)}{(3.40\,L)}\left(62.364\frac{L\cdot mmHg}{mol\cdot K}\right)(298.15\,K) = 168\,mmHg$$

29. (13.6 g·cm^{-3}) (200 mm) (g) = (2.89 g·cm^{-3}) (h) (g) where h is the height of the bromoform column, and g is the acceleration of gravity. Thus,

$$h=\left(\frac{13.6}{2.89}\right)(200\,\text{mm})=941\,\text{mm}=94.1\,\text{cm}$$

30. Both P and n are constant, so $T_2/T_1 = V_2/V_1$, and $T_2 = (V_2/V_1)\ T_1 = \left(\frac{3.00\,\text{L}}{3.75\,\text{L}}\right)$ (308.15 K)

= 246.52 K

31. Since the temperature is above 273 K, one mole must occupy a volume greater than 22.4 L at 1.00 atm, so the number of moles of gas that occupies 22.4 L must be less than 1. Only (a) and (b) are possible answers, and (b) is most likely, as the temperature is not every far above 273 K. To calculate the number of moles of gas, use the ideal gas law.

$$n=\frac{PV}{RT}=\frac{(1.00\,\text{atm})(22.4\,\text{L})}{(0.082058\,\text{L}\cdot\text{atm}\cdot\text{mol}^{-1}\text{K}^{-1})(303.2\,\text{K})}=0.900\,\text{mol}$$

32. $P_{total} = 753.2\ \text{mmHg} = P_{H_2} + P_{H_2}O = P_{H_2} + 25.8$

Thus $P_{H_2} = 727.4\,\text{mmHg}$. Since the number of moles of H_2 is constant,

$$V_2 = \left(\frac{T_2}{T_1}\right)\left(\frac{P_1}{P_2}\right)V_1 = \left(\frac{293.2\,\text{K}}{299.6\text{K}}\right)\left(\frac{727.4\,\text{mmHg}}{760.0\,\text{mmHg}}\right)(94.6\,\text{mL})=88.6\,\text{mL}$$

33. We have, δ = MP/RT. The mol. wt. of $C1_2(g)$ is 70.91 g·mol^{-1}. Hence

$$\delta = \frac{(70.91\,\text{g}/\text{mol})(1.60\,\text{atm})}{(0.082058\,\text{L}\cdot\text{atm}\cdot\text{mol}^{-1}\text{K}^{-1})(323.2\,\text{K})} = 4.28\ \text{g}\cdot\text{L}^{-1}$$

34. $P_{total} = P_{gas} + P_{H_2O} = P_{gas} + 21.7 = 758.3\,\text{mmHg}$ so $P_{gas} = 736.6$ mmHg

$$\therefore\ M = \frac{(0.4153\,\text{g})}{(0.13\,\text{L}}\left(62.36\frac{\text{L}\cdot\text{mmHg}}{\text{mol}\cdot\text{K}}\right)\frac{(296.7\,\text{K})}{(736.6\,\text{mmHg})} = 80\,\text{g}\cdot\text{mol}^{-1}$$

35. M = δRT/P

$$M = \frac{(1.22\,\text{gL}^{-1})(0.08206\,\text{L}\cdot\text{atm}\cdot\text{mol}^{-1}\text{K}^{-1})(300\,\text{K})}{(1.00\ \text{atm})} = 30.0\text{g mol}^{-1}$$

Of the gases listed, only ethane, C_2H_6, has a molecular weight of 30.

36. If the total number of moles of gas is 5, there are 2 moles of Ar. Hence, $X_{Ar} = 2/5$.

$\therefore\ P_{Ar} = P_{total}X_{Ar} = (2/5)\ P_{total}$

37. The equation for the reaction tells us that No. moles H_2 formed = $\frac{3}{2}$ (No. moles Al used)

$$n_{Al} = \frac{1.7093\,g}{26.9815\,g\,mol^{-1}} = 6.335\times10^{-2}\,mol$$

$$n_{H_2} = (3/2)(6.335\times10^{-2}) = 9.503\times10^{-2}\,mol$$

$$V = nRT/P = \frac{(9.503\times10^{-2}\,mol)}{(738.5\,mmHg)}(62.364\,L\cdot mmHg\,mol^{-1}K^{-1})(298.0\,K) = 2.391\,L$$

38. We know that the molar volume is 22.4 L at 273 K and 1.00 atm. If we halve the pressure, the volume will double to 44.8 L. Increasing the temperature will further increase the molar volume, as $V/n = RT/P$. At 313 K and 0.500 atm, V/n = (0.082058) (313.2)/0.500 = 51.4 L. Option (b) is ruled out because at a pressure of 0.25 atm and 273 K the molar volume is 4(22.4 L) = 89.6 L.

39. Since all gases are at the same temperature and pressure, the molar ratios are the same as the volume ratios. Only (4/5) (60.0 mL) = 48.0 mL of NH_3 are required to react completely with 60.0 mL of O_2, so there is excess ammonia. Thus the O_2 is the limiting reagent, and the volume of $H_2O(g)$ formed is

(6/5) (volume O_2 used) = (6/5) (60.0 mL) = 72.0 mL.

40. Mol. wts: SO_2 = 64, O_2 = 32. For 100–g samples of each,

$$n_{O_2} = 3.13\,mol,\quad n_{SO_2} = 1.56\,mol,\quad n_{total} = 4.69\,mol$$

$$X_{SO_2} = \frac{1.56}{4.69} = 0.333, \text{ and}$$

Now, P_{total} = 600 mm Hg.

$$\therefore\quad P_{SO_2} = P_{total}\,X_{SO2} = (0.333)(600\,mmHg) = 200\,mmHg$$

41. Burning 0.20 mol of C_xH_y will yield 0.20x mol of CO_2, by the conservation of C atoms. Since 0.20x = 0.80, ⇒ x = 4. Burning 0.20 mol of C_xH_y will yield (1/2) (0.20y) mol of H_2O, by the conservation of H atoms, as each H_2O molecule has 2 H atoms. Thus,

(1/2) (0.20y) = 0.10y= 1.0 mol H_2O, and ∴ y = 10

The hydrocarbon is therefore butane, C_4H_{10}.

42. The average molecular speed is proportional to $(T/M)^{1/2}$. Thus light molecules at high temperatures move fastest. The mean speed is independent of the size of the sample. The lightest molecule of those listed is He, and the highest temperature of those given is 560 K. Hence, just calculate T/M for choices (a) and (d) for comparison. For (a), M_{N_2} = 28.0, and T/M = 560/28.0 = 20. For (d), M_{He} = 4.0, and T/M = 140/4.0 = 35. Choices (c) and (e) are immediately eliminated as the temperature is lower and the molecular weight greater than the sample of choice (a). For option (b), the molecular weight of Ne is 5 times that of He, but 500 K is less than 5 time 140 K. Thus T/M, and therefore the mean speed, is greatest in sample (d).

43. A real gas most closely approaches the behaviour of an ideal gas at low pressure and high temperature.

44. $$\frac{\text{rate of effusion of } O_2}{\text{rate of effusion of He}} = \left(\frac{4.0}{32.0}\right)^{1/2} = 0.35$$

45. At 0.5 atm, P_1V_1 = (0.5 atm) (2.0 L) = 1.0 L·atm

At 100 atm, P_2V_2 = (100 atm) (2.0 L) = 1.3 L·atm

An increase in PV with increasing pressure is due to the fact that the volume of the gas molecules is a significant fraction of the volume of the container. Option (c) is a correct statement, but the increasing force of attraction causes PV to decrease with increasing pressure. If the molecules dimerize, the number of moles of gas will decrease, and PV will decrease significantly.

46. $$\frac{\text{effusion rate of } N_2 \text{ at T}}{\text{effusion rate of } SO_2 \text{ at 323}} = \left[\frac{(T/28.01)}{(323.2/64.06)}\right]^{1/2} = 1.625$$

Squaring the two terms on the right yields

$$\frac{64.06\ T}{(28.01)\ (323.2)} = (1.625)^2 = 2.6406$$

$$\text{or } 7.076 \times 10^{-3}\ T = 2.6406 \quad \therefore\ T = 373.2\ K$$

47. The first four quantities listed all increase with temperature at constant volume. The density depends on the number of moles of gas and the volume. As these are both constant, the density does not change.

48. The rate of effusion is inversely proportional to the time it takes for the gas to effuse, that is, the less the time, the faster the rate.

$$\frac{\text{rate of effusion of gas X}}{\text{rate of effusion of } H_2} = \left(\frac{2.0}{M_X}\right)^{1/2} = \frac{(1/130)}{(1/26)} = \frac{1}{5}$$

Thus

$$\frac{2.0}{M_X} = \frac{1}{25} \quad \therefore\ M_X = (2.0)\ (25) = 50$$

49. $u_{rms} = (3RT/M)^{1/2}$

50. u is proportional to $T^{1/2}$. Thus,

$$\frac{\text{u at 375 K}}{\text{u at 250 K}} = \left(\frac{375}{250}\right)^{1/2} = 1.22$$

51. At a specified temperature, the gas with the highest molecular weight has the lowest mean speed. Of the gases listed, CO_2 has the highest molecular weight.

52. The number of collisions of Ar atoms with the walls of the container depends on the partial pressure of the Ar. If the temperature decreases at constant volume, the pressure decreases and the number of collisions decreases. If the average K.E. increases, the temperature has increased, and the number of collisions increases. If the volume decreases, the pressure increases, and the number of collisions increases. Adding CO_2 does not change the partial pressure of the Ar, and hence does not affect the number of collisions of Ar atoms with the walls.

53. The reason that PV for a real gas is *less* than PV for an ideal gas is the non-zero force of attraction between the molecules.

54. $$\frac{\text{rate of effusion of } NH_3}{\text{rate of effusion of X}} = \left(\frac{M_X}{17.03}\right)^{1/2} = 3.32$$

Thus,

$M_X/17.03 = (3.32)^2 = 11.02$

$M_X = (17.03)\ (11.02) = 187.7 = 188 \text{ g mol}^{-1}$

55. As these are monatomic gases, they have only translational kinetic energy. $E_{trans} = \frac{3}{2}RT$, and depends on T only. Gases at the same T have the same molar translational kinetic energy.

56. The total translational kinetic energy, $E_{trans} = \frac{3}{2}nRT$.

Helium		Argon

Thus, $$\frac{3}{2}(0.30)RT = \frac{3}{2}(0.40)R(400)$$

The $\frac{3}{2}$ and R factors cancel. $\therefore\ T = \frac{(0.40)}{(0.30)}(400\text{K}) = 533\text{ K}$

57. Moles of He $= \frac{0.4\text{g}}{4\text{g mol}^{-1}} = 0.1$ mole

Moles of O_2 $= \frac{1.6\text{g}}{32\text{g mol}^{-1}} = 0.05$ mole

Moles of N_2 $= \frac{1.4\text{g}}{28\text{g mol}^{-1}} = 0.05$ mole

Total no. of moles $= 0.1 + 0.05 + 0.05 = 0.2$ mole

$\therefore$ PV $= nRT$

or P $= \frac{nRT}{V}$

$$= \frac{(0.3\,\text{mole})\,(0.082\,\text{L atm K}^{-1}\text{mol}^{-1})(300\,\text{K})}{10\,\text{L}} = 0.492 \text{ atm}$$

58. The reaction is

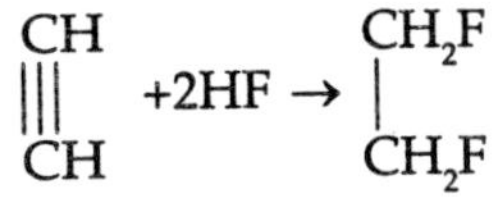

It is evident that 1 mole of C_2H_2 combines with 2 moles of HF to form 1 mole of $C_2H_4F_2$. Thus after the reaction

No. of moles of C_2H_2 $= 1 - 1 = 0$ mole

No. of moles of HF $= 5 - 2 = 3$ mole

No. of moles of $C_2H_4F_2$ $=$ 1 mole

Therefore total no. of moles after the completion of reaction

$= 0 + 3 + 1 = 4$ mole

According to gas equation,

$$PV = nRT$$

$$\text{or } P = \frac{nRT}{V}$$

$$= \frac{(4.\text{mole})\,(0.0821\,\text{L atm K}^{-1}\,\text{mol}^{-1})\,(273\text{ K})}{10\text{ L}} = 8.95 \text{ atm}$$

59. Mass of cylinder = 14.8 kg.

∴ Mass of gas in the cylinder when filled

= 29.0 – 14.8 = 14.2 kg.

Mass of gas after usage

= 23.2 – 14.8 = 8.4 kg.

Molar mass of n-Butane

$C_4H_{10} = 4 \times 12 + 10 \times 1 = 58$ g = 0.058 kg.

$$\text{Initial no. of moles, } n_1 = \frac{14.2}{0.058} = 245$$

$$\text{Final no. of moles, } n_2 = \frac{8.4}{0.058} = 145$$

Assuming normal usage condition to be 300 K and 1 atm pressure.

Volume of 1 mole at 273.15K = 22.4 litres = $0.0224\ m^3$

Volume of 145 moles at 1 atm and 300 K

$$= 0.0224 \times \frac{300}{273.115} \times 145 = 3.56\,m^3$$

Final pressure also depends directly on number of moles

$$\frac{n_1}{n_2} = \frac{3.5\,\text{atm}}{P}$$

$$\therefore \qquad P = \frac{3.5 \times n_2}{n_1} = \frac{3.5 \times 145}{245}$$

= 2.07 atm pressure (*i.e.* actual pressure)

Now pressure shown would be (2.07 – 1) = **1.07 atm.**

60. According to gas equation,

$$PV = nRT \text{ or } n = \frac{PV}{RT}$$

$$= \frac{(6\,\text{atm})\,(3\,\text{L})}{(0.0821\,\text{L atm mol}^{-1}\,\text{K}^{-1})\,(300\,\text{K})} = 0.731\,\text{mole}$$

Number of moles of unknown gas

$$= 0.731 - 0.7 = 0.031 \text{ mole}$$

Now,

$$\frac{r_1}{r_2} = \frac{0.7}{0.031}$$

and

$$\frac{r_1}{r_2} = \sqrt{\frac{M_2}{M_1}}$$

or

$$M_2 = \left(\frac{r_1}{r_2}\right)^2 M_1$$

$$= \left(\frac{0.7}{0.031}\right)^2 (2\,\text{g mol}^{-1}) = \mathbf{1033\ g\ mol^{-1}}$$

61. van der Waals' equation for n moles of gas may be written as :

$$\left(P + \frac{an^2}{V^2}\right)(V - nb) = nRT$$

On substitution of given values, we get

$$\left[11 + \frac{a(2)^2}{(4)^2}\right][4 - 2 \times 0.05] = (2)(0.082)(300)$$

$$\left(11 + \frac{a}{4}\right) = \frac{2 \times 0.082 \times 300}{4 - 0.1} = 12.61$$

or $\quad a = 4(12.61-11)$

$= \mathbf{6.46\ atm\ L^2\ mol^{-2}}$

62. Given that the tube is 3 m long.

Assuming the distance from HCl inlet end of the tube at which NH_2Cl will first appear $= x$ m

Then, $$\frac{\text{rate of diffusion of HCl } (= r_1)}{\text{rate of diffusion of } NH_3 (= r_2)} = \frac{x}{3-x} \quad \ldots (1)$$

Moreover, from Graham's Law of Diffusion,

$$\frac{r_1}{r_2} = \sqrt{\frac{M_2}{M_1}} = \sqrt{\frac{\text{mol. mass of } NH_3}{\text{mol. mass of HCl}}} = \sqrt{\frac{17}{36.5}} = 0.68 \quad \ldots (2)$$

From equations (1) and (2), we get

$$\frac{x}{3-x} = 0.68 \Rightarrow x = 1.214\,\text{m}$$

63. Formula used $v_{rms} = \sqrt{\frac{3RT}{M}}$ or $T = \frac{V_{rms}^2 M}{3R}$

where v_{rms} = Root-mean-square velocity = 100 ms^{-1} (given)

R = Gas constant = 8.314 $JK^{-1}\ mol^{-1}$

M = molar mass of gas = 2×10^{-3} kg mol^{-1} for H_2 gas

and T = Temperature (to be determined)

$$\therefore \quad T = \frac{(100\,ms^{-1})^2\ (2\times10^{-3}\,kg\,mol^{-1})}{3\times8.314(JK^{-1}\,mol^{-1})}$$

$$= 0.8\ K\ \frac{kg\,m^2\,s^{-1}}{J} \quad (\because 1J = 1\ kg\ m^2\ s^{-1})$$

$\therefore \quad T = 0.8$ K

64. We know that $v_{mp} = \sqrt{\frac{2RT}{M}} = \left(\frac{2RT}{M}\right)^{1/2}$

Hence $$\frac{(V_{mp})_{O_2}}{(v_{mp})_{H_2}} = \frac{T_{O_2}/M_{O_2}}{T_{H_2}/M_{H_2}} = 1$$

$$\Rightarrow \quad \frac{T_{O_2}}{M_{O_2}} = \frac{T_{H_2}}{M_{H_2}}$$

where T stands for temperature and M stands for molar mass

$$\Rightarrow \qquad T_{O_2} = T_{H_2} \times \frac{M_{O_2}}{M_{H_2}} = 298\text{K} \times \frac{32\,\text{g mol}^{-1}}{2\,\text{g mol}^{-1}} = 4768 \text{ K}$$

65. Average speed $v_{avg} = \sqrt{\frac{8KT}{\pi m}}$ . . . (1)

$\therefore \qquad v_{avg}\text{ (He)} = \sqrt{\frac{8K\,(330)}{\pi\,(4)}}$ . . . (2)

and $\qquad v_{avg}\,(N_2) = \sqrt{\frac{8KT}{\pi\,(28)}}$ . . . (3)

Because given that average speeds are same, so equating the right-hand sides of equations (2) and (3), we get

$$\sqrt{\frac{8K\,(330)}{\pi\,(4)}} = \sqrt{\frac{8KT}{\pi(28)}}$$

$$\Rightarrow \qquad T = \left(\frac{28}{4}\right)(330) = 2310\,\text{K}.$$

66. Root mean square speed $v_{rms} = \sqrt{\frac{3KT}{m}}$

Here $\sqrt{\frac{3KT_{He}}{4}} = \sqrt{\frac{3K\,(200)}{2}} \Rightarrow T_{He} = 2\,(200) = 400$ K.

67. Most probable speed, $v_{mP} = \sqrt{\frac{2KT}{m}}$

Here, $\sqrt{\frac{2KT}{m}} = 2\sqrt{\frac{2K\,(293)}{m}}$

$\Rightarrow \qquad T = 2^2\,(293) = 4 \times 293 = 1172$ K

68. $\frac{r\text{(Ne)}}{r\text{(He)}} = \sqrt{\frac{M\text{ (He)}}{M\text{ (Ne)}}} = \sqrt{\frac{4}{20}} = \frac{1}{\sqrt{5}}$. $\therefore$ Ratio is $1:\sqrt{5}$

69. According to Van der Walls' equation

$$\left(P + \frac{a\,n^2}{V^2}\right)(V - nb) = n\,RT$$

Hence, Pressure $P = \dfrac{n\,RT}{V - nb} - \dfrac{a\,n^2}{V^2}$

Substituting the values we have:

$$= \frac{(10\text{ moles})(8.314\text{ N m k}^{-1}\text{ mol}^{-1})(298\text{ K})}{(25\times10^{-3}\text{ m}^3)-(10\text{mol})(0.037\times10^{-3}\text{ m}^3\text{ mol}^{-1})} - \frac{(10.0\text{ mol})^2(0.417\text{ N m}^4\text{ mol}^{-2})}{(25\times10^{-3}\text{ m}^{-3})^2}$$

$= 9.39 \times 10^5$ Nm^{-2}

70. The pressure exerted by a column of water h_w whose density is d_w, is gd_wh_w. A similar result holds for mercury, and because the pressures are equal you can equate these expressions.

$$\therefore\ gd_wh_w = gd_{Hg}h_{Hg}$$

Rearranging gives

$$\frac{h_w}{h_{Hg}} = \frac{d_{Hg}}{d_w}$$

This says that the height of the liquid column is inversely proportional to density. After substituting values into the above equation,

$$\frac{h_w}{760.00\text{mm}} = \frac{13.596\text{g cm}^{-3}}{0.99987\text{g cm}^{-3}}$$

Hence,

$$h_w = 760.00\text{ mm} \times \frac{13.596}{0.99987} = 1.0334 \times 10^4\text{ mm} = 10.334\text{ m}$$

71. We have $V_i = 12.0$ dm^3 $P_i = 98.9$ kPa

$V_f = ?$ $P_f = 119.0$ kPa

} Temperature and moles remain constant.

Here P and V vary but temperature and moles are constant.

Application of Boyle's law gives

$$V_f = V_i \times \frac{P_1}{P_f} = 12.0\text{dm}^3 \times \frac{98.9\text{ kPa}}{119.0\text{ kPa}} = 9.97\text{dm}^3$$

Note that the pressure on the gas increases, so the gas is compressed (gives a smaller volume), and the ratio of pressures is less than 1.

72. In asking for the mass of oxygen, we are in effect asking for moles of gas, n, because mass and moles are directly related. The problem gives P, V, and T, so you can use the ideal gas law to solve for n. The proper value to use for R depends on the units of P and V. Here they are in atmospheres and liters, respectively, so you use $R = 0.0821$ L · atm/(K · mol). Once you have solved for n, convert to mass by using the molar mass of oxygen.

The data given in the problem are :

Variable	Value
P	15.7 atm
V	50.0L
T	(21 + 273) K = 294 K
n	?

Solving the ideal gas law for n gives

$$n = \frac{PV}{RT}$$

Substituting the data gives

$$n = \frac{15.7\ \cancel{\text{atm}} \times 50.0\ \cancel{\text{L}}}{0.0821\ \cancel{\text{L}} \cdot \cancel{\text{atm}}\ \text{K}^{-1} \cdot \text{mol}^{-1} \times 294\,\text{K}} = 32.5\ \text{mol}$$

and converting moles to mass of oxygen yields

$$32.5\ \cancel{\text{mol}\,\text{O}_2} \times \frac{320\,\text{g}\,\text{O}_2}{1\ \cancel{\text{mol}\,\text{O}_2}} = 1.04 \times 10^3\ \text{g}\,\text{O}_2$$

73. The density of a gas in g/L is numerically equal to the mass of 1L. therefore, using the ideal gas law to calculate the moles, then the mass, of 1L of O_2 to obtain its density.

The data given here are :

Variable	Value
P	0.850 atm
V	1L (exact value)
T	298 K
n	?

Therefore,

$$n = \frac{PV}{RT} = \frac{0.850\ \cancel{\text{atm}} \times 1\ \cancel{\text{L}}}{0.0821\ \cancel{\text{L}} \cdot \cancel{\text{atm}}\ \text{K}^{-1} \cdot \text{mol}^{-1} \times 298\ \cancel{\text{K}}} = 0.0347\ \text{mol}$$

Now convert mol O_2 to grams

$$0.0347\ \cancel{\text{mol}\,\text{O}_2} \times \frac{32.0\,\text{g}\,\text{O}_2}{1\ \cancel{\text{mol}\,\text{O}_2}} = 1.11\,\text{g}\,\text{O}_2$$

Therefore, the density of O_2 at 298 K and 0.850 atm is 1.11 g.L^{-1}

74. One can calculate the moles of vapor from the ideal gas law and then calculate the molar mass by dividing mass by moles. Note that the temperature of the vapour is the same as the temperature of the boiling water and that the pressure of the vapor equals the barometric pressure.

The following data is given for the vapor.

Variable	Value
P	$733 \text{ mmHg} \times \frac{1\text{ atm}}{760 \text{ mmHg}} = 0.964 \text{ atm}$
V	200.0 mL = 0.2000 L
T	372 K
n	?

From the ideal gas law, $PV = nRT$,

$$n = \frac{PV}{RT} = \frac{0.964 \text{ atm} \times 0.2000 \text{ L}}{0.0821 \text{ L} \cdot \text{atm K}^{-1} \cdot \text{mol}^{-1}) \times 372 \text{ K}}$$

$$n = 0.00631 \text{ mol}$$

Dividing the mass of the vapor by moles gives the mass per mole (the molar mass).

$$\text{Molar mass} = \frac{\text{grams vapor}}{\text{moles vapor}} = \frac{0.970 \text{ g}}{0.00631 \text{ mol}} = 154 \text{ g mol}^{-1}$$

Thus, the molecular weight is 154 amu.

75. The chemical equation relates the moles of NaN_3 to moles of N_2. Use the ideal gas law to relate the volume of N_2 to moles of N_2 and hence moles of NaN_3. Then obtain the grams of NaN_3 from the moles of NaN_3.

The available data is:

Variable	Value
P	$748 \text{ mmHg} \times \frac{1\text{ atm}}{760 \text{ mmHg}} = 0.984 \text{ atm}$
V	75.0L
T	298 K
n	?

Note that the pressure was converted from mmHg to atmospheres and the temperature from degrees Celcius to kelvins. Rearrange the ideal gas law to obtain n.

$$n = \frac{PV}{RT}$$

Substituting from the data gives the moles of N_2.

$$n = \frac{0.984 \text{ atm} \times 75.0 \text{ L}}{0.0821 \text{ L} \cdot \text{atm K}^{-1} \cdot \text{mol}^{-1} \times 298 \text{ K}} = 3.02 \text{ mol}$$

From the moles of N_2, you use the chemical equation to obtain moles of NaN_3.

$$3.02 \text{ mol } N_2 \times \frac{6 \text{ mol } NaN_3}{9 \text{ mol } N_2} = 2.01 \text{ mol } NaN_3$$

Now you can calculate the grams of NaN_3. The molecular weight of NaN_3 is 65.01 amu.

$$2.01\ \text{mol NaN}_3 \times \frac{65.01\,\text{g NaN}_3}{1\ \text{mol NaN}_3} = 131\,\text{g NaN}_3$$

76. The gas collected is hydrogen mixed with water vapor. To obtain the amount of hydrogen, first find its partial pressure in the mixture, using Dalton's law (Step 1). Then calculate the moles of hydrogen from the ideal gas law (Step 2). Finally obtain the mass of hydrogen from the moles of hydrogen (Step 3).

Step 1 : The vapor pressure of water at 292K is 16.5 mmHg. Dalton's law of partial pressures, says that the total gas pressure equals the partial pressure of hydrogen, and partial pressure of water.

i.e. $P = P_{H_2} + P_{H_2O.}$

Substituting and solving for the partial pressure of hydrogen,

$$P_{H_2} = P - P_{H_2O} = (769 - 16.5)\ \text{mmHg} = 752\ \text{mmHg}$$

Step 2 : Now use the ideal gas law to find the moles of hydrogen collected. The data :

Variable	Value
P	$752\ \text{mmHg} \times \frac{1\,\text{atm}}{760\ \text{mmHg}} = 0.989\,\text{atm}$
V	156 mL = 0.156 L
T	293 K
n	?

From the ideal gas law, $PV = nRT$, you have

$$n = \frac{PV}{RT} = \frac{0.989\ \text{atm} \times 0.156\ \text{L}}{0.0821\ \text{L}\cdot\text{atm}\ \text{K}^{-1}\cdot\text{mol}^{-1} \times 293\ \text{K}} = 0.00641\ \text{mol}$$

Step 3 : Convert moles of H_2 to grams of H_2.

$$0.00641\ \text{mol H}_2 \times \frac{2.02\,\text{g H}_2}{1\ \text{mol H}_2} = 0.0129\,\text{g H}_2$$

77. Let S_1 be the solubility of the gas at partial pressure P_1, and let S_2 be the solubility at partial pressure P_2. Then Henry's law for both pressures is,

$$S_1 = k_H P_1 \qquad \text{(i)}$$

$$S_2 = k_H P_2 \qquad \text{(ii)}$$

Dividing the second equation by the first,

$$\frac{S_2}{S_1} = \frac{k_H P_2}{k_H P_1} \quad \text{or} \quad \frac{S_2}{S_1} = \frac{P_2}{P_1}$$

Now, At 1.0 atm partial pressure of acetylene (P_1), the solubility S_1 is 27 g C_2H_2 per liter of acetone. For a partial pressure of 12 atm (P_2).

$$\frac{S_2}{27\,g\,C_2H_2\,L(\text{acetone})} = \frac{12\,\cancel{atm}}{1.0\,\cancel{atm}}$$

$\therefore S_2 = 12 \times 27 = 324 = 3.24 \times 10^2 g.L^{-1}$ (acetylene)

78. First rearrange the van der Waals equation to give P in terms of the other variables.

$$P = \frac{nRT}{V\text{-}nb} - \frac{n^2a}{V^2}$$

Given, $R = 0.08206$ L · atm K^{-1} · mol^{-1}, $T = 273.2$ K, $V = 22.41$ L, $a = 6.865$ L^2 · atm mol^{-2}, and $b = 0.05679$ L mol^{-1}.

$$P = \frac{1.000\,\cancel{mol} \times 0.08206\,\cancel{L} \cdot atm\,\cancel{K^{-1}} \cdot \cancel{mol^{-1}} \times 273.2\,\cancel{K}}{22.41\,\cancel{L} - 1.000\,\cancel{mol} \times 0.05679\,\cancel{L\,mol^{-1}}} - \frac{(1.000\,\cancel{mol})^2 \times 6.865\,\cancel{L^2} \cdot atm\,\cancel{mol^{-2}}}{(22.41\,\cancel{L})^2}$$

$= 1.003$ **atm** $- 0.014$ atm $= 0.989$ atm

79. Assuming ideal gas behaviour,

so $\quad PV = nRT$

Thus, $$n = \frac{PV}{RT} = \frac{\left(10^{-5}\,mmHg \times \dfrac{1\,atm}{760\,mmHg}\right)\left(10^{-3}\,cm^3 \times \dfrac{1L}{1000\,cm^3}\right)}{(0.0821\,L\,atm\,K^{-1}\,mol^{-1})\,(298\,K)}$$

$= 5.3 \times 10^{-16}$ moles

Since, 1 more of each gas contains 6.023×10^{23} molecules,

Hence, number of gas molecules left

$= (5.3 \times 10^{-16})\,(6..023 \times 10^{23}) = 3.192 \times 10^8$

80. Given $\quad P_1 = 298$ K $\qquad T_1 = 298$ K $\qquad T_2 = 1808$ K

$P_2 = ?$

Here,

$$\therefore \quad \frac{P_1}{T_2} = \frac{P_2}{T_2}$$

$$\Rightarrow \quad P_2 = \frac{P_1}{T_1} \times T_2 = \frac{200}{298} \times 1808 = 1213.4 \text{ atm}$$

Since $\quad P_2 > 700$ atm.

Hence the tank will blow up.

81. From Graham's Law of Diffusion

$$\frac{r_1}{r_2} = \sqrt{\frac{M_2}{M_1}}$$

Here, molar mass of $N_2 = M_2 = 28$ g mol^{-1}

and molar mass of He = M_1 = 4 g mol^{-1}

$$\therefore \quad \frac{r_1}{r_2} = \sqrt{\frac{28}{4}} = \sqrt{7} = 2.65$$

82. Assuming ethane (C_2H_6) will behave as an ideal gas

So $\quad PV = nRT$

Thus $\quad T = \dfrac{PV}{nR}$

$$= \frac{(10\,\text{atm})\left(10^{-3}\,\text{m}^3 \times \dfrac{1\text{L}}{10^{-3}\text{m}^3}\right)}{\left(\dfrac{5\times10^{-3}\,\text{kg}}{30\,\text{kg mol}^{-1}} \times (0.0821\,\text{L atm K}^{-1}\,\text{mol}^{-1})\right)} = 730.9\text{ K}$$

83. Volume of vapour = 100 cm^3 – 10 cm^3 = 90 cm^3

Once again, the measurements are converted to S1 units with the exception of the mass which is left as grams.

p = 100 600 Pa, V = 90 × 10^{-6} m^3, m = 0.180 g

M = ?, R = 8.31 JK^{-1} mol^{-1} and T = 353 K

Insert these values into the ideal gas equation :

$$pV = \frac{mRT}{M}$$

$$100\,600 \times 90 \times 10^{-6} = \frac{0.180}{M} \times 8.31 \times 353$$

Rearranging :
$$M = \frac{0.180 \times 8.31 \times 353}{100600 \times 90 \times 10^{-6}} = 58$$

84. Theoretical vapour density of

$$N_2O_4 = \frac{28+64}{2} = 46$$

N_2O_4 dissociates as:

$$N_2O_4 \rightleftharpoons 2NO_2$$

	N_2O_4	$2NO_2$
Initial conc.	1	0
Final conc.	$1-x$	$2x$

Total no. of moles = $1-x+2x = 1+x$

and $\quad$ d = 30·6 (given)

x being degree of dissociation. For NO_2 (mol wt = 14 + 32 = 46)

$$x = \frac{D-6}{d} = \frac{46-30{\cdot}6}{30{\cdot}6} = 50{\cdot}32\%.$$

85. At constant pressure

$$U_1 = \sqrt{\frac{3RT}{M}}$$

$$U_2 = \sqrt{\frac{3R\,(2T)}{M}}$$

$$\frac{U_1}{U_2} = \sqrt{\frac{3RT}{M} \times \frac{M}{3R \times 2T}} = \frac{1}{\sqrt{2}} = 0.707$$

86. Since pressure, volume and temperature all three parameters are changing, the process could be polytropic.

Chapter 4 : STOICHIOMETRY

1. *Step 1:* Write down equals sign (=)
Step 2: To the right of the equals sign, write down the units you want the answer to be in. Examination of the problem reveals that you want your answer in "liters of whiskey", so you have

$$= \text{liters of whiskey}$$

Step 3: Begin the problem with the item you are limited to. In this case, it is 50. moles of carbon—no more, no less, Place this 50 moles of carbon over 1.

$$\frac{50 \text{ moles of carbon}}{1} = \text{liters of whiskey}$$

Step 4: Get rid of the units "moles of carbon" by placing them in the denominator of the next factor. What do you know about moles of carbon? There are 2 moles of carbon in each moles of ethanol.

$$\frac{50 \text{ moles of carbon}}{1} \times \frac{1 \text{ mol ethanol}}{2 \text{ moles of carbon}} = \text{liters of whiskey}$$

Step 5: Continue in this fashion, getting rid of unwanted units until you are left in the units you despite. At the point, stop and to the calculations.

$$\frac{50 \text{ moles of carbon}}{1} \times \frac{1 \text{ mole of ethanol}}{2 \text{ moles of carbon}} \times \frac{46 \text{ gms of ethanol}}{1 \text{ mole of ethanol}} \times \frac{1 \text{ ml. of ethanol}}{0.79 \text{ gms of ethanol}} \times$$

$$\frac{200 \text{ mL whiskey}}{92 \text{ ml. of ethanol}} \times \frac{1 \text{ L whiskey}}{10^3 \text{ mL whiskey}} = 3.2 \text{ L whiskey}$$

2. First change the grams of each element to moles. You end up with 0.100 mole of carbon 0.200 mole of hydrogen, and 0.200 mole of chlorine. This represent a 1 carbon: 2 hydrogen : 2 chlorine molar ratio.

3. Begin balancing equations by typing suitable coefficients that will give the same number of atoms of each elements on both side of the equation. Remember to change coefficients, not subscripts.

$$2NH_3 \rightarrow 1N_2$$

Step 2: Look for elements that appear only once on each side of the equation and with equal numbers of atoms on each side. The formulas containing these elements must have the same coefficients.

$$CuO \rightarrow Cu$$

Step 3: Look for elements that appear only once on each side of the equation but in unequal numbers of atoms. Balance these elements.

$$\underline{2NH_3} \rightarrow \underline{3H_2O}$$

Step 4: Balance elements that appear in two or more formulas on the same side of the equation.
Step 5: Double check you balanced equation and be sure the coefficients are the lowest possible whole numbers.

$$2NH_3 + 3CuO \rightarrow 3Cu + N_2 + 3H_2O$$
$$2 + 3 + 3 + 1 + 3 = 12$$

(Be sure to include the unwritten 1 that is in front of N_2)

4. Begin by writing a balanced equation:

$$3BaCl_{2(aq)} + 2K_3AsO_{4(aq)} \rightarrow Ba_3(AsO_4)_{2(s)} + 6KCl_{(aq)}$$

You may want to write the net-ionic equation:

$$3Ba^{2+}_{(aq)} + 2AsO_4^{3-}{}_{(aq)} \rightarrow Ba_2(AsO_4)_{2(s)}$$

Next realize that this problem is a limiting-reactant problem. That is, one of the two reactants will run out first, and when that happens, the reaction will stop. You need to determine which one of the reactions will run out first. To do this, you need to be able to compare then on a 1:1 basis. But their coefficient are different, so you need to relate both reactants to a common product, say $Ba_2(AsO_4)_2$. Set the problem up like this:

$$\frac{0.600 \text{ moles } BaCl_2}{1} \times \frac{1 \text{ mole } Ba_3(AsO_4)_2}{3 \text{ moles } BaCl_2} = 0.200 \text{ mole } Ba_3(AsO_4)_2$$

$$\frac{0.250 \text{ mole } K_3\ AsO_4}{1} \times \frac{1 \text{ mole } Ba_3(AsO_4)}{2 \text{ mole } K_3\ AsO_4} = 0.125 \text{ mole } Ba_3(AsO_4)_2$$

Given the two amount of starting materials, you discover that you can make a maximum of 0.125 mole of **$Ba_3(AsO_4)_2$**, because at that point you will have exhausted you supply of K_3AsO_4.

5. Begin by writing a balanced equation. Remember that all Group II carbonates decompose to yield the metallic oxide plus carbon dioxide gas.

$$CaCO_3(s) \rightarrow CaO(s) + CO_2(g)$$

According to your balanced equation, any loss of mass during the experiment would have to have come from the carbon dioxide gas leaving the test tube. 4,400 grams of CO_2 gas correspond to 0.100 mole. Because all of the calcium carbonate decomposed, and the calcium carbonate and carbon dioxide gas are in a 1:1 molar ratio, you must originally have had 0.1000 mole of calcium carbonate, or 10.01 grams. The calcium carbonate and test tube weighted 30.08 grams, so if you get rid of the calcium carbonate, you are left with 20.07 grams for the empty test tube.

6. For challenge all the grams to moles:

$$32.0 \text{ grams of } O_2 = 1.00 \text{ mole}$$
$$32.0 \text{ grams of } CH_4 = 2.00 \text{ moles}$$
$$32.0 \text{ grams of } SO_2 = 0.500 \text{ mole}$$

Mole fraction of oxygen gas:

$$\text{Mole fraction of oxygen gas } \frac{1 \text{ mole } O_2}{3.50 \text{ total moles}} = 0.286$$

7. To solve this problem, multiply the percentage of each isotope by its atomic mass and add those products.

$$(0.900 \times 20.0) + (0.100 \times 22.0) = 20.2 \text{ atomic mass of elements X.}$$

8. According to the information given, you have 0.50 mole of element X (50.g/100 g · mole^{-1} = 0.50 mole). For the oxygen, remember that you will use 16 g mole^{-1} of the atomic weight giving you 2.0 moles of oxygen atoms. A 0.50:2.0 molar ratio is the same as a 1:4 molar ratio so the answer is XO_4.

9. From the information provided, you know that the oxide X_2O_7 contains 76.8% X and 23.2% oxygen by weight. If you had 100.0 g of the oxide, you would have 76.8 g of X and 23.2 g of O (or 23.2/16.0 moles of O atoms). Because for each mole of O in the oxide you have 2/7 mole of X, you have in effect (23.2/16.0) × 2/7 mole of X. Since the units of atomic mass are g/moles, the setup is:

$$\frac{76.8}{\left(\frac{23.2}{16.0}\right)\times\left(\frac{2}{7}\right)}$$

Finding the solution is unnecessary for selecting an answer, however, here is the solution for your information.

$$\frac{76.8\text{ g X}}{\left(\frac{23.2\text{ g }O_2}{16.0\text{ g }O_2/\text{mol }O_2}\right)\times\left(\frac{2\text{ mole X}}{7\text{ mol }O_2}\right)} = 186\text{ g mol}^{-1}$$

(The element is rhenium.)

10. 30% H_2O in the hydrate sample represents

$$\frac{\text{mass of hydrate} - \text{mass of anhydride}}{\text{mass of hydrate}} \times 100\%$$

In a problem like this, I like to make up some *easy* fictitious numbers that I can use to fit the scenarios and see how the various changes effect the final outcome. Let's say the mass of the hydrate is 10 g and the mass of the anhydride is 7 g. This would translate as

$$\frac{10\text{ g} - 7\text{ g}}{10\text{ g}} \times 100\% = 30\%\ H_2O$$

In examining choice A, the original mass of the hydrate, would not change, however, because rust will not evaporate, the final mass of the anhydride would be higher that expected—let's say 8 g. Substituting this value into the formulas for % water would give

$$\frac{10\text{ g} - 8\text{ g}}{10\text{ g}} \times 100\% = 20\%\ H_2O$$

which is less than the theoretical value of 30%. This is in the direction of the student's experimental results, and since we are looking for the choice that is NOT the cause, we can rule out A as an answer.

In choice B, the mass of the hydrate would not change, but the mass of the anhydride would be higher than it should be. Let's estimate the anhydride at 8 g again.

$$\frac{10\text{ g} - 8\text{ g}}{10\text{ g}} \times 100\% = 20\%\ H_2O$$

In choice C, because all masses are being measured on a *consistently* wrong balance, the faultiness does not matter in the final answer.

$$\frac{11.0\,g - 7.7\,g}{11.0\,g} \times 100\% = 30\%\ H_2O$$

In choice D, the original sample is wet. The freshman chemist weighed out 10 g of the hydrate, but more weight is lost in the heating process than expected, making the final mass of the anhydride lower than expected, say 6 g. Using he equation for % H_2O shows

$$\frac{10\,g - 6\,g}{10\,g} \times 100\% = 40\%\ H_2O$$

which is higher than the theoretical value of 30% and in line with the reasoning that the could NOT have caused the error.

11. Begin with a balanced equation:

$$C_4H_{10} + 13/2\ O_2 \rightarrow 4CO_2 + 5H_2O$$

Next, set up the equation in factor-label fashion:

$$\frac{100\,g.C_4H_{10}}{1} \times \frac{1\text{ mole }C_4H_{10}}{58.14\,g.C_4H_{10}} \times \frac{5\text{ mole }H_2O}{1\text{ mole }C_4H_{10}} \times \frac{18.02\,g\,H_2O}{1\text{ mole }H_2O} = g\ H_2O$$

12. All of the numbers are multiplies of 19.00 (fluorine). Use the law of multiple proportions.

13. Number of moles of $Ca(NO_3)_2$ $= \dfrac{7.0\,g}{164\,g\,mol^{-1}}$

Number of moles of water $= \dfrac{100\,g}{18\,g\,mole^{-1}}$

The ionization of $Ca(NO_3)_2$ may be represented as:

$$Ca(NO_3)_2 \rightleftharpoons Ca^{++} + 2NO_3^{+}$$

No. of atoms $\dfrac{7.0}{164}(1 - 0.7)$ $\quad \dfrac{7.0}{164}(0.7)$ $\quad \dfrac{7.0}{164}(2 \times 0.7)$

Total no. of moles in the solution,

$$n_{total} = n_{water} + n_{Ca^{2+}} + n_{NO_3}$$

$$= \frac{100}{18} + \frac{7.0}{164}(0.3 + 0.7 + 1.4) = 5.658$$

$$\therefore \text{ Mole fraction of water} = \frac{5.556}{5.658}$$

According to Raoult's law

$$p \times {}_1p_{1^\circ} = \frac{5.556}{5.658} \times 760\text{ mm Hg} = \mathbf{7463\ mm\ Hg}$$

14. We know that,

$$\Delta T_t = K_f \times \text{molatity}$$

or $$\text{molality} = \frac{\Delta T_f}{K_f} = \frac{273 - 272.70}{1.86}$$

$$= \frac{0.30}{1.86} = 0.1613$$

According to Raoult's law

$$\frac{p - p_s}{p} = \frac{n}{N} \text{ (For dilute soln.)}$$

$$\frac{23.51 - p_s}{23.51} = \frac{m}{1000/18}$$

$$23.51 - p_s = \frac{0.1613 \times 23.51 \times 18}{1000} = 0.068$$

or $$p_s = 23.51 - 0.068 = \mathbf{23.44\ mm\ Hg.}$$

15. Since, $$M = V \times D$$

$\therefore$ $$\text{Mass of benzene} = 56.8\ \text{cm}^3 \times 0.889\ \text{g cm}^{-3}$$
$$= 50.4952\ \text{g}$$

$$\text{Moles of benzene} = \frac{50.4952}{78} = 0.6474$$

According to Raoult's law,

$$\frac{p - p_s}{p} = \frac{n}{N} \text{ (for very dilute solutions)}$$

Substituting the values we get,

$$\frac{100 - 98.88}{100} = \frac{n}{0.6474}$$

$$\therefore n = \frac{1.12 \times 0.6474}{100} = 0.007251$$

$$\text{Molality} = \frac{\text{No. of moles of solute}}{\text{Mass of solvent in kg}} = \frac{0.007251}{50.4952 \times 10^{-3}}$$

Who know that

$$\Delta T_1 = K_f \times \text{molality}$$

or $$K_f = \frac{\Delta T_f}{\text{molatity}} = \frac{0.73}{0.144} = \mathbf{5.07}$$

16. Since

$$\Delta T_b = K_b \times \text{molality}$$

Here, $\Delta T = 373.34 - 373$ (B.P. of water)

$$= 0.34 = 0.51 \times \frac{12}{m \times 0.1}$$

$$\text{or,} \quad m = \frac{0.51 \times 12}{0.34 \times 0.1} = \mathbf{180\ g\ mol^{-1}}$$

17. We know that,

$\Delta T_f = K_f \times$ molality. Here $\Delta T = 278.51–278.03=0.48$,

Let m be the molecular mass of the compound.

$$\therefore \quad 0.48 = 5.12\left[\frac{0.643}{m} \times \frac{1}{50 \times 0.879 \times 10^{-3}}\right]$$

$$\text{or} \quad m = \frac{5.12 \times 0.643}{0.48 \times 50 \times 0.879 \times 10^{-3}} = \mathbf{156\ g\ mol^{-1}}$$

18. Molecular weight of benzole acid, C_6H_5COOH

$72 + 12 + 32 + 6 = 122$

$$2C_6H_6COOH \rightleftharpoons (C_6H_5COOH)_2$$

$$1 - \alpha \qquad\qquad \frac{\alpha}{2}$$

moles after dissociation

$$= 1 - \alpha + \alpha/2 = 1 - \alpha/2$$

$\because \quad \alpha = 0.84$

$$\text{moles after association} = 1 - \frac{0.84}{2} = 0.58$$

$\because \quad \Delta T_b = K_b \times$ molality

$$= (2.3)\ \frac{0.61 \times 1000}{122 \times 50} = 0.23$$

$$\Delta T_{obs} = \Delta T_{normal} \times \text{moles after dissociation}$$

$$= 0.23 \times 0.58 = \mathbf{0.133}$$

$$\text{B.P.} = 319.2 + 0.133 = 319.33\ \text{K}$$

19. Let w g of ethylene glycol be added.
Molecular mass of ethylene glycol,

$$\begin{array}{l} CH_2OH \\ | \\ CH_2OH \end{array} = 62\ g\ mol^{-1}$$

We know that,

$$\Delta T_f = K_f \text{ molality}$$

Here, $\Delta T_f = 273–267$

$$= 6 = 1.85\left[\frac{w}{62}\times\frac{1}{4}\right]$$

or $$w = \frac{6\times 62\times 4}{1.85} = \mathbf{804\ g}$$

20. $$\frac{W_1}{m_1} = \frac{W_2}{m_2}$$

$$\frac{1.2}{m_1} = \frac{7.2}{180}$$

$$\underset{\text{(observed)}}{m_1} = \frac{1.2\times 180}{7.2} = 30$$

Vant Hoff's factor, $$i = \frac{M_c}{M_o} = \frac{58.5}{30} = \mathbf{1.95}\ (\because \text{ molecular weight of NaCl} = 58.5)$$

21. Mass of solution = volume × Density
$= 1000\ cm^3 \times 1.005\ g\ cm^{-3} = 1005\ g$
Mass of solvent – 1005 – 15 = 990 g

No. of moles of solute (n) $= \dfrac{15\ g}{150\ g\ mol^{-1}} = 0.1$ mole

No. of moles of solvent (N) $= \dfrac{990\ g}{150\ g\ mol^{-1}} = 55$ mole

Mole fraction of solute $= \dfrac{n}{n+N} = \dfrac{0.1}{0.1+55} = \mathbf{0.0018}$

22. Mass of given H_2SO_4 in 1 litre
$= 1000 \times$ strength
$= 1000\ mL \times 1.84\ g\ mL^{-1} \times \dfrac{93}{100} = 1711.2\ g$
Mole of 1 litre solution = density × volume
$= 1.84\ mL^{-1} \times 1000\ mL = 1840\ g$
$\therefore$ Mass of solvent = 1840×1711.2 g
$= 128.8\ g = 0.1288\ kg$

$$\text{Molality} = \frac{\dfrac{\text{Mass of solute}}{\text{Mol. mass of solute}}}{\text{Mass of solvent in kg}} = \frac{\dfrac{1711.2\ g}{98\ g\ mol^{-1}}}{0.1288\ kg} = \mathbf{135.56\ m}$$

23. We know that, $PV = nRT = \frac{w}{m} RT \quad m = \frac{wRT}{PV}$

Here, $w = 0.2$g, $T = 285$K, $P = 570/760$ atm, $V = 100 \times 10^{-3}$L

Substituting the values, $m = \dfrac{(0.2\text{ g})(0.0821\text{ L atm K}^{-1}\text{ mol}^{-1})(285\text{ K})}{(570/760)\text{ atm }(100 \times 10^{-3}\text{ L})}$

$= \mathbf{62.40\ g\ mol^{-1}}$

24. We know that,

$$PV = nRT$$

Here, $V = 0.1$L, $T = 300$K

$$\therefore \quad P \times 0.1\text{L} = \left(\frac{0.85\text{ g}}{85\text{ g mol}^{-1}}\right)(0.082\text{ L atm K}^{-1}\text{ mol}^{-1}).\ (300\text{ K})$$

or $\quad P = 2.46$ atm

$NaNO_3$ dissociates as:

$$NaNO_3 \rightleftharpoons Na^+ + NO_3$$

1	0	0	(Initially)
$1 - \alpha$	α	α	(At equil.)

Total no. of moles at equal

$= 1 - \alpha + \alpha + \alpha = 1 + \alpha$

(But $\alpha = 90\% = 0.9$)

$\therefore \quad 1 + \alpha = (1 + 0.9) = 1.9$ mole

Therefore, osmotic pressure

$= 1.9 \times 2.46$ atm $=$ **4.674 atm.**

25. (a) The mass of water required to fill the pycnometer at 293K is

35.552 g – 25.601 g = 9.951 g.

The volume of water, and hence the volume capacity of he pycnometer, is

$$\text{no. cm}^3 = 9.951\text{ g} \times \frac{1.0000\text{ cm}^3}{0.99823\text{ g}} = 9.970\text{ cm}^3$$

(b) The mass of methanol required to fill the pycnometer at 293 K is

33.490 g – 25.601 g = 7.889 g

The density of methanol at 293 K is

$$d = \frac{m}{V} = \frac{7.889\text{ g}}{9.969\text{ cm}^3} = 0.78\text{ g cm}^{-3}.$$

26. (i) Use the formula for the volume of a sphere.

$$V = \frac{4}{3}\pi r^3 = \frac{4}{3}\pi\ (3.14)(0.635)^3\text{ cm}^3 -= 1.07\text{ cm}^3$$

(ii) Use density as the conversion factor between volume and mass, followed by percent composition.

$$\text{no. g carbon} = 1.07\ \text{cm}^3 \times \frac{7.75\ \text{g steel}}{1.00\ \text{cm}^3\ \text{steel}} \times \frac{0.25\ \text{g carbon}}{100.0\ \text{g steel}}$$

$$= 0.021\ \text{g carbon}$$

(iii) The mass of carbon from part (ii) must be converted to mol C and then, through the Avogadro constant, to number of C atoms.

$$\text{no. C atoms} = 0.021\ \text{g C} \times \frac{1\ \text{mol C}}{12.0\ \text{g C}} \times \frac{6.02 \times 10^{23}\ \text{C atoms}}{1\ \text{mol C}}$$

$$= 1.1 \times 10^{21}\ \text{C atoms}$$

(iv) Use the result of part 3 and percent natural abundance as conversion factor.

$$\text{no. C-13 atoms} = 1.1 \times 10^{21}\ \text{C atoms} \times \frac{1.11\ \text{C-13 atoms}}{100\ \text{C atoms}} = 1.2 \times 10^{19}\ \text{C–13 atoms.}$$

(v) First convert the number of C-13 atoms found in part (iv) to the number of moles of C-13. From the definition of the mole, the mass of one mole of C-13 is 13.00335 g. This provides us with a factor to convert from moles to grams.

$$\text{no. g C-13} = 1.2 \times 10^{19}\ \text{C-13 atoms} \times \frac{1\ \text{mol C-13}}{6.02 \times 10^{23}\ \text{atoms}} \times \frac{13.0\ \text{g C-13}}{1\ \text{mol C-13}}$$

$$= 2.6 \times 10^{-1}\ \text{g C-13}$$

27. 1. The 1.647 g mass corresponds to the *anhydrous* compounds $CuSO_4$. The loss of mass (2.574 – 1.647) g = 0.927 g is the mass of water of hydration in the original sample. Convert the masses of the anhydrous compound and of the water to number of moles,

$$\text{no. mol } CuSO_4 = 1.647\ \text{g } CuSO_4 \times \frac{1\ \text{mol } CuSO_4}{159.6\ \text{g } CuSO_4} = 0.01032\ \text{mol } CuSO_4$$

$$\text{no. mol } H_2O = 0.927\ \text{g } H_2O \times \frac{1\ \text{mol } H_2O}{18.02\ \text{g } H_2O} = 0.05144\ \text{mol } H_2O$$

Now, write a formula based on these molar amounts $(CuSO_4)_{0.01032} \cdot (H_2O)_{0.05144}$ and reduce to the simplest formula,

$$(CuSO_4)_{\frac{0.01032}{0.01032}} \cdot (H_2O)_{\frac{0.05144}{0.01032}} = (CuSO_4) \cdot (H_2O)_{4.98} = CuSO_4 \cdot 5H_2O$$

2. The loss of water occurring at 413 K is (2.574 – 1.833) = 0.741 g. The mass of water *remaining* at 413 K is the *total* loss of water (0.927 g) minus the loss that occurs at 413 K (0.741 g): 0.927 – 0.741 = 0.186 g H_2O. Now express this amount of water of water on a mole basis.

$$\text{no. mol } H_2O = 0.186\ \text{g } H_2O \times \frac{1\ \text{mol } H_2O}{18.02\ \text{g } H_2O} = 0.01032\ \text{mol } H_2O$$

The number of moles of $CuSO_4$ was determined in part 1. It is 0.01032 mol. The $CuSO_4$ and the H_2O are in a 1:1 mol ratio. The formula of the hydrate obtained at 413 K is $CuSO_4 \cdot H_2O$

3. Recall that the 1.647 g anhydrous $CuSO_4$ sample corresponds to 0.01032 mol. The 0.01032 mol Cu present in this sample appears in the 0.812 g of copper oxide residue. The mass of Cu, its percentage, and the percent O_2 in copper oxide are

$$\text{no. g Cu} = 0.01032 \text{ mol Cu} \times \frac{63.546 \text{ g Cu}}{1 \text{ mol Cu}} = 0.6558 \text{ g Cu}$$

$$\% \text{ Cu} = \frac{0.6558 \text{ g Cu}}{0.812 \text{ g cpd.}} \times 100 = 80.8\% \quad \%O_2 = 100.0 - 80.8 = 19.2\%$$

We can readily determine the empirical formula from the percentage composition

$$\text{no. mol Cu} = 80.8 \text{ g Cu} \times \frac{1 \text{ mol Cu}}{63.55 \text{ g Cu}} = 1.27 \text{ mol Cu}$$

$$\text{no. mol O} = 19.2 \text{ g O} \times \frac{1 \text{ mol O}}{16.00 \text{ g O}} = 1.20 \text{ mol O}$$

The empirical formula is $Cu_{1.27}O_{1.20} = Cu_{1.06}O = CuO$.

28. (i) The series of conversion required here is

$$\text{g } Na_2CO_3 \rightarrow \text{mol } Na_2CO_3 \rightarrow \text{mol } NaNO_2 \rightarrow \text{g } NaNO_2$$

These conversion can be done on a step-by-step basis or in single setup,

$$\text{no. g Na/NO}_2 = 748 \text{ g } Na_2CO_3 \times \frac{1 \text{ mol } Na_2CO_3}{106.0 \text{ g } Na_2CO_3} \times \frac{4 \text{ mol } NaNO_2}{2 \text{ mol } Na_2CO_3} \times \frac{69.00 \text{ g } NaNO_2}{1 \text{ mol } NaNO_2}$$

$$(\text{g } Na_2CO_3 \quad \rightarrow \text{mol } Na_2CO_3 \quad \rightarrow \text{and } NaNO_2 \quad \rightarrow \text{g } NaNO_2)$$

$$= 974 \text{ g } NaNO_2$$

(ii) The problem in best broken down into *three* smaller problems,

(a) Which is the limiting reagent? $O_2(g)$ is stated to be in excess, We must choose between Na_2CO_3 and NO.

Assuming Na_2CO_3 is the limiting reagent:

$$\text{no. mol } NaNO_2 = 0.225 \text{ L} \times \frac{1.50 \text{ mol } Na_2CO_3}{1 \text{ L}} \times \frac{4 \text{ mol } NaNO_2}{2 \text{ mol } Na_2CO_3}$$

$$= 0.675 \text{ mol } NaNO_2$$

Assuming NO is the limiting regant:

$$\text{no mol } NaNO_2 = 22.1 \text{ g NO} \times \frac{1 \text{ mol NO}}{30.01 \text{ g NO}} \times \frac{4 \text{ mol } NaNO_2}{4 \text{ mol NO}}$$

$$= 0.736 \text{ mol Na/NO}_2$$

Since 0.675 mol $NaNO_2$ is the smaller of the two results, Na_2CO_3 is the limiting reagent.

(b) What is the theoretical yield of $NaNO_2$? Simply convert the amount of $NaNO_2$ determined in (a) to a mass in grams.

$$\text{no. g } NaNO_2 = 0.675 \text{ mol } NaNO_2 \times \frac{69.00 \text{ g } NaNO_2}{1 \text{ mol } NaNO_2} = 16.6 \text{ g } NaNO_3$$

(c) What is the actual yield of $NaNO_2$?

$$\text{actual yield} = \frac{\% \text{ yield} \times \text{theoretical yield}}{100} = \frac{95 \times 46.6 \text{ g}}{100}$$

$$= 44.3 \text{ g } NaNO_2$$

29. $5Fe^{2+}(aq) + MnO_4^-(aq) + 8H^+(aq)$

$$\rightarrow 5Fe^{3+}(aq) + Mn^{2+}(aq) + 4H_2O(l)$$

$$\text{moles of } MnO_4^- = \text{volume in dm}^3 \times \text{conc}$$

$$= 0.0210 \times 0.0200 = 4.20 \times 10^{-4}$$

From the equation:

$$\text{moles of } Fe^{2+} = 5 \times \text{moles of } MnO_4^- = 2.10 \times 10^{-5}$$

Therefore:

$$[MnO_4^-(aq)] = \frac{\text{moles of } MnO_4^-}{\text{volume of } MnO_4^- \text{ (in dm}^3)}$$

$$= \frac{2.10 \times 10^{-3}}{0.0250} = 0.0840 \text{ mol dm}^{-3}$$

30. Determine the no. mol CrO_4^{2-} that reacts.

$$\text{no. mol } CrO_4^{2-} = 100.0 \text{ L} \times \frac{0.0148 \text{ mol } CrO_4^{2-}}{1 \text{ L}} = 1.48 \text{ mol } CrO_4^{2-}$$

Determine the no. mol $S_2O_4^{2-}$ that reacts.

$$\text{no. mol } S_2O_4^{2-} = 1.48 \text{ mol } CrO_4^{2-} = \frac{3 \text{ mol } S_2O_4^{2-}}{2 \text{ mol } CrO_4^{2-}} = 2.22 \text{ mol } S_2O_4^{2-}$$

Determine the mass of $Na_2S_2O_4$

$$\text{no. g } Na_2S_2O_4 = 2.52 \text{ mo } S_2O_4^{2-} \times \frac{1 \text{ mol } Na_2S_2O_4}{1 \text{ mol } S_2O_4^{2-}} \times \frac{173.1 \text{ g } Na_2S_2O_4}{1 \text{ mol } Na_2S_2O_4}$$

$$= 3.87 \, Na_2S_2O_4$$

31. (i) First we will determine the number of moles of $CO_2(g)$ produced. Thus we follow this with an application of the ideal gas equation.

$$\text{no. mol } CO_2 = 100 \text{ g } C_8H_{18} \times \frac{1 \, 1 C_8H_{18}}{114.2 \text{ g } C_8H_{18}} \times \frac{16 \text{ mol } CO_2}{2 \text{ mol } C_8H_{18}} = 7.01 \text{ mol } CO_2$$

$$V = \frac{nRT}{P} = \frac{7.01 \text{ mol} \times 0.08206 \text{ L atm mol}^{-1} \text{ K}^{-1} \times 295 \text{ K}}{(745/760) \text{ atm}} = 173 \text{ L } CO_2$$

(ii) This calculation is based solely on the chemical equation and does not require us to use any of the gas laws.

$$\text{no. g } O_2 = 100 \text{ g } C_8H_{18} \times \frac{1 \text{ mol } C_8H_{18}}{114.2 \text{ g } C_8H_{18}} \times \frac{25 \text{ mol } O_2}{2 \text{ mol } C_8H_{18}} \times \frac{32.00 \text{ g } O_2}{1 \text{ mol } O_2}$$

$$= 3.50 \text{ g } O_2$$

(iii) This question can be answered in two ways. One method is to consider the partial volumes and then the masses of the several gases present in a 100 L sample of air. These individual masses can be converted to percentages A second method uses the given values.

Here, In one mole of air, with a mass of 28.96 g, there is present 0.2095 mol of O_2, weighing

$$0.2095 \text{ mol } O_2 \times \frac{31.999 \text{ g } O_2}{1 \text{ mol } O_2} = 6.704 \text{ g } O_2$$

The mass percent of O_2 is

$$\frac{6.704 \text{ g } O_2}{28.96 \text{ g air}} \times 100 = 23.15\% \ O_2$$

(iv) Base the calculation on the combustion of 100 g C_8H_{18}. The required mass of O_2 is 350 g (from part ii). Use the percent composition of air from part C to determine the mass of air containing 350 g O_2

$$\text{no. g air} = 350 \text{ g } O_2 \times \frac{100.\text{ g air}}{23.15 \text{ g } O_2} = 1.51 \times 10^3 \text{ g air}$$

The air/octane ratio, by mass, is

$$\frac{1.51 \times 10^3 \text{ g air}}{100.\text{ g } C_8H_{18}} = 15.1$$

32. To work out how many moles of atoms are in a particular mas of substance, use the equation:

$$\text{amount in moles} = \frac{\text{mass in grams}}{\text{mass of one mole (in grams)}}$$

$$\therefore \quad \text{moles of silicon} = \frac{56 \text{ g}}{28 \text{ g}} = 2 \text{ moles}$$

33. Mass in grams = amount in moles × mass of 1 mole

mass of iron = 0.25 mol × 56 g = 14 g

34. The sample with the largest number of atoms will be the sample with the largest number of moles of atoms, because a mole contains 6.022×10^{23} atoms. Since all samples weigh 1 gram, and

$$\text{No. moles} = \frac{\text{mass in grams}}{\text{formula wt. in g mol}^{-1}}$$

the substance with the smallest formula weight will be the one for which the number of moles is largest. Both O_2 and N_2 are diatomic, and the number of moles of atoms is twice the number of moles of the molecules. Of the three solids (Ni, Ca, and B), boron has the smallest atomic weight. The molecular weight of N_2 is 28.0134, so the number of moles of N atoms is

$$2 \times \left(\frac{1.00 \text{ g}}{28.0134 \text{ g mol}^{-1}} \right) = \frac{1.00}{14.0067} \text{ mol of N atoms}$$

The number of moles of B atoms is larger, it is (1.00/10.81) mol.

Note that you can answer this question without using Avogadro's number or actually calculating the number of moles or the number of atoms of any of these substances. You are asked for a comparison, and need only find out which of these 5 elements has the smallest atomic weight.

35. There are two factors to be considered: the number of atoms per molecule, and the molecular weight of the substance. There are 3 atoms per molecule for CO_2, 8 for C_2H_6, 26 for C_8H_{18}, 2 for LiF, and 14 for B_4H_{10}. The molecular weights are 44 for CO_2, 30 for C_2H_6, 114 for C_8H_{18}, 25.9 for LiF, and 53.3 for B_4H_{10}. The number of atoms is directly proportional to the number of atoms of atom, which is 3(1/44) = 0.068 for CO_2, and 2(1/25.9) = 0.077 for LiF. The other have many more atoms per molecule, and therefore a larger number of moles of atoms.

36 In $(NH_4)_2SO_4$, $\dfrac{\text{No. mol of } O_2 \text{ atoms}}{\text{No. mol of H atoms}} = \dfrac{4}{8} = \dfrac{1}{2}$

37. In $(NH_4)_3PO_4$, $\dfrac{\text{No. mol of } O_2 \text{ atoms}}{\text{No. mol of H atoms}} = \dfrac{4}{12} = \dfrac{1}{3}$

Thus,

$$\begin{aligned} \text{No. moles } O_2 \text{ atoms} &= (1/3) \text{ (No. moles H atoms)} \\ &= (1/3)\ (3.18) = 1.06 \text{ mol} \end{aligned}$$

38. The atomic weight of Cu is 63.546 g mol^{-1}. Thus the number of moles of Cu is

$$\frac{3.782 \text{ g}}{63.546 \text{ g mol}^{-1}} = 5.9516 \times 10^{-2} \text{ mol Cu}$$

In $CuSO_4 \cdot 5H_2O$, $\dfrac{\text{No. mol } O_2}{\text{No. mol Cu}}$ Note that there are 4 0 atoms per SO_4^{2-} ion, and one O_2 atom per H_2O.

$$\begin{aligned} \text{Thus, No. mol } O_2 \text{ in sample} &= 9(\text{No. mol Cu}) = 9(5.9516 \times 10^{-2}) \\ &= 5.356_4 \times 10^{-1} \text{ mol 0 atoms} \\ \text{mass of } O_2 &= (5.356_4 \text{ mol})\ (15.9994 \text{ g} \cdot \text{mol}^{-1}) = 8.570 \text{ g } O_2 \end{aligned}$$

39. A 1.1596 g sample of this compound contains 1.000 g of X. It must therefore contain 0.1596 g of O_2.

$$\text{No. moles } O_2 = \frac{0.1596 \text{ g}}{15.9994 \text{ g mol}^{-1}} = 9.975_4 \times 10^{-3} \text{ mol } O_2$$

In X_2O_3, $\dfrac{\text{No. moles X}}{\text{No. moles } O_2} = \dfrac{2}{3}$

Hence, No. mol X = (2/3) $(9.9754 \times 10^{-3}) = 6.650 \times 10^{-3}$ mol X

$$\frac{\text{wt. in grams of } X}{\text{atomic wt. of } X} = \text{No. mol X}$$

$$\text{at. wt. of X} = \frac{\text{wt. in g of } X}{\text{No. mol of } X} = \frac{1.000}{6.650 \times 10^{-3}} = 150.4 \text{ g} \cdot \text{mol}^{-1}$$

Element X is samarium, Sm, Z = 62. The compound is Sm_2O_3.

40. The number of moles of Ag in a 1.00-g sample is:

$$\frac{1.00 \text{ g}}{107.868 \text{ g mol}^{-1}} = 9.27 \times 10^{-3} \text{ mol Ag}$$

$$\text{In } Ag_2S, \frac{\text{No. mol Ag}}{\text{No. mol } Ag_2S} = \frac{2}{1}$$

Thus, No. mol Ag_2S = (1/2) (No. mol Ag) = 4.635×10^{-3} mol Ag_2S. Formula weight of Ag_2S = 2(107.87) + 32.06 = 247.80 g mol^{-1}

The weight of Ag_2S needed to yield 1.00 g of Ag is therefore

$$(4.635 \times 10^{-3} \text{ mol}) (247.80 \text{ g} \cdot \text{mol}^{-1}) = 1.149 \text{ g} = 1.15 \text{ g}$$

Let x = the weight of ore needed. Then the weight of Ag_2S is 1.34% of x, or $0.0134x = 1.149$ g, and x = 85.7 g of ore

Alternative method: Let x = wt. of ore that must be processed. Then,

wt. of Ag_2S in ore = $0.0134x$

How many moles of Ag_2S are in x grams of ore? As the formula weight of Ag_2S is 247.80 g/mol.

$$\text{No. moles } Ag_2S \text{ in ore} = \frac{0.0134\, x\text{g}}{247.80 \text{ g mol}^{-1}}$$

$$\text{No. moles Ag} = 2(\text{No. moles } Ag_2S) = \frac{2(0.0134x)}{247.80}$$

$$\text{No. grams Ag} = (\text{No. moles Ag}) (\text{at. wt. in g mol}^{-1})$$

$$= \frac{2(0.0134\,x)(107.87)}{247.80}$$

$$= 1.167 \times 10^{-2}\, x = 1.00 \text{ g of Ag}$$

Hence $$x = \frac{1.00}{1.167 \times 10^{-2}} = 87.5 \text{ go of ore are needed.}$$

41. On the right-hand side there are 5 × 2 + 6 = 16 mol of O_2 atoms. Therefore there must be 16 mol of O_2 atoms, or g mol of O_2 molecules on the left.

42. There must be 2 moles of C atoms on the right, therefore the coefficient of CO_2 is 2. There must be 6 moles of H atoms on the right, therefore the coefficient of H_2O is 3. The ratio of the coefficients of CO_2 to H_2O is therefore 2/3.

43. Mol. wt. C_2H_5OH = 2(12.011) + 6(1.0079) + 15.9994 = 46.07.

Mass 1.2 mol of C_2H_5OH = (1.2 mol) (46.07 g mol^{-1}) = 55.3 g

$$\text{Density} = \frac{\text{mass in g}}{\text{volume in mL}} = 0.7893 \text{ g mL}^{-1} \text{ Hence,}$$

$$\text{volume in mL} = \frac{\text{mass in g}}{0.7893 \text{ g mL}^{-1}} = \frac{55.3 \text{ g}}{0.7893 \text{ g mL}^{-1}} = 70 \text{ mL}$$

44. Mol. wt. camphor = 10(12.011) + 16(1.0079) + 15.9994

= 152.24 g · mol^{-1}

25.0 mg = 25.0 × 10^{-3} g, or 2.50 × 10^{-2} g

$$\text{No. mol camphor} = \frac{2.5 \times 10^{-2} \text{ g}}{152.24 \text{ g mol}^{-1}} = 1.64 \times 10^{-4} \text{ mol}$$

Each molecule of camphor contains 27 atoms (10C, 16H, 1 O_2).

No. atoms = (1.64 × 10^{-4} mol) (6.022 × 10^{23} molecules/mol)

(27 atoms/molecule)

= 2.67 × 10^{21} atoms in 25.0 mg of camphor

45. Mol. wt. = [3(12) + 4(1) + (16)]n = 56n. There is no need for more figures because the molecular weight given has a large uncertainty. Thus 56n ~ 170, and n is exactly 3. The value of n must be an exact integer; it is the molecular weight that is uncertain. The compound is therefore, $C_9H_{12}O_3$, and its exact molecular weight is 9(12.011) + 12(1.0079) + 3(15.9994) = 168.19 g mol^{-1}.

170 ± 5 means between 165 and 175. The value 168.19 is within these limits.

46. Consider a 100.00-g sample of this compound. It contains 65.03 g Ag. 15.68 g Cr, and 19.29 g O_2.

$$\text{No. moles Ag} = \frac{65.03 \text{ g}}{107.868 \text{ g mol}^{-1}} = 0.6029 \text{ mol Ag}$$

$$\text{No. moles Cr} = \frac{15.68 \text{ g}}{51.996 \text{ g mol}^{-1}} = 0.3016 \text{ mol Cr}$$

$$\text{No. moles } O_2 = \frac{19.29 \text{ g}}{15.9994 \text{ g mol}^{-1}} = 1.206 \text{ mol } O_2$$

$$\text{Molar ration: } \frac{\text{No. moles Ag}}{\text{No. moles Cr}} = \frac{0.6029}{0.3016} = 1.999 = 2.00$$

$$\frac{\text{No. moles } O_2}{\text{No. moles Cr}} = \frac{1.206}{0.3016} = 3.999 = 4.00$$

The simplest formula of this compound is Ag_2CrO_4.

47. In one mole of cortisone, the weight of C is 21(12.011) g, or 2.52.23 g C per mole of cortisone. By the definition or a weight percentage.

$$\frac{\text{wt. of C in 1 mol}}{\text{Mol. wt. of cortisone}} = 0.699 \text{ g}$$

Thus,

$$\text{Mol. wt. of cortisone} = \frac{252.23}{0.6998} = 360.4 \text{ g mol}^{-1}$$

48. Mol. wt. benzoic acid = 7(12.011) + 6(1.0079) + 2(15.9994)
= 122.12 g mol^{-1}

$$\text{No. mol } C_6H_5COOH = \frac{24.4 \text{ g}}{122.12 \text{ g mol}^{-1}} = 0.200 \text{ mol}$$

The mass of 70 mL of CH_3OH is (70 mL) (0.791) g · mL^{-1}) = 55.4 g
Mol. wt. CH_3OH = 12.011 + 4(1.0079) + 15.9994
= 32.04 g · mol^{-1}

$$\text{No. mol } CH_3OH = \frac{55.4 \text{ g}}{32.04 \text{ g mol}^{-1}} = 1.73 \text{ mol } CH_3OH$$

Benzoic acid and methanol react in a 1:1 molar ratio, in this mixture there is excess CH_3OH. Thus benzoic acid the limiting reagent. The maximum amount of methyl benzoate that can be produced is 0.200 mol.

Mol. wt. $C_6H_5COOCH_3$ = 8(12.011) + 8(1.0079) + 2(15.9994)
= 136.15 mol^{-1}

maximum yield methyl benzoate = (0.200 mol) 136.15 g mol^{-1})
= 27.23 g

The actual yield of methyl benzoate was 21.6 g. Thus,

$$\text{yield} = \left(\frac{21.6 \text{ g}}{27.23 \text{ g}}\right) \times 100 = 79.3\%$$

49. The atomic number of Rb is 37. Hence ^{85}Rb has (65 – 37) = 48 neutrons. The atomic number of Sr is 38. Since 38 + 48 = 86, ^{86}Sr also has 48 neutrons.

50. Since $N_A = 6.0 \times 10^{23}$, 2.0×10^{23} atoms is 1/3 mol of atoms.
No. moles Be atoms = (3.0 g)/(9.0 g · mol^{-1}) = 1/3 mol Be
No. moles O atoms = 2(8.0 g)/(32.0 g · mol^{-1}) = 1/2 mol O
as there are 2 atoms of O per molecule of O. The other choices are 2/3 mol of C atoms, 3 mol He atoms, and 1 mol of F atoms.

51. Tellurium, Te, is atomic number 52. To form an anion with a charge of –2 requires the addition of 2 electrons. Since the neutral atom has 52 electrons, there are 54 in the Te^{2-} ion.

52. Nitrogen dioxide, NO_2 is one of the few gases that are intensely colored. The red-brown color of NO_2 is distinctive.

53. Laughing gas, used as an anesthetic, is dinitrogen oxide, N_2O (also called nitrous oxide). Inhaling small amounts of N_2O causes a giddy sensation. It must be mixed with oxygen when it is administered or death by suffocation results.

54. Note that the question asks only for an approximate value of the natural abundance. We can therefore use the mass number as the mass of each isotope. One can say immediately that the percentage of ^{63}Cu must be greater than 50%, because the atomic weight of Cu is significantly less than 64, and a 50/50 mixture of masses 63 and 65 would result in an atomic weight of 64. Thus we need only consider the two choices (d) and (e). If you try (d) first you find 0.70(63) + 0.30(65) = 63.6, very close to the actual atomic weight. It is clear that the natural abundance of the ^{63}Cu isotope is approximately 70%. If you try (e) first, you find 0.900 (63) + 0.10(65) = 63.2, much further from the actual atomic weight. To be certain, though, you would have to do the calculation for both 70% and 90% if you had tried 90% first.

55. The net ionic equation for the reaction between yttrium metal and dilute hydrochloric acid is:

$$Y(s) + 3H^+(aq) \rightarrow Y^3(aq) + \frac{3}{2}H_2(g)$$

or $$2Y(s) + 6H^+(aq) \rightarrow 2Y^{3+}(aq) + 3H_2(g)$$

Hence, the molar ratio of yttrium: H_2 = 1(3/2) = 2:3

56. Molar mass of glycine

$$= (14 \times 1) + (1 \times 5) + (12 \times 2) + (16 \times 2) = 75 \text{ gm}$$

As $$M = \frac{n_a}{V}$$

$\therefore$ $$0.01 = \frac{x/75}{100 \times 1000} \Rightarrow x = 0.075 \text{ gm}$$

57. (a) As $N_1V_1 = N_2V_2$

$\therefore$ $(N_1 \times 180)$ of conc. H_2SO_4 = (6.61 × 1000) of dil. H_2SO_4

$\Rightarrow$ $$N_1 = \frac{6.61 \times 1000}{180} = 36.72 \text{ N}$$

Thus, the normality of conc. H_2SO_4 = 36.72 N (i)

(b) For H_2SO_4, Eq. Wt = $\frac{\text{mol. wt.}}{2}$

Thus, 1 eq. L^{-1} = 0.5 L^{-1} (for H_2SO_4)

As Normality of dil H_2SO_4 = 6.61 given

So, the molarity of dil. $H_2SO_4 = \frac{6.61}{2} = 3.305 \text{ mol } L^{-1}$

$= 3.305$ M (ii)

(c) From Eq. wt. of H_2SO_4 (= 49) and its normality (= 6.61 N), we can calculate the weight of H_2SO_4 actually contained in solution as

= 6.61 × 49 = 323.89 gm

Now, weight of 180 mL of conc. H_2SO_4 (having sp gr. = 1.84)

= 180 × 1.84 = 331.2 gm

$\therefore$ %H_2SO_4 by weight = $\frac{323.89}{331.2} \times 100 = 97.79 \approx 97.8\%$ (iii)

(d) Weight of H_2SO_4 in 1 L of dil. Solution = 49 × 6.61 = 323.89 gm (I)

Moreover, from the sp. gr. (= 1.198), volume (1000 mL) and y (let % by weight in dil. Solution), the weight of H_2SO_4 in 1 L of dil. Solution

$$= 1000 \times 1.198 \times 0.01\ y \qquad \text{(II)}$$

Equating (I) and (II), we get

$$323.89 = 1000 \times 1.198 \times 0.1\ y \Rightarrow y = 27.035\% \approx 27.00\% \qquad \text{(iv)}$$

(e) Weight of H_2SO_4 in 1 L of dil. Solution = 323.89 gm (from I)
Weight of 1 L of dil. solution = 1000 × 1.198 = 1198 gm
∴ Weight of water = 1198 – 323.85 = 874.11 gm.
From answer of part (b), molarity of dil. H_2SO_4 = 3.305 mol L^{-1} so L of solution contains 3.305 moles of H_2SO_4.

$$\text{Thus, molality} = \frac{\text{moles of } H_2SO_4}{\text{weight in kg of solvent } (H_2O)}$$

$$= \frac{3.305}{874.11/1000} = 3.781 \text{ m} \qquad \text{(v)}$$

58. *Step 1.* Write the balanced reaction for the reaction:

$$CH_4(g) + 2O_2(g) \rightarrow CO_2(g) + 2H_2O(l)$$

Step 2. Convert the equation to volumes:
Since: 1 mol CH_4 reacts with 2 mol O_2
1 volume CH_4 reacts with 2 volumes O_2

Step 3. Use the simple whole number ratio of volumes to work out the actual volume used:
600 cm^3 CH_4 reacts with 2 × 600 cm^3 O_2
So the volume of oxygen used is 1200 cm^3.

59. Molar mass of glucose $C_6H_{12}O_6$ is 12 × 6 + 1 × 12 + 6 × 6 = 180 gm. Suppose, x gm of glucose is required to prepare 2 M soln.

Here, M = 2 molar, V = 1 litre, $n_a = x/180$

Since $M = \frac{n_a}{V}$ ∴ $2 = \frac{x/180}{1}$

⇒ x = 360 gm.

Thus, for preparing 2 M solution of glucose, 360 gm of glucose is dissolved in enough water to produce 1.0 L solution
It is wrong to dissolve the solution in 1 L of water.

60. The oxidation number of oxygen is almost always –2.
The sum of all the known oxidation numbers (due to oxygen) is 7 × (–2) = –14
The sum overall of the oxidation numbers must be the charge on the ion = –2.
So, the sum of the oxidation numbers of the two chromium atoms = –2 – (–14) = +12
This means that each chromium atom has an oxidation number of +6, which explains why the dichromate ion is more accurately referred to as the dichromate (VI) ion.

61. *Step 1.* Write the balance equation for the reaction:

$$S(s) + O_2(g) \rightarrow SO_2(g)$$

Step 2. Convert the equation to amounts:
1 mol S gives 1 mol SO_2
Note: In this example we do not need to know the number of moles of O_2.

Step 3. Work out the amount being used (1 tonne is 1,00,000 g):
100 tonnes of coal contains 0.5 per cent S.
so the mass of S = 0.5 tonnes of S = 0.5 tonnes = 500 000 g.

$$\text{Moles of S in 0.5 tonnes} = \frac{\text{mass (in grams)}}{\text{mass of one mole (in grams)}}$$

$$= \frac{500000 \text{ g}}{32 \text{ g}} = 15625 \text{ mol.}$$

Step 4. Scale the amounts in the equation:
1 mol S producers 1 mol SO_2
15625 mol S produces 15625 mol SO_2

Step 5. Calculate the volume of gas:
1 mole of gas occupies 24 dm^3 at 298 K and 101 KPa.
Volume of SO_2 = 15625 × 24 = 375 000 dm^3

62. Write the chemical equation for the combustion reaction and calculate the theoretical yield of MgO.

$$2 \text{ Mg(s)} + O_2\text{(g)} \rightarrow 2 \text{ MgO(s)}$$

$$\text{no. g MgO} = 0.200 \text{ g Mg} \times \frac{1 \text{ mol Mg}}{24.30 \text{ g Mg}} \times \frac{2 \text{ mol MgO}}{2 \text{ mol Mg}} \times \frac{40.30 \text{ g MgO}}{1 \text{ mol MgO}}$$

$$= 0.322 \text{ g MgO}$$

We should have obtained 0.332 g MgO but obtained only 0.315 g of product. The product cannot be pure MgO.

63. Applying the law of conservation of mass. According to this law, the total mass remain constant before during a chemical reaction; that is,
Mass of substance before reaction = mass of substances after reaction
From the law of conservation of mass,
Mass of mercury + mass of oxygen = mass of red-orange residue
Substituting, you obtain

$$2.53 \text{ grams} + \text{mass of oxygen} = 2.73 \text{ grams}$$

or,

$$\text{Mass of oxygen} = (2.73 - 2.53) \text{ grams} = 0.20 \text{ grams}$$

The mass of oxygen collected when the red-orange residue decomposes equals the mass of oxygen that originally reacted (0.20 grams).

64. Since 1 mg = 10^{-3} g, you can write.

$$243 \text{ mg} \times \frac{10^{-3} \text{ g}}{1 \text{ mg}} = 2.43 \times 10^{-1} \text{ g}$$

Then, because the prefix *kilo-* means 10^3, you write

$$1 \text{ kg} = 10^3 \text{ g}$$

and

$$2.43 \times 10^{-1} \text{ g} \times \frac{1 \text{ kg}}{10^3 \text{ g}} = 2.34 \times 10^{-4} \text{ g}$$

Note, however, that you can do the two conversions in one step, as follows:

$$2.43 \text{ mg} \times \underbrace{\frac{10^{-3}\text{ g}}{1\text{ mg}}}_{\text{converts mg to g}} \times \underbrace{\frac{1\text{ kg}}{10^{3}\text{ g}}}_{\text{converts g to kg}} = 2.43 \times 10^{-4} \text{ kg}$$

65. Convert cubic kilometers to cubic decimeters, which are equivalent to liters. Do this in two steps, first cubic kilometers to cubic meters, then cubic meters to cubic decimeters. (The steps may be combined.) For a conversion such as that for cubic kilometers to cubic meters, note that 1 km = 10^3 m, so the factor is $(10^3 \text{ m}/1 \text{ km})^3$

$$1.35 \times 10^{9} \text{ km}^3 \times \underbrace{\left(\frac{10^{3}\text{ m}}{1\text{ km}}\right)}_{\text{converts km}^3\text{ to m}^3} \times \underbrace{\left(\frac{1\text{ dm}}{10^{-1}\text{ m}}\right)}_{\text{converts m}^3\text{ to dm}^3} = 1.35 \times 10^{21} \text{ dm}^3$$

66. Taking the average
Multiply each isotopic mass by its fractional abundance, then sum:

49.9461 amu × 0.0435 = 2.17 amu
51.9405 amu × 0.8379 = 43.52 amu
52.94087 amu × 0.0950 = 5.03 amu
53.9389 amu × 0.0236 = 1.27 amu
Total = 51.99 amu

The atomic weight of chromium is 51.99 amu.

67. (a) Because a compound is neutral, the sum of positive and negative charges equals zero. Consider the ionic compound $CaCl_2$, which consists of one Ca^{2+} ion and two Cl^- ions. The sum of the charges is:

$$\underbrace{1\times(+2)}_{\text{Positive charge}} + \underbrace{2\times(-1)}_{\text{Negative charge}} = 0$$

Note that the number of calcium ions in $CaCl_2$ equals the magnitude of charge on the chloride ion (1), whereas the number of chloride ions in $CaCl_2$ equals the magnitude of charge on the calcium ion (2). In general, you use the magnitude of charge on one ion to obtain the subscript for the other ion. You may need to simplify the formula you obtain this way so that it expresses the simplest ratio of ions.
You can achieve electrical neutrality by taking as many cations as there are units of charge on the anion and as many anions as there units of charge on the cation. Two Cr^{3+} ions have a total charge of 6+, and three O^{2-} ions have a total charge of 6–, giving the combination a net charge of zero. The simplest ratio Cr^{3+} to O^{2-} is 2 : 3, and the formula is Cr_2O_3. Note that the charge (without its sign) on one ion becomes the subscript of the other ion.

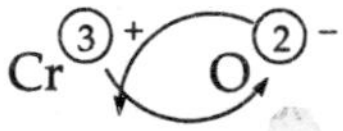

or Cr_2O_3

68. Because each mole of substance contains Avogadro's number of units of that substance, dividing the mass of a substance in one mole (molar mass) by Avogadro's number gives the mass of that unit of substance. Thus,

(a) The atomic weight of Cl is 35.5 amu, so the molar mass of Cl is 35.5 g mol^{-1}, Dividing 35.5 g (per mole) by 6.02 × 1023 (Avogadro's number) gives the mass of one atom.

$$\text{Mass of a Cl atom} = \frac{35.5\text{ g}}{6.02 \times 10^{23}} = 5.90 \times 10^{-23}\text{ g}$$

(b) The molecular weight of HCl equals the AW of H plus the AW of Cl or 1.01 amu + 35.5 amu = 36.5 amu. Therefore, 1 mol HCl contains 36.5 g HCl and

$$\text{Mass of an HCl molecule} = \frac{36.5\text{ g}}{6.02 \times 10^{23}} = 6.06 \times 10^{-23}\text{ g}$$

69. Use the formula weight to write the factor that converts from mol ZnI_2 to g ZnI_2. The molar mass of ZnI_2 is 319 g mol^{-1}. (The formula weight is 319 amu, which is obtained by summing the atomic weights ion the formula.) Therefore,

$$0.0654\text{ mol ZnI}_2 \times \frac{319\text{ g ZnI}_2}{1\text{ mol ZnI}_2} = 20.9\text{ g ZnI}_2$$

70. The molar mass of $PbCrO_4$ is 323 g mol^{-1}. That is,

1 mol $PbCrO_4$ = 323 g $PbCrO_4$

Therefore,

$$45.6\text{ g PbCrO}_4 \times \frac{1\text{ mol PbCrO}_4}{323\text{ g PbCrO}_4} = 0.141\text{ mol PbCrO}_4$$

71. The number of molecules in a sample is related to moles of compound (1 mol HCl = 6.02 × 10^{23} HCl molecules). Therefore, if you first convert grams HCl to moles, then you can convert moles to number of molecules.

Here is the calculation:

$$3.46\text{ g HCl} \times \frac{1\text{ mol HCl}}{36.5\text{ g HCl}} \times \frac{6.02 \times 10^{23}\text{ HCl molecules}}{1\text{ mol HCl}} = 5.71 \times 10^{22}\text{ HCl molecules}$$

Note how the units in the numerator of a factor are canceled by the units in the denominator of the following factor.

72. To calculate mass percentage, you need the mass of an element in a given mass of compound. You can get this information by interpreting the formula in molar terms and then converting moles to masses, using a table of atomic weights. Thus, 1 and mol CH_2O has a mass of 30.0 g and contains 1 mol C (12.0 g), 2 mol H (2 × 1.01 g), and 1 mol O (16.0 g). You divide each mass of element by the molar mass, then multiply by 100, to obtain the mass percentage.

Thus we have,

$$\%\text{ C} = \frac{12.0\text{ g}}{30.0\text{ g}} \times 100\% = 40.0\%$$

$$\% \text{ H} = \frac{2 \times 1.01 \text{ g}}{30.0 \text{ g}} \times \%\% = 6.73\%$$

You can calculate the percentage of O in the same way, but it can also be found subtracting the percentage of C and H from 100%.

$$\% \text{ O} = 100\% - (40.0\% + 6.73\%) = 53.3\%$$

73. You first convert the mass of CO_2 to moles of CO_2. Then you convert this to moles of C, noting that 1 mol C produces 1 mol CO_2. Finally, you convert to mass of C. Similarly, you convert the mass of H_2O to mol H_2O, then to mol H, and finally to mass of H. (Remember that 1 mol H_2O produces 2 mol H.) Once you have the masses of C and H, you can calculate the mass percentages. Subtract from 100% to get %O.
Thus the grams of C is:

$$6.21 \times 10^{-3} \text{ g } CO_2 \times \frac{1 \text{ mol } CO_2}{44.0 \text{ g } CO_2} \times \frac{1 \text{ mol C}}{1 \text{ mol } CO_2} \times \frac{12.0 \text{ g C}}{1 \text{ mol C}}$$

$$= 1.69 \times 10^{-3} \text{ g C (or 1.69 mg C)}$$

For hydrogen, you note that 1 mol H_2O yield 2 mol H_2, so you write

$$2.54 \times 10^{-3} \text{ g } H_2O \times \frac{1 \text{ mol } H_2O}{18.0 \text{ g } H_2O} \times \frac{2 \text{ mol } H_2}{1 \text{ mol } H_2O} \times \frac{1.01 \text{ g } H_2}{1 \text{ mol } H_2}$$

$$= 2.85 \times 10^{-4} \text{ g } H_2 \text{ (or 0.285 mg } H_2)$$

You can now calculate the mass percentages of C and H_2 in acetic acid.

$$\text{Mass \% C} = \frac{1.69 \text{ mg}}{4.24 \text{ mg}} \times 100\% = 39.9\%$$

$$\text{Mass \% } H_2 = \frac{0.285 \text{ mg}}{4.24 \text{ mg}} \times 100\% = 6.72\%$$

You find the mass percentage of oxygen by subtracting the sum of these percentages from 100%:

$$\text{Mass \% O} = 100\% - (39.9\% + 6.72\%) = 53.4\%$$

Thus, the percentage composition of acetic acid is 39.9% C, 6.7% H, and 53.4% O.

74. Which is the limiting reactant? To answer, this, you take each reactants in turn and ask how much product (H_2) would be obtained if each were totally consumed. The reactant that gives the smaller amount of product is the limiting reactant.

So, you obtain the amount of product actually obtained from the amount of limiting reactant.
Thus,

$$0.30 \text{ mol Zn} \times \frac{1 \text{ mol } H_2}{1 \text{ mol Zn}} = 0.30 \text{ mol } H_2$$

$$0.52 \text{ mol HCl} \times \frac{1 \text{ mol } H_2}{2 \text{ mol HCl}} = 0.26 \text{ } H_2$$

Hence hydrochloric acid must be the limiting reactant and that some zinc most be left uncomposed. (Zinc is the excess reactant.)

Since HCl is the limiting reactant, the amount of H_2 produced must be 0.26 mol.

75. (i) How much acetic acid is produced?

To determine which reactant is limiting, you can convert grams of each reactant (20.0 g CH_3CHO and 10.0 g O_2) to moles of product, $HC_2H_3O_2$. Acetaledyhde has a molar mass of 44.1 g/mol, and oxygen has a molar mass of 32.0 g mol^{-1}.

$$20.0 \text{ g } CH_3CHO \times \frac{1 \text{ mol } \cancel{CH_3CHO}}{44.1 \text{ g } \cancel{CH_3CHO}} \times \frac{2 \text{ mol } HC_2H_3O_2}{2 \text{ mol } \cancel{CH_3CHO}} = 0.454 \text{ mol } HC_2H_2O_3$$

$$10.0 \text{ g } O_2 \times \frac{1 \cancel{\text{mol } O_2}}{32.0 \cancel{\text{g } O_2}} \times \frac{2 \text{ mol } HC_2H_3O_2}{1 \cancel{\text{mol } O_2}} = 0.625 \text{ mol } HC_2H_3O_2$$

Thus, acetaldehyde, CH_3CHO, is the limiting reactant, so 0.454 mol $HC_2H_3O_2$ was produced.

Now, convert 0.454 mol $HC_2H_3HO_2$ to grams of $HC_2H_3O_2$.

$$0.454 \text{ mol } HC_2H_3O_2 \times \frac{60.1 \text{ g } HC_2H_3O_2}{1 \text{ mol } \cancel{HC_2H_3O_2}} = 27.3\ HC_2H_3O_2$$

(ii) To find how much the excess reactant (oxygen) was left over? You convert the moles of acetic acid to grams of oxygen (the quantity of oxygen needed to produce this amount of acetic acid).

$$0.454 \cancel{\text{mol } HC_2H_3O_2} \times \frac{1 \cancel{\text{mol } O_2}}{2 \text{ mol } \cancel{HC_2H_3O_2}} \times \frac{32.0 \text{ g } O_2}{1 \cancel{\text{mol } O_2}} = 7.26 \text{ g } O_2$$

So, the quantity remaining is

(10.0×7.26) g O_2 = 2.7 g O_2

76. To calculate the molarity, you need the moles of solute. Therefore, you first convert grams $NaNO_3$ to moles. The molarity equals the moles of solute divided by the liters of solution. You find that 0.38 g $NaNO_3$ is 4.47×10^{-3} mol $NaNO_3$; the last significant figure is underlined. The volume of solution is 50.0 mL, or 50.0×10^{-3} L, so the molarity is

$$\text{Molarity} = \frac{4.47 \times 10^{-3} \text{ mol } NaO_3}{50.00 \times 10^{-3} \text{ L soln}} = 0.089\ M\ NaNO_3$$

77. You first need to convert grams NaOH to moles NaOH, because molarity relates moles of solute to volume of solution .The, you convert moles NaOH to liters of solution, using the molarity as a conversion factor. Here, 0.150 *M* means that 1 L of solution contains 0.150 moles of solute, so the conversion factor is:

$$\underbrace{\frac{1 \text{ L soln}}{0.150 \text{ mol NaOH}}}_{\substack{\text{Converts} \\ \text{mol NaOH to L soln}}}$$

Now,
The molar mass of NaOH is 40.0 g/mol.

$$\therefore \quad 0.184 \text{ g NaOH} \times \frac{1 \text{ mol NaOH}}{40.0 \text{ g NaOH}} \times \frac{1 \text{ L soln}}{0.150 \text{ mol NaOH}}$$

$$= 3.07 \times 10^{-2} \text{ L soln (or 30.7 mL)}$$

You need to add 30.7 mL of 0.150 *M* NaOH solution to the reaction vessel.

78. All of the lead in the water solution is precipitated as lead(II) sulfate, $PbSO_4$. The mass percentage of Pb in $PbSO_4$ is obtained by dividing the molar mass of Pb by the molar mass of $PbSO_4$, then multiplying by 100%.

$$\therefore \quad \%Pb = \frac{207.2 \text{ g/mol}}{303.3 \text{ g/mol}} \times 100\% = 68.32\%$$

So, the 1.000-L sample of water contains

Amount Pb in sample = 229.8 mg $PbSO_4$ × 0.6832 = 157.0 mg Pb

The water sample contains 157.0 mg Pb per liter.

79. Here we, convert from 35.0 mL (or 35.0 × 10^{-3}) H_2SO_4 solution to moles H_2SO_4 (using the molarity of H_2SO_4), then to moles (from the chemical equation). Finally, we convert this to volume of NaOH solution (using the molarity of NaOH).
Thus,

$$35.0 \times 10^{-3} \text{ L } H_2SO_4 \text{ soln} \times \frac{0.175 \text{ mol } H_2SO_4}{1 \text{ L } H_2O_4 \text{ soln}} \times \frac{2 \text{ mol NaOH}}{1 \text{ mol } H_2SO_4} \times \frac{1 \text{ L NaOH soln}}{0.250 \text{ mol NaOH}}$$

$$= 4.90 \times 10^{-2} \text{ L NaOH soln (or 49.0 mL NaOH soln)}$$

Thus, 35.0 mL of 0.175 *M* sulfuric acid solution reacts with 49.0 mL of 0.250 M sodium hydroxide solution.

80. You first convert mass of glucose to moles, because the molality equals moles of solute (glucose) divided by mass of solvent (water) in kilograms.

(i) The moles of glucose (MW = 180.2 amu) in 5.67 g are found as follows:

$$5.67 \text{ g } C_6H_{12}O_6 \times \frac{1 \text{ mol } C_6H_{12}O_6}{180.2 \text{ g } C_6H_{12}O_6} = 0.0325 \text{ mol } C_6H_{12}O_6$$

The mass of water is 25.2 g, or 25.2 × 10^{-3} kg.

$$\text{Molality} = \frac{0.0315 \text{ mol } C_6H_{12}O_6}{25.2 \times 10^{-3} \text{ kg solvent}} = 1.25 \, m \; C_6H_{12}O_6$$

(ii) It is found that 5.67 g of glucose equals 0.0315 of glucose. The moles of water in the solution are

$$25.2 \text{ g } H_2O \times \frac{1 \text{ mol } H_2O}{18.0 \text{ g } H_2O} = 1.40 \text{ mol } H_2O$$

Hence, the total moles of solution are
1.40 mol + 0.0315 mol = 1.432 mol
We retain an extra figure in the answer for further computation). Finally, we get

$$\text{Mole fraction glucose} = \frac{0.0315 \text{ mol}}{1.432 \text{ mol}} = 0.0220$$

$$\text{Mole fraction water} = \frac{1.40 \text{ mol}}{1.432 \text{ mol}} = 0.978$$

The sum of the mole fractions is 1.000.

(iii) A 0.120 *m* of glucose solution contains 0.120 mol of glucose in 1.00 kg of water. After converting 1.00 kg H2O to moles, you can calculate the mole fraction using its definition.
The moles of H_2O in 1.00 kg of water is

$$1.00 \times 103 \ \cancel{\text{g H2O}} \times \frac{1 \text{ mol } H_2O}{18.0 \ \cancel{\text{g } H_2O}} = 55.6 \ H_2O$$

Hence,

$$\text{Mole fraction glucose} = \frac{0.120 \text{ mol}}{(0.120 + 55.6) \text{ mol}} = 0.00215$$

$$\text{Mole fraction water} = \frac{55.6 \text{ mol}}{(0.120 + 55.6) \text{ mol}} = 0.998$$

(iv) One mole of solution contains 0.150 mol of glucose and 0.850 mol of water. So, convert moles of solvent (H_2O) to kilograms, then calculate the molality.
The mass of this amount of water is

$$0.850 \ \cancel{\text{mol } H_2O} \times \frac{18.0 \text{ g } H_2O}{1 \ \cancel{\text{mol } H_2O}} = 15.3 \text{ g } H_2O \ (= 0.0153 \text{ kg } H_2O)$$

The molality of glucose, $C_6H_{12}O_6$, in the solution is

$$\text{Molality of } C_6H_{12}O_6 = \frac{0.150 \text{ mol } C_6H_{12}O_6}{0.0153 \text{ kg solvent}} = 9.80 \ m \ C_6H_{12}O_6$$

81. The molality tells that the solution contains 0.273 mol KCl in 1 kg of water takes the mass of solution that contains 1 kg of water (the mass of 0.273 mol KCl plus 1.000×10^3 g H_2O) and use the density to convert this quantity of solution to volume. Molarity equals moles of solute (KCl) divided by volume of solution in liters.
The mass of potassium chloride in this quantity of solution is

$$0.273 \ \cancel{\text{mol KCl}} \times \frac{74.6 \text{ g KCl}}{1 \ \cancel{\text{mol KCl}}} = 20.4 \text{ g KCl}$$

The total mass of solution equals the mass of water plus the mass of potassium chloride.

$$1.000 \times 10^3 \text{ g} + 20.4 \text{ g} = 1.020 \times 10^3 \text{ g}$$

The volume of solution equals the mass divided by the density of the solution.

$$\text{Volume of solution} = \frac{1.020 \times 10^3 \text{ g}}{1.011 \times 10^3 \text{ g/L}} = 1.009 \text{ L}$$

Hence, the molarity of the solution is

$$\frac{0.273 \text{ mol KCl}}{1.009 \text{ L solution}} = 0.271\ M \text{ KCl}$$

Note that the molarity and the molality of this solution are approximately equal. This happens in cases where the solutions and dilute and the density is about 1 g/mL.

82. Mass of solution = density × volume

$$= 1.252 \text{ g/mL} \times 1.000 \times 10^3 \text{ mL}$$
$$= 1.252 \times 10^3 \text{ g}$$

The mass of lead nitrate is

$$0.907 \text{ mol Pb(NO}_3)_2 \times \frac{331.2 \text{ g Pb(NO}_3)_2}{1 \text{ mol Pb(NO}_3)_2} = 3.00 \times 10^2 \text{ g Pb(NO}_3)_2$$

This mass of water is

$$\text{Mass of } H_2O = \text{mass of solution} - \text{mass of Pb(NO}_3)_2$$
$$= 1.252 \times 10^3 \text{ g} - 3.00 \times 10^2 \text{ g}$$
$$= 9.52 \times 10^2 \text{ g } (= 0.952 \text{ kg})$$

Hence, the molality of lead nitrate in this solution is

$$\frac{0.907 \text{ mol Pb(NO}_3)_2}{0.952 \text{ kg solvent}} = 0.953\ m \text{ Pb(NO}_3)_2$$

83. As KOH can donate one hydroxide,

∴ normality $_{KOH}$ = molarity $_{KOH}$

Here, $N_T = 0.1354\ N$

$V_T = 42.20$ mL

and $V_S = 50.00$ mL

Let $N_S = N_{H_2SO_4}$

∴ $(0.13545 \text{ N}) \times (42.20 \text{ mL}) = (N_{H_2SO_4}) \times (50.00 \text{ mL})$

$$\Rightarrow \quad N_{H_2SO_4} = \frac{0.1354 \times 42.20}{50.00} = 0.1143 \text{ N}$$

Now, $\text{molarity}_{H_2SO_4} = \frac{1}{2} \text{normality}_{H_2SO_4}$

$$\Rightarrow \quad M_{H_2SO_2} = \frac{0.1143}{2} = 0.571 \text{ M.}$$

Chapter 5 : SOLUTIONS

1. This problem can be easily solved using the factor-label method:

$$\frac{1.00 \text{ liter solution}}{1} \times \frac{1000 \text{ mL solution}}{1 \text{ liter solution}} \times \frac{2.00 \text{ g solution}}{1 \text{ mL solution}} \times \frac{10.0 \text{ g sucrose}}{100.0 \text{ g solution}}$$

$$= 2.00 \times 10^2 \text{ g sucrose}$$

2. This problem can be easily solved using the factor-label method:

$$\frac{500 \text{ mL (2.00 M soln)}}{1} \times \frac{1 \text{ liter (2.00 M soln)}}{1000 \text{ mL (2.00 M soln)}} \times \frac{2.00 \text{ moles } HNO_3}{1 \text{ liter (2.00 M soln)}}$$

$$\times \frac{63 \text{ g } HNO_3}{1 \text{ mole } HNO_3} \times \frac{100. \text{ g } 50.0\% \text{ soln}}{50.0 \text{ g } HNO_3} \times \frac{1 \text{ mL } 50.0\% \text{ soln}}{2.00 \text{ g } 50.0\% \text{ soln}}$$

$$= 63.0 \text{ mL of a } 50.0\% \text{ soln}$$

3. This problem can be solved using the factor-label method:

$$\frac{9.80 \text{ g } H_2SO_4}{200. \text{ mL sol'n}} \times \frac{1000. \text{ mL sol'n}}{1 \text{ liter sol'n}} \times \frac{1 \text{ mole } H_2SO_4}{98.1 \text{ g } H_2SO_4}$$

$$\times \frac{2 \text{ g-eq } H_2SO_4}{1 \text{ mole } H_2SO_4} = 1.00 \text{ g·eq liter}^{-1} = 1.00 \text{ N}$$

4. This problem can be solved using the factor-label method:

$$\frac{0.500 \text{ liter soln}}{1} \times \frac{3.00 \text{ g-eq}}{1 \text{ liter soln}} = 1.50 \text{ g} \cdot \text{eq}$$

5. This problem can be solved using the factor-label method:

$$\frac{10.0 \text{ moles NaCl}}{1000. \text{g } H_2O} \times \frac{58.50 \text{ g NaCl}}{1 \text{ mole NaCl}} = \frac{585 \text{ g NaCl}}{1000. \text{ g } H_2O}$$

The question is asking for the parts of NaCl per *total* solution (solute solvent).

$$\text{which is } \frac{585 \text{ g NaCl}}{1000. \text{ g } H_2O + 585 \text{ g NaCl}} = \frac{585}{1585} \simeq 0.3$$

6. This problem can be solved using the factor-label method:

$$\frac{0.592° \text{C}}{1} \times \frac{1 \text{ mole solute}}{1.86 °\text{C} \cdot \text{kg } H_2O} \times \frac{186 \text{ g } H_2O}{5.92 \text{ g solute}} \times \frac{1 \text{ kg } H_2O}{1000 \text{ g } H_2O} = 0.0100 \text{ mol g}^{-1} = 100. \text{ g mol}^{-1}$$

7. This problem can be solved using the factor-label method: Here, $\Delta T_b = 102 - 100 = 2°C$

$$\frac{520\ \text{g H}_2\text{O}}{1} \times \frac{1\ \text{kg H}_2\text{O}}{1000\ \text{g H}_2\text{O}} \times \frac{1\ \text{mole C}_3\text{H}_3(\text{OH})_3}{0.52°\ \text{C}\cdot\text{kg H}_2\text{O}} \times \frac{92.1\ \text{g C}_3\text{H}_5(\text{OH})_3}{1\ \text{mole C}_3\text{H}_3(\text{OH})_3} \times \frac{2.00°\ \text{C}}{1}$$

$$= 184\ \text{g C}_3\text{H}_5(\text{OH})_3$$

8. Begin with the equation
$MW = gRT/\pi V$

$$= \frac{0.010\ \text{g}}{1} \times \frac{0.0821\ \text{liter atm}}{\text{mole}\cdot\text{K}} \times \frac{293\ \text{K}}{1} \times \frac{1}{0.0821\ \text{torr}} \times \frac{1}{293\ \text{mL}} \times \frac{760\ \text{torr}}{1\ \text{atm}} \times \frac{100\ \text{mL}}{1\ \text{liter}}$$

$= 7.6 \times 10^4$ g/mol

9. This problem can be solved using the factor-label method

$$\frac{10.0\ \text{moles NH}_3}{1\ \text{kg H}_2\text{O}} \times \frac{1\ \text{kg H}_2\text{O}}{1000\ \text{g H}_2\text{O}} \times \frac{18.02\ \text{g H}_2\text{O}}{1\ \text{mole H}_2\text{O}} = \frac{0.180\ \text{mole NH}_3}{1.00\ \text{mole N}_2\text{O}}$$

total number of moles = 0.180 mole NH_3 + 1.00 mole H_2O = 1.18 moles.

$$\therefore \text{ mole fraction of water} = \frac{1.00\ \text{mole H}_2\text{O}}{1.18\ \text{mole}} \approx 0.85$$

10. Determine k by using $C = kP$ (Henry's law).

$$k = \frac{\text{concentration N}_2}{\text{pressure N}_2} = \frac{6.0 \times 10^{-4}\ \text{M}}{1.00\ \text{atm}} = 6.0 \times 10^{-4}\ \text{M}\cdot\text{atm}^{-1}$$

To solve the problem

$$P = 0.80 \times 3.0\ \text{atm} = 2.4\ \text{atm}$$

$$C = kP = \frac{6.0 \times 10^{-4}\ \text{moles}}{\text{liter}\cdot\text{atm}} \times \frac{2.4\ \text{atm}}{1} = 1.4 \times 10^{-3}\ \text{M}$$

11. Here,

$$P_1 = X_1 P_1^\circ$$

$$X_1 = \frac{P_1}{P_1^\circ} = \frac{4.50\ \text{mmHg}}{1000.\ \text{mmHg}} = 0.450$$

The mole fraction of the solute is

$$1.000 - X_1 = 1.000 - 0.450 = 0.550$$

12. In 0.040 F $Sr(NO_3)_2$,
$Sr^{2+} = 0.040\ M$; $[NO_3^-] = 0.080$ M

13. In 0.10 m $NiCl_2$, the molality of Ni^{2+} is 0.10 m and of Cl^- is 0.20 m, so that the *total* concentration of all solute particles is 0.30 m. The larger the total concentration of all solute particles, the lower the freezing point. Sucrose and glucose are nonelectrolytes, and the number of ions per formula unit for $CuSO_4$ and NH_4NO_3 is 2.

14. Mo. wt. CH_3COCH_3 is 58.08 g mol^{-1}. Consider 100.00 g of this solution. It contains 10.00 g of acetone and 90.00 g of H_2O.

$$\text{No. moles acetone} = \frac{10.00 \text{ g}}{58.08 \text{ g mol}^{-1}} = 0.1722$$

$$\text{No. moles } H_2O = \frac{90.00 \text{ g}}{18.015 \text{ g mol}^{-1}} = 4.996$$

$$X_{\text{acetone}} = \frac{0.1722}{0.1722 + 4.996} = \frac{0.1722}{5.168} = 0.3332 \text{ or } 3.332\%$$

15. Consider one liter of this solution. It weights 986.7 g. Of this, 98.67 g are acetone. The molecular weight of CH_3COCH_3 is 58.08 g mol^{-1}.

$$\text{No. moles acetone per liter of solution} = \frac{98.67 \text{ g}}{58.08 \text{ g mol}^{-1}} = 1.699$$

Hence the solution is 1.699 M in acetone.

16. The formula wt. of $CuSO_4 \cdot 5H_2O$ is 249.68 g mol^{-1}.

$$\text{No. moles } CuSO_4 \cdot 5H_2O = \frac{10.00 \text{ g}}{249.68 \text{ g mol}^{-1}} = 0.04005 \text{ mol}$$

in 0.5000 L. Thus the molarity is
0.04005 mol/0.5000 L = 0.08010 M

17. $K_f = 1.86$ for H_2O. $\Delta T_f = K_f\, m = 0.14 = 1.86\, m$, so that m is 0.075 molal.

18. As $P_b = P^\circ_B X_B = (93.4 \text{ mmHg})\, X_B$, we must calculate the mole fraction of benzene to find the vapor pressure of the solution. The mol. wt. of C_6H_6 is 78.114 g mol^{-1}.

$$\text{No. mol } C_6H_6 = \frac{250.0 \text{ g}}{78.114 \text{ g mol}^{-1}} = 3.200 \text{ mol}$$

$$\text{No. mol 1,2-benzanthracene} = \frac{18.2632 \text{ g}}{228.29 \text{ g mol}^{-1}} = 0.08000 \text{ mol}$$

$$X_B = \frac{3.200}{3.200 + 0.0800} = 0.9756$$

$$P_B = (93.4 \text{ mmHg})\,(0.9756) = 91.1 \text{ mmHg}$$

19. The number of milimoles of HNO_3 is the same before and after the dilution. Let V = the final volume before and after the dilution. Let V = the final volume the 0.500 F HNO_3.

$$(25.00 \text{ mL})\left(6.00\frac{\text{m mol}}{\text{mL}\cdot}\right) = (V)\left(0.500\frac{\text{m mol}}{\text{mL}}\right)$$

and V = 300 mL, If the volumes are additive, we must add 275 mL of water to 25.00 mL of 6.00 F HNO_3, to prepare 300 mL of 0.500 F HNO_3.

20. There are two NH_4^+ ions per $(NH_4)_2SO_4$ formula unit.
21. The mol. wt. of CH_3CH_2OH is 46.069 g mol^{-1}. Consider exactly 100 g of solution. It contains 1.00 g or ethanol and 88.00 g of H_2O.

$$\text{No. mol } CH_3CH_2OH = \frac{12.00 \text{ g}}{46.069 \text{ g mol}^{-1}} = 0.2605$$

$$\text{molality} = \frac{0.2605 \text{ mol}}{0.08800 \text{ kg}} = 2.960\ m$$

22. $\Delta T_f = 5.48 - 4.33 = 1.15°$. For C_6H_6, $K_f = 5.1$ (From Table). Thus $m = 1.15/5.1 = 0.2255 = 0.23$. Let n be the no. of moles of solute in 0.4000 kg of C_6H_6.

Then $0.2255 = \dfrac{\text{n moles solute}}{0.4000 \text{ kg}}$, and n is 0.0902 mol. As the mass of solute is 20.5461 g, the

approximate molecular weight of solute is $\dfrac{20.5461 \text{ g}}{0.0902 \text{ mol}} = 228 \text{ g mol}^{-1}$. The formula weight of C_3H_2 is 38, so $38n \sim 228$, and $n = 6$. Thus the correct molecular formula is $C_{18}H_{12}$.

23. For $CaCl_2$, there are 3 ions per formula unit. For an ideal solution, $\Delta T_b = vK_b m = 3(0.51)(1.00) = 1.53°$. Solution of electrolytes deviate from identity more than solutions of nonelectrolytes, and the actual boiling point elevation will be somewhat less than = 1.53°.
24. Let V be the volume, in mL, of the 0.0500 F solution. Then (V) (0.0500 F) = (500.0 mL). (0.0200 F), and V = 200 mL,

25. $$\text{No. mol } C_6H_6 = \frac{64.05 \text{ g}}{78.11 \text{ g mol}^{-1}} = 0.8200 \text{ mol}$$

$$\text{No. mol toluene} = \frac{106.26 \text{ g}}{92.14 \text{ g mol}^{-1}} = 1.153 \text{ mol}$$

$$X_T^{\text{liq}} = \frac{1.153}{1.973} = 0.5844;\ X_B^{\text{liq}} = 1 - 0.5844 = 0.4156$$

$$P_B = (160.0 \text{ mmHg})\ (0.4156) = 66.49 \text{ mmHg}$$

$$P_T = (50.0 \text{ mmHg})\ (0.5844) = 29.2 \text{ mmHg}$$

$$P_{\text{total}} = 66.49 + 29.2 = 95.7 \text{ mmHg},\ K_T^{\text{vap}} = \frac{29.2}{95.7} = 0.305$$

26. The formula weight of $KMnO_4$ is 158.034 g/mol. In 1 L we need 0.1000 mol or 15.8034 g, and in 0.25000 L we need ¼ of 0.1000 mol or (15.8034/4) = 3.951 g.

27. $$\pi = cRT = \left(1.0 \times 10^{-3} \frac{\text{mol}}{\text{L}}\right)\left(0.08206 \frac{\text{L} \cdot \text{atm}}{\text{mol} \cdot \text{K}}\right) (298 \text{ K})$$

$$= 2.44 \times 10^{-3} \text{ atm} = (2.44 \times 10^{-3} \text{ atm}) \left(760 \frac{\text{mmHg}}{\text{atm}}\right)$$

$$= 18.6 \text{ mmHg} = 19 \text{ mmHg}$$

28. $[NO_3^-] = \dfrac{(20.00\text{ mL})(2)(0.100\text{ M}) + (30.00\text{ mL})(0.400\ M)}{50.00\text{ mL}}$

$$= \frac{(4.00 + 12.0)\text{ m mol}}{50.00\text{ mL}} = 0.320\ M$$

28. The solubility of a compound with a molecular weight of many thousands is usually quite small, so that the vapor pressure lowering, boiling point elevation, and freezing point depression are too small to be measured accurately. Only the osmotic pressure is large enough for an accurate determination of the molecular weight.

30. The formula wts. of KCl and K_2CO_3 are, respectively, 74.551 and 138.204 g mol^{-1}.

$$\text{No. mol KCl} = \frac{1.864\text{ g}}{74.551\text{ g mol}^{-1}} = 0.02500\text{ mol}$$

$$\text{No. mol } K_2CO_3 = \frac{8.293\text{ g}}{138.204\text{ g mol}^{-1}} = 0.06001\text{ mol}$$

There are 2 moles of K^+ ions per mole of K_2CO_3. The final volume of solution is 0.5000 L. Thus,

$$[K^+] = \frac{(0.02500 + 2(0.06001))\text{ mol}}{0.5000\text{ L}} = 0.2900\ M$$

31. $P_{CS_2(\text{ideal})} = (512\text{ mmHg})\ (0.25) = 128\text{ mmHg}$

$$P_{A(\text{ideal})} = (344\text{ mmHg})\ (0.75) = 258\text{ mmHg}$$

Thus, if the solution were ideal, its total vapor pressure would be 128 + 258 = 386 mmHg. As the actual vapr pressure is 600 mmHg, the system exhibits positive deviations from ideality. Options (a) and (d) are only true for ideal solutions, and options (c) and (e) are true for systems that exhibit negative deviations from Raoult's law. Only option (b) is correct for positive deviations.

32. No. of m mol Pb^{2+} added $= (20.00\text{ mL})\left(0.100\dfrac{\text{mmol}}{\text{mL}}\right) = 2.000$ m mol.

No. of m mol SO_4^{2-} added $= (30.00\text{ mL})\left(0.100\dfrac{\text{mmol}}{\text{mL}}\right) = 3.000$ m mol.

Since Pb^{2+} and SO_4^{2-} ions react in a 1 : 1 molar ratio to form $PbSO_4$, 2.000 m mol Pb^{2+} ions react with 2.000 m mol SO_4^{2-} ions to precipitate 2.000 m mol $PbSO_4$, leaving 1.000 m mol SO_4^{2-} ions in excess. The formula weight of $PbSO_4$ is 303.26 g · mol^{-1}. The maximum mass of $PbSO_4$ that can be precipitated is, therefore,

$$(2.000 \times 10^{-3}\text{ mol})\ (303.26\text{ g} \cdot \text{mol}^{-})= 0.6065\text{ g}$$

33. $[Na^+] = \dfrac{(30.00\text{ mL})(0.120\text{ M}) + (70.00\text{ mL})(0.150\text{ M})(2)}{100.00\text{ mL}}$

$= 0.246\ M$

34. Acetic acid, CH_3COOH, is a weak electrolyte. Casium hydroxide, CaOH, calcium hydroxide, $Ca(OH)_2$, and potassium hydroxide, KOH, are all strong bases. Methyl alcohol, CH_3OH, is a nonelectrolyte.

35. The formula weight of Ag_2SO_4 is 311.79 g mol^{-1}. Thus the number of moles of dissolved Ag_2SO_4 is

$$\frac{0.57 \text{ g}}{311.79 \text{ g mol}^{-1}} = 1.83 \times 10^{-3} \text{ mol} = 1.8 \times 10^{-3} \text{ mol in } 100.00 \text{ mL.}$$

Ten times as much Ag_2SO_4 can be dissolved in 1 L as in 100.00 mL, so 1.83×10^{-2} mol of Ag_2SO_4 dissolve per liter.

$$\frac{\text{moles Ag}^+ \text{ ion in solution}}{\text{moles Ag}_2\text{SO}_4 \text{ dissolved}} = \frac{2}{1}$$

The number of moles of Ag^+ ions per liter is $2(1.83 \times 10^{-2})$, and

$$[Ag^+] = 0.037 \text{ M}$$

36. Benzene, C_6H_6, is a nonpolar organic molecule that cannot hydrogen bond to water. Because it is nonpolar, the force of attraction between two H_2O molecules is very much greater than the force of attraction between C_6H_6 and H_2O, and C_6H_6 does not mix with H_2O. Ammonium dichromate, $(NH_4)_2Cr_2O_7$, is a sult, a soluble, strong electrolyte. Ethanol, CH_3CH_2OH, hydrogen bonds to H_2O, and mixes with it in all proportions. Methylamine, CH_3NH_2, is a weak base that is very soluble in water, and oxalic acid, $H_2C_2O_4$, is also very soluble in water.

37. No. m mol OH^- = $(2)\left(0.1080 \frac{\text{mmol}}{\text{mL}}\right)$ (25.00 mL) = 5.400 m mol Remember that $Ba(OH)_2$ is a strong base, and that there are 2 OH^- ions per $Ba(OH)_2$ formula unit. The net ionic equation for the reaction is $H^+(aq) + OH^-(aq) \rightarrow H_2O$. The NO_3^- and Ba^{2+} ions take no part in the reaction, they are only spectator ions. Since H^+ and OH^- ions react in a 1 : 1 molar ratio, we need 5.400 m mol of H^+ ions for the reaction.

$$\text{volume HNO}_3 \text{ required} = \frac{5.400 \text{ mmol}}{0.1250 \text{ mmol mL}^{-1}} = 43.20 \text{ mL}$$

38. Sodium chloride, NaCl, is an ionic crystalline solid. It is not soluble in nonpolar liquids. The only nonpolar liquid of those listed is carbon tetrachloride.

39. The reaction that occurs when these two solutions are mixed is

$$H^+(aq) + OH^-(aq) \rightleftharpoons H_2O$$

No. of m mol H^+ added = (80.00 mL) (0.200 *M*) = 16.0 m mol

No. of m mol OH^- added = (120.00 mL) (0.150 *M*) = 18.0 m mol

As H^+ and OH^- react in a 1 : 1 molar ratio, there are 2.00 m mol or OH^- in excess. Thus there are 16.0 m mol Cl^-, 18.0 m mol K^+, and 2.00 m mol OH^- in a total volume of 200.00 mL. Hence, $[Cl^-]$ = 16.0 m mol/200.00 mL = 0.0800 *M*; $[K^+]$ = 0.0900 *M*; and $[OH^-]$ = 0.0100 *M*. This is identical with a solution that is 0.0800 *F* KCl and 0.0100 *F* KOH.

40. Let V = volume, in Ml, of each solution used. Then the volume of the final solution is 2V mL.

$$\text{No. of m mol } Ba^{2+} \text{ added} = (\text{V mL})\left(0.200\frac{\text{mmol}}{\text{mL}}\right) = 0.200 \text{ V m mol}$$

$$\text{No. of m mol } CrO_4^{2-} \text{ added} = (\text{V ml})\left(0.120\frac{\text{mmol}}{\text{mL}}\right) = 0.120 \text{ V m mol}$$

Since Ba^{2+} and CrO_4^{2-} must react in a 1 : 1 molar ratio to produce an electrically neutral compound, there are 0.080 V m mol Ba^{2+} in excess, in 2V mL.

$$[Ba^{2+}] = \frac{0.80V}{2V}\frac{\text{mmol}}{\text{mL}} = 0.040\ M$$

41. The reaction is $Pb^{2+}(aq) + 2Cl^{-}(aq) \rightleftharpoons PbCl_2$

No. of m mol Pb^{2+} added = (40.00 mL) (0.100 M) = 4.00 m mol

No. of m mol Cl^{-} added = (60.00 mL) (0.300 M) = 18.0 m mol

The limiting reagent is Pb^{2+}, as only 8.00 m mol of Cl^{-} are needed to precipitate completely the 4.00 m mol of Pb^{2+}. Thus a maximum of 4.00 m mol of $PbCl_2$ could be precipitated. The formula weight of $PbCl_2$ is 278.1 g · mol^{-1}. The maximum mass of $PbCl_2$ that could be precipitated is therefore,

$$(4.00 \times 10^{-3} \text{ mol})(278.1 \text{ g} \cdot \text{mol}^{-1}) = 1.112 \text{ g}$$

$$\text{Percentage yield} = \frac{\text{actual yield}}{\text{maximum yield}} = \left(\frac{1.068 \text{ g}}{1.112 \text{ g}}\right) \times 100 = 96.0\%$$

42. Of the 5 solutions listed, the only one containing a cation that reacts with Cl^{-} to form an insoluble precipitate is mercury (I) nitrate. The reaction that occurs is

$$Hg_2^{2+}(aq) + 2Cl^{-}(aq) \rightleftharpoons Hg_2Cl_2\downarrow.$$

43. Many weak and nonelectrolytes are very soluble in water. While a large number of strong electrolytes are very soluble in water, there are a significant number of insoluble salts.

44. The reaction that occurs is

$$H^{+}(aq) + OH^{-}(aq) \rightleftharpoons H_2O$$

As the volumes are equal and the concentration of KOH is twice that of HCl, there will be OH^{-} in excess. Thus only options (b) and (c) need to be considered. These two options differ only in the $[K^{+}]$, so focus on that. Mixing equal volumes doubles that volume. Since K^{+} ions take no part in the reaction, the concentration of K^{+} must be halved when the volume is doubled. As $[K^{+}]$ is 0.400 M initially, it must be 0.200 M in the final solution.

45. The reaction that occurs is

$$Fe^{3+}(aq) + 3OH^{-}(aq) \rightleftharpoons Fe(OH)_3$$

Thus $Fe3^{+}$and OH^{-} react in a 1 : 3 molar ratio. In (a) there is excess OH^{-} and only 1.00 m mol of $Fe(OH)_3$ can be precipitated, as there is only 1.00 m mol of Fe^{3+} added. In (b) there is excess OH^{-} and only 2.00 m mol of $Fe(OH)_3$ can be precipitated. In (c) the solutions are mixed in a 1 : 3 molar ratio, neither reagent is in excess, and 2.5 m mol of $Fe(OH)_3$ can be precipitated. In (d) and (e) the limiting reagent is OH^{-}. The maximum number of moles of $Fe(OH)_3$ that can be precipitated is 1/3 the number of moles of OH^{-} added, or 1.80 m mol for (d) and 1.20 m mol for (e).

46. Strontium hydroxide and sulfuric acid react in a 1 : 1 molar ratio. The net ionic equation for the reaction is

$$Sr^{2+} + 2OH^- + H^+(aq) + HSO_4^- \rightleftharpoons SrSO_4\downarrow + 2H_2O$$

No. of m mol H_2SO_4 added = (34.57 mL) $\left(0.0452 \dfrac{\text{mmol}}{\text{mL}}\right)$ = 1.563 m mol.

Let x = formality of the $Sr(OH)_2$ solution. Then,
No. of m mol $Sr(OH)_2$ added = 50.00 x = (34.57 mL) (0.0452 F) and x = 0.0313 F

47. $2La^{3+}(aq) + 3C_2O_4^{2-}(aq) \rightleftharpoons La_2(C_2O_4)_3\downarrow$.
No. of m mol La^{3+} added = (30.00 mL) (0.0860 M) = 2.58 m mol
No. of m mol $C_2O_4^{2-}$ added = (20.00 mL) (0.114 M) = 2.28 m mol
The limiting reagent is the $C_2O_4^{2-}$. There is excess La^{3+} present, as only (2/3) (2.28 m mol) of La^{3+} are needed to precipitate all the $C_2O_4^{2-}$,

$$\frac{\text{No. mol } La_2(C_2O_4)_3 \text{ precipitated}}{\text{No. mol } C_2O_4^{2-} \text{ added}} = \frac{1}{3}$$

The maximum amount of $La_2(C_2O_4)_3$ precipitate is therefore,
(1/3) (2.28 m mol) = 0.760 m mol

48. Barium carbonate, $BaCO_3$ precipitates when $BaCl_2$ and Na_2CO_3 are mixed. Silver chloride, AgCl, precipitates when $BaCl_2$ and $AgNO_3$ are mixed. Barium sulfate, $BaSO_4$, precipitates when $BaCl_2$ and $(NH_4)_2SO_4$ are mixed. Lead chloride, $PbCl_2$, precipitates when $BaCl_2$ and $Pb(NO_3)_2$ are mixed.

49. Precipitates of $PbCl_2$, $PbBr_2$, and PbI_2 will form when HCl, HBr, and HI, respectively, are added to $Pb(OH)_2$. A typical reaction is

$$Pb(OH)_2(s) + 2H^+(aq) + 2Br^-(aq) \rightarrow 2H_2O + PbBr_2$$

Precipitates of $SrSO_4$ and $PbSO_4$ will form when H_2SO_4 is added to $Sr(OH)_2$ or $Pb(OH)_2$. As nitrates are soluble, no precipitate can be formed when nitric acid is added to any insoluble hydroxide.

50. (i) The cinnamaldehyde molecule features a benzene ring and hydro-carbon side chain. We would not expect hydrogen bonding to occur in cinnamaldehyde (all H atoms are bonded to C atoms). These facts suggest that the molecule is much more similar to benzene than to water. We should expect cinnamaldehyde to dissolve in benzene but to be rather insoluble in water.

(ii) no. mol C_9H_8O = 1.00 g $C_9H_8O \times \dfrac{1 \text{ mol } C_9H_8O}{132.2 \text{ g } C_9H_8O}$ = 0.00756 mol C_9H_8O

$$\text{mol freaction } C_6H_6 = \frac{0.6401 \text{ mol } C_6H_6}{(0.6401 + 0.0076) \text{ mol}} = 0.9883$$

The partial pressure of benzene above the solution is

$$P_{benz} = X_{benz}\, P^{\circ}_{benz} = 0.9883 \times 95.1 \text{ mmHg} = 94.0 \text{ mmHg}$$

(iii) Our main task is to determine the molality concentration of this solution. We know the number of moles of solute from part (ii)

$$\therefore \quad \text{molality} = \frac{0.00756 \text{ mol } C_9H_8O}{0.05000 \text{ kg } C_6H_6} = 0.151\ m$$

Now, use the solution molality (0.51 m) and the given freezing point depression constant of benzene (K_f = 5.12), to obtain the freezing point depression,

$$\therefore \quad \Delta T_f = \frac{5.12 \text{ K kg solvent}}{\text{mol solute}} \times \frac{0.151 \text{ mole solute}}{\text{kg solvent}} = 0.773 \text{ K}$$

That is

The freezing point of the solution is 0.773 K below that of benzene ∴ f.p. = 5.533 K – 0.773 K = 4.760 K.

(iv) The solubility is approximately 1 g of cinnamaldehyde in 700 g water, and 700 g of water occupies a volume of essentially 700 mL. The solution volume will be very close to that of the water—700 mL = 0.7 L. We have already determined the number of moles of cinnamaldehyde in 1.00 g (part (ii)); it is 0.00756 mol C_9H_8O. The *molarity* of the saturated water solution is 0.00756 mol C_9H_8O/0.7L = 0.01 M C_9H_8O. Now, we can calculate the osmotic pressure by using equation.

$$\pi = MRT = 0.01 \text{ mol L}^{-1} \times 0.0821 \text{ L atom mol}^{-1} \text{ K}^{-1} \times 298 \text{ K} = 0.2 \text{ atm}$$

51. As $\Delta T_f = K_f m_B$

where $$m_B = \frac{\text{moles of solute (ethylene glycol)}}{\text{kg of solvent (water)}}$$

For water, assuming density = 1 ⇒ 10 L = 10 kg

$$\text{moles of ethylone glycol} = \frac{\text{weight}}{\text{Molecular weight}} = \frac{\text{Weight}}{62 \text{gm}}$$

Given $\Delta T_f = 10$ K

$K_f = 1.86$ K kg mol^{-1}

$$\therefore \quad 10 \text{ K} = \frac{(1.86 \text{ K kg mol}^{-1}) \times (\text{weight}/62 \text{ gm mol})}{(10 \text{ kg})}$$

$$\Rightarrow \quad \text{weight} = \frac{10 \times 10 \times 62}{1.86} = 3334 \text{ gm} = 3.334 \text{ kg}.$$

52. As $\Delta T_f = K_f m_B$

or $$\Delta T_f = K_f \frac{w_B / M_B}{w_A} \quad \therefore \quad M_B = \frac{k_f w_B}{\Delta T_f w_A}$$

Here $w_B = 28$ gm, $w_A = 750$ gm $= \frac{750}{1000}$ kg,

$\Delta T_f = 5.4$ K and $K_f = 30$ K kg mol^{-1}

Thus, $$m_B = \frac{30 \times 28}{5.4 \times 750/1000} = 207.4 \text{ gm mol}^{-1}$$

53. As we are considering aqueous solution, so water is the pure solvent and thus $T_b^* = 373$ K $\therefore$ $\Delta T_b = 373.1 - 373 = 0.1$ K. Also $T_f^* = 273$ K

Moreover, as same solution is used, its molality will be same.

$\therefore$ $$\Delta T_b = K_b m_B \text{ and } \Delta T_f = K_f m_B$$

Hence, $$\frac{\Delta T_b}{K_b} = \frac{\Delta T_f}{K_f} \Rightarrow \Delta T_f = \Delta T_b \frac{K_f}{K_b} \text{ (i)}$$

Now $$K_b = \frac{RT_b^{*2}}{\Delta H_{vap}} \text{ and } K_f = \frac{RT_f^{*2}}{\Delta H_{fus}}$$

With these values of K_b and K_f, equation (i) becomes:

$$\Delta T_f = \Delta T_b \frac{K_f}{K_b} = 0.1 \times \frac{RT_f^{*2} / \Delta H_{fus}}{RT_b^{*2} / \Delta H_{vap}}$$

$$= 0.1 \frac{(273)^2}{(373)^2} \times \frac{\Delta H_{vap}}{\Delta H_{fus}} \Rightarrow \boxed{\Delta T_f = 0.054 \frac{\Delta H_{vap}}{\Delta H_{fus}}}$$

where ΔT_f = Depression in freezing point of solvent; ΔH_{fus} = Enthalpy of fusion of pure solvent (water) ΔH_{vap} = Enthalpy of vaporization of pure solvent (water)

54. As $\pi = [B]RT = \frac{w/M}{V} RT$

$\therefore$ $$w = \frac{\pi MV}{RT}$$

$$= \frac{(4.82 \text{ atm})(342 \text{ g mol}^{-1}) \times (1 \text{ L})}{(0.082 \text{ L atm K}^{-1} \text{ mol}^{-1})(298 \text{ K})} = 67.5 \text{ gm}$$

or weight of sucrose = 67.5 gm

$\therefore$ weight of water in 1015 gm of sucrose solution = 1015 – 67.5

= 947.5 gm = 947.5×10^{-3} kg.

Now, moles of sucrose solution = $\frac{w}{M} = \frac{67.5}{342} = 0.1974$.

Hence molality of the sucrose solution = $\frac{0.1974}{947.5 \times 10^{-3}} = 0.208$ m.

55. As $\pi = [B]\, RT$

or $$\pi V = n_B RT$$

Suppose initially volume was V and finally it was diluted by dilution factor a. Therefore final volume is aV.

At 283'K : $\frac{500}{760} \times V = n_B \times 0.0821 \times 283$ (1)

At 298 K : $\frac{105.3}{760} \times a \times V = n_B \times 0.0821 \times 298$ (2)

Dividing (2) by (1), we get

$$\frac{105.3\,a}{500} = \frac{298}{283} \Rightarrow a = \frac{298}{283} \times \frac{500}{105.3} = 5$$

Thus the solution was diluted 5 times.

56. Isotonic solutions exert the same osmotic pressure at a given temperature For isotonic solutions at the same temperature (T is constant), let $(n_B)_1$ and $(n_B)_2$ moles of solute are present per litre (V is constant) of solution. R and π and already constant.

Thus $$(n_B)_1 = \frac{\pi V}{RT} \text{ and } (n_B)_2 = \frac{\pi V}{RT}$$

$$\therefore \quad (n_B)_1 = (n_B)_2 \text{ or } \frac{w_1}{M_1} = \frac{w_2}{M_2}$$

If subscript 1 indicates sucrose and subscript 2 indicates urea, then,

w_1 = 10g/100 mL; M_1 = 342 g mol^{-1};

w_2 = 1.754 g/100 mL; M = ?

$$\therefore \quad \frac{10\text{ g}/100\text{ mL}}{342\text{ g moL}^{-1}} = \frac{1.754\text{ g}/100\text{ mL}}{M_1}$$

$$\Rightarrow \quad M_2 = 1.754 \times \frac{342}{10} = 60 \text{ g mol}^{-1}.$$

57. As $\pi = [B]_l RT$.At T= 300 K, π = 3 atm

$$\therefore \quad [B]_t = \frac{3}{300\,R}$$

At $$T = 400\text{ K}, [B]_f = \frac{[B]}{2} = \frac{3}{300\,R \times 2}$$

$$\therefore \quad \pi_f = [B\}_f RT$$

$$= \frac{3}{300\,R \times 2} \times R \times 400 = 2 \text{ atm}$$

58. $\pi = [B\}\, RT$

Here T = 300 K

Since osmotic pressure is a colligative property, thus

$$[B] = [B]_{sucrose} + [B]_{glucose} + [B]_{urea}$$

$$= \frac{0.17/342}{50/100} + \frac{0.18/180}{50/1000} + \frac{0.06/60}{50/100}$$

$$= 20\left[\frac{0.17}{342} + \frac{0.18}{180} + \frac{0.06}{60}\right] = 0.05\,M \qquad (3)$$

Since $R = 0.0821$ L atm K^{-1} mol^{-1} (4)

$\therefore$ $\pi = (0.05 \text{ mol L}^{-1})(0.0821 \text{ L atm K}^{-1} \text{ mol}^{-1})(300 \text{ K})$

$\pi =$ **1.232 atm.**

59. *Step 1: Determination of Osmotic pressure in atm.* The osmotic pressure given in terms of the height of a liquid column, needs it conversion into an equivalent height of a mercury column.

As density of mercury = 13.6 g cm^{-3}.

$\therefore (h_1 d_1)$ for mercury column = $(h_2 d_2)$ for liquid column

$\Rightarrow$ $h_1 \times 13.6 = 2.4 \times 0.88$

$$\Rightarrow \quad h_1 = \frac{2.4 \times 0.88}{13.6} \; 0.155 \text{ mm} = \frac{0.155}{760} \text{ atm} = 204 \times 10^{-4} \text{ atm}$$

Thus, $\pi = 204 \times 10^{-4}$ atm. (1)

Step2 Determine of molecular weight

As $\pi = [B]\, RT$

Here $$[B] = \frac{0.25/M}{100/1000} = \frac{2.5}{M}$$

$R = 0.0821$ L atm K^{-1} mol^{-1}

$T = 310$ K

and $\pi = 2.04 \times 10^{-4}$ atm from (1) above

$$\therefore \quad \pi = \frac{2.5}{M} RT \;\Rightarrow\; M = \frac{2.5}{\pi} RT$$

$$\Rightarrow \quad M = \frac{2.5}{2.04 \times 10^{-4}} \times 0.0821 \times 310 = 3.1 \times 10^5 \quad (2)$$

60. Molality equals moles of solute (moles of substance) divided by kilograms of solvent (H_2O). By substituting values for molality and kilograms H_2O, solve for moles of substance. The molar mass of the substance equals mass of substance (0.131 g) divided by moles of substance. The molecular weight (in amu) has the same numerical value as molar mass in g mol^{-1}.

The molality of an aqueous solution is

$$\text{Molality} = \frac{\text{moles of substance}}{\text{kgH}_2\text{O}}$$

Hence, for the given solution,

$$0.056 \frac{\text{mol}}{\text{kg H}_2\text{O}} = \frac{\text{moles of substance}}{25.4 \times 10^{-3} \text{ kg H}_2\text{O}}$$

Rearranging this equation,

$$\text{Moles of substance} = 0.056 \frac{\text{mol}}{\cancel{\text{kgH}_2\text{O}}} \times 25.4 \times 10^{-3} \cancel{\text{kgH}_2\text{O}} = 1.42 \times 10^{-3} \text{ mol}$$

(An extra digit is retained for further computation.) The molar mass of substance equals

$$\text{Molar mass} = \frac{0.131 \text{ g}}{1.42 \times 10^{-3} \text{ mol}} = 92.3 \text{ g mol}^{-1}$$

You round the answer to two significant figures. The molecular weight is 92 amu.

61. From the melting point of the solution, calculate the freezing point depression, ΔT_f then solve the equation $\Delta T_f = K_f c_m$ for the molality, m. Use the method of the preceding example to obtain the molecular weight, given the mass of the sample and the molality of the solution.

(a) The freezing point lowering is

$$\Delta T_f = 4525 - 449 \doteq 3.5 \text{ K}$$

so the molality of the solution is

$$\frac{\Delta T_f}{K_f} = \frac{3.5}{40 \text{ km}} = 0.088\ m$$

the moles of the compound that are dissolved in 78.1 mg of is

$$0.088 \text{ mol kg}^{-1} \times 78.1 \times 10^{-5} \text{ kg} = 6.9 \times 10^{-6} \text{ mol}$$

The molar mass of the compound is

$$M_m = \frac{1.07 \times 10^{-3} \text{ g}}{6.9 \times 10^{-5} \text{ mol}} = 1.6 \times 10^2 \text{g mol}^{-1}$$

∴ The molecular on weight is 160 amu.

62. Obtain the molecular weight of $(C_6H_{10}O_5)_{200}$, and use it to obtain the molarity of the starch solution. Substitute into the formula for the osmotic pressure.

The molecular weight of $(C_6H_{10}O_5)_{200}$ is 32,400 amu. The number of moles in 0.798 g of starch is

$$0.798 \text{ g stach} \times \frac{1 \text{ mol starch}}{32{,}400 \text{ g strach}} = 2.46 \times 10^{-5} \text{ mol starch}$$

The molarity of the solution is

$$\frac{2.46 \times 10^{-5} \text{ mol}}{0.1000 \text{ L solution}} = 2.46 \times 10^{-4} \text{ mol starch}$$

and the osmotic pressure at 298 K is

$$\pi = MRT$$

$$= 2.46 \times 10^{-4} \text{ mol/L} \times 0.0821 \text{ K} \cdot \text{atm (K} \cdot \text{mol)} \times 298 \text{ K}$$

$$= 6.02 \times 10^{-3} \text{ atm} = 6.02 \times 10^{-3} \text{ atm} \times \frac{760 \text{ mmHg}}{1 \text{ atm}}$$

$$= \mathbf{4.58\ mmHg}$$

Chapter 6 : THERMODYNAMICS

1. Note, this is a Hess's law problem, which basically requires that you rearrange the three reactions listed if necessary (remembering to reverse the sign of $\Delta H°$ if you reverse the reaction) and then add up the reactions and the $\Delta H°$'s for the answer.

Looking at the first reaction, you notice that $H_2(g)$ is on the wrong side. This requires that you reverse the reaction, thus changing the sign of $\Delta H°$. In the second reaction, C is on the correct side of the equation, so the second equation should be left as is. In the third reaction, $C_2H_2(g)$ is on the wrong side of the equation, so the reaction and the sign of $\Delta H°$ must be reversed. At this point the set-up should look like

$$H_2O(l) \rightarrow H_2(g) + 12O_2(g) \qquad \Delta H° = +286 \text{ kJ}$$
$$CO2(g) \rightarrow C(s) + O_2(g) \qquad \Delta H° = +394 \text{ kJ}$$
$$C_2H_2(g) + 5/2\ O_2(g) \rightarrow 2CO_2(g) + H_2O(l) \qquad \Delta H° = -1300 \text{ kJ}$$

Before you add up the three reactions, note that there is no $CO_2(g)$ in the final reaction. The 2 $CO_2(g)$ in the third reaction must be able to cancel with two $CO_2(g)$'s in the second reaction. In order to get 2 $CO_2(g)$ in the second reaction, multiply everything in the second equation by 2, which yields

$$\cancel{H_2O(l)} \rightarrow H_2(g) + \cancel{1/2O_2(g)} \qquad \Delta H° = +286 \text{ kJ}$$
$$\cancel{2CO_2(g)} \rightarrow 2C(s) + \cancel{2O_2(g)} \qquad \Delta H° = +788 \text{ k}$$
$$C_2H_2(g) + 5/\cancel{2O_2(g)} \rightarrow \cancel{2CO_2(g)} + \cancel{H_2O(l)} \qquad \Delta H° = -1300 \text{ kJ}$$
$$\overline{C_2H_2(g) \rightarrow H_2(g) + 2C(s)} \qquad \overline{\Delta H° = -26 \text{ kJ}}$$

2. The structural diagram of the reaction is,

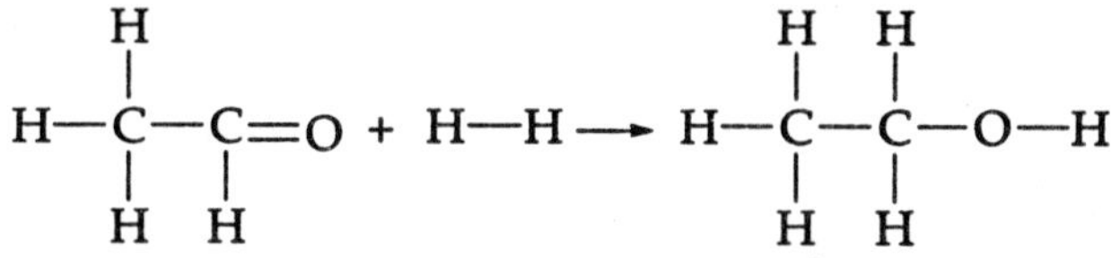

There are three steps you need to take to do this problem:

Step 1: Decide which bonds need to be broken on the reactant side of the reaction. Add up all the bond energies for the bonds that are broken. Call this subtotal ΔH_1. Assign ΔH_1 a positive value because energy is required when bonds are broken. In the example given, a C==O and a H—H bond need to be broken. This becomes x kJ mol^{-1} + 435 kJ mol^{-1}, or $\Delta H_1 = 435 + x$ kJ mole^{-1}

Step 2: Decide which bonds need to be formed on the product side of the reaction. Add up all of the bond energies that are formed. Call this subtotal ΔH_2. Assign ΔH_2 a negative value because energy is released when bonds are formed. In the example given, a C—H, a C—O and a O—H bond need to be formed. This becomes (414 + 351 + 464 = 1229) kJ mol^{-1}. Remember to assign a negative sign, which makes $\Delta H_2 = -1229$ kJ mol^{-1}.

Step 3: Apply Hess's law: $\Delta H° = \Delta H_1 + \Delta H_2$. You know that $\Delta H°$ is –71 kJ mol^{-1} so Hess's law becomes

$$-71 \text{ kJ mol}^{-1} = 435 \text{ kJ mol}^{-1} + x \text{ kJ mol}^{-1} - 1229 \text{ kJ mol}^{-1}$$
$$x = 723 \text{ kJ mol}^{-1}.$$

which represents the bond energy of the C=O bond.

3. Begin this problem by realizing that $\Delta H°$ of combustion for acetic acid represents the amount of heat released when acetic acid burns in oxygen gas. The products are carbon dioxide and water. According to the following balanced equation.

$$CH_3COOH((l)) + 2O_2(g) \rightarrow 2CO_2(g) + 2H_2O((l))$$

Because, $\Delta H° = \Sigma H^{\circ}_{f\,products} - \Sigma H^{\circ}_{f\,reactants}$ you can substitute at this point to give

$$\Delta H° = 2(-397.1 \text{ kJ mol}^{-1}) + 2(-250.8 \text{ kJ mol}^{-1}) - (-5015 \text{ kJ mol}^{-1})$$
$$= -794.2 \text{ kJ mol}^{-1}$$

4. Begin this problem by drawing a structural diagram:

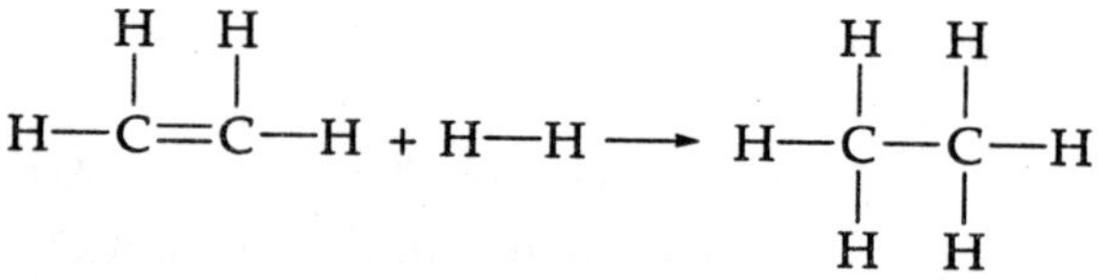

There are three steps you need to take to do this problem.

Step 1: Decide which bonds need to be broken on the reactant side of the reaction. Add up al the bond energies for the bonds that are broken. Call the subtotal ΔH_1, and assign it a positive value because when energy is required, bonds are broken. In the example given, a C==C and a H—H bond need to be broken. This becomes 614 kJ mol^{-1} + 436 kJ mol^{-1} = ΔH_1 = 1050 kJ mol^{-1}.

Step 2: Decide which bonds need to be formed on the product side of the reaction. Add up all the bond energies for the bonds that are formed. Call this subtotal ΔH_2. Assign ΔH_2 a negative value because when energy is released, bonds are formed. In the example given, two C—H bonds and a C—C bond need to be formed. This becomes (2 × 413 kJ mol^{-1}) + 348 kJ mol^{-1}, or 1174 kJ mol^{-1}. Remember to assign a negative sign, which makes ΔH_2 = – 1174 kJ mol^{-1}.

Step 3: Apply Hess's law: $\Delta H° = \Delta H_1 + \Delta H_2$
This becomes 1050 kJ mol^{-1} + (–1174 kJ mol^{-1}) = –124 kJ mol^{-1}.

5. Examine the first reaction and realize that $SO_2(g)$ needs to be on the reactant side. Reverse the equation and change the sign of $\Delta H°$. When you examine the second reaction, you notice that $SO_3(g)$ is on the correct side, so there is no need to reverse this equation. At this point, your two reactions can be added together.

$SO_2(g) \rightarrow S(s) + O_2(g)$	$\Delta H° = 295 \text{ kJ mol}^{-1}$
$S(s) + 3/2O_2(g) \rightarrow SO_3(g)$	$\Delta H° = -395 \text{ kJ mol}^{-1}$
$SO_2(g) + 1/2O_2(g) \rightarrow SO_3(g)$	$\Delta H° = 100 \text{ kJ mol}^{-1}$

But before concluding that this is your answer, note that the question asks for $\Delta H°$ in terms of 2 moles of SO_2(g). Doubling the $\Delta H°$ gives the answer, –200 kJ mol^{-1}.

6. The first law of thermodynamics states that $\Delta E = q + w$. Since $w = -P_{ext} \cdot \Delta V$, the equation can be stated as:

$$\Delta E = \Delta H - P_{ext} \cdot \Delta V$$
$$\Delta E = 418 \text{ J} - (2.00 \text{ atmospheres} \cdot 3 \text{ liters}. 101 \text{ J.L}^{-1}.\text{atm}^{-1}) = -188 \text{ J}$$

7. $w = -P\Delta V = -(1.00 \text{ atm})(8.00 \text{ liters}) = -8.00 \text{ L} \cdot \text{atm}$
Because the gas was expanding, w is negative (work was being done by the system).

8. Δn = moles of gaseous product – moles of gaseous reactants
$= 1 - 0 = 1$
Use the equation $\Delta H = \Delta E + \Delta n \cdot RT$

$$= \frac{62.30 \text{ kJ}}{\text{mole}} \times \frac{1000 \text{ J}}{1 \text{ kJ}} + [1 \cdot (8.314 \text{ J/mole} \cdot \text{K}) \cdot (298 \text{ K})]$$

9. $\Sigma S^\circ = \Sigma\Delta S^\circ_{\text{products}} - \Sigma\Delta S^\circ_{\text{reactants}}$

$$\Delta S^\circ = [3(33.0) + 51.0] - [2(28.0) + 3(27.0)] = 13.0 \text{ JK}^{-1}$$

10. This problem requires us to use the Gibbs-Helmholtz equation:
$$\Delta G^\circ = \Delta H^\circ - T\Delta S^\circ$$
Step 1: Using the given information, calculate ΔH°.

$$\Delta H^\circ = \Sigma\Delta H^\circ_{\text{products}} - \Sigma\Delta H^\circ_{\text{reactants}}$$
$$= [2(-100.0)] - [2(-218.0) + 2(-279.0)] = 830.0 \text{ kJ mol}^{-1}$$

Step 2: Calculate ΔS°.

$$\Delta S^\circ = \Sigma\Delta S^\circ_{\text{products}} - \Sigma\Delta S^\circ_{\text{reactants}}$$
$$= [2(91.0) + (3(205.0)] - [2(70.0) + 2(248.0)]$$
$$= 797.0 - 636.0 = 161.0 \text{ J mole}^{-1} \text{ K}^{-1} = 0.161 \text{ kJ mol}^{-1} \cdot \text{K}^{-1}$$

Step 3: Substitute into the Gibbs-Helmholtz equation.
$$\Sigma G^\circ = \Delta H^\circ - T\Delta S^\circ$$

$$\frac{830.0 \text{ kJ}}{1 \text{ mole}} - \frac{298 \text{ K} \cdot (0.161 \text{ kJ})}{\text{mole} \cdot \text{K}} = 782.0 \text{ kJ mol}^{-1}$$

11. $\Delta G^\circ = \Sigma\Delta G^\circ_{\text{products}} - \Sigma\Delta G^\circ_{\text{reactants}}$
$= [2(-237.00) + (-394.00) - (-51.00)$
$= -817.00 \text{ kJ mol}^{-1}$

12. At equilibrium, $\Delta G = 0 = \Delta G^\circ + 2.303 \text{ RT} \log \text{K}$ (at equilibrium $Q = K$).
$$\Delta G^\circ = -2.303(8.314 \text{ J} \cdot \text{K}^{-1})(298 \text{ K})(\log 1.0 \times 10^{-3})$$
Rounding
$$\sim -2.3(8.3)(300)(-3.0) = 17181 \text{ K} \approx 17 \text{ kJ}$$

13. Entropy is a measure of the randomness or disorder of a system. The greater the disorder of a system, the greater its entropy.

In $H_2O(g) \rightarrow H_2O(l)$, the reaction is going from a disordered state (g) to a more ordered state (l); low entropy, $\Delta S < 0$.

In $2 \text{ HCl}(g) \rightarrow H_2(g) + Cl_2(g)$, the change in energy will very small since there are two moles of gas molecules on each side of the equation.

In $SiO_2(s) \rightarrow Si(s) + O_2(g)$, the system is becoming more disordered, apparent from the presence of gas molecules on the product side; high entropy, $\Delta S > 0$.

14. At equilibrium Hg((*l*)) ↔ Hg(*g*), which represents the condition of boiling, and at equilibrium $\Delta G^\circ = 0$. The word normal in the question refers to conditions at 1 atm of pressure, which is reflected in the notation for standardized conditions for ΔS° and ΔH°. Therefore, using the Gibbs-Helmholtz equation, $\Delta G^\circ = \Delta H^\circ - T\Delta S^\circ$, we can substitute 0 for ΔG° and solve for *T*.

$$T = \frac{\Delta H^\circ}{\Delta S^\circ} = \frac{63.000 \text{ J} \cdot \cancel{\text{mol}^{-1}}}{100 \text{ J} \cdot \text{K}^{-1} \cdot \cancel{\text{mol}^{-1}}} = 630 \text{ K}$$

15. To solve this problem, use the Gibbs-Helmholtz equation:

$$\Delta G^\circ = \Delta H^\circ - T\Delta S^\circ$$

Step 1: Solve for ΔH°. realize that you will have to use Hess's law to determine ΔH°. Be sure to multiply through the stepwise equations to achieve the lowest common denominator (6), and reverse equations where necessary.

$\cancel{3Fe_2O_3(s)} + 9CO(g) \rightarrow 6Fe(s) + 9CO_2(g)$	-81 kJ mol^{-1}
$\cancel{2Fe_2O_4(s)} + CO_2(g) \rightarrow \cancel{3Fe_2O_3(s)} + CO(g)$	61 kJ mol^{-1}
$6FeO(s) + 2CO_2(g) \rightarrow \cancel{2Fe_3O_4(s)} + 2CO(g)$	-76 kJ mol^{-1}
$6FeO(s) + 6CO(g) \rightarrow 6Fe(s) + 6CO_2(g)$	-96 kJ mol^{-1}

$$\Delta H^\circ = \frac{-96 \text{ kJ mol}^{-1}}{6} = -16 \text{ kJ mol}^{-1}$$

Step 2: Solve for ΔS°.

$$FeO(s) + CO(g) \rightarrow Fe(s) + CO_2(g)$$

$$\Delta S^\circ = \Sigma \Delta S^\circ_{\text{products}} - \Sigma \Delta S^\circ_{\text{reactants}}$$

$$= (27.0 + 214.0) - (61.0 + 190.0) = -10.0 \text{ J} \cdot \text{K}^{-1} \text{ mole}^{-1}$$

Step 3: Substitute ΔS° and ΔH° into the Gibbs-Helmholtz equation.

$$\Delta G^\circ = \Delta H^\circ - T\Delta S^\circ$$

$$= -16 \text{ kJ mole}^{-1} - 298 \text{ K}(-0.0100 \text{ kJ} \cdot \text{K}^{-1} \cdot \text{mol}^{-1})$$

$$\approx -13 \text{ kJ mole}^{-1}$$

16. Realize that you will need to use the equation

$$\Delta G = \Delta G^\circ + RT \ln Q$$

Step 1: Solve for the reaction quotient, *Q*.

$$Q = \frac{(P_{HF})^2}{(P_{H_2})(P_{P_2})} = \frac{(1.00)^2}{(0.50)(2.00)} = 1.00$$

$$\ln 1.00 = 0$$

Step 2: Substitute into the equation.

$$\Delta G = \Delta G^\circ + RT \ln Q$$

$$= -546{,}000 \text{ J} + (8.3148 \text{ J} \cdot \text{K}^{-1} \cdot \text{mol}^{-1}) \cdot 773 \text{ K}(0)$$

$$= -546 \text{ kJ mol}^{-1}$$

17. Examine the Gibbs-Helmholtz equation, $\Delta G^\circ = \Delta H^\circ - T\Delta S^\circ$, too see the mathematical relationships of negative ΔH°'s and ΔS°'s. (Refer to text book)

18. Use the equation

$$\Delta S = 2.303 C_p \log \frac{T_2}{T_1}$$

where Cp represents the heat capacity

$$(C_p = \Delta H/\Delta T = 290{,}700 \text{ J}/2907 \text{ K} = 100.0 \text{ JK}^{-1}).$$

Substituting into the equation yields

$$\Delta S = 2.303 C_p \log \frac{T_2}{T_1}$$

$$= 2.303\left(100.0 \frac{\text{J}}{\text{K}}\right) \log \frac{3230 \text{ K}}{323 \text{ K}} = 230 \text{ JK}^{-1}$$

19. Work done by an ideal gas when it expands isothermally and reversible is given by

$$W = n\, RT \ln (V_2/V_1)$$

$$= -\ (10 \text{ mol})(0.8206 \text{ dm3 atm. K}^{-1} \text{ mol}^{-1})\ (298 \text{ K}) \ln \left(\frac{50 \text{ dm}^3}{5 \text{ dm}^3}\right)$$

$$= -563.07 \text{ dm}^3 \text{ atm.}$$

$$= -563.07 \text{ dm}^3 \text{ atm } \frac{8.314 \text{ JK}^{-1} \text{ mol}^{-1}}{0.08206 \text{ dm}^3 \text{ atm K}^{-1} \text{ mol}^{-1}}$$

$\Rightarrow$ $$W = -57048.2 \approx -57 \text{ kJ}$$

20. Work done (W) $= -n\ RT \ln (P_1/P_2)$

$\Rightarrow$ $$W = -(1 \text{ mol})(8.314 \text{ JK}^{-1} \text{ mol}^{-1})(298 \text{ K}) \ln \left(\frac{5 \text{ atm}}{1 \text{ atm}}\right)$$

$$= -3987.5 \text{ J} \qquad \ldots\ldots \text{(A)}$$

Let m' be the mass which can be lifted through a height of 10metres. Work done in lifting the mass $= m'gh = (m' \text{ kg})\ (9.81 \text{ ms}^{-2})\ (10 \text{ m})$ ..... (B)

From equation (A) and (B) we get,

$98.1\ m'\ (\text{kg } m^2 \text{ s}^{-2}) = 3987.5 \text{ J}$

(Note. $1 \text{ J} = \text{kg m}^2 \text{ s}^{-2}$)

$\Rightarrow$ $$m' = \frac{3987.5}{98.1} = 40.65 \text{ kg} \ \Rightarrow\ m' = 40.65 \text{ kg}$$

21. As the expansion is isothermal and reversible, so $\Delta E = 0$. From First Law of thermodynamics, $q = \Delta E - W \Rightarrow q = -W$

Given

$\therefore$ $$W = -q = -41840 \text{ J}$$

As the ideal gas is udner the conditions of S.T.P., so $V_1 = 22.4 \text{ dm}^3$

Since $$W = -n\ RT \ln (V_2/V_1)$$

$\Rightarrow$ $$-41840 = -(10 \text{ moles}) (8.314 \text{ JK}^{-1} \text{ mol}^{-1}) (273 \text{ K}) \ln\left(\frac{V_2}{22.4 \text{ dm}^3}\right)$$

$\Rightarrow$ $\ln V_2 - \ln (22.4) = 1.84$

$\Rightarrow$ $V_2 = 14.15 \text{ dm}^3$

22. $W = -P(V_2 - V_1)$
$= -(4 \text{ atm}) (25 \text{ dm}^3 - 5 \text{ dm}^3)$
$= -80 \text{ dm}^3 \text{ atm}$

$$= -80 \text{ dm}^3 \text{ atm} \times \frac{8.314 \text{ JK}^{-1} \text{ mol}^{-1}}{0.08206 \text{ dm}^3 \text{K}^{-1} \text{ mol}^{-1}} = -8105.3 \text{ J}$$

From the first law of thermodynamics,

$\Delta E = q + W$
$= 500 \text{ J} + (-8105.3 \text{ J}) = -7605.3 \text{ J}$

23. Let the initial pressure = P_1
and Final pressure = P_2

Given $P_2 = 5 P_1;\ \gamma = 1.66$

Since $\boxed{T_2/T_1 = (P_2/P_1)^{(\gamma-1)/\gamma}}$

$\therefore$ $T_2 = T_1(P_2/P_1)^{(\gamma-1)/\gamma}$

$$T_2 = (289 \text{ K}) \left[\frac{5p_1}{p_1}\right]^{\frac{(1.66-1)}{1.66}} = (298 \text{ K}) (5)^{0.397}$$

$\Rightarrow$ $\ln T_2 = \ln (298) + 0.397 \ln (5) = 6.336 \therefore T_2 = \text{anti} \ln (6.336)$

$\Rightarrow$ $T_2 = 564.5 \text{ K}$

24. At constant pressure,

$$\boxed{(\Delta S)_p = n\, C_p \ln (T_2/T_1)}$$

$$= (5 \text{ moles})\left(\frac{5}{2} \times 8.314 \text{ JK}^{-1} \text{ mol}^{-1}\right) \ln\left(\frac{1000 \text{ K}}{500 \text{ K}}\right)$$

$\Rightarrow$ $(\Delta S)_p = 72.04 \text{ JK}^{-1}$

25. $\boxed{\Delta S = n[C_V \ln (T_2/T_1) + \text{R} \ln (V_2/V_1)]}$

$$\therefore \quad \Delta S = \left[\frac{3}{2}(8.314 \text{ JK}^{-1} \text{ mol}^{-1}) \ln\left(\frac{323 \text{ K}}{298 \text{ K}}\right) + (8.314 \text{ JK}^{-1} \text{ mol}^{-1}) \ln\left(\frac{2 \text{ dm}^3}{1 \text{ dm}^3}\right)\right]$$

$\Delta S = 67.68 \text{ JK}^{-1}$.

26. Per mole of CH_4:

$$CH_4 + 2O_2 \longrightarrow CO_2 + 2H_2O$$
4(C–H) 2(O–O) 2(C=O) 4(HO)

$\therefore$ $4(418) + 2(498) - 2(741) - 4(464) = -670$ KJ mol^{-1}.

$\therefore$ Pergram $\Delta H = \dfrac{-670}{16} = -42$ KJ g^{-1}

Per mole of C_8H_{18}:

$$C_8H_{18} \quad + \quad \frac{25}{2}O_2 \longrightarrow 8CO_2 \quad + \quad 9H_2O$$

$$8(C\text{–}H),\ 7(C\text{–}C) \quad \frac{25}{2}(O\text{–}O) \quad 16(C{=}O) \quad 18(H{=}O)$$

or $\Delta H = 18(418) + 7(343) + \dfrac{25}{2}(498) - 16(741) - 18(464) = -4200$ KJ mol^{-1}

Per gram $\Delta H = \dfrac{-8458}{114\,g} = -36.8$ KJ g^{-1}

$\therefore$ Methane produces more heat per gram.

27. The bond energy is the energy needed to dissociate gaseous H_2 into separated atoms.

$$H_2 \rightarrow 2H$$

ΔH for this reaction is twice ΔH_f for mol of H.

$$\Delta H = 2(218 \text{ kJ mol}^{-1}) = 436 \text{ kJ mol}^{-1}$$

28. The left-hand side of a formation reaction can have only elements in their standard states. Option (a) is wrong because CO is not an element. Option (b) is wrong because the standard states of hydrogen and oxygen are H_2 and O_2, and not H(g) or O(g). Option (c) its wrong because H_2O is not an element.

29. Only for an element in its standard state is $\Delta H°$ equal to zero. The standard state of Br_2 is the liquid, not the gas. The standard state of nitrogen is $N_2(g)$, and of carbon is graphite.

30. Krypton is a monatomic gas. The Law of Dulong and Petit applies to solid elements only.

31. mass Ca needed $= \dfrac{(1000 \text{ kJ})(40.08 \text{ g mol}^{-1})}{(635.1 \text{ kJ mol}^{-1})} = 62.11$ g

32. Remember that the standard heat oc combustion produces liquid water. To obtain the desired equation we must combine the following two equations:

$$C_3H_8(g) + 5\ O_2(g) \rightarrow 3\ CO_2(g) + 4H_2O(l) \qquad \Delta H^{\circ}_{comb} = -2220.1 \text{ kJ}$$

$$H_2O(l) \rightarrow H_2O(g) \qquad \Delta H^{\circ}_{vap} = +44.0 \text{ kJ}$$

Thus,

$$\Delta H° = -2220.1 + 4(44.0) = -2044.1 \text{ kJ}$$

33. $\Delta E° = \Delta H_f^{\circ}(CO_2, g) + 2\Delta H_f^{\circ}(H_2O, g) - \Delta H_f^{\circ}(CH_4, g)$

$= -393.51 + 2(-241.82) - (-74.81)$

Note that -285.83 is ΔH_f° for liquid H_2O, not gaseous.

34. Only elements in their standard states can be on the left-hand side of the formation reaction. Options (a) and (b) have compounds on the left-hand side. In both (b) and (c) hydrogen is not in its standard state.

35. $C_4H_{10}(g) + \frac{13}{2}O_2\,(g) \rightarrow 4\ CO_2(g) + 5\ H_2O(\ell)$

There are 13 O atoms on each side of the equation.

36. Neon is a monatomic as, so that $C_V = \frac{3}{2}R$ and $p = \frac{5}{2}R = 20.8\ J \cdot mol^{-1}K^{-1}$. Since $q_p = nC_p\Delta T$,

$\Delta T = (100.0\ J)/(20.8\ J \cdot K^{-1}) = 4.8°$

37. To obtain the desired equation from the two that are given, multiply the second equation by 3 {to have $3O_2(g)$ on the left} and subtract twice the first equation (so that we have $2Fe_2O_3$ on the right, and 4Fe on the left).

38. $w = -P_{ext}\Delta V = -(1.50\ atm)\ (-6.00\ L) = 9.00\ L \cdot atm$

$$w = (9.00\ L \cdot atm)\ (101.325\ J \cdot L^{-1}\ atm^{-1}) = 912\ J$$
$$\Delta E = q + w = 100\ J + 912\ J = 1012\ J$$

39. $\Delta H° = \Delta H_f°(N_2H_4, (\ell)) + \Delta H_f°(H_2O, (\ell)) - \Delta H_f°(N_2O, g)$

$$-317.25 = \Delta H_f°(N_2H_4, (\ell)) + (-285.83) - 82.05$$
$$\Delta H_f°(N_2H_4, (\ell)) = -137.25 + 285.83 + 82.05$$

40. Argon is monatomic, so $C_p = 20.8\ J \cdot mol^{-1}\ K^{-1}$ $q_p = nC_p\Delta T = (0.250\ mol)\ (20.8\ J \cdot mol^{-1}\ K^{-1})\ (16\ K) = 83.2\ J$

41. $B(s) + \frac{3}{4}(g) \rightarrow \frac{1}{2}B_2O_3(s)$

By the definition of heat of combustion, only 1 mol of $B(s)$ is burned, and this yields one half mole of B_2O_3.

42. The first law states: $\Delta E = q + w$. For an adiabatic process, $q = 0$, therefore $\Delta E = w$.

43. The molecular weight of C_2H_4 is 28.054 $g \cdot mol^{-1}$. The 0.105-g sample contains. 3.743×10^{-3} mol C_2H_4.

$$\Delta E = \frac{-C_V\,\Delta T}{n} = \frac{-(2.47\ kJ\,K^{-1})\,(2.14\ K)}{(3.743 \times 10^{-3}\ mol)} = -1410$$

44. The standard state of Br_2 is the liquid. From text books we obtain $\Delta H_f°\ (Br_2, g) = 30.907\ kJ\ mol^{-1}$.

$$\text{No. mol } Br_2 = \frac{10.00\ g}{159.808\ g\,mol^{-1}} = 0.062575\ mol$$

To vaporize 10.00 g of liquid bromine therefore requires (0.062575 ml) $(30.907\ kJ\ mol^{-1}) = 1.934$ kJ

45. $w = -P_{ext}\Delta V = -(2.00\ atm)(3.40\ L) = -6.80\ L \cdot atm$

$$w = (-6.80\ L \cdot atm)\ (101.325\ J \cdot L^{-1}\ atm^{-1}) = -689\ J$$
$$\Delta E = q + w = 400 - 689 = -289\ J$$

46. The equation for the combustion of Al is $Al(s) + \frac{3}{4}O_2(g) \rightarrow \frac{1}{2}Al_2O_3(s)$ $\Delta H° = -837.8$ kJ · mol^{-1}
Burning 1 mol Al yields 837.8 kJ, hence we must combust

$$(1046 \text{ kJ})/(837.8 \text{ kJ} \cdot \text{mol}^{-1}) = 1.249 \text{ mol Al}$$

to realise 1046 kJ. The reaction of 1.249 mol Al is not one of the options, but that reaction will produce 0.624 mol of Al_2O_3, which is option (b), the correct answer.

47. We must calculate $\Delta H°$ for the reaction

$$CCl_4(g) \rightarrow C(g) + 4Cl(g)$$

The standard state of CCl_4 is the liquid, so $\Delta H_f^\circ(CCl_4, g)$ is the same as ΔH_f° (CCl_4). The series of reactions needed is
(1) $C(s) + 2Cl_2(g) \rightarrow CCl_4(l)$
(2) $CCl_4(l) \rightarrow CCl_4(g)$
(3) $C(s) \rightarrow C(g)$ and
(2) $Cl_2(g) \rightarrow 2Cl(g)$

48. On the right-hand side of the equation there are 6 C—H and 1 C—C single bond. Thus the bond energy of the right side is $6\varepsilon_{C-H} + \varepsilon_{C-C}$. On the left-hand side of the equation there are 4 C—H bonds, 1 C=C double bond, and 1 H—H bond. Thus the bond energy of the left-hand side is $4\,\varepsilon_{C-H} + \varepsilon_{C=C} + \varepsilon_{H-H}$. When the reaction occurs, the bonds on the left side must be broken, expending that bond energy, and the bonds on the right side are formed, releasing that bond energy. Thus,

$$\Delta H° = 4\,\varepsilon_{C-H} + \varepsilon_{C=C} + \varepsilon_{H-H} - (6\,\varepsilon_{C-H} + \varepsilon_{C-C})$$
$$= \varepsilon_{C=C} + \varepsilon_{H-H} - 2\,\varepsilon_{C-H} - \varepsilon_{C-C}$$

49. Add the first equation given to twice the second equation to obtain the desired equation. Thus,

$$\Delta H° = -26.8 + 2(16.5) = +6.2 \text{ kJ}$$

50. $\Delta S° = S°(C_2H_6) - 2S°(H_2) - 2S°(C, gr)$
$= 229.59 - 3(130.57) - 2(5.740) = -173.7 \text{ JK}^{-1}$

51. $\Delta G° = \Delta H° - T\Delta S° = -197.7 - (298.15)(-0.1880)$
$= -197.7 + 56.05 = -141.6$ kJ

52. For an ideal gas, E is a function only of temperature. Therefore at constant temperature $\Delta E = 0$. The entropy increases ($\Delta S > 0$) because the final volume is larger than the initial volume and the molecules have more freedom of motion in the final state.

53. $\Delta G° = \Delta G_f^\circ(H_2) + \Delta G_f^\circ(I_2) - 2\Delta G_f^\circ(HI)$
$= 0 + 0 - 2(1.7) = -3.4 \text{ kJ} = -3400 \text{ J}$

$$\Delta G° = -3400 \text{ J} = -RT(\ln K_p) = -(8.3145)(298.15)(\ln K_p)$$

$$\ln K_p = \frac{3400}{(8.3145)(298.15)} = 1.37$$

$$K_p = e^{1.37} = 3.9$$

54. Liquid water and water vapour are in equilibrium at 373 K and 1-atm pressure. Therefore, $\Delta G = 0 = \Delta H - T\Delta S$. Hence, $\Delta H = T\Delta S$ for this phase equilibrium.

55. This is the formation reaction for propane as there are only elements in their standard states on the left-hand side. $\Delta G° = \Delta G_f^\circ(C_3H_8) = -23{,}490 \text{ J} = -RT(\ln K_{eq})$

$$\ln K_{eq} = (23{,}490)/(8.1345)(298.15) = 9.48$$
$$K_{eq} = e^{9.48} = 1.3 \times 10^4$$

56. Since the equilibrium constant is less than 1, $\Delta G°$ must be positive. This means that options (a) and (b) are both wrong. To determine the sign of ΔG we must know the relative magnitudes of the reaction quotient, Q, and K_{eq}.

$$Q = P_C P_B / P_A = (1.0)(0.50)/(2.0) = 0.25$$

Since $Q > K_{eq}$, the reaction proceeds to the left; therefore ΔG must be positive, that is, $\Delta G > 0$. You can also use the formula $\Delta G = RT \ln (Q/K)$ to calculate ΔG, that is not necessary to answer the question.

57. The side of tan equation with more moles of gas is the side with higher entropy. For $\Delta S°$ to be negative, there must be more moles of gas on the left side than on the right. This is only true for option (c).

58. Since $\Delta G = \Delta G° + RT(1\ nQ)$, if $Q = 1$, $1\ nQ = 0$, and $\Delta G = \Delta G°$.

59. $Q = P^2_{ICl}/P_{Cl_2} = (0.800)^2/(2.1 \times 10^{-3}) = 304.8$

$$\Delta G = \Delta G° + RT\ (1\ nQ) = -10.9 + (8.3145 \times 10^{-3})\ (298.15) \ln (304.8)$$
$$= -10.9 + (2.4790)(5.1795) = -10.9 + 14.2 = 3.3$$

60. $\Delta G° = \Delta H° - T\Delta S° = 33.2 - (298.15)\ (-0.0609)$

$$= 33.2 + 18.16 = 51.4 \text{ kJ}$$

61. For reaction to be spontaneous, ΔG must be negative. Since $\Delta G = \Delta H - T\Delta S$, if $\Delta H > 0$, then ΔS must also be positive in order for $\Delta G < 0$.

62. Trouton's rule states that at the normal boiling point, $\Delta S_{vap} \sim 88\ \text{J} \cdot \text{mol}^{-1}\ \text{K}^{-1}$. Thus, for HCl,

$$88\ \text{J} \cdot \text{mol}^{-1}\text{K}^{-1} = \frac{\Delta H_{vap}}{T_{bp}} = \frac{16{,}200}{T_{bp}} \quad \text{so that } T_{bp} = 184\ \text{K}$$

63. The standard state of bromine is liquid Br_2.

Thus
$$\Delta G° = \frac{1}{2}\Delta G_f^\circ(Br_2, g) + 0 - \Delta G_f^\circ(HBr, g)$$
$$= (0.5)\ (3.144) - (-53.43) = 55.00$$

64. $\Delta H° = 2\Delta H_f^\circ\ (NH_3) = 2(-46.11) = -92.22$ kJ

$$\Delta S° = 2S°\ (NH_3, g) - 3S°\ (H_2, g) - S°(N_2, g)$$
$$= 2(192.34) - 3(130.57) - 191.50 = -198.53\ \text{J} \cdot \text{K}^{-1}$$
$$\Delta G° = \Delta H° - T\Delta S° = -92.22 - (373.15)(-0.19853) = -18.14 \text{ kJ})$$

65. $\Delta H_{vap} = \Delta H_f^\circ(CCl_4, g) - \Delta H_f^\circ(CCl_4, l)$

$$= -102.9 - (-135.44) = 32.5 \text{ kJ}$$
$$\Delta S_{vap} = S°(CCl_4, g) - S°(CCl_4, l) = 309.74 - 216.4 = 93.3\ \text{J K}^{-1}$$

There are the values at 298 K, and are not the values at the normal bp, although the differences are not large. Using these values we obtain

$$T_{bp} = \frac{\Delta H_{vap}}{\Delta S_{vap}} = \frac{32{,}500\ \text{J}}{93.3\ \text{J/K}} = 348\ \text{K}$$

66. $\Delta G^{\circ}_{800} = -40{,}000 \text{ J} = -R \text{ (800 K) (ln } K_{eq})$

$$\ln K_{eq} = (40{,}000)/(8.1345)(800) = 6.014$$

$$K_{eq} = e^{6.014} = 409 = 4.1 \times 10^2$$

67. Decomposition are endothermic, so $\Delta H > 0$. There is more freedom of motion, that is, more molecular disorder, for the two separate molecules, PCl_3 and Cl_2, than when all the atoms are bound together as PCl_5. Therefore $\Delta S > 0$ for this reaction.

68. For these figures, there is no need to do a calculation.
Since ΔH° is negative and $-T\Delta S^{\circ}$is negative (absolute temperature cannot be negative), Then $\Delta G^{\circ} = \Delta H^{\circ} - T\Delta S^{\circ}$ must also be negative. Therefore, the reaction is spontaneous.

69. $\Delta G^{\circ} = \Delta H^{\circ} - T\Delta S^{\circ}$

$= +176 - 298\ (+284 \times 10^{-3})$

[Remember: kJ must be used through out.]

$$= +91 \text{ kJ mol}^{-1}$$

Since the free energy change is positive, the reaction cannot be spontaneous at 298 K.

70. Assuming that both ΔH and ΔS are constants, the reaction becomes spontaneous when $\Delta G = 0$.
That is $0 = \Delta H - \Delta S$

$$0 = +184 - (T \times 25 \times 10^{-3})$$

[Note: energy is measured in kJ through out.]

Therefore: $$T = \frac{184}{25 \times 10^3} = 7360 \text{ K}$$

71. The enthalpy of formation of source can be calculated with help of following equation:

$$12\ C(s) + 11\ H_2(g) + \frac{11}{2} O_2(g) \rightarrow C_{12}H_{22}O_{11}(s)$$

Let, enthalpy of formation of Sucrose = a kJ
Given, $C(s) + O_2(g) \rightarrow CO_2(g);\ \Delta H = -393.5 \text{ kJ}$ (i)

$H_2(g) + \frac{1}{2} O_2(g) \rightarrow H_2O(l);\ \Delta H = -286.2 \text{ kJ}$ (ii)

and $C_{12}H_{22}O_{11}(s) + 12\ O2(g) \rightarrow 12\ CO_2(g) + 11\ H_2O(l);\ \Delta H = -5644.2 \text{ kJ}$

From the above equations, we have

$$-5644.2 = [12 \times (-393.5) + 11 \times (286.2)] - [a + 0]$$

$\Rightarrow$ $a = -2226.00$ kJ = Enthalpy of formation of Sucrose.

72. The reaction is hydrogenation of ethane,

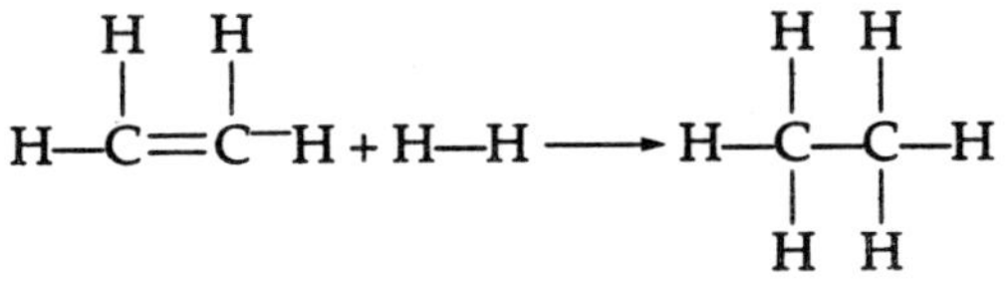

In this reaction,

Bonds broken are:
(i) One C=C bond and
(ii) One H—H bond
and Bonds Formed are:
(i) One C—C bond and
(ii) Two new C—H bonds
Now Energy required to break the bonds
= (Bond energy of C=C bond) + (Bond energy of H—H bond)
= 606 + 434.7 = 1040 kJ
And Energy released in bond formation
(Bond energy of C—C bond) + (2 × Bond energy of C—H bond) = 346.9 + 2 × 413.8 = 1174.5

Thus, Enthalpy change = Energy required to break the bonds – Energy released in the formation bonds

$$= 1040 - 1174.5 = -134.5 \text{ kJ}$$

A negative value of enthalpy change indicates that the reaction is exothermic.

73. From the energy cycle we conclude:

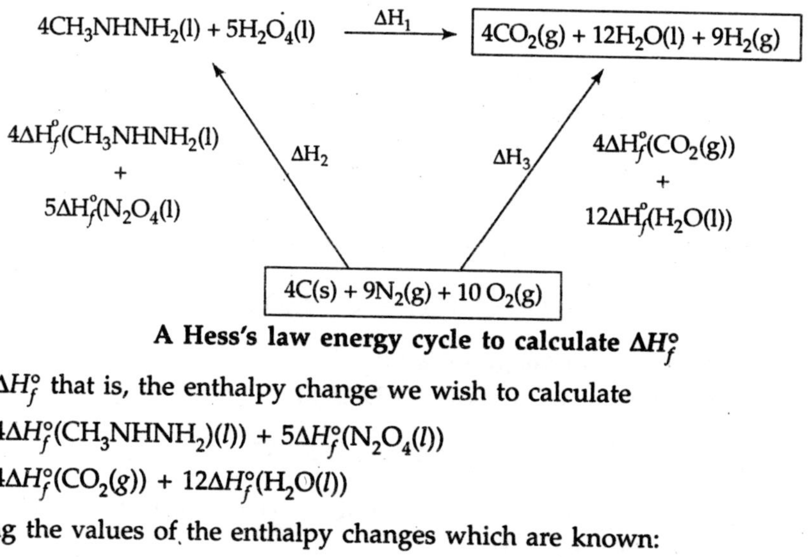

A Hess's law energy cycle to calculate ΔH_f°

$\Delta H_1 = \Delta H_f^\circ$ that is, the enthalpy change we wish to calculate

$\Delta H_2 = 4\Delta H_f^\circ(CH_3NHNH_2)(l)) + 5\Delta H_f^\circ(N_2O_4(l))$

$\Delta H_3 = 4\Delta H_f^\circ(CO_2(g)) + 12\Delta H_f^\circ(H_2O(l))$

Inserting the values of the enthalpy changes which are known:

$\Delta H_1 = \Delta H_3 - \Delta H_2$

$\Delta H_f^\circ = 4\Delta H_f^\circ\ (CO_2(g)) + 12\Delta H_f^\circ(H_2O(l))$

$\quad - [4\ \Delta H_f^\circ\ (CH_3NHNH_2(l)) + 4\Delta H_f^\circ\ (N_2O_4(l))]$

$\Delta H_f^\circ = 4(-393) + 12(-286) - [4(+54) + 5(-20)]$

$\quad = -5004 - [+116] = -5120 \text{ kJ mol}^{-1}$

74. Step 1. Write out the enthalpy change you have been asked to calculate—in this case $\Delta H_f^\circ(CH_3OH(l))$:

$$C(s) + 2H_2(g) + \frac{1}{2}O_2(g) \rightarrow CH_3OH(g)$$

Step 2. Construct an energy cycle with two alternative routes. It help if you always put the enthalpy change you want to calculate along the top of the cycle and call it ΔH_1.

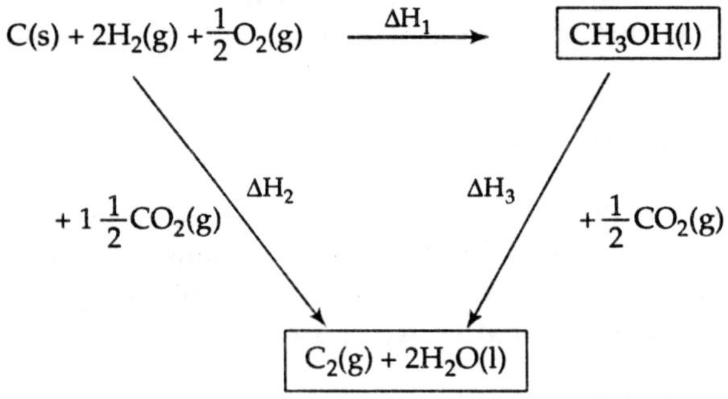

A Hess' law energy cycle to calculate ΔH_f° ($CH_3OH(l)$)

Step 3. Write out the enthalpy changes for the two routes:

$$\Delta H_1 + \Delta H_3 = \Delta H_2$$

Step 4. Decide what enthalpy changes are represented by ΔH_1, ΔH_2 and ΔH_3. It is a good idea to do this on the energy cycle diagram first.

$$\Delta H_1 = \Delta H_f^\circ(CH_3OH(l))$$

that is, the enthalpy change we wish to calculate

$$\Delta H_2 = \Delta H_f^\circ(C(s)) + 2\Delta H_f^\circ(H_2(g))$$

$$\Delta H_3 = \Delta H_f^\circ(CH_3OH(l))$$

Step 5. Rearrange the equation in step 3 and insert the values of the enthalpy changes which are known:

$$\Delta H_1 = \Delta H_2 - \Delta H_3$$

$$\Delta H_f^\circ(CH_3OH(l)) = \Delta H_f^\circ(C(s) + 2\Delta H_f^\circ(H_2(g)) - \Delta H_f^\circ(CH_3OH(l))$$

$$\Delta H_f^\circ(CH_3OH) = -393 + 2(-286) - (-726)$$

$$= -239 \text{ kJ mol}^{-1}$$

Thus the standard enthalpy change of formation of methanol is –239 kJ mol^{-1}

75. **(i)** The molar volume of a gas STP is:

$$\text{no. mol gas} = 1.00 \text{ L} \times \frac{1 \text{ mol gas}}{22.4 \text{ L}} = 0.0446 \text{ mol gas } (n_{tot.})$$

But, the fraction of a gas in a mixture is the same as its volume fraction i.e. $\frac{n_{CO}}{n_{tot.}} = \frac{V_{CO}}{V_{tot.}}$

Thus, if there is 55 L CO for every 100 L of the gaseous mixture, there musty also be 55 mol CO for every 100 mol of the mixture. As a result, the molar amounts of CO and H_2 must be

$$n_{CO} = 0.0446 \text{ mol gas} \times \frac{55 \text{ mol CO}}{100 \text{ mol gas}} = 0.0245 \text{ mol CO}$$

$$n_{H_2} = 0.0446 \text{ mol gas} \times \frac{33 \text{ mol } H_2}{100 \text{ mol gas}} = 0.0147 \text{ mol } H_2$$

(ii) If the equation for the chemical reaction on which an enthalpy calculation is based is not given, writing this equation is always a good starting point. For the combustion of $H_2(g)$.

$$H_2(g) + \frac{1}{2}O_2(g) \rightarrow H_2O(l) \qquad \Delta H° = ?$$

Once you have written this equation, you should immediately recognize that it also represents the formation of $H_2O(l)$ from its elements. $\Delta H°$ for this reaction is the tabulated value of the enthalpy of formation of one mole of $H_2O(l)$: $\Delta H_f°[H_2O(l)] = -285.8 \text{ kJ mol}^{-1} H_2O(l)$. That is, we can write

$$H_2(g) + \frac{1}{2}O_2(g) \rightarrow H_2O(l) \qquad \Delta H° = -285.8 \text{ kJ}.$$

$$\therefore \qquad \text{no. kJ} = 0.0147 \text{ mol } H_2 \times \frac{-285.8 \text{ kJ}}{1 \text{ mol } H_2} = -4.20 \text{ kJ}$$

(iii) Again we can start by writing the equation for the combustion of one mole of $CO(g)$ to form one mole of $CO_2(g)$.

$$CO(g) + \frac{1}{2}O_2(g) \rightarrow CO_2(g) \qquad \Delta H° = ?$$

Thus time, however the enthalpy change is not the enthalpy of formation of $CO_2(g)$, since the $CO_2(g)$ is not being formed from its elements. (O_2 is an element but not CO.) Here we have

$$\Delta H°_{rxn} = \Delta H_f°[CO_2(g) - \Delta H_f°[CO(g) = -393.5 \text{ kJ} - (-110.5 \text{ kJ})$$
$$= -283.0 \text{ kJ}$$

Now we can write

$$CO(g) + \frac{1}{2}O_2(g) \rightarrow CO_2(g) \qquad \Delta H° = -283.0 \text{ kJ}$$

Finally,

$$\text{No. kJ} = 0.0245 \text{ mol CO} \times \frac{-283.0 \text{ kJ}}{1 \text{ mol CO}} = -6.93 \text{ kJ}$$

(iv) In parts (ii) and (iii), we calculated the heats of combustion of the two combustible gases in 1.00 L (STP) of the synthesis gas. Wee need simply to add them together.

The heat of combustion of 1.00 L (STP) = –4.20 kJ – 6.93 kJ = –11.13 kJ of synthesis gas.

(v) We use, in the equation below $q_{water} = +11.13$ kJ.

$$q_{water} = \text{mass water} \times \text{sp. ht. Water} \times \Delta T \qquad (Q = m.s.t)$$

$$11.13 \text{ kJ} = 11{,}130 \text{ J} = 1.00 \times 10^3 \text{ g} \times \frac{4.184 \text{ J}}{\text{g °C}} \times \Delta T$$

$$\Delta T = \frac{11{,}130 \text{ J}}{1.00 \times 10^3 \text{ g} \times 4.184 \text{ J g}^{-1} \text{ °C}^{-1}} = 2.66\text{°C}$$

Final temp. = initial temp. + ΔT = 298 K + 2.66°C $\simeq$ 300.7 K

76. (i) The only *quantitative* relationship that includes a heat of vaporization is the Clausius–Clapeyron equation. If we have two vapor pressure values and their corresponding temperatures, we can solve the equation for ΔH_{vap}. Whenever we are given a normal boiling point we automatically have one value—P = 760 mmHg at the boiling point temperature (386.7 K). A second value is given at 298 K—14.4 mmHg.

$$\ln\frac{P_2}{P_1} = \frac{\Delta H_{vap}}{R}\left(\frac{1}{T_1} - \frac{1}{T_2}\right) \text{ and } \ln\frac{P_2}{P_1} = \frac{\Delta H_{vap}}{R} \times \left(\frac{T_2 - T_1}{T_2 T_1}\right)$$

$$\Delta H_{vap} = \frac{RT_2T_1}{T_2 - T_1} \times \ln(P_2/P_1).$$

$$\Delta H_{vap} = \frac{8.314 \text{ J mol}^{-1} \text{ K}^{-1} \times 386.7 \text{ K} \times 298.2 \text{ K}}{(386.7 - 298.2) \text{ K}} \times \ln(760/14.4)$$

ΔH_{vap} = 1.08 × 10^4 J mol^{-1} × 3.97 = 4.29 × 10^4 J/mol^{-1} = 42.9 kJ mol^{-1}

(ii) ΔH_{subl} is simply the sum of ΔH_{fusion} (given) and ΔH_{vap} (from part (i)).

$\Delta H_{subl} = \Delta H_{vap} + \Delta H_{fusion}$ = 42.9 kJ mol^{-1} + 12.66 kJ/mol^{-1} = 55.56 kJ/mol^{-1}

77. (i) Our first thought should be whether any significant amount of $CH_3OH(g)$ will be present at equilibrium. For this, we need a value of the equilibrium K, or at least, of the standard free energy change, $\Delta G°$.

$$\Delta G° = \Delta G_f°[CH_3OH(g)] - \Delta G_f°[CO(g)]$$
$$= -162.0 \text{ kJ mol}^{-1} - (-137.2 \text{ mol}^{-1}) = -24.8 \text{ kJ mol}^{-1}$$

The *negative* value of $\Delta G°$ signifies that the conversion of $CO(g)$ and $H_2(g)$ in their standard states to $CH_3OH(g)$ in its standard state is indeed a spontaneous reaction. We should expect the forward reaction to proceed to a considerable extend before equilibrium is reached.

(ii) Here we use the equation

$$\Delta G° = -RT \ln K \quad \text{and} \quad K = \frac{-\Delta G°}{RT}$$

$$\therefore \quad \ln K = \frac{24.8 \times 10^3 \text{ J mol}^{-1}}{8.314 \text{ J mol}^{-1} \text{ K}^{-1} \times 298.15 \text{ K}} = 10.0$$

or K = antiln 10.0 = 2.2 × 10^4

(iii) We need to recognize that the thermodynamic equilibrium constant for this reaction is the same as K_p, that is,

$$K = \frac{a_{CH_3OH(g)}}{(a_{CO(g)})(a_{H_2(g)})^2} = \frac{P_{CH_3OH(g)}}{(P_{CO(g)})(P_{H_2(g)})^2} = K_p$$

The question then is one of finding K_p at 500. J, and for this we use equation

$$K = K_p = K_c\ (RT)^{\Delta n_{gas}} = 14.5 \times (0.8206 \times 500.)^{-2} = 8.61 \times 10^{-3}$$

Where, number of moles of gas, $\Delta n_{gas} = -2$.

(iv) The quantities in question here are related through the van't Hoff equation. The unknown is $\Delta H°$.

Here,

$$K_2 = 8.61 \times 10^{-3} \qquad T_2 = 500 \text{ K}$$
$$K_1 = 2.2 \times 10^4 \qquad T_1 = 298.15 \text{ K} \qquad \Delta H° = ?$$

$$\ln \frac{8.61 \times 10^{-3}}{2.2 \times 10^4} = \frac{\Delta H°}{8.314 \text{ J mol}^{-1} \text{ K}^{-1}} \left(\frac{1}{298.15 \text{ K}} - \frac{1}{500. \text{ K}} \right) = -14.8$$

$$-14.8 = \frac{\Delta H°}{8.314 \text{ J mol}^{-1} \text{ K}^{-1}} (3.35 \times 10^{-3} - 2.00 \times 10^{-3}) \text{ K}^{-1}$$

$$\Delta H° = \frac{-14.8 \times 9.314}{1.35 \times 10^{-3}} \text{ J mol}^{-1} = -9.11 \times 10^4 \text{ J mol}^{-1} = -91.1 \text{ kJ mol}^{-1}$$

78.

Cycle	A *to* B	B *to* C	C *to* A
I	Isochoric	Isobaric	Isothermal
II	Isobaric	Isochoric	Isothermal
III	Isothermal	Isobaric	Isochoric

79. $P_2 = P_1 \left(\frac{V_1}{V_2} \right)^{\gamma} = 1 \left(\frac{4.8}{10} \right)^{1.41}$

$= 8.30$ atm

If it is adiabatic expansion, final temperature, T_2

$$T_2 = \frac{P_2 V_2}{nR} = \frac{8.3 \times 10}{2 \times 0.0821} = 505 \text{ K}$$

80. (i) Isothermal irreversible work in given by

$$dW = P_{ext}\ (V_2 - V_1)$$
$$P_{ext} = 1 \text{ atm} = 1.013 \times 10^5 \text{ N m}^{-2}$$
$$dW = 1.013 \times 10^5\ (24.6 - 2.46) \times 10^{-3}$$
$$= 22.4278 \times 10^2 \text{ Joules mole}^{-1} = 22\ 42.9 \text{ Joules mol}^{-1}$$

(ii) The reversible work done by one mole of gas when it expands from 2.46×10^{-3} to 24.6×10^{-3} m^3

$$= 2.303\ RT \log \frac{V_2}{V_1} = 2.303 \times 8.314 \times 300 \log 10$$

$$= 5744 \text{ J mol}^{-1}$$

81. The internal energy of an ideal gas depends only on temperature therefore.

$$\Delta E = 0$$

Now,

$$q = -w$$

$$w = -nRT \ln\frac{V_2}{V_1} = -1 \times 8.314 \times 300 \text{ K} \ln\frac{4.56}{2.28}$$

$$= -1728.85 \text{ J} = -1.728 \text{ kJ}$$

$\therefore$ $$q = 1728.85 \text{ J} = +1.728 \text{ kJ}$$

$\therefore$ $$W = -1.728 \text{ kJ};\ q = 1.728 \text{ kJ}$$

$\therefore$ $$\Delta H = \Delta E + \Delta(PV) = \Delta E + \Delta(RT) = 0 + 0 = 0$$

($\because$ T temperature is constant)

82. Work done in isothermal reversible compression is given by.

$$W = nRT \times 2.303 \log\frac{P_2}{P_1}$$

$$= \frac{16}{32} \times 8.314 \times 300 \text{ K} \times 2.303 \log\frac{1.01325 \times 10^5}{1.01325 \times 10^3}$$

$$= 0.5 \times 8.314 \times 300 \times 2.303 \log 10^2$$

$$= 0.5 \times 8.314 \times 300 \times 2.303 \times 2 \log 10$$

$$= 5744.14 \text{ Joules.}$$

or $$= 5.744 \text{ k Joules.}$$

83. Work done in isothermal expansion is given by

$$W = nRT\ 2.303 \log\frac{V_2}{V_1}$$

$$= 2 \times 8.314 \times 298 \times 2.303 \log\frac{20}{10}$$

$$= 3435.26 \text{ Joules}$$

$$= 3.435 \text{ kJ.}$$

84. The production of gastric juice from body fluids is a slow process and hence the change may be considered a reversible one. The work done for the process can be calculated by using equation.

$W = -2.303\ nRT \log (C_1/C_2)$. From this case

C_1 = H^+ ions concentration in body fluid = $[H^+]_1$ = 10^{-pH} = 3.98×10^{-8} mol dm^{-3}

C_2 = H^+ ions concentration is the gastric juice produced = $[H^+]_2$ = 10^{-pH} = 10^{-1} = 0.10 mol dm^{-3}

n = Mol of H^+ ions present in 0.1 M of 3 dm^3 gastric juice = 0.3 mol

T = Body temperature = 310 K

$$w = -2.303 \times 0.3 \times 8.314 \text{ JK}^{-1} \text{ mol}^{-1} \times 310 \text{ K} \log\frac{3.98 \times 10^{-8}}{0.1}$$

$$= 11400 \text{ J} = 11.4 \text{ kJ.}$$

The positive sign with work done implies that the work is done on the system. This is obvious because te change is from a low concentration (pH = 7) to high concentration (pH = 1)

85. The maximum work done in a reversible manner

(i) When a gas expands isothermally and reversibly

$$W_{max} = nRT \ln \frac{V_2}{V_1}$$

$$n = 1, T = 300 \text{ K}, V_2 = 20 \text{ dm}^3, V_1 = 10 \text{ dm}^3$$

$$W_{max} = -(1 \text{ mol}) \times (8.314 \text{ JK}^{-1} \text{ mol}^{-1}) \times (300 \text{ K})$$

$$\times 2.303 \log\left(\frac{20}{10}\right)$$

$$= -1.729 \text{ kJ}$$

(ii) In adiabatic expansion gas

$$w = nC_V (T_2 - T_1)$$

$$C_V = \frac{5}{2}R; \text{ Cp}, m = \frac{7}{2}R, T_1 = 300 \text{ K}$$

$$(T_1V_1)^{\gamma-1} = T_2V_2^{\gamma-1}$$

$$\left(\frac{T_2}{T_1}\right) = \left(\frac{V_1}{V_2}\right)^{\gamma-1}$$

$$T_2 = \left(\frac{10}{20}\right)^{7/5-1} \times 300 \qquad \left(\because \gamma = \frac{C_p}{C_v} = \frac{7}{5}\right)$$

$$= 227.4 \text{ K}$$

$$w = (1 \text{ mol}) \times \left(\frac{5}{3} \times 8.314 \text{ JK}^{-1} \text{ mol}^{-1}\right) \times (2227.4 - 300) \text{ K}$$

$$= -1.509 \text{ kJ}$$

86. Let 1 – 2 represent the adiabatic process. During any process

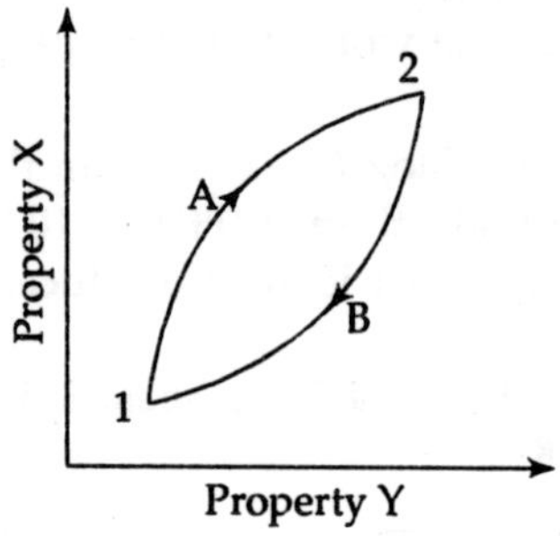

$$Q_{1-2} = W_{1-2} + U_2 - U_1$$

In case of adiabatic process,

$$Q_{1-2} = 0$$

Hence $\quad W_{1-2} = (U_2 - U_1) = 0$

But $\quad W_{1-2} = -5000$ J as work is done on the system

Hence $\quad -5000 + (U_2 - U_1) = 0$

or $\quad U_1 - U_2 = -500$ J

Let the system return from 2 to 1 by a non adiabtatic process represented by B. During this process.

$$Q_{2-1} = W_{2-1} + U_1 - U_2$$

In this $\quad Q_{2-1} = +1000$ since heat is added to the system

$$U_1 - U_2 = -5000 \text{ J}$$

Hence $\quad 1000 = W_{2-1} - 5000$

or $\quad W_{2-1} = 1000 + 5000$

$$= 6000 \text{ J}$$

87. In case of irreversible process,

$$ds > \frac{dQ}{T}$$

or $\quad \int T\, ds > Q$

88. Air is to be considered as an ideal gas with ratio of specific heats $\gamma = 1.4$. Changes in kinetic and potential energies are to be neglected. Also there is no heat transfer between the air in the main and the air in the vessel. Consider first the control volume in Fig. below.

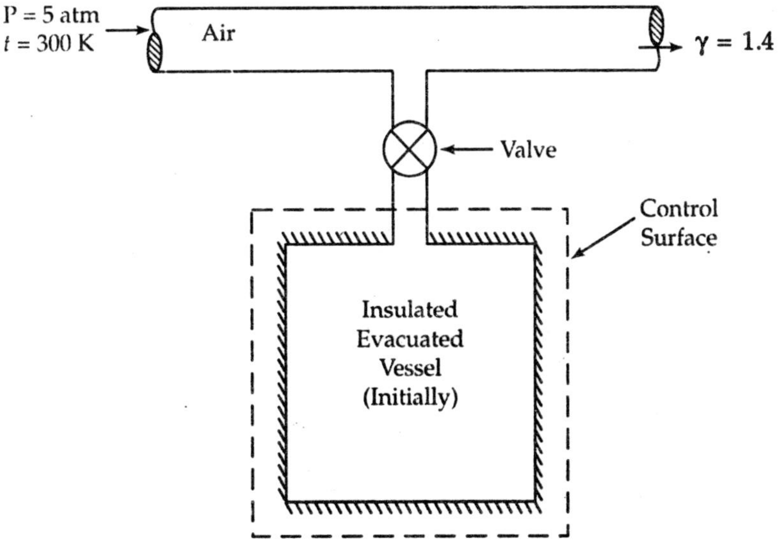

or $\quad C_v(T_2 - T_1) = RT_1$

$\therefore \quad (T_2 - T_1) = \dfrac{RT_1}{C_v} = (\gamma - 1)T_1$

$$= (1.4 - 1) \times 300 = 120$$

Thus the temperature rises by 120 K.

89. In a reversible isothermal process undergone by a gas

$$\int_1^2 pdV = p_1V_1 \ln\frac{V_2}{V_1}$$

$$Q = p_1V_2 \ln\frac{V_2}{V_1}$$

$$U_2 - U_1 = p_1V_1 \ln\frac{V_2}{V_1}$$

Hence in a reversible isothermal process, work transfer is equal to heat transfer.

90. During isobaric heating of an ideal gas

$$Q = C_p(T_2 - T_1)$$
$$W = p(V_2 - V_1).$$

91. In case of isothermal process $p_1V_1 = p_2V_2$ and change ion entropy is given

$$R \ln\frac{V_2}{V_1} = R \ln\frac{P_1}{P_2}$$

92. Availability $= \Delta Q - T_0\, \Delta s$

$$= 1000 - 300 \times \left(\frac{1000}{600}\right)$$

$$= 500 \text{ kJ}$$

93. The sum of internal energy and pressure volume product (i.e. $u + pv$) is known as enthalpy.

94. We have the relations for an ideal gas a given below:

$$\left(\frac{\partial u}{\partial v}\right)_T = 0 \quad \text{and} \quad \left(\frac{\partial u}{\partial p}\right)_T = 0$$

Also $$\left(\frac{\partial u}{\partial T}\right)_p = C_p \quad \text{and} \quad \left(\frac{\partial u}{\partial T}\right)_v = C_v$$

From the above relations it is evident that internal energy is a function of temperature only.

95. As the gas is compressed suddenly it will undergo adiabatic compression. Hence

$$\left(\frac{T_1}{T_2}\right)^{\gamma} = \left(\frac{p_2}{p_1}\right)^{1-\gamma}$$

Here, $$T_1 = 273° \text{ k}$$
$$P_2 = 20\, p_1$$

$$\therefore \quad \left(\frac{273}{T_2}\right)^{1.42} = \left(\frac{20\, p_1}{p_1}\right)^{1-1.42}$$

$$= (20)^{-0.42}$$

$$\therefore \quad T_2 = 662.2 \text{ k}$$

96. The integrated form of Clapeyron—Clausius equation is

$$2.303 \log \frac{P_2}{P_1} = \frac{\Delta H_v}{R} \left(\frac{T_2 - T_1}{T_1 T_2} \right)$$

Let T_2 be the boiling point at 750 mm pressure,

$$\therefore \qquad 2.303 \log_c \frac{750}{760} = \frac{369.86 \times 74}{8.314}$$

$$\left(\frac{T_2 - 306.5}{306.5 \times T_2} \right)$$

Solving $\qquad T_2 = 306.0$ K.

97. Total heat supplied to the engine

= heat rejected + work done

= 142 + 98

= 240 K Joule min

Let Q_1 kJ min^{-1} to be the heat taken the engine from the reservoir at 600K.

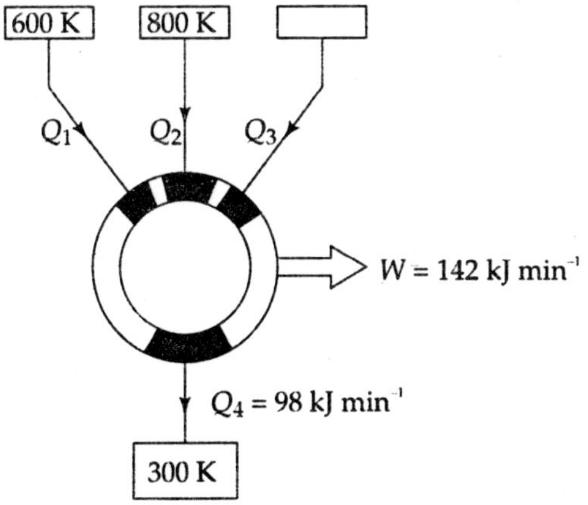

Then heat taken from the reservoir at 1000°K = $Q_3 = 0.6\ Q_1$ kJ/min

Also heat taken from reservoir at

$$800\text{K} = Q_s = 240 - Q_1 - 0.6\ Q_1$$

$$= (240 - 1.6\ Q_1) \text{ kJ min}^{-1}$$

As the engine is reversible, hence the change in entropy must be zero. Hence

$$\frac{Q_1}{600} + \frac{Q_2}{800} + \frac{Q_3}{1000} - \frac{Q_4}{300} = 0$$

or

$$\frac{Q_1}{600} + \frac{200 - 1.6\,Q_1}{800} + \frac{0.6\,Q_2}{1000} - \frac{98}{300} = 0$$

Solving, $\qquad Q_1 = 100$ kJ min^{-1}, (for reservoir at 600K)

Also $Q_2 = 240 - 1.6 \times 100$
$= 80$ kJ min^{-1} (for reservoir at 800K)

and $Q_3 = 0.6\ Q_1$
$= 60$ kJ min^{-1} (for reservoir at 1000K).

98. $\boxed{\Delta S = n\ C_p \ln (T_2/T_1) + nR \ln (P_1/P_2)}$

For an ideal gas,

$$C_P - C_V = R \Rightarrow C_P = C_V + R = 5/2R = \frac{5}{2} \times 8.314 = 20.785^{-1}\ \text{mol}^{-1}$$

$$\therefore\ \Delta S = 5 \times (20.785\ \text{JK}^{-1}\ \text{mol}^{-1}) \ln\left(\frac{298\ \text{K}}{323\ \text{K}}\right) + 5 \times (8.134\ \text{JK}^{-1}\ \text{mol}^{-1}) \ln\left(\frac{760\ \text{mm Hg}}{380\ \text{mm Hg}}\right)$$

or $\Delta S = 20.44$ JK^{-1}

99. $\Delta S = n\ R \ln (V_2/V_1)$

$$= (1\ \text{mol})\ (8.314\ \text{J K}^{-1}\ \text{mol}^{-1}) \ln\left(\frac{10}{1}\right) = 19.14\ \text{J K}^{-1}$$

$\Rightarrow$ $\Delta S = 19.14$ JK^{-1}.

100. The maximum efficiency of an engine working between temperature T_2 and T_2 is given by

$$\boxed{\eta = \frac{T_2 - T_1}{T_2}}$$

$$\therefore \qquad \eta = \frac{373 - 293}{373} = 0.215 = 21.5\%$$

101. Given, $\eta = 80\% = 0.8$

$T_1 = 350$ K

As $$\eta = \frac{T_2 - T_1}{T_2} = 1 - \frac{T_1}{T_2}$$

$$\Rightarrow \qquad \frac{T_1}{T_2} = 1 - \eta \quad \Rightarrow \quad \boxed{T_2 = \frac{T_1}{1-\eta}}$$

$$\Rightarrow \qquad T_2 = \frac{350}{1-0.8} = \frac{350}{0.2} = 1750\ \text{K}.$$

102. In this case, $W = -P(V_2 - V_1)$

$\Rightarrow$ $W = -(5\ \text{atm})\ (5\ \text{dm}^3 - 1\ \text{dm}^3) = -20\ \text{dm}^3$ atm

$$\Rightarrow \qquad W = -20\ \text{dm}^3\ \text{atm} \times \frac{8.314\ \text{JK}^{-1}\ \text{mol}^{-1}}{0.08206\ \text{dm}^3\ \text{atm K}^{-1}\ \text{mol}^{-1}} = -2026\ \text{J}$$

103. The combustion of ethane can be represented as

$$C_2H_6(g) + \frac{7}{2}O_2(g) \rightarrow 2\ CO_2(g) + 3H_2O(l)$$

The enthalpy of combustion for ethane is thus given by

$$\Delta H^\circ = [(2\ \Delta H^\circ_{CO_2(g)} + 3\ \Delta H^\circ_{H_2O(l)}) - \Delta H^\circ_{C_2H_6(g)}$$
$$= [(2 \times -393.5 + 3 \times -285.8) - (-84.6)]\ kJ\ mol^{-1}$$
$$= -787 - 857.4 + 84.6 = -1559.8\ kJ\ mol^{-1}$$

104. We want to calculate the enthalpy of neutralization of acetic acid according to the following equation:

$CH_3COOH\ (aq) + OH^-\ (aq) \rightarrow CH_3COO^-(aq) + H_2O(l)$

Let, $\Delta H^\circ = x$ kJ mol^{-1}

Adding reactions (i) and (ii), we get

$$\Delta H^\circ = -57.32 + 1.89$$

$\Rightarrow$ $x = -55.43$ kJ mol^{-1} = Enthalpy of neutralization of Acetic acid.

105. Given $C(s) + 2H_2(g) \rightarrow CH_4(g)$ $\Delta H = -74.85$ kJ^{-1} mol^{-1}

$T = 298$ K,

now $\Delta ng = 1 - 2 = -1$

Moreover $R = 8.314$ J k^{-1} mol^{-1}

As $\boxed{\Delta H = \Delta E - \Delta n_g RT}$

$\therefore$
$$\Delta E = \Delta H - \Delta n_g RT$$
$$= (-74850\ J^{-1}\ mol^{-1}) = -\{(-1) \times (8.314\ J\ k^{-1}\ mol^{-1}) \times (298\ K)\}$$
$$= -72{,}372\ J^{-1}\ mol^{-1} = -72.37\ kJ\ mol^{-1}.$$

106. The C—H bond energy in methane is defined as one fourth of the energy required to break the four C—H bonds

$$CH_4(g) \rightarrow C(g) + 4H(g)$$

To determine the enthalpy change of this reaction the data are combined in the following manner:

$C(s) \rightarrow C(g)$	$\Delta H_{sub} = 717.7$ kJ mol^{-1}
$2H_2(g) \rightarrow 4H(g)$	$2\Delta(H_2) = 2(435.5)$ kJ mol^{-1}
$CH_4(g) \rightarrow C(s) + 2H(g)$	$-\Delta H_f = 74.8$ kJ mol^{-1}
$CH_4(g) \rightarrow C(g) + 4H(g)$	$\Delta H = 1663.5$ kJ mol^{-1}

The C—H bond energy is one-fourth of the energy required to break all four bonds, ≈ 416 kJ mol^{-1}

107. We have, $N_2 + 3H_2 \rightarrow 2NH_3$

Note that there are three N—H bonds in each NH3 molecule,

$\therefore$ $3H(g) + N(g) \rightarrow NH_3(g)$	$-3\Delta H_f(H—N) = -3(389) = -1167$ kJ mol^{-1}
$\frac{1}{2}N_2(g) \rightarrow N(g)$	$\frac{1}{2}\Delta H_f(N_2) = \frac{1}{2}(946) = +473$ kJ mol^{-1}
$\frac{1}{2}H_2(g) \rightarrow 3H(g)$	$\frac{1}{2}\Delta H_f(H_2) = \frac{3}{2}(436) = +654$ kJ mol^{-1}
$\frac{1}{2}N_2(g) + \frac{1}{2}H_2(g) \rightarrow NH_3(g)$	$\Delta H_f = -40$ kJ mol^{-1}.

108. The calculation involves converting grams of NH_3 to moles of NH_3 and then to kilojoules of heat.

Grams of NH_3 → moles of NH_3 → kilojoules of heat

From the given equation the production of 2 mol NH_3 is accompanied by q_p = –91.8 kJ. 9.07 kg = 9.07 × 105 g Hence,

$$9.07 \times 10^3 \text{ g NH}_3 \times \frac{1 \text{ mol NH}_3}{17.0 \text{ g H}_3} \times \frac{-91.8 \text{ kJ}}{2 \text{ mol NH}_3} = -2.45 \times 10^6 \text{ kJ}$$

Thus, 2.45 × 105 kJ of heat evolves.

109. Calculate ΔG_f° from ΔG_f° values, similar to the way you calculate ΔH° from ΔH_f°. Write the balanced equation with values of ΔG_f° multiplies by stoiochiometric coefficients below each formula.

$$C_2H_5OH(l) + 3O_2(g) \rightarrow 2CO_2(g) + 3H_2O(g)$$

ΔG_f°: –174.8 0 2(–394.4) 3(–228.6) kJ

$$\therefore \quad \Delta G^\circ = \Sigma n\, \Delta G_f^\circ(\text{products}) - \Sigma m \Delta G_f^\circ(\text{reactants})$$
$$= [2(-394.4) + 3(-228.6)] - [(-174.8)] \text{ kJ}$$
$$= -1299.8 \text{ kJ}$$

110. Calculate ΔH° and ΔG°; then interpret the signs of these quantities. Give the interpretation of the sign of ΔH. As long as ΔG° is not close to zero, you can interpret its sign as follows: a negative ΔG° means the reactants tend to go mostly to products; a positive ΔG° means the reactants tend to go mostly to products; a positive ΔG° means that the reaction is nonspontaneous as written. When the value of ΔG° is close to zero, the reaction given an equilibrium mixture.

Now, The problem is set up as follows:

	$2KClO_3(s) \rightarrow$	$2KCl(s)$	$+ 3O_2(g)$
ΔH_f°:	2 × (–391.2)	2 × (–436.7)	0 kJ
ΔG_f°:	2 × (–289.9)	2 × (–408.8)	0 kJ

Thus,

$$\Delta H^\circ = [2 \times (-436.7) - 2 \times (-391.2)] \text{ kJ} = -91.0 \text{ kJ}$$
$$\Delta G^\circ = [2 \times (-408.8) - 2 \times (-289.9)] \text{ kJ} = -237.8 \text{ kJ}$$

Hence,

The reaction is exothermic, liberating 91.0 kJ of heat. The large negative value for ΔG° indicates that the equilibrium composition is mostly potassium chloride and oxygen.

111. The heat released by the reaction is absorbed by the calorimeter and its contents. Let q_{rnx} be the quantity of heat from the reaction mixture, and let C_{cal} be the capacity of the calorimeter and contents. The quantity of heat absorbed by the calorimeter is $C_{cal}\Delta t$. This will have the same magnitude as q_{rxn} but the opposite sign: $q_{rxn} = -C_{cal}\Delta t$. Substituting values of C_{cal} and Δt, you obtain the heat q_{rxn} from the sample of C(0.562 g). The factor (q_{rxn}/0.562 g C) converts.

grams C → kJ heat

The ΔH of reaction is the heat obtained from 1 mol C.

The heat from the graphite samples is

$$q_{rxn} = -C_{cal}\Delta t = -20.7 \text{ kJ K}^{-1} \times (298.89 - 298) \text{ K}^{-1}$$
$$= -20.7 \text{ kJ} \times 0.89 = -18.4 \text{ kJ}$$

The negative sign indicates the reaction is exothermic as expected for a combustion. The factor to convert grams C to kJ heat is –18.4 kJ/0.562 gC
The conversion of 1 mol C to kJ heat for 1 mol (ΔH) is

$$1 \text{ mol C} \times \frac{12 \text{ g C}}{1 \text{ mol C}} \times \frac{-18.4 \text{ kJ}}{0.562 \text{ g C}} = -3.9 \times 10^2 \text{ kJ}$$

Thus, $C(\text{graphite}) + O_2(g) \rightarrow CO_2(g)$; $\Delta H = -3.9 \times 10^2$ kJ

112. Here we have to multiply Equations 1, 2, and 3 by factors so that when we add the three equations we obtain the desired equation for the formation of WC(*s*). To obtain these factors, compare Equations 1, 2, and 3 in turn with the desired equation. For instance, note that Equation 1 has 2W(*s*) on the left side, whereas the desired equation has W(*s*). Therefore we multiply Equation 1 (and its ΔH) by $\frac{1}{2}$.

Multiplying Equation 1 by $\frac{1}{2}$ we get,

$$W(s) + \frac{3}{4}O_2(g) \rightarrow WO_3(s); \frac{5}{2}O_2(g)\Delta H = \frac{1}{2} \times (-1680.6 \text{ kJ}) = -840.3 \text{ kJ}$$

Compare Equation 2 with the desired equation. Both have C(graphite) on the left side; therefore, leave Equation 2 as it is. Now, compare Equation 3 with the desired equation. Equation 3 has 2WC(*s*) on the left side, whereas the desired equation has WC(*s*) on the right side. Hence, you reverse Equation 3 and multiply it (and its ΔH) by $\frac{1}{2}$.

$$WO_3(s) + CO_2(g) \rightarrow WC(s) + \frac{5}{2}O_2(g); \Delta H = -\frac{1}{2} \times (-2391.6 \text{ kJ}) = 1195.8.\text{kJ}$$

Note that the ΔH is obtained by multiplying the values for Equation 3 by $-\frac{1}{2}$. Now these three equations and the corresponding ΔH's are added together.

$$W(s) + \frac{3}{2}O_2(g) \rightarrow WO_3(s) \qquad \Delta H = -840.3 \text{ kJ}$$
$$C(\text{graphite}) + O_2(g) \rightarrow C_2(g) \qquad \Delta H = -393.5 \text{ kJ}$$
$$WO_3(s) + CO_2(g) \rightarrow WC(s) + \frac{5}{2}O_2(g) \qquad \Delta H = 1195.8 \text{ kJ}$$
$$W(s) + C(\text{graphite}) \rightarrow WC(s) \qquad \Delta H = -38.0 \text{ kJ}$$

113. Write the values of ΔH_f° under the formulas in the equation, multiplying them by the coefficients in the equation, calculate ΔH° by subtracting values for the reactants from values for the products.

Thus, $4NH_3(g) + 5O_2(g) \rightarrow 4NO(g) + 6H_2O(g)$

4(–45.9) 5(0) 4(90.3) 6(–241.8) (kJ)

(ignore the catalyst because its only effects is to speed up the reaction)
Hence,

$$\Delta H^\circ = \Sigma n\ \Delta H_f^\circ \text{ (products} - \Sigma H_f^\circ \text{ (reactants)}$$
$$= [4\ \Delta H_f^\circ(NO) + 6\ \Delta H_f^\circ(H_2O)] - [4\ \Delta H_f^\circ(NH_3) + 5\ \Delta H_f^\circ(O_2)]$$
$$= [4(90.3) + 6(-241.8)]\ kJ - [4 - 45.9) + 5(0)]\ kJ$$
$$= -960\ J$$

114. When liquid CCl_4 evaporations, it absorbs heat: ΔH_{vap} = 430 kJ/mol (43.0 × 103 J/mol) at 298 K. the entropy change, ΔS, is

$$\Delta S = \frac{\Delta H_{vap}}{T} = \frac{43.0 \times 10^3\ J\,mol^{-1}}{298\ K} = 144\ J\ mol^{-1} \cdot K^{-1}$$

In other words, 1 mol of carbon tetrachloride increases in entropy by 144 JK^{-1} when it vaporizes. The entropy of 1 mol of the vapor equals the entropy of 1 mol of liquid (214 JK^{-1}) plus 144 JK^{-1}.

$$\text{Entropy vapor} = (214 + 144) J\ mol^{-1} \cdot K^{-1} = 358\ J\ mol^{-1} \cdot K^{-1}$$

115. (i) A molecule (glucose) breaks into smaller molecules (C_2H_5OH and CO_2). Moreover, this results in a gas being released. You predict that ΔS° for this reaction is positive. That is, the entropy increases.

(ii) In this reaction, the moles of gas decrease (by 3 mol), which would decrease the entropy. You predict that the entropy should decrease. That is, ΔS° is negative

(iii) Because there is no change in the number of moles of gas, you cannot predict the sign of ΔS° from the rules given.

116. It is convenient to put the standard entropy values multiplied by stoichiometric coefficients below the formulas in the balance equation.

$2NH_3(g) + CO_2(g) \rightarrow NH_2CONH_2(aq) + H_2O(l)$

S°: 2 × 193 214 174 70

Here we, calculate the entropy change by subtracting the entropy of the reactants from the entropy of the products.

$$\Delta S^\circ = \Sigma nS^\circ(\text{products}) - \Sigma mS^\circ(\text{reactants}) = [(174 + 70) - (2 \times 193 + 214)]\ JK^{-1} = -356\ JK^{-1}$$

117. Calculate ΔH° and ΔS°, then substitute values into $\Delta H^\circ - T\Delta S^\circ$ to obtain ΔG°.
Write the balanced equation and place below each formula the values of ΔH_f° and S° multiplies by stoichiometric coefficients.

	$N_2(g)$ +	$3H_2(g)$	$\rightarrow$	$2NH_3(g)$
ΔH_f°:	0	0		2 × (–45.9) kJ
S°:	191.5	3 × 130.6		2 × 193 JK^{-1}

You calculate ΔH° and ΔS° by taking values for products and subtracting values for reactants.

$$\Delta H^\circ = \Sigma n\Delta H_f^\circ(\text{products}) - \Sigma m\Delta H_f^\circ(\text{reactants}) = [2 \times (-45.9) - 0]\ kJ = -91.8\ kJ$$
$$\Delta S^\circ = \Sigma nS^\circ(\text{products}) - \Sigma mS^\circ\ (\text{reactants}) = [2 \times 193 - (191.5 + 3 \times 130.6)]\ JK^{-1} = -197\ JK^{-1}$$

You have substitute into the equation for ΔG° in terms of ΔH° and ΔS°. Note that you substitute ΔS° in units of kJ K^{-1}.

$$\Delta G^\circ = \Delta H^\circ - T\Delta S^\circ = -91.8\ kJ - (298\ K)(-0.197\ kJ\ K^{-1}) = -33.1\ kJ$$

Chapter 7 : CHEMICAL KINETICS

1. Begin this problem by writing a balanced expiation representing the reaction.

$$CH_3CHO(g) \rightarrow CH_4(g) + CO(g)$$

Next, write a rate expression,

$$\text{rate} = k(\text{conc. } CH_3CHO)^2$$

Because you know the rate and the concentration of CH_3CHO, solve for k, the rate-specific constant.

$$k = \frac{\text{rate}}{(\text{conc. } CH_3CHO)^2} \rightarrow \frac{0.01 \text{ mol liter}^{-1} \cdot \text{sec}^{-1}}{(0.01 \text{ mol liter}^{-1})} = 1.0 \times 10^3 \text{ liters mol}^{-1} \cdot \text{sec}^{-1}$$

Finally, substitute the rate-specific constant and the new concentration into the rate expression.

$$\text{rate} = \frac{1.0 \times 10^3 \text{ liter}}{1 \text{ mole} \cdot \text{sec}} \times \left(\frac{0.50 \text{ mol}}{1 \text{ liter}}\right)^2 = 2.5 \times 10^2 \text{ moles liter}^{-1} \cdot \text{sec}^{-1}$$

2. Let X be what needs to be done to [B]

$$\frac{\text{rate}_{new}}{\text{rate}_{old}} = \frac{1}{2} = \frac{k([A]/2)^2 \cdot X[B]}{k[A]^2 \cdot [B]} = \frac{X}{4}$$

$$X = 2$$

3. With all other factors held constant, decreasing the number of molecules decreases the chance of collision. Adding an accelerating catalyst has no effect on the *rate* of collisions. It lowers the activation energy, thereby increasing the chance for *effective* molecular collisions. Furthermore, a increases the rate of production.

4. The concentration of the reactant is increased by a factor of 4, from 0.10 M to 0-.40 M. If the reaction is second order, the rate will then increase by a factor of $4^2 = 16$.

$$\frac{16}{1} \times \frac{0.50 \text{ mole}}{1 \text{ liter} \cdot \text{sec}} = \frac{8.0 \text{ moles}}{1 \text{ liter} \cdot \text{sec}}$$

5. Given a reaction mechanism, the order with respect to each reactant is its coefficient in the chemical equation for that stop. The slowest step is the rate-determining step, so

rate = $k[X]^3[Y]^2$

$$k = \frac{\text{rate}}{[X]^2\,[Y]^2} = \frac{0.45 \text{ mole liter}^{-1} \cdot \text{min}^{-1}}{\left(\frac{3.0 \text{ moles}}{5.0 \text{ liters}}\right)^3 \left(\frac{2.0 \text{ moles}}{5.0 \text{ liters}}\right)^2}$$

6. Examine experiments 2 and 3, wherein [X] is held constant. Note that as [Y] doubles (from 1.5 to 3.0), the rate does not change. Hence, the rate is independent of [Y] and the order is 0 for Y.

Now examine experiments 1 and 3, wherein [Y] is held constant. Note that as [X] doubles, the rate is increased by a factor of 4. In this case, the rate is proportional to the square of the concentration of the reactant. This is a second order reactant.

Combining these reactant orders in a rate equation gives

$$\text{rate} = k[X]^2[Y]^0 = k[X]^2$$

7. The activation energy to the amount of energy that the reactant must absorb from the system in order to react. In the reaction diagram, the reactant begin at *A*. The reactants must absorb the energy from *A* to *B* in order to form the activated complex. The energy necessary to achieve this activated complex is the distance from *A* to *B* n the diagram and is mathematically the difference (*B* –*A*).

8. The enthalpy of the reaction ΔH, is the difference between the enthalpies of the products and the enthalpies of the reactants.

$$\Delta H = H_{\text{products}} - H_{\text{reactants}}$$

The products are represented at point *C* and the reactants are represented at point *A*, so the change in enthalpy is *C* – *A*.

9. The rate is determined by the slowest step. Only when given the complete mechanisms *in steps* can you identify the slowest step and use the coefficients of the reactants as the orders in the rate equation.

10. To find the overall equation when given the reaction mechanism, simply add the given reactions of problem 9 i.e.

$$2\,A + 3\,C \rightarrow 4\,D + 5\,E \text{ slow}$$

$$A + 2\,D \rightleftharpoons 2A + 2\,E \text{ fast}$$

$$2\,E + C \rightarrow 3\,D + 2\,C \text{ fast}$$

There is a total of 3 A's on the left and 2 A's on the right: balance of 1 *A* on the left. There is a total of 4 C's on the left and 2 C's on the right; balance of 2 C's on the left. There is a total of 2 D's on the left and 7 D's on the right: balance of 5 D's on the right. There is a total of 2 E's on the left and 7 E's on the right: balance of 5 E's on the right.

$A + 2\,C \rightarrow 5\,D + 5\,E$

11. At $t = t_{1/2}$ $[R]_t = [R]_0/2$
and at $t_{99.9}$, $[R]_t = 0.001\,[R]_0$

$$\therefore \quad t_{99.9} = \frac{2.303}{k}\log\left(\frac{[R]_0}{0.001[R]_0}\right) = \frac{2.303 \times 3}{k} \qquad \text{(i)}$$

and

$$t_{1/2} = \frac{2.303}{k}\log\left(\frac{[R]}{[R]_0/2}\right) = \frac{0.693}{k} \qquad \text{(ii)}$$

Now

$$\frac{t_{99.9}}{t_{1/2}} = \frac{2.303 \times 3}{k} \times \frac{k}{0.693} = 9.97 \simeq 10$$

12. Given $t_{1/2} = 50$ minutes

$$\therefore \quad k = \frac{0.693}{t_{1/2}} = \frac{0.693}{50} = 1.386 \times 10^{-2} \text{ minute}^{-1}$$

Since

$$k = \frac{2.303}{t}\log\left(\frac{[R]_0}{[R]_t}\right)$$

$$\therefore \qquad \log([R]_t/[R]_0) = \frac{-kt}{2.303} = \frac{-1.386 \times 10^{-2} \times 100}{2.303}$$

$$\Rightarrow \qquad [R]_t/[R]_0 = 0.25$$

13. Given that V is the number of mL of $KMnO_4$ required to decompose a definite volume of H_2O_2 solution, hence it will corresponds to the undercomposed H_2O_2.
The volume of $KMnO_4$, at zero time corresponds to the initial concentration $[R]_0$ and the volume used after time t, corresponds to $[R]_t$ at that time. Putting these values, in rate expression for first order reaction

$$k = \frac{2.303}{t}\log\frac{[R]_o}{[R]_t}, \text{ we get}$$

When t = 10 min., $k = \frac{2.303}{10} \log(25/20) = 2.23 \times 10^{-2}\ \text{min}^{-1}$

When t = 20 min., $k = \frac{2.303}{20} \log(25/15.7) = 2.32 \times 10^{-2}\ \text{min}^{-1}$

When t = 30 min., $k = \frac{2.303}{30} \log(25/12.5) = 2.25 \times 10^{-2}\ \text{min}^{-1}$

When t = 40 min., $k = \frac{2.303}{40} \log(25/9.6) = 2.39 \times 10^{-2}\ \text{min}^{-1}$

The constancy of k show that the decomposition of H_2O_2 in aqueous solution is a first order reaction.

14. For first order gas equation,

$$k = \frac{2.303}{t}\log\frac{[R]_0}{[R]_t}$$

$$\therefore \qquad \frac{[R]_0}{[R]_t} = \text{antilog}\left(\frac{kt}{2.303}\right)$$

$$= \text{antilog}\left[(2.2 \times 10^{-5}\ \text{sec}^{-1})\frac{(90 \times 60\ \text{sec})}{2.303}\right] = 1.126$$

Now, the % of SO_2Cl_2 decomposed $= \left\{\frac{[R]_0 - [R]_t}{[R]_0}\right\} \times 100 = \left\{1 - \frac{[R]_t}{[R]_0}\right\} \times 100$

$$= 11.19 = \mathbf{11.2\%}$$

15. (i) Given in t = 50 min, reaction is 40% complete, so$[R]_t$ = 60% of $[R]_0 = 0.6\ [R]_0$

Now for a first order reaction, rate constant $(k) = \frac{2.303}{t}\log\frac{[R]_0}{[R]_t}$

$$\Rightarrow \qquad k = \frac{2.303}{t}\log\left\{\frac{[R]_0}{0.6\,[R]_0}\right\} = \mathbf{0.0102\ min^{-1}}.$$

(ii) When the reaction is 80% complete, $[R]_t$ = 20% of $[R]_0$ = 0.2 $[R]_0$

thus, time required, $(t) = \frac{2.303}{k}\log\left\{\frac{[R]_0}{0.2[R]_0}\right\} = \frac{2.303}{0.0102}\log\left\{\frac{1}{0.2}\right\}$

$= \mathbf{157.8\ min}$

16. Here $a = 0.5\ M,\ b = 1\ M,\ T_a = 50$ sec, $T_b = 25$ sec

∴ Using, $T_b/T_a = (a/b)^{n-1}$; we get,

$$25/50 = \left(\frac{0.5}{1}\right)^{n-1}$$

⇒ $(0.5)^1 = (0.5)^{n-1}$

⇒ $1 = n - 1$

⇒ $n = 2$. Hence the reaction is of second order.

17. Rate of reaction, $r = k[NO]^2[O_2]$ (given)

∴ $r = [2.0 \times 10^{-6}\ (s^{-1}\ mol^{-2}\ L^2)]\ [0.04\ (mol\ L^{-1})]^2\ (0.1\ mol\ L^{-1}]$

$= 6.4 \times 10^{-10}\ mol\ L^{-1}\ sec^{-1}$

18. The action of increasing $[H_2]$ stimulates a shift of the equilibrium condition "to the right". "However, only a portion of the added H_2 is consumed in this reaction. When a new equilibrium is established, there will be more H_2 than present originally. The amount of NH_3 will also the greater, but the amount of N_2 will be *smaller*. Some of the original N_2 must be consumed in converting some of the added H_2 to NH_3.

19. Adding or removing substances will affect an equilibrium condition only if the produces changes in concentration terms appearing in an equilibrium constant expression. An additional small quality of $CaCO_3$ does not affect the reaction volume, and there is no concentration term for $CaCO_3$(s) in the equilibrium constant expression. Adding more $CaCO_3$(s) will not affect the equilibrium condition.

20. Raising the temperature favors the endothermic reaction, the reverse reaction above. To favor the forward (exothermic) reaction requires that we lower the temperature. Therefore, an equilibrium mixture would have a higher concentration of SO_3 at lower temperatures; the conversion of SO_2 to SO_3 is favored at *low* temperature.

21. The reaction of a total of 4 mol of hydrogen and nitrogen would produce 2 mol of ammonia. In a given volume fewer moles of gas exert less pressure, thus, as required by Le Chatelier's principle, if additional pressure is applied the equilibrium will shift to the right, in the direction of fewer moles of gas.

22. Since the equation states that 2 mol of gaseous reactant(s) would produce 2 mol of gaseous product(s), changing the total pressure will not shift the equilibrium at all.

23. In the given reaction, the catalyst does not effect the equilibrium.

24. The Cl_2 causes a shift to the left, producing heat and thus raising the temperature.

25. The ratio $\frac{[Product]}{[Reactant]}$ is maximum when $K = 10^{10}$ and thus, reaction goes farthest to completion when $K = 10^{10}$.

26. For $A(g) \rightleftharpoons B(g)$; $K_C = \dfrac{[B]}{[A]} = 1.1. \Rightarrow [B] > [A]$

27. (i) Here, the number of PAN molecules drops by about one-half in the first 30 min (from 5.0×10^{14} to 2.6 x. 10^{14} to 1.3×10^{14}). In the next 30 min the number drops by exactly one-half (from 2.6×10^{14} to 1.3×10^{14}). Between 10.0 and 40.0 min and between 20.0 and 50.0 min the number of PAN molecules/L also drops by essentially one-half. The half-life has a *constant* value of 30. min. The reaction must be first order.

(ii) Since the reaction is **first order**, we can use the expression $k = 0.693/t_{1/2} = 0.693/30.$ min $= 0.023$ min^{-1}.

(iii) From the half-life at 273 K we can get the rate constant

$$k = \frac{0.693}{t_{1/2}} = \frac{0.693}{(35 \times 60)\text{min}} = 3.3 \times 10^{-4} \text{ min}^{-1}$$

Now we have values of k at two different temperatures 273 and 298 K and we can use the Arrhenius equation to solve for E_a.

$$\ln \frac{k_2}{k_1} = \frac{E_a}{R}\left(\frac{1}{T_1} - \frac{1}{T_2}\right)$$

$$\ln \frac{0.023}{3.3 \times 10^{-4}} = 4.24 = \frac{E_a}{8.314}\left(\frac{1}{273} - \frac{1}{298}\right) \; (0.00366 - 0.00336)$$

$$E_a = \frac{8.314 \times 4.24}{0.00030} = 1.2 \times 10^5 \text{ J } (1.2 \times 10^2 \text{ kJ})$$

28. If concentrations for given in moles per liter and time is minutes, then the rate of reaction of any order is in units of mol $\cdot$ L^{-1} min^{-1}

29. mol $\cdot$ L^{-1} min^{-1} = rate = k_3 (mol $\cdot$ L^{-1})^3 for a third order reaction. Hence, k_3 will have units of L^2 mol^{-1} min^{-1} so that both sides of the equation have the same units.

30. The order of a reaction can be zero, fractional or integral. The order indicates nothing about the molecularity unless it is an elementary process. Order is temperature independent.

31. This is just stoichiometry. Every time an A molecule is used up, two B molecules are used up. Thus, B disappears twice as fast as A.

$$-d[B]/dt = -2(d[A]/dt) = 2(2.6 \times 10^{-2} \text{ M} \cdot \text{s}^{-1})$$
$$= 5.2 \times 10^{-2} \text{ M} \cdot \text{s}^{-1}$$

32. No. mol C formed = No. mol A used = 1/2 No. mol B used. Since the partial pressure of a gas is directly proportional to the number of moles at constant T and V. when

$$P_C = 0.20 \text{ atm}, \quad P_A = (0.60 - 0.20) = 0.40 \text{ atm}$$
$$P_B = 0.80 - 2(0.20) = 0.40 \text{ atm}$$

Since this is an elementary process,

$$\text{rate} = kP_A P_B^2$$

Initially:

$$\text{rate} = k(0.60)\,(0.80)^2$$

Then $P_C = 0.20$ atm: rate $= k(0.40)\,(0.40)^2$

Ratio of rates = 0.40 $(0.40)^2/(0.60)(0.80)^2 = \left(\frac{2}{3}\right)\left(\frac{1}{4}\right) = \frac{1}{6}$

33. The rate is directly proportional to $[Cl_2]$ since doubling the $[Cl_2]$ doubles the rate, Hence, doubling [NO] must increase the rate of a factor of 4. This means the order of the reaction with respect to NO. is 2.

34. Comparing Runs 1 and 2, we see that keeping $[I^-]_o$ constant while doubling $[S_2O_8^{2-}]_0$ doubles the reaction rate, since it takes only half as long (22 s compared to 44 s) for the blue color to appear. This means that the order of the reaction with respect to $S_2O_8^{2-}$ is 1/(Note that this name conclusion can be drawn using some other combination of runs.]

35. A comparison of Runs 4 and 1 shows that doubling $[I^-]_O$ while keeping $[S_2O_8^{2-}]_n$ constant halve the time (a reaction is therefore first order with respect to $[I^-]$, Comparing Runs 1 and 3 shows that simultaneously doubling $[I^-]_o$ and having $(S_2O_8^{2-}]_o$ leaves the rate unchanged ($t \cong 44$ s in both runs). Hence, rate = $k[I^-]\,[S_2O_8^{2-}]$. The overall order of the reaction is 2.

36. Compare Run 4 with this run, in which $[S_2O_8^{2-}]_o$ is also 0.0400 M, but $[I^-]_o$ is three times higher, the rate will be 3 times higher and the time required for the blue color to appear will be 1/3 of the time for Run 4.

$$\text{time} = (1/3)\ (88.0) = 29.3 \text{ s}$$

37. $\dfrac{d[C]}{dt} = \dfrac{-3d[A]}{dt} = -\dfrac{3}{2}\dfrac{d[B]}{dt}$

From the stoichiometry of the reaction we see that C is formed 3 times as fast as A is used up and 3/2 times as fat as B is used up.

38. Option (a) is wrong because there is no relationship between the magnitude of the activation energy and the sign of ΔH. Option (e) is wrong because for some reactions with complex mechanisms there can be several slow steps.

39. Here, $T_1 = 318$ K, and $k_1 = 6.2 \times 10^{-4}$
$T_2 = 328$ K, and $k_2 = 2.1 \times 10^{-3}$

$$\therefore \quad \ln\left(\frac{6.2\times10^{-4}}{2.1\times10^{-3}}\right) = \frac{-E_{act}}{R}\left[\frac{1}{318} - \frac{1}{328}\right]$$

$$-1.22 = \frac{-E_{act}}{R}\left[\frac{10}{(318)(328)}\right]$$

$$E_{act}/R = \frac{(1.22)(318)(328)}{10} = 1.27 \times 10^4$$

$E_{act} = (8.3145 \text{ J} \cdot \text{mol}^{-1}\text{ K}^{-1})(1.27 \times 10^4 \text{ K}) = 1.06 \times 10^5 \text{ J mol}^{-1} = 1.1 \times 10^2 \text{ kJ mol}^{-1}$

40. The rate law must be determined by experiment it cannot, in general, be determined from the overall reaction stoichiometry.

41. All the other choices change only $[NO_2]$. The value of k does not depend on $[NO_2]$, but on temperature, solvent, and the concentration of any catalyst.

42. We have, $E_{a_f} - E_{a_b} = \Delta H$

If $E_{a_f} - E_{a_b}$, then ΔH must be zero.

43. The other three options involve *strong* acids; acetic acid is a weak acid. $[H^+] = 1.3 \times 10^{-3}$ M in 0.10 F HOAc. Thus, in the solution of (c), rate = $k[H^+]$ [TA] = $k(1.3 \times 10^{-3})$ (0.15) = $(1.95 \times 10^{-4})\ k$

The next lowest rate is in (d), where $[H^+]$ [TA] = 6.0×10^{-3}.

44. Sodium acetate, NaOAc, is a basic salt. Adding OAc^- decreases the $[H^+]$ and therefore decreases the rate. It does not affect k, which is a constant at constant T.

45. The value of k increases with increasing T and thus the rate increases.

46. Comparing the relative rates for A and B shows that the rate in B doubles when $[BrO_3\ (aq)]$ doubles. Note that the other concentrations remain the same. Therefore,

$$\text{rate} \propto [HrO_3^-(aq)]$$

Comparing the relative rates for B and C shows that $[BrO_3\ (aq)]$ and $[H^+(aq)]$ are the same in both experiments while $[I^-(aq)]$ doubles in C. This doubles experiments while $[I^-(aq)]$ doubles in C. This double the rate in C. Therefore,

$$\text{rate} \propto [I^-(aq)]$$

Comparing the relative rates of B and D shows that, while the other two concentration stay the same, doubling $[H^+(aq)]$ quadruples the rate in D. Therefore,

$$\text{rate} \propto [H^+(aq)]^2$$

Combining the three reaction rates gives:

$$\text{rate} = k[BrO_3^-(aq)] \times [I^-(aq)] \times [H^+(aq))]^2$$

So, the rate equation is:

$$\text{rate} = k[BrO_3^-(aq)][I^-(aq)][H^+(aq)]^2$$

The reaction is **first order** with respect to $[BrO_3^-(aq)]$, because in the rate equation the power in which the concentration is raised is 1.

It is also **first order** with respect to $[I^-(aq)]$, because the power to which this concentration is raised is 1.

The reaction is second order with respect to $[H^+\ (aq)]$, because in the rate equation the power to which this concentration is raised is 2.

The overall order of the reaction is the sum of the individual orders, which gives 1 + 1 +2 = 4.

Hence, the reaction is **fourth order**.

47. Comparing experiments 1 and 3, the concentration of NO(g) doubles, which quadruples the rate of reaction.

The concentration of $O_2(g)$ is constant.

Therefore: $\quad \text{rate} \propto [NO(g)]^2$

So, the reaction is **second order** with respect to NO(g).

48. The rate of disappearance of a reactant X is $-\Delta[X]/\Delta t$, the rate of formation of a products Y is $\Delta[Y]/\Delta t$. You can express the rate of the reaction in terms of either rate expression. These rate expression can be equated, if you first divide each by the coefficient of the corresponding substance in the chemical equation

Now, $\quad$ Rate of formation of $NO_2F = \dfrac{\Delta[NO_2F]}{\Delta t}$

and Rate of reaction of $F_2 = -\frac{\Delta[F_2]}{\Delta t}$

You divide each rate by corresponding coefficient and then equate them:

$$\frac{1}{2}\frac{\Delta[NO_2F]}{\Delta t} = -\frac{\Delta[F_2]}{\Delta t}$$

49. You calculate a Δ quantity in the rate expression by taking the final value minus the initial value.

$$\text{Average rate of decomposition of } N_2O_5 = -\frac{\Delta[N_2O_5]}{\Delta t}$$

$$\frac{(0.93 - 1.24)\times 10^{-2}\ M}{(1200 - 600)\ S} = \frac{-0.31\times 10^{-2}\ M}{600\ S} = -\ 5.2 \times 10^{-6}\ M/S$$

Note that this rate is twice the rate of formation of O_2 in the same time interval

50. You assume that the rate law has the following from

$$\text{Rate} = k[H_2O_2]^m\ [I^-]^n[H^+]^p$$

and you determine the reaction orders (exponents m, n and p) by comparing two rate experiments in which all concentrations of reactants but one are held constant.

(a) Comparing Experiment 1 and Experiment 2, you see the when the H_2O_2 concentration is doubled (with other concentrations constant), the rate is doubled,

that $m = 1$ The reaction is first order in H_2O_2.

To solve a problem such as this in a general way, you approach it algebraically. You write the rate law algebraically for two experiments. (The subscripts denote the experiments).

$$(\text{Rate})_1 = k[H_2O_2]_1^m\ [I^-]_1^n\ [H^+]_1^p$$
$$(\text{Rate})_2 = k[H_2O_2]_2^m\ [I^-]_2^n\ [H^+]_2^p$$

Now you divide the second equation by the first.

$$\frac{(\text{Rate})_2}{(\text{Rate})_1} = \frac{k[H_2O_2]_2^m\ [I^-]_2^n\ [H^+]_2^p}{k[H_2O_2]_1^m\ [I^-]_1^n\ [H^+]_1^p}$$

The rate constant cancels. Grouping the terms. You obtain

$$\frac{(\text{Rate})_2}{(\text{Rate})_1} = \left(\frac{[H_2O_2]_2}{[H_2O_2]_1}\right)^m \left(\frac{[I^-]_2}{[I^-]_1}\right)^n \left(\frac{[H^+]_2}{[H^+]_1}\right)^p$$

Now you substitute values from Experiment 1 and Experiment 2. (In a ratio the unit cancel and can be omitted.)

$$\frac{2.30\times \cancel{10^{-6}}}{1.15\times \cancel{10^{-6}}} = \left[\frac{0.020}{0.010}\right]^m \left[\frac{\cancel{0.010}}{\cancel{0.010}}\right]^n \left[\frac{\cancel{0.00050}}{\cancel{0.0050}}\right]^p$$

This gives $2 = 2^m$, from which you obtain $m = 1$. That is doubling the H_2O_2 concentration doubles the rate.

The exponents n and p are obtained in the same way Comparing Experiment 1 and Experiment 3, you see that doubling the I^- concentration (with the other concentrations constant) doubles the rate. Therefore, $n = 1$ (the reactions is first order in I^-). Finally comparing Experiment 1 and Experiment 4, you see the doubling the H^+ concentration (holding other concentrations constant) has no effect on the rate. Therefore, $p = 0$ (the reaction is zero order in H^+). Because $[H^+]^0 = 1$, the rate law is

Rate = $k[H_2O_2][I^-]$

The reaction orders with respect to H_2O_2, I^-, H^+, are 1, 1 and 0, respectively Note that the orders are *not* related to the coefficients of the overall equation.

(b) You calculate the rate constant by substituting value from any of the experiments into the rate law. Using Experiment 1, you obtain

$$1.15 \times 10^{-6} \frac{\cancel{\text{mol}}}{\cancel{\text{L}} \cdot \text{s}} = k \times 0.010 \frac{\cancel{\text{mol}}}{\cancel{\text{L}}} \times 0.010 \frac{\text{mol}}{\text{L}}$$

$$k = \frac{1.15 \times 10^{-6}\ \text{s}^{-1}}{0.010 \times 0.010 \times \text{mol L}^{-1}} = 1.2 \times 10^{-2}\ \text{L mol}^{-1} \cdot \text{s}^{-1}$$

51. You substitute $k = 2.20 \times 10^{-5}\ \text{s}^{-1}$ into the equation relating k and $t_{1/2}$

$$t_{1/2} = \frac{0.693}{k} = \frac{0.693}{2.20 \times 10^{-5}\ \text{s}^{-1}} = 3.15 \times 10^4\ \text{s}$$

and the half-life is the time required for one half (50.0%) of the SO_2Cl_2 to decompose. This is 3.15×10^4 s, or 8.75 hr. After another half-time one half of the remaining SO_2Cl_2 decompose. The total decomposed is $\frac{1}{2} + \left(\frac{1}{2} \times \frac{1}{2}\right) = \frac{3}{4}$ or 75%. The time required is two half lives, or 2×8.75 hr = 17.5 hr.

52. Substituting the data given in the problem statement into the equations and then solve, for E_a. Use the same equations, but substitute for k_1, T_1, T_2, and E_a obtained and solve for k_2. Thus,

(i) $$\log \frac{3.5 \times 10^{-3}}{2.7 \times 10^{-4}} = \frac{E_a}{2.303 \times 8.31\ \text{J} \cdot \text{mol}^{-1}\text{K}^{-1}} \left(\frac{1}{600\ \text{K}} - \frac{1}{650\ \text{K}}\right)$$

$$\text{or } \log 1.30 \times 10^1 = 1.11 = \frac{E_a}{2.303 \times 8.31\ \text{J} \cdot \text{mol}^{-1}} \times (1.28 \times 10^{-4})$$

Hence,

$$E_a = \frac{1.11 \times 2.303 \times 8.31\ \text{J mol}^{-1}}{1.28 \times 10^{-4}} = \mathbf{1.66 \times 10^5\ J \cdot mol^{-1}}$$

(ii) Substitute $E_a = 1.66$ <x. $10^5\ \text{J} \cdot \text{mol}^{-1}$ and
$k_1 = 2.7 \times 10^{-4}\ \text{L} \cdot \text{mol}^{-1} \cdot \text{s}^{-1}$ ($T_1 = 600$ K)
k_2 = unknown as yet ($T_2 = 700$ K)

You get

$$\log \frac{k_2}{2.7 \times 10^{-4}\ \mathrm{L/(mol \cdot s)}} = \frac{1.66 \times 10^5\ \mathrm{J \cdot mol^{-1}}}{2.303 \times 8.31\ \mathrm{J \cdot mol^{-1} \cdot K^{-1}}} \times \left(\frac{1}{600\ \mathrm{K}} - \frac{1}{700\ \mathrm{K}} \right) = 2.07$$

Taking antilogarithms,

$$\frac{k_2}{2.7 \times 10^{-4}\ \mathrm{L \cdot mol^{-1} \cdot s^{-1}}} = 10^{2.07} = 1.2 \times 10^2$$

Hence,

$$k_2 = (1.2 \times 10^2) \times (2.7 \times 10^{-4})\ \mathrm{L \cdot mol^{-1} \cdot s^{-1}} = 3.2 \times 10^{-2}\ \mathrm{L \cdot mol^{-1} \cdot s^{-1}}$$

Chapter 8 : Equilibrium

1. Being by writing the balanced equation in equilibrium.

$$Cl_2(g) \leftrightarrow 2Cl(g)$$

Next, write an equilibrium expression

$$K_c = \frac{[Cl]^2}{[Cl_2]}$$

Then create a chart that outlines the initial and final concentration for the various species.

Species	*Initial Concentration*	*Final Concentration*
Cl_2	$\frac{6.0 \text{ moles}}{3.0 \text{ liters}} = 2.0$ M	6.0 moles – (0.5)(6.0) = 3.0 moles $\frac{3.0 \text{ moles}}{3.0 \text{ liters}} = 1.0$ M
Cl	$\frac{0 \text{ moles}}{3.0 \text{ liters}} = 0$ M	$\frac{3.0 \text{ moles } \cancel{Cl_2 \text{ dissociated}}}{1} \times \frac{2 \text{ moles Cl}}{1 \cancel{\text{ mole } Cl_2}}$ = 6.0 moles Cl at equation $\frac{6.0 \text{ moles Cl}}{3.0 \text{ liters}} = 2.0$ M

Finally, substitute the concentrations (at equilibrium) into the equilibrium expression.

$$K_c = \frac{[Cl]^2}{[Cl_2]} = \frac{(2.0)^2}{1.0} = 4.0$$

2. Write the balanced equation at equilibrium

$$N_2(g) + O(g) \leftrightarrow 2NO(g)$$

Write the equilibrium expression.

$$K_{eq} = \frac{[NO]^2}{[N_2][O_2]} = 4.00$$

Create a chart that shows the initial concentrations, the final concentrations, and the changes in concentration. Let x represent the concentration (M) of either N_2 or O_2 (their concentration are in a 1:1 molar ratio) that is transformed through the reaction into NO.

Species	*Initial Concentration*	*Change in Concentration*	*Final Concentration*
N_2	6/2 = 3.00 M	$-x$	$3.00 - x$
O_2	6/2 = 3.00 M	$-x$	$3.00 - x$
NO	0/2 = 0 M	$+2x$	$2x$

Take the concentrations at equilibrium and substitute them into the equilibrium expression.

$$K_{eq} = \frac{[NO]^2}{[N_2][O_2]} = \frac{(2x)^2}{(3.00 - x)^2} = 4.00$$

Solve for x by taking the square root of both sides.

$$\frac{2x}{3.00 - x} = 2.00$$

$$2x = 6.0 - 2.00\ x$$
$$x = 1.50$$

Plug the value for x into the expression of the equilibrium concentration for NO.

$$[NO] = 2x = 2(1.50) = 3.00\ M$$

Note: You could also solve *this problem* using equilibrium partial pressure of the gases:

$$K_p = \frac{P_{NO^2}}{(P_{N_2})(P_{O_2})} = 4.00$$

Use P = CRT, where C represent the molar concentration of the gas, $\frac{n}{v}$

Try this approach to confirm that [NO] = 3.00 M

3. Write the balanced equilibrium equation.

$$C(s) + CO_2(s) \leftrightarrow 2\ CO(g)$$

Write the equilibrium expression.

$$K_p = \frac{(P_{CO})^2}{P_{CO_2}} = 0.50$$

Express the two unknowns, pressure of CO and pressure of CO_2, in terms of a single unknown, pressure of CO.

$$P_{total} = P_{CO} + P_{CO_2} = 3.5\text{ atm}$$

$$P_{CO_2} = 3.5\text{ atm} - P_{CO}$$

Rewrite the equilibrium expression in terms of the single unknown.

$$K_p = 0.50 = \frac{(P_{CO})^2}{3.5 - P_{CO}}$$

Rewrite this relationship in terms of the quadratic equation so that you can solve for the unknown x, the pressure of the CO.

$$x^2 = 1.75 - 0.50\ x$$

Putting this equation into the standard form, $ax^2 + bx + c = 0$, you get

$$x^2 + 0.50\ x - 1.75 = 0$$

Use the quadratic equation to solve for x

$$x = \frac{-0.50 \pm \sqrt{(0.50)^2 - 4(1)(-1.75)}}{2(1)}$$

4. Write the balanced equation in a state of equilibrium.

$$HC_2H_3O_{2(aq)} \rightleftharpoons H^+_{(aq)} + C_2H_3O^-_{2(aq)}$$

Write the equilibrium expression.

$$K_a = \frac{[H^+][C_2H_3O_2^-]}{[HC_2H_3O_2]}$$

Use the pH of the solution to determine $[H^+]$

$$pH = -\log[H^+] = 2.00$$
$$H^+ = 10^{-2.00} = 0.0100 \text{ M}$$

Determine $[C_2H_3O_2^-]$
The molar ratio of $[H^+]$ to $[C_2H_3O_2^-]$ is 1:1, $[C_2H_3O_2^-] = 0.0100$ M also
Substitute the concentrations into the equilibrium expression.

$$K_a = \frac{[H^+][C_2H_3O_2^-]}{[HC_2H_3O_2]} = \frac{(0.0100)^2}{0.99} \approx 1.00 \times 10^{-4}$$

5. Write the generic formula for % dissociation.

$$\% \text{ dissociation} = \frac{\text{part}}{\text{whole}} \times 100\% = \frac{M_{HC_2H_3O_2 \text{ dissociated}}}{M_{HC_2H_3O_2 \text{ available}}} \times 100\%$$

Substitute the known information into the generic equation and solve.

$$\% \text{ dissociation} = \frac{0.0100}{0.99} \times 100\% \approx 1.00\text{‰}$$

6. The problem involves the concept of multiple equilibria. The dissociation constants given in the example are related to the following reactions:

$$H_3PO_4(aq) \rightleftharpoons H^+(aq) + H_2OPO_4^-(aq) \qquad K_1 = 7.0 \times 10^{-3}$$
$$H_2PO_4^-(aq) \rightleftharpoons H^+(aq) + HPO_4^{2-}(aq) \qquad K_2 = 6.0 \times 10^{-8}$$
$$HPO_4^{2-}(aq) \rightleftharpoons H^+(aq) + PO_4^{3-}(aq) \qquad K_3 = 5.0 \times 10^{-13}$$

For multiple equilibria dissociation constants (such as polyprotic acids), K for the overall reaction is the product of the equilibrium constants for the individual reactions. Therefore,

$$K = K_1 \times K_2 \times K_3 = \frac{(H^+][H_2PO_4^-]}{[H_3PO_4]} \times \frac{[H^+][HPO_4^{2-}]}{[H_2PO_4^-]}$$

$$\times \frac{[H^+][PO_4^{3-}]}{[HPO_4^{2-}]} = \frac{[H^+][PO_4^{3-}]}{[H_2PO_4]} = 210 \times 10^{-24} = 2.10 \times 10^{-22}$$

which is the equilibrium constant for the sum of three individual reactions:

$$H_3PO_4(aq) \rightleftharpoons 3H^+(aq) + PO_4^{3-}(aq)$$

7. Get a picture of the solution in equilibrium. Ammonia (NH_3) is a weak base. NH_3 reacts with water in accordance with the following equilibrium equation:

$$NH_{3(aq)} + H_2O_{(l)} \rightleftharpoons NH^+_{4(aq)} + OH^-_{(aq)}$$

Ammonium choloride is soluble in water. Therefore, the concentration NH_4^+ and that of Cl^- are both 0.20 M.
Write an equilibrium expression.

$$K_b = \frac{[NH_4^+][OH^-]}{[NH_3]} = 1.8 \times 10^{-5}$$

Set up a chart showing initial concentrations of the species and final concentrations at equilibrium (the Cl^- does not contribute to the pH). Let x represent the portion of NH_3 that eventually converts to NH_4^+, x will also represent the amount by which the concentration of NH_4^+ increases. As OH^- and HN_4^+ are at equilibrium (1:1 molar ratio), the con. of OH will increase by amount x.

Species	Initial Concentration	Final Concentration
NH_3	0.35 M	$0.35 \times x$
NH_4^+	0.20 M	$0.20 + x$
OH^-	–M	$-x$

Substitute these chart values into the equilibrium expression.

$$K_b = \frac{[NH_4^+][OH^-]}{[NH_3]} = 1.8 \times 10^{-5} = \frac{(0.20 + x)(x)}{(0.35 - x)}$$

Since, K_b <<< 0.20, $K_b = \dfrac{(0.20)(x)}{(0.35)} = 1.8 \times 10^{-5}$ could be used to solve for x. Therefore C is the correct expression.

8. Write the equation for the dissociation of copper (II) iodate.

$$Cu(IO_3)_{2(s)} \rightleftharpoons Cu^{2+}_{(aq)} + 2IO^-_{3(aq)}$$

Write down the concentrations during the process of dissociation

3.3×10^{-3} M $Cu(IO_3)_2 \rightarrow 3.3 \times 10^{-3}$ M Cu^{2+}(aq) + 2 (3.3×10^{-3} M) O_3^-(aq)

$\therefore$ $[Cu^{2+}] = 3.3 \times 10^{-3}$ M

and $[IO_3^-] = 6.6 \times 10^{-3}$ M

Write the equilibrium expression.

$$K_{sp} = [Cu^{2+}][IO_3^-]^2$$

Substitute the equilibrium concentration into the K_{sp} expression.

$$K_{sp} = (3.3 \times 10^{-3})\ (6.6 \times 10^{-3}) = 1.4 \times 10^{-7}$$

9. Write the equilibrium equation for the dissociation of lead iodide.

$$PbI_2(s) \leftrightarrow Pb^{2+}(aq) + 2I^-(aq)$$

Write the equilibrium expression.

$$K_{sp} = [Pb^{2+}][I^-]^2 = 1.08 \times 10^{-7}$$

Set up a chart the express initial and final concentrations (at equilibrium) of the Pb^{2+}(aq) and I^-(aq).

Species	*Initial Concentration*	*Final Concentration at Equilibrium*
Pb^{2+}	0 M	x
I^-	0 M	$2x$

Substitute the equilibrium concentration of the ions into the equilibrium expression.

$$K_{sp} = [Pb^{2+}][I^{-2}] = (x)(2x)^2 = 1.08 \times 10^{-7}$$
$$4x^3 = 1.08 \times 10^{-7}$$
$$x^3 = 27 \times 10^{-9}$$
$$x = 3.0 \times 10^{-3}$$

10. Recognize that this problem in the involving the ion product, Q. We calculate Q in the same manner as K_{sp}, except that we use initial concentrations of the species instead of equilibrium concentrations. We then compare the value of Q to that of K_{sp}.

If $Q < K_{sp}$ — no precipitate.
If $Q = K_{sp}$ — no precipitate.
If $Q > K_{sp}$ — a precipitate forms.

At this point you can rule out choices B, C, and D because they do not make sense. If you have forgotten how to do the problem mathematically, you should guess now you have a 50% chance of getting the answer right.

Realize that this problem involves a possible double displacement, the possible precipitate being either KNO_3 or $SrCrO_4$. Rule out the KNO_3 since most nitrates are soluble and that you were provided with the K_{sp} for $SrCrO_4$. To answer these questions, you must know you solubility rules!

Write the net ionic equation.

$$Sr^{+2}(aq) + CrO_4^{2-} \rightarrow SrCrO_{4(s)}$$

Write the equilibrium expression,

$$K_{sp} = [Sr^{2+}][CrO_4^{2}]$$

Determine the initial concentration of the ions that may from the precipitate in the mixed solution

$$Sr^{2+}(aq): \frac{0.075 \text{ liter}}{1} \times \frac{0.10 \text{ mole}}{1 \text{ liter}} = 7.5 \times 10^{-3} \text{ mole}$$

$$CrO_4^{2}:(aq): \frac{0.075 \text{ liter}}{1} \times \frac{0.05 \text{ mole}}{1 \text{ liter}} = 3.8 \times 10^{-3} \text{ mole}$$

The total liters of solution = 0.075 + 0.075 = 0.15 liter. Therefore,

$$[Sr^{2+}] = \frac{7.5 \times 10^{-3} \text{ mole}}{0.15 \text{ liter}} = 0.050 \text{ M}$$

$$[CrO_4^{2}] = \frac{3.8 \times 10^{-3} \text{ mole}}{0.15 \text{ liter}} = 0.025 \text{ M}$$

Now let us, determine Q, the ion product, we have,

$$Q = [Sr^{2+}][CrO_4^{2-}] = (0.05)\ (0.025) = 1.3 \times 10^{-3}$$

Therefore, since $Q(1.3 \times 10^{-3}) > K_{sp}\ (7.1 \times 10^{-4})$, a precipitate will form.

11. $NH_4HS(s) \rightleftharpoons NH_3(g) + H_2S(g)$

0.5 atm   0 atm (Initially)

(0.5 + x) atm x atm (At equilibrium)

$$K_p = \frac{p_{NH_3} \times p_{H_2S}}{p_{NH_4HS}}$$

$$= p_{NH_3} \times p_{H_2S} \qquad \ldots\ldots (1)$$

∵ NH_4HS is solid its p_{NH_4} may be taken 1

Total number of moles at equilibrium

$$= 0.5 + x + x = 0.5 + 2x$$

∴ $$p_{NH_3} = \frac{0.5+x}{0.5+2x}P, \; p_{H_2S} = \frac{x}{0.5+2x}P$$

Given that $K_p = 0.11$

Putting these values in eq. (1), we get

$$0.11 = \left(\frac{0.5+x}{0.5+2x}P\right)\left(\frac{x}{0.5+2x}P\right)$$

or $$0.11 = \frac{x^2+0.5x}{4x^2+2x+0.25}\,p^2$$

Neglecting x^2 and taking $P = 1$ atm, we get

$$0.11 = \frac{0.5x}{2x+0.25}$$

or $$0.22x + 0.0275 = 0.5\,x$$

or $$x = \frac{0.0275}{0.28} = 0.098$$

∴ Pressure of $NH_3(g) = 0.5 + 0.098$

$= \mathbf{0.598\ atm}$

Pressure of $H_2S(g) = \mathbf{0.098\ atm}$

12. $N_2O_4(g) \rightleftharpoons 2NO_2(g)$

0.28 atm   1.1 atm   (At equilibrium)

$$K_p = \frac{[P_{NO_2}]^2}{[P_{N_2O_4}]} = \frac{(1.1\,\text{atm})^2}{0.28\,\text{atm}} = 4.32\text{ atm}$$

When the volume of the container is doubled, the pressure is believed to be one half.

Now, $N_2O_4(g) \rightleftharpoons 2NO_2(g)$

	N_2O_4	NO_2
Initial pressure	$\frac{0.28\,\text{atm}}{2}$	$\frac{1.1\,\text{atm}}{2}$
At equilibrium	$0.14 - x$	$0.55 + 2x$

$$\therefore \qquad K_p = \frac{(P_{NO_2})^2}{P_{N_2O_4}} = \frac{(0.55 + 2x)^2 \text{ atm}}{0.14 - x}$$

or $$4.32 \text{ atm} = \frac{(0.55 + 2x)^2 \text{ atm}}{0.14 - x}$$

or $$4x^2 + 6.52x - 0.3023 = 0$$

or $$x = \frac{-6.52 \pm \sqrt{(6.52)^2 - 4(4)(-0.3023)}}{2 \times 4}$$

or $$x = 0.045$$

$$\therefore \qquad P_{N_2O_4} = 0.14 - 0.045 = \mathbf{0.095 \text{ atm}}$$

$$P_{NO_2} = 0.55 + 2 \times 0.045 = \mathbf{0.64 \text{ atm}}$$

13. $N_2O_4(g) \rightleftharpoons 2NO_2(g)$

1	0	(Initially)
$1 - \alpha$	2α	(At equil.)

Total no. of moles = $1 - \alpha + 2\alpha = 1 + \alpha$

$$P_{N_2O_4} = \frac{1-\alpha}{1+\alpha}(0.5), P_{NO_2} = \frac{2\alpha}{1+\alpha}(0.5)$$

Now, $$K_P = \frac{(P_{NO_2})^2}{P_{N_2O_4}} = \frac{4\alpha^2}{1-\alpha^2} \quad (0.5)$$

or $$0.66 = \frac{2\alpha^2}{1-\alpha^2}$$

or $$\alpha^2 = \frac{33}{133}$$

or $$\alpha = 0.5$$

Thus, $$P_{N_2O_4} = \frac{1-\alpha}{1-\alpha}(0.5)$$

$$= \frac{1 \ 0.5}{1+0.5}(0.5) = \mathbf{0.166}$$

$$P_{NO_2} = \frac{2\alpha}{1+\alpha}(0.5)$$

$$= \frac{2 \times 0.5}{1+0.5}(0.5) = \mathbf{0.33}$$

14. For the reaction,

$$2H_2S(g) \rightleftharpoons 2H_2(g) + S_2(g)$$
$$1 \qquad 0.2 \qquad 0.8 \text{ (At equil.)}$$

$\therefore$
$$[H_2S] = \frac{1}{2} = 0.5,\ [H_2] = \frac{0.2}{2} = 0.1,$$

$$[S_2] = \frac{0.8}{2} = 0.4$$

$$K_c = \frac{[H_2]^2[S_2]}{[H_2S]^2}$$

$$= \frac{(0.1)^2(0.4)}{(0.5)^2}$$

$$= \mathbf{0.016}$$

15. $2H_2S(g) \rightleftharpoons 2H_2(g) + S_2(g)$

$$0.1 - x \qquad x \qquad x/2 \quad \text{(At equil.)}$$

$$V = 0.4 \text{ L}$$

$$K_c = \frac{[H_2]^2[S_2]}{[H_2S]^2} = \frac{\left[\frac{x}{V}\right]^2\left[\frac{x}{2V}\right]}{\left[\frac{0.1-x}{V}\right]^2}$$

$$K_c = \frac{x^3}{2V(0.1-x)^2}$$

Since x is very very small, hence we may take $0.1 - x = 0.1$

or
$$1.0 \times 10^{-6} = \frac{x^3}{(2)(0.4)(0.1)^2}$$

or
$$x^3 = 1.0 \times 10^{-6} \times 2 \times 0.4 \times 0.1 \times 0.1$$
$$= 8 \times 10^{-9}$$
$$x = 2 \times 10^{-3}$$

$$\% \text{ dissociation} = \frac{2 \times 10^{-3} \times 100}{0.1} = \mathbf{2\%}$$

16. $S_8(g) \rightleftharpoons 4S_2(g)$

$$1 \qquad 0 \qquad \text{(Initially)}$$
$$1 - \alpha \qquad 4\alpha \qquad \text{(At equilibrium)}$$

$$K_p = \frac{P_{S_2}^4}{P_{S_8}} = \frac{(4\alpha)^4}{(1-\alpha)} \text{ atm}^3$$

$$= \frac{(4 \times 0.29)^4}{1-0.29} \text{ atm}^3 \qquad (\because \alpha = 29\% = 0.29)$$

$$= \mathbf{266.8 \text{ atm}^3}$$

17. According to Grahm's law of diffusion,

$$\frac{r_1}{r_2} = \sqrt{\frac{M_2}{M_1}}$$

$\therefore$ $$\frac{r}{r_{Kr}} = \sqrt{\frac{M_{Kr}}{M_{eff}}}$$

or $$1.16 = \sqrt{\frac{84}{M_{eff}}}$$

or $$M_{eff} = \frac{84}{(1.16)^2} = 63.42$$

Now, $$Cl_2 \rightleftharpoons 2Cl$$

	1	0	(Initially)
	$1-x$	$2x$	(At equilibrium)

$\therefore$ $$M_{eff} = \frac{(2x)(35.5)+(1-x)(71)}{1+x}$$

or $$62.42 = \frac{71x + 71 - 71x}{1+x}$$

or $$1 + x = \frac{71}{62.42} = 1.137$$

or $$\mathbf{x = 0.137}$$

18.

H_2	+ I_2	$\rightleftharpoons$	$2HI$	
1	2		3	(Initially)
$(1-x)$	$(2-x)$		$(3+2x)$	(At equil.)

Since $$K_c = \frac{[HI]^2}{[H_2][I_2]}$$

$$45.9 = \frac{(3+2x)^2}{(1-x)(2-x)}$$

$$41.9x^2 - 125.7x - 82.8 = 0$$

$$x = \frac{125.7 \pm \sqrt{(125.7)^2 - (4)(41.9)(-82.8)}}{2 \times 41.9}$$

$$x = 0.684$$

Thus at equilibrium

$[H_2] = 1 - x = 1 - 0.684 =$ **0.316 mol · L⁻¹**

$[I_2] = 2 - x = 2 - 0.684 =$ **1.316 mol · L⁻¹**

$[HI] = \mathbf{3 + 2x} = 3 + 2 \times 0.684 =$ **4.368 mol · L⁻¹**

19. $N_2(g) \quad + \quad O_2(g) \rightleftharpoons 2NO(g)$

$\quad x \qquad (0.982 - x) \qquad 0.018$

∵ At equilibrium the mole % of NO (1.8)

∴ Mole of $N_2 + O_2 = 1.000 - 0.018 = 0.982$ mole

Let moles of N_2 be x

∴ moles of $O_2 = 0.982 - x$

Now,

$$K = \frac{[NO]^2}{[N_2][O_2]}$$

$$2.1 \times 10^{-3} = \frac{(0.018)^2}{(x)(0.982 - x)}$$

$$x^2 - 0.982x + 0.154 = 0$$

or $\quad x = 0.196$

∴ At equilibrium

$N_2 = 0.196$ mol

$O_2 = 0.982 - 0.196$

$= 0.786$ mol

∵ 0.018 mol NO is formed at equilibrium

∴ 0.009 mol each N_2 and O_2 are reacted.

∴ Initially $N_2 = 0.196 + 0.009$

= **0.205 mole**

$O_2 = 0.786 + 0.009$

= **0.795 mole**

20. We know that

$$K_p = K_C(RT)^{\Delta n}$$

For the equilibrium,

$$N_2 + 3H_2 \rightleftharpoons 2NH_3$$

$K_c = 0.5,\ T = 400\ K,\ \Delta n = 2 - (1 + 3) = -2$

$R = 0.082$ L atm K^{-} mol^{-1}

∴ $K_p = (0.5)\ (0.08 \times 400)^{-2}$

$= \dfrac{0.5}{(32.8)^2} =$ **4.65 × 10⁻⁴ atm⁻²**

21. $H_2 \quad + \quad Br_2 \rightleftharpoons 2HBr$

H_2	Br_2	$2HBr$	
0.6	0.2	0	(Initially)
$(0.6 - x)$	$(0.2 - x)$	$2x$	(At equil.)

$$K = \frac{[HBr]^2}{[H_2][Br_2]}$$

$$5 \times 10^8 = \frac{2x^2}{(0.6 - x)(0.2 - x)}$$

or $\quad (0.6 - x)(0.2 - x) = 4x^2$

(Here $4x^2/5 \times 10^8$ is considered as negligible w.r.t x^2)

or $\quad (x - 0.6)(x - 0.2) = 0$

$$x = 0.2 \text{ or } 0.6$$

Since $[Br_2]$ is taken 0.2 mol L^{-1} hence x can not be 0.6. Therefore, $x = 0.2$

$\therefore \quad [H_2] = 0.6 - 0.2 = \mathbf{0.4\ mol\ L^{-1}}$ (approx.)

$[Br_2] = 0.2 - 0.2 = 0.0$ or **negligible**

$[HBr] = 2 \times 0.2 = \mathbf{0.4\ mol\ L^{-1}}$

22. Dissociation of PCl_5 may be represented as:

$PCl_5 \rightleftharpoons PCl_3 + Cl_2$

PCl_5	PCl_3	Cl_2	
0.1	0	0	(Initially)
$0.1 - \alpha$	α	α	(At requil.)

Total no. of moles at equilibrium

$$= 0.1 - \alpha + \alpha + \alpha$$
$$= (0.1 + \alpha)$$

According to gas equation,

$$P_{eq} V = nRT$$

1.0 atm $\times 8.0$ L $= (0.1 + \alpha)$ mol

$(0.082$ L atm K^{-1} $mol^{-1})$ $(540$ K$)$

$$(0.1 + \alpha) = \frac{1 \times 8}{0.082 \times 540} = 0.18$$

or $\quad \alpha = 0.18 - 0.1 = 0.08$

$$P_{PCl_5} = \frac{0.1 - \alpha}{0.1 + \alpha} P, \quad Pcl_3 = P_{PCl_2} = \frac{\alpha}{0.1 + \alpha} P$$

$$= \frac{0.1 - 0.08}{0.1 + 0.08} P = \frac{0.08}{0.1 + 0.08} P$$

$$= \frac{0.02\,P}{0.18} = 0.111\,P = \frac{0.08}{0.18} = .444\,P$$

$$K_P = \frac{P_{PCl_3} \times P_{Cl_2}}{P_{PCl_5}}$$

$$= \frac{0.444 \times 0.444 \times 1\,\text{atm}^2}{0.111 \times 1\,\text{atm}} = \mathbf{1.78\ atm}$$

$\because \quad K_p = K_c(RT)^{\Delta n}$

or $\quad K_C = \dfrac{K_P}{(RT)^{\Delta n}}$

$$= \frac{1.78\,\text{atm}}{[(0.082\,\text{L atm K}^{-1}\,\text{mol}^{-1})(540\,\text{K})]^1} = \mathbf{0.04\ mol\ L^{-1}}$$

23. Let initial moles of AB_2 be 1.
Therefore

	$2AB_2(g)$	$\rightleftharpoons$ $2AB(g)$	+ $B_2(g)$
Initial moles	1	0	0
Moles at equilibrium	$1 - x$	x	$x/2$

Where x is the degree of dissociation
Total no. of moles at equilibrium

$$1 - x + x + \frac{x}{2} = 1 + \frac{x}{2}$$

$\because$ Total pressure $= P$

$$P_{AB_2} = \frac{1-x}{1\ (x/2)}P;\ P_{AB} = \frac{x}{1+(x/2)}P;\quad P_{B_2} = \frac{x/2}{1+(x/2)}P$$

$$\therefore K_P = \frac{P_{AB}^2\, P_{B_2}}{P_{AB_2}^2}$$

$$\text{or } K_P = \frac{\left(\dfrac{(x/2)}{1+(x/2)}\right)P \cdot \left(\dfrac{x}{1+(x/2)}\right)P^2}{\left(\dfrac{1-x}{1+(x/2}\right)^2 P^2} = \frac{x}{2} \times \frac{x^2 \times P}{\left(1+\frac{x}{2}\right)(1-x)^2}$$

Since x is very small w.r.t. 1 hence

$$K_P = \frac{x^3}{2}P$$

$$\text{or} \quad x = \sqrt[3]{\frac{2Kp}{P}}$$

24. $PCl_5 \rightleftharpoons PCl_3 + Cl_2$

1	0	0	(Initially)
$1-\alpha$	α	α	(At equl.)

Total no. of moles at equil.

$$= 1 - \alpha + \alpha + \alpha = 1 + \alpha$$

But the degree of dissociation is 10% i.e. $\alpha = 0.1$

Therefore

$$p_{PCl_5} = \frac{1-\alpha}{1+\alpha} \times 4; \quad \text{and} \quad p_{PCl_3} = p_{Cl_2} = \frac{\alpha}{1+\alpha} \times 4$$

$$= \frac{1-0.1}{1\ 0.1} \times 4 = \frac{36}{11} \qquad = \frac{0.1 \times 4}{1+0.1} = \frac{4}{11}$$

$$\therefore K_P = \frac{p_{PCl_3} \times p_{Cl_2}}{p_{PCl_5}} = \frac{\frac{4}{11} \times \frac{4}{11}}{\frac{36}{11}} = 0.04 \text{ atm}$$

Let at P atm, degree of dissociation be 20% i.e. $\alpha = 0.2$

$$K_P = \frac{p_{PCl_3} \times p_{Cl_2}}{p_{PCl_5}} = \frac{\alpha^2 P}{1-\alpha^2}$$

$$0.04 = \frac{(0.2)^3 (P)}{1-\alpha^2}$$

$$\text{or} \quad P = \frac{0.04 \times 0.96}{0.04} = \mathbf{0.96\ atm}$$

25. We have, $K_p = K_c(RT)^{\Delta n}$

Here $\Delta n = 2 - (1 + 3) = -2$

Given $T = 673$ K

$R = 0.0821$ dm^3 K^{-1} mol^{-1}

and $K_c = 0.495$

$\because$ $K_p = K_c(RT)^{\Delta n}$

$\therefore$ $K_p = (0.495)\ (0.0821 \times 673)^{-2} = 1.62 \times 10^{-4}$

26. For the equilibria

$$H_2(g) + I_2(g) \rightleftharpoons 2HI(g),\ K_c = \frac{[HI]^2}{[H_2][I_2]} = 50$$

	H_2	I_2	$2HI$
Initial conc.	2	4	0
Equilibrium conc.	$(2-x)$	$(4-x)$	$(2x)$

Hence, $K_c = \frac{(2x)^2}{(2-x)(4-x)} \frac{4x^2}{x^2-6x+8} = 50$

$\Rightarrow$ $46x^2 - 300x + 400 = 0$

$\Rightarrow$ $x = \frac{300 + \sqrt{90000 - 4 \times 46 \times 400}}{2 \times 46} = 1.86 \text{ (or } 4.65)$

Therefore moles of HI formed at equilibrium = $2x = 2 \times 1.86 = 3.72$.

27. Let x moles of ester are formed at equilibrium.

	$CH_3COOH_{(l)}$ +	$C_2H_5OH_{(l)}$ $\rightleftharpoons$	$CH_3COO\,C_2H_{5(l)}$ +	$H_2O_{(l)}$
Initial molar concentration	$1.0/V$	$0.5/V$	0	0
Molar concentrations at equilibrium	$\left(\frac{1-x}{V}\right)$	$\left(\frac{0.5-x}{V}\right)$	x/V	x/V

$\therefore$ $K_c = \frac{[CH_3COOC_2H_5][H_2O]}{[CH_3COOH][C_2H_5OH]}$

$\Rightarrow$ $4 = \frac{(x/V)(x/V)}{(1-x/V)(0.5-x/V)}$

$\Rightarrow$ $4 = \frac{x^2}{(1-x)(0.5-x)} \Rightarrow x = 0.425$ mole.

28. (i) To combine the given equation we must use (a) as written and reverse (b), The values of ΔH and the values of K_c are calculated as below:

(a) $CO(g) + H_2O(g) \rightleftharpoons CO_2(g) + H_2(g)$

$\Delta H = -40.$ kJ $K_c = 1.4$

(b) $CH_4(g) + H_2O(g) \rightleftharpoons CO(g) + 3H_2(g)$

$\Delta H + 230$ kJ $K_c = 1/190.$

net: $CH_4(g) + 2H_2O(g) \rightleftharpoons CO_2(g) + 4H_2(g)$

$\Delta H = +190.$ kJ $K_c = 1.4/190$

$= 7.4 \times 10^{-3}$

(ii) Here we need to use the expression $K_p = K_c(RT)^{\Delta n_{(aq)}}$, with a value of $\Delta n_{gas} = 2$(5 moles of gases on the right minus 3 moles on the left).

$$k_p = K_c(RT)^{\Delta n_{(aq)}} = 7.4 \times 10^{-3}\,(0.082 \times 1000)^2 = 50.$$

(iii) In the tabulation below, x represents the change in the number of moles of CH_4 and of CO_2. In accordance with the stoichiometry of the reaction, changes in the numbers of moles of H_2O and H_2 are $2x$ and $4x$, respectively. Also increases in amounts of substances are denoted by positive signs ($+x$, $+4x$) and decreases in amounts by negative signs ($-x$, $-2x$).

the reaction:	$CH_4(g)$	+	$2H_2O(g)$	$\rightleftharpoons$	$CO_2(g)$	+	$4H_2(g)$
initial amts:	1.00 mol		1.00 mol		0.0 mol		0.0 mol
changes:	$-x$ mol		$-2x$ mol		$+x$ mol		$+4x$ mol
equal. amts:	$(1.00 - x)$ mol		$(1.00 - 2x)$ mol		x mol		$4x$ mol
equil. concens.M:	$(1.00 - x)/10.0$		$(1.00 - 2x)/10.0$		$x/10.0$		$4x/10.0$

$$K_c = \frac{[CO_2][H_2]^4}{[CH_4][H_2O]^2} = \frac{\left(\frac{x}{10.0}\right)\left(\frac{4x}{10.0}\right)^4}{\left(\frac{1.00-x}{10.0}\right)\left(\frac{1.00-2x}{10.0}\right)} = \frac{x(4x)^4}{100(1.00-x)(1.00-2x)^2}$$

$$= 7.4 \times 10^{-3}$$

$$256x^5 = 0.74\ [(1.00 - x)(1.00 - 2x)^2]$$

$$256x^5 = 0.74[(1.00 - x)\ (1.00 - 2x)^2] = 0$$

Solving we get $4x = 0.92$

or $x = 0.23$

(iv) To describe these effect qualitatively means simply to indicate whether the amount of $H_2(g)$ would increase, decrease, or remain uncharged when equilibrium is reestablished. We can do this most simply by applying Le Chatelier's principle

The amount of $H_2(g)$ will

(a) Add a catalyst ---------- no change

(b) Add 24.0 g CH_4 ---------- increase

(c) Add 10.0 g CO_2 ---------- decrease

(d) Remove 6.00 g H_2O ---------- decrease

(e) Raise the temperature to 1200 K ---------- increases

(f) Transfer to mixture to a 15.0-L vessel ---------- increases

(g) Add 10.0 g He(g) at constant volume ---------- no change

29. A solubility product is the equilibrium constant for the equilibrium between a very slightly electrolyte and the saturated equeous solution of that solid. The solid is always written on the left side of the equation, and the ions of which the solid is composed are on the right side. No other species but the solid and its ions appear in the equation. The equilibrium constant for the reaction given in (c) is the inverse of K_{sp} for AgCl.

30. Only if $\Delta n_{gas} = 0$ does $K_p = K_c$.

For (a), $\Delta n_{gas} = 2 + 1 - 2 = 1$. For (b), $\Delta n_{gas} = 2 - 1 - 3 = -2$.

For (c), $\Delta n_{gas} = 2 - 1 - 1 = 0$. For (d), $\Delta n_{gas} = 2 - 1 = 1$

Iodine, a solid, does not appear in K_{eq} for (d). For (*e*) $\Delta n_{gas} = 1 + 1 - 1 = 1$.

31. $K_p = P_{CO_2}$ The pressure of CO_2 is a constant at constant temperature, independent of the amount of either solid, as long as both solids are present.

32. The equilibrium constant for $PCl_3 + Cl_2 \rightleftharpoons PCl_5$ is the inverse of K_{eq} for $PCl_5 \rightleftharpoons PCl_3 + Cl_2$. Thus

$$K_p = \frac{1}{0.0024} = 4.2 \times 10^2$$

33. When a system is not at equilibrium, Q and K_{eq} have different numerical values. The reaction quotient is zero when none of the products of the reaction are present but there are reactants present.
34. Doubling the volume means halving each pressure. The reaction will shift to the side with more moles of gas to restore equilibrium. Only in (d) are there more moles of gas on the right than on the left.
35. An equilibrium constant is a constant (invariant) at constant temperature.
36. An acidity constant is the equilibrium constant for the proton-transfer reaction between a weak acid and water.
37. As each pressure is halved, the reaction shifts to the side with more moles of gas, that is, to the left. More Cl_2 is formed.
38. In (a), $\Delta n_{gas} = 0$. Both sides have the same number of moles of gas. Decreasing the volume increases each pressure by the same factor, but as K_p is dimension less, Q is unaffected if each pressure increases by the same factor. As Q remain equal to K_p, no shift occurs.
39. Only increasing the temperature increases K_p for the decomposition reaction, Increasing the temperature for any endothermic reaction drives the reaction to the right, to absorb heat.
40. $K_c = \dfrac{[H_2O][CO]}{[CO_2][H_2]} = 3.24$

Let $x = [H_2O] = [CO]$ at equilibrium. Then,
$H_2 = [CO_2] = 0.800 - x$, as the volume is 1.00 L. As the reaction proceeds to form H_2O and CO, both H_2 and CO_2 are used up. Thus

$$\frac{x^2}{(0.800 - x)^2} = 3.24$$

Take the square root of both sides of this equation.

$$\frac{x}{0.800 - x} = 1.80$$

Thus

$$x = (1.80)(0.800) - 1.80x \text{ or } 2.80x = 1.44$$

and

$$x = 0.514 \text{ mol of } H_2O(g) \text{ and } CO(g) \text{ formed}$$

41. $K_c = [C]^2/[B]^2$

$$[C] = \frac{0.40 \text{ mol}}{2.00 \text{ L}} = 0.20 \text{ M} \quad [B] = \frac{0.80 \text{ mol}}{2.00 \text{ L}} = 0.40\ M$$

Thus,

$$K_c = \frac{(0.20)^2}{(0.40)^2} = (1/2)^2 = ¼ = 0.25$$

42. $[NO_2] = \dfrac{0.120\text{ mol}}{4.00\text{ L}} = 3.00 \times 10^{-2}\ M$

$[O_2] = \dfrac{0.640\text{ mol}}{4.00\text{ L}} = 1.6 \times 10^{-1}\ M$

$[NO] = \dfrac{0.080\text{ mol}}{4.00\text{ L}} = 2.00 \times 10^{-2}\ M$

$$K_c = \frac{[NO_2]^2}{[O_2][NO]^2} = \frac{(3.00 \times 10^{-2}\ M)^2}{(1.60 \times 10^{-1}\ M)(2.0 \times 10^{-2}\ M)^2} = \frac{9.00 \times 10^{-4}}{6.4 \times 10^{-5}}\ M^{-1}$$
$$= 1.4 \times 10\ M^{-1} = 1.4\ L \cdot mol^{-1}$$

43. $K_p = \dfrac{P_{HI}^2}{P_{H_2} P_{I_2}} = 9.0$

For each gas, $P = n(RT/V)$. As K_p has no units, the (RT/V) factors all cancel, and

$$K_p = \frac{n_{HI}^2}{n_{H_2} n_{I_2}} = \frac{(0.060)^2}{(0.40)(x)}$$

Thus,

$$x = \frac{(0.60)^2}{(9.0)(0.40)} = 1.0 \times 10^{-1} = 0.10$$

44. First, write down the stoichiomeric equation. At the start of the reaction there is no product. But once equilibrium has been reached, the reaction vessel contains ethanoic acid, ethanol, ethyl ethanoate and water.

From the equation, every mole of ethanoic and requires 1 mole of ethanol and makes 1 mole of ethyl ethanoate and 1 mole of water. This means that if the yield at equilibrium is x moles of ethyl ethanoate, then x moles of each of ethanol and ethanoic acid most have reacted. This leaves behind $(1.0 - x)$ moles of each of ethanoic acid and ethanol.

	CH_3COOH +	C_2H_3OH ⇌	$CH_3COOC_2H_5$ +	H_2O
As start of reaction moles^{-1}	1.0	1.0	0	0
At equilibrium moles^{-1}	$1.0 - x$	$1.0 - x$	x	x

The equilibrium constant is: $K_c = \dfrac{[CH_3COOC_2H_5][H_2O]}{[CH_3COOH][C_2H_5OH]}$

The concentration of each substance must be substituted into the above expression, so the moles of equilibrium must be converted to concentrations.

$$\text{Concentrations is in mol dm}^{-3} = \frac{\text{moles}}{\text{volume (dm)}^3}$$

So be careful to substitute concentrations into the expression for K_c, *not* the number of moles.

Assuming that the volume of the esterification mixture is $V\text{dm}^3$, this gives:

$$K_c = 4.0 = \frac{\left(\frac{x}{V}\right)\left(\frac{x}{V}\right)}{\left(\frac{1.0-x}{V}\right)\left(\frac{1.0-x}{V}\right)}$$

Cancelling v gives. $4.0 = \dfrac{x^2}{(1.0-x)^2}$

Taking the square root of both sides gives: $2.0 = \dfrac{x}{1.0-x}$

So, $2.0 = 2.0x - x$

That is, $x = 0.67$ moles

45. The problem is essentially one in stoichiometry. It involves initial, or *starting*, amounts of reactants. These amounts *change* as reaction occurs. Later, the reaction comes to *equilibrium*, and you analyze the reaction mixture for the amount of one of the reactants or products. It is convenient to solve this problem by first setting up a table in which you write the starting change, and equilibrium value of each substance under the balanced equation. This way you can easily see what you have to calculate. The given equation is :

$$CO_2(g) + 3H_2(g) \rightleftharpoons CH_4(g) + H_2O(g)$$

You fill in the starting amounts from the values given in the problem statement. You are not given explicit values for the changes that occur, so you let x be the molar change. That is, each product increases by x moles multiplied by the coefficient of the substances in the balanced equation. Reactants decreases by x moles multiple by the corresponding coefficient. The decrease is indicated by a negative sign Equilibrium values are equal to starting values plus the changes,

Using the information given in the problem, you set up the following table:

Amount (mol)	**$CO(g)$ +**	**$3H_2(g)$**	**$\rightleftharpoons$ $CH_4(g)$**	**+ $H_2O(g)$**
Starting	1.0	3.0	0	0
Change	$-x$	$-3x$	$+x$	$+x$
Equilibrium	$1.0 - x$	$3.0 - 3x$	x	$x = 0.387$

The problem statement gives the equilibrium amount for H_2O. Thus tells you that $x = 0.387$ mol. You calculate equilibrium amounts for other substances from the expressions given in the table, using this value of x.

Equilibrium amount CO $= (1.0 - x)$ mol
$= (1.0 - 0.387)$ mol
$= 0.613$ mol

Equilibrium amount $H_2 = (3.0 - 3x)$ mol
$= (3.0 - 3 \times 0.387)$ mol
$= 1.839$ mol

Equilibrium amount $CH_4 = x$ mol $= 0.387$ mol

Therefore, the amounts of substances in the equilibrium mixture are **0.613 mol CO, 1.839 mol H_2, 0.387 mol CH_4, and 0.387 mol H_2O.**

46. You will first need to calculate the molar concentrations of substances, because the equilibrium constant is stated in terms of these quantities. The problem statement gives the starting amount of HI and the equilibrium amount of I_2. It is convenient to set up a table of starting, change, and equilibrium concentrations,. You use the equilibrium values to calculate the equilibrium constant K_c.

To obtain the concentrations of HI and I_2, you divide the amounts by the volume of the reaction vessel (5.00 L).

$$\text{Starting concentration of HI} = \frac{4.00 \text{ mol}}{5.00 \text{ L}} = 0.800\ M$$

$$\text{Equilibrium concentration} = \frac{0.442 \text{ mol}}{5.00 \text{ L}} = 0.0884\ M$$

From these values, you set up the following table:

Concentration (*M*)	$2HI(g)$	$\rightleftharpoons$	$H_2(g)$	+	$I_2(g)$
Starting	0.800		0		0
Change	$-2x$		x		0
Equilibrium	$0.800 - 2x$		x		$x = 0.0884$

The equilibrium concentrations of substances can be evaluated from the expressions given in the last line of this table. You know that the equilibrium concentrations of I_2 is 0.0884 *M* and that this equals x. Therefore,

$$[HI] = (0.800 - 2x)\ M = (0.800 - 2 \times 0.0884)\ M = 0.623\ M$$
$$[H_2] = x = 0.0884\ M$$

Now you substitute into the equilibrium constant expression for the reaction. From the chemical equation, you write

$$K_c = \frac{[H_2][I_2]}{[HI]^2}$$

Substituting,

$$K_c = \frac{(0.0884)(0.0884)}{(0.623)^2} = 0.0201$$

47. The solution of an equilibrium problem involves these steps. In Step 1, you use the given information to set up a table of starting, change, and equilibrium concentration. In this problem, you are given starting amounts of reactants. You use these to obtain the starting concentrations. Express the change concentrations in terms of x, In Step 2, you substitute the equilibrium concentrations in the table into the equilibrium constant expression and equate this to the value of the equilibrium constant. In Step 3, you solve this equilibrium constant equation. This example, given an equation that is perfect square, taking the square root of it gives a linear equation, which is easy to solve. (More generally, an equilibrium equation involves solving an nth-order equation for its roots. For $n = 2$, you can use the quadratic formula, for $n > 2$, it is time to think about using a computer).

Now, The *starting concentrations* of CO and H_2O are

$$[CO] = [H_2O] = \frac{1.00 \text{ mol}}{50.0 \text{ L}} = 0.0200 \text{ mol/L}$$

The starting concentrations of the products, CO_2 and H_2, are 0. The *changes* in concentrations when the mixture goes to equilibrium are not given. However, you can write them all in terms of a single unknown. If you let x be the moles of CO_2 formed per liter, then the moles of H_2 formed per liter is also x. Similarly, x moles each of CO and H_2O are consumed. You write the changes for CO and H_2O as $-x$. You obtain the *equilibrium concentration* by adding the change in concentrations to be starting concentrations, as shown in the following table:

Concentrations (*M*)	**CO(*g*)**	+	**H_2O(*g*)**	$\rightleftharpoons$	**CO_2(*g*)**	+	**H_2(*g*)**
Starting	0.0200		0.0200		0		0
Change	$-x$		$-x$		$+x$		$+x$
Equilibrium	$0.0200 - x$		$0.0200 - x$		x		x

You then substitute the values for the equilibrium concentrations into the equilibrium constant equation,

$$K_c = \frac{[CO_2][H_2]}{[CO][H_2O]}$$

and you get

$$0.58 = \frac{(x)(x)}{(0.0200 - x)(0.200 - x)}$$

or

$$0.58 = \frac{x^2}{(0.0200 - x)^2}$$

You now solve this equilibrium equation for the value of x. Note that the right hand side is a perfect square. If you take the square root of both sides. You get

$$\pm 0.76 = \frac{x}{0.0200 - x}$$

We have written ± to indicate that you should consider both positive and negative values, because both are mathematically possible. Rearranging the equation gives

$$x = \frac{0.0200 \times 0.76}{1.76} = 0.0086, \text{ and } x = \frac{-0.0200 \times 0.76}{0.24} = -0.063$$

You can dismiss the negative value as physically impossible (x can only be positive, it represents the concentration of CO_2 formed). If you substitute for x in the last line of the table, the equilibrium concentration are 0.0114 *M* CO. 0.0114 *M* H_2O, 0.0086 *M* CO_2, and 0.0086 *M* H_2. To find the moles of each substance in the 50.0-L vessel, you multiply the concentrations by the volume of the vessel, For example, the amount of CO is

$0.0114 \text{ mol L}^{-1} \times 50.0 \text{ L} = 0.570 \text{ mol}$

The equilibrium composition of the reaction mixture is 0.570 mol CO, 0.570 mol H_2O 0.43 mol CO_2 and 0.43 mol H_2.

48. *Step 1* The concentrations of substances is as follows:

Concentrations (*M*)	$H_2(g)$	+	$I_2(g)$	$\rightleftharpoons$	$2HI(g)$
Starting	1.00		2.00		0
Change	$-x$		$-x$		$2x$
Equilibrium	$1.00 - x$		$2.00 - x$		$2x$

Note that the changes in concentrations equal x multiplied by the coefficient of that substance in the balanced chemical equation. The change is negative for a reactant and positive for a product Equilibrium concentrations equal starting concentrations plus the changes in concentrations.

Step 2 Substituting into the equilibrium constant equation,

$$K_c = \frac{[HI]^2}{[H_2][I_2]}$$

you get

$$49.7 = \frac{(2x)^2}{(1.00 - x)(2.00 - x)}$$

Step 3 Because the right-hand side is not a perfect square, you must use the quadratic formula to solve for : x. The equation rearranges to give

$$(1.00 - x)\ (2.00 - x) = (2x)^2/49.7 = 0.0805x^2$$

or

$$0.920x^2 - 3.00x + 2.00 = 0$$

Hence,

$$x = \frac{3.00 \pm \sqrt{9.00 - 7.36}}{1.84} = 1.63 \pm 0.70$$

There are two mathematical solutions to a quadratic equation. You obtains one solution by taking the upper (positive) sign in ± and the other by taking the lower (negative) sign. You get

$$x = 2.33 \text{ and } x = 0.93$$

However, $x = 2.33$ gives a negative value to $1.00 - x$ (the equilibrium concentration of H_2), which is physically impossible. Only $x = 0.93$ remains. You substitute this value of x into the last line of the table in Step 1 to get the equilibrium concentrations, and then multiply these by the volume of the vessel (1.00 L) to get the amounts of substances. The last line of the table is rewritten

Concentration (*M*)	$H_2(g)$	+	$I_2(g)$	$\rightleftharpoons$	$2HI(g)$
Equilibrium	$1.00 - x = 0.07$		$2.00 - x = 1.07$		$2x = 1.86$

The equilibrium composition is **0.07 mol H_2, 1.07 mol I_2, and 1.86 mol HI.**

49. Substituting numerical values into this equation,

$$\log K = \frac{\Delta G^\circ}{-2.303\,RT} = \frac{-13.6 \times 10^3}{-2.303 \times 8.31 \times 298} = 2.38$$

Hence,

$$K = \text{antilog } 2.38 = 10^{2.38} = 2.4 \times 10^2$$

Note that although the value of K indicates that products predominate at equilibrium, K is only moderately large. You would expect that the composition could be easily shifted toward reactants if you could remove either NH_3 or CO_2 (according to Le Chatelier's principle). This is what happens when urea is used as fertilizer. As NH_2 is used up, more NH_3 is produced by the decomposition of urea.

50. You must first calculate ΔG_f° for the ionic equation. Note that K_{sp} equals K, the thermodynamic equilibrium constant.

You first calculate ΔG°. Writing ΔG_f° values below the formulas in the equation gives

$$\begin{array}{llll} & AgCl(s) & \rightleftharpoons Ag^+(aq) & + Cl(aq) \\ \Delta G_f^\circ & -109.7 & 77.1 & -131.2\ kJ \end{array}$$

Hence,

$$\Delta G^\circ = [(77.1 - 131.2) - (-109.7)]\ kJ$$
$$= 55.6\ kJ\ (\text{or } 55.6 \times 10^3\ J)$$

You now substitute numerical values into the equation relating log K and ΔG°.

$$\log K = \frac{\Delta G^\circ}{-2.303\,RT} = \frac{55.6 \times 10^3}{-2.303 \times 8.31 \times 298} = -9.75$$

Therefore,

$$K = \text{antilog}\ (-9.75) = 10^{-9.75} = \mathbf{1.8 \times 10^{-10}}$$

51. (i) Calculate ΔH° and ΔS° at 298 K using standard enthalptes of formation and standard entropies. Then substitute into the equation for ΔG_T°. (b) Use the value of ΔG_T° to find K (= K_p)

ΔH° and ΔS° from these given value we get

$$\Delta H^\circ = [(-635.1 - 393.5) - (-1206.9)]\ kJ = 178.3\ kJ$$
$$\Delta S^\circ = [(38.2 + 213.7) - (92.9)]\ J \cdot K^{-1} = 159.0\ J \cdot K^{-1}$$

Now you substitute ΔH°, ΔS° (= 0.1590 k J $\cdot$ K^{-1}), and T = (1273 K) into the equation for ΔG_T°.

$$\Delta G_T^\circ = \Delta H^\circ - T\Delta S^\circ = 178.3\ kJ - (1273\ K)\ (0.1590\ k\ J \cdot K^{-1}) = -24.1\ kJ$$

(ii) Because ΔG_T° is negative, the reaction should be spontaneous at 1273 K and 1 atm. Substitute the value of ΔG° at 1273 K, which equals – 24.1 × 10^3 J, into the equation relating log K and ΔG°

$$\log K = \frac{\Delta G^\circ}{-RT\,2.303} = \frac{24.1 \times 10^3}{-8.31 \times 1273 \times 2.303} = 0.989$$

$$K = K_p = \text{antilog}\ 0.989 = 9.75$$

$K_p = P_{CO_2}$ so the partial pressure of CO_2 is 9.75 atm.

52. $2XO + O_2 \rightleftharpoons 2XO_2 \quad K = \dfrac{[XO_2]^2}{[XO]^2\,[O_2]} = 1.0 \times 10^{-4}\ L \cdot mol^{-1}$

The equation $XO + \frac{1}{2}O \rightleftharpoons XO_2$ would have an equilibrium constant with units $(L \cdot mol^{-1})^{\frac{1}{2}}$. The sequence of construction of the concentrations table is shown by the circled numbers.

	Initial	*produced*	*used up*	*equilibrium*
[XO]	1.0 ①		$2x$ ④	$1.0 - 2x \simeq 1.0$ ⑤
[O_2]	2.0 ①		x ④	$2.0 - x \simeq 2.0$ ⑤
[XO_2]	0.0 ①	$2x$ ③		$2x$ ②

①The initial concentrations are given in the statement of the problem; the [XO_2] by implication②Letting the equilibrium concentration of XO_2 equal $2x$ allows the deduction ③that $2x$ mol · L^{-1} was produced by the reaction.④The concentrations of XO and O_2 used up, $2x$ and x, are in the ratio of the coefficients in the balanced chemical equation. ⑤ Knowledge of the initial concentrations and how much is used up in the reaction allows calculation of the equilibrium concentrations. The latter are approximated as 1.0 and 2.0 mol · L^{-1}, subject to later checking. These values are substituted in the equilibrium constant expression.

$$K = \frac{(2x)^2}{(1.0)^2\,(2.0)} = 1.0 \times 10^{-4}$$

$x = 7.1 \times 10^{-3}$ [The approximations made, e.g. $1.0 - 0.014 \simeq 1.0$, are within 1.4% of the true value, which is acceptable).

$$[XO_2] = 2x = 1.4 \times 10^{-2} \text{ mol} \cdot L^{-1} \qquad [XO] = 1.0 - 0.014 \simeq 1.0 \text{ mol} \cdot L^{-1}$$

$$[O_2] = 2.0 - 0.0071 \simeq 2.0 \text{ mol} \cdot L^{-1}$$

53. $6HCHO \rightarrow C_6H_{12}O_6$; $K = \dfrac{[C_6H_{12}O_6]}{[HCHO]^6} = \dfrac{1.00}{[HCHO]^6} = 6 \times 10^{22}$

$$[HCHO]^6 = \frac{1.00}{6 \times 10^{22}} \qquad [HCHO] = 1.6 \times 10^{-4} \text{ M}$$

54. At equilibrium, $\Delta G = 0$. The larger $\Delta G°$, the less likely the reaction is spontaneous, and the smaller the value of the equilibrium constant.

55. Hence, $\Delta n = 1$

and, $\Delta H° = \Delta E° + \Delta(PV) = \Delta E° + (\Delta n)RT = \Delta E° + RT = -376000 + (8.314 \times 300) = -373505$

also, $\Delta G° = \Delta H° - T\Delta S° = -373505 - (300 \times 408) = -495905$

$$-\Delta G° = 2.30\, RT \log K$$

$$\therefore \qquad \log K = \frac{+495905}{(2.30)(8.314)(300)} = 86.4 \text{ or } K = 3 \times 10^{86}$$

56. (a) The ions in solution are Na^+, which does not bydrolyze, and CN^-, which does. CN^- is the conjugate base of HCN and should form a basic solution.

$$CN^- + H_2O \rightleftharpoons HCN + OH^-.$$

(b) Neither K^+ nor Cl^- undergoes hydrolysis KCl(aq) is neutral—pH = 7.

(c) Both NH_4^+ and CN^- hydrolyze in aqueous solution, one to produce H_3O^+ and the other, OH^-. With values from Tables from text books we can write

$$NH_4^+ + H_2O \rightleftharpoons NH_3 + H_3O^+ \quad K_a = 5.7 \times 10^{-10}$$

$$CN^- + H_2O \rightleftharpoons HCN + OH \quad K_6 = 2.5 \times 10^{-5}$$

Because K_b is larger than K_a, we should expect CN^- to hydrolyze to a greater extent than NH_4^+. This means that $[OH^-] > [H_3O^+]$ and the solution will be basic.

57. $\Delta G° = \Delta G_f°(CaO) + \Delta G_f°(CO_2) - \Delta G_f°(CaCO_3) = -(603) + (-394) - (-1127) = 130 \text{ KJ} = 130000\text{J}$

$$\Delta G° = -2.30\ RT \log K$$

$$\log K = \frac{\Delta G°}{2.30\ RT} = \frac{-130000}{2.30(8.314)(298)} = -22.8$$

$$K = 1.6 \times 10^{-23} \text{ atm} = P(CO_2)$$

Chapter 9 : Ionic Equilibria

1. $[H^+][OH^-] = 10^{-14}$. Substituting gives $(5.00 \times 10^{-3})(x) = 10^{-14}$. Solving for x yields $x = 2.00 \times 10^{-12}$.
2. The conjugate acid of a base if formed when the base acquires a proton from the acid. In this reaction, water acts as an acid because it donates a proton to the ammonia molecule. The ammonium ion (NH_4^+) is the conjugate acid of ammonia (NH_3), a base, which receives a proton from water. The hydroxide ion (OH^-) is the conjugate base.
3. For a series of oxyacids of the same structure that differ only in the halogen, the acid strength increases with the electronegativity of the halogen. Because the electronegativity of the halogens increases as we move up the column, the order at this point would be HClO > HBrO > HIO. For a series of oxyacids containing the same halogen, the H—O bond polarity, and hence the acid strength, increases with the oxidation state of the halogen. Therefore, in the series HClO, $HClO_2$, and $HClO_3$, the oxidation states of the chlorine are +1, +3, and +5 respectively. And thus the correct order for decreasing acid strength is $HClO_3 > HClO_2 > HClO > HBrO > HIO$.
4. Brønsted acids donate protons (H^+). In the equation, both H_2O and HCO_3^- donate H+.
5. Any anion derived from a weak acid acts as a base in a water solution. The carbonate polyatomic anion, CO_3^{2-}, is derived from the weak acid carbonic acid, H_2CO_3. There are no common basic cations.
6. To write a net ionic equation for an acid-base reaction between two solutions, use the following three steps:
   1. Determine the nature of the principal species in both solutions. $KHCO_3$ would ionize to produce K^+ and HCO_3^-. HBr would ionize to produce H^+ and Br^-.
   2. Determine which species take part into acid-base reaction. The bicarbonate anion (HCO_3^-) is basic, and the H^+ from the hydrobromic acid is acidic.
   3. Write a balanced net ionic equation. Because H_2CO_3 is a weak acid, it would tend to remain as carbonic acid, especially in the presence of a strong acid like HBr. KBr is an ionic solid, soluble in water; therefore, it would exist as separate ions known as spectator ions, which are *not* written in the net ionic equation.
7. All hydroxides of the Group I metals are strong bases. The hydroxides of the heavier Group II metals (Ca, Sr, and Ba) are also strong bases. $Mg(OH)_2$ is not very soluble in water, yielding relatively little $OH^-(aq)$.
8. First, determine the volume of the mixture.

   0.600 liter + 0.400 liter = 1.000 liter

   Next, determine the concentration of each acid.

$$\text{HCl:} \quad \frac{0.600 \cancel{\text{liter}} \times 1.0 \times 10^{-3}\ \text{M}}{1.00\ \text{liter}} = 0.000600\ \text{M}$$

$$\text{HNO}_3\text{:} \quad \frac{0.400 \cancel{\text{liter}} \times 1.0 \times 10^{-3}\ \text{M}}{1.00\ \text{liter}} = 0.000400\ \text{M}$$

Because both acids are strong (and monoprotic), the H^+ concentration is equal to the concentration of the acid. Therefore, $[H^+] = 6.00 \times 10^{-4}$ M + 4.00×10^{-4} M = 1.00×10^{-3} M, and pH = $-\log[H^+]$ = 3.00.

9. Write a balanced reaction.

$$2\ HCl + Ba(OH)_2 \rightarrow BaCl_2 + 2\ H_2O$$

Calculate the number of moles of H^+.

$$\frac{0.500\ \text{liter}}{1} \times \frac{0.0200\ \text{mole}}{1\ \text{liter}} = 1.00 \times 10^{-2}\ \text{mole}\ H^+.$$

Calculate the number of moles of OH^-.

There should be twice as many moles of OH^- as moles of $Ba(OH)_2$.

$$\frac{2\ \text{moles}\ OH^-}{1\ \text{mole}\ Ba(OH)_2} \times \frac{0.100\ \text{liter}}{1} \times \frac{0.100\ \text{mole}\ Ba(OH)_2}{1\ \text{liter}} = 0.0200\ \text{mole}\ OH^-$$

Write the net ionic equation.

$$H^+ + OH^- \rightarrow H_2O$$

Because every mole of H^+ uses 1 mole of OH^-, calculate the number of moles of excess H^+ or OH^-.

$$2.00 \times 10^{-2}\ \text{mole}\ OH^- - 1.00 \times 10^{-2}\ \text{mole}\ H^+$$
$$= 1.00 \times 10^{-2}\ \text{mole}\ OH^-\ \text{excess}$$

What is the approximate pH in the final solution?

$$pOH = -\log[OH^-] = -\log[1.00 \times 10^{-2}] = 2$$
$$pH = 14.00 - pOH = 14.00 - 2.00 = 12.00$$

Another way to do this step, if you could use a calculator, would be

$$[OH^-] = \frac{1.00 \times 10^{-2}\ \text{mole}}{0.600\ \text{liter}} = 0.0167\ M$$

$$pOH = 1.778$$
$$pH = 12.222$$

10. Calculate the number of moles of H^+ in 250 mL of a HNO_3 solution which has a pH of 2.00. HNO_3, is a monoprotic acid. pH = $-\log[H^+]$ so 2.00 = $-\log[H^+]$ and $[H^+] = 1.00 \times 10^{-2}$ M.

$$\text{mole}\ H^+ = \frac{1.00 \times 10^{-2}\ \text{mole}\ H^+}{1\ \text{liter}} \times \frac{0.25\ \text{liter}}{1} = 2.5 \times 10^{-3}\ \text{mole}\ H^+$$

Determine the number of milliliters of concentrated HNO_3 solution that is needed.

$$\frac{2.5 \times 10^{-3}\ \text{mole}\ H^2}{1} \times \frac{1000\ \text{mL sol'n}}{2.00\ \text{mole}\ H^+} = 1.3\ \text{mL soln}$$

11. The gram-equivalent weight (GEW) of a base is the mass of the base (in grams) that will provide 1 mole of hydroxide ions in a reaction or that will react with 1 mole of H^+ ions. This problem can do done by sing the factor-label method.

$$\frac{121.64\text{ g Sr(OH)}_2}{1\text{ mole Sr(OH)}_2} \times \frac{1\text{ mole Sr(OH)}}{2\text{ equiv. Sr(OH)}_2} = \frac{60.82\text{ g Sr(OH)}_2}{1\text{ equiv.Sr(OH)}_2}$$

Note: Had the reaction been

$$Sr(OH)_{2(aq)} + H^+_{(aq)} \rightarrow Sr(OH)^+_{(aq)} + H_2O(1)$$

there would have been only 1 equivalent per mole. In other words, you must know the specific reaction in order to use the concept of equations.

12. From the acidity constant table in text book we see that of the five acids listed, only sulfurous acid, H_2SO_3, has an acidity constant larger than 1.8×10^{-4}, the K_a of formic acid, HCOOH.

13. The reaction between amide ion, NH_2^-, and H_2O goes to completion:

$$NH_2^+ + H_2O \xrightarrow{100\%} NH_3(aq) + OH^-(aq)$$

Ammonia, NH_3, and nitrite ion, NO_2^- are weak bases and undergo proton-transfer reactions with H_2O that proceed to the right only to a slight extent. Ammonium ion, NH_4^+, is a weak acid, and the proton-transfer reaction with H_2O also proceeds only to a slight extent. The nitrate ion, NO_3^-, does not react with water in dilute solution.

14. The pH increases when a base is added. The only basic substance of the 5 choices is K_2CO_3, a basic salt. The carbonate ion, CO_3^{2-}, is a base and reacts with acetic acid as follows:

$$CO_3^{2-}(aq) + H_3COOH(aq) \rightleftharpoons HCO_3^-(aq) + CH_3COO^-(aq)$$

15. If pH = 10.82 then pOH = 14.00 – 10.82 = 3.18 $[OH^-]$ = antilog (–3.18) = $10^{-3.18}$ = 6.6×10^{-4} M.

16. The conjugate base of H_3O^+ is H_2O. The conjugate acid of OH^- is H_2O.

17. Water can act either as a proton donor or as a proton acceptor. The two anions, acetate ion, CH_3COO^-, and sulfide ion, S^{2-}, are bases. Ammonium ion, NH_4^+, and benzoic acid, C_6H_5COOH, are both acids.

18. $[H_3O^+] = [A^-]$ = antilog(–5.35) = 4.47×10^{-6} M

$$[HA] = 0.050 - 4.47 \times 10^{-6} = 0.050\text{ M}$$

$$K_a = \frac{[H_3O^+][A^-]}{[HA]} = \frac{(4.47 \times 10^{-6})^2}{5.0 \times 10^{-2}} = 4.0 \times 10^{-10}$$

19. Formic acid, HCOOH, is a weak acid and must be written in molecular form. Potassium hydroxide, KOH, is a strong base, 100% ions in dilute aqueous solution. The potassium ion is a spectator ion and taken no part in the proton-transfer reaction between the proton donor, HCOOH, and the proton acceptor, OH^-.

20. Acidity and basicity constants always apply to the proton-transfer reaction with H_2O. The basicity constants for HS^-, $K_b(HS^-)$, is the equilibrium constant for the reaction which HS^- accepts a proton from H_2O.

21. The acidity constant for HS^-, $K_a(HS^-)$, is the equilibrium constant for the reaction in which HS^- donates a proton to water.

22. The solution having the highest pH is the most basic. The solutions listed in (a), (b), (c), and

(d) are all acidic. Solutions of NaCl are neutral pH = 7.

23. The solution having the lowest pH is the most acidic. Both perchloric acid, $HClO_4$, and hydrochloric acid, are strong acids, but the $[H_3O^+] = 2.0$ M in (a) and only 0.020 M in (c).

24. Potassium oxide, K_2O, reacts with H_2O to form OH^-

$$K_2O(s) + H_2O \xrightarrow{100\%} 2K^+ (aq) + 2\ OH^-(aq)$$

25. Acetic acid, CH_3COOH, is a weak acid and must be written in molecular form. Sodium carbonate is a salt, a strong electrolyte, 100% ions in aqueous solution. A proton-transfer reaction between the donor, CH_3COOH, and the acceptor, CO_3^{2-} occurs. The sodium ions are by stander (spectator) ions and do not appear in the equation.

26. If $[OH^-] = 4.8 \times 10^{-3}$, pOH = $-\log(4.8 \times 10^{-3}) = 2.32$
Thus, pH = 14.00 – pOH = 14.00 – 2.32 = 11.68

27. $K_a(HCOOH) = 1.8 \times 10^{-4} = \dfrac{[H_3O^+][HCOO^-]}{[HCOOH]}$

Let $x = [H_3O^+] = [HCOO^-]$
since these two ions are formed in a 1 : 1 molar ratio by the proton-transfer reaction between HCOOH and H_2O. Then the concentration of HCOOH molecules is less than the initial value by the amount x:
$[HCOOH] = 0.20 - x$
Assume $x = < 10\%$ of 0.20. Then, $0.20 - x \sim 0.20$ and

$$1.84 \times 10^{-4} = \frac{x^2}{0.20}$$

$$x^2 = 3.6 \times 10^{-5} = 36 \times 10^{-6} \text{ and } x = 6.0 \times 10^{-3}$$

Check: Is $6.0 \times 10^{-3} < 10\%$ of 0.20? Yes, $6.0 \times 10^{-3} < 2.0 \times 10^{-2}$
Hence, $[H_3O^+] = 6.0 \times 10^{-3}$ M and pH = $-\log(6.0 \times 10^{-3}) = 2.22$

28. The $[H_3O^+]$ in a solution containing both a strong acid and a weak acid is essentially just the value contributed by the strong acid. The proton-transfer reaction between H_2S and H_2O proceeds to only a very slight extent, and is repressed by the H_3O^+ ions from the HCl. Thus, $[H_3O^+] = 0.500$ M, and pH = $-\log(0.500) = 0.30$

29. The highest pH means the most basic. Ammonium chloride, NH_4Cl, is an acidic salt, and KNO_3 is a neutral salt. The other three solutions are all basic. To determine the most basic, we must compare the K_b values of the bases.

$$K_b(CN^-) = K_w/K_a(HCN) = (1.00 \times 10^{-14})/(4.9 \times 10^{-10}) = 2.0 \times 10^{-5}$$
$$K_b(OAc^-) = K_w/K_a(HOAc) = (1.00 \times 10^{-14})/(1.8 \times 10^{-5}) = 5.6 \times 10^{-10}$$
$$K_b(CH_3NH_2) = K_w/K_a(CH_3NH_3^+) = (1.00 \times 10^{-14})/(2.4 \times 10^{-11}) = 4.2 \times 10^{-4}$$

Thus methylamine has the largest basicity constant of these three bases, and solution (a) has the highest pH.

30. Since we are only asked about α, the degree of dissociation, use the Ostwald dilution law.

$$K_a = \frac{\alpha^2 C}{(1-\alpha)} = \frac{\alpha^2\ (0.048)}{(1-\alpha)} = 6.3 \times 10^{-5}$$

Assume $\alpha < 100\%$ of 1. Then, $(1 - \alpha) \sim 1$ and
$\alpha^2(0.048) = 6.3 \times 10^{-5}$ or $\alpha^2 = 1.3 \times 10^{-3} = 13 \times 10^{-4}$

$$\alpha = 3.6 \times 10^{-2} = 3.6\%$$

Check: Is $3.6 \times 10^{-2} < 10\%$ of 1? Yes, $3.6 \times 10^{-2} < 1.0 \times 10^{-1}$. Thus the degree of dissociation of 0.048 *F* C_6H_5COOH is 3.6%.

31. $CH_3COOH(aq) + H_2O \rightleftharpoons CH_3COO^-(aq) + H_3O^+(aq)$

$$K_a = \frac{[H_3O^+][CH_3COO^-]}{[CH_3COOH]} = 1.8 \times 10^{-5}$$

Let $x = [CH_3COO^-] = [H_3O^+]$
Then $[CH_3COOH] = 0.080 - x$
The equilibrium constant equation is

$$1.8 \times 10^{-5} = \frac{x^2}{0.080 - x}$$

Assume $x < 10\%$ of 0.080. Then, $0.080 - x \sim 0.080$, and $x^2 = (0.080)(1.8 \times 10^{-5})$
$= 1.44 \times 10^{-6}$
$\therefore x = 1.2 \times 10^{-3}$
Check: Is $1.2 \times 10^{-3} < 10\%$ of 0.080? Yes, $1.2 \times 10^{-3} < 8.0 \times 10^{-3}$ Thus, $[H_3O^+]$
$= 1.2 \times 10^{-3}$ and $pH = -\log(1.2 \times 10^{-3}) = 2.92$

32. Both NaI and $NaNO_3$ solutions are neutral. The other three solutions are all basic, as their anions are bases. The most basic solution is the one for which for K_b for the anion base is the largest. Since $K_b = K_w/K_a$, where K_a is the acidity constant of the conjugate acid, the most basic anion is the one whose conjugate acid has the smallest K_a value. The three conjugate acids of F^-, CN^-, and OAc^-, are respectively, HF, HCN, and HOAc. Of these, the one with the smallest K_a is HCN. Thus CN^- is the strongest base of the three, and the most basic solution is the NaCN.

33. This solution contains both a strong base and a weak base. The total $[OH^-]$ in the solution is essentially just that contributed by the strong base, NaOH. The proton-transfer reaction between NH_3 and H_2O proceeds to only a small extent, and is represented by the OH^- from the NaOH. Thus, $[OH^-] = 0.100$ *M*, pOH = 1 and pH = 14 – 1 = 13.

34. $K_a(HONO) = 4.5 \times 10^{-4}$

$$pK_a = -\log(4.5 \times 10^{-4}) = 4 - \log(4.5) = 4 - 0.65 = 3.35$$

35. Methanol, CH_3OH, is an extremely weak acid. If the proton bonded to the oxygen atom is lost, the anion remaining is CH_3O^-, methoxide ion, a very string base.

36. An equilibrium constant that is very much larger than 1, as this one is, indicates that the position of equilibrium is far to the right. This means that CN^- has a greater tendency to accept a proton than ONO^- does, so that CN^- is a stronger base than ONO^-. It also means that HCN is a weaker acid than HONO, so that statement (b) is false. Statements (a) and (d) are correct, but the magnitude of the equilibrium constant for the reaction shown has nothing to do with either of those statements. The conjugate base of HONO and the conjugate acid of CN^- are known from the definitions of those quantities. There is no way to tell whether HONO will react with F^- solely from the magnitude of K_{eq} for the reaction given.

37. The pK_a of the weak acid should be within ± 1 of 3.4. This is true only for HNO_2 of the options given.

38. Of the weak acids in the options given, only NH_4^+ has a pKa close to 9.0; pKa (NH_4^+) = 9.2.

39. The titration reaction produces $HCOO^-$ ions. Equal volumes of acid and base are mixed, doubling the volume, so the $[HCOO^-]$ is half of 0.100 M.

40. pH = pK_a + log(ratio); 5.24 = 4.74 + log(ratio) Hence log(ratio) = 0.50 and ratio = or 3 : 1

41. No. mmol OH^- used = (29.38 mL) (0.1025 M) = 3.011.

∴ No. mmol C_6H_5COOH originally = 3.011 = No. mmol $C_6H_5COO^-$ formed during the titration.

∴ The final volume of solution is 69.38 mL. Hence

$$[C_6H_5COO^-] = \frac{3.011 \text{ mmol}}{69.38 \text{ mL}} = 0.04340\ M$$

42. Nitrous acid is a weak acid and must be written in molecular form. Potassium hydroxide is a strong electrolyte and is 100% ions.

43. No. mmol OH^- at start = (50.00 mL) (0.0100 M) (2)
= 1.00 mmol

We therefore need 1.00 mmol of H_3O^+, or 100.00 mL of 0.0100 F HCl. The volume at the equivalence point is 150.00 mL, and the solution contains 0.500 mmol Ba^{2+} and 1.00 mmol

Cl^- so that its formality is $\dfrac{0.500 \text{ mmol}}{150.00 \text{ mL}} = 3.33 \times 10^{-3}\ F\ BaCl_2$.

44. Only a 1 : 1 $HOAc/OAc^-$ buffer has pH = 4.75, since 4.75 = pK_a for acetic acid. The flask contains 10.0 mmol of acetic acid as (100.00 mL) (0.100 F) = 10.0. Thus we must and 10.0 mmol of OAc^- ions. Option (b) provides 10.0 mmol of OAc^- as (25.00 mL) (0.200 F) {2 OAc^- ions/$Ba(OAc)_2$}.

45. One-half of 38.62 mL is 19.31 mL, so that the halfway point in the titration occurs when 19.31 mL of base have been added. At the halfway point pH = pKa, and pKa for acetic acid is 4.7.

46. The solution must have both the weak base, NH_3, and its conjugate weak acid, NH_4^+, to be buffer. Option (a) mixes 5.00 mmol OH^- with 5.00 mmol NH_4^+. They react to produce 5.00 mmol NH_3, and there is no excess NH_4^+ remaining after the reaction with OH^-. In option (b) only 2.50 mmol OH^- are added to 5.00 mmol NH_4^+. The reaction produces 2.50 mmol NH_3 and there are 2.50 mmol NH_4^+ in excess. This is therefore a 1 : 1 NH_3/NH_4^+ buffer. In the solutions of options (c) and (d) there is no NH_3 at all, just the weak acid NH_4^+ and a strong acid, H_3O^+.

47. This is a $C_6H_5COO^-/C_6H_5COOH$ buffer. The pKa of benzoic acid is 4.20 = $-\log(6.3 \times 10^{-5})$. For this buffer

$$\text{pH} = \text{pKa} + \log\left(\frac{[C_6H_5COO^-]}{[C_6H_5COOH]}\right) = 4.20 + \log\left(\frac{2}{1}\right) = 4.50$$

48. We have mixed 10.0 mmol OH^- with 20.0 mmol of HCOOH. The reaction $OH^- + HCOOH \rightleftharpoons H_2O + HCOO^-$ produces 10.0 mmol of $HCOO^-$ ions, and uses up 10.0 mmol of HCOOH, leaving 10.0 mmol HCOOH in excess. This is therefore a 1 : 1 $HCOO^-$/HCOOH buffer, with pH = pK_a of HCOOH, or 3.75. For HCOOH, Ka = 1.8×10^{-4} and $pK_a = -\log(1.8 \times 10^{-4}) = 3.75$.

49. The titration reaction is

$$CH_3NH_2(aq) + H_3O^+(aq) \rightleftharpoons CH_3NH_3^+(aq) + H_2O$$

No. mmol H_3O^+ added = No. mmol $CH_3NH_3^+$ formed

$$= (39.26 \text{ mL})\ (0.150\ M) = 5.889 \text{ mmol}$$

Volume of solution at the equivalence point = 79.62 mL

$$CH_3NH_3^+(aq) + H_2O \rightleftharpoons CH_3NH_2(aq) + H_3O^+(aq)$$

Let, $x = [CH_3NH_2] = [H_3O^+]$ at the equivalence point

Then

$$[CH_3NH_3^+] = \frac{5.889 \text{ mmol}}{79.26 \text{ mL}} - x = 0.07430 - x$$

$$K_a(CH_3NH_3^+) = 2.4 \times 10^{-11} \text{ (From Tables).}$$

$$\text{or, } \frac{x^2}{0.07430 - x} = 2.4 \times 10^{-11}$$

Assume $x < 10\%$ of 0.7430. Then, $0.07430 - x \sim 0.07430$ and

$$x^2 = (2.4 \times 10^{-11})\ (7.430 \times 10^{-2}) = 1.78 \times 10^{-12}$$

$$x = [H_3O^+] = 1.34 \times 10^{-6}\ M$$

Check: Is $1.3 \times 10^{-6} < 10\%$ of 0.743? Yes, $1.3 \times 10^{-6} < 7.43 \times 10^{-3}$. The pH if therefore $-\log(1.34 \times 10^{-6}) = 5.87$

50. No. mmol H_3O^+ added = 7.50

No. mmol OH^- added = 4.50

Excess H_3O^+ = 3.00 mmol in a volume of 120.00 mL. $[H_3O^+] = \dfrac{3.00 \text{ mmol}}{120.00 \text{ mL}} = 0.025\ M$ and

pH = $-\log(0.025) = 1.60$

51. A solution of CO_2 in water is carbonic acid, for which $K_a = 4.3 \times 10^{-7}$

$$CO_2(aq) + 2H_2O \rightleftharpoons H_3O^+(aq) + HCO_3^-(aq)$$

$$x = [H_3O^+] = [HCO_3^-]$$

Then, $x^2/0.050 = 4.3 \times 10^{-7}$ or $x^2 = 2.15 \times 10^{-8}$

$$\therefore x = 1.47 \times 10^{-4} = [H_3O^+] \text{ and pH} = -\log(1.47 \times 10^{-4}) = 3.83$$

52. A buffer must contain a weak base and its conjugate weak acid. Aniline is a weak base; its conjugate weak acid is anilinium ion. Aniline reacts with H_3O^+ to produce anilinium ion. In option (d) 0.100 mol H_3O^+ from the HCl react s with 0.100 mol aniline to produce 0.100 mol anilinium ion, leaving 0.100 mol aniline in excess. The solution of option (d) is therefore a 1 : 1. In (a) there is no excess aniline after the reaction. In (b) there is no anilinium ion, two bases have been added. In (a) and (b) there is no weak base or weak acid.

53. Since the pH is greater than 7 in the titration of a weak acid versus a strong base, only options (a) and (b) are possible. The titration of HA versus OH^- produces the weak base A^-. For this weak base, $K_b = K_w/K_a = 2.5 \times 10^{-9}$
No. mmol A^- formed = (28.45 mL)(0.106 *M*) = 2.900 mmol
Volume of solution at the equivalence point = 53.54 mL
Let

$$x = [HA] = [OH^-] \text{ at the equivalence point}$$

$$[A^-] = \frac{2.900 \text{ mmol}}{53.54 \text{ mL}} - x = 0.5416 - x$$

$$K_b = 2.5 \times 10^{-9} = \frac{x^2}{0.05416 - x}$$

Assume $x < 10\%$ of 0.05416. Then $0.05416 - x \sim 0.05416$ and

$$x^2 = (0.05416)(2.5 \times 10^{-9}) = 1.35 \times 10^{-10}$$

$$x = 1.16 \times 10^{-5} = 1.2 \times 10^{-5}$$

Check: Is $1.2 \times 10^{-5} <$ 10% of 0.05416? Yes, $1.2 \times 10^{-5} < 5.4 \times 10^{-3}$
Hence $[OH^-] = 1.2 \times 10^{-5}$ *M*

pOH = $-\log(1.16 \times 10^{-5})$ = 4.93 and pH = 14.00 – 4.93 = 9.07. Thymolphthalein does not change color until after the equivalence point in this titration, so that phenolphthalein is the only suitable indicator.

54. $B + H_2O \rightleftharpoons BH^+ + OH^-$

$$K_b = 2 \times 10^{-5} = \frac{[BH^+][OH^-]}{[B]}$$

If $[B] = [BH^+]$ then $[OH^-] = 2 \times 10^{-5}$ and $[H_3O^+] = 5 \times 10^{-10}$
PH = $-\log(5 \times 10^{-10})$ = 9.3

55. Barium hydroxide and sulfuric acid are soluble string electrolytes and are written in ionic form. There are two products, H_2O and insoluble $BaSO_4$, both of which are written in molecular form.

56. No. mmol H_2X at start = (20.00 mL) (0.200 *F*) = 4.00
No. mol OH^- to reach second equivalence point = 8.00
The volume of base required, V_b, in mL is therefore

$$V_b = \frac{8.00 \text{ mmol}}{0.250 \text{ M}} = 32.0 \text{ mL}$$

Volume of solution at equivalence point = 52.0 mL As the titration reaction produces 4.00 mmol of X^{2-} from the original 4.00 mmol of H_2X,

$$[X^{2-}] = \frac{4.00 \text{ mmol}}{52.0 \text{ mL}} = 0.0769\ M$$

57. $H_2S(aq) + H_2O \rightleftharpoons HS^-(aq) + H_3O^+(aq)$
Since K_2 for H_2S is so much smaller than K_1 we can neglect the second stage of ionization relative to the first.

Let $x = [H_3O^+] = [HS^-]$ in a saturated solution of H_2S

$$K_a(H_2S) = 1 \times 10^{-7} = \frac{[H_3O^+][HS^-]}{[H_2S]} = \frac{x^2}{0.10}$$

Thus $x^2 = 1.0 \times 10^{-8}$ and $x = 1.0 \times 10^{-4}\ M$

58. From the preceeding question we have

$[H_3O^+] = [HS^-] = 1 \times 10^{-4}\ M$ in a saturated H_2S solution,

so that $pH = -\log(1.0 \times 10^{-4}) = 4.0$

59. As $K_1 >> K_2$ for H_2S, $[HS^-] = [H_3O^+]$ in a solution of H_2S in water. Using the second ionization constant,

$$K_2 = \frac{[H_3O^+][S^{2-}]}{[HS^-]} = 1.3 \times 10^{-13}$$

$$[S^{2-}] = K_2 = 1.3 \times 10^{-13}\ M$$

60. $H_3O^+(aq)\ OAc^-(aq) \rightleftharpoons HOAc(aq) + H_2O$

Since HOAc is a weak acid, this reaction goes virtually to completion. We added 60.0 mmol of OAc^- ions and 20.0 mmol of H_3O^+. As they react in a 1 : 1 molar ratio, 20.0 mmol OAc^- react with 20.0 mmol H_3O^+ to produce 20.0 mol HOAc, leaving 40.0 mmol OAc^- in excess. Thus the $[HOAc]/[OAC^-]$ ratio is 20.0/40.0 = 1/2.

$$K_a(HOAc) = 1.8 \times 10^{-5} = \frac{[H_3O^+][OAc^-]}{[HOAc]}$$

and $[H_3O^+] = (1.8 \times 10^{-5})(1/2) = 9.0 \times 10^{-6}\ M$

61. The weak acid of the conjugate pair used to prepare a buffer must have a pK_a within ±1 pH unit of the desired pH. The only weak acid of those given that has a pK_a with ±1 of 3.2 of formic acid, for which $pK_a = 3.75$.

62. A mixture of KCl and HCl has a neither a weak base nor a weak acid and is not a buffer, but it is acidic.

63. There are two weak bases in this solution, CO_3^{2-} and CH_3COO^- ions, but no weak acid. It is not a buffer, but it is basic, with pH > 7.

64. This is a 1 : 2 $HOAc/OAc^-$ buffer. Since pK_a of HOAc is 4.75, the pH of this buffer is between 4 and 7.

65. This solution contains a strong base and a weak base, but no weak acid, so it is not a buffer.

66. H_3O^+ and OH^- have been mixed in a 1 : 1 molar ratio. They react to form H_2O, and the solution has pH = 7.

67. $H_3O^+(aq) + OAc^-(aq) \rightleftharpoons HOAc(aq) + H_2O$

This reaction goes virtually to completion as HOAc is a weak acid. We have added 25.0 mmol OAc^- ions and 50.0 mmol H_3O^+ ions. Thus 25.0 mmol OAc^- react with 25.0 mmol H_3O^+ to produce 25.0 mmol HOAc, leaving 25.0 mmol H_3O^+ in excess. The solution contains a strong acid, H_3O^+, and a weak acid, HOAc, but no weak base, and is therefore not a buffer. It is however, acidic, with pH < 7.

68. The mercurous ion is diatomic, Hg_2^{2+}.

69. $$K_{eq} = \frac{[HX][M^+]}{[H^+]} = \frac{[HX][M^+][X^-]}{[H^+][X^-]} = \frac{K_{sp}\,(MX)}{K_a\,(HX)}$$

70. $Mn(OH)_2(s) \rightleftharpoons Mn^{2+}(aq) = 2OH^-(aq)$

Let s = molar solubility of $Mn(OH)_2$ in water

Then $[Mn^{2+}] = s$ and $[OH^-] = 2s$

$K_{sp} = [Mn^{2+}]\,[OH^-]^2 = (s)\,(2s)^2 = 4s^3$

71. Option (a) is incorrect because KCl is a soluble salt. At the instant of mixing,

$$[Ba^{2+}] = \frac{(50.00\text{ mL})\,(0.050\,M)}{(100.00\text{ mL})} = 0.025\,M$$

However, the reaction $Ba^{2+}(aq) + SO_4^{2-}(aq) \rightleftharpoons BaSO_4\downarrow$ goes virtually to completion so that the $[Ba^{2+}]$ in the final solution is much less than 0.025 M. The $[K^+]$ in the mixture is

$$\frac{(50.00\text{ mL})\,(0.10\,M)\,(2)}{(100.00\text{ mL})} = 0.10\text{ M}$$

72. All of the choices are electrolytes in which the molar ratio of cation: anion is 1 : 1. For this type of electrolyte, $s = (K_{sp})^{1/2}$, and the one with largest K_{sp} has the largest molar solubility in water.

$K_{sp}(SrCO_3) = 9.4 \times 10^{-10}$

$K_{sp}(PbSO_4) = 1.8 \times 10^{-8}$

$K_{sp}(AgCl) = 1.8 \times 10^{-10}$

$K_{sp}(MnS) = 5 \times 10^{-15}$

$K_{sp}(FeS) = 5 \times 10^{-18}$

73. $[Ba^{2+}] = K_{sp}(BaSO_4)/[SO_4^{2-}]$

$= (1.1 \times 10^{-10})/(5.5 \times 10^{-4}) = 2.0 \times 10^{-7}\,M$

74. $AgBr(s) \rightleftharpoons Ag^+(aq) + Br^-(aq)$

Let z = molar solubility of AgBr in 0.20 F KBr

Then $[Ag^+] = z$ and $[Br^-] = 0.20 + z$

$K_{sp} = [Ag^+]\,[Br^-] = (z)\,(0.20 + z) = 5.0 \times 10^{-13}$

Assume $z <$ 10% of 0.20. Then $(0.20 + z) \sim 0.20$, and

$z = (5.0 \times 10^{-13})/(0.20) = 2.5 \times 10^{-12}$

Since $2.5 \times 10^{-22} <<$ 10% of 0.20, $K_{sp} = 0.20z$

75. The relationship between molar solubility and K_{sp} is different for different types of electrolytes. For 1:1 electrolytes, $K_{sp} = s^2$. There are three 1:1 electrolytes among the options. The K_{sp} value of these 3 are:

AgI 1.5×10^{-16} $SrSO_4$ 2.8×10^{-7} MnS 5.0×10^{-15}

Of these three, $SrSO_4$ has by far the largest K_{sp} and therefore the largest molar solubility. There are two 1:2 electrolytes, for which $K_{sp} = 4s^3$.

$K_{sp}(PbI_2) = 8.7 \times 10^{-9}$ and $K_{sp}\{Zn(OH)_2\} = 4.5 \times 10^{-17}$

So that PbI_2 has the larger molar solubility of these two. The question therefore really comes down to: Which has the larger molar solubility in water, $SrSO_4$ or PbI_2?

For $SrSO_4$, $s = (K_{sp})^{1/2} = (28 \times 10^{-8})^{1/2} = 5.3 \times 10^{-4}\ M$

For PbI_2, $s = (K_{sp}/4)^{1/3} = (2.2)^{1/3} \times 10^{-3} = 1.3 \times 10^{-3}\ M$

Hence PbI_2 has the largest molar solubility of the 5 salts.

76. $$[K^+] = \frac{(20.00\ \text{mL})(0.100\ M)(2)}{100.00\ \text{mL}} = 0.040\ M$$

$$[NO_3^-] = \frac{(80.00\ \text{mL})(0.120\ M)(2)}{100.00\ \text{mL}} = 0.192\ M$$

Since $CaCO_3$ is an insoluble electrolyte, it precipitates from this solution, and the $[Ca^{2+}]$ and $[CO_3^{2-}]$ are both considerably smaller than at the instant of mixing.

No. mmol Ca^{2+} added = (80.00 mL) (0.120 M) = 9.60

No. mmol CO_3^{2-} added = (20.00 mL) (0.100 M) = 2.00

As there is excess Ca^{2+}, virtually all the CO_3^{2-} is precipitated, and there are 9.60 – 2.00 = 7.60 mmol of Ca^{2+} left in 100.00 mL of solution, so that $[Ca^{2+}] = 0.076\ M$

77. $[Pb^{2+}] = K_{sp}(PbSO_4)/[SO_4^{2-}] = 1.7 \times 10^{-8}/(0.10)$
$= 1.7 \times 10^{-7}\ M$

78. For any 1:1 electrolyte in pure water, $K_{sp} = s^2$. Therefore $s = (K_{sp})^{1/2} = (1.7 \times 10^{-8})^{1/2} = 1.3 \times 10^{-4}\ M$.

79. Calcium carbonate reacts with any strong acid:

$$CaCO_3(s) + 2H^+(aq) \rightleftharpoons H_2O + CO_2\uparrow + Ca^{2+}(aq)$$

As the CO_2 escapes into the air, a sample of $CaCO_3$ can be dissolved completely by adding sufficient acid.

80. At the instant of mixing,

$$[Ca^{2+}] = \frac{(60.00\ \text{mL})(0.010\ M)}{100.00\ \text{mL}} = 6.0 \times 10^{-3}\ M$$

$$[CrO_4^{2-}] = \frac{(40.00\ \text{mL})(0.025\ M)}{100.00\ \text{mL}} = 1.0 \times 10^{-2}\ M$$

$$Q = [Ca^{2+}]\,[CrO_4^{2-}] = 6.0 \times 10^{-5} < K_{sp} = 7.1 \times 10^{-4}$$

Since $Q < K_{sp}$, no $CaCrO_4$ will precipitate.

$$[NO_3^-] = \frac{(2)(0.010\ M)(60.00\ \text{mL})}{100.00\ \text{mL}} = 0.012\ M$$

No KNO_3 will precipitate because KNO_3 is a soluble salt.

81. For a 1 : 2 electrolyte, $K_{sp} = 4s^3$, where s is the molar solubility in water. Therefore

$$s^3 = (7.9 \times 10^{-10})/4 = 2.0 \times 10^{-10} = 200 \times 10^{-12}$$

$$s = 5.8 \times 10^{-4}\ M$$

82. $SrF_2(s) \rightleftharpoons Sr^{2+}(aq) + 2F^-(aq)$

Let x = molar solubility of SrF_2 in 0.2 F NaF

Then $[Sr^{2+}] = x$ and $[F^-] = 0.10 + 2x$

$$[Sr^{2+}][F^-]^2 = (x)(0.10 + 2x)^2 = 7.9 \times 10^{-10}$$

Assume $2x < 10\%$ of 0.10. Then, $(0.10 + 2x) \sim 0.10$

$$x(0.10)^2 = 7.9 \times 10^{-10} \text{ and } x = 7.9 \times 10^{-8}\ M$$

Check: Is $2(7.9 \times 10^{-8}) < 10\%$ of 0.10? Yes, $1.8 \times 10^{-7} < 1.0 \times 10^{-2}$

Thus the molar solubility of SrF_2 in 0.10 F NaF is 7.9×10^{-8}.

83. Let y – molar solubility of SrF_2 in 0.10 F $Sr(NO_3)_2$. Then, $[Sr^{2+}] = 0.10\ y + y$ and $[F^-] = 2y$.

$$K_{sp} = [Sr^{2+}][F^-]^2 = 7.9 \times 10^{-10} = (0.10 + y)(2y)^2$$

Assume $y < 10\%$ of 0.10. Then, $(0.10 + y) \sim 0.10$

$$(0.10)(4y^2) = 7.9 \times 10^{-10} \text{ and } y^2 = (7.9 \times 10^{-10})/(0.40)$$

$$y = 4.4 \times 10^{-5}\ M$$

84. $$K_{eq} = \frac{[Pb^{2+}][HSO_4^-]}{[H^+]} = \frac{[Pb^{2+}][SO_4^{2-}][HSO_4^-]}{[H^+][SO_4^{2-}]}$$

$$= K_{sp}(PbSO_4)/K_a(HSO_4^-) = 1.8 \times 10^{-6}$$

85. The anion I^- has virtually no tendency to accept a proton in dilute aqueous solution because HI is a strong acid, 100% dissociated into ions. The anions of all the other salts are bases, and react with H_3O^+ to form their conjugate weak acids.

86. $[Pb^{2+}] = 2.3 \times 10^{-2}\ M$ and $[Br^-] = 4.6 \times 10^{-2}\ M$

$$K_{sp} = [PB^{2+}][Br^-]^2 = (2.3 \times 10^{-2})(4.6 \times 10^{-2})^2$$

$$= (2.3)(4.6)^2 \times 10^{-6} = 4.9 \times 10^{-5}$$

87. $[Ca^{2+}] = K_{sp}/[F^-]^2 = K_{sp}/(0.080)^2$

88. If pH = 8.30 then pOH = 14.00 – 8.30 = 5.70

$$[OH^-] = \text{antilog}(-5.70) = 2.0 \times 10^{-6}\ M$$

$$Zn(OH)_2\ (s) \rightleftharpoons Zn^{2+}(aq) + 2\ OH^-(aq)$$

If $[OH^-] = 2.0 \times 10^{-6}\ M$ then $[Zn^{2+}] = 1.0 \times 10^{-6}\ M$

$$K_{sp} = [Zn^{2+}][OH^-]^2 = (1.0 \times 10^{-6})(2.0 \times 10^{-6})^2 = 4.0 \times 10^{-18}$$

89. If 99.9% is precipitated, then 0.1% is still in solution. Therefore

$$[PB^{2+}] = (0.001)(0.120\ M) = 1.2 \times 10^{-4}\ M$$

$$[CrO_4^{2-}] = K_{sp}(PbCrO_4)/[Pb^{2+}] = \frac{1.8 \times 10^{-14}}{1.2 \times 10^{-4}} = 1.5 \times 10^{-10}\ M$$

90. $MgF_2(s) \rightleftharpoons Mg^{2+}(aq) + 2F^-(aq)$

Let

z = molar solubility of MgF_2 in 0.10 F $Mg(NO_3)_2$

Then

$$[Mg^{2+}] = 0.10 + z \text{ and } [F^-] = 2z$$

$$K_{sp} = [Mg^{2+}]\,[F^-]^2 = (0.10 + z)\,(2z)^2$$

Assume $z < 10\%$ of 0.10. Then

$$(0.10)\,(4z^2) = K_{sp} \text{ and } z = (K_{sp}/0.40)^{1/2}$$

91. No. mmol Pb^{2+} added = (20.00 mL) (0.100 *M*) = 2.00
No. mmol $C_2O_4^{2-}$ added = (30.00 mL) (0.150 *M*) = 4.50
There are 2.50 mmol $C_2O_4^{2-}$ in excess in 50.00 mL solution.

Hence $$[C_2O_4^{2-}] = \frac{2.50 \text{ mmol}}{50.00 \text{ mL}} = 5.00 \times 10^{-2}\ M$$

92. No. mmol Ca^{2+} added = (60.00 mL) (0.100 *M*) = 6.00
No. mmol CO_3^{2-} added = (40.00 mL) (0.125 *M*) = 5.00

$$Ca^{2+}(aq) + CO_3^{2-}(aq) \rightleftharpoons CaCO_3\downarrow$$

Since Ca^{2+} and CO_3^{2-} react in a 1:1 molar ratio, there is 1.00 mmol Ca^{2+} in excess. Let x = $[CO_3^{2-}]$ at equilibrium

Then $$[Ca^{2+}] = \frac{1.00 \text{ mmol}}{100.00 \text{ mL}} + x = 0.0100 + x$$

$$K_{sP} = [Ca^{2+}][CO_3^{2-}] = 5 \times 10^{-9} = (x)\,(0.0100 + x)$$

Assume $x < 10\%$ of 0.0100. Then $0.0100 + x \sim 0.0100$ and

$$x = 5 \times 10^{-7}\ M$$

Since $5 \times 10^{-7} < 1.0 \times 10^{-3}$. Thus, $[CO_3^{2-}] = 5 \times 10^{-7}\ M$

93. $$K_{eq} = \frac{[OH^-]^2}{[S^{2-}]} = \frac{[Cu^{2+}][OH^-]^2}{[Cu^{2+}][S^{2-}]} = \frac{K_{sp}\,(Cu(OH)_2)}{K_{sp}\,(CuS)}$$

$$= \frac{2 \times 10^{-19}}{8.7 \times 10^{-36}} = 2 \times 10^{+16}$$

94. $ZnCO_3(s) + 2H^+(aq) \rightleftharpoons Zn^{2+}(aq) + H_2O + CO_2\uparrow$

95. (i) An important point for you to understand about acid-base titration is that, in the titration of a *strong* acid by a strong base, all the H_3O^+ to be neutralized is present in the solution initially. In the titration of a *weak* and by a strong base, the H_3O^+ to be neutralized is *not* just what is present in the solution initially. That is, think of the neutralization reaction in this way: as the H_3O^+ present initially is converted to H_2O more acid ionizes, until eventually all of it is ionized and neutralized, thus with oxalic acid this means that ionization occurs through the second step, yielding a final solution of $Na_2C_2O_4$.

no. mol OH^- consumed

$$= 28.77 \text{ mL} \times \frac{1 \text{ L}}{1000 \text{ mL}} \times \frac{0.3650 \text{ mol NaOH}}{1 \text{ L}} \times \frac{1 \text{ mol OH}}{1 \text{ mol NaOH}}$$

$$= 0.01050 \text{ mol OH}^-$$

$$\text{no. mol neutr.} = 0.01050 \text{ mol OH}^- \times \frac{1 \text{ mol } H_2C_2O_4}{2 \text{ mol OH}^-} = 0.005250 \text{ mol } H_2C_2O_4$$

$$\text{molarity} = \frac{0.005250 \text{ mol } H_2C_2O_4}{0.00500 \text{ L}} = 1.05 \text{ M } H_2C_2O_4$$

(ii) Here we should (a) write the equation for the *first* ionization step of oxalic acid, (b) describe the ionization equilibrium in a "standard" format, (c) substitute known data into this format, and (d) solve for a numerical value of K_{a_1}. Let us work with the saturated aqueous solution, for which we now know the total solution molarity from part 1) and the pH. Note that for pH = 0.67; log $[H_3O^+] = -0.67$; and $[H_3O^+] = 0.21$ M

	$H_2C_2O_4 + H_2O$	$\rightleftharpoons H_3O^+$	$+ HC_2O_4^-$
placed in solution:	1.05 M	—	—
changes:	–0.21 M	+0.21 M	+0.21 M
after first ionization:	0.84 M	0.21 M	0.21 M

$$K_{a_1} = \frac{[H_3O^+][HC_2O_4^-]}{[H_2C_2O_4]} = \frac{(0.21)(0.21)}{(0.84)} = 5.2 \times 10^{-2}$$

(iii) At the point where complete neutralization occurs we have present a solution of sodium oxalate, with neither unreacted oxalic acid nor excess NaOH present. The solution volume if the 5.00 mL of acid + 100 mL, of water + 28.77 mL of NaOH(aq) = 134 mL. [Note that the 100, mL water used to dilute the 5.00-mL acid sample has no effect on the volume of NaOH (aq) required for the titration, but it does affect the concentrations of the species present in the solution.] The amount of $H_2C_2O_4$ that is neutralized is 0.005250 mol, and one $C_2O_4^{2-}$ ion is produced for every H_2C_2O molecule that is neutralized by reaction.

$$\text{no. mol of } C_2O_4^{2-} = 0.005250 \text{ mol } H_2C_2O_4 \times \frac{1 \text{ mol } C_2O_4^{2-}}{1 \text{ mol } H_2C_2O_4}$$

$$= 0.005250 \text{ mol } C2O_4^{2-}$$

$$[C_2O_4^{2}] = \frac{0.0055250 \text{ mol } C_2O_4^{2-}}{0.134 \text{ L}} = 0.0392 \text{ M}$$

(iv) The oxalate ion in solution hydrolyzes.

$$C_2O_4^{2-} + H_2O \rightarrow HC_2O_4^- + OH^-$$

Since hydrolysis produces OH^-, the solution is basic.

96. (i) This is a straightforward titration of a weak acid by a strong base. The neutralisation reaction to the first equivalence point is

$$HOC_6H_4COOH + OH^- \rightarrow HOC_6H_4COO^- + H_2O$$

$$\text{no. mmol } HOC_6H_4COOH = 16.24 \text{ mL} \times \frac{0.0200 \text{ mmol OH}^-}{\text{mL}} \times \frac{1 \text{ mmol } HOC_6H_4COOH}{1 \text{ mmol OH}^-}$$

$$= 3.25 \text{ mmol } HOC_6H_4COOH$$

$$\text{molarity of } HOC_6H_4COOH = \frac{0.325 \text{ mmol } HOC_6H_4COOH}{25.00 \text{ mL}} = 0.0130 \text{ M}$$

(ii) The key to this question is in recognizing the special significance of the equivalence point (16.24 mL) and the *half-neutralization* point (8.12 mL). At the half-neutralization point, since $[HOC_6H_4COO^-] = [HOC_6H_4COO]$, then pH = pK_{a_1}. Recall that this conclusion is based on the condition

$$pH = pK_{a_1} + \log \frac{[HOC_6H_4COO^-]}{[HOC_6H_4COOH]} = pK_{a1} + \log 1 = pK_{a_1}$$

$$pK_{a_1} = 4.57; \quad \log pK_{a_1} = -4.57; \quad pK_{a_1} = \text{antilog } -4.57 = 2.7 \times 10^{-5}$$

To establish pK_{a_2}, note that the anion present at the first equivalence point, HOC_6H_4COO–, can ionize further as an acid or it can ionize (hydrolyze) as a base. For a solution containing such anions we have established the expression: pH = $\frac{1}{2}(pK_{a_1} + pK_{a_2})$.

The pH at the first equivalence point is 7.02, and the value of pK_a, from above is 4.57.

$$pK_{a_2} = (2 \times pH) - pK_{a_1} = (2 \times 7.02) - 4.57 = 9.47$$

$$pK_{a_2} = \text{antilog } -9.47 = 3.4 \times 10^{-10}$$

(iii) This point should lie within the range in which the solution being titrated acts as a buffer solution, and we can calculate its pH with the "buffer equation" [Henderson-Hasselbalch equation]. First, however, we have to work out the details of the stoichiometry of the neutralization to this point. The no. mmol of weak acid present initially is 25.00 mL × 0.0130 mmol HOC_6H_4COOH/mL = 0.325 mmol. The amount of OH– added through this point in the titration, in milimoles, is 10.00 mL × 0.0200 mmol OH^- mL^{-1} = 0.200 OH^-.

	HOC_6H_4COOH +	OH^- →	$HOC_6H_4COO^-$ + H_2O
initially present:	0.325 mmol	—	—
add:		0.200 mol	
changes:	–0.200 mmol	–0.200 mmol	+0.200 mmol
after reaction:	0.125 mmol	~0	0.200 mmol
concens	0.125 mmol/35.00 mL = 3.57×10^{-3} M		0.200 mmol/35.00 mL = 5.71×10^{-3} M

$$pH = pK_{a_1} + \log \frac{[HOC_6H_4COO^-]}{[HOC_6H_4COOH]} = 4.57 + \log \frac{5.71 \times 10^{-3}}{3.57 \times 10^{-3}}$$

$$= 4.57 + 0.20 = 4.77$$

97. (i) We have, $Ca(OH)_2(s) \rightleftharpoons Ca^{2+}(aq) + 2\,OH^-(aq)$

The maximum pH will be found in the solution with the highest $[OH^-]$. This will be a saturated solution of $Ca(OH)_2$. We must calculate $[OH^-]$ in this saturated solution. Let us set $[Ca^{2+}] = S$ and $[OH^-] = 2S$.

$$K_{sp} = [Ca^{2+}][OH^-]^2 = (S)(2S)^2 = 4S^3 = 5.5 \times 10^{-6}$$

$$S^3 = 1.4 \times 10^{-6} \qquad S = 0.011 \text{ M} \qquad [OH^-] = 2S = 0.022 \text{ M}$$

$$pOH = -\log[OH^-] = -\log 0.022 = 1.66$$

$$pH = 14.00 - pOH = 14.00 - 1.66 = 12.34$$

The maximum pH attainable in an aqueous solution of $Ca(OH)_2$, is 12.34

(ii) Note that NH_4Cl is the salt of a weak base and a strong acid. This salt should hydrolyze to produce an acidic solution (that is, NH_4^+ can ionize as an acid). $CaOH)_2$, a base, should be more soluble in an acidic solution.

Also, with such a large value K, we can conclude that equilibrium in the dissolving reaction is displaced far to the right. $[Ca_2^+]$ should be much larger in this solution than in pure water saturated with $Ca(OH)_2$.

(iii) Now we can proceed

$$Ca(OH)_2(s) + CO_3^{2-}(aq) \rightleftharpoons CaCO_3(s) + 2\,OH^-$$

	CO_3^{2-}	OH^-
Initial concentrations	1.0 M	—
changes	$-x$ M	$+ 2x$ M
at equilibrium:	$(1.0 - x)$ M	$2x$ M

$$K = \frac{[OH^-]}{[CO_3^{2-}]} = \frac{(2x)^2}{(1.0 - x)} = 2.0 \times 10^3$$

Now assume that x is very small compared to 1.0. Notice that K is quite a large quantity (2.0×10^3). Equilibrium will be displaced for to the right; x will be very nearly equal to 1.0. We can obtain an exact solution with the quadrate formula, however.

$$4x^2 + (2.0 \times 10^3)\,x - \mathbf{2.0 \times 10^3 = 0}$$

$$x = \frac{-2.0\times10^3 \pm \sqrt{4.0\times10^6 + 3.2\times10^4}}{2\times4} = 0.998 \approx 1$$

Thy hydroxide ion concentration $[OH^-] = 2x \approx 2$ M

$$pOH = -\log 2 = -0.3 \text{ and } pH = 14.00 - pOH = 14.3$$

Reaction (A) goes essentially to completion.

98. For CH_3COOH

$$CH_3COOH + H_2O \rightleftharpoons CH_3COO^- + H_3O^+$$

	CH_3COOH	CH_3COO^-	H_3O^+
Initial conc.	C	0	0
Equilibrium conc.	$C - C\alpha$	$C\alpha$	$C\alpha$

Where C is initial conc. in mol L^{-1} and α is degree of dissociation.

Now, $$K_a = \frac{[CH_3COO^-][H_3O^+]}{[CH_3COOH]} = \frac{(C\alpha)(C\alpha)}{(C - C\alpha)} = \frac{C\alpha^2}{1-\alpha}$$

Assuming $$\alpha << 1 \Rightarrow K_a = C\alpha^2 \Rightarrow \alpha = \sqrt{\frac{K_a}{C}}$$

Thus, $$\alpha = \sqrt{\frac{1.8\times10^{-5}}{0.1}} = 0.0134$$

Now $[H_3O^+] = C\alpha = 0.1 \times 0.0134 = 1.34 \times 10^{-3}$

so $pH = -\log[H_3O^+] = -\log(1.34 \times 10^3) = 2.87.$

99. In water, ethanoic acid forms an equilibrium mixture very quickly:

$$CH_3COOH(aq) \rightleftharpoons H^+(aq) + CH_3COOH^-(aq)$$

The 0.100 mole dm^{-3} is the concentration of CH_3COOH before any has dissociated.

Given that only 1.3 percent of ethanoic acid molecules dissociate, then the concentration of ethanoic acid that has dissociated must be 1.3 per cent of 0.100 mol dm^{-3}, which is 1.3×10^{-3} mol dm^{-3}. From the equation mole of $CH_3COOH(aq)$ that dissociates gives 1 mole of $H^+(aq)$. So, [H+(aq)] must be 1.3×10^{-3} mol dm^{-3}.

Substituting this value into the equation for pH gives:

$$pH = -\log10\ (1.3 \times 10^{-3}) = 2.9$$

100. Since ethanoic acid is a weak acid, it forms an equilibrium mixture in water.

$$CH_3COOH(aq) \rightleftharpoons CH3COO^-(aq) + H^+(aq)$$

Therefore, the expression for the acid dissociation constant is:

$$K_a = \frac{[CH_3COO(aq)][H^+(aq)]}{[CH_3COOH(aq)]}$$

From the equation, every mole of CH_3COOH that dissociates given 1 mole of CH_3COO^- and 1 mole of $H^+(aq)$.

So, if $[H^+(aq)]$ at equilibrium is x, then it follows that $[CH_3COO^-(aq)]$ must also be x.

It also follows that $[CH_3COOH(aq)]$ must be $(0.01 - x)$, since some of the ethanoic acid has dissociated.

This can be summarised as follows:

	$CH_3COOH \rightleftharpoons$ undissociated ethanoic acid	$CH_3COO(aq)$ +	$H^+(aq)$
At start before equilibrium mol dm^{-3}	0.10	0	0
At equilibrium mol dm^{-3}	$0.10 - x$	x	x

These values can be substituted in the equation for the acid dissociation constant. Therefore,

$$1.8 \times 10^{-5} = \frac{x^2}{0.10 - x}$$

The concentration of the $H^+(aq)$ is very small compared with that of the undissociated acid, therefore the concentration of the undissociated acid is very early 0.10. So we can assume that $(0.10 - x)$ is approximately 0.10. Using the assumption, we can write the equation as:

$$1.8 \times 10^{-5} = \frac{x^2}{0.10}$$

Which can be rearranged to give:

$$x^2 = 1.8 \times 10^{-6}$$

Therefore:

$$x = (1.8 \times 10^{-6}) = 1.34 \times 10^{-3} \text{ (ignoring significant figures)}$$

Since x is the $H^+(aq)$ concentration, we can substitute this value of x into the equation for pH.

$$\therefore pH = -\log[H^+(aq)]$$
$$= -\log(1.34 \times 10^{-3}) = 2.9$$

101. Assume the excess calcium carbonate is shaken with distilled water until equilibrium is attained. The solubility of calcium carbonate corresponds to the concentration of the aqueous calcium ions (for the aqueous carbonate ions) in the equilibrium solution:

$$CaCO_3(s) \rightleftharpoons Ca^{2+}(aq) + CO_3^{2-}(aq)$$

At start (before shaking) mol dm^{-3}	0	0
At equilibrium mol dm^{-3}	6.9 × 10–5	6.9×10^{-5}

$$K_{sp}(CaCO_3) = [Ca^{2+}(aq)]\,[CO_3^{2-}(aq)]$$
$$= (6.9 \times 10^{-5})\,(6.9 \times 10^{-5})$$
$$= 4.8 \times 10^{-9}\ mol^2\ dm^{-6}$$

102. (i) Let s be the solubility of barium chromate (VI).

$$BaCrO_4(s) \rightleftharpoons Ba^{2+}(aq) + CrO_4^{2-}(aq)$$

At equilibrium: s s

Therefore: $K_{sp} = [Ba^{2+}(aq)]\,[CrO_4^{2-}(aq)] = s^2 = 2 \times 10^{-10}.\ mol^2\ dm^{-6}$

This gives: $s = 1.4 \times 10^{-5}\ mol\ dm^{-3}$.

(ii) In this calculation, it is necessary to consider dissolving barium chromate (VI) in aqueous potassium chromate (VI):

$$BaCrO_4(s) \rightleftharpoons Ba^{2+}(aq) = C_2O_4^{2-}(aq)$$

At start mol dm^{-3}:	0	1.0
At equilibrium mol dm^{-3}:	s	(1.0 + s)

Barium chromate (VI) is sparingly soluble in water Therefore s is small compared with 1.0, and so (s + 1.0) is approximately equal to 1.0.

Now: $K_{sp} = [Ba^{2-}(aq)]\,[CrO_4^{2-}(aq)]$

$$2 \times 10^{-10} = s(1.0)$$

The gives $s = 2 \times 10^{-10}\ mol\ dm^{-3}$.

Note that the solubility of barium chromate (VI) is very much smaller in potassium chromate(VI) than in water along. This is the common ion effect

103. Use the ionic product of water:

$$K_w = [H^+(aq)]\,[OH^-(aq)]$$

Substitute into this equation the hydrogen ion concentration and the numerical value for the ionic product:

$$1.0 \times 10^{-14} = 2.5 \times 10^{-8} \times [OH^-(aq)]$$

Rearranging this equation gives:

$$[OH^-(aq)] = \frac{1.0 \times 10^{-14}}{2.5 \times 10^{-8}}$$
$$= 4.0 \times 10^{-7}\ mol\ dm^{-1}$$

104. Potassium hydroxide is a strong base that fully dissociates into its constituent aqueous ions.

$$KOH(aq) \rightarrow K^+(aq) + OH^-(aq)$$

The stoichiometry of the equation shows that in 0.150 mol dm^{-3} aqueous potassium hydroxide, $[OH^-(aq)]$ = 0.150 mol dm^{-3}. Using the ionic product of water gives:

$$[H^+(aq)] = \frac{K_w}{[OH^-(aq)]} = \frac{1.00 \times 10^{-14}}{0.150} = 6.67 \times 10^{-14} \text{ mol dm}^{-3}$$

So: $$pH = -\log_{10}(6.67 \times 10^{-14}) = 13.2$$

105. Aqueous barium hydroxide is a strong base. It fully dissociates in water to give $Ba^{2+}(aq)$ and $OH^-(aq)$.

$$Ba(OH)_2(aq) \rightarrow Ba^{2+}(aq) + 2OH^-(aq)$$

1 mol of barium hydroxide contains 2 mol of hydroxide ions, so:

$$[OH-(aq)] = 2 \times [(Ba(OH)_2(aq)] = 0.400 \text{ mol dm}^{-3}$$

$$POH = -\log_{10}(0.400) = 0.40$$

Using the relationship between pOH and pH gives:

$$PH = 14 - pOH = 13.60$$

106. $$NH_3(aq) + H_2O(l) \rightleftharpoons NH^*(aq) + OH^-(aq)$$

At start:	0.10 mol dm^{-3}	0 mol dm^{-3}	0 mol dm^{-3}
At equilibrium:	0.10 – x mol dm^{-3}	x mol dm^{-3}	x mol dm^{-3}

Using the expression for K_b: $1.8 \times 10^{-5} = \dfrac{x \times x}{0.10 - x}$

Ammonia is a weak base, so the equilibrium lies to the left. Therefore, x is small compared to 0.10, so:

$$1.8 \times 10^{-5} = \frac{x \times x}{0.10}$$

Solving this equation gives: $x = 1.34 \times 10^{-3}$ mol dm^{-3}

∴ $$pOH = -\log_{10}(.00134) = 2.9$$

Since x is $[OH^-(aq)]$, using the expression pOH + pH = 14 gives the pH = 11.1.

107. Let the solubility of silver chloride in water is S mol L^{-1}, then

$$AgCl(s) \rightleftharpoons Ag^+_{(aq)} + Cl^-_{(aq)};\ K_{sp} = [Ag^+]\,[Cl^-]$$

As $$S = [Ag^+] = [Cl^-].$$

Thus, $$K_{sp} = S \times S = S^2$$

⇒ $$S' = \sqrt{K_{sp}} = \sqrt{1.6 \times 10^{-10}} = 1.3 \times 10^{-5} \text{ mol L}^{-1}$$

Let the solubility of silver chloride in the presence of 0.1 M NaCl is S'. As NaCl is a strong electrolytes so it will be completely ionized in solution.

$$NaCl(s) \rightarrow Na^+(aq) + Cl^-(aq)$$

Thus, $$[NaCl] = [Na^+] = [Cl^-] = 0.1 \text{ mol L}^{-1}$$

We assume that Cl^- ions added by NaCl greatly exceeds the conc. of Cl^- ions that arise from AgCl.

$\therefore$ $$K_{sp} = [Ag^+][Cl^-]$$
$$= (S')(0.1 + S') = (S')(0.1)$$

$\Rightarrow$ $$S' = \frac{K_{sp}}{0.10} = \frac{1.6 \times 10^{-10}}{0.10} \approx 2 \times 10^{-9} \text{ mol L}^{-1}$$

108. $S = (K_{sp})^{1/3} = \left(\frac{9 \times 10^{-12}}{4}\right) = 1.3 \times 10^{-4} \text{ mol.dm}^{-3}$

109. (i) Let the solubility of $M_g(OH)_2$ in water = S then

$$Mg(OH)_2 \rightleftharpoons Mg^{2+} + 2\,OH^-;\ K_{sp} = [Mg^{2+}][OH^-]^2$$

Here $[Mg^{2+}] = S$ and $[OH^-] = 2S$

$\therefore$ $K_{sp} = (S)(2S)^2 = 4\,S^3 \Rightarrow S = (K_{sp}/4)^{1/3}$

$\Rightarrow$ $$S = \left(\frac{1.2 \times 10^{-11}}{4}\right) = 1.442 \times 10^{-4} \text{ mol dm}^{-3}.$$

110. Since $K_{sp}(AgI) < K_{sp}(AgCl)$ hence AgI is less double and it will precipitate out first. Solubility product expression for AgI is

$$K_{sp}(AgI) = [Ag^+][I^-] \qquad \text{(i)}$$

Hence, the concentration of Ag^+ ions at which AgI will start precipitation is

$$\left[Ag^+ = \frac{K_{sp}(AgCl)}{[I^-]}\right] \qquad \text{(ii)}$$

Similarly, the concentration of Ag^+ ions at which AgCl will start precipitation is

$$[Ag^+] = \frac{K_{sp}(AgCl)}{[Cl^-]} \qquad \text{(iii)}$$

when AgCl starts to precipitate, I^- present in solution

$$\frac{[I^-]}{[Cl^-]} = \frac{K_{sp}(AgI)}{K_{sp}(AgCl)} = \frac{1.5 \times 10^{-16}}{1.56 \times 10^{-10}} = 9.6 \times 10^{-7}$$

Hence, AgCl will start precipitating out when the $[I^-]$ is approximately one million part of $[Cl^-]$, or the precipitation of I^- as AgI is almost complete.

111. Molarity of H_2SO_4 solution $= \frac{0.049/98}{200/1000} = 2.5 \times 10^{-3}$ M

$$\text{As } H_2SO_4 \rightarrow 2H^+_{aq} + SO_4^{2-}$$

Hence, $[H_3O^+] = 2[H_2SO_4] = 2 \times 2.5 \times 10^{-3} = 5 \times 10^{-3}\ M$

Now, $pH = -\log[H_3O^+] = -\log(5 \times 10^{-3}) = 2.3$

112. Total volume of solution = 150 + 150 = 300 mL

Total milligram equivalents of excess HCl

$= 150 \times 0.2 - 150 \times 0.1 = 30 - 15 = 15$ gm equivalents.

Hence, Normality of HCl = $\frac{15}{300} = \frac{1}{20}$ N

As HCl is strong acid so $HCl + H_2O \rightarrow H_3O^+ + Cl^-$

Thus, $[HCl] = [H_3O^+] = \frac{1}{20} N = \frac{1}{20} M$

$\therefore$ $pH = -\log [H_3O^+] = \left(\frac{1}{20}\right) = 1.301$

113. Before mixing, pH = 5 $\therefore$ $[H_3O^+] = 10^{-5}$ M

pH = 3 $\therefore$ $[H_3O^+] = 10^{-3}$ M

After mixing, total value of the solution becomes twice,

Therefore, $$[H_3O^+] = \frac{10^{-5} + 10^{-3}}{2} = 5.05 \times 10^{-4}\ M$$

$\therefore$ $pH = -\log [H_3O^+] = -\log(5.05 \times 10^{-4}) = 3.297$

114. For weak acid

$$HX + H_2O \rightleftharpoons H_3O^+ + X^-$$

	HX	H_3O^+	X^-
Initial conc.	c	0	0
Equilibrium conc.	$c - c\alpha$	$c\alpha$	$c\alpha$

where initial conc. = c mol/L and degree of dissociation = α

Now $$K_a = \frac{[H_3O^+][X^-]}{[HX]} = \frac{(C\alpha)(C\alpha)}{C - C\alpha} = \frac{C\alpha^2}{1-\alpha}$$

Assuming $\alpha << 1 \Rightarrow K_a = C\alpha^2$

$\Rightarrow$ $$\alpha = \sqrt{\frac{K_a}{c}} = \sqrt{\frac{7.3 \times 10^{-6}}{0.23}} = 5.63 \times 10^{-3}$$

Now $[H_3O^+] = c\ \alpha = 0.23 \times 5.63 \times 10^{-3} = 1.296 \times 10^{-3}$

So $pH = -\log [H_3O^+] = -\log (1.296 \times 10^{-3}) = 2.887$

115. As the HCl solution is very dilute with sources of H_3O^+ viz. HCl and H_3O must be considered.

Now $[H_3O^+]$ from HCl = 10^{-8} M (because $[H_2O^+]$ = [HCl] as strong acid (HCl ionizes almost completely)

Let $[H_3O^+]$ from H_2O = x M so $[OH^-] = x$ M

Consider the ionization of water

$$2\ H_2O(l) \rightleftharpoons H_3O^+(aq) + OH^-(aq)$$

conc. at Equilibrium. $(10^{-8} + x)$ x

since $K_w = [H_3O^+]\ [OH^-] = 10^{-14}$

$\therefore$ $(10^{-8} + x)\ (x) = 10^{-14}$

Solving the quadratic we get $x = 9.5 \times 10^{-8}$.

Thus total $[H_3O^+] = 10^{-8} + x = 10^{-8} + x = 10^{-8} + 9.5 \times 10^{-8} = 10.5 \times 10^{-8}$

And $pH = -\log[H_3O^+] = -\log(10.5 \times 10^{-8}) = 6.98$

116. Given $K_a = 1.75 \times 10^{-5}$

$\therefore$ $pK_a = -\log K_a = -\log (1.75 \times 10^{-5}) = 4.757$

Now $[CH_3COOH] = \left(\frac{1}{60} \times \frac{1000}{100}\right) = 0.1667 \text{ mol dm}^{-3}$

And $[CH_3COONa] = \left(\frac{1.5}{82} \times \frac{1000}{100}\right) = 0.1829 \text{ mol dm}^{-3}$

From Henderson's equation

$$PH = pK_a + \log \frac{[Salt]}{[Acid]}$$

$$\Rightarrow \quad pH = pK_a + \log \frac{[CH_3COONa]}{CH_3COOH}$$

$$= 4.757 + \log\left(\frac{(0.1829)}{(0.1667)}\right)$$

$$\Rightarrow \quad pH = 4.797 = 4.78$$

117. The NH_3(aq) will be in equilibrium with its conjugate acid:

Conc.	$NH_3(aq)$ + $H_2O(l)$	$\rightleftharpoons$ $NH_4^+(aq)$	+ $OH^-(aq)$
At start:	0.10 mol dm^{-3}	0.20 mol dm^{-3} $NH^+(aq)$	0 mol dm^{-3}
At equilibrium:	$0.10 - x \text{ mol dm}^{-3}$	$0.20 + x \text{ mol dm}^{-3}$	$x \text{ mol dm}^{3}$

Since $NH_3(aq)$ is a weak base, the equilibrium lies almost completely on the left. Therefore, x is very small compared to 0.10 and 0.20.

At equilibrium: $[NH_3(aq)] = 0.10 \text{ mol dm}^{-3}$

$[NH_4^+(aq)] = 0.20 \text{ mol dm}^{-3}$

$[OH^-(aq)] = x \text{ mol dm}^{-3}$

Substituting these values into the expression for the base dissociation constant gives:

$$K_b = 1.8 \times 10^{-5} = \frac{0.20 \times x}{0.10}$$

Solving for x gives.

$x = 9 \times 10^{-6} \text{ mol dm}^{-3}$

So $pOH = 5.0$ and $pH = 9.0$.

118. Let x be the concentration of the H^+(aq). We the have:

Conc.	$CH_3COOH(aq)$ $\rightleftharpoons$	$CH_3COO^-(aq)$ +	$H^+(aq)$
At start	0.10	0.20	0
At equilibrium	$(0.10 - x)$	$(0.20 + x)$	x

We can assume that x is very small compared with 0.10 and 0.20, since ethanoic acid is a weak acid. Therefore, at equilibrium:

$[CH_3COOH(aq)] = 0.10 \text{ mol dm}^{-3}$
$[CH_3COO^-(aq)] = 0.20 \text{ mol dm}^{-3}$

Substituting these values into the expression for the K_a for ethanoic acid:

$$K_a = \frac{[CH_3COO^-][H^+]}{[CH_3COOH]} = 2x$$

Now: $K_a = 1.8 \times 10^{-5}$

So: $x = 9.0 \times 10^{-6} \text{ mol dm}^{-3}$

$$pH = -\log_{10} [H^+(aq)]$$
$$= -\log_{10} (9.0 \times 10^{-6}) = 5.0$$

The pH of the buffer solution is therefore 5.0.

119. Given $K_b = 1.85 \times 10^{-5}$

$$pK_b = -\log K_b = -\log (1.85 \times 10^{-5}) = 4.73$$

As $pH = 11.45$

$\therefore$ $pOH = 14 - pH = 14 - 11.45 = 2.55$

Since $$pOH = pK_b + \log \frac{[NH_4Cl]}{[NH_3]}$$

$\therefore$ $$2.55 = 4.73 + \log \frac{[NH_4Cl]}{(0.05)}$$

$\Rightarrow$ $[NH_4Cl] = 0.00033\ M$

Hence, number of grams of NH_4Cl in 1 litre buffer solution.

$$= 0.00033 \times 53.5$$
$$= 0.0177 \text{ gms} = 1.77 \times 10^{-2} \text{ gms.}$$

120. **(i)** $$pH = pK_a + \log \frac{[CH_3COONa]}{[CH_3COOH]}$$

$$= -\log (1.75 \times 10^{-5}) + \log\left(\frac{0.01}{0.1}\right)$$

$$pH = 3.757 = 3.76$$

(ii) Amount of HCl added = 0.2 ml of 1 *M* HCl
= 0.2 millimole = 0.0002 moles

As HCl is a strong acid so it will be completely dissociated,. Thus, the amount of H+ ions added will be 0.002 moles. The addition of H^+ ions will result in the reaction

$$CH_3COO^- + H^+ \rightarrow CH_3COOH$$

Hence, the amount of salt will decrease and that of the acid will increase by 0.0002 moles.

$\therefore$ After adding HCl

$[CH_3COOH] = 0.1 + 0.0002 = 0.1002 \text{ mol dm}^{-3}$

and $[CH_3COONa] = 0.01 - 0.002 = 0.0098 \text{ mol dm}^{-3}$

Now $$pH = pKa + \log \frac{[CH_3\ COO\ Na]}{[CH_3COOH]}$$

$$= -\log (1.75 \times 10^{-5}) + \left(\frac{(0.0098)}{(0.1002)}\right)$$

$\Rightarrow$ $pH = 3.747 = 3.75$

(iii) As sodium hydroxide is a strong base, so it will be completely dissociated. Thus, the amount of OH^- ions added will be = 0.002 moles.

The addition of the OH^- will result in the reaction

$$CH_3COOH + OH^- \rightarrow CH_3COO^- + H_2O$$

Hence, the amount of acetic acid will decrease and that of sodium acetate will increase by 0.002 moles.

$\therefore$ After adding NaOH

$$[CH_3COOH] = 0.1 - 0.002 = 0.098 \text{ mol dm}^{-3}$$

And $[CH_3COONa] = 0.01 + 0.002 = 0.012 \text{ mol dm}^{-3}$

Now $$pH = pK_a + \log \frac{[CH_3COONa]}{[CH_3COOH]}$$

$$= -\log (1.75 \times 10^{-5}) + \log \left(\frac{0.012}{0.098}\right)$$

$\Rightarrow$ $pH = 3.845$

121. We have $Mg(OH)_2 \rightleftharpoons Mg^{2+} + 2OH^-$

$\therefore K_{sp} = [Mg^{2+}][OH^-]^2 = 1.2 \times 10^{-11}$

Any OH^- concentration higher than the contained in a saturated solution would cause precipitation. Hence, the solution must be at the point of attaining equilibrium, and the concentrations of ions in solution must be no greater than those required to satisfy the solubility product constant. In this solution $[Mg^{2+}] = 0.10\ M$

$$[Mg^{2+}][OH^-]^2 = 1.2 \times 10^{-11}$$

$$[OH^-]^2 = \frac{1.2 \times 10^{-11}}{0.10} = 1.2 \times 10^{-10} \quad [OH^-] = 1.1 \times 10^{-5} \quad pOH = 4.96 \quad pH = 9.04$$

122. Let solubility of $PbBr_2$ be s mol litre^{-1}

$$PbBr_2(s) \rightleftharpoons PbBr_2(aq) \rightleftharpoons Pb^{2+} + 2Br^-$$

$$\frac{s \times 80}{100} \qquad \frac{2s \times 80}{100}$$ $\because$ Ionisation of $PbBr_2(s) = 80\%$

$\therefore$ $$K_{sp} = [Pb^{2+}][Br^-]^2$$

or, $$8 \times 10^{-5} = \left[\frac{s \times 80}{100}\right]\left[\frac{2s \times 80}{100}\right]^2$$

$\therefore$ $s = 0.034 \text{ mol litre}^{-1}$ ($\because$ M.wt. of $PbBr_2 = 367$)

$$s = 0.034 \times 367 \text{ g litre}^{-1}$$

$$s = 12.48 \text{ g litre}^{-1}$$

123. For, $$CaSO_4 \rightleftharpoons Ca^{2+} + SO_4^{2-}$$

$$[Ca^{2+}][SO_4^{2-}] = K_{sp}$$

Let $[SO_4^{2-}] = a$, just sufficient to precipitate $CaSO_4$ from a solution having $[Ca^{2+}] = 0.005$ M

Then, $[0.005]\,[a] = 2.4 \times 10^{-5} \Rightarrow a = \dfrac{2.4 \times 10^{-5}}{0.005}$

or $[SO_4^{2-}] = 4.8 \times 10^{-3}$ mol litre^{-1}

124. $MgCl_2 + 2NaOH \rightarrow Mg(OH)_2 + 2NaCl$

	$MgCl_2$	$2NaOH$	$Mg(OH)_2$	$2NaCl$
mm before reaction	10	20	0	0
after reaction	0	0	10	20

Thus, 10 *m* mole of $Mg(OH)_2$ are formed. The product of $[Mg^{2+}]\,[OH^-]^2$ is therefore $\left[\dfrac{10}{200}\right] \times \left[\dfrac{20}{200}\right]^2 = 5 \times 10^{-4}$ which is more than K_{sp} of $Mg(OH)_2$. Now solubility (*s*) of $Mg(OH)_2$ can be derived by

$$K_{sp} = s^3$$

$$\therefore \quad s = (K_{sp})^{1/3} = (1.2 \times 10^{-11})^{1/3} = 1.4 \times 10^{-4}$$

$$\therefore \quad [OH^-] = 2s = 2.8 \times 10^{-4}$$

125. $H_2S + H_2O \rightleftharpoons H_3O^+ + HS^- \quad K = \dfrac{[H_3O^+][S^{2-}]}{[H_2S]} = 1 \times 10^{-21}$

$HS^- + H_2O \rightleftharpoons H_3O^+ + S^{2-}$

In 0.10 *M* H_2S nd 0.15 M H_3O^+, the concentrations are effectively $[H_3O^+] = 0.15$ and $[H_2S] = 0.10$

$$K = \frac{(0.15)^2\,[S^{2-}]}{(0.10)} = 1 \times 10^{-21} \qquad [S^{2-}] = \frac{1 \times 10^{-22}}{0.0225} \approx 4 \times 10^{-21}$$

Now, $Ag_2S \rightleftharpoons 2Ag^+ + S^{2-}$ and $K_{sp} = [Ag^+]^2[S^{2-}] = 1.6 \times 10^{-49}$

$$[Ag^+]^2 = \frac{1.6 \times 10^{-49}}{(4 \times 10^{-21})} = 4 \times 10^{-29} \qquad \text{hence } [Ag^+] = 6 \times 10^{-15}\ M$$

126. For the equilibrium

$$NH_4OH \rightleftharpoons NH_4^+ + OH^-$$

$$K = \frac{[NH_4^+][OH^-]}{[NH_4OH]} = 1.8 \times 10^{-5}$$

Let $x = [NH_4^+]$ ormed by the disociation of NH_4OH.

$$x = [NH_4^+] = [OH^-]$$

$$[NH_4OH] = 0.10 - x$$

Assuming $x << 0.10$

$$\frac{x^2}{0.10} = 1.8 \times 10^{-5}$$

$$x = [OH^-] = .3 \times 10^{-3}\ M$$

127. Let us express all quantities as their equivalent in cm^3 of a 0.02 mole dm^{-3} solution

	Initial	Final
	Ac^- = negligible	$Ac^- \equiv$ NaOH added
		$= 10\ cm^3$ of 0.05 mole dm^{-3}
		$= 25\ cm^3$ of 0.02 mole dm^{-3}
	HA 35	HAc = Initial HAc – Ac^- produced
		$= 35 - 25 = 10.$

Thus in the final solution

$[Ac^-] = 25\ cm^3$ of 0.02 mole dm^{-3} solution diluted to 45 cm^3. [HAc] $\equiv$ 10 cm^3 of 0.02 mole dm^{-3} solution diluted to 45 cm^3.

$$\therefore \quad \frac{[Ac^-]}{[HAc]} = \frac{25}{10} = 2.5$$

$$pH = 4.74 + \log \frac{[Ac^-]}{[HAc]}$$

$$pH = 4.74 + \log 2.5$$

$$\text{or} \quad pH = 5.14$$

128. We know, $[HCHO_2] = 1.0$, and $[H^+]$ must be 1.0×10^{-4}, so we can solve for $[HCO_2^-]$ in the ionization constant expression

$$\frac{[H^+][HCO_2^-]}{[HCHO_2]} = \frac{[1.0 \times 10^{-4}][HCO_2^-]}{[1.0]} = 2.1 \times 10^{-4}.$$

$$[HCO_2^-] = 2.1.$$

And if the number of moles per litre must be 2.1, the number of moles be added to 200 ml is $\frac{2.1}{4} = 0.53$

129. pH = 2.70; If $[H^+] = A \times 10^{-3}$, log A must be 0.30, so $[H^+] = 2.0 \times 10^{-3}$. From the hydrolysis equilibrium we see that if $[H^+] = 2.0 \times 10^{-3}$, $[Al(OH)^{2+}]$ must also 2.0×10^{-3}, and $[Al^{3+}]$ would be $1.0 - 0.002 \approx 1.0$ so

$$\frac{[Al(OH)^{2+}][H^+]}{[Al^{3+}]} = \frac{[2.0 \times 10^{-3}][2.0 \times 10^{-3}]}{[1.0]}$$

$$= 4.0 \times 10^{-6}$$

130. (i) Let us calculate the product of the concentration of Ag^+ and $C_2H_3O_2^-$ and compare it to $K_{sp} = (2 \times 10^{-3})$

$$\text{conc. } Ag^+ \times \text{conc. } C_3H_2O_2^- = (1 \times 10^{-2})(2 \times 10^{-2})$$
$$= 2 \times 10^{-4}.$$

Since The concentration product is less than K_{sp} equilibrium is not established, and no silver acetate forms.

(ii) As the precipitate begins to form, the concentration product, $[Ag^+]$ × conc. $[C_2H_3O_2^-]$, must equal the equilibrium value 2×10^{-3}, Neglecting dilution effects, the concentration of $C_2H_3O_2^-$ at that point must be 0.020 *M*. Therefore $[Ag^+] \times 0.020 = 2 \times 10^{-3}$.

$$[Ag^+] = 0.10\ M$$

This is the minimum concentration of Ag^+ at which a precipitate will form. At higher Ag^- ion concentrations precipitation will occur and sufficient $C_2H_2O_2^-$ ions will be removed from solution to maintain the concentration product at 2×10^{-3}.

131. Ammonium Chloride is a strong electrolyte and may be assumed to be completely ionized.

$$\therefore \quad [NH_4^+] = 1000 \text{ ml} \times \left(\frac{0.020 \text{ mole } NH_4^-}{100 \text{ ml}}\right)$$

$$= 0.20 \text{ mole L}^{-1}$$

$$NH_3 + H_2O \rightleftharpoons NH_4^+ + OH^-$$

Conc. $\quad 0.15\ M \quad 0.02$

For this reaction the equilibrium constant K_{eq} is given by

$$K_{eq} = \frac{[NH_4^+][OH^-]}{[NH_3]}$$

$$1.8 \times 10^{-5} = \frac{[0.02][OH^-]}{[0.15]}$$

$$[OH^-] = 1.4 \times 10^{-5} \text{ M.}$$

132.

	V *(litre)*	*molarity (mole litre⁻¹)*	*= number of moles*
NaOH	0.40	0.10	4.0×10^{-3}
HCl	0.010	0.45	$4. \times 10^{-3}$

Both the substances are completely ionized in solution, therefore, 4.0×10^{-3} mole of each neutralizes the other. The total volume of the solution is 0040 + 0.010 = 0.050 litre which contains 0.5×10^{-3} mole each of H^+ and Cl^- ions.

$$[H^+] = \frac{0.5 \times 10^{-5} \text{ mole}}{0.050 \text{ litre}}$$

$$[H^+] = 1 \times 10^{-2} \text{ molar}$$

$$pH = -\log [H^+]$$

$$pH = -\log [10^{-2}]$$

$$PH = 2.$$

133. Let $x = [H_3O^+]_{eq}$

	HOOCH + H_2O ⇌	H_3O^+ +	$CHOO^-$	
initial	0.20	0	0	M
change	$-x$	x	x	M
equil	$0.20 - x$	x	x	M

Since K_a is small and the equilibrium lies to the left, assume $x << 0.20$ and thus $0.20 - x = 0.20$

$$K_a = \frac{[H_3O^+]_{eq}[CHOO^-]_{eq}}{[HOOCH]_{eq}} = \frac{(x)(x)}{(0.20\ x)} = \frac{(x)(x)}{(0.20)} = 1.76 \times 10^{-4}$$

$$x = [H_3O^+]_{eq} = 5.9 \times 10^{-3}\ M$$

134. Let $x = [OH^-]_{eq}$

	NH_3	+ H_2O $\rightleftharpoons$	NH_{4+}	+ OH^-	
Initial Conc.	0.100		0	0	M
Change in Conc.	$-x$		x	x	M
AT equilibrium	$100 - x$		x	x	M

Since K_b is very small and the equilibrium lies for to the left, assume that $x << 0.100\ M$ and $0.100 - x \simeq 0.100\ M$.

$$K_b = \frac{[NH_4^+]_{eq}\,[OH^-]_{eq}}{[NH_3]_{eq}} = \frac{(x)(x)}{(0.100-x)} \cong \frac{(x)(x)}{(0.100)} = 1.76 \times 10^{-5}\ \text{M}$$

$\therefore \quad x^2 = 1.76 \times 10^{-6} \Rightarrow x = 1.3 \times 10^{-3}$

$\therefore \quad pOH = -\log[OH^-] = -\log(1.3 \times 10^{-3}) = 2.876$

and $\quad PH = pK_w - pOH = 14.000 - 2.876 = 11.124$

135. Given $T_1 = 925$ K,

$T_2 = 1000$ K,

$(K_P)_{925} = 18.5$,

$(K_P)_{1000} = ?$ and $\Delta H^\circ = -71.09$ kJ mol^{-1}

Now, we have,

ln $$\ln\frac{(K_p)_{1000}}{(K_p)_{925}} = \frac{\Delta H^\circ}{R}\left[\frac{T_2 - T_1}{T_2 T_1}\right]$$

$$\Rightarrow \quad \ln\frac{(K_p)_{1000}}{18.5} = \frac{-71.09 \times 10^3\ \text{J mol}^{-1}}{8.314\ \text{J K}^{-1}\ \text{mol}^{-1}}\left[\frac{(1000-925)\ \text{K}}{(925\ K)(1000\ \text{K})}\right]$$

$$\Rightarrow \quad (K_P)_{1000} = 9.2486 = 9.25$$

136. A 0.56 mole dm^{-3} solution of K^+OH^- will contain 0.56 mole dm^{-3} of OH^-

$\therefore \quad [OH^-] = 5.6 \times 10^{-1}$ mole dm^{-3}

Hence $\quad pOH = -\log[OH^-]$

$pOH = -\log(5.6 \times 10^{-1})$

$= -(-1 + 0.75) = 0.25$

$pH + pOH = 14$

$pH = 14 - 0.25$

$pH = 13.75.$

137. Since $[H^+][OH^-] = 1.00 \times 10^{-14}$ ..... (1)

A mass balance on chloride ion combined with the charge balance gives the proton condition,

$$[OH^-] = [H^+] - 1.00 \times 10^{-7} \quad \text{..... (2)}$$

Since the solution is so dilute, $[OH^-]$ is not negligible compared to 1.00×10^{-7} and the equations must, therefore, be solved exactly. Substitute the value of $[OH^-]$ from Equation (2) in (1) we get

$$[H^+]^2 - 1.00 \times 10^{-7}[H^+] - 1.00 \times 10^{-14} = 0 \quad \text{..... (3)}$$

On solving Equation (3) for the positive root we get

$$[H^+] = 1.62 \times 10^{-7}$$

$$\therefore \quad pH = 6.79.$$

138. $pH = \dfrac{1}{[H_3O^+]} = \log\dfrac{1}{3\times10^{-3}} = \log\left(\dfrac{10^3}{3}\right) = 3 - \log 3$

or, $\quad pH = 3 - 0.48 = 2.52$

$\quad pH = 2.52$

$$pH = 6.5 = \log\frac{1}{(H_3O^+)}$$

139. (i) $\therefore [H_3O] = 10^{-pH} = 10^{-6.5}$

or $[H_3O^+] = 10^{-7} \times 10^{0.5}$

or $[H_3O^+] = 3 \times 10^{-7}$ g ion L^{-1} ($\because$ $10^{0.5} \approx 3$.)

(ii) Since $\quad [H_3O^+][OH^-] = 1 \times 10^{-14}$

$$\therefore \quad [OH^-] = \frac{1\times10^{-14}}{3\times10^{-7}} \approx 3 \times 10^{-8}$$

or $\quad [OH^-] = 3.0 \times 10^{-8}$ g ion L^{-1}

140. $[OH^-] = 1.5 \times 10^{-1}$

$\therefore \log [OH^-] = \log 1.5 + \log 10^{-1}$

$= 0.2 - 1 = -0.8$

$\therefore pOH = 0.8$

Thus, $\quad pH = 14 - 0.8 = 13.2$

141. Acetic acid ionizes as,

$$CH_3COOH + H_2O \rightleftharpoons CH_3COO^- + H_3O^+$$

$$\text{K-ionization} = \frac{[H_3O^+][CH_3COO^-]}{[CH_3COOH]}$$

Concentration of acetic acid is 0.2 mole L^{-1}

Let $\quad x = [H_3O^+]$ then

$0.2 - x = [CH_3COOH]$

$\therefore \quad 1.8 \times 10^{-5} = \dfrac{x \times x}{0.2 - x}$

or $\quad x = 1.9 \times 10^{-3}$

$\therefore \quad [H_3O^+] = 1.9 \times 10^{-3}$.g ion L^{-1}

Let y by the fraction of acetic acid ionized,

$\therefore \quad 0.2y = 1.9 \times 10^{-3}$

$$y = \frac{1.9 \times 10^{-3}}{0.2} = 9.5 \times 10^{-3} = 0.0095$$

or $y = 0.95$ percent.

142. pH = – log [H$^+$] = 10.6

$\log [H^+] = -10.6$
$= 0.4 - 11$
$=$ antilog 0.4×10 antilog (-11)
$\therefore [H^+] = 2.5 \times 10^{-11}$

143. Since 9.6 cm^{-3} of 0.1 mole dm^{-3} NaOH will completely neutralize an equivalent quantity of the strong acid, HCl.

$$HCl + (Na^+)\ OH^- = H_2O + (Na^+)\ Cl^-$$

Since 10 cm^3 of 0.1 mole dm^{-3} HCl was initially present, a quantity equivalent to 0.4 cm^3 of 0.1 mole dm^{-3} HCl will remain unneutralized in the solution whose total volume is now 19.6 cm^3.

$\therefore$ Concentration of HCl remaining neutralized

$= 0.4/19.61 \times 0.1$ mole dm^{-3}
$= 2.04 \times 10^{-3}$ mole dm^{-3}

[H$^+$] in 1 mole dm^{-3} HCl = 1 mole dm^{-3}

$\therefore$ [H$^+$] in 2.04×10^{-3} mole dm^{-3} HCl

$= 2.04 \times 10^{-3}$ mole dm^{-3}

$pH = -\log [H^+] = -\log (2.04 \times 10^{-3})$

or $pH = 2.69$

144. $AgCl(s) \rightleftharpoons Ag^+ + Cl^-$

$K_{sp} = [Ag^+]\,[Cl^-]$

$[Ag^+] = [Cl^-] = C$

Mol. wt of AgCl = 107.87 + 35.453 = 143.323

$\therefore$ [Cl] = 0.2474

$$\therefore\ C = \frac{1.94 \times 10^{-3}}{1.43 \times 10^{2}} = 1.36 \times 10^{-5}\ mL^{-1}$$

$K_{sp} = (1.36 \times 10^{-5})^2 = 1.85 \times 10^{-10}$
$pK_{sp} = -\log K_{sp} = -\log (1.85 \times 10^{-10})$
$pK_{sp} = -[0.27 + (-10.00)] = -(-9.73)$
$pK_{sp} = 9.73.$

145. $$pOH = pK_b + \log \frac{[NH_4^+]}{[NH_3]}$$

$$\therefore \quad \frac{[NH_4^+]}{[NH_3]} = 1.00$$

$pOH = 4.75 + 0.000$ ($\because$ log 1 = 0)

The original pOH = 4.75. The pOH after addition of NaOH cannot be less than 3.75

$$pOH = 3.75 = 4.75 + \log \frac{[NH_4^+]}{[NH_3]}$$

$$\log \frac{[NH_4^+]}{[NH_3]} = -1.00$$

$$\therefore \quad \frac{[NH_4^+]}{[NH_3]} = 0.10.$$

Hence NaOH can be added until the ratio of $[NH_4^+]$ to $[NH_3]$ 0.10. Initially $[NH_4^+]$ + $[NH_3]$ = 0.200 Although the reaction with OH^- converts NH_4^+ into NH_3, the sum of the concentrations remains 0.200

$$[NH_4^+] + [NH_3] = 0.200$$

$$[NH_4^+] = 0.10\ [NH_3]$$

$$\therefore \quad 0.10\ [NH_3] + [NH_3] = 0.200$$

$$1.10\ [NH_3] = 0.200$$

$$[NH_3] = 0.182\ M$$

Hence $[NH_4^+] = 0.018\ M$

Assuming no change in volume, 0.182 – 0.100 = 0.082 moles of NaOH can be added without changing the pOH by more than 1.00 pOH unit.

146. (i) Given, $K_a = 2.0 \times 10^{-3}$

$$\therefore \log K_a = \log (2 \times 10^{-5}) = \log 2 + \log 10^{-5}$$

$$= 0.3010 - 5 = -4.6990$$

$$\therefore \quad pKa = -\log K_a = -(-4.6990) = 4.6990$$

Approximating to second place of decimal

pKa of the acid = 4.7

(ii) HCl being strong, degree of dissociation $\alpha = 1$

Given concentration of HCl, [HCl] = 0.0001 = 1.0×10^{-4}

$$\therefore \quad 1.0 \times 10^{-4} = [H^+] = [Cl^-]$$

$$\text{or } \log [H^+] = \log (1 \times 10^{-4}) = 0 - 4.0 = -4.0$$

$$\therefore \quad pH = -\log [H^+] = -(-4.0) = 4.0$$

(iii) Molarity of NaOH solution $= \dfrac{\text{mass of solute}}{\text{Molar mass}} \times \dfrac{1000}{\text{Vol. in cm}^2}$

$$\therefore \quad [NaOH] = \frac{0.4}{40} = 0.01 = 1 \times 10^{-2}$$

NaOH is strong base. Therefore $\alpha = 1.0$ or $[OH^-] = [NaOH]$

Thus $[OH^-] = 1 \times 10^{-2}$

$$\therefore \log [OH^-] = \log (1 \times 10^{-2}) = 0 - 2.0 = -2.0$$

$$pOH = -\log [OH^-] = -(-2.0) = 2.0.$$

147. (i) $pH = -\log [H^+] = 5.2$

$$\log [H^+] = -5.2 = -5 - 1 + 1 - 0.2 = (\bar{6} + 0.8)$$

$\Rightarrow$ $[H^+] = 6.31 \times 10^{-6}$

(ii) $pKw = -\log Kw = 13.02$

or $\log Kw = -13.02 = -13 - 1 + 1 - 0.02 = (\bar{14} + 0.98)$

$\therefore$ $Kw = 9.55 \times 10^{-14}$.

148. (i) From the given data

$$[CH_3COOH] = 0.5\ M;\ [CH_3COONa] = 0.5\ M$$
$$pKa = 4.75$$

Substituting these values in the Henderson-Hasselbalch equation for buffers,

$$PH = pK_a + \log \frac{[CH_3COONa]}{[CH_3COOH]}$$

$$= 4.75 + \log \frac{0.5}{0.5} = 4.75 + \log 1.0$$

$$pH = 4.75 + 0 = 4.75$$

(ii) Substituting $[CH_3COONa] = 0.6\ M$ and $[CH_3COOH] = 0.5\ M$ in the above equation,

$$pH = 4.75 + \log \frac{0.6}{0.5} = 4.75 + \log 1.2 = 4.75 + 0.08 = 4.83$$

149. Substituting [Salt] 0.5 M; [acid] 0.05 and pH 5.5 from the given data in the equation

$$pH = pK_a + \log \frac{[Salt]}{[Acid]},\ \text{we get}$$

$$5.5 = pK_a + \log \frac{0.50}{0.05} = pK_a + \log 10$$

$\Rightarrow$ $5.5 = pK_a + 1.0$

or $pK_a = 5.5 - 1.0 = 4.5$

$pK_a = -\log K_a = 4.5$

$\log K_a = -4.5 = -4 - 1 + 1 - 0.5 = (\bar{5} + 0.5)$

Hence $K_a = 3.162 \times 10^{-5}$.

150. Let $x = [H_3O^+]$

	$2H_2O$ $\rightleftharpoons$	H_3O^+	+ OH^-	
initial	–	0	0	M
change	$-2x$	x	x	M
equil	–	x	x	M

Now, $K_w = [H_3O^+][OH^-] = (x)(x) = 1.00 \times 10^{-14}$

$x = [H_3O^+] = 1.00 \times 10^{-7}\ M$

151. $pH = -\log(6.38 \times 10^{-6}) = -\log(6.38) - \log(10^{-6}) = -0.805 + 6.000 = 5.195$

152. Silver chloride (AgCl) will precipitate provided the product $[Ag^+][Cl^-]$ exceeds 2.8×10^{-10} mole$^2 \cdot L^{-2}$.

$$[Ag^+] \text{ in } Ag\,NO_3 = 1.00 \times 10^{-3}\ M;\ [Cl^-] \text{ in } NaCL = 1.00 \times 10^{-4}\ M$$

The product of these two quantities (1.00×10^{-7} mol$^2 \cdot$ L^{-2}) is much large than 2.8×10^{-10} mol$^2 \cdot$ L^{-2}. Therefore some AgCl must precipitates until the condition

$$[Ag^+] = [Cl^-] = 2.8 \times 10^{-10} \text{ mol}^2 \cdot L^{-2} \text{ is reached.}$$

153. The ionic reactions occurring in the solution are

$$PbCl_2 \rightleftharpoons Pb^2 + 2Cl^-$$

$$CuCl \rightleftharpoons Cu^+ + Cl^-$$

lead chloride precipitates when

$$[Pb^{2+}] = [Cl^-]^2 = Ksp = 1.6 \times 10^{-5} \text{ mol}^3 \cdot L^{-3}$$

In this case

$$[Pb^{2+}] = 0.0010\ M$$

Therefore, $PbCl_2$ precipitates when

$$[Cl^-] = \sqrt{\frac{1.6 \times 10^{-5}}{0.00100}} = 1.3 \times 10^{-1}\ M$$

Cuprous chloride precipitates when

$$[Cu^{2+}] = [Cl^-] = 3.2 \times 10^{-7} \text{ mol}^3 \cdot L^{-3}$$

In this case,

$$[Cu^+] = 0.010\ M$$

Therefore, CuCl precipitates when

$$[Cl^-] = \frac{3.2 \times 10^{-7}}{0.010} = 3.2 \times 10^{-5}\ M$$

It is obvious therefore that CuCl would precipitate first as $[Cl^-]$ of 3.2×10^{-5} is reached earlier than $[Cl^-]$ of 1.3×10^{-1} M of $PbCl_2$.

154. Thallous chloride undergoes dissociation as under:

$$TlCl(s) \rightleftharpoons Tl^+\ Cl^-$$

$$K_{sp} = [Tl^+]\ [Cl^-] = 1.7 \times 10^{-4}$$

If the solubility of TlCl = x

$$[Tl^-] = [Cl^-] = x$$

$\therefore$ $$x^2 = 1.7 \times 10^{-4}$$

or $$x = \sqrt{1.7 \times 10^{-4}} = 1.3 \times 10^{-2} \text{ mol} \cdot L^{-1}$$

155. We have in solution,

$$C_2H_5COO\,Ag \rightleftharpoons Ag^+ + C_2H_5COO^-$$

$\therefore$ $$K_{sp} = [Ag^+]\ [C_2H_5COO^-] = x^2 \text{ (say)} = 49 \times 10^{-4}$$

Since the ethanoate and silver ions are produced in equal quantity in solution when the salt dissolves, the solubility of silver ethanoate is

$\therefore$ $$x = \sqrt{(49 \times 10^{-4})}$$

$$= 7 \times 10^{-2} \text{ mole dm}^{-3}$$

156. pH = 5.20, ∴ pOH = 14 – 5.20 = 8.80

$$pOH = 8.80 = -\log [OH^-]$$

$$[OH^-] = 1.6 \times 10^{-9}\ M = [NH_3] \text{ in buffer solution}$$

$$k = \frac{[NH_4^+][OH^-]}{[NH_4OH]}$$

$$[OH^-] = 1.6 \times 10^{-9}\ M;$$

Now, $[NH_4^+] = 0.10\ M$ in NH_4CL

Let $[NH_4OH] = x$

$$\therefore \quad \frac{(1.6 \times 10^{-9})(0.10)}{x} = 1.8 \times 10^{-5} \text{ mole L}^{-1}$$

$x = [NH_4OH] = 8.9 \times 10^{-6}\ M$. This is same as $[NH_3]$. Thus total concentration $= 8.9 \times 10^{-6} + 1.6 \times 10^{-9}$ mole of NH_3.

$$= 8.9016 \times 10^{-6} \text{ mole of } NH_3.$$

157. Since $K_h = \frac{K_w}{K_{HAc}} = \frac{[HAc][OH^-]}{[Ac^-]}$

Let x be the number of moles of acetic acid formed, then

$$[HAc] = x,\ [OH^-] = x$$

$$[Ac^-] = 0.10 - x$$

Assuming $x << 0.10$

$$\therefore \quad \frac{x^2}{0.10} = \frac{10^{-14}}{1.9 \times 10^{-5}}$$

$$x^2 = 5.3 \times 10^{-11} \text{ mole}^2 \text{ L}^{-2}$$

$$x = 7.3 \times 10^{-6} \text{ M}$$

Solving we get, pOH = 5.14

∴ pH = 14.0 – 5.14 = 8.86

158. Here degree of hydrolysis, $x = \frac{0.023}{100} = 2.3 \times 10^{-4}$

Concentration of the salt, $c = 0.01\ M$

$$K_h = x^2c\ (2.3 \times 10^{-4})^2 \times 0.01 = 5.29 \times 10^{-10}\ M$$

From the reaction

$$CH_3COO^- + H_2O \rightleftharpoons CH_3COOH = OH^-$$

$$[OH^-] = c.x = 0.01 \times 2.3 \times 10^{-4} = 2.3 \times 10^{-6}\ M.$$

159. (i) To start with, when $f = 0$, we have 50 cm^3 of 01 N acetic acid ($K_a = 1.85 \times 10^{-5}$). Applying Ostwald's dilution law,

$$a = \sqrt{\frac{K_a}{c}} \text{ and } [H_3O^+] = c\alpha = \sqrt{K_a c}$$

Substituting $c = 0.1$ and $K_a = 1.85 \times 10^{-5}$ in (i)

$$[H_3O^+] = \sqrt{1.85 \times 10^{-5} \times 0.1} = 1.36 \times 10^{-3}$$

Hence at $f = 0$. $pH = -\log (1.36 \times 10^{-3})$

$$= -(1.36) + \log 10^{-3}) = -(0.13 - 3.00)$$

or $pH = 2.87$

(ii) At $f = 05$ (the mid point) we will having a solution which has equal concentration acetic acid and sodium acetate which is formed by the neutralization of the acid by NaOH.
As such, $pH = pKa = -\log Ka = -\log (1.85 \times 10^{-5})$

$$= -(0.27 - 5.00) = 4.73$$

(iii) At $f = 1.0$ neutralisation is complete. Only sodium acetate will be present with concentration $0.1/2 = 0.05\ M$ because 50 cm^3 of 0.1 M acetic acid will require 50cm^3 of 0.1 M. Due to hydrolysis (of this salt) the p increases. The pH of the solution at this stage is given by Eqn. 31.19.

$$pH = \frac{1}{2}(pK_W + pK_a + \log c)$$

Putting $pK_W = 14.0$, $pK_b = 4.73$ and $c = 0.05\ M$, we get

$$pH = \frac{1}{2}(14.0 + 4.73 + \log (5 \times 10^{-2})$$

$$= \frac{1}{2}(14.0 + 4.73 - 1.30) \quad (\because \log 5 \times 10^{-2} = -1.30)$$

$$= \frac{17.43}{2} = \boxed{8.72}$$

160. The centre of the range may not be calculated by averaging the pH values, since pHs are exponential inform. The centre of the range can be determined by a averaging the corresponding concentrations of H^+.

(i) For, $pH = 3.1$

$$\log [H^+] = -3.1 = 0.9 - 4.0$$

$\therefore$ $[H^+] = 7.9 \times 10^{-4} = 79.0 \times 10^{-5}$

(ii) For, $pH = 4.5$

$$\log [H^+] = 4.5 = 0.5 - 5$$

$\therefore$ $[H^+] = 3.2 \times 10^{-5}$

Average $$[H^+] = \frac{79 \times 10^{-5} + 3.2 \times 10^{-5}}{2}$$

$$= 4.1 \times 10^{-4}$$

Now, $K = [H^+]$ = ionization constant of indicator

or $K = 4.1 \times 10^{-4}$

161. (Molecular Weight of $NaC_2H_3O_2$ = 82.03 g mol^{-1}; $K_a = 1.8 \times 10^{-5}$)

$$[CH_3COOH] = 0.150\ M$$

$$[CH_3COONa] = (10.3\ g\ NaC_2H_3O_2/82.03\ g\ mol^{-1} \times 0.500\ L = 0.251\ M$$

$$pK_a = -\log (1.8 \times 10^{-5}) = 4.75$$

$$pH = pK_a + \log \frac{[CH_3COONa]}{[CH_3COOH]} = 4.75 + \log(0.251/0.150)$$

$$pH = 4.75 + \log (1.67) = 4.75 + 0.22 = 4.97$$

162. Hasselbalch Henderson Equation can be used to calculate the pH of the formic acid buffer system:

$$pH = pK_a + \log[HCOO^-]/[HCOOH]$$
$$= 3.7^- + \log (0.12/0.15) = 3.6$$

Addition of 0.02 M HCl, a strong acid, converts 0.02 M formate ion to formic Acid. The new concentration of the buffer components are $[HCOO^-]$ 0.10 M and $[HCOOH] = 0.17$ M. The final pH is

$$pH = pK_a + \log [HCOO^-]/[HCOOH]$$
$$= 3.7 + \log (0.10/0.17) = 3.5$$

163. (i) When 0.01 mole HCl (strong acid) is added, 0.01 mole of H_2O^+ ions$^-$ will be added to the solution. These long will combine with CH_3COO^- to produce 0.01 mole of CH_4COOH. As a net result of this $[CH_3COOH] = [CH_3COONa] = 1.00 - 0.01 = 0.99$. Substituting these values in Henderson-Hasselbatch equation

$$pH = pK_a + \log \frac{[CH_3COONa]}{[CH_3COOH]}$$

$$= 4.74 + \log \frac{0.99}{1.01} = 4.74 - 0.01 = 4.73$$

Decrease in pH change in this case = 0.01

(ii) When 0.01 NaOH is added, the base will combine with acid thereby, increasing the concentration of the salt from 1.00 to 1.01 mol dm^{-3} and decreasing the concetration of acid from 1.00 to 0.99 mol dm^{-3}

From the above equation

$$pH = 4.74 + \log \frac{1.01}{0.99} = 4.74 + 0.01 = 4.74$$

Hence in the case increase in pH = 0.01

164.

	$CaSO_4(s)$ $\rightleftharpoons$	$Ca^{2+}(aq)$ +	$SO_4^{2-}(aq)$
Initial conc.	Ignore	0	0
x disolves (solubility $x = 4.9 \times 10^{-3}$ M.)			
at equilibrium concentration are		x	x

$$K_{sp} = [Ca^{2+}[SO_4^{2-}]\ ^2 = (4.9 \times 10^{-3})^2 = 2.4 \times 10^{-5}$$

165. Writing the dissociation equation for Lead II Iodide

$$PbI_2 \rightleftharpoons Pb^{+2} + 2I^-$$

The solubility product is given by

$$K_{ap} = [Pb^{+2}]\ [I^-]^2$$

For, the molar solubility

Let x = molar solubility

The equilibrium concentration of Lead (II) ion and Iodide ion are as follows:

According to the balanced dissociation equation for every pBI_2 that dissolves one Pb^{2+} will be produced

If, $[Pb^{2+}] = x$

For every PBI_2 that dissolves two Iodine ions are formed

$$\therefore [I^-] = 2x$$

Substituting the equilibrium concentration and K_{sp} into the equation for it, we get

$$K_{sp} = 1.4 \times 10^{-8} = (x)(2x)^2$$

Solving for x

$$K_{sp} = 1.4 \times 10^{-8} = 4x^2$$

$$1.4 \times 10^{-8}/4 = x^3 = 3.5 \times 10^{-9}$$

cube root $(3.5 \times 10^{-9}) = x = 1.55 \times 10^{-3}$ M

166. The equation and the equilibrium expression can be written as,

$$Ag_2CrO_4(s) \rightleftharpoons 2Ag^+(aq) + CrO_4^{2-}(aq)$$

$$K_{sp} = [Ag^+]^2[CrO_4^{2-}]$$

Let "x" be the number of moles of silver chromate that dissolves is every litre of solution (its solubility).

The Concentrations at various stages can be written as

	$Ag_2CrO_4(s)$	$Ag^+(aq)$	$CrO^{2-}(aq)$
Initial Concentration	All solid	0	0
Change in Concentration	$-x$ dissolves	$+2x$	$+x$
Equilibrium Concentration	Less solid	$2x$	x

Substituting the equilibrium amounts and the K_{sp} into the equilibrium expression and solving for x, we get

$$1.1 \times 10^{-12} = [2x]^2[x]$$

$$x = 6.50 \times 10^{-5} \text{ M}$$

167. $PbCl_2(s) \rightleftharpoons Pb^{2+}(aq) + 2\,Cl^-(aq)$

$$[Pb^{2+}] = x$$

$$[Cl^-] = 2x$$

$$K_{sp} = [Pb^{2+}][Cl^-]^2$$

$$x(2x)^2 = 4x^3 = 1.7 \times 10^{-5}$$

$$x^3 = (1.7 \times 10^{-5})/4$$

$$x = (4.25 \times 10^{-6})^{1/3}$$

$$x = 1.6 \times 10^{-2} \text{ M}$$

168. $PbCl_2(s) \rightleftharpoons PbCl_2 \rightarrow Pb^{2+}(aq) + 2Cl^-(aq)$

Initial Concentration	0	1.00
Change in concentration	$+x$	$+2x$
equilibrium	$+x$	$1.0 + 2x$

$$K_{sp} = [Pb^{2+}][Cl^-] = 1.7 \times 10^{-5}$$

$$= x(2x)^2 = 4x^3 = 1.7 \times 10^{-5}$$

$$x^3 = 1.7 \times 10^{-5})/4$$
$$x = (4.3 \times 10^{-6})^{1/3}$$
$$x = 1.6 \times 10^{-2} \text{ M.}$$

If no HCl or any other chloride had been present we would have determined the solubility to be 1.6×10^{-2} M. The solubility is significantly reduced due to *common ion effect.*

169. Writing the equation and the equilibrium equation for the dissolving of barium suphate

$$BaSO_4(s) \rightleftharpoons Ba^2 + (aq) + SO_4^{2-}(aq)$$
$$K_{sp} = [Ba^{2+}]\,[SO_4^{2-}]$$

Let "x" represent the barium sulfate that dissolves in the "sodium sulfate solution expressed in moles per litre.

	$BaSO_4(s)$	$Ba^{2+}(aq)$	$SO_4^{2-}(aq)$
Initial Concentration	All solid	0	0.020 M (from Na_2SO_4)
Change in Concentration	$-x$ dissolves	$+ x$	$+x$
Equilibrium Concentration	Less solid	x	0.020 M + x

Substituting into the equilibrium expression and solving for x. The assumption that since x is going to the very small (the solubility is reduced in the presence of a common ion), the term "0.020 + x" is the same as 0.020," (one can leave x in the term and use the quadratic equation of the method of successive approximations to solve for x, but it will not improve the significance of the answer.)

$$1.1 \times 10^{-10} = [x]\,[0.20 + x] = [x]\,[0.20]$$
$$x = 5.5 \times 10^{-9} \text{ M}$$

170. A strong acid (or base) is fully ionized in solution, so you can use to formula acid, molar concentration to determine the H^+ (or OH^-) ion concentration (Any contribution to this ion concentration from the self-ionization of water is negligible). K_w relates the H^+ and OH^- ion concentration; therefore, you use this relation to obtain the OH^- concentration, given the H^+ concentration, or vice versa.

(i) Every mole of HNO_3 contributes one mole of H^+ ion, so the H^+ concentration is 0.15 *M*. The OH^- concentration is obtained from the equation for K_w.

$$K_w = [H^+][OH^-]$$
$$1.0 \times 10^{-14} = 0.15 \times [OH^-]$$

so

$$[OH^-] = \frac{1.0 \times 10^{-14}}{0.15} = 6.7 \times 10^{-14}$$

The concentration of OH^- in 0.15 *M* HNO_3 is **6.7 × 10$^{M\text{->}14}$ *M*.**

(ii) $Ca(OH_2)(s) \rightleftharpoons Ca^{2+}(aq) + 2OH^-(aq)$

Every mole of $Ca(OH)_2$ than dissolves yields two moles of OH^-. Therefore, 0.010 *M* $Ca(OH)_2$ contains 2 × 0.010 *M* OH^- = **0.020 *M* OH^-**. The H^+ concentration is obtained from

$$K_w = [H^+][OH^-]$$
$$1.0 \times 10^{-14} = [H^+] \times 0.020$$

so

$$[H^+] = \frac{1.0 \times 10^{-14}}{0.020} = 5.0 \times 10^{-13}$$

The hydrogen-ion concentration is **5.0×10^{-13} *M*.**

171. pH is the negative logarithm of the molar hydrogen ion concentration. Determine the log by using either an electronic calculator or log tables. On a calculator, either the H^+ concentration, press *log*, and change the sign of the result. You should write the pH with the same number of decimal places as there are significant figures in $[H^+]$

$$pH = -\log[H^+] = -\log(2.9 \times 10^{-4}) = \mathbf{3.54}$$

172. The hydrogen-ion concentration is

$$[H^+] = \text{antilog}(-pH) = 10^{-pH}$$

On some calculators, you obtain the antilogarithm by taking the *inverse* of the logarithm. On other, you use a 10^x key

The result is

$$[H^+] = \text{antilog}(-7.40) = 10^{-7.40} = \mathbf{4.0 \times 10^{-8}}\ M$$

Because the pH was given to two decimal places, write the hydrogen-ion concentration with two signification figures.

173. It is important to realize the when we say the solution is 0.012 *M*, this refers to *how the solution is prepared*. The solution is made up by adding 0.012 mol of substance to enough water to give a liter of solution. once the solution is prepared, some molecules ionize, so the actual concentration is some what less than 0.02 *M*. To solve for K_a, you follow the three steps outlined is some what less than 0.012 *M*. To solve for K_a.

Step 1: Abberviate the formula for nicotinic acid as HNic. Then in 1 L of solution of HNic concentration is 0.012 mol and has [H] and [HNic] 0 mol before ionization. The H^+ concentration at the start is that from the self-ionization of water. It is usually much smaller than that the obtained from the acid (unless the solution is extremely dilue or K_a is quite small), so you can write $[H^+] = \sim 0$ (meaning approximately zero). If x mole HNic ionizes, x mol each of H^+ and Nic^- is formed leaving $(0.012 - x)$ mol HNic in solution.

Now,

Concentration (*M*)	**HNic(aq)** $\rightleftharpoons$	**H^+(aq)** +	**Nic^-(aq)**
Starting	0.012	~0	0
Change	$-x$	$+x$	$+x$
Equilibrium	$0.012 - x$	x	x

Thus, the molar concentration of HNic, H^+, and Nic^- at equilibrium are $(0.012 - x)$, x, and x, respectively

Step 2: The equilibrium constant equation is:

$$K_a = \frac{[H^+][Nic^-]}{[HNic]}$$

When you substitute the expression for the equilibrium concentrations, you get:

$$K_a = \frac{x^2}{(0.02 - x)}$$

Step 3: The value of x equals the numerical value of the molar hydrogen-ion concentration, and can be obtained from the pH or the solution.

$$x = [H^+] = \text{antilog}\ (-pH) = \text{antilog}\ (-3.39) = 4.1 \times 10^{-4} = 0.0041$$

You can substitute this value of x into the equation obtained in Step 2. Note first however that

$$0.012 - x = 0.0012 - 0.0041 = 0.01159 = 0.012 \text{ (to two significant figure)}$$

This means that the concentration of undissolved acid is equal to the original concentration of the acid within the precision of the data. (We will make use of this type of observation in later problem solving) Therefore,

$$K_a = \frac{x^2}{(0.012 - x)} = \frac{x^2}{0.012} = \frac{(0.00041)^2}{0.012} = \mathbf{1.4 \times 10^{-4}}$$

To obtain the degree of ionization, note that x mol out of 0.012 mol of nicotinic acid ionizes. Hence,

$$\text{Degree of ionization} = \frac{x}{0.012} = \frac{0.00041}{0.012} = \mathbf{0.034}$$

The percent ionization is obtained by multiplying this by 100, which gives 3.4%

174. (i) *Step 1:* At the start (before ionization) the concentration of nicotinic acid, HNic, is 0.10 *M* and that of its conjugate base, Nic^-, is 0. The concentration of H^+ is essentially zero (~0), assuming that the contribution from the self-ionization of water can be neglected. In 1 L of solution, the nicotinic acid ionizes to give x mol H^+ and $x$mol Nic^-, leaving $(0.10 - x)$ mol of nicotinic acid. Thus,

Concentration (*M*)	**HNic(aq)**	$\rightleftharpoons$ **H^+(aq)**	**+ Nic^-(aq)**
Starting	0.10	~0	0
Change	$-x$	$+x$	$+x$
Equilibrium	$-0.10 - x$	x	x

The equilibrium concentration of HNic, H^+, and Nic^- are $(0.10 - x)$, x, and x, respectively.

Step 2: Now substitute these concentration and the value of K_a into the equilibrium constant equation for acid ionization.

$$\frac{[H^+][Nic^-]}{[HNic]} = K_a$$

You get

$$\frac{x^2}{(0.10 - x)} = 1.4 \times 10^{-3}$$

Step 3 Now solve that equation for x. This is actually a quadratic equation, but it can be simplified so that the value of x is easily found. Because the acid-ionization constant is small, the value of x is small. Assume that x is much smaller than 0.10, so that

$$0.10 - x = 0.10$$

You will need to check that this assumption is valid after you obtain a value for x. The equilibrium-constant equation becomes

$$\frac{x^2}{0.10} \approx 1.4 \times 10^{-5}$$

or

$$x^2 = 1.4 \times 10^{-5} \times 0.10 = 1.4 \times 10^{-6}$$

$$\therefore \quad x \approx 1.2 \times 10^{-3} = 0.0012$$

At this point, you should check to make sure that the assumption that $0.10 - x \approx 0.10$ is valid. You substitute the value obtained for x into $0.10 - x$.

$$0.10 - x = 0.10 - 0.0012 = 0.10 \text{ (to two significant figures)}$$

The assumption is indeed valid.

Now you can substitute the value of x into the last line of table written is Step 1 to find the concentration of species. The concentration of nicotinic acid, hydrogen ion, and nicotinate.**10 *M*, 0.0012 *M*, and 0.0012 *M*,** respectively.

(ii) The pH of the solution is

$$\text{pH} = -\log[\text{H}^+] = -\log(0.0012) = \mathbf{2.92}$$

The degree of ionization equals the amount per line of nicotinic acid that ionizes (x = 0.0012) divided by the total amount per liter of nicotinic acid initially present (0.10). Thus, the degree of (ionization of 0.0012/0.10) = **0.012.**

175. The molar mass of $HC_9H_7O_4$ is 180.2 g. From this you find that an aspirin tablet contains 0.00180 mol of the acid. a Hence, the concentration of the aspirin solution is 0.00180 mol/ 0.500 L, or 0.0035 *M*.

Abbreviate the formula for acetylsalicylic acid as HAcs and let x be the amount of H^+ ion formed in 1 L of solution. The amount of acetylsalicylate ion formed is also x mol; the amount o nonionized acetylasalicylic acid is $(0.0036 - x)$ mol. Thus,

Concentration (*M*)	**HAcs(aq)**	**⇌ H^+(aq)**	**+ Acs^-(aq)**
Starting	0.0036	~0	0
Change	$-x$	$+x$	$+x$
Equilibrium	$0.0036 - x$	x	x

If you substitute the equilibrium concentrations and the value of K_a into the equilibrium constant equation, you get

$$\frac{[\text{H}^+][\text{Acs}^-]}{[\text{HAcs}]} = K_a$$

$$\frac{x^2}{0.0036} = 3.3. \times 10^{-4}$$

Note that the concentration of acid divided K_a is $0.0036/3.3 \times 10^{-4} = 11$, which is considerably smaller than 100. Thus, you can expect that x cannot be neglected compared with 0.0036.

$$\therefore \quad x^2 = (0.0036 - x) \times 3.3 \times 10^{-4} = 1.2 \times 10^{-6} - 3.3 \times 10^{-4}\, x$$

or

$$x^2 + 3.3 \times 10^{-4}\, x - 1.2 \times 10^{-6} = 0;$$ **solving for x we have,**

the quadratic formula,

$$x = \frac{-b \pm \sqrt{b^2 - 4ac}}{2a} = \frac{-3.3 \times 10^{-4} \pm \sqrt{(3.3 \times 10^{-4})^2 + 4(1.2 \times 10^{-6})}}{2}$$

$$= \frac{-33 \times 10^{-4} \pm 2.2 \times 10^{-3}}{2}$$

If you take the lower sign in ±, you will get a negative value for x. But x equals the concentration of H^+ ion, so it must be positive. Therefore you ignore the negative solution. Taking the upper sign, you get

$$x = [H^+] = 9.4 \times 10^{-4}$$

Now,

$$pH = -\log[H^+] = -\log(9.4 \times 10^{-4}) = \mathbf{3.03}$$

176. The calculation for the ionization of a weak base parallels that used with weak acids you write the equation, make a table of concentrations (Step 1), set up the equilibrium-constant equation for K_b (Step 2), and solve for $x = [OH^-]$ (step 3). Assume the self-ionization of water of water can be neglected. You obtain $[H^+]$, and then the pH, by solving $K_w = [H^+][OH^-]$

Abbreviate the formula for morphine as Mor. Morphine ionizes by picking up a proton from water (as does ammonia).

$$\text{Mor(q)} + H_2O \rightleftharpoons \text{HMor}^+\text{(aq)} + OH^-\text{(aq)}$$

Step 1: Summarize the concentrations ina table. For the change values, note that the morphine in 1 L of solution ionizes to give x mol $HMor^+$ and x mol OH^-.

Concentration (*M*)	**Mor(aq) + $H_2O(l)$**	$\rightleftharpoons$ **$HMor^+$(aq)**	**+ OH^-(aq)**
Starting	0.0075	0	~0
Change	$-x$	$+x$	$+x$
Equilibrium	$0.0075 - x$	x	x

Step 2: Substituting into the equilibrium equation

$$\frac{[HMor^+][OH^-]}{[Mor]} = K_b \Rightarrow \frac{x^2}{0.0075 - x} = 1.6 \times 10^{-6}$$

Step 3: If you assume that x is small enough to neglect compared with 0.0075, you have

$$0.0075 - x \approx 0.0075$$

and you can write the equilibrium-constant equation as

$$\frac{x^2}{0.0075} \approx 1.6 \times 10^{-6}$$

Hence, $x^2 \simeq 1.6 \times 10^{-6} \times 0.0075 \simeq 1.2 \times 10^{-8}$

$$x = [OH^-] = 1.1 \times 10^{-4}$$

You can now calculate the hydrogen-ion concentration. The product of the hydrogen-ion concentration and the hydroxide-ion concentration equals 1.0×10^{-14} at 298 K.

$$[H^+][OH^-] = 1.0 \times 10^{-14}$$

Substituting into this equation gives

$$[H^+] \times 1.1 \times 10^{-4} = 1.0 \times 10^{-14}$$

Hence,

$$[H^+] = \frac{1.0 \times 10^{-14}}{1.1 \times 10^{-4}} = 9.1 \times 10^{-11}$$

The pH of the solution is

$$pH = -\log [H^+] = -\log(9.1 \times 10^{-11}) = 10.04$$

Note that you could also calculate the pH using the formula pH + pOH = 14.00. You would first calculate the pOH; then pH = 14.00 – pOH.

177. This is acid-ionization problem,

Here, starting concentrations are $[HC_2H_3O_2]$ = 0.10 *M*, $[H^+]$ = 0.010 m (from HCl), and $[C_2H_3O_2^-]$ = 0. The acetic acid ionizes to give an additional x mol L^{-1} of H^+ and x mol L^{-1} of $C_2H_3O_2^-$. The table is

Concentration (*M*)	$HC_2H_3O_2(aq)$ $\rightleftharpoons$	$H+(aq)$ +	$C_2H_3O_2^-(aq)$
Starting	0.10	0.010	0
Change	$-x$	$+x$	$+x$
Equilibrium	$0.10 - x$	$0.010 + x$	x

Now, you substitute into the equilibrium-constant equation

$$\frac{[H^+][C_2H_3O_2^-]}{[HC_2H_3O_2]} = K_a$$

obtaining

$$\frac{(0.010 + x)}{0.10 - x} = 1.7 \times 10^{-5} \text{ (i)}$$

Assume that x is small compared with 0.010. Then

$$0.010 + x \approx 0.010$$

$$0.10 - x \approx 0.10$$

The equation (i) becomes

$$\frac{0.010x}{0.10} \approx 1.7 \times 10^{-5}$$

Solving for x, you get

$$x = 1.7 \times 10^{-5} \times \frac{0.10}{0.010} = 1.7 \times 10^{-4}$$

Check that x can indeed be neglected in $0.010 + x$ and $0.10 - x$.) The degree of ionization of $HC_2H_3O_2$ is $x/0.10 = 0.0017$.

178. The buffer contains a base and its conjguate acid iin equilibrium. The equation is

$$NH_3(aq) + H_2O(l) \rightleftharpoons NH_4^+(q) + OH^-(aq)$$

To obtain the concentration of H+ or OH^- in this equilibrium mixture, you need the starting concentrations in the solution obtained by mixing the NH_3 and NH_4Cl solutions.

For this, you calculate the moles of NH_3 and moles of NH_4^+ added to the buffer solutions, then divide by the total volume of buffer.

Now, $$\text{Molarity } NH_3 = \frac{\text{moles } NH_3}{\text{liters } NH_3 \text{ solution}}$$

Note that 60. mL (or 0.060 L) of .100 M NH_3 is added to make up buffer. So

$$\text{Moles } NH_3 \text{ added} = \text{molarity } NH_3 \times \text{liters } NH_3 \text{ solution}$$

$$= \frac{\text{mol}}{\text{L}} \times 0.060 \text{ L} = 0.0060 \text{ mol } NH_3$$

In the same way, you find that you have added 0.0040 mol NH_4^+ (from NH_4Cl). Assume that the total volume of buffer equals the sum of the volumes of the two solutions.

Total volume of buffer = 60. ML + 40. ML = 100. ML (0.100 L)

Therefore, the concentrations of base and conjugate acid are

$$[NH_3] = \frac{0.0060 \text{ mol}}{0.100 \text{ L}} = 0.060\ M$$

$$[NH_4^+] = \left(\frac{0.0040 \text{ mol}}{1.100}\right) = 0.040\ M$$

Now,

Concentration (*M*)	$NH_3(aq)$ + $H_2O(l)$	$\rightleftharpoons$ $NH_4^+(aq)$ +	$OH^-(aq)$
Starting	0.060	0.040	0
Change	$-x$	$+x$	$+x$
Equilibrium	$0.060 - x$	$0.040 + x$	x

You substitute the equilibrium concentrations into the equilibrium-constant equation.

$$\frac{[NH_4^+][OH^-]}{[NH_3]} = K_b$$

$$\frac{(0.040 + x)}{0.060 - x} = 1.8 \times 10^{-5}$$

To solve this equation, you assume that x is small compared with 0.040 and 0060. So this equation becomes

$$\frac{0.40x}{0.060} = 1.8 \times 10^{-5}$$

Therefore $x = (0.060/0.040) \times (1.8 \times 10^{-5}) = 2.7 \times 10^{-5}$. (Check that x can be neglected in $0.040 + x$ and $0.060 - x$). Thus, the hydroxide-ion concentration is 12.7×10^{-5} M. The pH of the buffer is

$$pH = 14.00 - pOH = 14.00 + \log(2.7 \times 10^{-5}) = 9.43$$

179. When hydrogen ion (from hydrochloric acid) is added to the buffer, it reacts with acetate ion.

$$H+(aq) + C_2H_3O_2^-(aq) \rightarrow HC_2H_3O_2(aq)$$

Because acetic acid is a weak acid, assume that as a first approximation the reaction goes to completion. Also assume that the acetic acid ionizes slightly.

Now, calculate the amounts of hydrogen ion, acetate ion, and acetic acid present in the solution before reaction. The molar amount of hydrogen ion will equal the molar amount of hydrochloric acid added, for 0.10 *M* HCl that 1 L HCl is equivalent to 0.10 mol HCl. Hence, 9.5 mL (=9.5×10^{-3} L) of hydrochloric acid,

$$= 9.5 \times 10^{-3}\ \text{L HCl} \times \frac{0.10\ \text{mol HCl}}{1\ \text{L HCl}} = 0.00095\ \text{mol HCl}$$

Hydrochloric acid is a strong acid, so it exists in solution as the ions H^+ and Cl^-. Therefore, the amount of H^+ added is 0.00095 mol.

The amount of acetate ion and acetate acid are found in a similar way. The buffer in contains 0.20 mol of acetate ion and 0.10 mol of acetic acid in 1 L of solution. The amounts in 75 mL (= 0.075 mL) of solution are:

$$0.075\ \text{L soln} \times \frac{0.20\ \text{mol}\ C_2H_3O_2^-}{1\ \text{L soln}} = 0.015\ \text{mol}\ C_2H_3O_2^-$$

and

$$0.075\ \text{L soln} \times \frac{0.10\ \text{mol}\ HC_2H_3O_2}{1\ \text{L soln}} = 0.0075\ \text{mol}\ HC_2H_3O_2$$

Assume that all of the hydrogen ion added (0.00095 mol) reacts with acetate ion. Therefore, 0.00095 mol of acetite acid is produced and 0.00095 mol of acetate ion is used up. Hence, after reaction you have

$$\text{Moles of acetate} = (0.015 - 0.00095)\ \text{mol}\ C_2H_3O_2^- = 0.014\ \text{mol}\ C_2H_3O_2^-$$

$$\text{Moles of acetic acid} = (0.0075 + 0.00095)\ \text{mol}\ HC_2H_3O_2 = 0.0085\ \text{mol}\ HC_2H_3O_2$$

Now, calculate the concentration of $HC_2H_3O_2$ and $C_2H_3O_2^-$ present the solution before equilibrium. The total volume of solution (buffer plus hydrochloric acid) is 75 mL + 9.5 mL, or 85 mL (0.085 L) Hence the concentrations are

$$[HC_2H_3O_2] = \frac{0.0085\ \text{mol}}{0.085\ \text{L}} = 0.10\ M$$

$$[HC_2H_3O_2^-] = \frac{0.014\ \text{mol}}{0.085\ \text{L}} = 0.16\ M$$

Thus we have the following

Concentration (*M*)	$HC_2H_3O_2(aq)$	$\rightleftharpoons$	$H^+(aq)$	+	$C_2H_3O_2^-(aq)$
Starting	0.10		~0		0.16
Change	$-x$		$+x$		$+x$
Equilibrium	$0.10 - x$		x		$0.17 + x$

The equilibrium-constant equation is

$$\frac{[H^+][C_2H_3O_2^-]}{[HC_2H_3O_2]} = K_a$$

Substituting, we get

$$\frac{x(0.16+x)}{0.10-x} = 1.7 \times 10^{-5}$$

Assume that x is small enough so that $0.16 + x \approx 0.16$ and $0.10 - x \simeq 0.10$, thus the equation becomes

$$\frac{x(0.16)}{0.10} = 1.7 \times 10^{-5}$$

or

$$x = 1.7 \times 10^{-5} \times \frac{0.10}{0.16} = 1.1 \times 10^{-5}$$

∴ The H^+ concentration is 1.0×10^{-5} *M*. The pH is

$$pH = -\log [H^+] = -\log(1.1 \times 10^{-5}) = 4.96$$

Adding HCl to Pure Water : If 9.5 mL of 0.01 *M* hydrochlric acid were added to 75 mL of pure water, the hydrogen-ion concentration would change to

$$[H^+] = \frac{\text{amount of } H^+ \text{ added}}{\text{total volume of solution}}$$

$$[H^+] = \frac{0.00095 \text{ mol } H^+}{0.085 \text{ L solution}} = 0.011\ M$$

(The total volume is 75 mL of water plus 9.5 mL ($\simeq$ 85 mL) HCl, assuming no change of volume on mixing). The pH is

$$pH = -\log [H^+] = -\log(0.011) = 1.96$$

180. The grams of residue form a liter of saturated solution equals the solubility in grams per liter. Convert this to molar solubility (solubility in moles per liter).

The solubility of calcium oxalate is 0.0061 g/L of solution, and the formula weight of CaC_2O_4 is 128 amu. Convert grams per liter to moles per liter:

$$\text{Molar solubility of } CaC_2O_4 = 0.0061\ \cancel{g\ CaC_2O_4}/L \times \frac{1 \text{ mol } CaC_2O_4}{128\ \cancel{g\ CaC_2O_4}}$$

$$= 4.8 \times 10^{-5} \text{ mol } CaC_2O_4/L$$

Now

Suppose you mix solid CaC_2O_4 in a liter of solution. Of this solid, 4.8×10^{-5} mol with dissolve to form 4.8×190^{-5} mol of each ion.

Concentration (*M*)	**$CaC_2O_4(s) \rightleftharpoons Ca^{2+}(aq) + C_2O_4^{2-}(aq)$**	
Starting	0	0
Change	$+4.8 \times 10^{-5}$	$+4.8 \times 10^{-5}$
∴ At Equilibrium	4.8×10^{-5}	4.8×10^{-5}

substituting the values into the equilibrium-constant equation we get

$$K_{sp} = [Ca^{2+}][C_2O_4^{2-}] = (4.8 \times 10^{-5})(4.8 \times 10^{-5}) = 2.3 \times 10^{-9}$$

181. (i) Consider the equation $PbI_2(s) \rightleftharpoons Pb^{2+}(aq) + 2I^-(aq)$

Here every mole of the compound (PbI_2) dissolves to form two moles of the anion (I^-).

Suppose the solid lead(II) iodide is mixed into 1 L of solution. So, 1.2×10^{-3} mol dissolves to form 1.2×10^{-3} mol Pb^{2+} and $2 \times (10^{-3})$ mol I^-.

Thus we have

Concentration (*M*)	**$PbL_2(s) \rightleftharpoons Pb^{2+}(aq)$**	**+ $2I^-(aq)$**
Starting	0	0
Change	$+1.2 \times 10^{-3}$	$+2 \times (1.2 \times 10^{-3})$
At Equilibrium	1.2×10^{-3}	$2 \times (1.2 \times 10^{-3})$

substituting the values into the equilibrium-constant equation: we get

$$K_{sp} = [Pb^{2+}]\,[I^-]^2 = (1.2 \times 10^{-3}) \times (2 \times 1.2 \times 10^{-3})^2 = 6.9 \times 10^{-9}$$

(ii) This problem is reverse of the preceding ones; instead of finding K_{sp} from the solubility, here we calculate solubility from the K_{sp} followed by the three steps of equilibrium problems, but since the molar solubility is not immediately known we assign if the value x. Obtain the concentration of each ion by multiplying x by the coefficient of the ion in the chemical equation. Next we calculate K_{sp} as a cubic in x. Then solve the equilibrium-constant equation for x, which will involve taking a cube root.

Let x be the molar solubility of CaF_2. When solid CaF_2 is mixed into a liter of solution, x mol dissolves, forming x mol Ca^{2+} and $2x$ mol F^-.

Concentration (*M*)	**$CaF_2(s) \rightleftharpoons Ca^{2+}(aq)$**	**+ $2F^-(aq)$**
Starting	0	0
Change	$+x$	$+2x$
At Equilibrium	x	$2x$

Substituting these values into the equilibrium-constant equation, we get

$$[Ca^{2+}]\,[F^-]^2 = K_{sp}$$

$$(x) \times (2x)^2 = 3.4 \times 10^{-11}$$

$$4x^3 = 3.4 \times 10^{-11}$$

$$\therefore \quad x = \left(\frac{3.4 \times 10^{-11}}{4}\right)^{\frac{1}{3}} = 2.0 \times 10^{-4}$$

The molar solubility is 2.0×10^{-4} mol CaF_2 per liter. To get the solubility in grams per liter, we covert, using the molar mass of CaF_2 (78.1 g/mol).

$$\text{Solubility} = 2.0 \times 10^{-4}\ \text{mol CaF}_2\text{/L} \times \frac{78.1\ \text{g CaF}_2}{1\ \text{mol CaF}_2} = 1.6 \times 10^{-2}\ \text{g CaF}_2\text{/L}$$

182. Suppose solid CaC_2O_4 is mixed into 1 L of 0.15 *M* $CaCl_2$ and that the molar solubility of the CaC_2O_4 is x *M*. At the start (before CaC_2O_4 dissolves), there is 0.15 mole Ca^{2+} in the solution. Of the solid CaC_2O_4, x mol dissolves to give x mol of additional Ca^{2+} and x mol $C_2O_4^{2-}$. Thus we have

Concentration (*M*)	**$CaC_2O_4(s) \rightleftharpoons Ca^{2+}(aq)$**	**+ $C_2O_4^{2-}(aq)$**
Starting	0.15	0
Change	$+x$	$+x$
At Equilibrium	$0.15 + x$	x

Substituting the values into the equilibrium-constant equation we get:

$$[Ca^{2+}]\,[C_2O_4^{\;2-}] = K_{sp}$$

$$(0.15 + x)x = 2.3 \times 10^{-9}$$

or $$x = \frac{2.3 \times 10^{-9}}{0.15 + x}$$

Because calcium oxalate is only slightly soluble, you might expect x to be negligible compare with 0.15. In that case,

$$0.15 + x \simeq 0.15$$

and the previous equation becomes

$$x \simeq \frac{2.3 \times 10^{-9}}{0.15} = 1.5 \times 10^{-8}$$

Note that x is indeed much smaller than 0.15, so your assumption was correct. Therefore, the molar solubility of calcium oxalate in 0.15 M $CaCl_2$ is 1.5×10^{-8} M:

183. Here we need to calculate Q_c the ion product, to compare it with K_{sp}, the solubility product. For Q_c we need to obtain the concentrations of ions in the solution just after mixing, but before precipitation, if nay occurs. First calculate the moles of each ion added. Then, using the total volume of solution (assumed to equal the sum of the two added solutions), calculate the molar concentration of each ion present. Finally, calculate Q_c compare it with K_{sp}.

The molar amount Ba^{2+} present in 50.0 mL (=0.0500 L) of 0.0010 M $BaCl_2$ is

$$\text{Amount of } Ba^{2+} = \frac{0.0010 \text{ mol } Ba^{2+}}{1 \text{ L soln}} \times 0.050 \text{ L soln} = 5.0 \times 10^{-5} \text{ mol } Ba^{2+}$$

The molar concentration of Ba^{2+} in the total solution equals the molar amount of Ba^{2+} divided by the total volume (0.0500 L $BaCl_2$ + 0.0500 L Na_2SO_4 = 0.1000 L).

$$[Ba^{2+}] = \frac{5.0 \times 10^{-5} \text{ mol}}{0.1000 \text{ L soln}} = 5.0 \times 10^{-4}\ M$$

Similarly, we have $[SO_4^{2-}] = 5.0 \times 10^{-5}\ M$

$$\therefore \quad Q_c = [Ba^{2+}]_i[SO_4^{2-}]_i = (5.0 \times 10^{-4}) \times (5.0 \times 10^{-5}) = 2.5 \times 10^{-8}$$

Because the ion product is greater than the solubility product constant (1.1×10^{-10}), you expect barium sulfate to precipitate.

184. (i) We can set up the stoichiometric calculation for the precipitation equation with initial and final amount of substance as:

Moles of substances	$Ba^{2+}(aq)$	+	$SO_4^{\;2-}(aq)$	→	$BaSO_4(s)$
Initial (before ppt.)	0.00050		0.000050		0
Final (after ppt.)	(0.00050 – 0.000050) = 0.00045		0		0.000050

For the solubility equilibrium we have,

Concentration (M)	$BaSO_4(s) \rightleftharpoons$	$Ba^{2+}(aq)$ +	$SO_4^{2-}(aq)$
Starting		0.00045	0

Change	$+x$	$+x$
Equilibrium	$0.00045 + x$	x

K_{sp} for $BaSO_4$ is 1.1×10^{-10}. The equilibrium-constant equation is

$$[Ba^{2+}]\,[SO_4^{2-}] = K_{sp}$$

$$(0.00045 + x)x = 1.1 \times 10^{-10}$$

If you assume that x can be neglected in comparison with 0.00045, this equation is easily solved for x.

$$x \simeq \frac{1.1 \times 10^{-10}}{0.00045} = 2.4 \times 10^{-7}$$

(ii) The SO_4^{2-} concentration is 2.4×10^{-7} *M*. The concentration of the ion before precipitation of $BaSO_4$ was 5.0×10^{-5} *M*, so the percentage of SO_4^{2-} remaining in solution is

$$\frac{2.4 \times 10^{-7}}{5.0 \times 10^{-5}} \times 100\% = \mathbf{0.48\%}$$

185. A string acid affects the solubility of a salt of a weak acid by providing hydrogen ion that reacts the anion of the salt in the solution. As the anion is removed by this reaction, more salt dissolves to replenish the anion concentration. You need to ask, which salt has the anion corresponding to the weaker acid?

Calcium carbonate gives he solubility equilibrium

$$CaCO_3(s) \rightleftharpoons Ca^{2+}(aq) + CO_3^{2-}(aq)$$

When a strong acid is added, the hydrogen ion reacts with carbonate ion, because it is conjugate to a weak acid (HCO_3^-).

$$H^+(aq) + CO_3^{2-}(aq) \rightleftharpoons HCO^{3-}(aq)$$

As carbonate ion is removed, calcium carbonate dissolves. Moreover, the hydrogen carbonate ion itself is removed in further reaction.

$$H^+(aq) + HCO_3^-(aq) \rightarrow H_2CO_3(aq) \rightarrow H_2O(l) + CO_2(g)$$

Bubbles of carbon dioxide gas appear as more calcium carbonate dissolves.

For calcium sulfate, the corresponding equilibria are

$$CaSO_4(s) \rightleftharpoons Ca^{2+}(aq) + SO_4^{2-}(aq)$$

$$H^+(aq) + SO_4^{2-}(aq) \rightleftharpoons HSO_4^-(aq)$$

Again, the anion of the insoluble salt is removed by reaction with hydrogen ion. You would except calcium sulfate to become more soluble in strong acid. However, HSO_4^- is a much stronger acid than HCO_3^-, as you can see by comparing acid-ionization constants. (The value K_{a2} for H_2CO_3 and K_a for HSO_4^- from Table are 4.8×10^{-11} and 1.1×10^{-2}, respectively.) Thus, calcium carbonate is much more soluble in acidic solution, whereas the solubility of calcium sulfate is only slightly affected.

186. First calculate the minimum S^{2-} ion concentration that must be present before each metal sulfide ZnS and PbS can precipitate. From, the S^{2-} ion concentrations calculate the corresponding H^+ ion concentrations. Now,

$$ZnS \rightleftharpoons Zn^{2+} + S^{2-}$$

$$\therefore \quad [Zn^{2+}]\,[S^{2-}] = K_{sp}$$

Substituting the values we get

$$(0.10)\ [S^{2-}] = 1.1 \times 10^{-21}$$

The minimum S^{2-} ion concentration needed to give a precipitate of ZnS is

$$\therefore \quad [S^{2-}] = \frac{1.1\times 10^{-21}}{0.10} = 1.1 \times 10^{-20}$$

Similarly, the S^{2-} ion concentration needed to precipitate from PbS is:

$$[S^{2-}] = \frac{K_{sp}\text{ (for PbS)}}{[Pb^{2+}]} = \frac{2.5\times 10^{-27}}{0.10} = 2.5 \times 10^{-26}$$

Substituting these minimum S^{2-} ion concentration into the equilibrium constant equation for the double ionization of H_2S,

$$\frac{[H^+]^2\,[S^{2-}]}{[H_2S]} = 1.1 \times 10^{-20}$$

we obtain the H^+ ion concentrations. For ZnS, we get

$$[H^+]^2 = 1.1 \times 10^{-20}\ \frac{[H_2S]}{[S^{2-}]} = 1.1 \times 10^{-20} \times \frac{0.10}{1.1\times 10^{-20}} = 0.10$$

or

$$[H^+] = .32$$

At any concentration H^+ ion equal to or less than 0.32 *M*, the S^{2-} concentration will be equal to or greater than that necessary to give a precipitate of ZnS. When the H^+ ion concentration is greater than 0.32 *M*, the Zn^{2+} ion will remain in solution.

For PbS :

$$[H^+]^2 = 1.1 \times 10^{-20} \times \frac{[H_2S]}{[S^{2-}]} = 1.1 \times 10^{-20} \times \frac{0.10}{2.5\times 10^{-26}} = 4.4 \times 10^4$$

$$[H^+] = 2.1 \times 10^2$$

At any H^+ ion concentration equal to or less than 2.1×10^2 *M*, PbS will precipitate. It is not possible to get an H^+ ion concentration this large, so you can expect PbS to precipitate regardless of pH.)

When saturated with H_2S any solution of 0.10 *M* Zn^{2+} and 0.10 *M* Pb^{2+} in which the H^+ ion concentration is greater than 0.32 *M* will give a precipitate of PbS but not of ZnS.

187. In 1L of solution, we initially have 0.010 mol $Ag^+(aq)$ from $AgNO_3$. This reacts to give 0.010 mol $Ag(NH_3)_2^+$, leaving (1.00 – 2 × 0.010) mol NH_3, which equals 0.98 mol NH_3

One liter of the solution contains 0.010 mol $Ag(NH_3)_2^+$ and 0.98 mol NH_3. The complex ion dissociates slightly, so that 1 L of solution contains *x* mol Ag^+.

Thus,

Concentration (*M*)	$Ag(NH_3)_2{}^+(aq)$	$\rightleftharpoons$ $Ag^+(aq)$	+ $2NH_3(aq)$
Starting	0.010	0	0.98
Change	$-x$	$+x$	$+2x$
Equilibrium	$0.010 - x$	x	$0.98 + 2x$

The formation constant is

$$K_f = \frac{[Ag(NH_3)_2^+]}{[Ag^+][NH_3)^2]} = 1.7 \times 10^7$$

From this we can write the dissociation constant, which corresponds to the reaction as

$$\frac{[Ag^+][NH_3]^2}{[Ag(NH_3)_2^+]} = K_d = \frac{1}{K_f}$$

Substituting the values into this equation we get

$$\frac{x(0.98+2x)^2}{(0.010-x)} = \frac{1}{1.7\times 10^7} = 5.9 \times 10^{-8}$$

It is small compared with .010,

Then, $$\frac{x(0.98)^2}{0.010} \simeq 5.9 \times 10^{-8}$$

$\Rightarrow$ $$x \simeq 5.9 \times 10^{-8} \times 0.010/(0.98)^2 = 6.1 \times 10^{-10}$$

The silver-ion concentration 6.1×10^{-10} *M* (over 10 million times smaller than its value in 0.010 *M* $AgNO_3$ that does not contain ammonia).

188. (i) To determine whether a precipitate should form, we calculate the ion product and compare it with K_{sp} for AgCl (1.8×10^{-10}).

$$\text{Ion product} = [Ag^+]_i\,[Cl^-]_i = (0.010)(0.010) = 1.0 \times 10^{-4}$$

This is greater than $K_{sp} = 1.8 \times 10^{-10}$, so a precipitate should form.

(ii) Now, calculate the concentration of $Ag^+(aq)$ in a solution containing 1.00 M NH_3. It is found that $[Ag^+]$ equals 6.1×10^{-10}.

Hence, Ion product $= [Ag^+]_i\,[Cl^-]_i = (6.1 \times 10^{-10})\,(0.010) = 6.1 \times 10^{-12}$

Because the ion product is smaller than $K_{sp} = 1.8 \times 10^{-10}$, no precipitate should form.

189. We have,

$$AgCl(s) \rightleftharpoons Ag^+(aq) + Cl^-(aq)$$
$$Ag^+(aq) + 2NH_3(aq) \rightleftharpoons Ag(NH_3)_2^+\,(aq)$$
$$\overline{AgCl(s) + 2NH_3(aq) \rightleftharpoons Ag(NH_3)_2{}^+(aq) + Cl^-\,(aq)}$$

The equilibrium constant for the overall equation is

$$K_c = \frac{[Ag(NH_3)_2^+][Cl^-]}{[NH_3]^2}$$

Because the overall equation is the sum of the solubility and complex-ion equilibria the equilibrium constant K_c equals the product of the equilibrium constants for the solubility and complex-ion equilibria.

$$K_c = K_{sp}K_f = \cancel{[Ag^+]}\,[Cl^-]\frac{[Ag(NH_3)_2^+]}{[\cancel{Ag^+}][NH_3]^2}$$

Substituting the values we get

∴ $K_{sp}K_f = (1.8 \times 10^{-10})\ (1.7 \times 10^{7}) = 3.1 \times 10^{-3}$

Thus,

Concentration (*M*)	$AgCl(s)$ +	$2NH_3(aq)$ ⇌	$Ag(NH_3)_2{}^{+}(aq)$ +	$Cl^-(aq)$
Starting		1.0	0	0
Change		$-2x$	$+x$	$+x$
Equilibrium		$1.0 - 2x$	x	x

Substituting these values into the equilibrium-constant equation we get

$$\frac{Ag(NH_3)_2^+][Cl^-]}{[NH_3]^2} = K_{sp}K_f$$

or
$$\frac{x^2}{(1.0 - 2x)^2} = 3.1 \times 10^{-3}$$

taking the square root of both side, gives

$$\frac{x}{1.0 - 2x} = 0.056$$

⇒
$$x = (0.056(1.0 - 2x) = 0.056 - 0.11x$$

or
$$x = \frac{0.056}{1.11} = 0.050$$

Note that the molar solubility of AgCl equals the molar concentration of silver in the solution. Because of the stability of $Ag(NH_3)_2^+$, most of the silver in solution will be in the form of this complex ion. Therefore, because x equals the concentration of $Ag(NH_3)_2^+$, the molar solubility of AgCl equals 0.050 *M*.

190. Consider $HClO_2$ in reaction (a). It is converted to ClO_2^- by losing a proton (H^+). $HClO_2$ must be an acid, and ClO_2^- is its conjugate base. H_2O accepts the proton lost by $HClO_2$. Then H_2O must be a base and H_3O^+, its conjugate acid. In reaction (b), H_2O acts as an acid and OH^- is its conjugate base. Reactions (c) and (d) illustrate another species that can either donate or accept a proton—an amphiprotic species. In this case the amphiprotic species is $H_2PO_4^-$

(a) $HClO_2 + H_2O \rightleftharpoons H_3O^+ + ClO_2$

acid(1) base(2) acid(2) base(1)

(b) $H_2O + OCl^- \rightleftharpoons HOCl + OH^-$

acid(1) base(2) acid(2) base(1)

(c) $H_2PO_4 + NH_3 \rightleftharpoons NH_4^+ + HPO_4^{-2}$

acid(1) base(2) acid(2) base(1)

(d) $HCl + H_2PO_4 \rightleftharpoons H_3PO_4^+ + Cl^{-1}$

acid(1) base(2) acid(2) base(1)

191. For the reaction $H_2C_2H_3O_2 + H_2O \rightleftharpoons H_3O+ + C_2H_3O_2^-$, Let $x = [H_3O^+] = [C_2H_3O_2^-]$ and for $HC_2H_3O_2$, molarity = *M*, let us make the assumption that $(M - x) \approx M$, we get the general expression

$$K_a = \frac{x^2}{M}; \quad x^2 = MK_a; \quad x = [H_3O^+] = \sqrt{MK_a}$$

$$\% \text{ ionization} = \frac{x}{M} \times 100 = \frac{\sqrt{MK_a}}{M} \times 100 = \sqrt{\frac{K_a}{M}} \times 100$$

192. $[N_2] = 5.0 \text{ mol}/1.5 \text{ L} = 3.3 \text{ mol L}^{-1}$ $\quad [O_2] = 7.0 \text{ mol}/1.5 \text{ L} = 4.7 \text{ mol L}^{-1}$

$[NO_2] = 0.10 \text{ mol}/1.5 \text{ L} = 0.067 \text{ mol L}^{-1}$

$$K = \frac{[NO_3]^2}{[N_2][O_2]^2} = \frac{(0.067)^2}{(3.3)(4.7)^2} = 6.1 \times 10^{-5}$$

193. Here,

	$NaOH$ +	CH_3COOH →	CH_3COONa +	H_2O
Millimole added	20×0.2	50×0.2	——	——
	= 4	= 10	0	0
Millimole after reaction	0	6	4	4

$\because \quad [\text{Molarity}] = \frac{\text{millimole}}{\text{Total volume}}$

$\therefore \quad [CH_3COOH] = \frac{6}{70}, \quad [CH_3COONa] = \frac{4}{70}$

$\therefore \quad pH = -\log 1.8 \times 10^{-5} + \frac{4/70}{6/70}$

$\therefore \quad pH = 4.5686$

194. K_w for H_2O at 298 $K = 10^{-14}$

$\therefore \quad [H^+][OH^-] = 10^{-14}$ $\quad (\because K_w = [H^+][OH^-])$

$\therefore \quad [H^+] = 10^{-7} M$ $\quad \therefore pH = 7$

Now K_w for H_2O at 333 $K = 9.62 \times 10^{-14}$

$\therefore \quad [H^+][OH^-] = 9.62 \times 10^{-14}$

For pure water $\quad [H^+] = [OH^-]$

$\therefore \quad [H^+]^2 = 9.62 \times 10^{-14}$

$\therefore \quad [H^+] = \sqrt{(9.62 \times 10^{-14})} = 3.10 \times 10^{-7} M$

$\therefore \quad pH = -\log H^+ = -\log 3.10 \times 10^{-7}$

$pH = 6.51$

Thus, pH of water becomes 6.51 at 333 K but the nature is neutral since calculation for pure water has been made, i.e., pH scale at 333 K becomes in between 0 to 13.02.

195. Here,

	$CHCl_2 \cdot COOH$ ⇌	$CHCl_2 \cdot COO^-$ +	H^+
Conc. before dissociation	0.2	–	–
Conc. after dissociation	$(0.2 - x)$	x	x

also, $\quad CHCl_2 \cdot COONa \rightarrow CHCl_2COO^- + Na^+$

$\qquad\qquad 0.1 \qquad 0.1$

Conc. on dissociation

For the dissociation of acid,

$$K_a = 5 \times 10^{-2} = \frac{[CHCl_2COO^-][H^+]}{[CHCl_2COOH]}$$

or $$0.05 = \frac{[0.1 + x][x]}{[0.2 - x]}$$

$\therefore$ $$x = 0.05$$

or $$[H^+] = 0.05$$

196. Whenever you have to decide whether a particular solution has buffer properties, the key question is, can you identity a component in the solution capable of neutralizing small added amounts of acid and a component that will neutralize added base? In the present case the components are NH_3 and NH_4^+, respectively.

$$\underset{\text{(small added amt.)}}{NH_3 + H_3O^+} \rightarrow NH_4^+ + H_2O$$

$$NH_4^+ + \underset{\text{(small added amt.)}}{OH} \rightarrow NH_3 + H_2O$$

In all aqueous solutions containing NH_3 and NH_4^+ we know that

$$NH_3 + H_2O \rightleftharpoons NH_4^+ OH^-$$

and

$$K_b = \frac{[NH_4^+][OH^-]}{[NH_3]} = 1.74 \times 10^{-5}$$

If approximately equal concentrations of NH_4^+ and NH_3 maintained in a solution, $[OH^-] \approx 1 \times 10^{-5}$ M; pOH $\approx$ 5; and pH $\approx$ 9. Thus, ammonia ammonium chloride solutions are alkaline (basic) buffer solutions.

197. If the $CaCO_3$ can decompose to yield CO_2 at 1.00 atm (to push back the air), it will do so.

$K_p = P(CO_2) = 1.00$ $\qquad$ $\log K_p = 7.282 - \frac{8500}{T}$

$$\therefore T = \frac{8500}{7.282} = 1167K$$

198. Here, $K = \frac{[PCl_3][Cl_2]}{[PCl_3]} = \frac{\left(\frac{0.32}{12}\right)\left(\frac{0.32}{12}\right)}{\left(\frac{0.21}{12}\right)} = 0.041$

199. Here, $AgCl(s) \rightleftharpoons Ag^+ + Cl^-$

$\therefore K_{sp} = [Ag^+][Cl^-]$ But, $[Ag^+] = [Cl^-] = 1.0 \times 10^{-5}$

$\therefore K_{sp} = (1.0 \times 10^{-5})(1.0 \times 10^{-5}) = 1.0 \times 10^{-10}$

200. Here, $A_2X_3 \rightarrow 2A^3 + 3X^{2-}$ $K_{sp} = [A^{3+}]^2[X^{2-}]^3 = 1.1 \times 10^{-23}$

If the solubility of A_2X_3 is y, then $[A^{3+}] = 2y$ and $[X^{2-}] = 3y$

$\therefore (2y^3)(3y)^2 = 108y^5 = 1.1 \times 10^{-23}$

$$y^5 = 1.0 \times 10^{-25}$$

or $y = 1.0 \times 10^{-5}$ mol.L^{-1}

201. For PbI_2, $PbI_2 \rightleftharpoons Pb^{2+} + 2I^-$

$\qquad\qquad\qquad s \quad 2s$

where s is solubility of PbI_2

$\therefore$ $$K_{sp} = 4s^3$$

$\therefore$ $$s^3 = \frac{K_{sp}}{4}$$

or $$s = \sqrt{\frac{K_{sp}}{4}} = \sqrt{\frac{4 \times 10^{-9}}{4}} = 10^{-3}$$

$\therefore$ $$[I^-] = 2s = 2 \times 10^{-3} \text{ mol.L}^{-1}$$

Now, for $Ag^+ + I^- \rightarrow AgI$

Meq. of Ag^+ = M eq. of I

$$N \times 10 = 2 \times 10^{-3} \times 25$$

$\therefore$ N or M $AgNO_3 = 5 \times 10^{-3}$

202. Here, $[Ag^+][Cl^-] = K_{sp}$

or $(4 \times 10^{-3})[Cl^-] = 1.0 \times 10^{-10}$

$\therefore$ $[Cl^-] = 2 \times 10^{-10}$

Hence, $[Cl^-]$ of 2×10^{-8} M must be exceeded before AgCl precipitates.

203. Here, $BaSO_4 \rightleftharpoons Ba^{2+} + SO_4^{2-}$ $\therefore$ $K_{sp} = [Ba^{2+}][SO_4^{2-}] = 1 \times 10^{-10}$

Let $[Ba^{2+}] = [SO_4^{2-}] = x$

$\therefore x^2 = 1 \times 10^{-10}$ or, $x = 1 \times 10^{-5}$

$= 1 \times 10^{-5}$ mol L^{-1} (0.450 L) (233 g mol^{-1}) = 0.01 g = 1 mg

204. Here, $[Cl^-] = 0.20$ M from $BaCl_2$

$\therefore$ $[Ag^+][Cl^-] = 1.0 \times 10^{-10}$

or, $$[Ag^+] = \frac{1.0 \times 10^{-10}}{0.20} = 5.0 \times 10^{-10}\ M$$

205. Here,

	$CaCl_2$	+	Na_2SO_4	$\rightarrow$ $CaSO_4$	+ $2NaCl$
Millimole added	0.02M		0.00054 M	0	0

Suppose 1 liter of both are mixed

$\therefore$ total volume = 2 liters

$\therefore \qquad [Ca^{2+}] = \frac{0.02}{2} \quad [SO_4^{2-}] = \frac{0.0004}{2}$

$\therefore \qquad = K_{sp}\ [Ca^{2+}]\ [SO_4^{2-}] = \frac{0.02}{2} \times \frac{0.0004}{2}\ (mol^2\ L^{-2})$

$= 2 \times 10^{-6}$

Thus, $[Ca^{2+}]\ [SO_4^{2-}]$ in solution $< K_{sp}$. (given) i.e. $2 \times 10^{-6} < 2.4 \times 10^{-5}$

$\therefore$ $CaSO_4$ will not precipitate.

Chapter 10 : Electro-Chemistry

1. Write the reaction that would occur at the cathode.

$$Cu^{2+}(aq) + 2e^- \rightarrow Cu(s)$$

This problem can be solved by using the factor-label method:
(Note all the conversion factors that you should be comfortable with)

$$\frac{96.487 \text{ amperes}}{1} \times \frac{100.0 \text{ minutes}}{1} \times \frac{60 \text{ seconds}}{1 \text{ minutes}}$$

$$\times \frac{1 \text{ coulomb}}{\text{ampere-second}} \times \frac{1 \text{ Faraday}}{96.487 \text{ coulombs}} \times \frac{1 \text{ mole } e^-}{1 \text{ Faraday}}$$

$$\times \frac{1 \text{ mole Cu}}{2 \text{ moles}} \times \frac{3.55 \text{ g Cu}}{1 \text{ mole Cu}} = 190.65 \text{ g Cu}$$

2. Notice that E°_{red} for silver is lower than E°_{red} for gold. This means that because silver is higher in the activity series, silver metal will reduce the gold ion.
Write the net cell reaction

$$Ag(s) + Au^+(aq) \rightleftharpoons Ag^+(aq) + Au(s)$$

Write the two half-reactions and include the E°_{red} mol E°_{ox} values.

ox; $Ag(s) \rightleftharpoons Ag^+(aq) + e^-$ $E^\circ_{ox} = -0.7991$ volt

red: $Au^+(aq) + e^- \rightleftharpoons Au(s)$ $E^\circ_{red} = 1.68$ volts

Hence, $Ag(s) + Au^-(aq) \rightleftharpoons Ag^+(Aq) + Au(s)$ $E^\circ + +0.88$ volt

Because the sign of E° is positive, the reactions will proceed spontaneously.

3. Decide what elements are undergoing oxidation and what elements are undergoing reduction.

ox: $Fe^{2+} \rightarrow Fe^{3+}$

red: $MnO_4^- \rightarrow Mn^{2+}$

Balance each half-reaction with respect to atoms and then charges.

ox: $Fe^{2+} \rightarrow Fe^{3+} + e^-$

Balance the reduction half-reaction, using water to balance the O's.

red: $MnO_4^- + Mn^{2+} + 4H_2O$

Balance the H atoms with H^+ ions.

red: $MnO_4^- + 8H^+ \rightarrow Mn^{2+} + 4H_2O$

Balance charges with electrons.

red: $MnO_4^- + 8H^+ + 5e^- \rightarrow Mn^{2+} + 4H_2O$

Equalize the number of electrons lost and gained. The were $5e^-$ gained in the reduction half-reaction, so there must be $5e^-$ lost in the oxidation half reaction

ox: $5Fe^{2+} \rightarrow 5Fe^{3+} + 5e^-$

Add the two half-reactions (cancel the electrons).

ox: $5Fe^{2+} \rightarrow 5Fe^{3+} + 5Fe^{-}$

red: $MnO_4^{-} + 8H^{+} + 5e^{-} \rightarrow Mn^{2+} + 4H_2O$

$5Fe^{2+} + MnO_4 + 8H^{-} \rightarrow 5Fe^{3+} + Mn^{2+} + 4H_2O$

4. The vertical lines represent phase boundaries. By convention, the anode is written first, at the left of the double vertical lines, following by the other components of the cell as they would appear in order from the anode to the cathode. The platinum is present to represent the presence of an inert anode. The two half reactions that occur are

anode; $H_2(g) \rightarrow 2H^{+}(aq) + 2e^{-}$ oxidation

Cathode : $Ag^{+}(aq) + e^{-} \rightarrow Ag(s)$ reduction

In adding the two half-reactions, multiply the reduction half-reaction by 2 so the electrons are in balance, giving the overall reaction.

$H_2(g) + 2Ag^{+}(aq) \rightarrow 2H^{+}(aq) + 2Ag(s)$

5. The more positive the E°_{red} value, the greater the tendency for the substance to be reduced, and conversely, the less likely it is to be oxidized. It would help in this example to reverse the last equation so it can be easily compared to the other E°_{red} values. The last equation becomes $Ni^{2+}(aq) + 2e^{-}(aq) \rightarrow Ni(s)$; $E^{\circ}_{red} = -0.25$ V.

The equation with the largest E°_{red} is $Cu^{2+}(aq) + e^{-} \rightarrow Cu^{+}(aq)$, $E^{\circ}_{red} = 0.34$ V. Thus $Cu^{2+}(aq)$ is the strongest oxidizing agent of those listed because it has the greatest tendency to be reduced Conversely, $Cu^{+}(aq)$ would be the weakest reducing agent. $Fe^{2+}(aq)$ would be the weakest oxidizing agent because it would be the most difficult species to reduce. Conversely, Fe(s) would be the strongest reducing agent.

6. The formula you need for this problem is $\Delta G^{\circ} = -n\,F\,E^{\circ}$. The Faraday constant, F, is equal to 9.65×10^4 joules · volt^{-1} · mole^{-1} · n is the number of electrons transferred between oxidizing and reducing agents in a balanced redox equation.

Write the balanced redox equation.

$Zn(s) + Cu^{2+}(aq) \rightarrow Zn^{2+}(aq) + Cu(s)$

Identity the variables needed for the equation.

$\Delta G^{\circ} = ?\; F = 9.65 \times 10^4$ joules · volt^{-1} · mole^{-1}

$n = 2, \quad E^{\circ} = 1.00$ volt

Substitute into the equation and solve

$$\Delta G^{\circ} = \frac{(2\ \text{moles}\ e^{-})}{1} \times \frac{9.65 \times 10^4\ \text{joules}}{\text{volt} \cdot \text{mole}\ e^{-}} \times \frac{1.00\ \text{volt}}{1}$$

$$= 1.93 \times 10^5\ \text{joules} = -193\ \text{kJ}$$

7 There are two types of cells: electrolytic (which requires a battery or external power source) and voltaic (which requires no battery or external power sources). The reaction in the diagram is voltaic and therefore spontaneous. In a voltaic cell the anode is the negative terminal and oxidation occurs at the anode. Remember Oxidation Is Losing electrons and (Anode is where Oxidation occurs).

$Zn(s) \rightarrow Zn^{2+}(aq) + 2e^{-}$

The cathode is the positive terminal in the voltaic cell, and reduction occurs at the cathode. Remember Reduction Is Gaining electrons and it occurs at the Cathode.

$$Ni^{2+}(aq) + 2e^- \rightarrow Ni(s)$$

8. Determine the oxidation and reduction half reaction and E°_{cell}

Ox:	$3[Zn(s) \rightarrow Zn^{2+}(aq) + 2e^-]$	$E^\circ_{ox} = +0.76$ V
red:	$2[Cr^{2+}(aq) + 3e^- \rightarrow Cr(s)]$	$E^\circ_{red} = -0.74$ V
	$3Zn(s) + 2Cr^{3+}(aq) \rightarrow 3Zn^{2+}(aq) + 2Cr(s)$	$E^\circ_{cell} = 0.02$ V

Use the equation

$$\ln K = \frac{n \cdot E^\circ_{cell}}{0.0257} = \frac{6 \times 0.02 \text{ V}}{0.0257} = 4.7$$

$$K = e^{4.7}$$

9. Reduction occurs at the cathode. you can eliminate choices (B) and (D), because these reaction are oxidation E°_{red} for $Fe^{2+}(aq)$ is –0.44 V, and E°_{red} for $Zn^{2+}(aq)$ is –0.76 V. Because $Fe^{2+}(aq)$ has the more positive E°_{red} of the two choices $Fe^{2+}(aq)$ is the more easily reduced and therefore plates out on the cathode.

10. Use the relationships

$$\Delta G^\circ = -nF\,E^\circ = \Delta H^\circ - T\Delta S^\circ$$

to derive the formula

$$E^\circ = \frac{\Delta H^\circ - T\Delta S^\circ}{-nF}$$

Next, take the given equation and break it down into the oxidation and reduction half reactions so that you can discover the value for n, the number of moles of electrons either lost or gained.

anode reaction (oxidation)

$$Pb(s) + SO_4^{2-} \rightarrow PbSO_4(s) + 2e^-$$

Cathode reaction (reduction)

$$PbO_2(s) + SO_4^{2-}(aq) + 4H^+(aq) + 2e^- \rightarrow PbSO_4(s) + 2H_2O(l)$$

$$Pb(s) + PbO_2(s) + 4H^+(aq) + 2SO_4^{2-}(aq) \rightarrow 2PbSO_4(s) + 2H_2O(l)$$

Now substitute all the known information into the derived equation.

$E^\circ = H^\circ - T\Delta S^\circ/-nF = (-315.9 \text{ kJ mol}^{-1} - 348 \text{ K } (0.2635 \text{ kJ K}^{-1} \cdot \text{mole}^{-1})/-2(96.487) \text{ kJ V}^{-1} \cdot \text{mole}^{-1})$

11. By the cell convention, the anode is on the left side of the cell, so this is where the oxidation occurs:

$$Co(s) \rightarrow Co^{2+}(aq) + 2e^-$$

Since $E^\circ_{red} = -0.28$ V for $Co^{2+}(aq) + 2e^- \rightleftharpoons Co(s)$) $\therefore$ $E^\circ_{oxid} = +0.28$ V

At the cathode, reduction takes place, so the half reaction must involve the gain of electrons:

$$Cr^{3+}(aq) + 3e^- \rightarrow Cr(s)$$

From the table of electrode potentials $E^\circ_{red} = -0.74$ V

Since $E^\circ_{cell} = E^\circ_{oxid} + E^\circ_{red}$ then:

$$E^\circ_{cell} = +0.28 + -(0.78) = -0.50 \text{ V}$$

12. The oxidation half reaction is:

$$Zn \rightarrow Zn^{2+} + 2e^-$$

for which $E^\circ_{oxid} = +0.76$ V

The reduction half reaction is:

$$I_2 + 2e^- \rightarrow 2I^-$$

for which $E^\circ_{red} = 0.54$ V

Using $E^\circ_{cell} = E^\circ_{oxid} + E^\circ_{red}$ gives

$$E^\circ_{cell} = +0.76 + 0.54 = +1.30 \text{ V}$$

13. Since $E^\circ_{cell} = E^\circ_{oxid} + E^\circ_{red}$

then: $E^\circ_{cell} = +1.18 - 0.40 = +0.78$V

From the two half reactions, it can be seen that 2 moles of electrons are transferred during the reaction.

So: $\Delta G^\circ = -nF E^\circ_{cell}$

$$\Delta G^\circ = -2 \times 96500 \times 0.78 \text{ J mol}^{-1}$$
$$= -150.5 \text{ kJ mol}^{-1}$$

14. From the overall cell reaction, the two half equations are:

Oxidation half reaction:

$$Fe^{2+}(aq) \rightarrow Fe^{3+}(aq) + e^- \qquad E^\circ_{oxid} = -0.77 \text{ V}$$

Reduction half equation:

$$Cr^{3+}(aq) + e^- \rightarrow Cr^{2+}(aq) \qquad E^\circ_{red} = 0.41 \text{ V}$$

Since the cell potential is given by $E^\circ_{cell} = E^\circ_{oxid} + E^\circ_{red}$ then:

$$E^\circ_{cell} = -0.77 - 0.41 = -1.18 \text{ V}$$

A negative cell potential predicts that the reaction should not take place spontaneously. The free energy change for this reaction can also be calculated to show that the free energy charge is positive.

$$\Delta G = -nF E^\circ_{cell}$$
$$= -1 \times 96500 \times (-1.18) \text{ J mol}^{-1}$$
$$= +114 \text{ kJ mol}^{-1}$$

The positive value of the free energy change is another indicator that the reaction will not take place spontaneously.

15. Concentration of Zn^{2+} (with 96% dissociation)

$$= 0.1 \times \frac{96}{100} = 96 \times 10^{-3} \text{ M}$$

The electrode reaction: $Zn^{2+}(aq) + 2e^- \rightleftharpoons Zn(s)$

According to the Nernst equation, the potential of the electrode is

$$E = E^\circ = \frac{RT}{nF}\ln\left(\frac{1}{Zn^{2+}}\right)$$

$$= (-0.76) - \frac{8.314 \times 298}{2 \times 96500}\ln\left[\frac{1}{(96 \times 10^{-3})}\right]$$

$$E = -0.79 \text{ volt}$$

16. The overall reaction is

$$Cu(s) + 2H^+ \rightleftharpoons Cu^{2+}(aq) + H_2(g)$$

The potential of the cell is given by

$$E_{cell} = E^\circ_{cell} - \frac{0.0591}{2}\log\frac{[Cu^{2+}]P_{H_2}}{[H^+]^2}$$

$$= (0.00 - 0.34) = \frac{0.0591}{2}\log\left[\frac{(0.50)(0.95)}{(0.01)^2}\right]$$

$$E_{cell} = -0.449 \text{ V}$$

17. Overall cell reaction:

$$Fe(s) + Cd^2(aq) \rightleftharpoons Fe^{2+}(aq) + Cd(s)$$

$\therefore$ $$E = E^\circ - \frac{0.0591}{2}\log\frac{[Fe^{2+}]}{[Cd^{2+}]}$$

$\Rightarrow$ $$-0.02 = 0.04 - \frac{0.0591}{2}\log\left(\frac{0.1}{x}\right)$$

$\Rightarrow$ $$x = 0.00093$$

or $$x = .001 \text{ M}$$

18. For the given reduction reaction

$$E^\circ = (-0.21) - (-0.06) = -0.15 \text{ V}$$

$\therefore$ $$\ln K_{eq} = \frac{nFE^\circ}{RT} = \frac{2 \times 96500 \times (-15)}{8.314 \times 298} = -11.68$$

$\Rightarrow$ $$K_{eq} = 8.42 \times 10^{-6}$$

19. Decreasing order of displacement power is

$$Mg\ (-2.37 \text{ V}) > Al(-1.66 \text{ V}) > Zn\ (-0.76 \text{ V}) > Fe(-0.44 \text{ V}) > Cu(+0.34 \text{ V})$$

Because lower the reduction potential, more the tendency to lose electrons and thus more will be the displacement power.

20. $E^\circ_{cell} = E^\circ_{Ag^+/Ag} - E^\circ_{Cu^{2+}/Cu} = 0.80 - 0.34 = 0.56$ V

$\therefore$ $$\ln K_{eq} = \frac{nFE^\circ}{RT} = \frac{2 \times 96500 \text{ C mol}^{-1} \times 0.56 \text{ V}}{8.314 \text{ JK}^{-1}\text{ mol}^{-1} \times 298 \text{ K}} = 43.62$$

$\Rightarrow$ $$K_{eq} = 8.5 \times 10^{18}$$

21. Here, $Cu^+(aq) + e^- \rightarrow Cu(s)$; $E^\circ(Cu^+, Cu) = +0.52$ V

$Cu^{2+}(aq) + e^- \rightarrow Cu^+(aq)$; $E^\circ(Cu^{2+}, Cu^+) = +0.16$ V

$$\therefore 2Cu^+(aq) \rightleftharpoons Cu(s) + Cu^{2+}(aq);\ E^\circ = (0.52 \text{ V}) = (0.16 \text{ V}) = +0.36 \text{ V}$$

Now $$\ln K = \frac{nFE^\circ}{RT}$$

$$= \frac{1 \times 96500 \times +0.36}{8.314 \times 298} = 14.022$$

$\Rightarrow$ $$K = 1.2 \times 10^6$$

Because of the large value of K, the equilibrium lies strongly towards the right and so Cu^+ disproportionales almost totally in solution.

22. For the given reaction

$$E° = E°(Sn^{2+}/Sn) - E°(Pb^{2+}/Pb)$$
$$= -0.14 - (-0.13) = -0.01 \text{ V}$$

$$\therefore \quad \ln K = \frac{nFE°}{RT} = \frac{2 \times 96500 \times (-0.01)}{8.314 \times 298} = 0.779$$

so $\quad K = 0.46.$

23. (a) The cell diagram is

Cd(s)	|	Cd^{2+}(aq)	||	Ag^+(aq)	|	Ag(s)
Anode		Electrolyte	Salt bridge	Electrolyte		Cathode

and $\quad E^\circ_{cell} = E^\circ_{cathode} - E^\circ_{anode} = 0.78 - (-0.40) = 1.18 \text{ V}$

24. The equation for the half-reactions are:

At anode $\quad Zn \rightarrow Zn^{2+} + 2e^-, \quad E° = +0.76$ V (1)

At cathode: $\quad Fe^{3+} + e^- \rightarrow Fe^{2+}, \quad E° = +0.77$ V (2)

To obtain the equation for the cell reaction, the equation for this $Fe^{2+} \mid Fe^{3+}$ half-reaction must be multiplies by 2(to obtain 2 mol electron) and combined with the $Zn \mid Zn^{2+}$ half reaction. If is to be noted that potential does not depend on the quantity of substance undergoing reaction so the value of the half-cell potential is not multiplied by 2. Thus, reaction at cathode is re written as:

At cathode: $\quad 2Fe^{3+} + 2e^- \rightarrow 2Fe^{2+}, \quad E° = +0.77$ V (3)

And now adding equations (1) and (3), we get desired cell reaction

$$Zn + 2Fe^{3+} \rightarrow 2Fe^{2+} + Zn^{2+}, \quad E° = +0.76 + 0.77 = +1.53 \text{ V}$$

A positive sign of E°_{cell} means that the reaction is spontaneous as written

25. As all the species are in their standard states, and because the potential of the standard hydrogen electrode (SHE) is zero. The total potential of the cell may be attributed to the copper half cell Thus, the standard electrode potential of the $Cu \mid Cu^{2+}$ half cell is +0.34 V.

26. $E^\circ_{cell} = E^\circ_{(Ag/AgCl)} - E^\circ_{(SCE)}$

$\Rightarrow \quad E^\circ_{(Ag/AgCl)} = E^\circ_{cell} - E^\circ_{(SCE)}$ (Standard Calomel electrode)

$$= (-0.022 \text{ V}) + (0.244 \text{ V}) = 0.222 \text{ V}.$$

27. The scheme of cell is:

$$Cu(s) \mid Cu^{2+}(0.01 \text{ M}) \parallel Cu^{2+}(0.1 \text{ M}) \mid Cu(s)$$

And emf of cell, $\quad E_{cell} = \frac{0.0591}{2}\log\left(\frac{0.1}{0.01}\right) = 0.0296 \text{ V}$

28. The given cell is concentration cell and its emf is given by:

$$E_{cell} = \frac{2.303\,RT}{nF}\log\left(\frac{C_2}{C_1}\right)$$

Here, $E_{cell} = 0.029$ V, $T = 291°$K, $R = 8.314$ JK^{-1} mol^{-1},

$F = 96{,}500$, $C_1 = 0.001$ N and $C_2 = 0.01$ N

Substituting these values in the above expression, we get

$$0.029 = \frac{2.303 \times 8.314 \times 291}{n \times 96500} \log\left(\frac{0.01}{0.001}\right)$$

$\Rightarrow$ $n = 2$

Hence, the valency of the mercurous ion is 2 and its formula is Hg_2^{2+}

29. 32 gm copper is deposited by 96,500 C

$\therefore$ 2 gm copper is deposited by $= \frac{96,500}{32} \times 2 = 6031.25$ C

As Quantity of electricity (Q) = current strength (I) × time (l)

$$\therefore \quad t = \frac{Q}{I} = \frac{6031.25\text{ C}}{5\text{ A}} = 1206.25\text{ s} = 20\text{ min, } 6.25\text{ s.}$$

30. The given cell is represented as:

$$Ag(s) \mid AgNO_3\ (0.018\text{ m}) \parallel AgNO_3(1.2\text{ m})\ Ag(s)$$

The EMF of the above cell is given by

$$E = \frac{RT}{nF} \ln \frac{(a_R)}{(a_L)}$$

where Activity of right electrode $(a_R) = \gamma_R m_R$
$= 1 \times 1.2$ m

And Activity of left electrode $(a_L) = \gamma_L\ m_L$
$= 1 \times 0.018$ m

Here, we have assumed that activity coefficients γ_R and γ_L are equal to unity for dilute solutions.

Thus, EMF of the cell $E = \frac{RT}{nF} \ln \frac{(1.2)}{(0.018)}$

$$\Rightarrow \quad E = \frac{2.303\,RT}{2F} \log \frac{(1.2)}{(0.018)} = \frac{0.0591}{2} \log \frac{(1.2)}{(0.018)} = 0.054\text{ volts.}$$

31. (i) We can simply return to the setup in part 1 and substitute the information we have available. We must solve for $E°$.

Oxidation $-E°$

Reduction: $E° = 0.401$ V (Given)

Ne: $E°_{cell} = -E° + 0.401\text{ V} = +2.73\text{ V}$

$E° = 0.401\text{ V} - 2.73\text{ V} = -2.33\text{ V}$

(ii) From the description fo the anode (oxidation) half-reaction, we should expect it to involve Al and OH^- as *reactions* and $[Al(OH)_4]^-$ as a product. The reduction half-reaction must involve $O_2(g)$ as a *reactant* and H_2O and/or OH^- as needed in complete and balance the equation (since the solution is basic). The following half equation meet these requirements.

First we must make use of the relationship between $\Delta G°$ and $E°_{cell}$ for a reaction.

$$\Delta G° = -nF\,E°_{cell} = -12\text{ mol } e^- \times 96{,}500\text{ C/mol } e^- \times 2.73\text{ V}$$

$$= -3.15 \times 10^6 \text{ V} \cdot \text{C} = -3.16 \times 10^6 \text{ J} = -3.16 \times 10^3 \text{ kJ}$$

Now,

$$\Delta G^\circ = 4\Delta G_f^\circ[\text{Al(OH)}_4]^- - 6\Delta G_f^\circ[\text{H}_2\text{O(l)}] - 4\Delta G_f^\circ[\text{OH}^-]$$

$$-3.16 \times 10^3 \text{ kJ} = 4\Delta G_f^\circ[\text{Ai(OH)}_4]^- - 6(-273 \text{ kJ}) - 4(-157 \text{ kJ})$$

$$\Delta G_f^\circ[\text{Al(OH)}_4]^- = \frac{-3.16 \times 10^3 - 1422 - 628}{4} = 1.30 \times 10^3 \text{ kJ mol}^{-1}$$

(iii) Think of this as you would an electrolysis calculation. The total amount of charge transferred is

$$10.0 \frac{\text{C}}{\text{s}} \times 4.0 \text{ h} \times \frac{60 \text{ min}}{1 \text{ h}} \times \frac{60 \text{ S}}{1 \text{ min}} = 1.44 \times 10^5 \text{ C}$$

This corresponds to

$$1.44 \times 10^5 \text{ C} \times \frac{1 \text{ mol } e^-}{96.485 \text{ C}} = 1.49 \text{ and } e^-$$

The quantity of Al that must be oxidized to produce this number of electrons is

$$\text{no. g Al} = 1.49 \text{ mol } e^- \times \frac{1 \text{ mol Al}}{3 \text{ mol } e^-} \times \frac{26.98 \text{ g Al}}{1 \text{ mol Al}} = 13.4 \text{ g Al}$$

32. Lead is below silver in the table of Standard Reduction Potentials, but above Ni. Thus it is a stronger reducing agent than Ag, but weaker than Ni. The other metals are stronger reducing agents than both Ag and Ni.

33. $\Delta E^\circ_{\text{cell}} = E^\circ_{\text{cathode}} - E^\circ_{\text{anode}} = -0.126 - = (-0.763)$

$= 0.763 - 0.126 = 0.637 \text{ V}$

34. anode: $\text{Zn}(s) \rightarrow \text{Zn}^{2+} + 2e^-$
cathode:$2\text{Ag}^+ + 2e^- \rightarrow 2\text{Ag}(s)$
net cell reaction:$\text{Zn}(s) + 2\text{Ag}^+ \rightarrow \text{Zn}^{2+} + 2\text{Ag}(s)$

35. Zinc is a stronger reducing agent than H_2, so H_2 will be evolved at the cathode. Nitrate ions cannot be oxidized at the anode; H_2O or OH^- will be oxidized and $O_2(g)$ will be evolved at the anode. Electrolyzing zinc nitrate is essentially electrolyzing water.

36. $\text{MnO}_4^- + 5e^- + 8\text{H}^+(\text{aq}) \rightarrow \text{Mn}^{2+} + 4\text{H}_2\text{O}$
Thus 5 faradays are needed to reduce 1 mol of MNO_4^- to Mn^{2+}

37. Cobaltic ion, Co^{3+}, is a powerful oxidizing agent. The value of E° for $\text{Co}^{3+} + e^- \rightarrow \text{Ce}^{2+}$ is 1.82 V. The other cations, Zn^{2+}, Fe^{2+}, and Mn^{2+} are week oxidizing agents, below H^+ in the Table of Standard Reduction Potentials. Option (a) is clearly wrong at Br^- can only serve as a reducing agent.

38. As $\Delta G = n\mathcal{F}\Delta E_{\text{cell}}$. If $\Delta E_{\text{cell}} > 0$, $\Delta G < 0$. Option (b) is wrong because the sign of ΔG cannot be determined solely from $\Delta E^\circ_{\text{cell}}$, the value of Q is also needed. Options (d) is wrong because K_{eq} is calculated from $\Delta E^\circ_{\text{cell}}$ and not from ΔE_{cell}. Option (c) is wrong because if $\Delta E^\circ_{\text{cell}} < 0$, the position of equilibrium is on the left, and $K_{eq} < 1$.

39. $\Delta E^\circ_{cell} = E^\circ_{right} - E^\circ_{left} = -0.403 - (-0.763) = +0.360$ V

$\Delta G^\circ = -nF\Delta E^\circ_{cell} = -2(96{,}485 \text{ C})(0.360 \text{ V}) = -69{,}470 \text{ J} = -69.47$ kJ.

40. anode: $Cd(s) \rightarrow Cd^{2+} + 2e^-$

cathode: $Cu^{2+} + 2e^- \rightarrow Cu(s)$

net cell reaction: $Cd(s) + Cu^{2+} \rightarrow Cd^{2+} + Cu(s)$

$Q = [Cd^{2+}]/[Cu^{2+}] = 1.00$

The way to increase the cell voltage is to decrease the reaction quotient Q. There are two ways to understand this. (1) A reaction proceeds to the eight if $Q < K_{eq}$. The smaller the value of q, the greater the tendency for the reaction to proceed spontaneously to the right, and the larger the cell voltage. (2) The term in the Nernst equation involving Q is $-(RT/nF)$ $(\ln Q)$. If $Q < 1$, $\ln Q$ is negative and this term is positive, making the cell voltage more positive than ΔE°_{cell}. The smaller Q is, the more negative $\ln Q$ is, and the larger (more positive) is the term $-(RT/nF)(\ln Q)$ in the Nernst equation for ΔE_{cell}.

Option (a) and (c) leave Q unchanged. Options (b) increase Q. Only option (d) decreases Q, from 1.00 to 0.100.

41. Metals that are weaker reducing agents than H_2 (below H_2 in the table) will plate out at the cathode during electrolysis, of the 5 listed. Cr and Cd are stronger reducing agents than H_2; Ag, Cu, and Hg are weaker.

42. Electrolysis of an aqueous solution of Li_2SO_4 is just the electrolysis of water. Both OH^- and H_2O are ensier to oxidize than SO_4^{2+}; Li is a stronger reducing agent then H_2 so that Li^+ is harder to reduce than H_2O.

43. Spontaneity cannot be predicted from the standard state conditions, but only from the actual values of ΔG and ΔE_{cell} with the concentrations used. For the spontaneous reaction $\Delta G < 0$, $Q < K_{eq}$, and $\Delta E_{cell} > Q$.

44. anode: $Zn(s) \rightarrow Zn^{2+} + 2e^-$

cathode: $Cu^{2+} + 2e^- \rightarrow Cu(s)$

net cell reaction: $Zn(s) + Cu^{2+} \rightarrow Zn^{2+} + Cu(s)$

$Q = [Zn^{2+}]/[Cu^{2+}]$

If $Q > 1$, $\ln Q > 0$, and $\Delta E_{cell} < \Delta E^\circ_{cell}$, because the term is the Nernst equation that depends on Q, $-(RT/nF)(\ln Q)$ is negative only if $[Zn^{2+}] > [Cu^{2+}]$ can $Q > 1$. Cell potentials are independent of the actual amount of material present, or the size of the electrode.

45. As two faradays are required to deposit 1 mol of Cu, 0.500 F is required to deposit 0.250 mol of Cu. No. coloumbs = $(96{,}485 \text{ C}/F)(0.500\ F) = (15.0 \text{ A})(t)$

$\therefore \quad t = 3216$ s

The time in minutes is (3216 s)/(60 s/min) = 53.6 min.

46. $\Delta E^\circ_{cell} = E^\circ_{cathode} - E^\circ_{anode} = 0.799 - (-0.403)$

$= 1.202$ V

$$\log K_{eq} = \frac{2(1.202)}{0.05916} = 40.6$$

47. In 70.2 min there are 4212 s. The charge in coulombs is (30.0 A) (4212 s) = 1.26×10^5 C. The number of faradays is $(1.26 \times 10^5\ \text{C})/(96{,}485\ \text{C}/F) = 1.31\ F$.

48. anode $\quad Co(s) \rightarrow Co^{2+} + 2e^-$

cathode: $\quad 2H^+(aq) + 2e^- \rightarrow H_2(g)$

net cell reaction: $\quad Co(s) + 2H^+(aq) \rightarrow Co^{2+} + H_2(g)$

$$Q = \frac{[Co^{2+}]P_{H_2}}{[H^+]^2}$$

A change that decrease the reaction quotient Q, increase the cell voltage. Increasing the $[H^+]$ decreases Q increases ΔE_{cell}. Options increases Q and therefore decreases ΔE_{cell}. ΔE_{cell} is independent of the amount of material present and depends only on the concentrations or partial pressure in Q.

49. anode reaction: $4OH^-(aq) \rightarrow O_2(g) + 4e^- + 2H_2O$ No. of coulombs passed = (1.50 h) (3600 s/h) (8.00 A) = 4.32×10^4 No. of faradays passed = $(4.32 \times 10^4\ \text{C})/(96.484\ \text{C}/F) = 4.48 \times 10^{-1}$. Since it requires $4F$ to produce 1 mol of O_2.

$$\text{No. mol } O_2 \text{ produced} = \frac{1}{4}(0.448) = 0.112.$$

50. anode: $\quad Ni(s) \rightarrow Ni^{2+} + 2e^-$

cathode: $\quad Cl_2(g) + 2e^- \rightarrow 2Cl^-$

net cell reaction: $\quad Ni(s) + Cl_2(g), \rightarrow 2Cl^- + Ni^{2+}$

$$Q = \frac{[Ni^{2+}][Cl^-]^2}{P_{Cl_2}} = \frac{(0.10)(0.40)^2}{0.50} = 3.2 \times 10^{-2}$$

51. The standard reduction potential for dilute nitric acid

$$NO_3^- + 4H^+(aq) + 3e^- \rightarrow NO^+\uparrow + 2H_2O$$

is 0.96 V. Dilute HNO_3 will, therefore, oxidize any metal whose cation has a reduction potential below +0.96 V, but cannot oxidize Au and Pt, which are relatively inert.

52. $Q = It$

$= (0.2\ \text{A})\ (3 \times 60 \times 60\ \text{s}) = 2160\ \text{C}$

Let thickness of silver deposited be (a) cm.

∴ Voume of silver deposited = $80\ \text{cm}^2 \times a\ \text{cm} = 80\ a\ \text{cm}^3$

and mass of Ag deposited = (Density of Ag) × (Volume of Ag deposited)

$= (10.5 \times 80\ a)\ \text{gm} = 840\ a\ \text{gm}$ (1)

Now $\quad Ag^+(aq) + e^- \rightarrow Ag(s)$

1 mol = 96,500 C $\quad$ 1 mol = 107.92 gm

∴ 96,500 C deposit Ag = 107.92 gm

so 2160 C will deposit Ag = $\frac{107.92}{96{,}500} \times 2160 = 2.42$ gm (2)

From equations (1) & (2), we have

$$840\ a\ \text{gm} = 2.42\ \text{gm}$$

$$\Rightarrow \qquad a = \frac{2.42}{840} = 2.88 \times 10^{-3}\ \text{cm.}$$

53. The Half-cell reactions are:

$$Zn^{2+}(aq) + 2e^- \rightleftharpoons Zn(s); \qquad E^\circ_{Zn^{2+}/Zn} = 0.76\ V$$

$$Ag^+(aq) + e^- \rightleftharpoons Ag(s); \qquad E^\circ_{Ag^+/Ag} = 0.8\ V$$

The overall cell reaction is:

$$Zn(s) + 2\ Ag^+(aq) \rightarrow Zn^{2+}(aq) + 2\ Ag(s)$$

The cell can be represented as:

$$Zn\ (s)\ |\ Zn^{2+}(aq)\ ||\ Ag^+(aq)\ |\ Ag\ (s)$$

Now

$$E^\circ_{cell} = E^\circ_{Ag^+/Ag} - E^\circ_{Zn^{2+}/Zn}$$
$$= 0.8 - (-0.76) = 1.56\ V$$

54. The cell notation gives the species involved in each half-reaction. Complete and balance each half-reaction, then multiply by factors so that when you add the half-reaction the electrons cancel. The result is the cell reaction.

(i) The half-cell reaction are

$$Tl(s) \rightarrow Tl^+(aq) + e^-$$
$$Sn^{2+}(aq) + 2e^- \rightarrow Sn(s)$$

Multiplying the anode reaction by 2 and then summing the half-cell reaction gives

$$2Tl(s) + Sn^+(aq) \rightarrow 2Tl^+(aq) + Sn(s)$$

(ii) The half-cell reaction are

$$Zn(s) \rightarrow Zn^{2+}(aq) + 2e^-$$
$$Fe^{3+}(aq) + e^- \rightarrow Fe^{2+}(aq)$$

and the cell reaction is

$$Zn(s) + 2Fe^{3+}(aq) \rightarrow Zn^{2+}(aq) + 2Fe^{2+}(aq)$$

55. The maximum work of a cell equals—nFE_{cell}. To get n, you split the cell reaction into half-reaction, multiplying each by a factor so that when added together they give the cell reaction. Then n equals the number of electrons in either half-reaction.

The half-reactions are

$$Hg_2^{2+}(aq) + 2e^- \rightleftharpoons 2Hg(l)$$
$$H_2(g) \rightleftharpoons 2H^+(aq) + 2e^-$$

n equals 2, and the maximum work for the reaction as written is

$$w_{max} = -nFE_{cell} = -2 \times 9.65 \times 10^4 C \times 0.650\ V$$
$$= -1.25 \times 10^5\ V \cdot C$$
$$= -1.25 \times 10^5\ J$$

(Remember that a joule is equal to a volt-coulomb. Also, note that the negative sign means work is done by the cell—that is, energy is lost by the cell.) For 0.500 g H_2, the maximum work is

$$0.500\ \text{g H}_2 \times \frac{1\ \text{mol H}_2}{2.02\ \text{g H}_2} \times \frac{-1.25 \times 10^5\ \text{J}}{1\ \text{mol H}_2} = \mathbf{-3.09 \times 10^4\ J}$$

Note that the conversion factor -1.25×10^5 J/1 mol H_2 is determined by the chemical equation as written. In the equation, 1 mol H_2 reacts, and the maximum work produced is -1.25×10^5 J.

56. (i) The half-reaction and corresponding electrode potentials are as follows:

$Fe^{3+}(aq) + e^- \rightarrow Fe^{2+}(aq)$ 0.77 V

$Cl_2(g) + 2e^- \rightarrow 2Cl^-(aq)$ 1.36 V

$H_2O(aq) + 2H^+(aq) + 2e^- \rightarrow 2H_2O(l)$ 1.78 V

The order by increasing oxidizing strength is Fe^{3+} (aq), $Cl_2(g)$, $H_2\,O_2$ (aq)

(ii) The half-reactions and corresponding electrode potentials are

$Al^{3+}(aq) + 3e^- \rightarrow Al(s)$ – 1.66 V

$2H^+(aq) + 2e^- \rightarrow H_2(g)$ 0.00 V

$Cu^{2+}(aq) + 2e^- \rightarrow Cu(s)$ 0.34 V

The order by increasing reducing strength is Cu(s), $H_2(g)$, Al(s).

57. The cell notation assumes that the anode (the oxidation half-cell) is on the left. Change the direction of this half-reaction and the sign of its electrode potential. (Assuming that the cell notation was written correctly, you change the direction of the half-reaction corresponding to the smaller, or more negative, electrode potential). Multiply the half-reaction (but not the electrode potentials) by factors so that when the half-reactions are added, the electrons cancel. The sum of the half-reactions is the cell reaction. Add the electrode potential to get the cell emf.

The reduction half-reactions and standard electrode potentials are

$Al^{3+}(aq) + 3e^- \rightarrow Al(s);$ $E^\circ_{Al} = -1.66$ V

$Fe^{2+}(aq) + 2e^- \rightarrow Fe(s);$ $E^\circ_{Fe} = -0.41$ V

Now, reverse the first half-reaction and its half cell potential to obtain

$Al(s) \rightarrow Al^{3+}(aq) + 3e^-;$ $-E^\circ_{Al} = 1.66$ V (i)

$Fe^{2+}(aq) + 2e^- \rightarrow Fe(s);$ $E^\circ_{Fe} = -0.41$ V (ii)

We obtain the cell emf by adding the half-cell potentials after balancing both (i) & (ii). Multiply the first half-reaction by 2 and the second half-reaction by 3, so that when the half-reaction are added to electrons cancel. The addition of half reactions is 1.25 V.

$2Al(s) \rightarrow 2Al^+(aq) + 6e^-$ $-E^\circ_{Al} = 1.66$ V

$3Fe^{2+}(aq) + 6e^- \rightarrow 3Fe(s)$ $E^\circ_{Fe} = -0.41$ V

$2Al(s) + 3Fe^{2+}(aq) \rightarrow 2Al^{3+}(aq) + 3Fe(s)$ $E^\circ_{Fe} - E^\circ_{Al} = 1.25$ V

Note that the cell emf is the standard electrode potential for the cathode minus the standard electrode potential for the anode. We could have calculated the cell emf from the formula

$$E^\circ_{cell} = E^\circ_{cathode} - E^\circ_{anode}$$

You obtain

$$E^\circ_{cell} - E^\circ_{Fe} - E^\circ_{Al} = -0.41 \text{ V} - (-1.66 \text{ V}) = 1.25 \text{ V}$$

58. The half reactions, corresponding half-cell potentials, and their sums are

$$Zn(s) \rightarrow Zn^{2+}(aq) + 2e^- \qquad -E^\circ = 0.76 \text{ V}$$

$$2Ag^+(aq) + 2e^- \rightarrow 2Ag(s) \qquad E^\circ = 0.80 \text{ V}$$

$$Zn(s) + 2Ag^+(aq) \rightarrow Zn^{2+}(aq) + 2Ag(s) \qquad E^\circ_{cell} = 1.56 \text{ V}$$

Here, each half-reaction involves two electrons; hence, $n = 2$ Also, $E^\circ_{cell} = 1.56$ V. and the faraday constant, F, is 9.65×10^4C

$$\therefore \quad \Delta G^\circ = -nFE^\circ_{cell} = -2 \times 9.65 \times 10^4 \text{ C} \times 1.56 \text{ V} = -3.01 \times 10^5 \text{ J} = -301 \text{ KJ}.$$

59. We have,

$$Zn(s) + Cl_2(g) \rightarrow Zn^{2+}(aq) + 2Cl^-(aq)$$

ΔG°_f 0   0   –147   2 × (–131) kJ

Hence,

$$\Delta G^\circ = \Sigma n\Delta G^\circ_f(\text{products}) - \Sigma n\Delta G^\circ_f(\text{reactants})$$
$$= [-147 + 2 \times (-131)] \text{ kJ}$$
$$= -409 \text{ kJ} = -4.09 \times 10^5 \text{ J}$$

Now,

$$Zn(s) \rightarrow Zn^{2+}(aq) + 2e^-$$
$$Cl_2(g) + 2e^- \rightarrow 2Cl^-(aq)$$

Each half reaction involves two electrons, so $n = 2$

But $\quad \Delta G^\circ = -nFE^\circ_{cell}$

or $\quad -4.09 \times 10^5 \text{ J} = -2 \times 9.65 \times 10^4 \text{ C} \times E^\circ_{cell}$

$$E^\circ_{cell} = 2.12 \text{ V}.$$

60. The reaction corresponds to the one for the voltaic cell. Note than $n = 2$. Substituting into the equation relating E°_{cell} and K gives

$$1.10 = \frac{0.0592}{2} \log K_c$$

Solving for $\log K_c$

$$\log K_c = 37.2$$

$$\therefore \quad K_c = \text{antilog } (37.2) = 1.6 \times 10^{37}$$

61. The cell reactions is

$$Zn(s) + Cu^{2+}(aq) \rightleftharpoons Zn^{2+}(aq) + Cu(s)$$

The number of electrons transferred is two, hence, $n = 2$. The reaction quotient is

$$Q = \frac{[Zn^{2+}]}{[Cu^{2+}]} = \frac{1.00 \times 10^{-5}}{0.100} = 1.00 \times 10^{-4}$$

The standard emf is 1.10 V, so the Nernst equation becomes

$$E_{cell} = E^{\circ}_{cell} - \frac{0.0592}{n} \log Q$$

$$= 1.10 - \frac{0.0592}{2} \log (1.0 \times 10^{-4})$$

$$= 1.10 - (0.12) = 1.22$$

The cell emf is **1.22 V**.

62. The electrode equation for copper says the 1 mol Cu is equivalent to 2 mol, e^-

But $$\text{g Cu} \rightarrow \text{mol Cu} \rightarrow \text{mol } e^- \rightarrow \text{coulombs (C)}$$

The current in amperes (A) equals the charge in coulombs divided by the time in sec. The coulombs of Cu required to deposit 404 mg of of copper is

$$0.404 \text{ g Cu} \times \frac{1 \text{ mol Cu}}{63.6 \text{ g Cu}} \times \frac{2 \text{ mol } e}{1 \text{ mol Cu}} \times \frac{9.65 \times 10^4 \text{ C}}{1 \text{ mol } e} = 1.226 \times 10^3 \text{ C}$$

The time lapse 5.00 h, equals 1.80×10^4 s. Thus,

$$\text{Current} = \frac{\text{charge}}{\text{time}} = \frac{1.226 \times 10^3 \text{ C}}{1.80 \times 10^4 \text{ g}} = \mathbf{6.81 \times 10^{-2} \text{ A}}$$

63. The current in amperes multiplied by the time in seconds equals the charge in coulombs. Convert the charge in coulombs to moles of electrons.

As, each mole of iodine produced requires 2 mol e^-, convert mol e^- to mol I_2 and then, mol I_2 to g I_2.

$$\text{Coulombs} \rightarrow \text{mol } e^- \rightarrow \text{mol } I_2 \rightarrow \text{g } I_2$$

When the current flows for 6.00×10^2 s (10.0 min) the amount of charge is

$$8.52 \times 10^{-3} \text{ A} \times 6.00 \times 10^2 \text{ s} = 5.11 \text{ C}$$

Note that two moles of electrons equivalent to one mole of I_2 Hence,

$$5.31 \text{ C} \times \frac{1 \text{ mol } e^-}{9.65 \times 10^4 \text{ C}} \times \frac{1 \text{ mol } I_2}{2 \text{ mol } e^-} \times \frac{254 \text{ g } I_2}{1 \text{ mol } I_2} = \mathbf{6.73 \times 10^{-3} \text{ g } I_2}.$$

64. The reaction may be given as:

$$Zn(s) + Ni^{2+}(aq) \rightleftharpoons Zn^{2+}(aq) + Ni(s)$$

The cell reaction may be written as

$$Zn(s) \mid Zn^{2+}(aq) \parallel Ni^{2+}(aq) \mid Ni(s)$$

The standard cell potential would be

$$E^{\circ}_{cell} = E^{\circ}_R - E^{\circ}_L = -0.24 \text{ V} - (-0.75 \text{ V}) = 0.51 \text{ V}$$

The equilibrium constant of the reaction would be,

$$\log K = \frac{nFE^\circ}{2.303\,RT}$$

$$= \frac{2 \times 96500 \times 0.51}{2.303 \times 8.314 \times 298} = 17.25$$

or $\qquad K = 1.78 \times 10^{17}$

Let x be the concentration of Ni^{2+} ions that have been reduced to Ni at equilibrium. It is obvious that

$$[Ni^{2+}] = 1.0 \times x$$

$$[Zn^{2+}] = x$$

Hence $$E = \frac{[Zn^{2+}]_{eq}}{[Ni^{2+}]_{eq}} = \frac{x}{1-x}$$

or

$$1.78 \times 10^{17} = \frac{x}{1-x}$$

$$x = \frac{1.78 \times 10^{17}}{1 + 1.78 \times 10^{17}}$$

$$[Ni^{++}] = 1 - x = 1 - \frac{1.78 \times 10^{17}}{1 + 1.78 \times 10^{17}}$$

$$= \frac{1}{1 + 1.78 \times 10^{17}} \simeq \frac{1}{1.78 \times 10^{17}} = 0.562 \times 10^{-17}$$

$$= \mathbf{5.62 \times 10^{-18}\ M}$$

65. The galvanic cell will be as:

$$Cu(s)/Cu^{++}(aq) \parallel Ag^{+}(aq)/Ag\ (s)$$

or $\qquad Cu(s) + 2Ag^{+}(aq) \rightarrow Cu^{2+}(aq) + 2Ag(s)$

$$E^\circ_{cell} = E^\circ_{RHE} - E^\circ_{LHE} = 0.799 - 0.337 = 0.462\ V$$

Now, according to Nernst's equation

$$E_{cell} = E^\circ_{cell} = \frac{0.0591}{n} \log \frac{[Product]}{[Reactant]}$$

$$0 = 0.462 - \frac{0.0591}{2} \log \frac{[Cu^{++}]}{[Ag^{+}]^2}$$

or $$\log \frac{[Cu^{++}]}{[Ag^{++}]^2} = \frac{0.642 \times 2}{0.0591}$$

$$\log \frac{0.01}{[Ag^+]^2} = \frac{0.462 \times 2}{0.0591}$$

or $$-2 - \log[Ag^+]^2 = 15.6$$

or $$\log[Ag^+] = -8.8 = \bar{9}.2$$

or $$[Ag^+] = \mathbf{1.52 \times 10^{-9}\ M}$$

66. The cell reaction of the chemical equation

$$Fe^{2+} + Ce^{4+} \rightleftharpoons Fe^{3+} + Ce^{3+}$$

may be written as:

$$Fe^{2+} \mid Fe^{3+} \parallel Ce^{4e} \mid Ce^{3+}$$

$\therefore$ $$E^\circ_{cell} = E^\circ_R - E^\circ_L$$

$$= 1.44 - 0.68 = 0.76\ V$$

$\because$ $$-\Delta G^\circ = nFE^\circ = 2.303\ RT \log K$$

or $$E^\circ = \frac{2.303\ RT}{nF} \log K = \frac{0.059}{n} \log K$$

or $$0.76 = \frac{0.059}{1} \log K$$

or $$\log K = \frac{0.76}{0.059} = 12.8813$$

$\therefore$ $$\mathbf{K = 7.61 \times 10^{12}}$$

67. According to Faraday's second law of electrolysis

$$\frac{M_1}{E_1} = \frac{M_2}{E_2}$$

or $$\frac{\text{Mass of Cu}}{\text{Eq. mass of Cu}} = \frac{\text{Mass of M}}{\text{Eq. mass of M}}$$

$$\frac{1.228}{31.77} = \frac{4.176}{\text{Eq. mass of M}}$$

or $$\text{Eq. mass of M} = \frac{4.176 \times 31.77}{1.228} = 108.04$$

Now according to Faraday's first law of electrolysis

$$m = Zit$$

$$4.176 = \frac{108.04 \times 1\ \text{amp} \times t}{96500}$$

or $$t = \frac{4.176 \times 96500}{108.04}$$

$$= \mathbf{3730\ s}$$

68. Since

$$W = Z\,I\,t$$

$$404 \times 10^{-3}\ g = \frac{63.5\ g\ C^{-1}}{2 \times 96500} \times I \times 5 \times 60 \times 60\ s$$

or
$$I = \frac{404 \times 10^{-3}\ g \times 2 \times 96500}{63.5\ g\ C^{-1} \times 5 \times 60 \times 60\ s}$$

$$= \mathbf{0.0682\ amp.}$$

69. let $[H^+]$ at cathode $= a$ mole L^{-1}

$\therefore$ $E_2 = E° + 0.059 \log a$ (i)

and $E_1 = E° + 0.059 \log 10^{-6}$ (ii)

Subtract (ii) from (i)

$$E_2 - E_1 = 0.059 (\log a - \log 10^{-6})$$

or $0.118 = 0.059 (\log a - \log 10^{-6})$

or $a = 10^{-4}$

$\therefore$ $H^+ = \mathbf{10^{-4}\ M}$

70. $Cu^{2+} + 2e^- \rightarrow Cu\ E° = +0.34$ V (1)

For $Cu(OH)_2$ $K_{sp} = 1.0 \times 10^{-19}$

$\therefore$ $[Cu^{2+}][OH^-]^2 = 1.0 \times 10^{-19}$

$\because$ pH of the solution is 14

$\therefore$ $[OH^-] = 1$ M

$\therefore$ $[Cu^{2+}] = 10^{-19}$

According to Nernst's equation for (1)

$$E = E° - \frac{0.059}{n} \log \frac{[Cu]}{[Cu^{++}]} = 0.34 - \frac{0.059}{2} \log \frac{1}{10^{-19}} = 0.34 - \frac{0.059}{2} \log 10^{19}$$

$$= 0.34 - \frac{0.059}{2} \times 19 = \mathbf{-0.22\ V}$$

71. (i) For the reaction,

$$NO_3^- + 2H^+ + e^- \rightarrow NO_2(g) + H_2O$$

$$E = E° - \frac{0.0591}{n} \log \frac{1}{[H^+]^2}$$

(The conc. of other species is unity)

$$= 0.78 - \frac{0.0591}{1} \log \frac{1}{(8)^2}$$

$$= 0.78 + 0.1062$$

$$= 0.886\ V$$

(ii) For neutral solution $[H^+] = 10^{-7}$ M

$$\therefore \quad E = E^\circ - \frac{0.0591}{n}\log\frac{1}{[H^+]^2}$$

$$= 0.78 - \frac{0.0591}{1}\log\frac{1}{(10^{-7})^2}$$

$$= 0.78 - 0.8274$$

$$= \mathbf{-0.047\ V}$$

72. The chemical reaction i:

$$2Fe^{3+} + 3I^- \rightleftharpoons 2Fe^{2+} + I_3^-$$

$$K_c = \frac{[Fe^{++}]^2\,[I_3^-]}{[Fe^{3+}]^2\,[I^-]^3}$$

The cell reaction may be written as:

$$I_3^- \mid I^- \parallel Fe^{3+} \mid Fe^{2+}$$

$$E_{cell} = E^\circ_{Fe^{3+}/Fe^{2+}} - E^\circ_{I_3^-/I^-}$$

$$= 0.77 - 0.54$$

$$= 0.23\ V$$

According to Nernst's equation

$$E_{cell} = E^\circ_{cell} - \frac{0.059}{n}\log K_c$$

At equilibrium

$$0 = 0.23 = \frac{0{,}059}{2}\log K_c$$

or $$\log K_c = \frac{0.23 \times 2}{0.059} = 7.797$$

Hence, $K_c = \mathbf{6.266 \times 10^7}$.

73. $H_2 \rightarrow 2H^+ + 2e^-$

$$E = E^\circ - \frac{0.059}{n}\log\frac{[H^+]^2}{[H_2]}$$

$$0.3 = 0\ \frac{0.059}{2}\log [H^+]^2$$

$$-\log\ [H^+] = \frac{0.3}{0.059} = 5.08$$

$$\therefore \quad pH = 5.08$$

Chapter 11 : Nuclear Chemistry

1. Think of the yield sign (arrow) as an equal sign. The superscript represents the mass number. The sum of the mass numbers on both sides of the arrow must be equal. The subscript represents the atomic number, and as with mass numbers, the sum of the numbers on both sides of the arrow must be equal.

   mass: 54 + 4 = 56 + 2(1)
   atomic number: 26 + 2 = 26 + 2(1)

2. We have

   mass: 65 + 1 = 64 + 2(1)
   atomic number: 29 + 0 = 29 + 2(0)

3. We have

   mass: 14 + 1 = 15 + 0
   atomic number: 7 + 1 = 8 + 0

4. Positron production occurs for nuclides that are below the zone of stability (those nuclides whose neutron/proton ratios are too small). The net effect of positron emission is to change a proton to a neutron. An example of positron emission would be

$$^{22}_{11}\text{Na} \rightarrow {}^{0}_{1}\text{e} + {}^{22}_{10}\text{Ne}$$

   Nuclides with to many neutrons per proton tend to undergo β-particle production. The net effect of β-particle production is to change a neutron to proton. Examples of β-particle production are

$$^{234}_{99}\text{Th} \rightarrow {}^{234}_{91}\text{Pa} + {}^{0}_{-1}\text{e}$$

$$^{131}_{53}\text{I} \rightarrow {}^{131}_{54}\text{Xe} + {}^{0}_{-1}\text{e}$$

5. An alpha particle is a helium nucleus.
6. Electron capture is a process by which one of the inner-orbital electrons is captured by the nucleus.
7. Fission is the process whereby a heavy nucleus splits into two nuclei (not including He nuclei) with smaller mass numbers.
8. A positron is a particle with the same mass as an electron but with the opposite charge. The net effect of positron production is to change a proton to a neutron.
9. The problem can be solved using the factor-label method.

$$\frac{28.850 \text{ years}}{1} \times \frac{1 \text{ half-life}}{5770 \text{ years}} \times = 5.00 \text{ half-lives}$$

   In 5.00 half-lives, the radioactive is reduced by

$$\left(\frac{1}{2}\right)^{5.00} = 0.0312 \times 100 = 3.12\%$$

10. Gamma rays (γ) have high penetrating power and are not deflected by electric or magnetic fields. Beta particles (β) have a lower ionizing power and greater penetrating power than alpha particles (α).

11. $t_{1/2} = 30$ min; $\quad R = \lambda N; \therefore N = \dfrac{R}{\lambda} = \dfrac{28 \text{ dis/min}}{0.693/30 \text{ min}} = 1200$

12. 90% decay corresponds to 10%, or 0.10, survival.

$$k = -\frac{(2.30)(\log [N]/[N]_o)}{t} = -\frac{(2.303)(\log 0.10)}{366 \text{ min}} = 6.29 \times 10^{-3} \text{ min}^{-1}$$

Then the half-life can be computed:

$$t_{1/2} = \frac{0.693}{k} = \frac{0.693}{6.29 \times 10^{-3} \text{ min}^{-1}} = 110 \text{ min}$$

13. The mass of carbon in the sample of old wood

$$= (7.32 \text{ g } CO_2)\left(\frac{1 \text{ mol } CO_2}{44.0 \text{ g}}\right)\left(\frac{1 \text{ mol C}}{\text{mol } CO_2}\right)\left(\frac{12.0 \text{ g C}}{\text{mol C}}\right) = 2.00 \text{ g C}$$

$$\therefore \quad \text{activity} = \frac{10.8 \text{ dis/min}}{2.00 \text{ g}} = 5.40 \text{ dis/min} \cdot \text{g}$$

$$= \frac{10.8 \text{ dis/min}}{2.00 \text{ g}} = 5.40 \text{ dis/min} \cdot \text{g}$$

Hence, $\quad (2.30)\left(\log \dfrac{R}{R_o}\right) = -\left(\dfrac{0.693}{t_{1/2}}\right)t = \dfrac{0.693 t}{5730 y}$

$$\therefore \quad t = (-2.30)\left(\log \frac{540}{15.3}\right)\left(\frac{5730 y}{0.693}\right) = 8610 y$$

14. Since $N/N_0 = m_0$, we have

$$(2.30)\left(\log \frac{m}{m_o}\right) = -\left(\frac{0.693}{t_{1/2}}\right)t = \left(\frac{0.693}{4.50 \times 10^9 y}\right)(3.00 \times 10^9 y) = -0.462$$

or $\quad \log(m/m_0) = -0.21$ hence $m/m_0 = 0.630 \quad \therefore m = m_0(0.630) = 1.26$ kg left

Also, of the 2.00 kg ^{238}U initially present, 2.00 kg – 1.26 kg = 0.74 kg disintegrated. For every U atom which disintegrates, one Pb atom is produced. The mass ratio is just 206/238. Hence,

$$(0.74 \text{ kg } ^{238}U)\left(\frac{206 \text{ g Pb}}{238 \text{ g U}}\right) = 0.64 \text{ kg Pb produced}$$

15. We have, $(2.30)\left(\log \dfrac{N}{N_0}\right) = \left(\dfrac{0.693}{t_{1/2}}\right)t = \dfrac{0.693 t}{5730 y}$

$$\therefore \qquad t = (-2.30)\left(\log\frac{2.80}{15.3}\right)\left(\frac{5730\ y}{0.693}\right) = 14000\ y$$

16. Here, $(2.303)\left(\log\dfrac{[A]}{[A]_0}\right) = -kt = \left(\dfrac{0.693}{t_{1/2}}\right)t$

or, $(2.303)\left(\log\dfrac{1070}{4280}\right) = \left(\dfrac{0.693}{t_{1/2}}\right)$ (200 min); $\therefore\ t_{1/2} = 100$ min

Alternatively, it is seen that the sample decomposed to one-fourth its original activity, which took two half-lives. Since two half-lives are 200 min, one half-life is 100 min.

17. Applying $2.303 \log \dfrac{N}{N_0} = \left(\dfrac{0.693}{t_{1/2}}\right) \cdot t$

we have,

$$(2.303)(\log 0.10) = \left(\frac{0.693}{2.62\ y}\right)t$$

or $t = 8.7\ y$

18. We have,

$$R = \left(\frac{0.693}{t_{1/2}}\right)n$$

$$\therefore \qquad t_{1/2} = \left(\frac{0.693}{A}\right)N = \left(\frac{0.693}{0.002\ s^{-1}}\right)\left(\frac{6.02 \times 10^{23}\ \text{atoms mol}^{-1}}{200\ \text{g mol}^{-1}}\right) = 1 \times 10^{24}\ s$$

$$= (1 \times 10^{24}\ s)\left(\frac{1\ h}{3600\ s}\right)\left(\frac{1\ \text{day}}{24\ h}\right)\left(\frac{1\ y}{365\ \text{days}}\right) = 3 \times 10^{16}\ y$$

19. We know that

$$\lambda = \frac{0.693}{t_{1/2}} = \frac{0.693}{22} = 31.5 \times 10^{-3}\ yr^{-1}$$

The decay of $^{222}_{89}Ac$ may be shown as:

$^{222}_{89}Ac$ — $-\beta$ → $^{227}_{90}Th$ 2%

$^{222}_{89}Ac$ — $-\alpha$ → $^{223}_{87}Fr$ 98%

Since both changes takes place at the same time, hence

$$t = \frac{2.303}{\lambda_1}\log\frac{100}{98} = \frac{2.303}{\lambda_2}\log\frac{100}{2}$$

where λ_1 and λ_2 are the decay constants for first and second change. On solving the above equation, we get

$$\lambda_2[2 - 1.99] = \lambda_1[2 - 0.30]$$

or $\lambda_2 = 170\ \lambda_1$

Since, $\lambda_1 + \lambda_2 = \lambda$

∴ $\lambda_1 + 170\lambda_1 = 31.5 \times 10^{-3}\ \text{yr}^{-1}$

or $\lambda_1 = \frac{31.5 \times 10^{-3}}{171}\ yr^{-1}$

$= 1.84 \times 10^{-4}\ \text{yr}^{-1}$

and $\lambda_2 = 170 \times 1.84 \times 10^{-4}\ \text{yr}^{-1}$

$= 3.13 \times 10^{-2}\ \text{yr}^{-1}$

20. Here first we convert gms to mol. Now, 23.8 g $^{238}U = \frac{23.8\ g}{238\ mol^{-1}} = 0.1$ mol

and 20.6 g $^{206}Pb = \frac{20.6\ g}{206\ mol^{-1}} = 0.1$ mol

Since 0.1 mol of ^{206}Pb (N_t) is formed the to the decay of 0.1 mol of ^{238}U hence the initial amount (N_0) of ^{238}U will be 0.1 + 0.1 = 0. 2 mol.
We know that

$$\lambda = \frac{0.693}{t_{1/2}} = \frac{0.693}{4.5 \times 10^9} = 1.54 \times 10^{-10}\ \text{yr}^{-1}$$

Also, $t = \frac{2.303}{\lambda}\log\frac{N_0}{N_t}$

$$= \frac{2.303}{1.54 \times 10^{-10}\ \text{yrs}}\log\frac{0.2}{0.1} = \frac{2.303 \times 0.30}{1.54 \times 10^{-10}}\text{yrs.} = 4.49 \times 10^9\ \text{yrs.}$$

21. Suppose x gms of mass is shipped.

$$\therefore \frac{N}{N_0} = \frac{\text{mass after 3 h}}{\text{mass shipped}} = \frac{10.0}{x}$$

and In $(N/N_0) = \frac{0.6931}{t_{1/2}}t = \frac{-(0.6931)(3.00\ h)}{(6.00\ h)} = -0.3466$

Thus, $N/N_0 = e^{-0.3466} = 0.707$

Hence (10.00 mg)/(x mg) = 0.707 and x = 14.1 mg shipped

22. Isobars are nuclides with the same mass number, A but different atomic number, Z.

23. An a-particle has $2n + 2p$. A triton has $2n$ but only a single p.

24. $^{25}_{13}Al \rightarrow ^{25}_{12}Mg^- + ^{0}_{1}\beta^+ + \nu$

After emitting a positron, the resulting atom has one less proton, one more neutron, but the same number of electrons as the atom that decayed. Of course $^{25}_{12}Mg^-$ will quickly lose the extra electron and become a neutral Mg atom.

25. For $^{13}_{7}N$, $n/p = 6/7 = 0.67$, too low for stability. At low Z, the most likely decay mode is β^+ emission.

26. For $^{25}_{11}Na$, $n/p = 1.27$, too high a Z = 11. Neutron-rich nuclides decay via $\bar{\beta}$ emission.

27. The plot of E_B/A versus A has a maximum close to $A \sim 60$. The nuclide $^{56}_{26}Fe$ has close to the maximum E_B/A value.

28. The injected activity is N_0, and the activity after 36.0 h is $N = 0.01\ \mu Ci$. The relation between them is $\ln (N/N_0) = -(0.6931)\ (t)/t_{1/2}$

$$= -(0.6931)\ (36.0\ h)/(6.0\ h) = -4.159$$

Thus, $N/N_0 = e^{-4.159} = 1.562 \times 10^{-2}$

Since $(0.01\ \mu Ci)/N_0 = 1.562 \times 10^{-2}$, $N_0 = 0.64\ \mu Ci$

30. The neutrino has no rest mass.

31. 1 amu = (1/2) (mass of one ^{12}C atom) The mass of one ^{12}C atom is $(12\ g)/N_A = (12 \times 10^{-3}\ kg)/N_A$.

1 amu = $(1/12)\ (12 \times 10^{-3}\ kg)/N_A = (1 \times 10^{-3})/N_A$ kg

32. $^{235}_{92}U + ^{1}_{0}n \rightarrow ^{146}_{57}La + ^{87}_{35}Br + 3(^{1}_{0}n)$

For the missing product, Z must be 92 – 57 = 35 and A must be (235 + 1) – (146 + 3) = 87

33. The target atom must have Z = 102 – 6 = 96 and

$$A = (154 + 4) - 12 = 146$$

$$\therefore \quad ^{146}_{96}Cm + ^{12}_{6}C \rightarrow ^{154}_{102}No + 4(^{1}_{0}n)$$

34. The only atom of those listed with an odd Z is $_{45}Rh$. Nuclei with odd Z have fewer stable isotopes than those with, even Z. Values of 20, 28, and 50 are magic numbers, so Ca, Ni, and Sn have many stable isotopes. Krypton, with Z = 36, has a 6 stable isotopes.

35. Because, $^{118}_{51}Sb \rightarrow ^{0}_{+1}\beta^+ + ^{118}_{50}Sn^- + \nu$; Hence product (daughter) nuclide is $^{118}_{51}Sb$.

36. Here, we have, $\dfrac{N}{N_0} = \dfrac{3.00 \times 10^8}{2.40 \times 10^9} = \dfrac{1}{8} = \left(\dfrac{1}{2}\right)^3$

which means that 3 half lives must pass for the activity to fall to 3.00×10^8 cpm.

$$t = 3t_{1/2} = 3(12.26) = 36.78 \text{ days}$$

Of course, this problem can also be done by direct substitution into the equation $\ln (N/N_0) = -(0.6931)\ (t)/t_{1/2}$, but it is a good idea to look first at the relation between N and N_0 and see if an integral number of half-lives are involved, so you can do the arithmetic in your head.

37. Here, $\dfrac{N}{N_0} = \dfrac{3.83}{15.3} = 0.2503 \approx \dfrac{1}{4}$

Since N/N_0 is very close to 1/4, essentially two half lives must have passed, and as two half lives are 11,460 years, the only possible answer is (c). If you are concerned about the third significant figure, then you must do all the calculations to 4 significant figures, and round to 3 figures at the end, as shown below, but for a multiple choices question that is unnecessary and should be avoided.

$$\ln(0.2503) = -1.385 = -(0.69315)\,(t)/(5730\text{ yr})$$

$$t = (1.385)\,(5730)/0.69315 = 11{,}449\text{ yr} \simeq 11{,}400\text{ yr to 3 significant}$$

figures.

38. When $^{126}_{88}\text{Ra}$ emits α-particle ($^{4}_{2}\text{He}^{2+}$), we get $^{222}_{86}\text{Rn}^{2-}$. The equation is,

$$^{126}_{88}\text{Ra} \rightarrow {}^{4}_{2}\text{He}^{2+} + {}^{222}_{86}\text{Rn}^{2-}$$

39. If the fraction remaining is $1/8 = (1/2)^3$, that means 3 half lives have elapsed. Thus, 96 min = $3t_{1/2}$, and

$$\therefore t_{1/2} = 32\text{ min}$$

40. Because particle eitted are β^-, β^- and α the equations are,

$$^{228}_{88}\text{Ra} \rightarrow {}^{0}_{-1}\beta^- + {}^{228}_{89}\text{Ac}^+ + \nu$$

$$^{228}_{89}\text{Ac} \rightarrow {}^{0}_{-1}\beta^- + {}^{228}_{90}\text{Th}^+ + \nu$$

$$^{128}_{90}\text{Th} \rightarrow {}^{4}_{2}\text{He}^{2+} + {}^{224}_{88}\text{Ra}^{2-}$$

Emitting a β^- particle increase Z by 1 ($n \rightarrow p^+ + \beta^-$). So emitting 2 β^- particles increases Z by 2. Since emitting an α-particle decreases Z by 2, the emission sequence β^-, β^-, α leaves Z unchanged. We start with on isotope of Ra and end with another.

41. (i) Polanium has an atomic number greater than 83, so $^{208}_{84}\text{Po}$ is radioactive, Bismuth 209 has 126 neutrons (a magic number), so $^{209}_{83}\textbf{Bi}$ is expected to be *stable*.

(ii) Of these two isotopes, $^{39}_{19}\text{K}$ has a magic number of neutrons (20), so $^{39}_{19}\text{K}$ is expected to be stable. The isotope $^{40}_{19}\text{K}$ has an odd number of protons (19) and an odd number of neutrons (21). Because stable odd–odd nuclei are rare, you might expect $^{40}_{19}\textbf{K}$ to be radioactive.

(iii) Of the two isotopes, $^{76}_{31}\text{Ga}$ lies farther from the center of the band of stability, so it is more likely to be radioactive. For this reason, you expect $^{76}_{31}\text{Ga}$ to be radioactive and $^{71}_{31}\text{Ga}$ to be *stable*.

42. (i) The atomic weight of calcium is 40.1 amu, so you expect calcium-40 to be a stable isotope. Calcium-47 has a mass number greater than that of the stable isotope, so you expect it to decay by *beta emission*. (This is the observed behavior of calcium-47.)

(ii) The atomic weight of aluminium is 28.0 amu, so you expect aluminium-27 to be a stable isotope. The mass number of aluminium-25 less than 27, so you expect aluminium-25 to decay by either positron emission or electron capture. *Positron emission* is actually observed.

43. The rate of decay in this sample is

$$\text{rate} = 1.7 \times 10^{-5}\ \text{Ci} \times \frac{3.7 \times 10^{10}\ \text{nuclei/s}}{1.0\ \text{Ci}} = 6.3 \times 10^{5}\ \text{nuclei/s}$$

The number of nuclei in this sample 1.0×10^{-3} g $^{99}_{43}$Tc is

$$1.0 \times 10^{-3}\ \text{g Tc-99} \times \frac{1\ \text{mol Tc-99}}{99\ \text{g Tc-99}} \times \frac{6.02 \times 10^{23}\ \text{Tc-99 mclei}}{1\ \text{mol Tc-99}} = 6.1 \times 10^{18}\ \text{Tc-99 nuclei}$$

The decay constant is

$$k = \frac{\text{rate}}{N_t} = \frac{6.3 \times 10^{5}\ \text{nuclei s}^{-1}}{6.1 \times 10^{18}\ \text{nuclei}} = 1.0 \times 10^{-13}\ \text{s}^{-1}$$

44. The conversion of half-life to seconds gives

$$12.3\ y \times \frac{365\ \text{d}}{1\ y} \times \frac{24\ \text{h}}{1\ \text{d}} \times \frac{60\ \text{min}}{1\ \text{h}} \times \frac{60\ \text{s}}{1\ \text{min}} = 3.88 \times 10^{8}\ \text{s}$$

Because $t_{1/2} = 0.693/k$, you solve this for k and substitute the half-life in seconds.

$$k = \frac{0.693}{t_{1/2}} = 1.79 \times 10^{-9}\ \text{s}^{-1}$$

The number of tritium nuclei in a sample containing 2.5×10^{-6} g of tritium,

$$\text{is} = 2.5 \times 10^{-6}\ \text{g H-3} \times \frac{1\ \text{mol H-3}}{3.02\ \text{g H-3}} \times \frac{6.02 \times 10^{23}\ \text{H-3 nuclei}}{1\ \text{mol H-3}} = 5.0 \times 10^{17}\ \text{H-3 nuclei}$$

Now substitute into the rate equation,

$$\text{Rate} = kN_t = 1.79 \times 10^{-9}\ \text{s}^{-1} \times 5.0 \times 10^{17}\ \text{nuclei} = 9.0 \times 10^{8}\ \text{nuclei s}^{-1}$$

Thus the activity of the sample is obtained by dividing the rate in disintegrations of nuclei per second by 3.70×10^{10} disintegrations of nuclei per second per curie.

$$\text{Activity} = \frac{9.0 \times 10^{8}\ \text{nuclei s}^{-1}}{3.70 \times 10^{10}\ \text{nuclei s}^{-1}\ (\text{s} \cdot \text{Ci})} = 0.024\ \text{Ci}$$

45. (i) Rate constant for first order reaction

$$k = \frac{0.693}{t_{1/2}} = \frac{0.093}{15} = 0.0462\ \text{min}^{-1}.$$

(ii) When reaction is 80% complete, $[R]_t$ = 20% of $[R]_0 = 0.2[R]_0$

$$\therefore \quad t = \frac{2.303}{k} \log \frac{[R]_0}{[R]_t} = \frac{2.303}{0.0462} \log \left\{ \frac{[R]_0}{0.2[R]_0} \right\} = 34.8\ \text{min.}$$

46. Since Rn is an alpha emitter hence

$$_{86}\text{Rn} \rightarrow {}_{84}\text{Po} + {}_{2}\text{He}^{4}$$

Rate of disintegration = k No

$$3.7 \times 10^{10}\ \text{dps} = 2 \times 10^{-6}\ \text{s}^{-1}.\ \text{No}$$

or $$No = \frac{3.7 \times 10^{10}}{2 \times 10^{-6}} = 1.85 \times 10^{16} \text{ atoms}$$

∴ Let amount left of Rn after 1 hr be N.

∵ $$t = \frac{2.303}{k} \log \frac{No}{N}$$

$$60 \times 60 \text{ s} = \frac{2.303}{2 \times 10^{-6} \text{ s}^{-1}} \log \frac{No}{N}$$

or $$\log \frac{No}{N} = \frac{60 \times 60 \times 2 \times 10^{-6}}{2.303}$$

$$= 3.126 \times 10^{-3} = 0.0031$$

∴ $$\frac{No}{N} = 1.0072$$

or $$N = \frac{No}{N} = \frac{1.85 \times 10^{16}}{1.0072}$$

or $$N = 1.837 \times 10^{6}$$

∴ No. of α-particles formed = No. of Rn atoms decay

$$= 1.85 \times 10^{16} - 1.837 \times 10^{16}$$

$$= 0.013 \times 10^{16}$$

$$\text{Energy} = 0.013 \times 10^{16} \times 5.5 \text{ MeV}$$

$$= 0.013 \times 10^{16} \times 5.5 \times 10^{6} \text{ eV}$$

$$= 0.0715 \times 10^{22} \text{ eV}$$

$$= 0.0515 \times 1022 \times 1.602 \times 10^{-19} \text{ J}$$

$$= 114.5 \text{ J}$$

Now mean life (t) = $\frac{1}{k}$

Let number of Rn atoms left after mean life (t) be N.

∴ $$t = \frac{2.303}{k} \log \frac{No}{N}$$

or $$\frac{1}{k} = \frac{2.303}{k} \log \frac{No}{N}$$

or $$\log \frac{No}{N} = \frac{1}{2.303} 0.4342 \Rightarrow \frac{No}{N} = 2.718$$

or $$N = \frac{No}{2.718} = \frac{1.85 \times 10^{16}}{2.718} = 0.6806 \times 10^{16}$$

∴No. of α-particles formed = No. of Rn atoms decay

$$= 1.85 \times 10^{16} - 0.6806 \times 10^{16}$$

$$= 1.1694 \times 10^{16}$$

$$\begin{aligned}\text{Energy} &= 1.1694 \times 10^{16} \times 5.5 \text{ MeV} \\ &= 1.1694 \times 10^{16} \times 5.5 \times 10^{6} \text{ eV} \\ &= 6.4317 \times 10^{22} \text{ eV} \\ &= 6.4317 \times 10^{22} \times 1.602 \times 10^{-19} \text{ J} = 1.03 \times 10^{4} \text{ J}\end{aligned}$$

47. It is clear that half life period ($t_{1/2}$) of radioactive element is 5 hours.

Since $N = \left(\frac{1}{2}\right)^n$

or $\frac{N_0}{8} = N_0\left(\frac{1}{2}\right)^n$

or $\frac{1}{8} = \left(\frac{1}{2}\right)^n$

or $n = 3$

$\because$ $n = \frac{\text{total time of disintegration}}{\text{half life period}}$

$\therefore$ Total time of disintegration

$$= n \times t_{1/2} = 3 \times 5 \text{ hours} = 15 \text{ hours}$$

48. $W_t = W_0 e^{-\lambda t}$

where W_t = mass of radioactive substance after time 't'

W_0 = mass of radioactive substance initially taken

λ = decay constant; t = time

$\because$ $\lambda = \frac{0.693}{t_{1/2}} = \frac{0.693}{28.1}$

$\therefore$ $W_t = 1\ \mu g \cdot e^{-(0.693 \times 20)/28.1}$

$= 1\ \mu g \cdot e^{-0.493}$

$\log_e W_t = -0.493$

or $\log_{10} W_t = \frac{0.493}{2.303} = -0.214$

or $W_t = 0.06\ \mu g$

49. Since,

$$\begin{aligned}\text{Average life} &= 1.44 \times \text{Half life time} \\ &= 1.44 \times 6930 \text{ days} \\ &= 9979.2 \text{ days}\end{aligned}$$

50. The mass of radioactive isotope left after 75 days i.e. 3 × half-life

$$= 4 \times \left(\frac{1}{2}\right)^3 = 0.5 \text{ g}$$

51. Number of α-particles emitted in 420 days

$= 2.24 \times 10^{13} \times 420 \times 24 \times 60$

$= 1.355 \times 10^{19}$

The volume of He is converted to STP

$$\frac{P_1 V_1}{T_1} = \frac{P_2 V_2}{T_2}$$

$$V_1 = \frac{750 \times 0.5 \times 273}{300 \times 760}$$

$= 0.449$ mL

Since 0.449 mL He at STP

$= 1.355 \times 10^{19}$ molecules

22400 mL He at STP

$$= \frac{1.355 \times 10^{19} \times 22400}{0.449}$$

$= 6.76 \times 10^{23}$

Hence Avogadro's constant $= 6.76 \times 10^{23}$

52. Since, $N = \text{No} \left(\frac{1}{2}\right)^n$

as $$\frac{1}{64} = \left(\frac{1}{2}\right)^6$$

$\Rightarrow$ $n = 6$

We know that, time $= t_{1/2} \times n$

or $$t_{1/2} = \frac{\text{Time}}{\text{n}} = \frac{2 \times 60 \text{ min}}{6}$$

$= 20$ min.

$$\lambda = \frac{0.693}{t_{1/2}} = \frac{0.693}{20 \text{ min}}$$

$= 0.03465 = 3.465 \times 10^{-2}\ \text{min}^{-1}$

53. Kinetic energy of α-particle $= \frac{1}{2} m_\alpha v_\alpha^2$

$$\text{Electrostatic energy} = \frac{\text{K(2e)(Ze)}}{\text{r}}$$

$\because$ $$\frac{1}{2} m_\alpha v_\alpha^2 = \frac{\text{K(2e)(Ze)}}{\text{r}}$$

$\therefore$ $$v_\alpha = \left[\frac{\text{2.K.(2e)(Ze)}}{\text{m}_\alpha (r)}\right]$$

$$= \left[\frac{(2)(9 \times 10^9)(2 \times 16 \times 10^{-19})(29 \times 16 \times 10^{-19})}{(4 \times 1674 \times 10^{-27})(10^{-13})} \right]^{\frac{1}{2}}$$

$= 6.318 \times 10^6 \text{ ms}^{-1}$

54. Number of moles of Cu in 0.635 g

$$= \frac{0.635 \text{ g}}{65.3 \text{ g mol}^{-1}}$$

$= 10^{-2}$ mol

$\because$ 1 mole copper contains 6.023×10^{23} atom of Cu

$\therefore$ 10^{-2} mole Cu contains $6.023 \times 10^{23} \times 10^{-2}$

$= 6.023 \times 10^{21}$ atoms of Cu.

55. We know that turpentine absorbs ozone only. When 100 mL ozonide oxygen at NTP is passed in turpentine the volume was reduced to 90 mL. It means there is 10 mL O_3 in ozonide oxygen.

Thus total volume of O_3 = 100 mL in 1 L

And total volume of O_2 = 900 mL in 1 L

Let the molecular mass of O_3 be x g mol^{-1}.

$\therefore$ Mass of 100 mL O_3 at N.T.P.

$$= \frac{x \times 100}{22400} = \frac{x}{224} \text{ g}$$

Mass of 900 mL O_2 at N.T.P. $= \dfrac{32 \times 900}{22400} = \dfrac{9}{7}$ g

But the mass of 1L ozonide oxygen at N.T.P. is 1.5 g

$\therefore$ $$\frac{x}{224} = 1.5$$

or $$x = 224 \left[1.5 - \frac{9}{7} \right]$$

$= 48$ g mol^{-1}

56. The reaction of iron and steam may be represented as:

$$3Fe + 4H_2O \rightarrow Fe_3O_4 + 4H_2$$

$\because$ 4 moles steam reacts with 3 moles Fe

$\therefore$ 1 mole (18 g) steam reacts with $\frac{3}{4}$ moles Fe

or $$\frac{3}{4} \text{ mole} \times 56 \text{ g mol}^{-1}$$

$= 42$ g Fe

57. In this process change in atomic mass = 234 – 206 = 28 a.m.u. Since by the emission of one α-particle atomic mass decreases by 4 units. Hence in this process emission of α-particles = 28/4 = 7.

If only α-particles emit then the atomic number of the product should be $90 - 2 \times 7 = 76$.
But here the atomic number of the product is 82 that is due to the emission of β-particles and their number is $82 - 76 = 6$.

58. We have,

$$^{198}_{79}\text{Au} \xrightarrow{-\beta} {}^{198}_{80}\text{Hg}$$

Amount of $^{198}_{79}$Au left after 260 hrs.

$$= \left(\frac{1}{2}\right)^4 = \frac{1}{16}\text{ g}$$

$\therefore$ $$\text{Amount og Hg} = 1 - \frac{1}{16} = \frac{15}{16}\text{ g} \cdot = 0.93\text{ g}$$

59. We know that

$$\lambda = \frac{0.693}{t_{1/2}} = \frac{0.693}{5770\text{ yrs}} = 1.2 \times 10^{-4}\text{ yr}^{-1}$$

Now, $$\lambda = \frac{2.303}{t}\log\frac{N_0}{N}$$

$$1.2 \times 10^{-4}\text{ yr}^{-1} = \frac{2.303}{11540\text{ yrs}}\log\frac{1}{N}$$

or $$\log\frac{1}{N} = \frac{(1.2 \times 10^{-4}\text{ yrs}^{-1})(11540\text{ yrs})}{2.303} = 0.6013$$

or $$N = \frac{1}{4} = 25\%.$$

Thus 25% of the sample will be left after 1154 yrs.

60. $\because$ 1 g atom contains 6.02×10^{23} atoms
$\therefore$ 0.1 g atom contains $0.1 \times 6.02 \times 10^{23}$ atoms
$= 6.02 \times 10^{22}$ atoms

$$\lambda = \frac{0.693}{t_{1/2}} = \frac{0.693}{5\text{ days}} = 0.1386\text{ day}^{-1}$$

$\because$ Half life period of the element is 5 days

$\therefore$ No. of atoms after 10 days $= \dfrac{6.02 \times 10^{22}}{2 \times 2} = 1.505 \times 10^{22}$

Now, number of atoms after 11 days $= N_0 e^{-\lambda t}$
$= 6.02 \times 10^{22}\, e^{-(0.1386)(11)} = 1.310 \times 10^{22}$

$\therefore$ Atoms decays during 11th day $= (1.505 - 1.310)\, 10^{22}$
$= 0.195 \times 10^{22}$ atoms

61. You substitute $k = 0.693/t_{1/2}$ into the equation for the number of nuclei in a sample after time t.

$$\log\frac{N_t}{N_0} = \frac{-kt}{2.303} = \frac{-0.693\,t}{2.303\,t_{1/2}}$$

Hence, $$t = \frac{2.303\,t_{1/2}}{0.693}\log\frac{N_0}{N_t}$$

To get N_0/N_t, you assume that the ratio of $^{14}_{6}C$ to $^{12}_{6}C$ in the atmosphere has remained constant. Then you can say that 1.00 g of total carbon from the living tree gave 15.3 disintegrations per minute. The ratio of the number of $^{14}_{6}C$ nuclei originally present to the number that existed at the time of dating equals the ratio of rates of disintegration. The ratio of rates of disintegration is

$$\frac{N_0}{N_t} = \frac{15.3}{7.0} = 2.2$$

Substituting this value of N_0/N_t and $t_{1/2} = 5730$ y into the equation gives

$$t = \frac{2.303_{1/2}}{0.693}\log\frac{N_0}{N_t} = \frac{2.303 \times 5730\, y}{0.693} \log 2.2 = 6.5 \times 10^3 \text{ y}$$

Thus, the date of the eruption was about 4500 B.C., 6500 years before the present.

62. The combined mass of the proton and neutron = 3.347551×10^{-27} kg
'Missing' mass = 0.03965×10^{-27} kg [$(3.347551 - 3.343586) \times 10^{-27}$ kg]
Energy released when the proton and neutron fuse = mc^2.
$= 0.003965 \times 10^{-27} \times (3.00 \times 10^8)^2$
$= 3.5685 \times 10^{-13}$ J
$= 3.57 \times 10^{-13}$

63. Find out how many half-lives 1.2 years is:
$21.2 \div 5.3 = 4$ half-lives
Thus,
2 g: $t_{1/2} = 5.3$ years → 1 g: $t_{1/2} = 5.3$ years → 0.5 g: $t_{1/2} = 5.3$ years → 0.25 g: $t_{1/2} = 5.3$ years → 0.125 g
You can see that the starting amount of the sample has been reduced four times by a half:

$$\left(\therefore\ 2 \times \frac{1}{2} \times \frac{1}{2} \times \frac{1}{2} \times \frac{1}{2} = 2 \times \left(\frac{1}{2}\right)^4 = 0.125\text{g}\right)$$

64. We have,
3000: $t_{1/2} = 6$ h → 1500: $t_{1/2} = 6$ h → 750: $t_{1/2} = 6$ h → 375: $t_{1/2} = 6$ h → 187.5: $t_{1/2} = 6$ h → 93.75 ≈ 94 c.p.m.

As shown, it takes 5 half-lives for activity to reduce to 94 c.p.m. Therefore, $6 \times 5 = 30$ hours.

Chapter 12 : Crystalography

1. In the body-centered cubic arrangement of K, the cell is a cube with one atom at each corner and one in center. Since a cube has eight corners, the atom at the center has eight nearest neighbours. Furthermore, since each corner is shared by eight cubes, a corner atom also as eight neighbours, which are the center atoms of the eight adjacent cubes. Alternatively, any given K atom can be visualized as being at the body-center of a unit cell. In fact, all atoms are in identical environments.

2. In MgO, both ions are doubly charged (Mg^{2+} and O^{2-}). All the other choices are crystals with singly charged ions. Thus, $U = \frac{N_A A Z_+ Z_-}{r_o}$, where, Z_+ and Z_- are the magnitudes of charges on Cations and anious, respectively, r_o is the equilibrium internuclear distance, A is a constant (the Madelung constant), that depends on crystal structure and N_A is Avagadro's number. Now, the product $Z_+ Z_-$ is 4 for MgO and 1 for all the other crystals. The product is the most important factor in determining lattice energies. It far outweighs variations due to differences in inter-ionic distance and madelung constant.

3. One eighth of each corner and one half of each face center are in the unit cell. There are 8 corners, and 2 faces with lattice points at the centers. Thus,

$$8(1/8) + 2(1/2) = 2$$

Therefore there are 2 lattice points per unit cell, and if there is a molecule at each lattice site, there are 2 molecules per unit cell.

4. No. X atoms per unit cell = (8 corners) $\left(\frac{1}{8}\text{ per corner}\right)$ = 1 No. Y atoms per unit cell = 1

The simplest formula is therefore XY.

5. The three crystals in which both ions are doubly charged are NiS, SrO, and BaS. Now, Ni is in the 4th period of the periodic table, while Sr and Ba are in the 5th and 6th periods, respectively, and are not transition metals. The O^{2-} ion is smaller than the S^{2-} ion.

For NiS, $r_c + r_a = 70 + 184 = 254$ pm

For SrO, $r_c + r_a + 113 + 140 = 253$ pm

As the interionic distance is considerably larger for BaS than for NiS of SrO, BaS has the smallest lattice energy of these 3. The Ni^{2+} ion is the only cation of the three that does not have rare gas electronic configuration and the S^{2-} anion is large and polarizable. Polarization of the anion by the cation increases the lattice energy, and this effect is maximised for small cations without rare gas configuration and for large anions. Because Ni^{2+} has partly filled d-orbitals and S^{2-} has low lying empty d-orbitals there is considerable covalent character to the NiS bond, so that the lattice energy of NiS is larger than that of SrO.

6. All the ions in the two compounds are singly charged and have rare gas electron configuration. Because Na^+ and F^- are much smaller than Rb^+ and Br^-, respectively, the internuclear distance in NaF is significantly smaller than that in RbBr. As a result, the lattice energy is larger for NaF, which is therefore more tightly bound and has a higher melting point.

7. In an fcc arrangement, the atoms make contact along the face diagonals. If the unit cell axial length is denoted a, the length a face diagonal is $a\sqrt{2}$ and the equal to four atomic radii. Thus

$$4x = a\sqrt{2} \quad \text{where } r \text{ is the atomic radius}$$

Setting r = 2.00 Å, we obtain a = $4\sqrt{2}$ Å = 5.657 Å. The volume of a unit cell is $a^3 = (5.657)^3 \times 10^{-24}$ cm^3 = 181 × 10^{-24} cm^3 = 1.81 × 10^{-22} cm^3 Recall that 1 Å = 10^{-8} cm, so that (1 Å)^3 = 10^{-24} cm^3.

8. The only cation of these compounds that does not have a rare gas electronic configuration is Cd^{2+}. The S^{2-} anion is large and polarizable. Thus the bond in CdS has the largest amount of covalent character of those listed.

9. The only crystal system which three unequal unit-cell axial lengths but three 90° cell angles is the ortho-cell rhombic.

10. Because Ni and Ti are in the 4th row of the iodic table, while Zr is in the 5th row, and Pt the 6th, the two smallest cations are Ti^{2+} and Ni^{2+}. As one goes across a transition metal series from left to right, ions of the same charge decrease slightly in size, as the *d* electrons being added do not shield each other very well, and the increasing nuclear charge pulls the electron cloud in more tightly.

11. The smallest of these anions is F^-, because F is in the second row of the periodic table, and F^- is singly charged. Remember that for isoelectronic anions, the more negative anions is I^-, because I is in the fifth period of the table. This leaves only sequences (b) and (d) to compare. As Br^- is larger than O2$^-$ (fourth versus second period), (d) must be the correct sequence.

12. Anion-anion contact occurs when the anions are very large compared to the cations. Of the compounds listed, NaI has the largest anion : cation radius ration.

13. Small, highly-charged ions have the largest heats of hydration. Cations are generally smaller than anions, and Al^{3+} is the most highly charged and the smallest of the ions listed.

14. Larger, more highly charged anions are more easily polarized than smaller anions. By far the largest ion of those listed is Te^{2-}.

15. The only compound of those given with both ions doubly charged is NiO, which therefore has the largest lattice energy and the highest melting point. Like SO_2, SeO_2 is covalent.

16. Electron affinities are difficult to measure, and are often calculated theoretically instead.

17. The total amounts of positive and negative charge in the unit cell must be equal, so the M^{3+} : X^- molar ratio must be 1 : 3. An M^{3+} at each corner means 1 M^{3+} ion per unit cell. An X^- at each body center means 1 X^- ion per unit cell (wrong molar ration) whereas an X^- at each face center means 6(1/2) = 3 X^- ions per unit cell, the correct molar ratio, consistent with the formula MX_3.

18. For the orthorhombic, tetragonal, and cubic systems, all three angles are 90°. For the rhombohedral system, all angles are equal, but are not, in general, 90°. In the hexagonal system, 2 angles are 90° but the third is 120°.

19. Metals form cations, which are smaller in size than the neutral atoms.

20. The packing fraction, or the fraction of the unit cell occupied by atoms is 74.0% for face-centered cubic, 68.0% for body-centered cubic, and 52.4% for simply cubic. Simple tetragonal or rhombohedral lattices have packing fractions less than an fcc cubic lattice.

21. As 1 Å = 1×10^{-8} cm, the volume of a unit cell is V = $(3.5238 \times 10^{-8}\ \text{cm})^3 = 4.3756 \times 10^{-23}\ \text{cm}^3$

There are 4 Ni atoms per unit cell. The mass of 4 Ni atoms is

$$\frac{(4\ \text{atoms})(58.69\ \text{g mol}^{-1})}{6.0221 \times 10^{23}\ \text{atoms mol}^{-1}} = 3.898 \times 10^{-22}\ \text{g}$$

Thus the density of Ni is

$$\frac{3.898 \times 10^{-22}\ \text{g}}{4.3756 \times 10^{-23}\ \text{cm}^3} = 8.909\ \text{g} \cdot \text{cm}^{-3}$$

22. The formula which related the radius of an atom (r) to the length of the side (s) of the unit cell for a face-centered cubic cell is $4r = s\sqrt{2}$

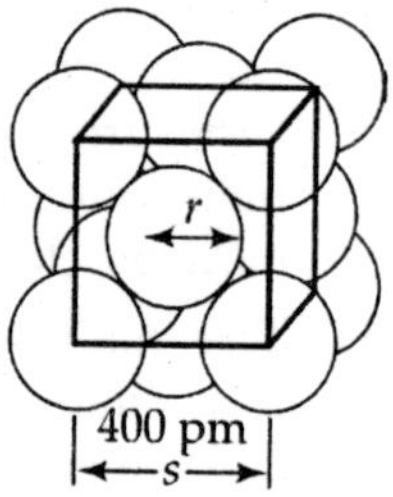

$$r = \frac{400.\text{pm}\sqrt{2}}{4} = 100.(1.414) = 141\ \text{pm}$$

23. First, calculate the mass of one cell, 100.pm on an edge:

$$\frac{200.\ \text{g}}{1\ \text{cm}^3} \times \left(\frac{100\ \text{cm}}{1\ \text{m}}\right)^3 \times \left(\frac{1\ \text{m}}{10^{12}\ \text{pm}}\right)^3 \times \left(\frac{100.\ \text{pm}}{1}\right)^3 = 2.00 \times 10^{-22}\ \text{g/cell}$$

Next, calculate the mass of one metal atom:

$$\frac{60.2\ \text{g}}{1\ \text{mole}} \times \frac{1\ \text{mole}}{6.02 \times 10^{23}\ \text{atoms}} = 10.0 \times 10^{-23} = 1.00 \times 10^{-22}\ \text{g/atom}$$

Finally, calculate the number of metal atoms in one cell:

$$\frac{2.00 \times 10^{-22}\ \text{g} \cdot \text{cell}^{-1}}{1.00 \times 10^{-22} \cdot \text{atoms}} = 2.00\ \text{atoms/cell}$$

24. From the edge length of a unit cell. We can calculate the unit-cell volume. Then, from the density, we can find the mass of the unit cell and hence the mass of a silver atom. The volume, V of the unit cell is obtained by cubing the length of an edge.

$$V = (4.086 \times 10^{-10}\ \text{m})^3 = 6.822 \times 10^{-29}\ \text{m}^3$$

The density, d, of silver in grams per cubic meter is

$$d = 10.50\frac{\text{g}}{\text{cm}^3} \times \left(\frac{1\ \text{cm}}{10^{-2}\ \text{m}}\right)^3 = 1.050 \times 10^7\ \text{g m}^{-3}$$

Density is mass per volume; hence, the mass of a unit cell equals the density times the volume of the unit cell.

$$m = dV$$
$$= 1.050 \times 10^7 \text{ g m}^{-3} \times 6.822 \times 10^{-29} \text{ m}^3$$
$$= 7.163 \times 10^{-22} \text{ g}$$

Because there are four atoms in a face-centered unit cell having one atom at each lattice point the mass of a silver atom is

$\therefore$ Mass of 1 Ag atom $= \dfrac{1}{4} \times 7.163 \times 10^{-22} \text{ g} = 1.791 \times 10^{-22} \text{ g}$

The known atomic weight of silver is 107.87 amu. Thus, the molar mass (Avogadro's number of atoms) is 107.87 g mol^{-1}, and Avogadro's number, N_A, is

$$N_A = \frac{107.87 \text{ g mole}^{-1}}{1.791 \times 10^{-22} \text{ g}} = 6.023 \times 10^{23} \text{ mol}^{-1}$$

25. Use Avogadro's number (6.022×10^{23} mol^{-1}) to convert the molar mass of platinum (195.08 g mol^{-1}) to the mass per atom.

$$\frac{195.08 \text{ g Pt}}{1 \text{ mol Pt}} \times \frac{1 \text{ mol Pt}}{6.022 \times 10^{23} \text{ Pt atoms}} = \frac{3.239 \times 10^{-22} \text{ g Pt}}{1 \text{ Pt atom}}$$

Because there are four atoms per unit cell, the mass per unit cell is calculated as follows:

$$\frac{3.239 \times 10^{-22} \text{ g Pt}}{1 \text{ Pt atom}} \times \frac{4 \text{ Pt atom}}{1 \text{ unit cell}} = \frac{1.296 \times 10^{-21} \text{ g}}{1 \text{ unit cell}}$$

The volume of the unit cell is

$$V = \frac{m}{d} = \frac{1.296 \times 10^{-21} \text{ g}}{21.45 \text{ g cm}^{-3}} = 6.022 \times 10^{-23} \text{ cm}^3$$

If the edge length of the unit cell is denoted as l, the volume if $V = l^3$. Hence, the edge length is

$$l = \sqrt[3]{V}$$
$$= \sqrt[3]{6.042 \times 10^{-23} \text{ cm}^3}$$
$$= 3.924 \times 10^{-8} \text{ cm (or } 3.924 \times 10^{-10} \text{ m)}$$

26. Since element has fcc structure hence there are 4 atoms in a unit cell (or $n = 4$), Atomic mass is 60 (or $M = 60$), $N_0 = 6.02 \times 10^{23}$ and $\rho = 6.23$ g cm^{-3}.

$\because$
$$\rho = \frac{n \times M}{V \times N_0}$$

or

$$V = \frac{n \times M}{\rho \times M_0} = \frac{(4)(60 \text{ g mol}^{-1})}{(6.23 \text{ g cm}^{-3})(6.02 \times 10^{23} \text{ mol}^{-1})}$$

$$= \frac{240 \times 10^{-23} \text{ cm}^{-3}}{37.5046}$$

$$= 6.4 \times 10^{-23} \text{ cm}^3 = 64 \times 10^{-24} \text{ cm}^3$$

Let l be the length of the edge of the unit cell

$\therefore$ $l^3 = V = 64 \times 10^{-24} \text{ cm}^3$

or $l = 4.0 \times 10^{-8}$ cm

27. Edge length (l) = 287 pm

$$= 2.87 \times 10^{-8} \text{ cm}$$

Volume of the unit cell = l^3

$$= (2.87 \times 10^{-8} \text{ cm})^3$$

$$= 2.364 \times 10^{-23} \text{ cm}^3$$

$$\text{Mass of unit cell} = \frac{2 \times 52}{6.023 \times 10^{23}} \text{ g}$$

$$= 1.7267 \times 10^{-22} \text{ g}$$

Now, $\text{Density} = \frac{\text{Mass}}{\text{Volume}}$

$$= \frac{1.7267 \times 10^{-22} \text{ g}}{2.364 \times 10^{-23} \text{ cm}^3}$$

$$= 7.3 \text{ g cm}^{-3}$$

For BCC lattice,

$$r = \frac{\sqrt{3}}{4} = \frac{\sqrt{3}}{4} \times 2.87 \times 10^{-8} \text{ cm.} = 1.243 \times 10^{-8} \text{ cm} = 1.24 \text{ Å}$$

28. Since, $\lambda = \frac{he}{eV}$

$$\lambda_{min} = \frac{12400}{V} \text{ Å}$$

or $V = \frac{12400}{\lambda_{min}} = \frac{12400}{1.54} = 8045.9 \text{ V} = 8.046 \text{ kV}$

29. In a BCC lattice the body diagonal is equal to $4r$.

$$4r = \sqrt{3a}$$

$$\Rightarrow \quad r = \frac{\sqrt{3}}{4} \times a = \frac{\sqrt{3}}{4} \times 4.29 \text{ Å} = 1.86 \text{ Å}$$

30. Volume (V) of the unit cell = 24×10^{-24} cm^3
Density of (ρ) of the element = 7.2 g cm^{-3}
Mass of the element = 200 g
Number of atoms (n) per unit cell = [The contribution of the 8 corners is $(1/8) \times 8 = 1$ and two atoms one of the diagonal is $2 \times 1 = 2$]
i.e. 1 + 2 = 3 atoms

$$\because \qquad \rho = \frac{n \times M}{V \times N_0} \quad \Rightarrow \quad N_0 = \frac{n \times M}{V \times 0}$$

$$\text{or} \qquad N_0 = \frac{(3)(200\text{ g})}{(24 \times 10^{-24}\text{ cm}^3)(7.2\text{ g cm}^{-3})}$$

$$= 3.47 \times 10^{24} \text{ atoms}$$

31. (i) Consider a corner gold atom in Fig. The closest distance to another corner atom is a. The distance to an atom at the center of a face is one-half the diagonal of that face, i.e.,

$$\frac{1}{2}(a\sqrt{2}) = \frac{a}{\sqrt{2}} < a$$

Thus the closest distance between atoms is $\frac{4.070}{\sqrt{2}} = 2.878$ Å

(ii) For the face-centered cubic structure, with 8 corners and 6 face centers,

$$\text{Mass per unit cell} = \frac{1}{8}(8m) + \frac{1}{2}(6m) = 4m$$

Where m is the mass of a single hold atom. 197.0 u. Converting to grams,

$$m = (197.0\text{ u})\left(\frac{1\text{g}}{6.023 \times 10^{23}\text{ u}}\right) = 3.27 \times 10^{-22}\text{ g}$$

$$\therefore \qquad \text{density} = \frac{4m}{a^3} = \frac{4(3.27 \times 10^{-22}\text{ g})}{(4.070 \times 10^{-8}\text{ cm})^3} = 19.4\text{ g cm}^{-3}$$

(iii) Since the atoms at closest distance are in contact in a close-packed structure, the closest distance between centers calculated in (i) $a/\sqrt{2}$, must equal the sum of the radii of the two spherical atoms, $2r$. Thus, $r = a/2^{3/2}$. From (ii), there are 4 gold atoms per unit cell. Then,

$$\text{Volume of 4 gold atoms} = 4\left(\frac{4}{3}\pi r^3\right) = 4\left(\frac{4}{3}\pi\right)\left(\frac{a}{2^{3/2}}\right)^3 = \frac{\pi a^3}{3\sqrt{2}}$$

and

$$\text{Packing fraction} = \frac{\text{volume of 4 gold atoms}}{\text{volume of unit cell}} = \frac{\pi a^3/3\sqrt{2}}{a^3} = \frac{\pi}{3\sqrt{2}} = 0.7405$$

32. Here, we calculate the mass first. Now,

$$m = (4 \text{ atom})\left(\frac{1 \text{ mol}}{6.02 \times 10^{23} \text{ atoms}}\right)\left(\frac{83.8 \times 10^{-3} \text{ kg}}{\text{mol}}\right) = 5.57 \times 10^{-25} \text{ kg}$$

$l = 0.559$ nm $\quad\therefore\quad V = (0.559 \times 10^{-9} \text{ m})^3 = 1.75 \times 10^{-28} \text{ m}^3$

Hence,
$$d = \frac{m}{V} = \frac{5.57 \times 10^{-2} \text{ kg}}{1.75 \times 10^{-19} \text{ m}^3} = 3190 \text{ kg m}^{-3}$$

33. (a) 2(2.32 Å) = 4.64 Å

(b) Using the Pythagoran theorem, $\sqrt{2}$ (2.32 Å) = 3.28 Å

34. With reference to Fig, two terminal carbon can be thought of as lying at A and B, while the central carbon is at O. Then

$$\overline{AB} = 2\overline{AP} = 2\left(\overline{AO}\sin\frac{\theta}{2}\right) = 2(1.54 \text{ Å})(\sin 54°44') = 2.51 \text{ Å}$$

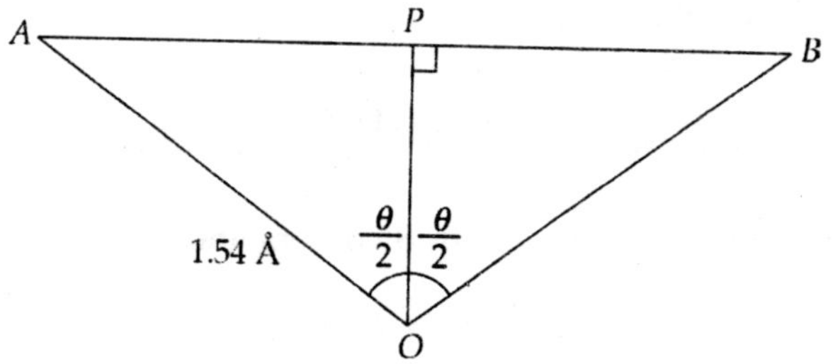

35. $V = \pi r^2 h = (3.14)\left(\frac{150 \text{ Å}}{2}\right)^2 (5000 \text{ Å}) = 8.83 \times 10^7 \text{ Å}^3 = (8.83 \times 10^7 \text{ Å}^3)\left(\frac{10^{-8} \text{ cm}}{\text{Å}}\right)^3$

$= 8.83 \times 10^{-17} \text{ cm}^3$

Now, specific volume = 0.75 cm^3g^{-1}.

$$\therefore \quad (8.83 \times 10^{-17} \text{ cm}^3)\left(\frac{1 \text{ g}}{0.75 \text{ cm}^3}\right) = 1.18 \times 10^{-16} \text{ g molecule}^{-1}$$

$$= (1.18 \times 10^{-16} \text{ g molecule}^{-1})\left(\frac{6.02 \times 10^{23} \text{ molecules}}{\text{mol}}\right) = 7.10 \times 10^7 \text{ g mol}^{-1}$$

36. We know that,

$$\text{density, } \rho = \frac{nm}{N_0 V} = \frac{nm}{N_0 (l)^3}$$

$$\therefore \quad \rho = \frac{4 \times 58.5 \text{ g mol}^{-1}}{(6.023 \times 10^{23} \text{ mol}^{-1})(0.564 \times 10^{-7} \text{ cm})^3}$$

where n = no. of molecules in a unit cell = 4
m = molecular mass (NaCl = 58.5)
N_0 = Avogadro's number (6.023×10^{23}/mol)
l = unit edge length = 0.564 nm = 0.564×10^{-7} cm^3
= 2.165 cm^{-3}

37. Figure (i) shows a schematic view of the unit cell, where the open circle represents Cl^- and the filled circle Cs^+. The circles are made small with respect to the unit cell length a to show more clearly the locations of the various represent Cl^- and the filled circle ions. Figure (ii) is a more realistic representation of the right triangle ABC, showing anion-cation-anion contact along the body diagonal AC.

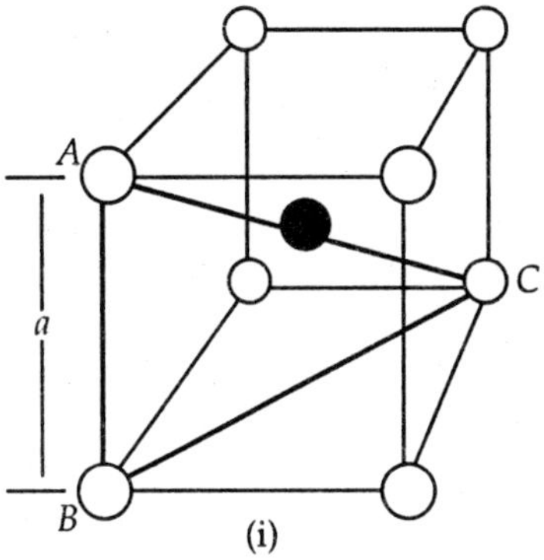

(i)

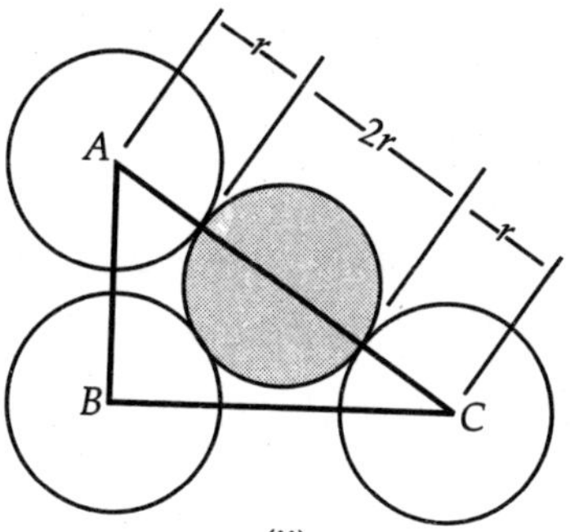

(ii)

Suppose that the closest CS^+-to-Cl^- distance is the sum of the ions radii of Cs^+ and Cl^-, or 1.69 + 1.81 = 3.50 Å. The distance is one-half the cube diagonal, or $a\sqrt{3/2}$.

$$\therefore \qquad \frac{a\sqrt{3}}{2} = 3.50 \text{ Å} \quad \text{so} \quad a = \frac{2(3.50 \text{ Å})}{\sqrt{3}} = 4.04 \text{ Å}$$

38. $$(1 \text{ CsCl unit})\left(\frac{168.4 \text{ g}}{6.02 \times 10^{23} \text{ pairs}}\right)\left(\frac{1 \text{ cm}^3}{3.988 \text{ g}}\right) = 7.014 \times 10^{-23} \text{ cm}^3$$

39. There are $\dfrac{1}{7.014 \times 10^{-23}} = 1.426 \times 10^{22}$ units cm^{-3} $\sqrt[3]{1.426 \times 10^{22}} = 2.42 \times 10^7$ units cm^{-1}

along the edge. The edge of each cube is the reciprocal of the number of units per centimeter:

$$\frac{1}{2.42 \times 10^7 \text{ units cm}^{-1}} = 4.125 \times 10^{-8} \text{ cm} = 4.125 \text{ Å}$$

40. The Cs-to-Cl distance, which is the distance from the corner to the center of a cube of edge 4.25 Å is

$$\frac{\sqrt{3}}{2}(4.125 \text{ Å}) = 3.572 \text{ Å}$$

41. Body-centered cubic: Body diagonal = $4r(\text{Fe}) = \sqrt{3}a$ thus $a =$ $(4/\sqrt{3})$ r(Fe) (See Fig.) Two atoms per unit cell (one at center; $\frac{1}{8}$ at each of 8 corners).

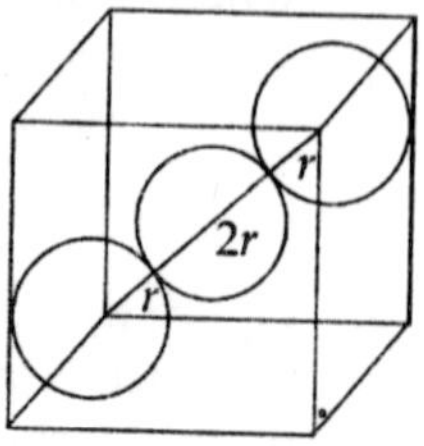

Face-centered cubic: Face diagonal = $4r(\text{Fe}) = \sqrt{2}a'$; $a' =$ $(2/\sqrt{2})\, r(\text{Fe})$

Four atoms per unit cell $\left(\frac{1}{8}\text{ at each of 8 corners; } \frac{1}{2}\text{ at each of six faces}\right)$

$$\therefore \qquad \text{Density ratio} = \frac{4/[(2\sqrt{2r}\,(\text{Fe})]^3}{2/[(4/\sqrt{3})r\,(\text{Fe})]^3} = \frac{4/(2^{9/2})}{2/(4^3/3^{3/2})} = 1.09$$

42. Here,

Body diagonal of little cube = $2r(\text{S}) + 2r(\text{Zn}) = \sqrt{3}(a/2)$

Face diagonal of little cube = $2r(\text{S}) = \sqrt{2}(a/2)$

$$\frac{a}{2} = \sqrt{2}\, r(\text{S}) = \frac{2}{\sqrt{3}}[r(\text{S}) + r(\text{Zn}]$$

$$\therefore \qquad r(\text{S}) = \sqrt{\frac{2}{3}}\,[r(\text{S}) + r(\text{Zn})]$$

$$\text{or} \quad \left(1 - \sqrt{\frac{2}{3}}\right)[r(\text{S})] = \sqrt{\frac{2}{3}}\,[r(\text{Zn})$$

$$r(\text{Zn})/r(\text{S}) = \frac{1-\sqrt{2/3}}{\sqrt{2/3}}\, 0.225$$

43. For Body centred cubic,

2 atoms per unit cell (one atom at center, $\frac{1}{8}$ at each of 8 corners)

Body diagonal = $\sqrt{3a} = 4r$; Volume of atoms = $2\left[\frac{4\pi}{3}\left(\frac{\sqrt{3}}{4}a\right)^3\right] = 0.680a^3$

Volume of unit cell = a^3

Packing factor = $(0.680a^3)/a^3 = 0.680$

and for simple cubic,

1 atom per unit cell

$r = a/2$ volume of atom $\therefore$ $V = \frac{4\pi}{3}\left(\frac{a}{2}\right)^3 = 0.524a^3$

Volume of unit cell = a^3 $\therefore$ Packing factor = $(0.524a^3)/a^3 = 0.524$

44. *A* atoms are present at the 8 corners of a cube, each shared by 8 cubes.

$\therefore$ The number of *A* atoms in a unit cell $= \frac{1}{8} \times 8 = 1$

B atoms are present at the centres of the 6 faces of the cube. Each face can be shared by 2 cubes.

$\therefore$ The number of *B* atoms in a unit cell $= \frac{1}{2} \times 6 = 3$

Hence the formula of the compound is AB_3.

45. *Q* atoms are present at the 8 corners of the cube, each shared by 8 cubes.

$\therefore$ The number of *Q* atoms in a unit cell $= \frac{1}{8} \times 8 = 1$

P atoms are present at the body centre.

$\therefore$ The number of *P* atoms in a unit cell = 1

Hence, the formula of the compound is *PQ*.

Since the packing of body centred close packing, the coordination number of *P* is 8 and that of *Q* is also 8.

46. *Y* atoms are present at the 8 corners of a cube, each shared by 8 cubes.

$\therefore$ Number of *Y* atoms in a unit cell $= \frac{1}{8} \times 8 = 1$

Now, there are 3 *X* atoms present at the centres of 3 alternate faces (since only alternate face has an *X* atom) and each face centred atom is shared by 2 cubes.

$\therefore$ Number of *X* atoms in a unit cell $= \frac{1}{2} \times 3 = \frac{3}{2}$

Therefore, the formula of the compound is $X_{3/2}Y$ or X_3Y_2.

47. We know that in a closest packed array of anions, there are two tetrahedral holes for each anion in the array.

$\therefore$ Number of tetrahedral holes per anion (sulphide ion) = 2

$\frac{1}{2}$ of these are occupied by zinc ions,

$\therefore$ Number of zinc ions per sulphide ion $= \frac{1}{2} \times 2 = 1$

Thus, the formula of zinc sulphide is ZnS.

48. We know that in a closest packed array of anions, number of octahedral holes per anion (oxide ion) = 1

Only $\frac{3}{2}$ of these holes are occupied by aluminium ions.

$\therefore$ Number of aluminuim ions per oxide ion $= \frac{3}{2} \times 1 = \frac{3}{2}$

Therefore, the formula of aluminium oxide is $Al_{2/3}O$. The simplest whole number ratio is 2 : 3. Hence, the formula of aluminum oxide is Al_2O_3.

49. Along each edge of the unit cell of NaCl, two Cl^- ions are in contact with one Na^+ ion. The edge length is equal to the radius of one Cl^- plus the diameter of Na^+, plus the radius of another Cl^-, i.e.,

$$\text{Edge length} = r_{Cl^-} + 2r_{Na^+} + r_{Cl^-}$$

$$= 2(r_{Na^+} + r_{Cl^-}) = 2(95 + 181) = 552 \text{ pm}$$

50. Since tungsten has BCC structure in a unit cell, atoms touch each other along the body diagonal.

We know that

For a cube, body diagonal = $\sqrt{3}$ (edge of the cube)

or $$4r = \sqrt{3}a$$

or $$r = \frac{\sqrt{3}}{4}a = \frac{\sqrt{3}}{4} \times 3.165 = 1.370 \text{ Å}$$

51. Edge length of a unit cell = 288 pm
= 288×10^{-10} cm

∴ Volume of a unit cell = $(288 \times 10^{-10})^3$
= 2.39×10^{-23} cm^3

Now, density of the unit cell = $\dfrac{\text{Mass the unit cell}}{\text{Volume of the unit cell}}$

∴ Mass of the unit cell = $7.2 \times 2.29 \times 10^{-23}$
= 1.72×10^{-22} g

We know that a BCC structure has 2 atoms in a unit cell.

∴ 1.72×10^{-22} g of the element contain 2 atoms.

Hence, 208 g of the element contains $\dfrac{2 \times 208}{1.72 \times 10^{-22}} = 2.42 \times 10^{24}$ atoms.

52. From the given data,

$M = 60$ g mol^{-1}
$a = 400$ pm $= 400 \times 10^{-10} = 4 \times 10^{-8}$ cm
$z = 4$(for a face-centered cubic unit cell)
$N_0 = 6.023 \times 1023$ mol^{-1}

We know that

$$\rho = \frac{zM}{a^3 N_0} = \frac{4 \times 60}{(4 \times 10^{-8})^3 \times 6.023 \times 10^{23}} = 6.230 \text{ g cm}^{-3}$$

∴ The density of the element is 6.230 g cm^{-3}.

53. From the given data,

a = 562 pm = 562×10^{-10} cm
ρ = 2.165 g cm^{-3}
M = 5.5 g mol^{-1}

We know that, $$\rho = \frac{z \times M}{a^3 N_0}$$

$$\therefore \quad z = \frac{\rho a^3 N_0}{M} = \frac{2.165 \times (562 \times 10^{-10})^3 \times 6.023 \times 10^{23}}{58.5} = 4$$

54. From the given data,

$$a = 200 \text{ pm} = 200 \times 10^{-10} = 2 \times 10^{-8} \text{ cm}$$
$$z = 4 \text{ (the structure if f.c.c.)}$$

$\because$ The mass of 24×10^{22} atoms = 200 g,

$\therefore$ The mass of 6.023×10^{23} atoms

$$= \frac{200 \times 6.023 \times 10^{23}}{24 \times 10^{22}}$$

$$= 50.19 \text{ g}$$

$\therefore$ $$M = 50.19 \text{ g mol}^{-1}$$

$\because$ $$\rho = \frac{zM}{a^3 N_0}$$

Substituting the proper values in the above expression,

$\therefore$ $$\rho = \frac{4 \times 50.19}{(2 \times 10^{-8})^3 \times 6.023 \times 10^{23}}$$

$$= 41.66 \text{ g cm}^{-3}$$

55. A cube has 4 (body) diagonals.

$\therefore$ The number of atoms present on the diagonals per unit cell

$$= 4 \times 2 = 8$$

These atoms are shared by one cube only and each of the atoms present on the 8 corners is shared by $\frac{1}{8}$ cube.

$\therefore$ Number of such atoms per unit cell

$$= \frac{1}{8} \times 8 = 1$$

Hence, total number of atoms per unit cell = 8 + 1 = 9

56. From the given data,

$$a = 305 \text{ pm} = 305 \times 10^{-10} \text{ cm}$$
$$\rho = 5.955 \text{ g cm}^{-3}$$
$$z = 2$$

We know that, $$\rho = \frac{z \times M}{a^3 N_0}$$

or $$M = \frac{\rho a^3 N_0}{z}$$

$$\therefore\ M = \frac{5.955 \times (305 \times 10^{-10})^3 \times 6.023 \times 10^{23}}{2} = 50.9$$

∴ The molar mass of the metal is 50.9 g mol^{-1}.

57. ∵ 2 Na^+ ions are replaced by 1 Sr^{2+} ion,

∴ Each Sr^{2+} ion causes one cation vacancy.

Hence, concentration of cation vacancies on being dopped with 10^{-3} mol% $SrCl_2$ = 10^{-3} mol%

$$= 10^{-3} \text{ mol\%} = \frac{10^{-3}}{100}$$

$$= \frac{10^{-3}}{100} \times 6.023 \times 1023 = 6.023 \times 10^{18} \text{ vacancies mol}^{-1}$$

58. From the given data,

z = 4(since the structure is FCC)

ρ = 7.2 g cm^{-3}

∵ Mass of 4.2832×10^{24} atoms = 208 g

∴ Mass of 6.023×10^{23} atoms

$$= \frac{208 \times 6.023 \times 10^{23}}{4.2832 \times 10^{24}} = 29.25 \text{ g}$$

∴ Molar mas of $A = M$ = 29.25 g mol^{-1}

$$\because\ \rho = \frac{zM}{a^3 N_0}$$

$$\Rightarrow a = \left(\frac{zM}{\rho N_0}\right)^{1/3} = \left(\frac{4 \times 29.55}{7.2 \times 6.023 \times 10^{23}}\right) = 29 \times 10^{-9} \text{ cm} = 290 \text{ pm}$$

Chapter 13 : Organic Chemistry

1. There are five isometric hexanes:

(a) $CH_3—CH_2—CH_2—CH_2—CH_2—CH_3$

n-hexane or hexane

(b) 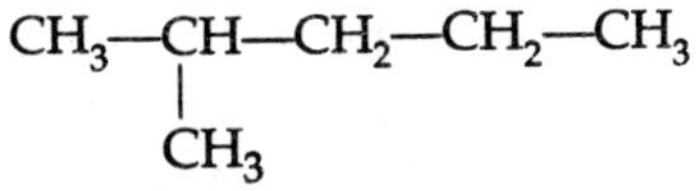

2-methylpentane

(c) 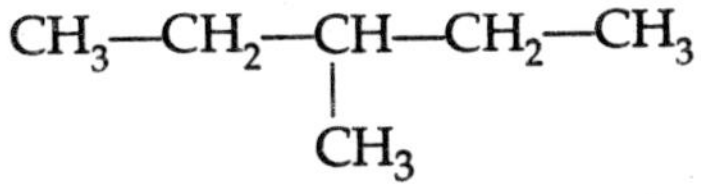

3-methylpentane

(d) 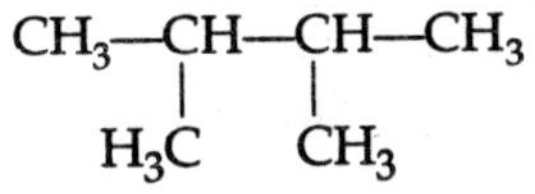

2, 3-dimethylbutane

(e) 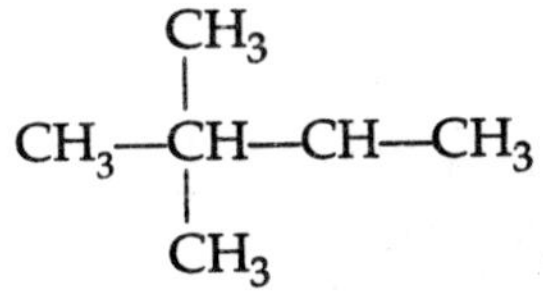

2, 2-dimethylbutane

2. Neither compound (a) nor (c) can exist as enantiomers. Neither has a carbon with four different substituents, not is there any other way that they are not superimpossible on their mirror images. The enantiomers of compounds (b), (d), and (e) are shown below:

(b) 

(d) 

(e) 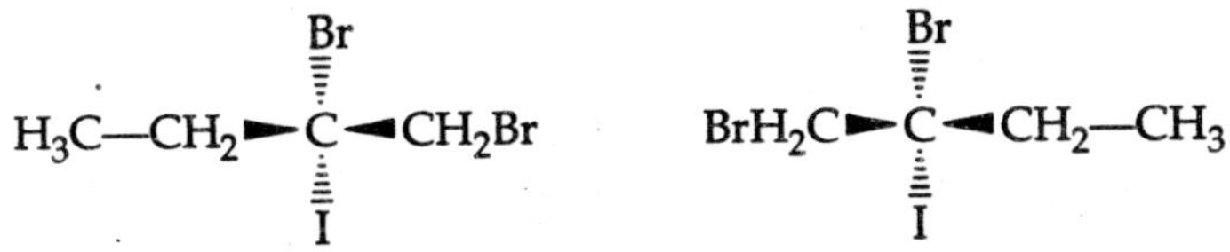

3. The order of increasing boiling point is
   (a) (iii) < (i) < (v) < (iv) < (ii)
4. (a) This is an oxidation (remember that MnO_4^- is a powerful oxidizing agent). Carbon atoms C-2 and C-3 are oxidized from –1 to the 0 state.
   (b) This is a reduction. The carbonyl carbon is reduced from the +2 to the –2 oxidation state.
   (c) This is a reduction. The double-bonded carbons (C-2 and C-3) are reduced from the –1 to –2 state.

(d) This is an oxidation. Atom C-1 is oxidized from the –state in the alcohal to the +1 state in the aldehyde.
(e) This is a reduction. The carbonyl carbon is reduced from the +3 to the –1 state, as the carbonxylic acid is reduced to the alcohal.

5. Going from left to right; the amino acids are named and categorized as:
(i) serine-polar, hydrophilic
(ii) valine-hydrophoic
(iii) histidine-hydrophilic
(iv) tyrosine-neither category
(v) glycine-neither category
(vi) cysteine-neither category

6. Polystyrene is an addition polymer. It is I.
(b) Dacron polyester is a condensation polymer. It is II.
(c) This is an addition polymer, the styrene, butadiene copolymer. It is I.

7. (a) cyclobutane
(b) methyl cylopropane
(c) 1-butane
(d) 2-methyl propene
(e) cis-2-butene
(f) trans-2-butene

8. (a) 1-pentene
(b) 2-methyl-2-butene
(c) trans-2-pentene
(d) cis-2-pentene
(e) 2-methyl-l-butane
(f) 3-methyl-1-butene

9. (a) cyclopentane
(b) methylcyclo butane
(c) 1, 1-dimethyl cyclopropane
(d) cis-1, 2-cyclopropane
(e) trans 1, 2-cyclopropane

10. (a) Ethyl alcohal because its molecules can form hydrogen bonds to each other. Methyl ether molecules have no hydrogens attached to oxygen.
(b) Heptane because it has a higher molecular weight. (Neither compound can form hydrogen bonds.)
(c) 1-Propanol because its molecules can form hydrogen bonds to each other. Acetone molecules have no hydrogen attached to oxygen.
(d) *Cis*-1, 2-dichloethane because its molecules have a higher dipole moment.
(e) Propionic acid because its molecules can form hydrogen bonds to each other.

11. (a) $CH_3\overset{\overset{CH_3}{|}}{C}{=}CHCH_2CH_3$ (major) + $CH_2{=}\overset{\overset{CH_3}{|}}{C}CH_2CH_2CH_3$

(b) $CH_2{=}\overset{\overset{CH_3}{|}}{C}CH_2CH_2CH_3$ (major) + $CH_3\overset{\overset{CH_3}{|}}{C}{=}CHCH_2CH_3$

(c) $CH_3\overset{\overset{CH_3}{|}}{C}{=}CHCH_2CH_3$ (major) + $CH_2{=}\overset{\overset{CH_3}{|}}{C}CH_2CH_2CH_3$

(d) 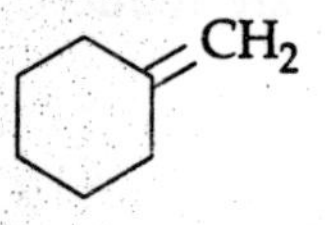

(e) 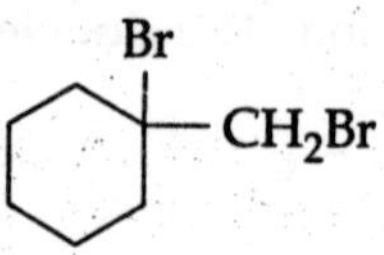

(f) 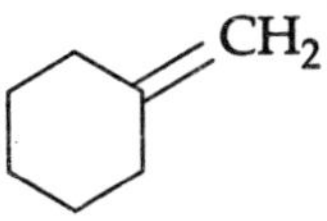
(g) 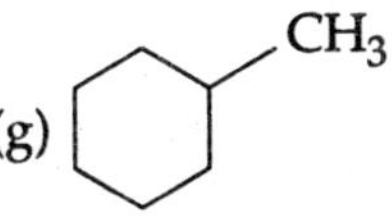
(h) 

12. (a) Enantiomers (b) Same
(c) Structural isomers (d) Diastereomers
(e) Same (f) Enatiomers
(g) Same (h) Disastereomers
(i) Same (j) Enantiomers
(k) Same (l) Enantiomers
(m) Same (n) Structural isomers
(o) Same (p) Disastereomers
(q) Enantiomers

13. (a) 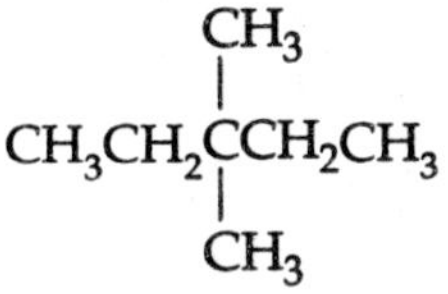
because most spherical of the three constitutional isomers, least vanderwaals attraction between the molecules

(b) 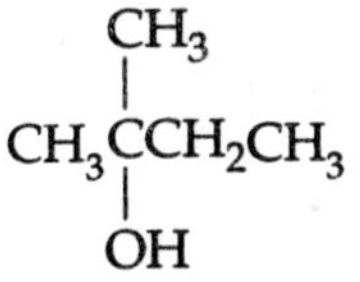
∵ same as (a)

(c) 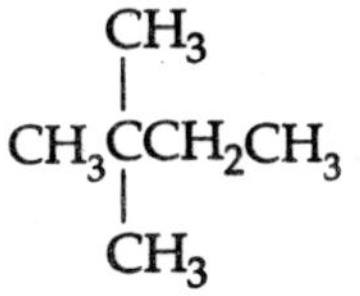
∵ same as (a)

(d) 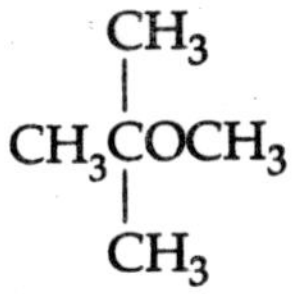
∵ least polar (the other two are alcohals and form hydrogen bonds)

(e) 
least polar

14. Following eight structrues are possible for C_7H_{16}.

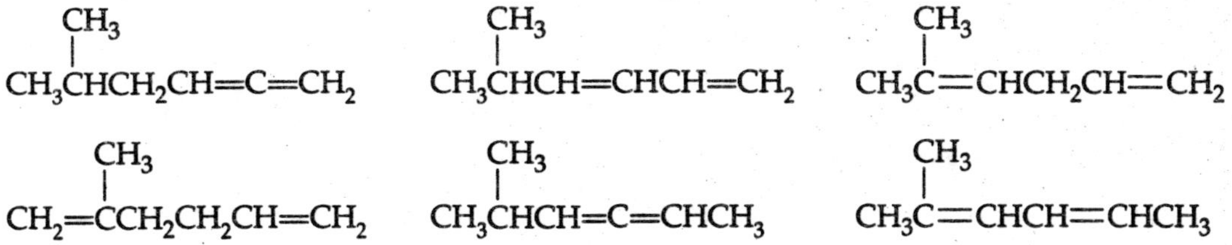

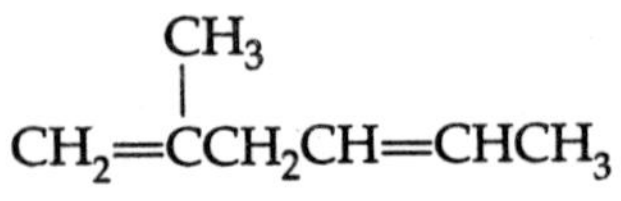

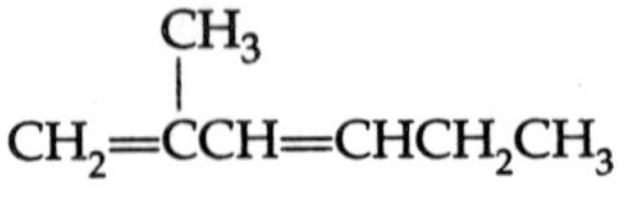

15. (a) No (b) No (c) No (d) No

(e) Yes

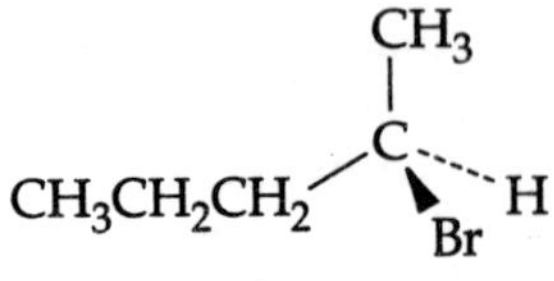

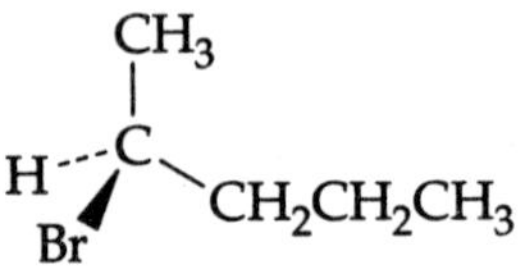

(f) Yes

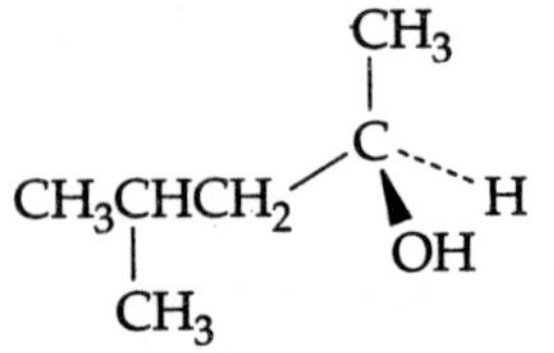

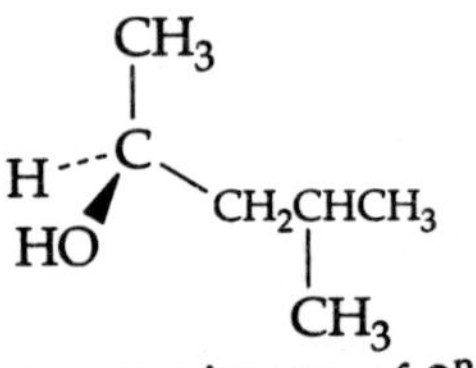

16. For a compound with n stereocenters, there can be a maximum of 2^n stereoisomers. 2, 3, 4-Tribromohexane has 3 stereocenters and thus a maximum of 8 stereoisomers. The number different isomers will be reduced if there are meso compounds.

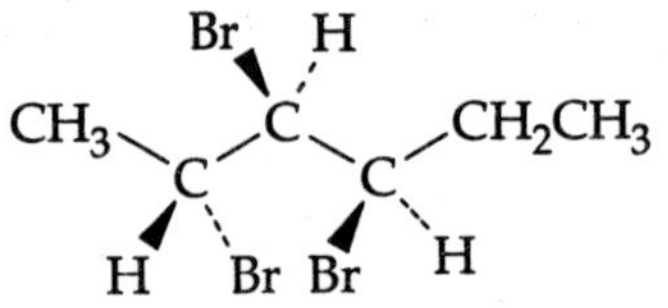

(2R,3R,4R)-2,3,4-tribromohexane

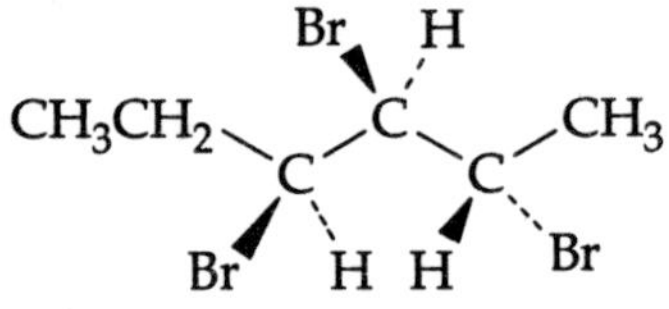

(2*S*,3*S*,4*S*)-2,3,4-tribromohexane

These two compounds are enantiomers and are diastereomers of the other six stereoisomers.

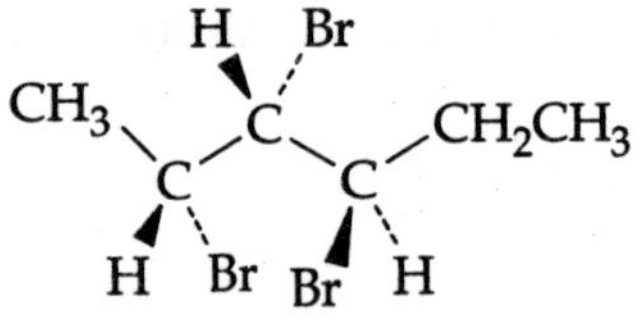

(2*R*,3*S*,4*R*)-2,3,4-tribromohexane

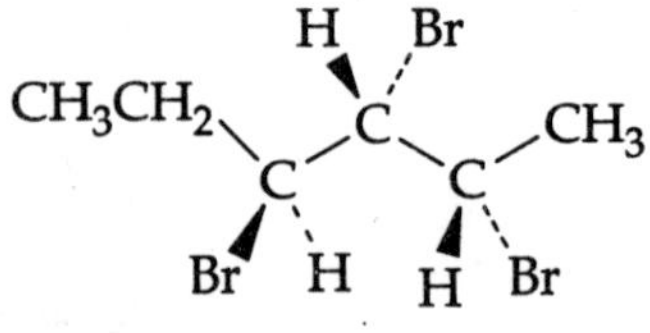

(2*R*,3*R*,4*S*)-2,3,4-tribromohexane

These two compounds are enantiomers and are diastereomers of the other six stereoisomer.

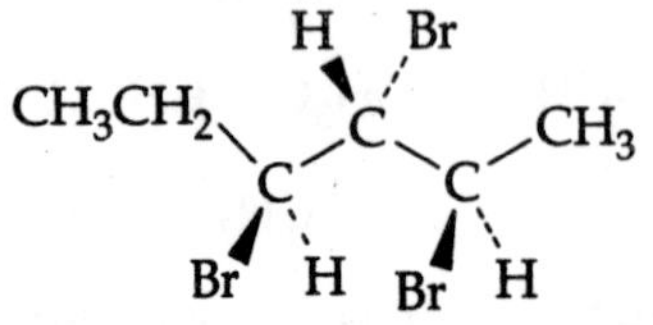

(2*R*,3*R*,4*S*)-2,3,4-tribromohexane

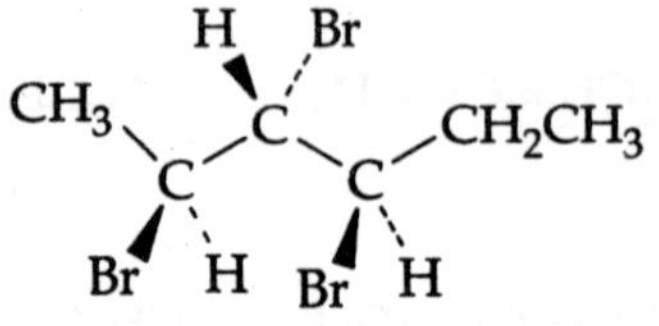

(2*S*,3*S*,4*R*)-2,3,4-tribromohexane

These two compounds are enantiomers and are diastereomers of the other six stereoisomer.

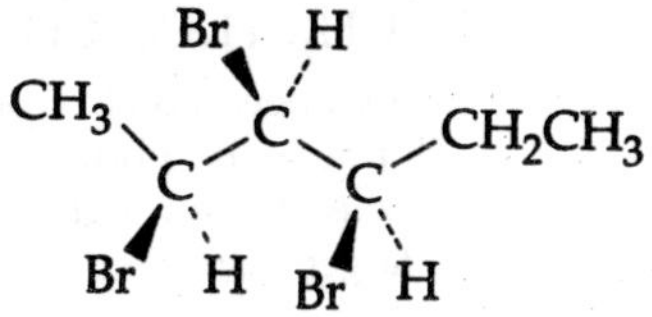

(2*S*,3*R*,4*R*)-2,3,4-tribromohexane

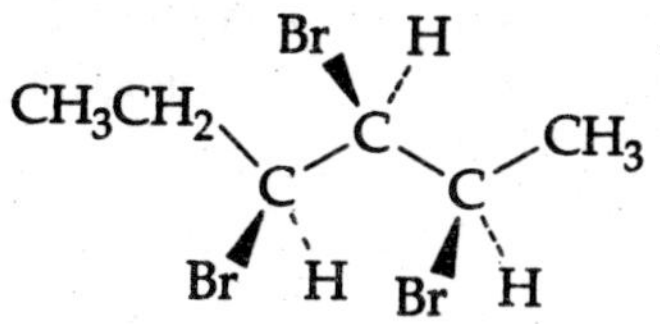

(2*S*,3*S*,4*S*)-2,3,4-tribromohexane

These two compounds are enantiomers and are diastereomers of the other six stereoisomer.

17. Three stereoisomers are possible for 1, 3-dimethyl cyclopentane.

(i)

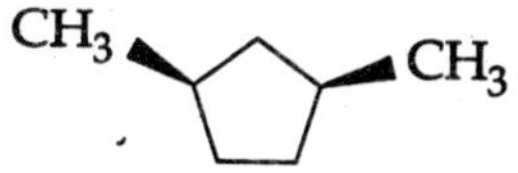

(1*R*, 3*S*)-1,3-dimethyl-cyclopentane

(1*R*, 2*R*)-1,3-dimethyl-cyclopentane

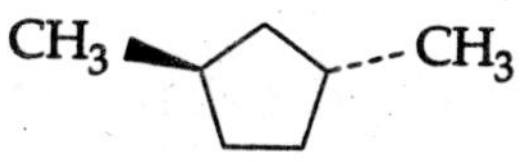

(1*S*, 3*S*)-1,3-dimethyl-cyclopentane

(ii) The possible stereoisomers are given below:

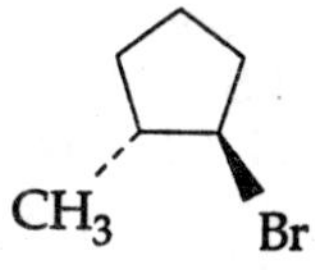

(1*R*, 2*R*)-1-bromo-2-methylcyclopentane

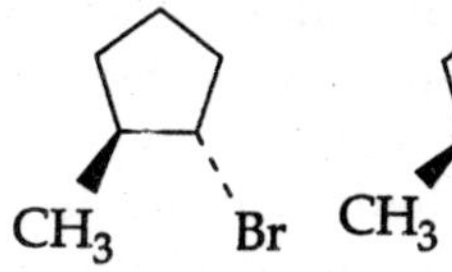

(1*S*, 2*S*)-1-bromo-2-methylcyclopentane

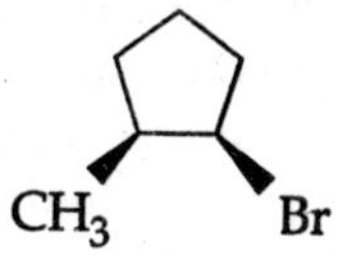

(1*R*, 2*S*)-1-bromo-2-methylcyclopentane

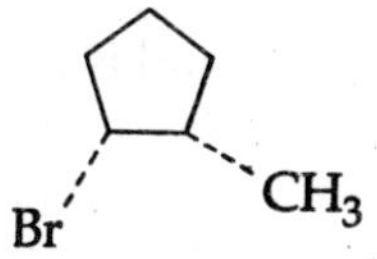

(1*S*, 2*R*)-1-bromo-2-methylcyclopentane

(iii) The different configurational isomers are:

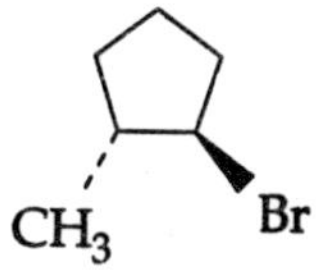

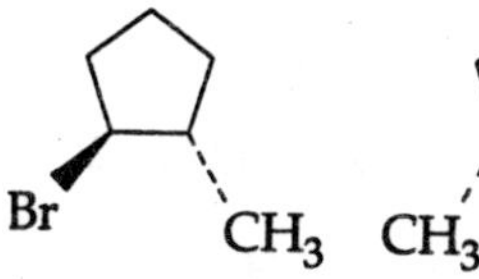

trans-1-bromo-2-methylcyclohexane

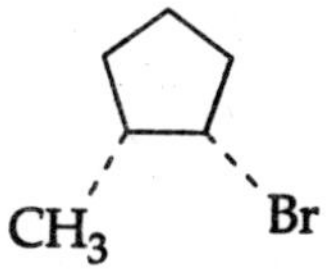

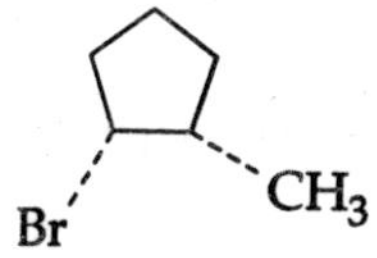

cis-1-bromo-2-methylcyclohexane

18. (a) $CH_3CH_2CH_2CH_2C(=O)O^-Na^+$ > $CH_3CH_2CH_2CH_2C(=O)OH$ > $CH_3CH_2C(=O)CH_2CH_3$

reason → ionic; covalent, but hydrogen bond donor as well as acceptor; only hydrogen bond acceptor

(b) $CH_3CH_2CH_2C(=O)OH$ > $CH_3CH_2CH_2CH_2OH$ > $CH_3CH_2C(=O)CH_2CH_3$

reason can hydrogen bond at the carbonyl group as well as the hydroxyl group; only one site for hydrogen bonding; only hydrogen bond acceptor

(c) $CH_3CH_2CH_2\overset{O}{\overset{\|}{C}}NH_2$ > $CH_3CH_2CH_2\overset{O}{\overset{\|}{C}}OCH_2CH_3$ > $CH_3CH_2CH_2\overset{O}{\overset{\|}{C}}O(CH_2)_4CH_3$

reason hydrogen bond donor as well as acceptor; hydrogen bond acceptor; hydrogen bond acceptor bit nonpolar portion of the molecule now significantly larger than the polar part

19. (a) 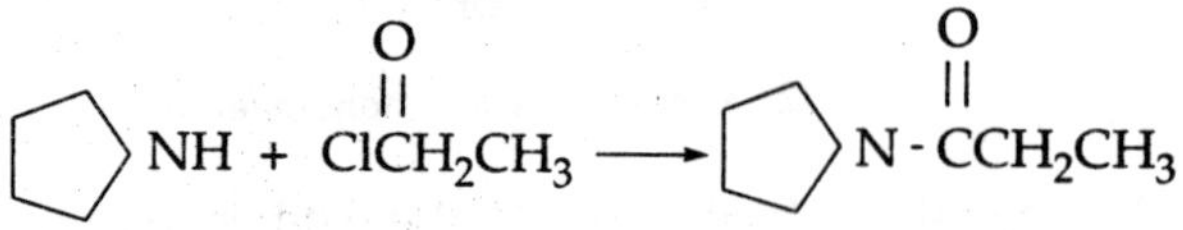

(b) 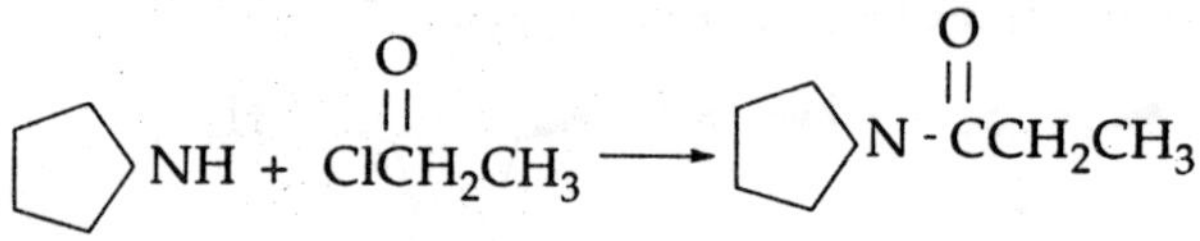

20. (a) For, $CH_3CH_2CH{=}CH_2 \xrightarrow{?} CH_3CH_2\underset{SH}{\underset{|}{C}}HCH_3$

The elements of hydrogen sulfide, H_2S have added to the double bond. The double bond is a nucleophile and is protonated by strong Bronsted acids. H_2S, with a pK_a~10.5, is not sufficiently acidic (hydrogen sulfide is similar to ethanethiol, CH_3CH_2SH, and can be expected to have a similar pK_a); a hydrogen halide such as HBr (pK_a – 8) is acidic. In the product that is formed, the bromine atom will be attached to an electropilic carbon atom and will leave when the hydrogen sulfide anion is used as nucleophile.

$$CH_3CH_2CH{=}CH_2 \xrightarrow{HBr} CH_3CH_2\underset{Br}{\underset{|}{C}}HCH_3$$

$$CH_3CH_2\underset{Br}{\underset{|}{C}}HCH_3 \xrightarrow{Na^+SH^-} CH_3CH_2\underset{SH}{\underset{|}{C}}HCH_3 + Na^+\ Br^-$$

(b) For, $CH_3CH{=}CHCH_3 \xrightarrow{?} CH_3\underset{Cl}{\underset{|}{C}}HCH_2CH_3$

The element of hydrogen chloride, HCl, have added to the double bond. Unlike H_2S, hydrogen chloride is a strong enough acid (pK_a – 7) to protonate the double bond. The cation produced by protonation will then react with the conjugate base of hydrogen chloride to give the desired product.

$$CH_3CH{=}CHCH_3 \xrightarrow{HCl} CH_3\overset{+}{C}H{-}\underset{H}{\underset{|}{C}}HCH_3 + Cl^- \rightarrow CH_3\underset{Cl}{\underset{|}{C}}HCH_2CH_3$$

(c) For, $CH_2{=}CH_2 \xrightarrow{?} CH_3CH_2\overset{+}{N}H_3\ \ I^-$

The element of the ammonium ion, NH_4^+, have added to the double bond. NH_4^+ is not sufficiently acidic (pK_a = 9.4) but HI (pK_a – 9) is acidic. Ammonia can then be used as nucleophile to react with the electrophilic carbon atom in ethyl iodide, the product of the first reaction.

$$CH_2{=}CH_2 \xrightarrow{HI} CH_3CH_2I \xrightarrow{NH_3} CH_3CH_2\overset{+}{N}H_3\ I^-$$

(d) For, $CH_3CH(Br)CH_2CH_2CH_3 \xrightarrow{?} CH_3CH(CN)CH_2CH_2CH_3$

Cyanide ion has substituted for bromide ion. This is a nucleophilic substitution reaction on 2-bromopentane, which requires only one species, NaCN.

$$CH_3CH(Br)CH_2CH_2CH_3 \xrightarrow{Na^+CN^-} CH_3CH(CN)CH_2CH_2CH_3 + Na^+Br^-$$

23. (a) (*S*)-3-bromo-2-methylpentane
(b) (*S*)-1-phenyl-1-butanol
(c) butane
(d) (*S*)-1,2-dibromopentane
(e) (*R*)-3-methyl-3-hexanol
(f) (*S*)-3-methyl-3-hexanol
(g) (2*R*, 3*S*)-2, 3-dibromobutane/meso-2, 3-dibromobutane
(h) (1*S*, 2*S*, 4*S*)-1, 2-dibromo-4-methylcyclohexane
(i) (1*R*, 2*R*)-2-bromcyclopentanol
(j) (2*R*, 3*S*)-2, 3-dichloropentane

24. (a) enantiomers — The molecule on the left has the *S* configuration, the one on the right, the *R* configuration.
(b) identical
(c) identical
(d) identical
(e) identical
(f) enantiomers — The molecule on the left has the *R* configuration; the other, the *S* configuration.
(g) enantiomers — The molecule on the left has the *R* configuration; the other, the *S* configuration. Note also that the molecule on the right is not the mirror image of the molecules on the left but is a conformer of the mirror image.
(h) constitutional isomers — (*R*)-3-bromo-1-chlorobutane and (2*R*, 3*S*)-2-bromo-3-chlorobutane
(i) identical — Both have the *R* configuration.
(j) conformers
(k) constitutional isomers — (*R*)-2-bromo-1-chlorobutane and (*S*)-3-bromo-1-chlorobutane
(l) conformers — The chlorine atoms is equatorial in the surface on the left and axial in the one on the right.

25. (a) 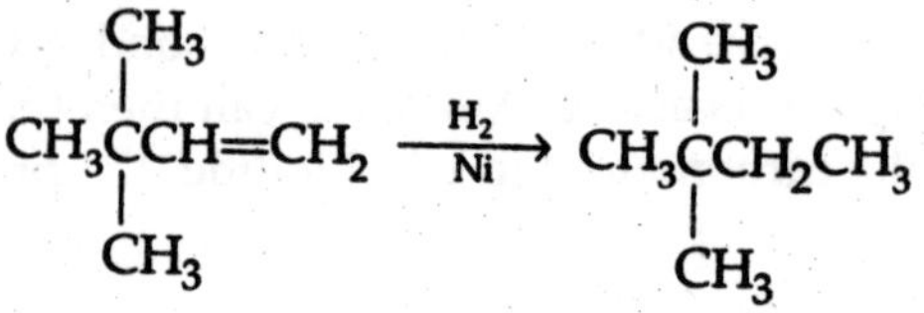

(b) 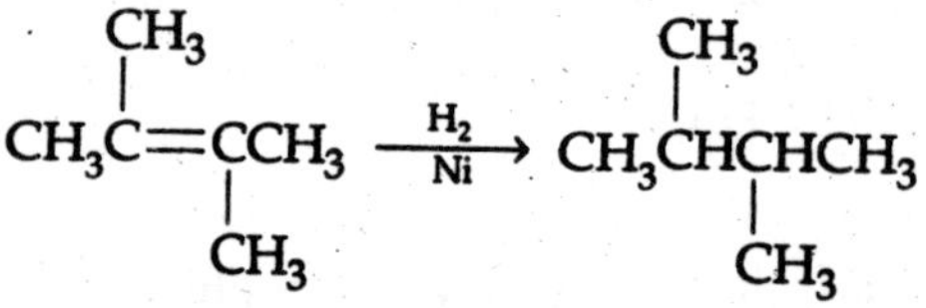

(c) 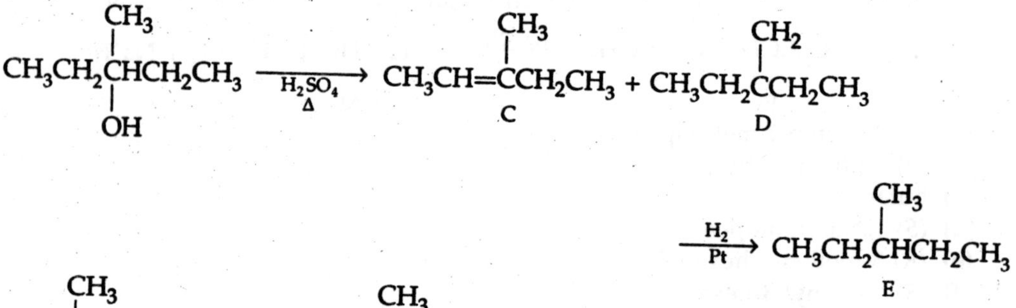

(d) 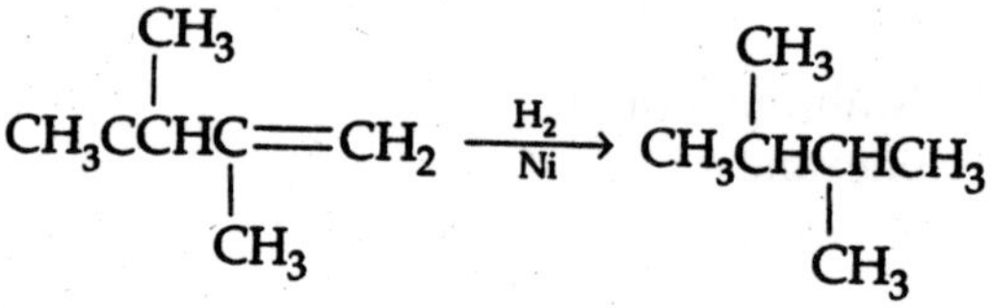

(e) 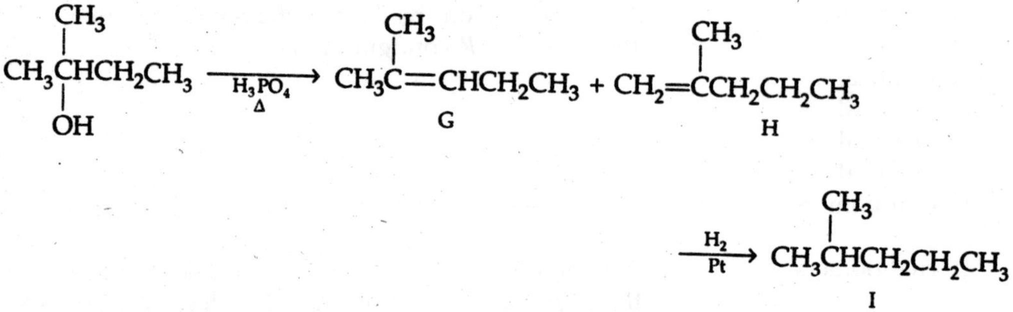

26. (a) $HC{\equiv}CH \xrightarrow{NaNH_2} H{-}C{\equiv}CNa \xrightarrow{C_6H_5CH_2Br} C_6H_5CH_2C{\equiv}CH$

$$C_6H_5CH_2C{\equiv}CH \xrightarrow{CH_3CH_2MgBr} C_2H_6 + C_6H_5CH_2C{\equiv}CMgBr$$

(b)

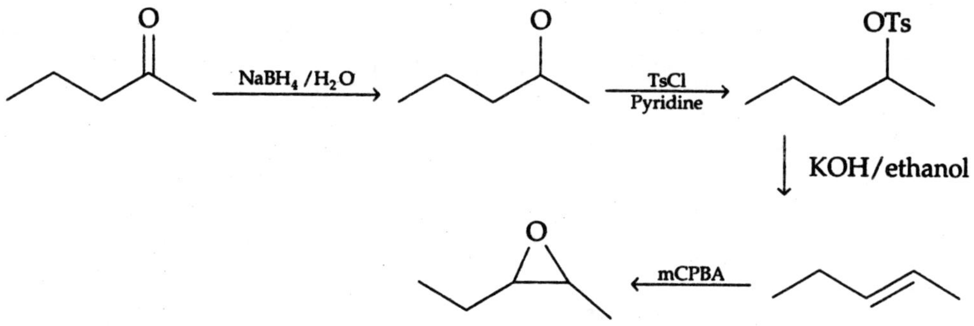
O
NaBH$_4$/H$_2$O
O
TsCl
Pyridine
OTs
KOH/ethanol
O
mCPBA

(c)

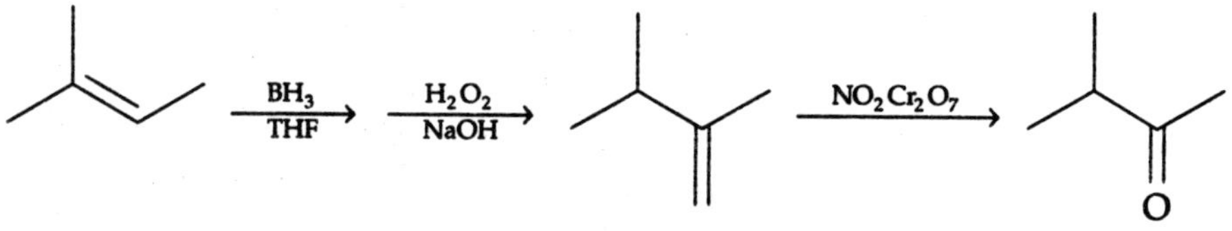
BH$_3$
THF
H$_2$O$_2$
NaOH
NO$_2$Cr$_2$O$_7$
O

27. (a)

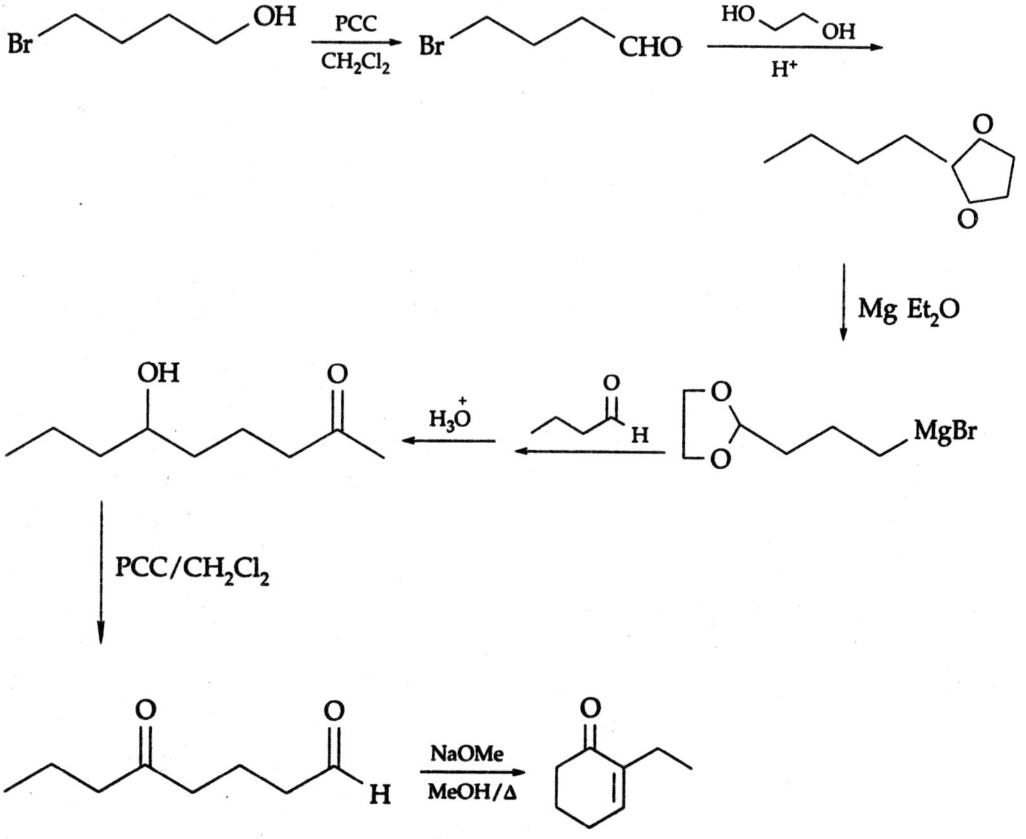
Br
OH
PCC
CH$_2$Cl$_2$
Br
CHO
HO
OH
H$^+$
O
O
Mg Et$_2$O
OH
O
H$_3\overset{+}{O}$
O
H
O
O
MgBr
PCC/CH$_2$Cl$_2$
O
O
H
NaOMe
MeOH/Δ
O

(b) 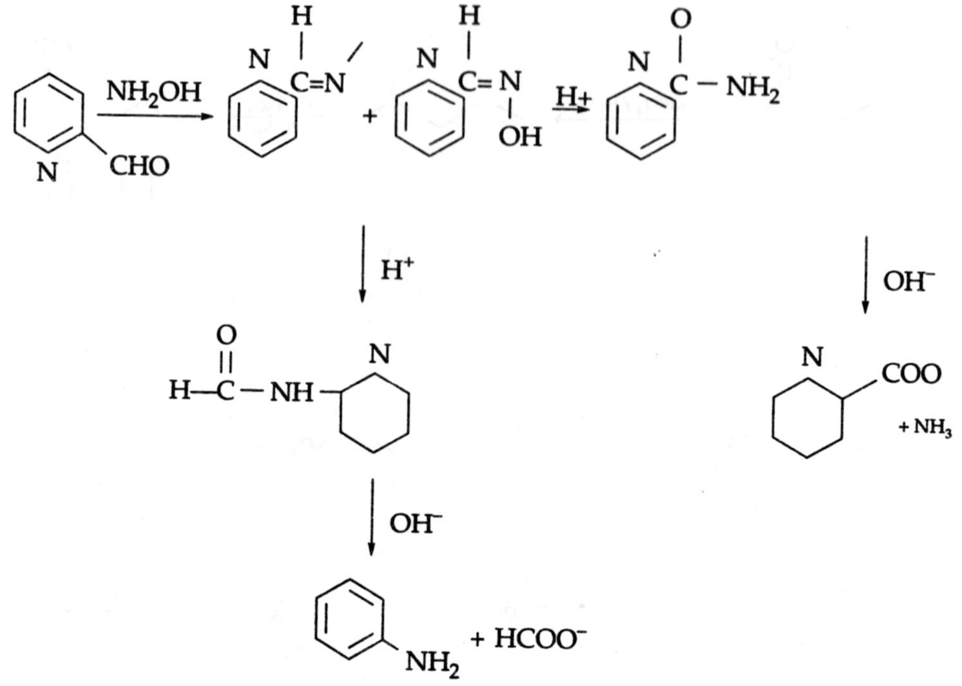

28. Compounds (b), (c) and (f) exhibit cis-trans isomerism. The isomers of compound (f) are shown below:

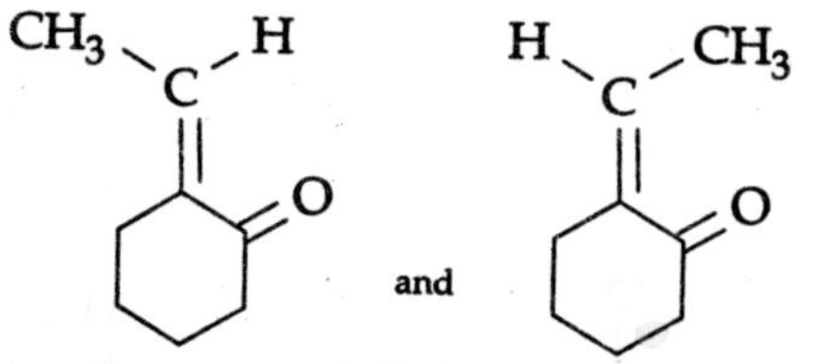

29. There are 3 isomers of dibromobenzene, shown here with their names:

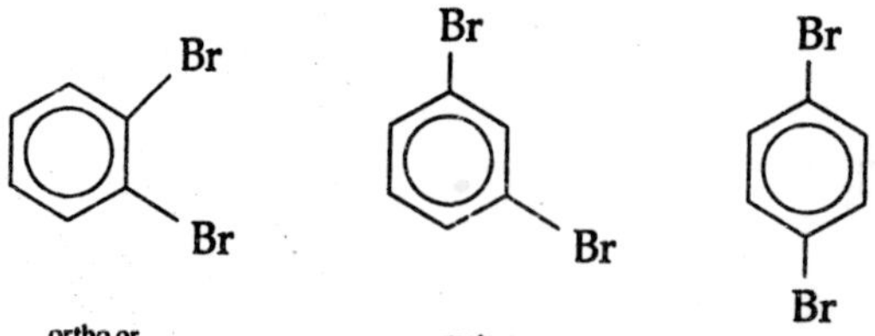

ortho or 1,2-dibromobenzene | meta or 1,3-dibromobenzene | para or 1,4-dibromobenzene

30. There are 6 isomers of dibromopyridine.

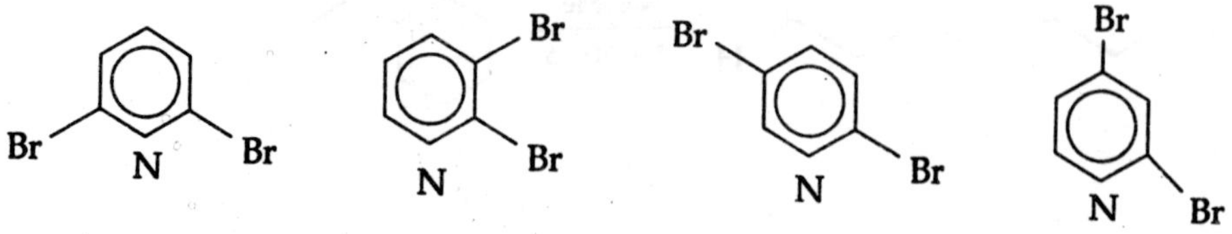

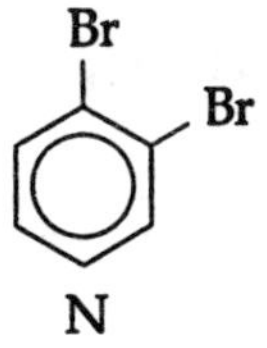
Br
Br
N

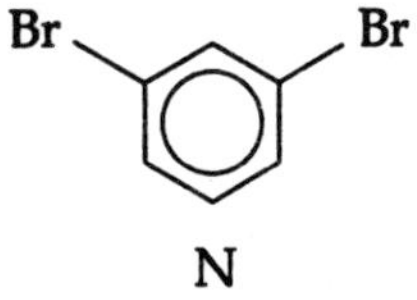
Br
Br
N

31. (i) 3-methyl-1-butanol.
(ii) 2-methylbutanol.
(iii) 2-ethyl-1-methyl 1-1-pentene.
(iv) benzamide
(v) 4-ethyl 1-2, 5, 5-trimethylheptane.
(vi) 3-pentanone.
(vii) ethyl isopropyl ether.
(viii) pentanoic acid.
(ix) N, N-dimethylaniline.

32. (a)

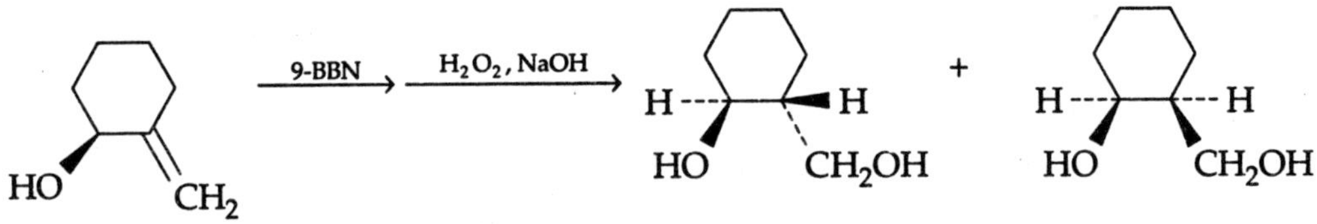
HO
CH2
9-BBN
H2O2, NaOH
H
H
HO
CH2OH
+
H
H
HO
CH2OH

Compound A, in which the large groups are trans to each other on the cyclohexane ring, is expected to be termodynamically more stable. Compound *B*, however, is formed by way of a sterically less hindered transitition state.

33. (a) $C_{13}H_{10}O \xrightarrow[\text{cat}]{H_2} C_{13}H_{14}O$ The corresponding saturated hydrocarbon would have the formula $C_{13}H_{28}O$.

$C_{13}H_{10}O$ and 9 units of unsaturation $\left(\dfrac{28-10}{2}=9\right)$.

After hydrogenation, 7 units of unsaturation remain $\left(\dfrac{28-14}{2}=7\right)$.

Two units of unsaturation (9 – 7 = 2) correspond to multiple bonds that are easily hydrogenated. Possible structures for Carlina-oxide are

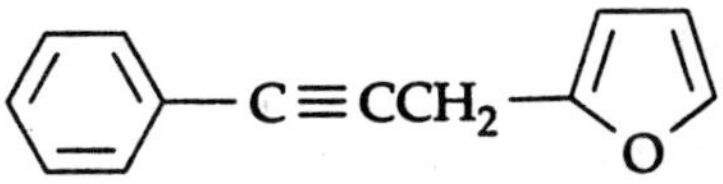
C≡CCH2
O

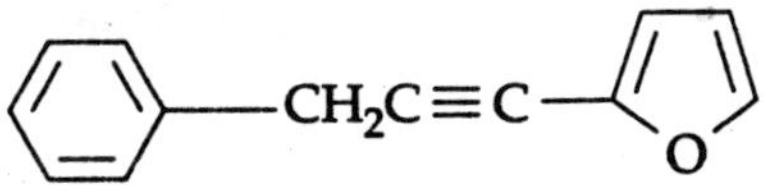
CH2C≡C
O

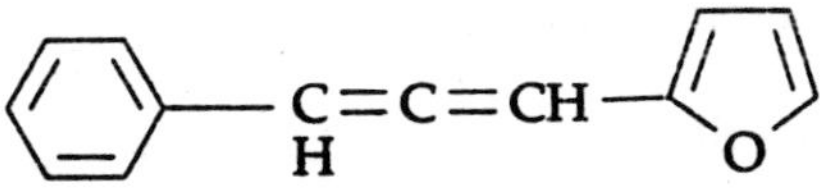
C=C=CH
H
O

34. (a)

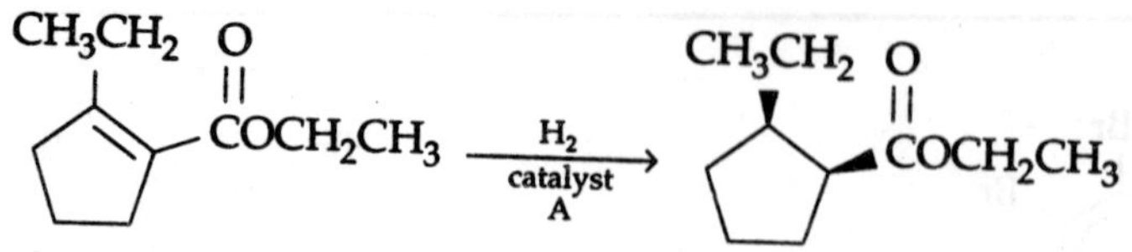

(b)

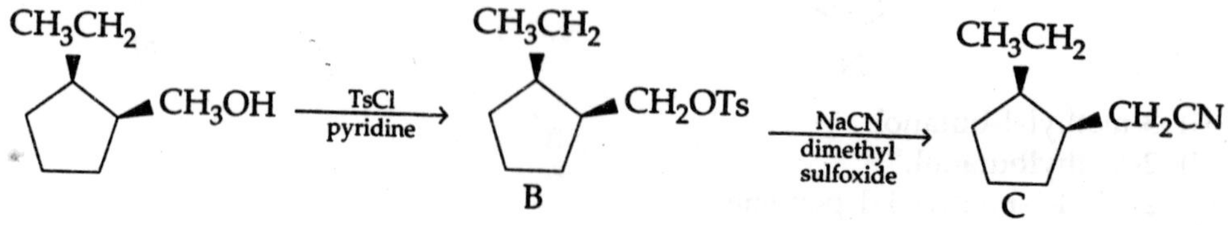

(c)

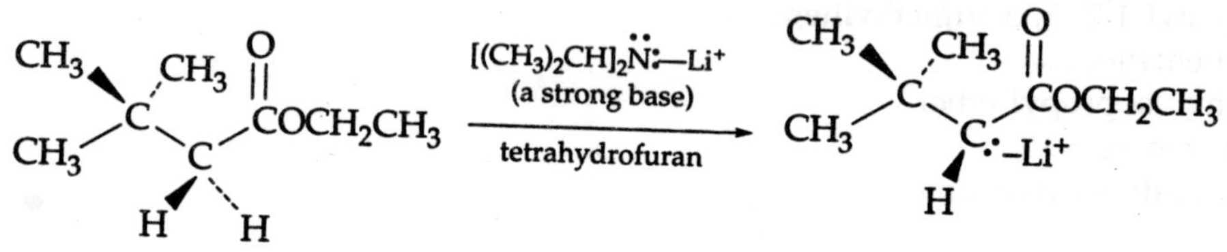

a good nuclephile

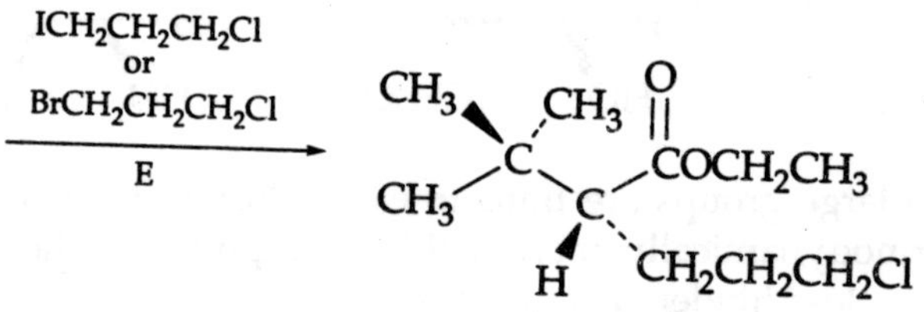

and enatiomer

35. (a)

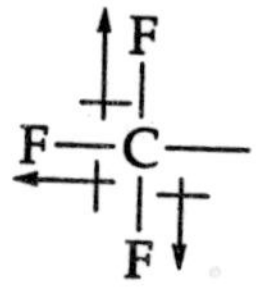

electron-withdrawing substituent; meta directing.

(b) $(CH_3)_3\overset{+}{P}$ positive charge on atom directly bonded to ring; meta.

(c)

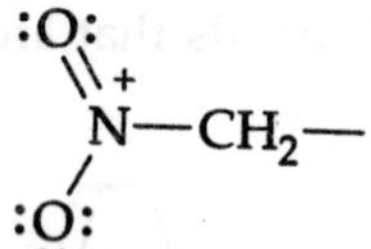

electron-withdrawing substituent (positively charged nitrogen atom closest to ring); meta directing.

(d) $(CH_3)_2\overset{+}{\ddot{S}}$— positive charge on atom directly bonded to ring; meta directing.

(e)

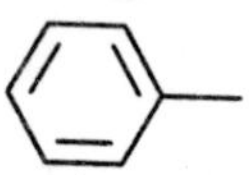

will stabilize a positive charge by resonance; orthopara directing.

(f) $CH_3CH{=}CH$— will stabilize a positive charge by resonance; orthopara directing.

36. (a) 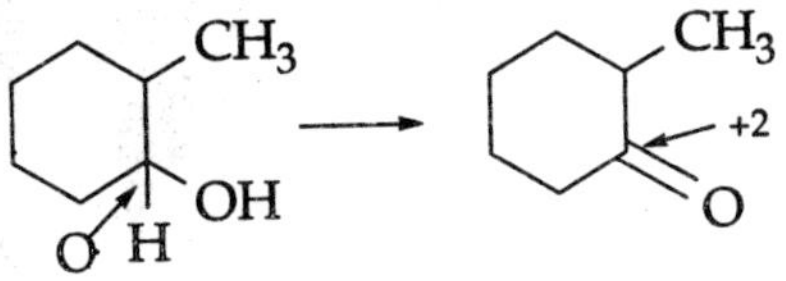

Due to loss of two hydrogens there is an increase in oxidation number ∴ oxidation and so an oxidising agent is needed.

(b) 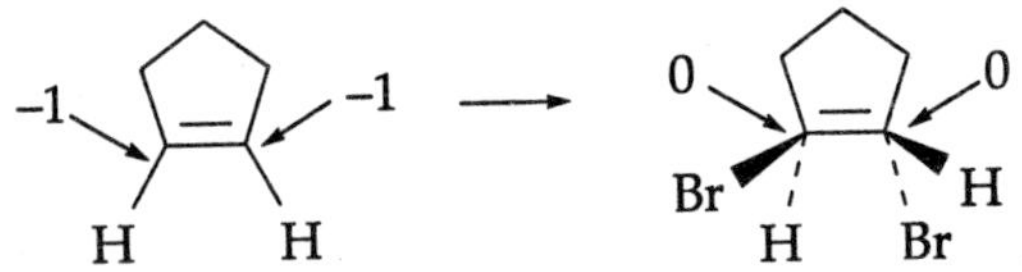

Due to an increase in oxidation number at each carbon it is oxidation and hence oxidizing agent is needed

(c) $CH_3CH_2Br \longrightarrow CH_3CH_2NH_2$
(−1) (−1)

As there is no change in oxidation state so it is not an oxidation-reduction reaction.

(d) 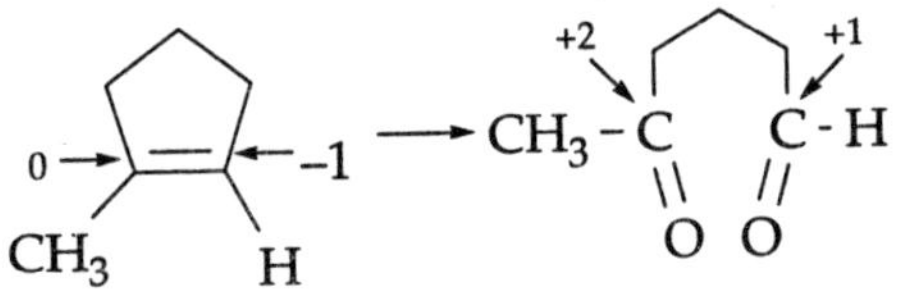

There is a gain of two oxygens thus increase in oxidation number at each carbon so it is oxidation and thus an agent is needed

(e) 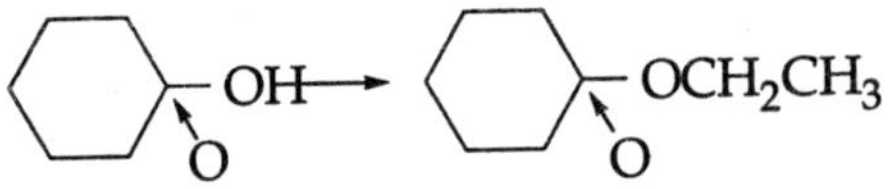

As there is no change in oxidation state so it is not an oxidation-reduction reaction

37. 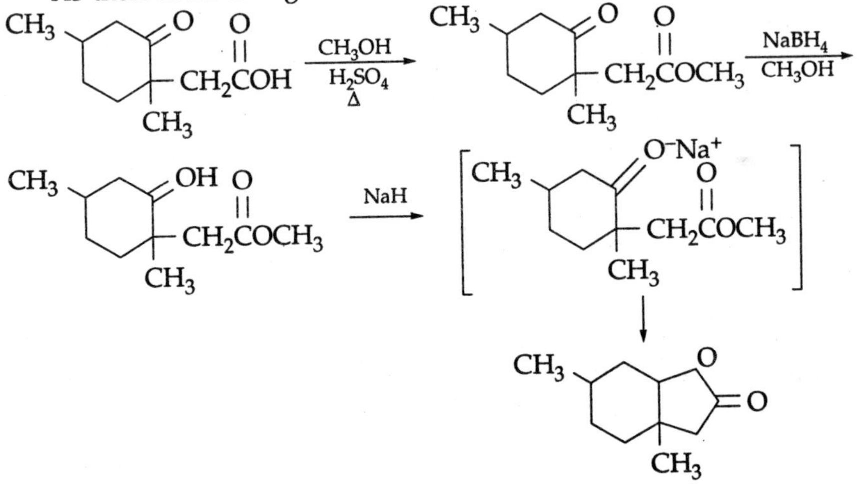

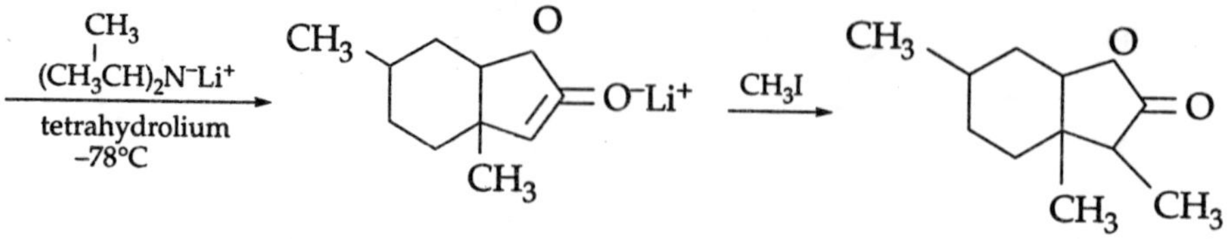
CH3
(CH3CH)2N-Li+
tetrahydrolium
–78°C
CH3
O
O-Li+
CH3
CH3I
CH3
O
O
CH3 CH3

38. (i) 1-chloro-3 phenyl propane
(ii) 3-iodo-3 phenyl hexane
(iii) 1, 2-dichloro-1 phenyl propane
(iv) 2-methyl-2 phenyl hexane
(v) 3-phenyl-1-propanol
(vi) 3-methyl-1-phenyl butane

39. (a) 1-bromo-1-methylcyclobutane
(b) ethylcyclopropane
(c) 1-ethyl-1-methylcyclopentane
(d) iodocyclopentane
(e) cyclobutanol
(f) bromocyclopropane

40. (a) 1-bromo-1-methylcyclopropane
(b) 2,5,5,5-tetramethylhexane
(c) 1, 1-dimethylcyclobutane
(d) 2-chloro-2-methylpentane
(e) isopropylcyclohexane or 2-cyclohexylpropane
(f) 1-ethyl-1-cyclopetanol
(g) 1-chloro-1-ethylcyclohexane
(h) 1-phenyl-1-cyclohexanol
(i) 3-bromo-3-ethyl-1-heptanol
(j) 6-tert-butyl-2,2-dichlorooctane or 6, 7-dichloro-3-ethyl-2, 2-dimethyloctane

41. (a) 3-hydroxy-2-methylcyclopropane
(b) 2-pentanone
(c) 4-chlorobenzaldehyde
(d) 4-heptyn-3-one propen-1-one
(e) (E)-1, 3-diphenyl-2-nonen-2-one
(f) 2,2-cdibromocyclohexanone
(g) trichloroethanal (trichloroacetaldehyde)
(h) 3,3-dimethylcyclohexane-carbaldehyde
(i) 2,4-pentanedione
(j) 2-ethyl-2-methylcyclobutanone
(k) *cis*-2-propylcyclopentane-carbaldehyde

42. (a) (i) 4-ethylbenzanilide
(b) (iii) methyl-2-bromo-4-methylpentanoate
(c) (v) ethyl2-aminobenzoate (2-methyladipic acid)
(d) (vii) 2-methylhexanedioic acid
(e) (vi) hexanioc anhydride

(f) (iv) (*E*)-4-octenoic acid
(g) (ii) (1*R*, 2*R*)-cyclobutanedicarboxylic acid

43. (a) 2,4-dichlorobenzoic acid
(b) 4-bromohexanoic acid
(c) 2-methylbutanedioic anhydride (2-methylsuccinic anhyride)
(d) 2-methylpentanenitrile
(e) ethyl (1*R*, 2*R*-2-ethylcyclohexane-carboxylate
(f) N, N-dimethypentanamide
(g) (*R*)-2-hydroxypentanoic acid
(h) 2-chloropentanoyl chloride

44. (a) cyclopentanecarbaldehyde
(b) cyclobutanone
(c) (*E*)-2-hexenal
(d) (*S*)-2-hydroxypropanal
(e) 2-methylcyclohexanone
(f) 5-methyl-2-hexanone
(g) 4-methyoxybenzaldehyde (anisaldehyde)
(h) 3-methyl-2-cyclohexanone
(i) (4*S*, 5*S*)-4-chloro-5-methyl-2-heptanone
(j) 4-phenylbutanal

45. (a) 3,4-dimethylnitrobenzene
(b) 4-methyl-1-phenylpentane
(c) 2-chloro-6-ethylnaphthalene
(d) 2,4-dinitrotoluene
(e) *m*-bromobenzoic acid
(f) methyl 3,5-dimethylbenzoate
(g) 1-(4-methylphenyl)-1-butanone
(h) *p*-bromobenzaldehyde 4-bromobenzaldehyde

46. (a) *trans*-2-ethyl-1-(N-methylamino)cyclobutane
(b) 2-methylpyridine
(c) triethylamine
(d) 2-amino-4-methylbenzoic acid
(e) 4-aminobutanoic acid
(f) 2,4,6-trichloroaniline
(g) 2-amino-5-methylhaxane
(h) *cis*-2-aminocyclohrxanol
(i) *N*,*N*-diethyl-4-nitroaniline

47.

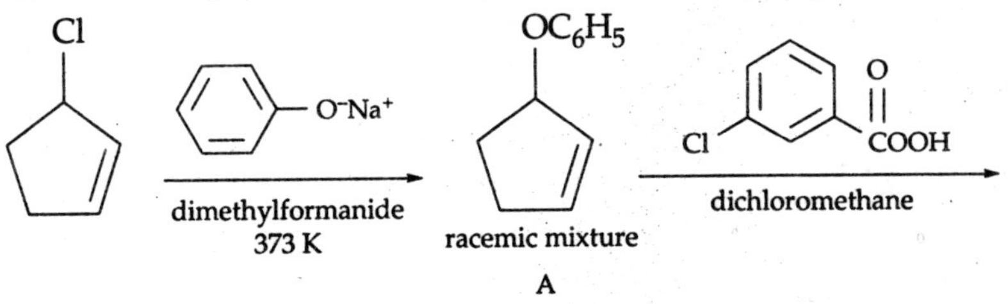

Cl
O⁻Na⁺
dimethylformanide
373 K
OC6H5
racemic mixture
A
O
Cl
COOH
dichloromethane

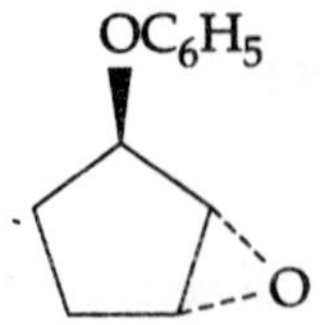

and enantiomer

B

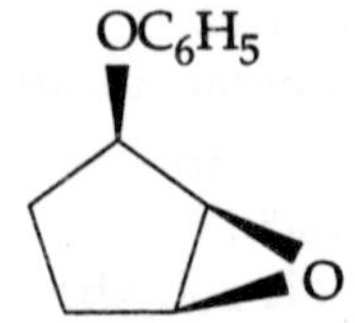

and enantiomer

C

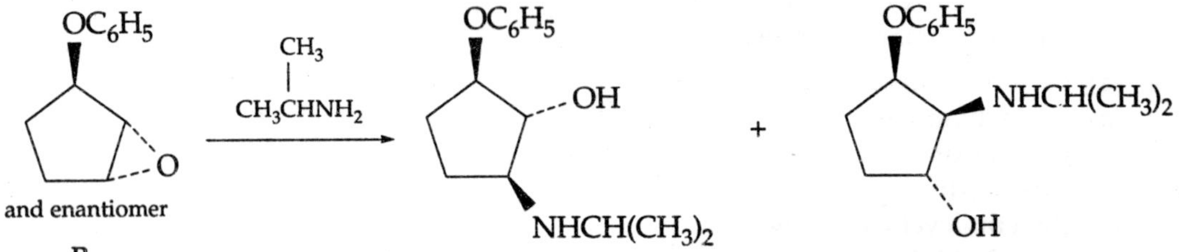
OC6H5
CH3
CH3CHNH2
OH
NHCH(CH3)2
+

and enantiomer B

and enantiomer D

and enantiomer (very minor product, bulky phemoxide group prevents the approach of amine to the carbon adjacent to the substituent.)

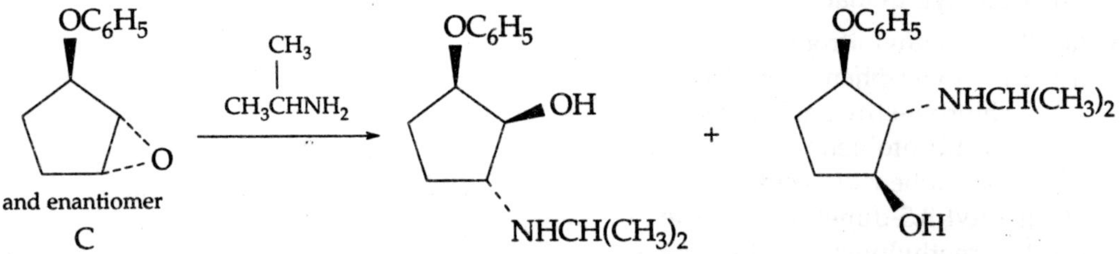
OC6H5
CH3
CH3CHNH2
OH
NHCH(CH3)2
+

and enantiomer C

and enantiomer E (major product)

and enantiomer (minor product)

48. (a) 2-methyl-1-propanamine isobutylamine
(b) *cis*-1-amino-2-ethylcyclopentane
(c) *N*,*N*-diethylbutanamine
(d) 3-cyclopropyl-1-propylamine
(e) 3-nitroaniline; *m*-nitroaniline
(f) 2,4-dibromoaniline

49. I. (a) 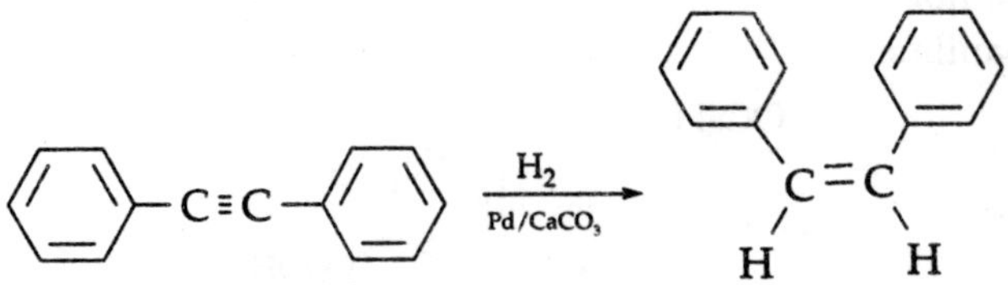
C≡C
H2
Pd/CaCO3
C=C
H
H

A

(b) $CH_2{=}CHBr \xrightarrow{HBr} CH_3CHBr_2$

B

(c) $CH_2{=}CHCH_2Br \xrightarrow[\text{carbon tetrachloride}]{Br_2} BrCH_2CHBrCH_2Br$

$\underline{C}$

(d) $C_6H_5{-}C{\equiv}CCH_2CH_3 \xrightarrow[NH_3\,(liq)]{Na}$ $C_6H_5(H)C{=}C(H)CH_2CH_3$ (trans)

$\underline{D}$

(e) 1-ethynylcyclohexanol (OH, $-C{\equiv}CH$) $\xrightarrow[\text{Pd/CaCO}_3\text{, quinoline}]{H_2}$ 1-vinylcyclohexanol (OH, $-CH{=}CH_2$)

$\underline{E}$

(f) $CH_3(CH_2)_7C{\equiv}C(CH_2)_7\overset{O}{\overset{\|}{C}}OH \xrightarrow[\text{carbon tetrachloride}]{O_3} \xrightarrow{H_2O} CH_3(CH_2)_7\overset{O}{\overset{\|}{C}}OH$

$\underline{G}$

$+ HO\overset{O}{\overset{\|}{C}}(CH_2)_7\overset{O}{\overset{\|}{C}}OH$

(g) $CH_3\overset{CH_3}{\overset{|}{C}}HCH_2C{\equiv}C\overset{CH_3}{\overset{|}{C}}HCH{=}CH\overset{O}{\overset{\|}{C}}OCH_3$ (trans: H and H on opposite sides) $\xrightarrow[\text{Pd/CaCO}_3\text{, quinoline}]{H_2}$

$\underline{H}$

$CH_3\overset{CH_3}{\overset{|}{C}}HCH_2CH{=}CH\overset{CH_3}{\overset{|}{C}}HCH{=}CH\overset{O}{\overset{\|}{C}}OCH_3$ (new C=C cis: both H on same side; C=C bearing $COCH_3$ trans)

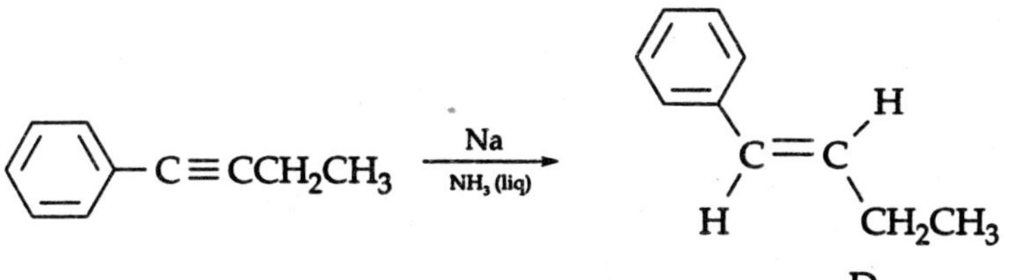

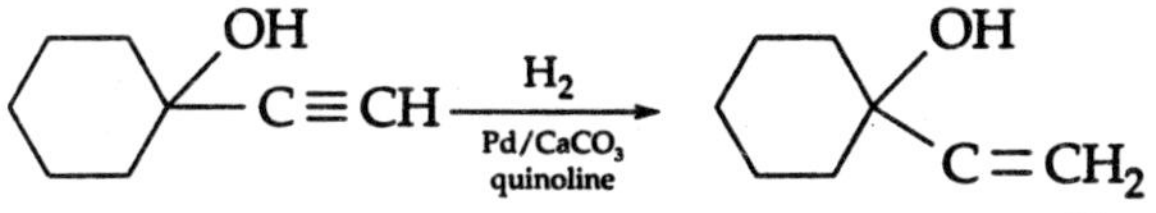

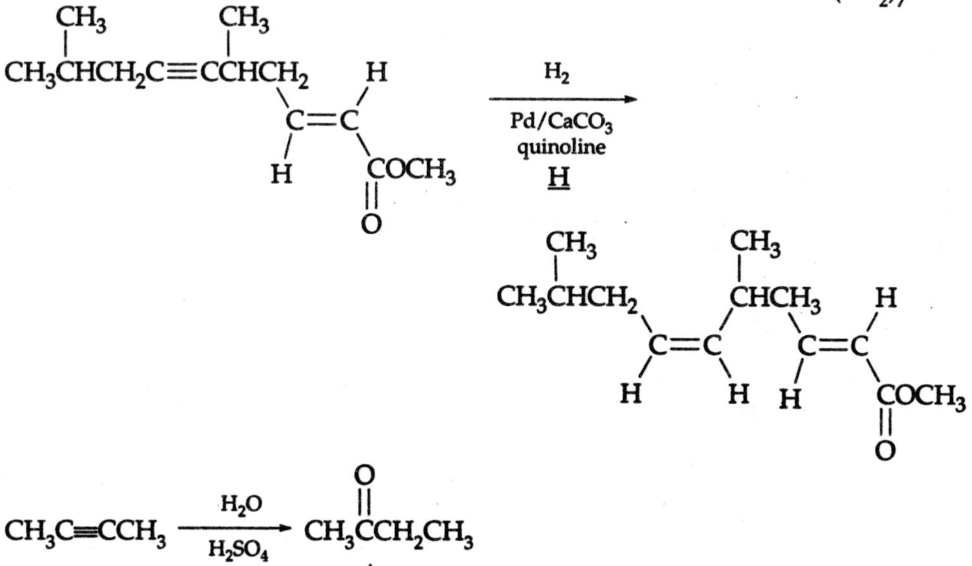

II. (a) $CH_3C{\equiv}CCH_3 \xrightarrow[H_2SO_4]{H_2O} CH_3\overset{O}{\overset{\|}{C}}CH_2CH_3$

$\underline{A}$

(b) tetrahydropyran-2-yl $-OCH_2C{\equiv}CH$ $\xrightarrow[NH_3(liq)]{NaNH_2}$ $\underline{B}$ $\xrightarrow{BrCH_2CH_2CH_2Cl}$ $\underline{C}$

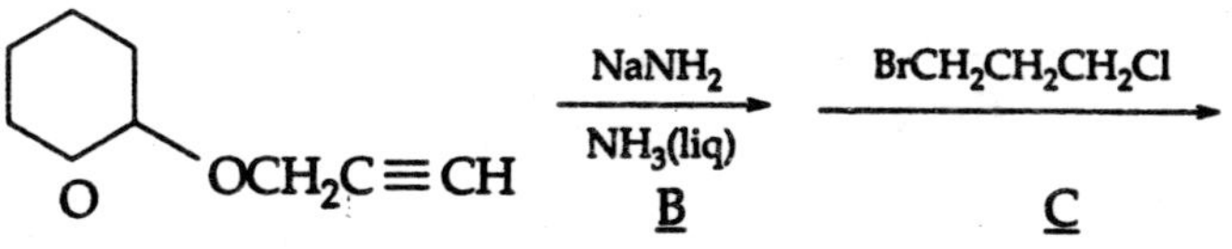

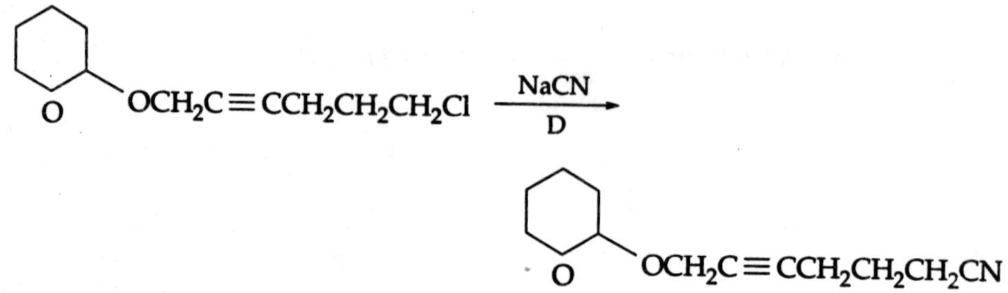

(c) $CH_2{=}CHCH_2CH_2Br \xrightarrow[NH_3(liq)]{HC{\equiv}C{:}\ Na^+} CH_2{=}CHCH_2CH_2C{\equiv}CH$

(d) 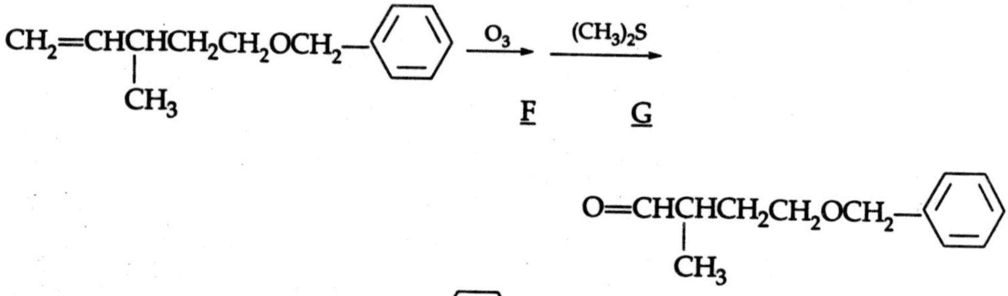

(e) 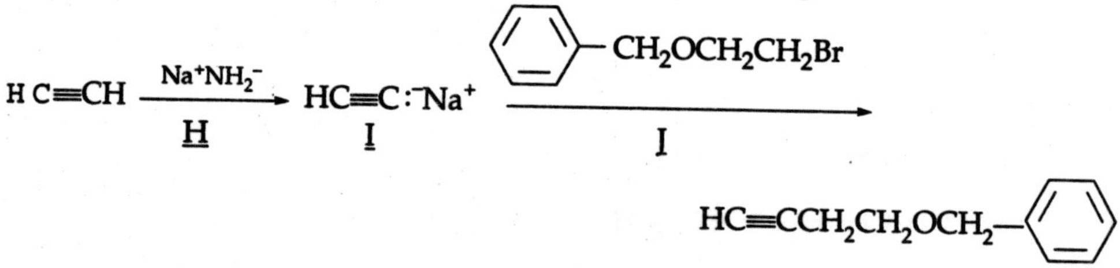

(f) 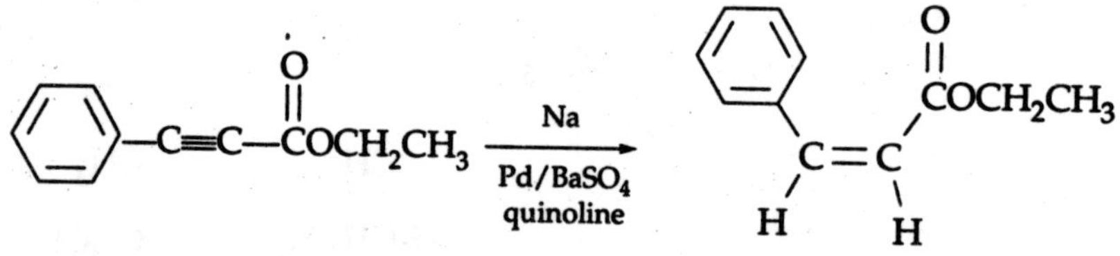

K

50. (a) $CH_3CH_2CH_2\overset{\overset{\large O}{||}}{C}CH_3 \xrightarrow[H_2O]{NaBH_4} CH_3CH_2CH_2\underset{\underset{\large OH}{|}}{C}HCH_3$

(b) 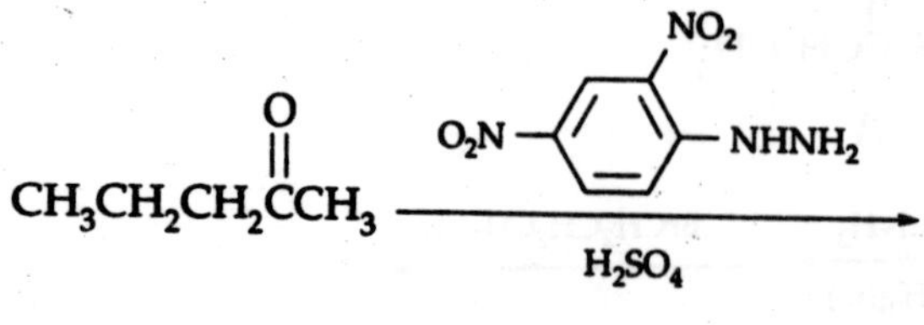

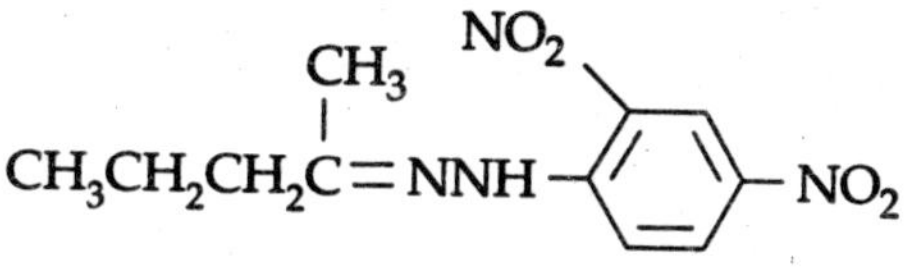

(c) 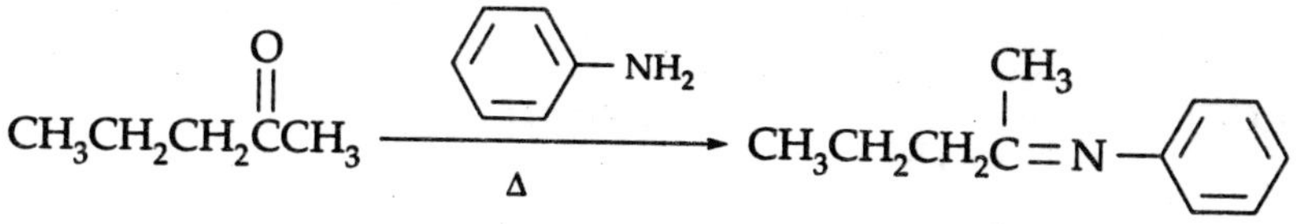

(d) 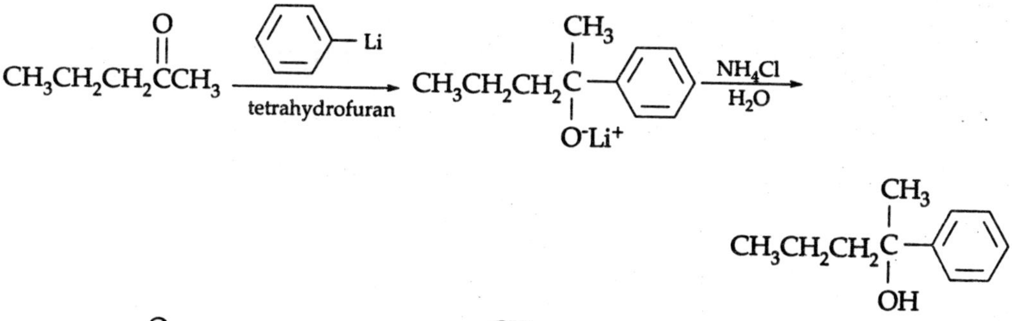

(e) 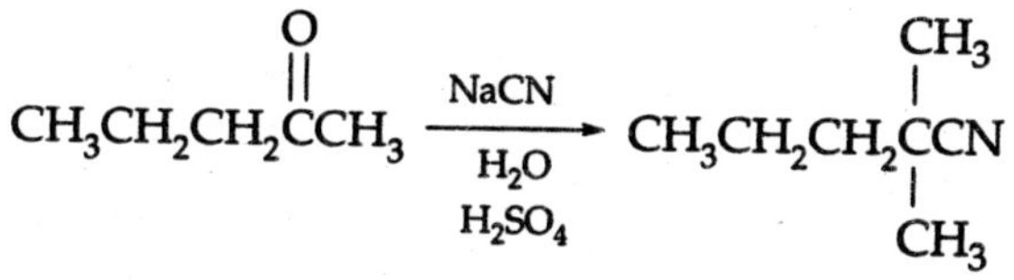

(f) 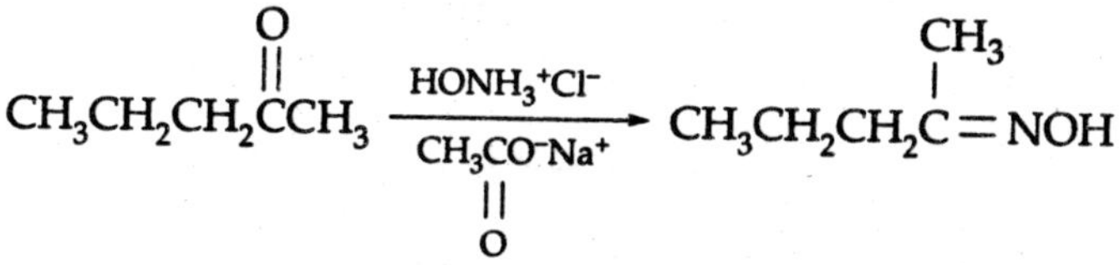

(g) 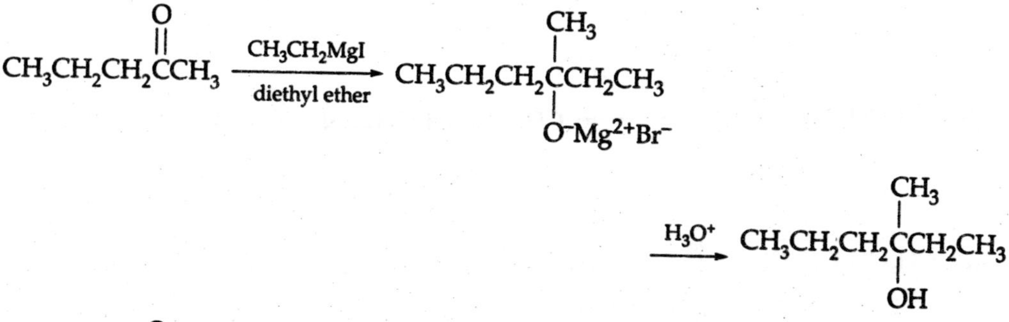

(h) 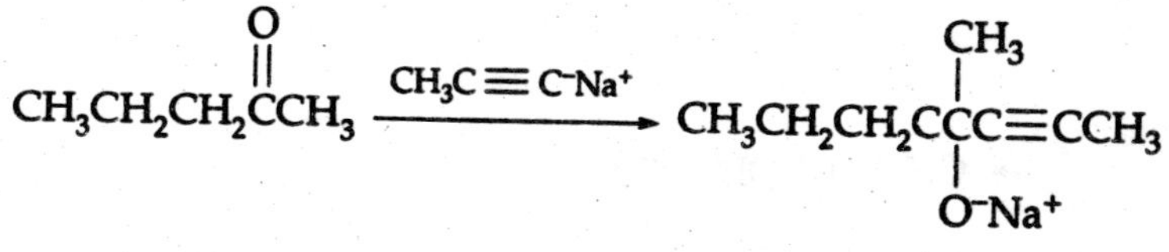

$$\xrightarrow{NH_4Cl} CH_3CH_2CH_2\underset{\underset{OH}{|}}{\overset{\overset{CH_3}{|}}{C}}C{\equiv}CCH_3$$

(i) $$CH_3CH_2CH_2\overset{\overset{O}{\|}}{C}CH_3 \xrightarrow[\text{diethylene glycol},\ \Delta]{H_2NH_2,\ KOH} CH_3CH_2CH_2CH_2CH_3$$

(j) $$CH_3CH_2CH_2\overset{\overset{O}{\|}}{C}CH_3 \xrightarrow[\Delta]{HCl,\ Zn(Hg)} CH_3CH_2CH_2CH_2CH_3$$

51. (a) $$CH_3CH_2CH_2\overset{\overset{O}{\|}}{C}H \xrightarrow[H_2O]{NaBH_4} CH_3CH_2CH_2CH_2OH$$

(b)

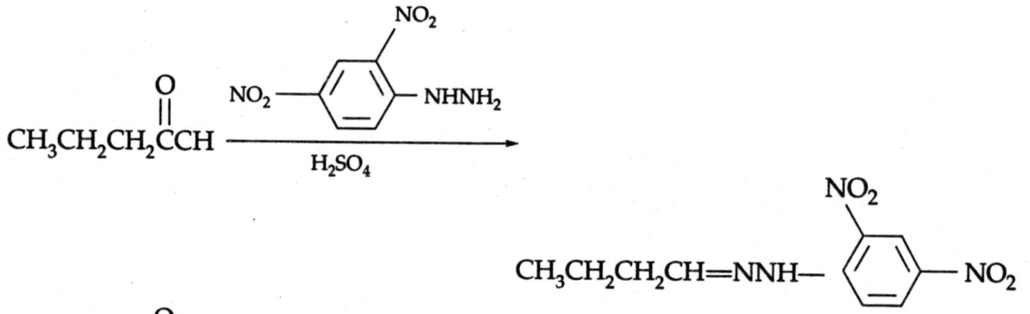

(c) $$CH_3CH_2CH_2\overset{\overset{O}{\|}}{C}H \xrightarrow{CH_3CH_2CH_2CH_2Li} \xrightarrow{H_3O^+} CH_3CH_2CH_2\underset{\underset{OH}{|}}{C}HCH_2CH_2CH_2CH_3$$

(d)

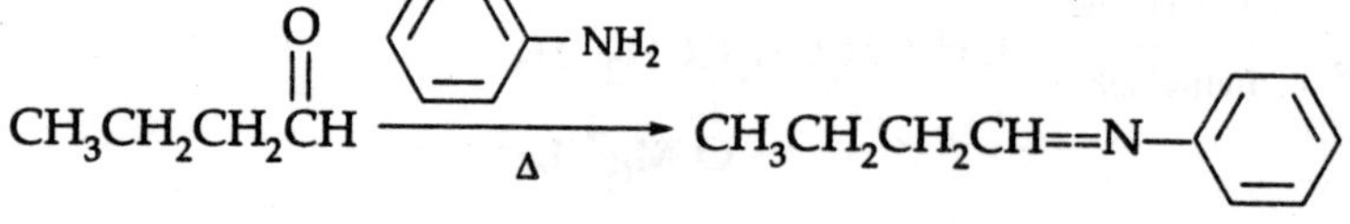

(e) $$CH_3CH_2CH_2\overset{\overset{O}{\|}}{C}H \xrightarrow[H_2O,\ H_2SO_4]{NaCN} CH_3CH_2CH_2\underset{\underset{OH}{|}}{C}HCN$$

(f) $$CH_3CH_2CH_2\overset{\overset{O}{\|}}{C}H \xrightarrow[H_3O^+,\ H_2O]{Na_2Cr_2O_7} CH_3CH_2\overset{\overset{O}{\|}}{C}OH + Cr^{+3}$$

(g) $$CH_3CH_2CH_2\overset{\displaystyle O}{\overset{\|}{C}}H \xrightarrow[CH_3\overset{\|}{\underset{O}{C}}O^-Na^+]{HONH_3{}^+Cl^-} CH_3CH_2CH_2CH{=}NOH$$

(h) $$CH_3CH_2CH_2\overset{\displaystyle O}{\overset{\|}{C}}H \xrightarrow[\text{diethyl ether}]{CH_3MgI} CH_3CH_2CH_2\underset{O^-Mg^{2+}I^-}{\underset{|}{C}}HCH_3 \xrightarrow{H_3O^+} CH_3CH_2CH_2\underset{OH}{\underset{|}{C}}HCH_3$$

(i) $$CH_3CH_2CH_2\overset{\displaystyle O}{\overset{\|}{C}}H \xrightarrow[\text{diethyl ether}]{LiAlH_4} \xrightarrow{H_3O^+} CH_3CH_2CH_2CH_2OH$$

(j) $$CH_3CH_2CH_2\overset{\displaystyle O}{\overset{\|}{C}}H \xrightarrow[\text{diethylene glycol}]{H_2NNH_2,\ KOH} CH_3CH_2CH_2CH_3$$

(k) $$CH_3CH_2CH_2\overset{\displaystyle O}{\overset{\|}{C}}H \xrightarrow[\Delta]{HCL,\ Zn(Hg)} CH_3CH_2CH_2CH_3$$

52. 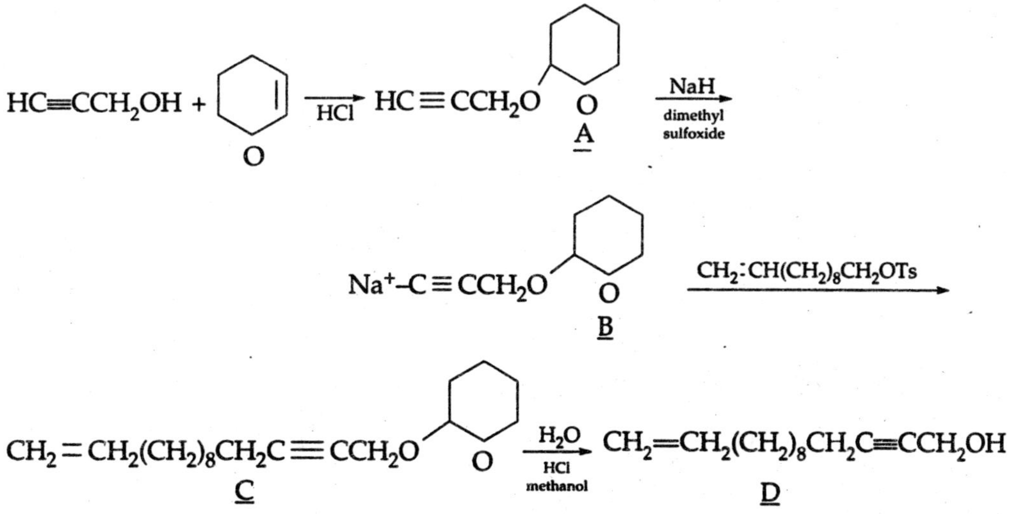

HC≡CCH₂OH +
O
HCl
HC≡CCH₂O
O
A
NaH
dimethyl sulfoxide
Na⁺-C≡CCH₂O
O
B
CH₂:CH(CH₂)₈CH₂OTs
CH₂=CH₂(CH₂)₈CH₂C≡CCH₂O
O
C
H₂O
HCl
methanol
CH₂=CH₂(CH₂)₈CH₂C≡CCH₂OH
D

53. (a) 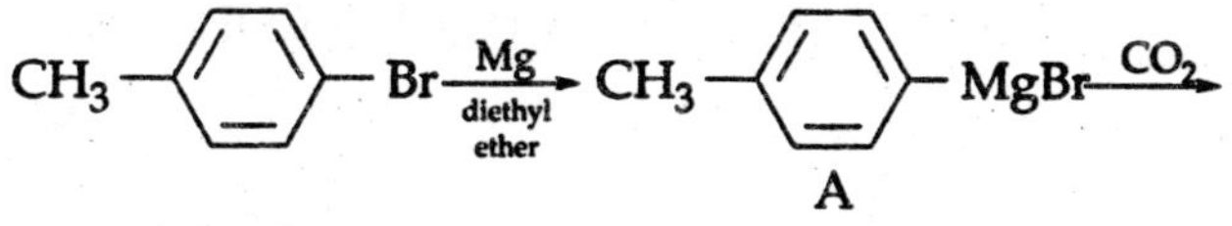

CH₃
Br
Mg
diethyl ether
CH₃
MgBr
CO₂
A

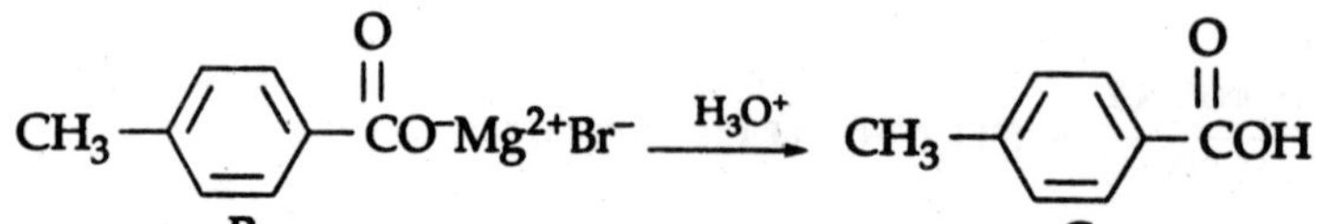

B C

(b) $CH_3CH_2CH_2CH(CH_3)CH_2CH_2OH \xrightarrow[H_2O,\ \Delta]{KMnO_4,\ NaOH} CH_3CH_2CH_2CH(CH_3)CH_2CO^-Na^+ \xrightarrow{H_3O^+}$

D

$CH_3CH_2CH_2CH(CH_3)CH_2COH$ (with C=O)

E

(c) 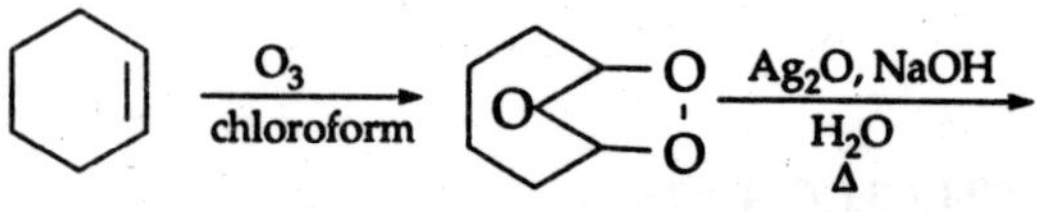

F

$Na^{+-}OC(CH_2)_4CO^-Na^+ \xrightarrow{H_3O^+} HOC(CH_2)_4COH$ (each C=O)

G H

(d) 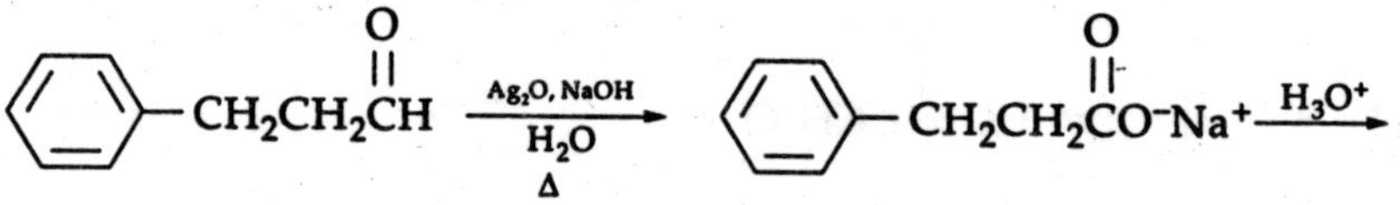

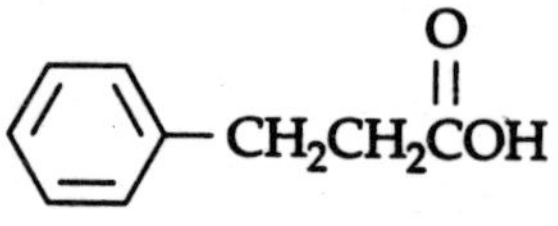

J

(e) $CH_3(CH_2)_5CH \xrightarrow[H_2SO_4,\ H_2O,\ \Delta]{CrO_3} CH_3(CH_2)_5COH$ (each C=O)

(f) $CH_3(CH_2)_9CH{=}CH_2 \xrightarrow[chloroform]{O_3} \xrightarrow{H_2O_2,\ NaOH} \xrightarrow{H_3O^+} CH_3(CH_2)_9COH + HCOH$ (each C=O)

L M

54. (a) 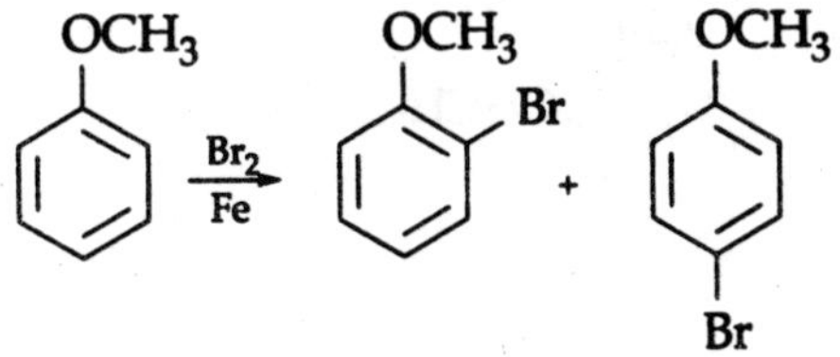

(b) 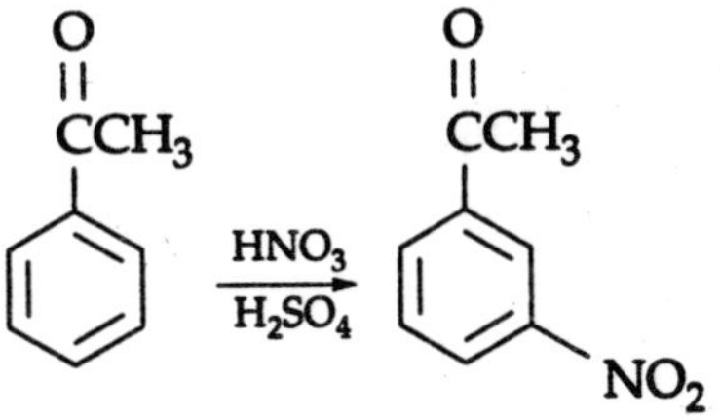

K

(c) 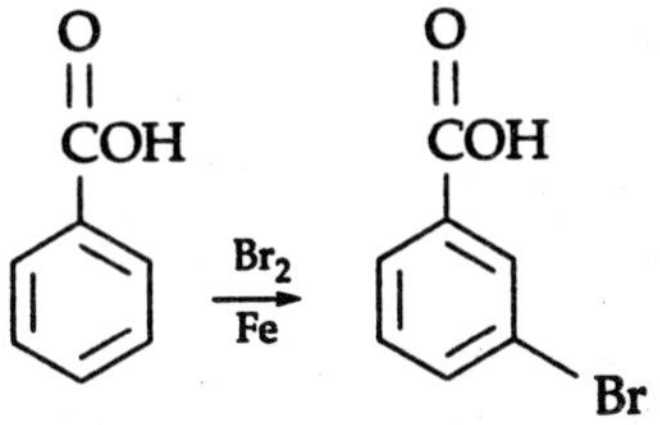

(d) 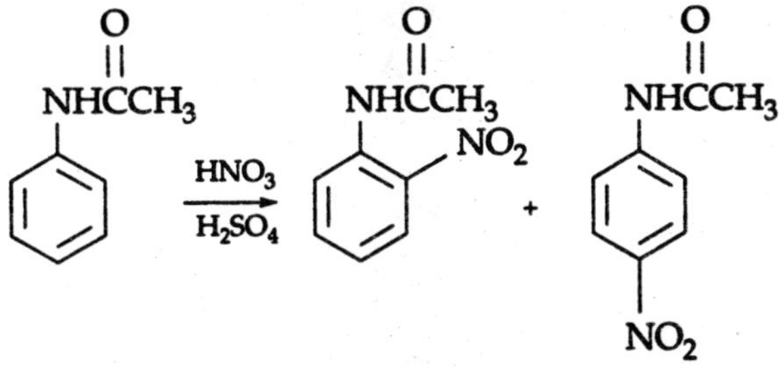

(e) 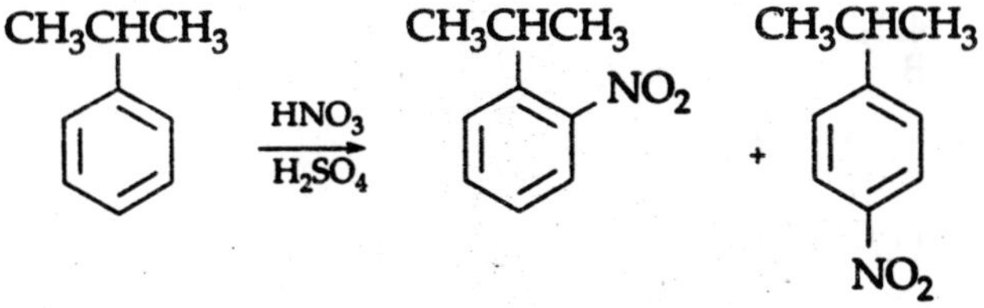

(f) 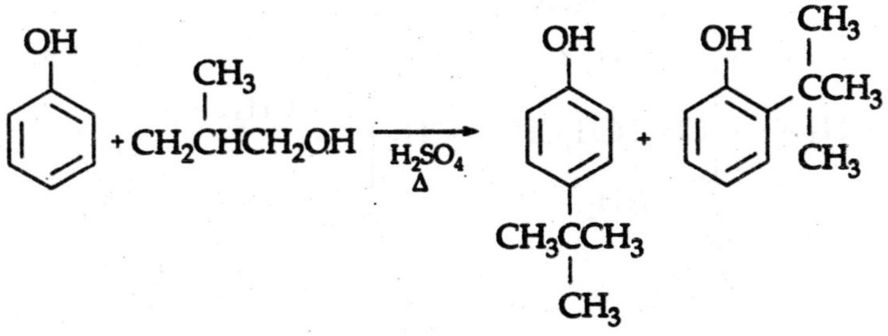

(g) Chlorobenzene $+ CH_3CH(CH_3)OH \xrightarrow[\Delta]{H_2SO_4}$ p-$ClC_6H_4CH(CH_3)_2$ (ring with Cl and CH_3CHCH_3) $+$ o-$ClC_6H_4CH(CH_3)CH_3$

(h) Benzene $+ CH_3CH_2CH{=}CH_2 \xrightarrow{H_2SO_4} C_6H_5CH(CH_3)CH_2CH_3$

(i) Benzene $+ CH_3C(CH_3)(Cl)CH_2CH_3 \xrightarrow[\Delta]{FeCl_3} C_6H_5C(CH_3)_2CH_2CH_3$

(j) Benzene $+ CH_3CH_2Br \xrightarrow{AlCl_3} C_6H_5CH_2CH_3$

55. (a) $C_6H_5CH{=}NC_6H_5 \xrightarrow[\text{diethyl ether}]{LiAlH_4} \xrightarrow{H_2O} C_6H_5CH_2NHC_6H_5$ (A)

(b) p-$HOOC{-}C_6H_4{-}CONH_2 \xrightarrow[H_2O]{Cl_2,\ NaOH} \xrightarrow{NH_3,\ pH\ 8}$ p-$Na^+\,{}^-OOC{-}C_6H_4{-}NH_2$ (B)

(c) $CH_3COCH{=}C(CH_3)CH_3 \xrightarrow[H_2O]{NH_3} CH_3COCH_2C(CH_3)(NH_2)CH_3$ (C)

(d) $CH_3C(CH_3)(NO_2)CH_2CH_2COOCH_3 \xrightarrow[\Delta]{H_2,\ Ni} \left[CH_3C(CH_3)(NH_2)CH_2CH_2COOCH_3\right] \longrightarrow$ D (5,5-dimethyl-pyrrolidin-2-one: ring with CH_3, CH_3, N–H, C=O) $+ CH_3OH$

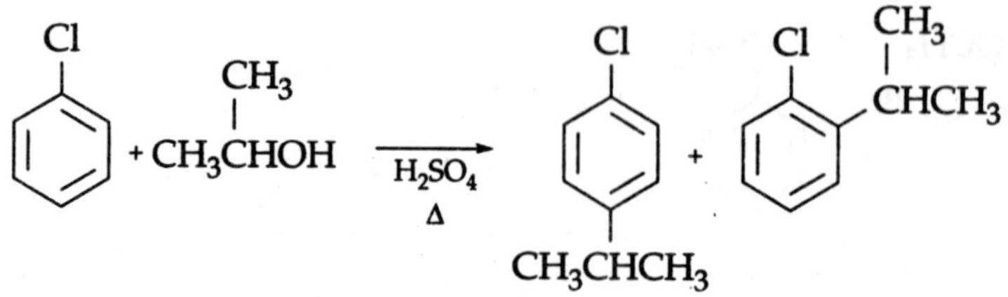

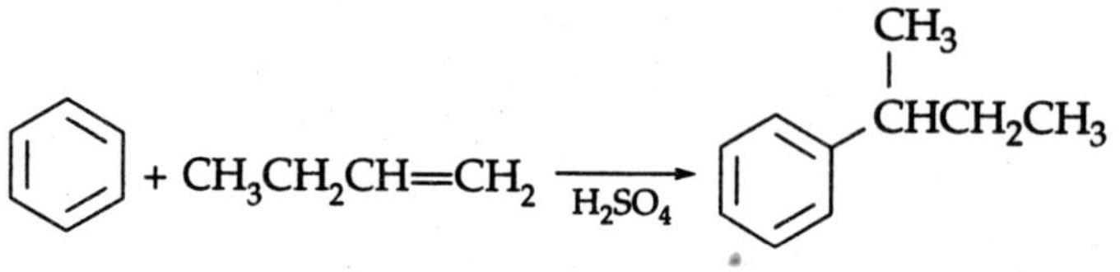

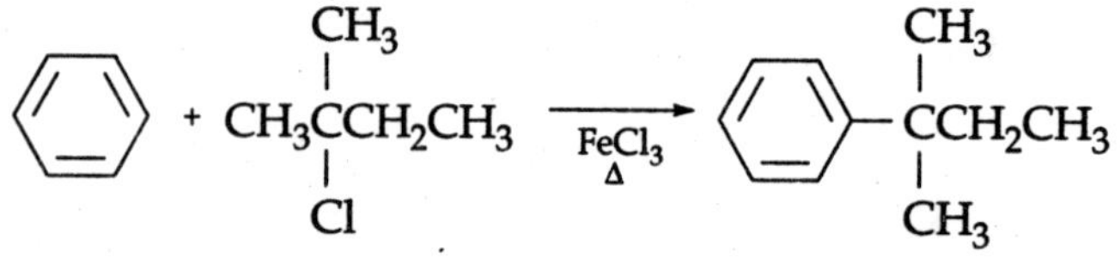

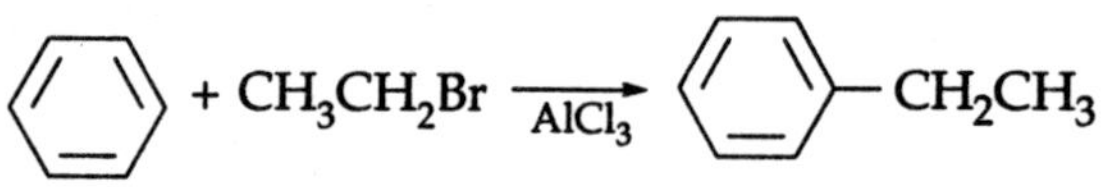

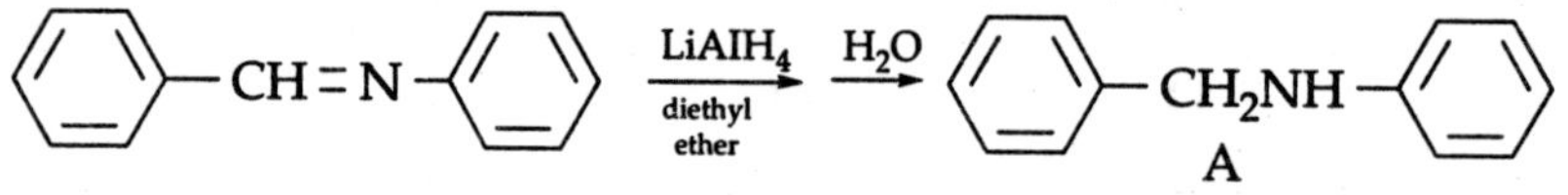

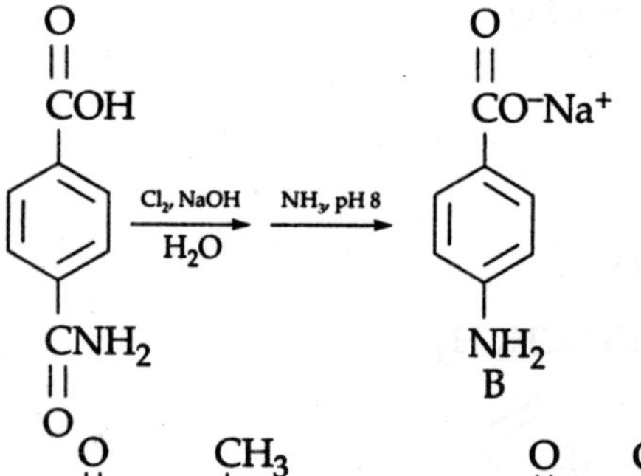

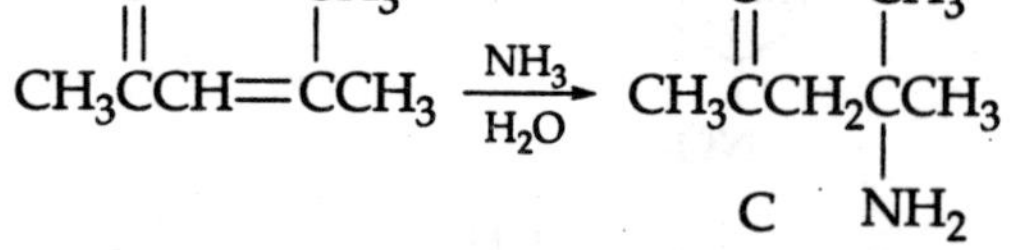

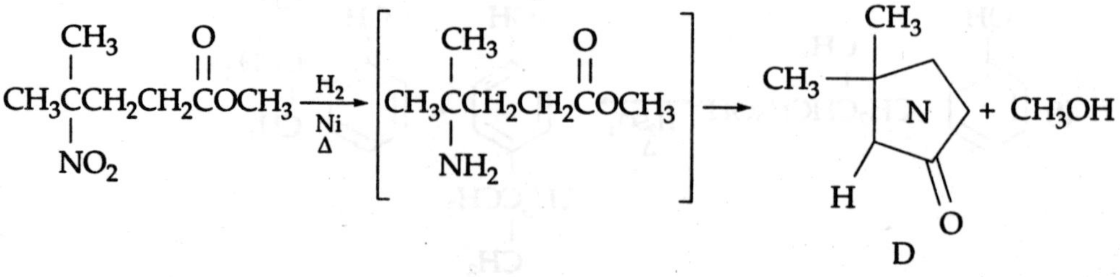

(e) 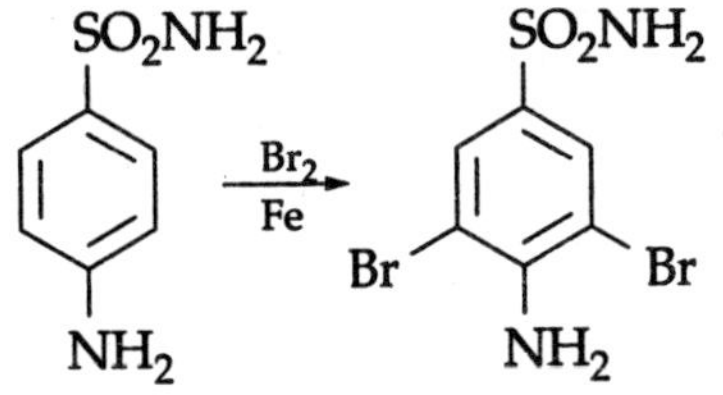

E

(f) $$CH_3CH_2\overset{O}{\overset{\|}{C}}CH_2CH_2CH_3 \xrightarrow[Na_2CO_3]{HO\overset{+}{N}H_3\ HSO_4^-}$$

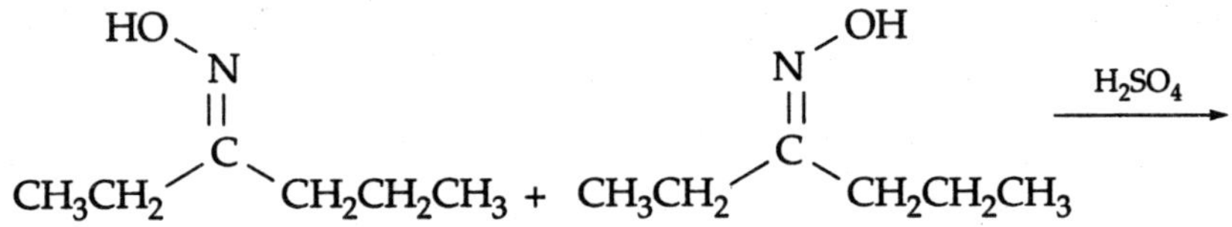

F G

$$\xrightarrow{H_2SO_4} CH_3CH_2\overset{O}{\overset{\|}{C}}NHCH_2CH_2CH_3 + CH_3CH_2CH_2\overset{O}{\overset{\|}{C}}NHCH_2CH_3$$

H I

(g) $$CH_3\overset{O}{\overset{\|}{C}}(CH_2)_4CH_3 \xrightarrow[Ni]{NH_3,\ H_2} \underset{NH_2}{CH_3\underset{|}{C}H(CH_2)_4CH_3}$$

J

(h) $$CH_3CH_2CH_2NO_2 \xrightarrow{\overset{O}{\overset{\|}{HCH}},\ NaOH} \underset{NO_2}{CH_3CH_2\underset{|}{C}HCH_2OH} \xrightarrow{H_2SO_4,\ Fe}$$

K

$$\underset{^+NH_3\ HSO_4^-}{CH_3CH_2\underset{|}{C}HCH_2OH} \xrightarrow{Ca(OH)_2} \underset{NH_2}{CH_3CH_2\underset{|}{C}HCH_2OH}$$

L M

56. (a) 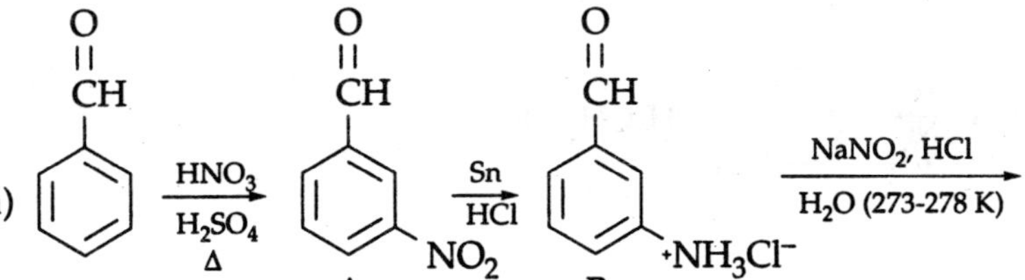

A B

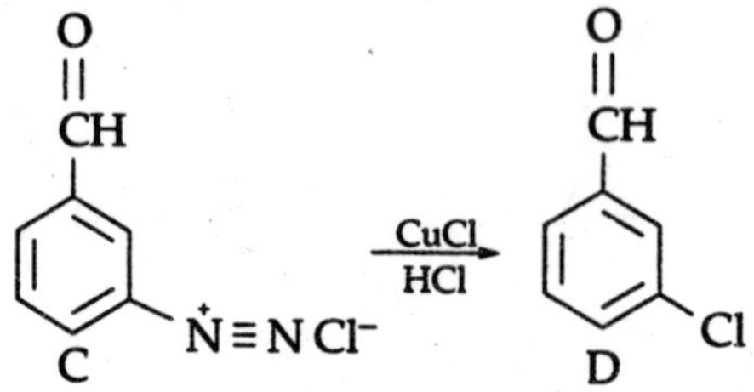

CuCl
HCl
C
D

(b)

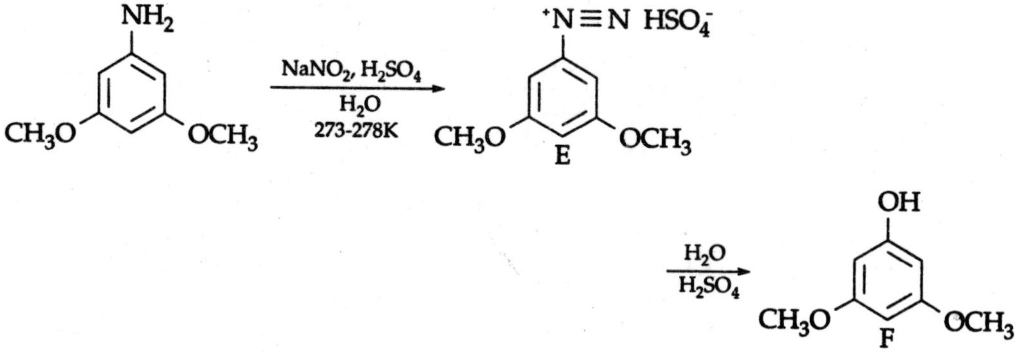

NaNO₂, H₂SO₄
H₂O
273-278K
E
H₂O
H₂SO₄
F

(c)

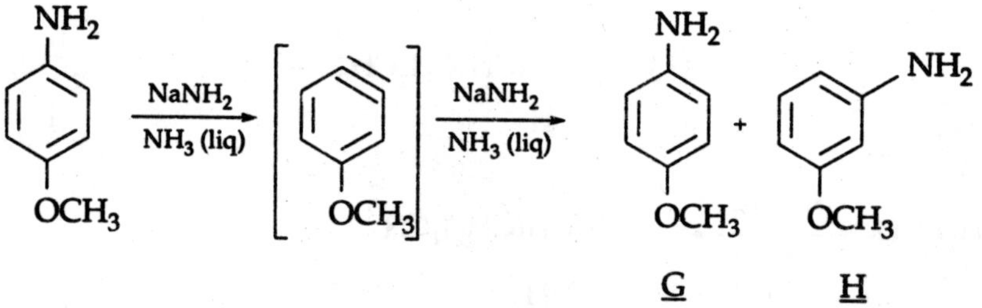

NaNH₂
NH₃ (liq)
NaNH₂
NH₃ (liq)
G
H

(d)

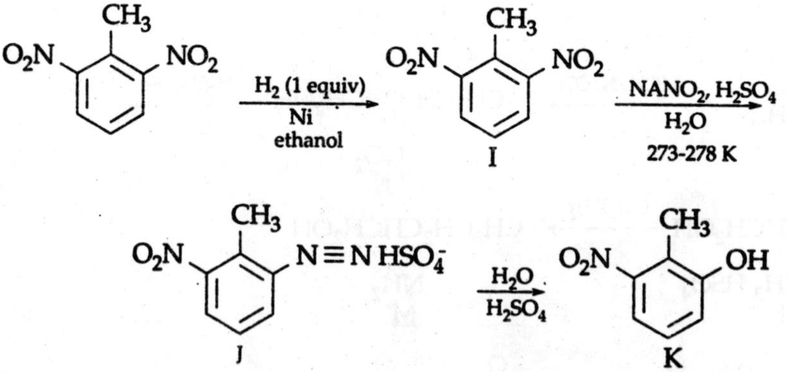

H₂ (1 equiv)
Ni
ethanol
I
NANO₂, H₂SO₄
H₂O
273-278 K
J
H₂O
H₂SO₄
K

(e)

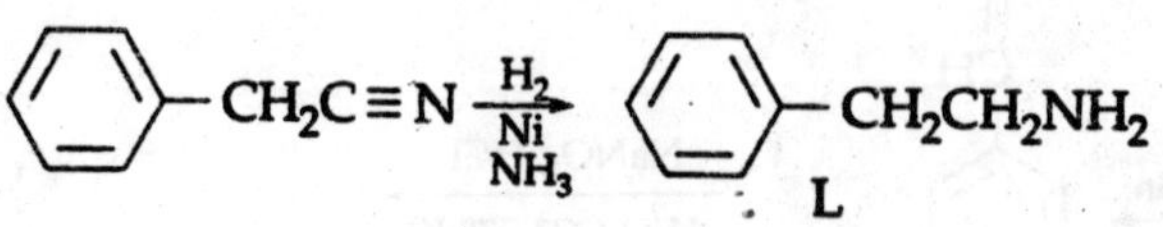

H₂
Ni
NH₃
L

(f) Cyclohexanone $\xrightarrow[\text{sodium acetate, } H_2O]{HO\overset{+}{N}H_3\ Cl^-}$ cyclohexanone oxime (NOH) **M** $\xrightarrow[\text{ethanol}]{Na}$ cyclohexylamine (NH_2) **N**

(g) 2-aminobenzoic acid (COOH, NH_2) $\xrightarrow[H_2O]{NANO_2,\ HCl}$ 2-carboxybenzenediazonium chloride ($\overset{+}{N}{\equiv}N\ Cl^-$) **O** $\xrightarrow{C_6H_5N(CH_3)CH_3}$ COOH–C_6H_4–N=N–C_6H_4–$N(CH_3)CH_3$ **P** methyl red

(h) C_6H_5–N(CH_3)–C_6H_5 $\xrightarrow[H_2O,\ 273\text{-}278\ K]{NaOH_2,\ HCl}$ C_6H_5–N(CH_3)–C_6H_4–NO **Q**

(i) C_6H_5–CH_2Br $\xrightarrow[\text{dimethyl-formamide}]{NaNO_2}$ C_6H_5–CH_2NO_2 **R**

(j) O_2N–C_6H_3(OH)–$COCH_2CH_3$ (C=O) $\xrightarrow[\text{Ni, ethyl ectate}]{H_2}$ H_2N–C_6H_3(OH)–$COCH_2CH_3$ (C=O) **S**

57. (a) C_6H_5–CH_2COOH $\xrightarrow{CH_3Li\ \text{(2 molar equiv)}}$ C_6H_5–CH_2–$C(O^-Li^+)_2CH_3$ $\xrightarrow{H_2O}$

C_6H_5–CH_2–$\overset{O}{\overset{\|}{C}}CH_3$ **A** $\xrightarrow[\text{diethyl ether}]{LiAlH_4}$ $\xrightarrow{H_3O^+}$ C_6H_5–CH_2–CH(OH)CH_3 **B**

(b) C_6H_5–CH_2–CH(OH)CH_3 **B** $\xrightarrow[\text{pyridine}]{TsCl}$ C_6H_5–CH_2–CH(OTs)CH_3 **C** $\xrightarrow{HCOOH}$

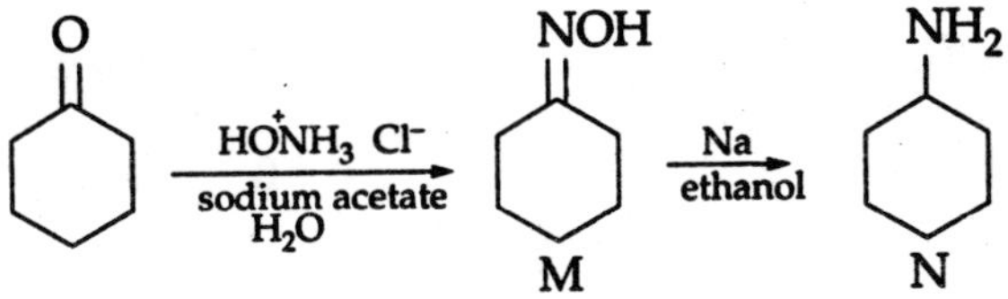

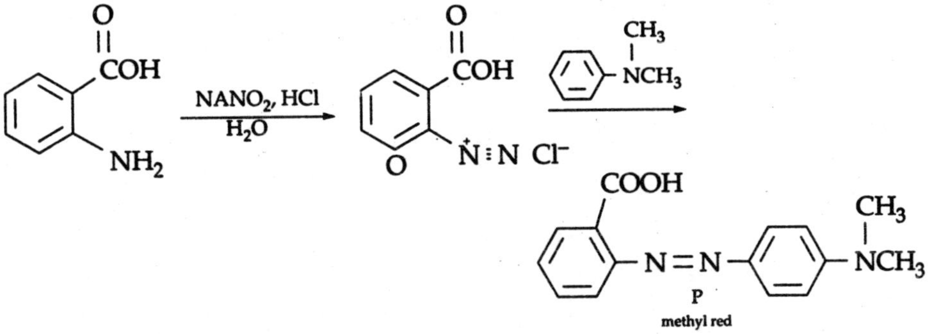

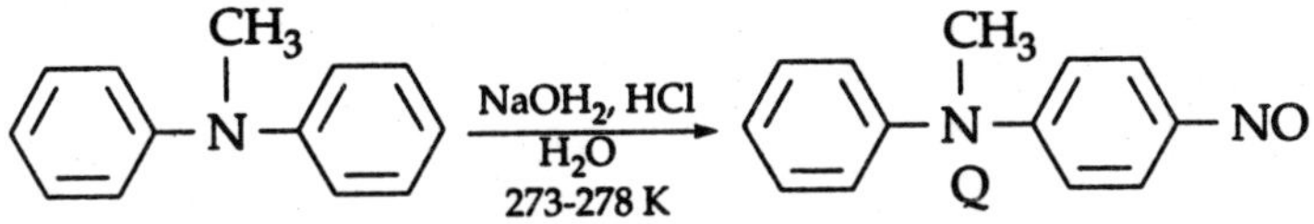

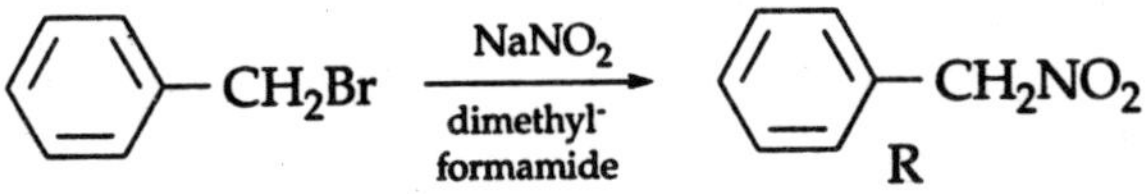

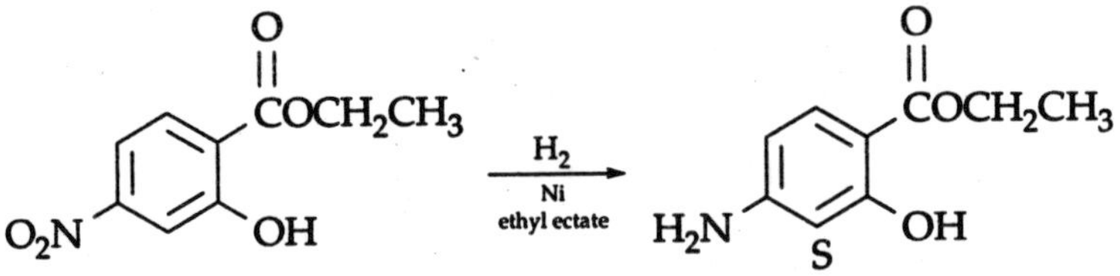

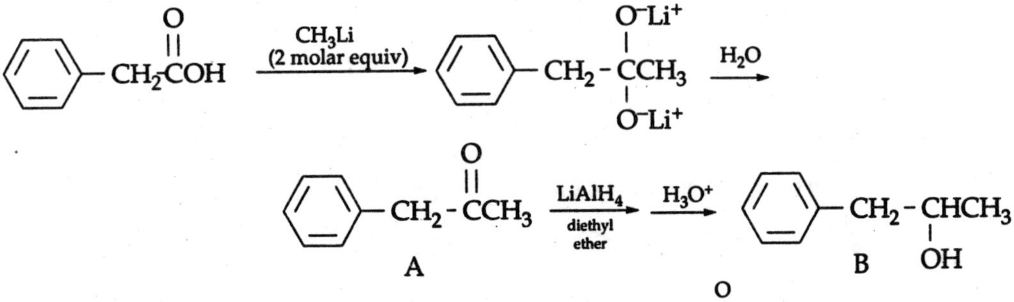

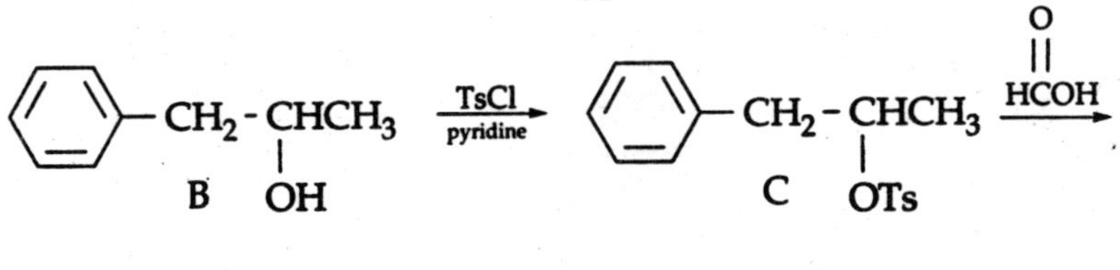

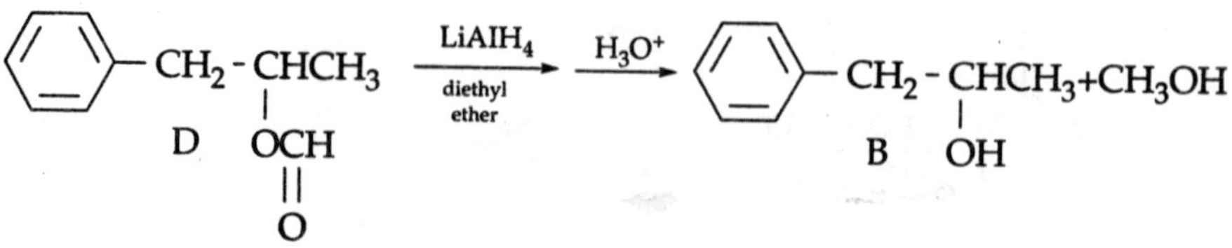
CH2 - CHCH3
D OCH
O
LiAlH4
diethyl
ether
H3O+
CH2 - CHCH3+CH3OH
B OH

(c) 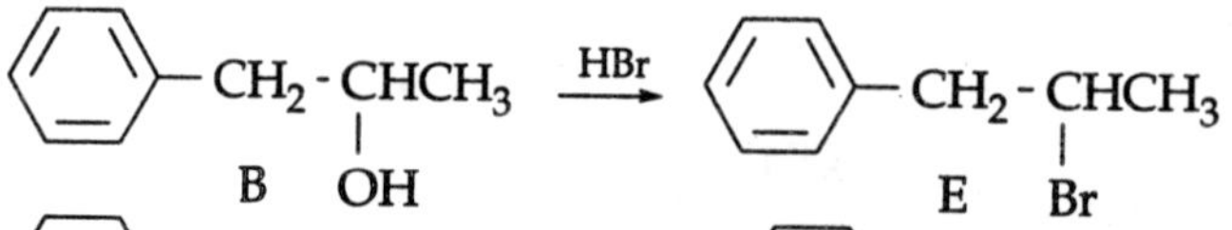
CH2 - CHCH3
B OH
HBr
CH2 - CHCH3
E Br

(d) 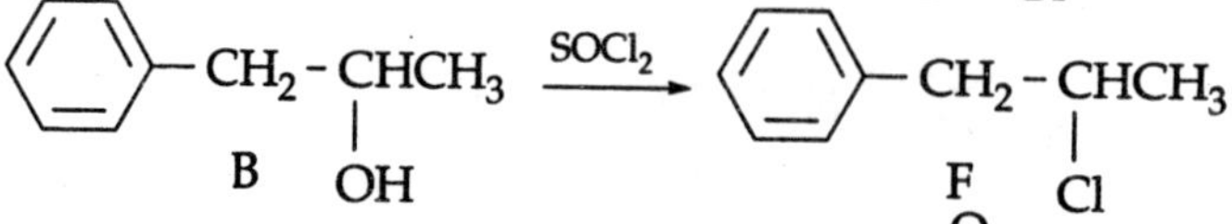
CH2 - CHCH3
B OH
SOCl2
CH2 - CHCH3
F Cl

(e) 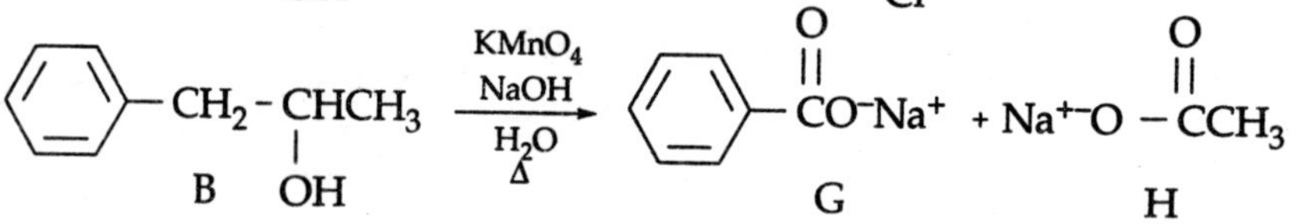
CH2 - CHCH3
B OH
KMnO4
NaOH
H2O
Δ
O
CO-Na+
G
+ Na+-O - CCH3
O
H

Most acidic Least acidic

58. (a) 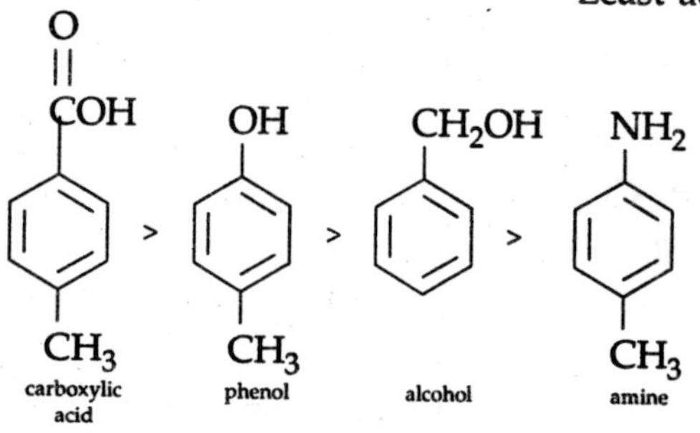
COH
O
OH
CH2OH
NH2
CH3
carboxylic acid
CH3
phenol
alcohol
CH3
amine

(b) 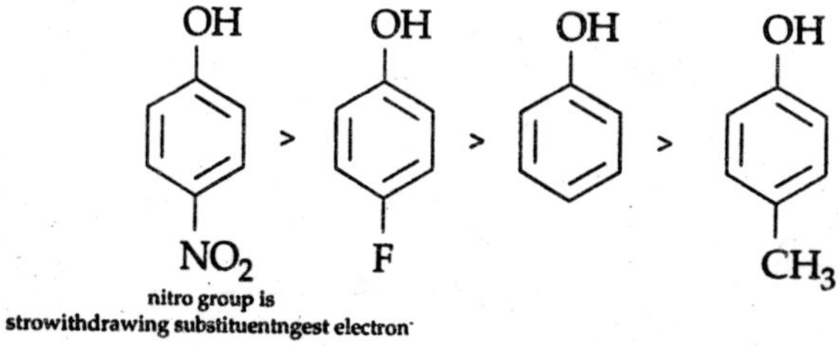
OH
OH
OH
OH
NO2
nitro group is
strowithdrawing substituentngest electron
F
CH3

(c) 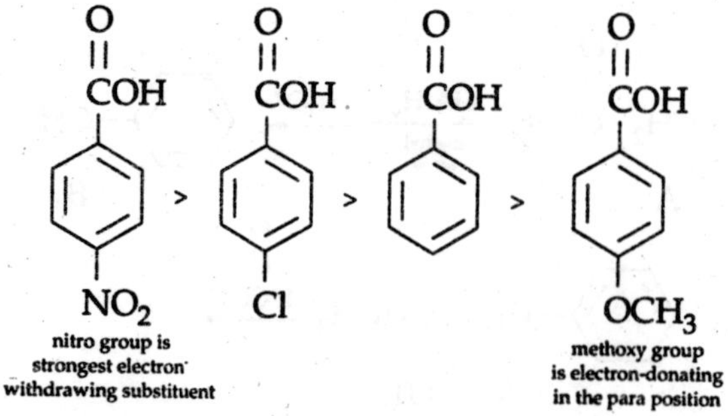
O
COH
O
COH
O
COH
O
COH
NO2
nitro group is
strongest electron
withdrawing substituent
Cl
OCH3
methoxy group
is electron-donating
in the para position

59. The linkage shown is an amide linkage. Glycogen, (b), is a starch and contains only C, H, and O. Of the other substances, (a), nylon, is a synthetic polyamide, (c), protein, is a naturally occurring polyamide (polypeptide), (d), valylglycine, is a dipeptide.

60. This is the only choice for which both carbons flanking that double bond have two different substituents. For the others, both substituents are the same on at least one of the carbons, thus precluding cis-trans isomerism.

61. This is a matter of definition.

62. (a) Compound (a) is methyl ethyl ether. Ethers tend to be less soluble than the other classes of compounds exemplified. Compound (b) is an alcohol, (c) a ketone, (d) an aldehyde. These molecules are sufficiently polar to be water soluble.

63. The longest chain has five carbons, so the compound must be a pentane. We number substituents so as to give the first one encountered the lowest possible number. In this case that means starting from the right. The second and third carbons in the chain then have methyl group substituents, leading to the name 2,3-dimethylpentane. The compound has been redrawn below, including the numbering, so that the name is immediately apparent.

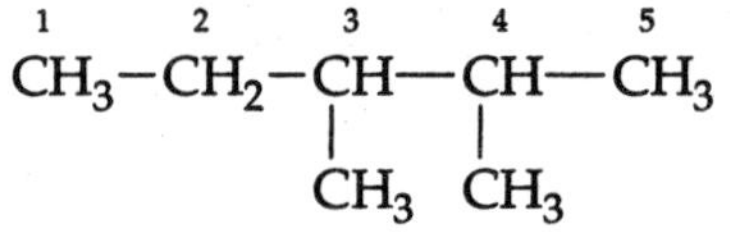

64. This is a definition. Hetero-is a combining form meaning "other than usual." In this context it means "other than carbon."A cyclic compound contains a ring of atoms.

65. None of the compounds has a tetrahedral carbon with four different substituents (a usually sufficient not necessary condition for optical isomerism), and none is non-superimposable on its mirror-image (the necessary condition).

66. The longest chain containing the double bond has four carbons; therefore the compound is a butene. The carbon chain is numbered so that the double-bonded carbons receive the lowest possible numbers, so this is a 1-butene; then the methylgroup is on the second carbons. The correct name is therefore 2-methyl-1-butene.

67. Note that what is given in parentheses has two—(CH_2CCl_2)—units, so that —(CH_2CCl_2)— -is really the basic repeating unit. The polymer has two connected carbons, one attached to two H's, one attached to two Cl's, and (a) is the only choice which also exhibits this property.

68. Compound II is the only alkene; alkenes have reactive double bonds and react with HBr.

69. The longest chain has five carbons; if you start numbering from either ethyl group the methyl substituent is on carbon number 3.

70. Compounds III and IV are both $C_5H_{10}O$.

71. Only trimethylamine, the compound in (c), has a nitrogen with three carbons attached to it. All three H atoms of NH_3 have been replaced by $—CH_3$.

72. Only compound (b) has a tetrahedral carbon bearing four different substituents.

73. The carboxyl group carbon, $-\overset{\overset{\displaystyle O}{||}}{C}-OH$ gets no contribution to its oxidation number from the methyl group to which it is attached. Of the other atoms in this group, the two oxygens have oxidation number -2 each, and the H has +1, so this C must have +3.

74. Only compounds I, III and IV exhibit the pattern of three alternating single and double bonds in a ring, which is the simplest way of depicting aromaticity.

75. This is a definition.

76. The structure of 2-butene is $CH_3-CH=CH-CH_3$. The addition of HBr to the double bond yields $CH_3-CH_2-CHBr-CH_2$. The carbon chain is numbered from the end closer to the Br, in order to give the lower numbering scheme (2-bromobutane, not 3(151) bromobutane).

77. If you forget that amylose is a starch, you might remember that nylon is a polyamide and that ribose is a monosaccharide. Compounds (c) is from its name, not a saccharide.

78. (i) (*A*) gives Lucas test within 5 minutes and thus, (*A*) is secondary alcohal.
(*A*) contains one O atom and thus, only one OH gp in (*A*). Thus, (*A*) is C_nH_{2n+1} OH.
∵ 1120 mL H_2 is given by the action of Na over 6 g (*A*).

$$\therefore \quad 11200 \text{ mL } H_2 \text{ is given by } \frac{6 \times 11200}{1120} = 60 \text{ g}$$

$$\therefore \quad \text{M. wt. of } (A) = 60$$

Since one mole of an alcohal having one OH gp gives 11200 mL H_2 at STP with Na.
Thus, $C_nH_{2n+1}OH = 60$

$\therefore \quad n = 3$

or (*A*) is C_3H_7OH; being secondary alcohal its formula is $CH_3{\cdot}CHOH{\cdot}CH_3$

(ii) $\underset{\substack{(A)\\ \text{(Propan-2-ol)}}}{CH_3{\cdot}CHOH{\cdot}CH_3} \xrightarrow{PBr_3} \underset{\substack{(B)\\ \text{2-bromopropane}}}{CH_3CHBr{\cdot}CH_3}$

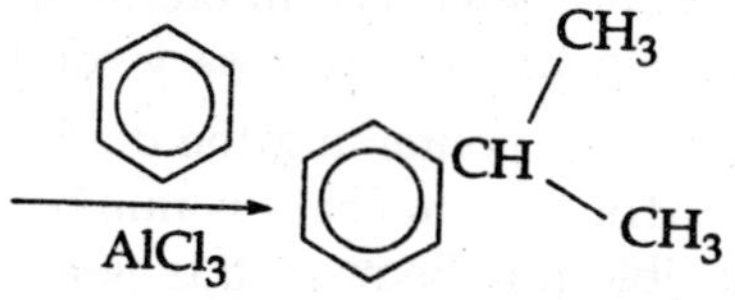

(C) 2-phenylpropane

This is Friedel Craft's reaction.

79. Compound (empirical formula CH_2) decolourizes Br_2 water, reacts with alk, $KMnO_4$ and adds one mole of H_2 and so (*B*) is alkene.
Compound (*B*) is obtained from (*A*) by the action of H_2SO_4 and so (*A*) is alcohal.
A has b.pt. 351 k so (*A*) is ethanol.

(i) $\underset{\substack{\text{Ethanol}\\ \text{(A)}}}{C_2H_5OH} \xrightarrow[-H_2O]{\text{conc. } H_2SO_4} \underset{\substack{\text{Ethene}\\ \text{(B)}}}{C_2H_4}$

(ii) $\underset{(B)}{C_2H_4} + Br_2 \text{ water} \rightarrow \text{decolourizes } (C_2H_4Br_2)$

$\underset{(B)}{C_2H_4} + KMnO_4 \rightarrow \text{decolourizes } \underset{\text{Pot. formate}}{(HCOOK)}$

(iii) $C_2H_4 + H_2 \rightarrow C_2H_6$

80. Given,

$A(\text{C, H, O}) \xrightarrow{\text{Na}}$ evolution of H_2 and so (A) is alcohal.

$A \xrightarrow{\text{O}} B \xrightarrow{\text{NaHSO}_3}$ Addition Product

$B \downarrow$ Tollen's regent

negative test

So, (B) is a carbonyl compound as it forms addition product with $NaHSO_3$. Also does not give silver mirror with Tollen's reagent and so, (B) is ketone.
Molecular weight of (A)

$$m = \frac{100 \times K \times w}{\Delta T \times W} = \frac{100 \times 18.58 \times 7.41}{1.85 \times 1000} = 74.1$$

Note: Although unit of K is not given, it is assumed in K mol^{-1} 100 g solvent since the given value of K stands of K mol^{-1} 100 g solvent.
Thus, (A) being R—OH

$\therefore$ M. wt. of R = M. wt. of (R—OH) – M. wt. of (—OH)
= 74.1 – 17 = 57

$\therefore$ $C_nH2_{2n+1} = 57$
$12n + 2n + 1 = 57$

$\therefore$ $n = 4$

Thus, (A) is C_4H_9OH
Since (A) gives ketone and thus, it is secondary alcohal, i.e., $CH_3{\cdot}CHOH{\cdot}CH_2{\cdot}CH_3$ or Butan-2-ol.

Thus we have,

(i) $CH_3{\cdot}CHOH{\cdot}CH_2{\cdot}CH_3 \xrightarrow{\text{Na}} CH_3{\cdot}CH(ONa){\cdot}CH_2{\cdot}CH_3 + 1/2\ H_2$

(A)

(ii)

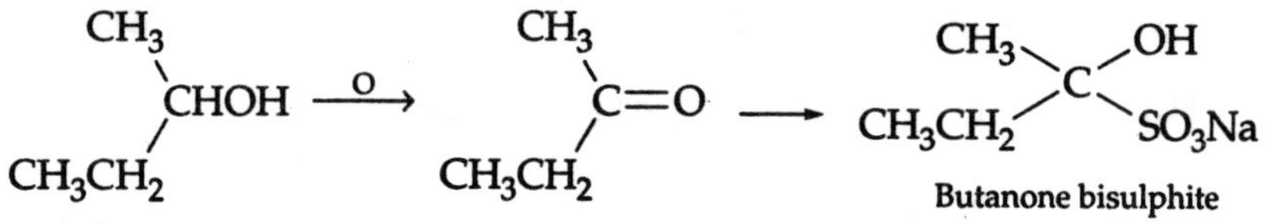

(A) (B) Butaone — Butanone bisulphite

81. (i) $ROH + CH_3MgI \rightarrow CH_4 + Mg(OR)(I)$

$\because$ 11.2 mL gas CH_4 is obtained by 0.037 g ROH

$\therefore$ 22400 mL gas is obtained by $\frac{0.037 \times 22400}{11.2} = 74$ g

$\therefore$ Molecular weight of $ROH = 74$

(ii) Since ROH on oxidation gives acid of same number of carbon atoms and thus, ROH is primary alcohol, i.e., C_nH_{2n+1} CH_2OH.

$\therefore$ $$12n + 2n + 1 + 12 + 2 + 16 + 1 = 74$$

$\Rightarrow$ $$n = 3$$

$\therefore$ Alcohol is $C_3H_7CH_2OH$

The alcohol on dehydration gives an alkene which on ozonolysis gives acetone as one product (for 3 carbon atoms) and thus, alkene is

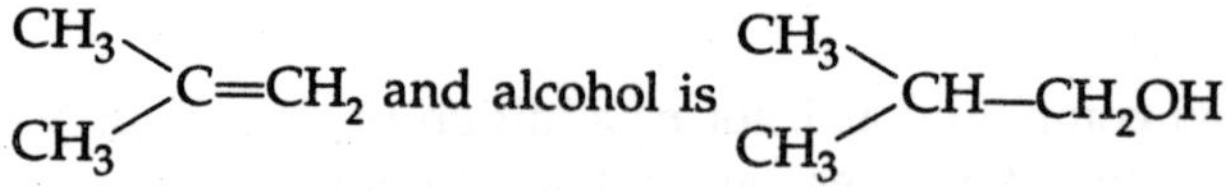

2-methyl-propanol

So, we have the reactions as given below:

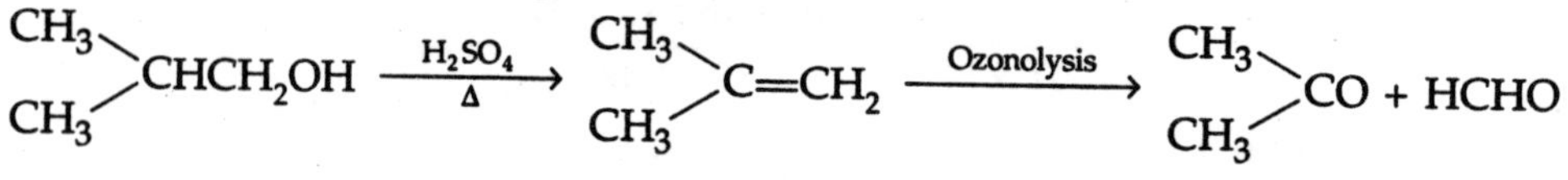

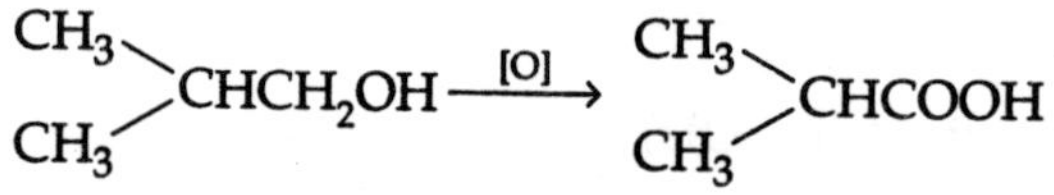

82. A decolourizes Br_2 in CCl_4 and absorbs H_2 and thus, (*A*) is unsaturated molecule having double or triple bond.

(*A*) gives a precipitate with ammonical Cu_2Cl_2 and thus, it has terminal alkyne linkage, i.e., triple bond at corner.

$$\text{Molecular weight of } (A) = \frac{22400 \times 1.49}{448} = 74.5$$

Since (*A*) has Cl and one$\equiv$bond at terminal so (*A*) may be $RCl{\cdot}C{\equiv}CH$

$\therefore$ $$R + 35.5 + 12 + 12 + 1 = 74.5$$

$\therefore$ $$R = 14$$

$\therefore$ $$R \text{ is } CH_2$$

Thus, (*A*) is $CH_2Cl{\cdot}C{\equiv}CH$, i.e., 3-chloropropyne

Hence we have the following reactions :

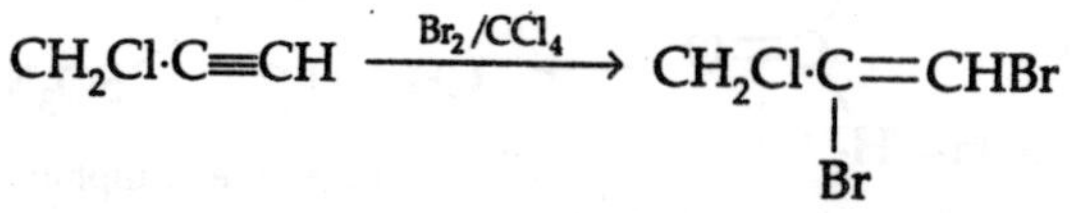

1,2-dibromo, 2-chloroprop-1-ene

$$\underset{(A)}{CH_2Cl{\cdot}C{\equiv}CH} \xrightarrow{H_2O} \underset{\text{1-chloropropane}}{CH_2Cl{\cdot}CH_2{-}CH_3}$$

$$\underset{(A)}{2CH_2Cl{\cdot}C{\equiv}CH} \xrightarrow[\text{or Am. } Cu_2Cl_3]{Cu_2O} \underset{\text{Red ppt.}}{2CH_2Cl{\equiv}C - Cu} + H_2O$$

83. (i) For empirical formula

Element	Percentage	relative no. of atoms	Simplest ratio
C	85.7	7.14	1
H	14.3	14.3	2

$\therefore$ Empirical formula is CH_2.

(ii) Now, 1 g of hydrocarbon uses 38.05 g of 5% solution Br_2. Since it is alkene, one mole of (*A*) uses 1 mole of Br_2.

$\therefore$ $\frac{38.05 \times 5}{100}$ Br_2 is used by 1 g of hydrocarbon

160 g Br_2 is used by $\frac{160 \times 100}{38.05 \times 5}$ of hydrocarbon

= 84 g of hydrocarbon

$\therefore$ Mol. wt. of hydrocarbon = 84

Empirical formula wt. of hydrocarbon, i.e., CH_2 = 14

$\therefore$ Molecular formula = $(CH_2)_6 = C_6H_{12}$

(iii) Because compound (C) is also prepared by the action $HgSO_4 + H_2SO_4$ on butyne-2, so (C) is $CH_3{\cdot}CH_2{\cdot}CO{\cdot}CH_3$

$$CH_3{-}C{\equiv}C \xrightarrow[HgSO_4\ +\ H_2SO_4]{H_2O} CH_3CH_2{\cdot}CO{\cdot}CH_3$$

But-2-yne 2-Butanone (C)

(iv) Since (*A*) C_6H_{12} on oxidation gives (C) and CH_3COOH and so, (*A*) is

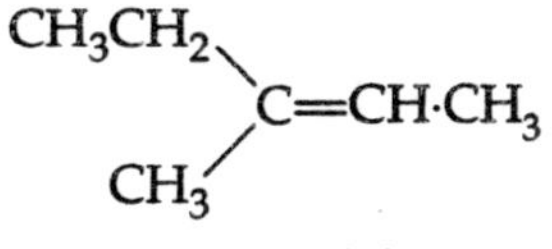

(A)

Thus, the reaction is:

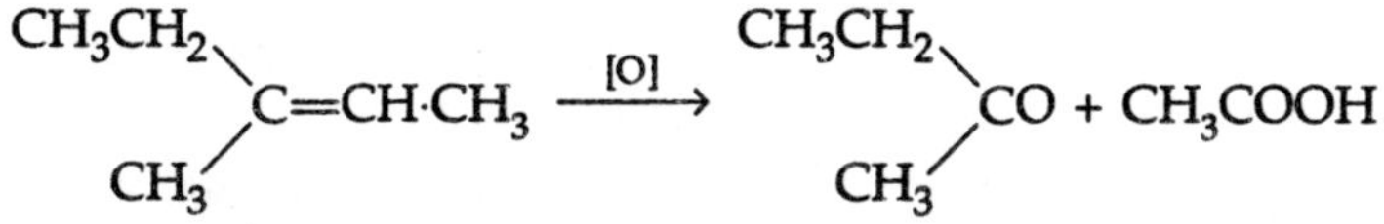

3-methylpent-2-ene (A) Butanone (C) Ethanoic acid (B)

84. Here we have, $C_2H_5OH + CH_3CHO = 0.535$ g

Let *a* g C_2H_5OH and *b* g CH_3CHO be present in mixture

$\therefore$ $a + b = 0.535$

Now the mixture reacts with Fehling solution to give a red precipitate. This suggests a characteristic reaction of an aldehyde, i.e.,

$$CH_3CHO + 2CuO \rightarrow CH_3COOH + Cu_2O$$

Fehling solution (2CuO); red ppt. (Cu_2O)

$\because$ 143.6 g Cu_2O is given by 44 g CH_3CHO

∴ 1.2 g Cu_2O is given by $\xrightarrow[143.6]{44 \times 1.2}$ = 0.368 g CH_3CHO

∴ $b = 0.368$ g

∴ % of $CH_3CHO = \frac{0.368}{0.535} \times 100 = 68.73\%$

85. The compound (x), C_5H_8O does not react with Lucas reagent appreciably at room temperature but gives precipitate with ammonical $AgNO_3$. Thus, (x) has terminal alkyne linkage as well as primary alcoholic gp.

Further, (x) on hydrogenation and then reacting with HI gives pentane and thus, (x) is straight chain compound.

Thus (x) may be

$CH_2 \cdot CH_2 \cdot C \equiv CH$ (with OH on the first carbon)

OH

Pent-4yn-1-ol

(x) with MeMgBr gives CH_4, thus, 1 mole of (x) will give two moles of CH_4.

$\underset{\substack{|\\ OH}}{CH_2} \cdot CH_2 \cdot \underset{(x)}{CH_2C \equiv CH} + 2CH_3MBr \rightarrow 2CH_4$

or 84 g (x) gives 2 × 22.4 litre CH_4

or 0.42 g (x) will give $\frac{2 \times 22. \times 0.42}{84} = 224$ mL CH_4

Hence, the given data is confirmed.

86. Given, Hexane + ethanol = 10 g

This mixture reacts with sodium to give H_2, which, only ethanol does so.

Therefore, $C_2H_5OH + Na \rightarrow C_2H_5ONa + \frac{1}{2}H_2$

Now Mole of H_2 formed $= \frac{PV}{RT} = \frac{760 \times 200}{760 \times 1000 \times 0.0821 \times 300}$

$= 8.13 \times 10^{-3}$ mole

(∵ 20 mL H_2 is formed S at 300 K and 760 mm pressure)

Now 1/2 mole H_2 is formed by = 46 g C_2H_5OH

∴ 8.13×10^{-3} is formed by = $46 \times 8.13 \times 10^{-3} \times 2$ g of C_2H_5OH

= 0.748 g

∴ % of ethanol in mixture $= \frac{0.748 \times 100}{10} = 7.48\%$

87. The Hydrocarbon (A) adds one mole of Br_2 and thus is an alkene.

(A) on ozonolysis forms CH_3COCH_3 and CH_3CHO and thus, (A) will be:

$(CH_3)_2C{=}CH{\cdot}CH_3$ (2-methylbut-2-ene)

Now, we check the given reactions:

(i) 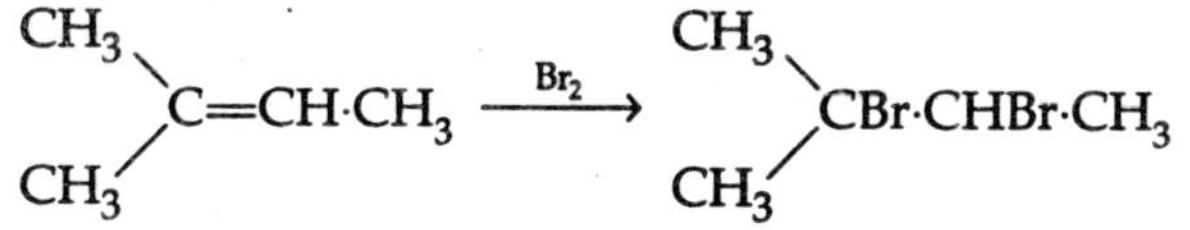

(A) 2,3-dibromo, 2-methylbutane

(ii) 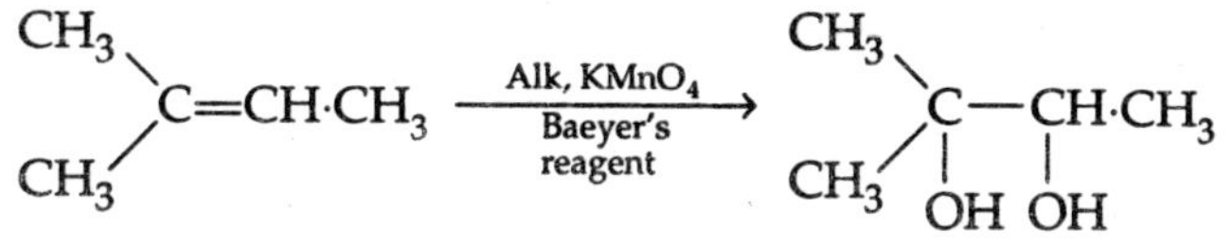

(A) 2-methylbutain-2,3-diol

(iii) 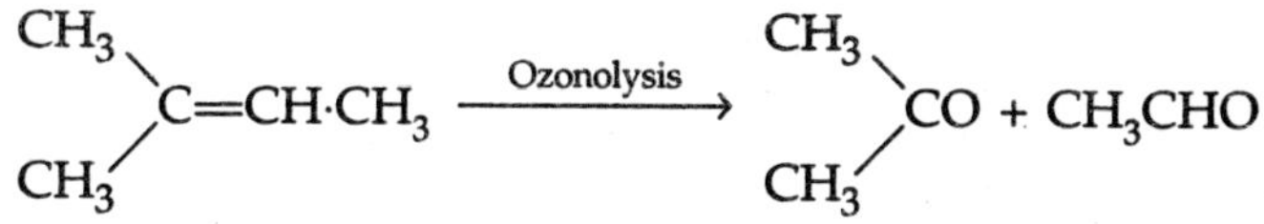

2-methylbut-2-ene(A) Propanone Ethanal

88. The alcohol is readily oxidised by dil. $KMnO_4$ and thus it is a, primary alcohol, i.e., *n*-butanol. Teritary alcohols are not oxidized by dil. $KMnO_4$.

$$CH_3CH_2CH_2CH_2OH \xrightarrow{[O]} CH_2CCH_2CH_2CHO$$

$$\downarrow [O]$$

$$CH_3CH_2CH_2COOK + MnO_2 + KOH$$

[brown ppt.]

89. (i)

Element	Percentage	Relative no. of atoms	Simplest ratio
C	88.89	7.41	1 or 2
H	11.11	11.11	1.49 3

∴ Empirical formula (*A*) is C_2H_3
Empirical formula weight = 27

(ii) Molecular weight of compound (*A*) can be determined by the fact that (*A*) is an alkyne as it gives white ppt. with ammonical $AgNO_3$ and on hydrogenation of (*A*), compound (*B*) is obtained with molecular weight 112.

$$A(\text{Alkyne}) \xrightarrow{2H_2} \text{Alkane}$$

∴ M. wt. of $A = 112 - 4 = 108$

(iii) $\because n = \frac{108}{27} = 4$ ∴ Molecular formula of (A) is $(C_2H_3)_4$

i.e., (*A*) is C_8H_{12}

(iv) Thus, the structure of compound (*A*), keeping in view of its molecular formula and alkyne nature may be given as:

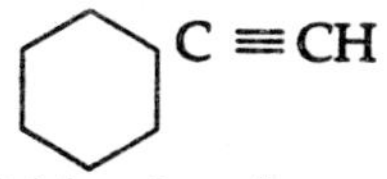

Cylohexylacetylene
(A)

(It gives white ppt. with ammonical silver nitrate)

Hence,

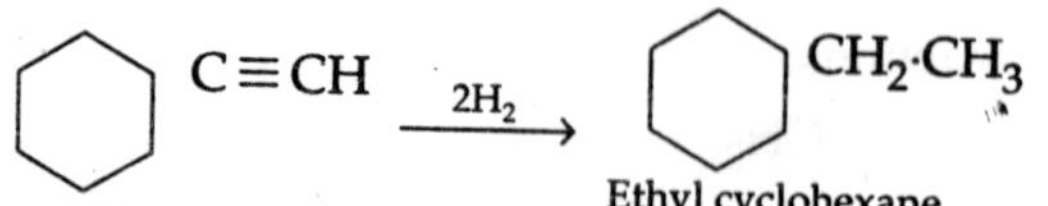

(A) Ethyl cyclohexane (B)

mol wt of B = 128

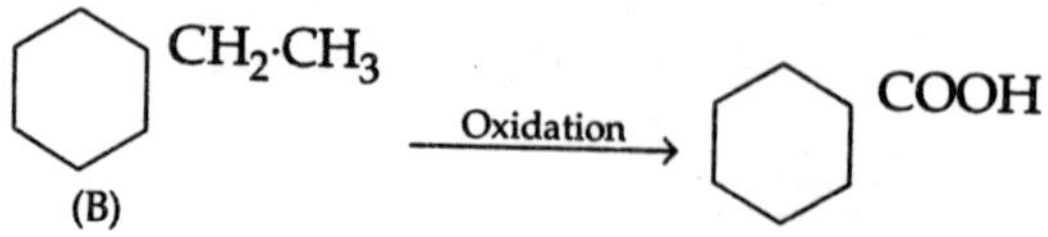

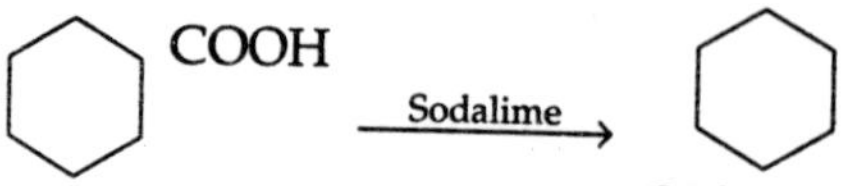

Cyelohexane

90. Compound C_8H_{12} adds up two mole of H_2 and thus, has two double bonds or on triple bonds; and
Ozonolysis of C_8H_{12} gives only butan-1, 4-dial and thus, C_8H_{12} seems to have closed ring of 8 carbon structure with two double bonds.

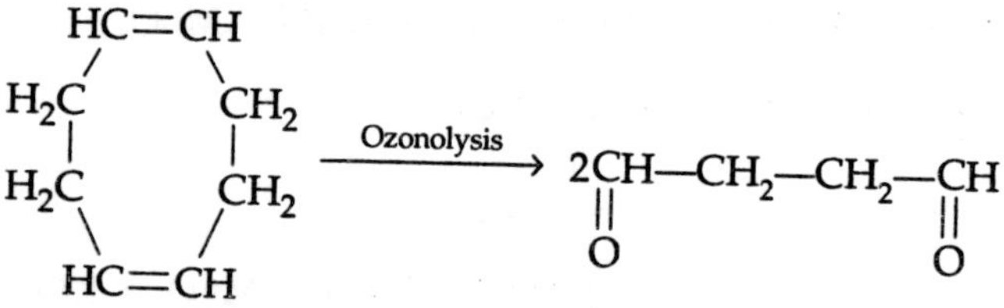

Cyclo.octa-1, 5 diene Butan-1, 4 dial

91. We have, $\Delta T = 452 - 440 = 12K$; $w = 0.0852$ g; $W = 2$ g; $K'_f = 40$

(i) Molecular weight of (A)

$$= \frac{1000 \times K' \times w}{\Delta T \times W}$$

$$= \frac{1000 \times 40 \times 0.0852}{12 \times 2} = 142$$

(ii) $\because$ 280 mL of (B) weigh = 0.375 g at STP

$\therefore$ 22400 mL of (B) weight $= \frac{0.375 \times 22400}{280} = 30$ g at STP

$\therefore$ Molecular weight of $B = 30$

(iii) Since $(A) \xrightarrow{Na} (B) + NaX$

$\therefore$ (B) is alkane. This is Wurtz reaction

$\therefore$ $C_nH_{2n+2} = 30$

$\Rightarrow$ $n = 2$

or (B) is C_2H_6 (Ethane)

(iv) Since (B) is obtained by (A), an alkyl halide, so (A) is CH_3X.

(v) Atomic weight of X = Molecular weight of CH_3X – Molecular weight of CH_3

$$= 142 - 15 = 127$$

$\therefore$ X is iodine.

(vi) Thus, (A) is CH_3I (methyl iodide) because,

$$\underset{(A)}{2CH_3I} + Na \rightarrow \underset{(B)}{C_2H_6} + 2NaI$$

92. Hydrocarbon (A) reacts with HBr to give mono bromo alkane (B), so (A) is alkene.

(A) gives same product on addition of HBr in absence and presence of peroxide and thus, (A) is symmetrical alkene.

Now 2 g Br_2 reacts with 0.7 g of (A)

$\therefore$ 160 g Br_2 reacts with $\frac{7.0 \times 160}{2} = 56$ g of (A)

$\therefore$ M.wt. of alkene = 56

or $C_nH_{2n} = 56$

$\therefore$ $n = 4$

Thus, alkene is C_4H_8 having structure $CH_3 \cdot CH{=}CH \cdot CH_3$

But-2-ene

(A)

$$\text{Now, } \underset{(A)}{CH_3{-}CH{=}CH{-}CH_3} \xrightarrow[\text{HBr}]{\text{Presence or absence of peroxide}} \underset{\text{2-bromobutane}}{CH_3CHBrCH_2CH_3}$$

93. (i) X forms semicarbazone and thus possesses carbonyl gp.

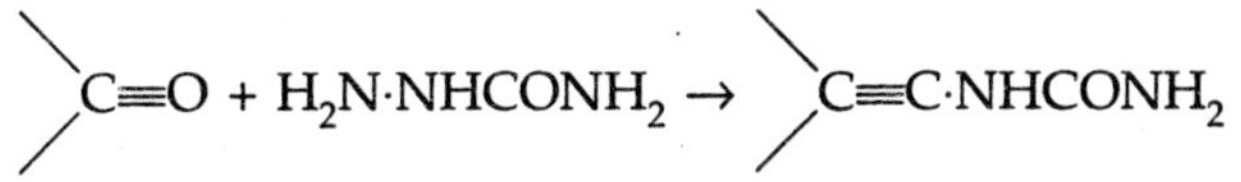

(ii) It does not give Tollen's reagent test and thus it is ketone.

(iii) It does not give iodoform test and thus it is not methyl ketone.

(iv) Keeping in view of the above facts and molecular formula; X is 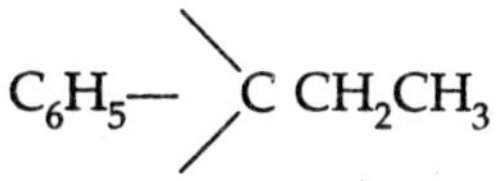

(v) This on reduction will give n-propyl benzene.

94. As alcohol is optically active, its structure should be 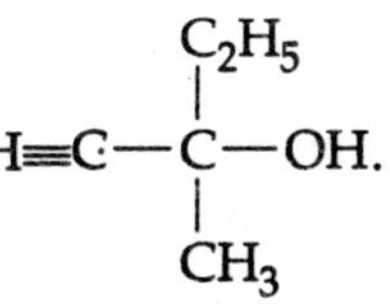.

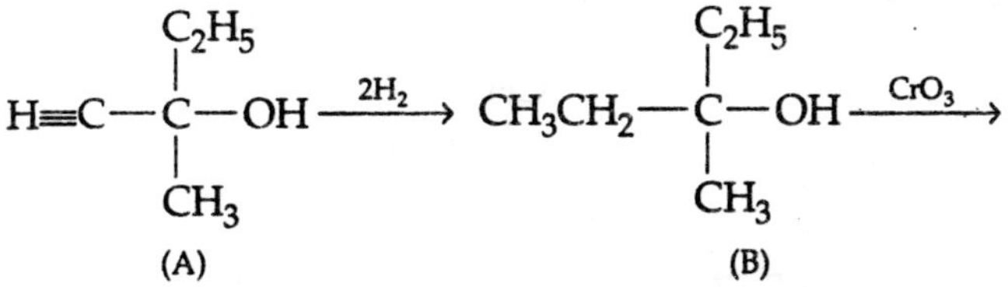

(A) (Optically active) (B) (Inactive)

Resistant to oxidation as it is tertiary alcohol.

95. $C_6H_5CHOH{\cdot}CH_3 \xrightarrow[\text{Indoform reaction}]{\text{(i) } I_2 + NaOH} C_6H_5COOH + CHI_3$

(ii) HOH (D) white (E) yellow

96.

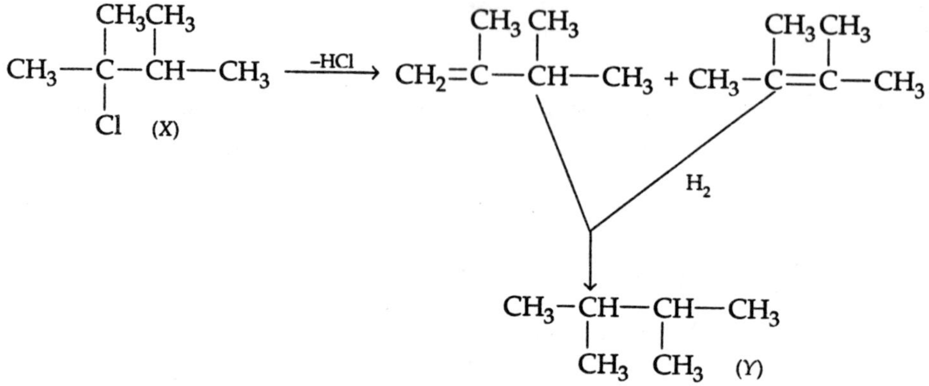

97. (i) We have, M. wt. of hydrocarbon $A = \dfrac{1.64 \times 22400}{448} = 82$

(ii)

Element	Percentage	Relative no. of atoms	Simplest ratio
C	87.8-	7.31	1 or 3
H	12.19	12.19	1.66 5

Thus, Empirical formula = C_3H_5

and Empirical formula wt. = 41

$\therefore$ Molecular formula = $(C_3H_5)_2 = C_6H_{10}$

(iii) Molecular formed if hydrocarbon (*A*) C_6H_{10} or C_nH_{2n-2} suggests that it contains either two double bonds or one triple bond.

(iv) The hydrocarbon reacts with H_2SO_4 in presence of $HgSO_4$ and thus, it is alkyne, i.e., has one triple bond.

(v) On hydrogenation it gives 2-methylpentane and thus, (*A*) has straight chain of 5 C atoms.

(vi) (A) does not react with ammonical $AgNO_3$ and thus, not a terminal alkyne.

(vii) Keeping in view of above facts the hydrocarbon can have only the structure

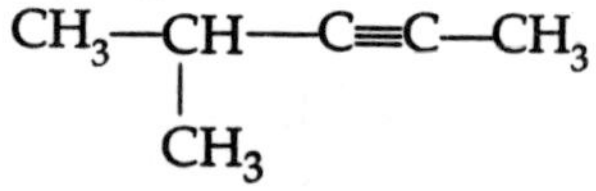

4-methylpent-2-yne

Thus the reactions of the hydrocarbon are as follows:

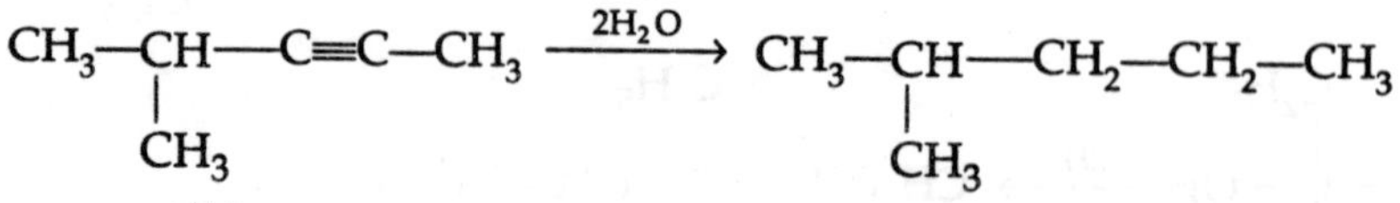

(A) 2-methylpentane

and,

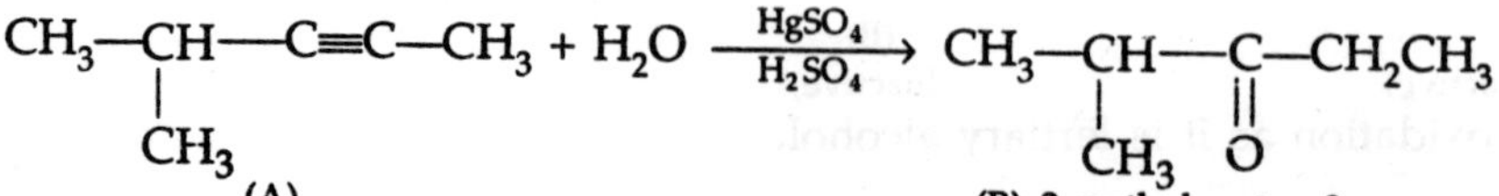

(A) (B) 2-methylpentan-3-one

or

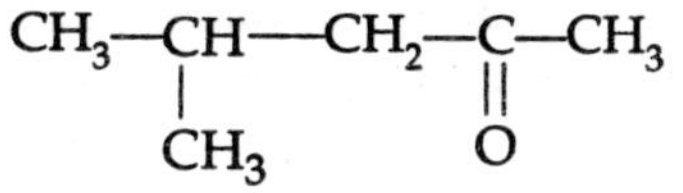

(B) 4-methylpentan-2-one

98. Here, the reaction is : $CH_3{\cdot}CHOH{\cdot}CH_3 \xrightarrow{H_2SO_4\ (conc.)} CH_3{\cdot}CH{=}CH_2$

isopropylalcohol — propene (X)

$\downarrow$ (HBr)

$CH_3{\cdot}CHBr{\cdot}CH_3$

2-bromopropane (Y)

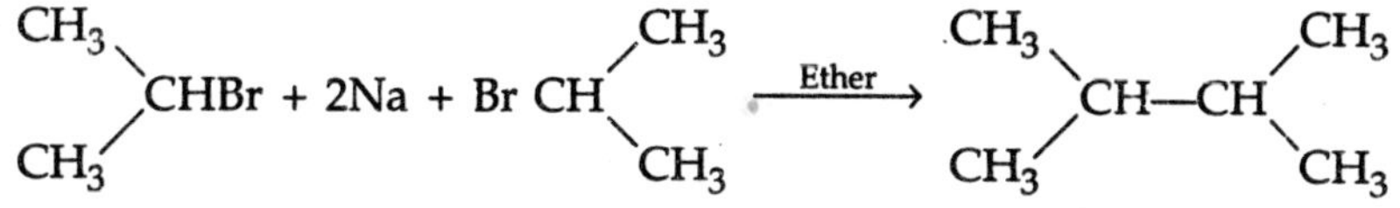

(Y) — 2,3-dimethylbutane (Z)

99. Here, we proceed with the reactions as:

(i) $C_7H_{15}Cl \xrightarrow[-HCl]{KOH} C_7H_{14} \xrightarrow{Ozonolysis}$ Acetone and butyraldehyde

(ii) Since, (*B*) C_7H_{14} gives on ozonolysis acetone $(CH_3)_2CO$ and butyraldehyde $(CH_3{\cdot}CH_2{\cdot}CH_2{\cdot}CHO)$; thus (*B*) should be 2 methylhex-2-ene; because

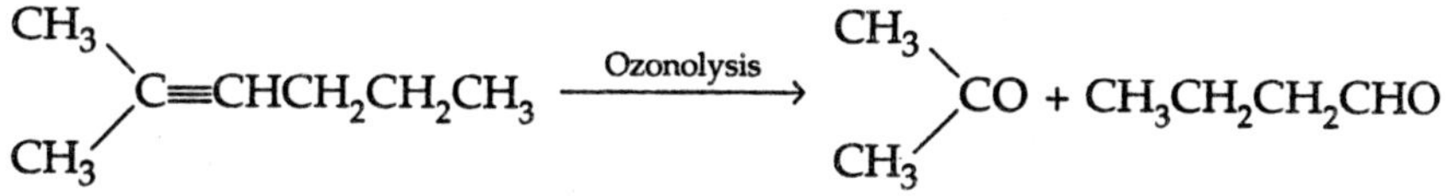

(*B*) 2-methylhex-2-ene — Acetone — Butanol

(iii) Since, (*B*) is formed by elimination of HCl from (*A*) and thus, (*A*) can be either I or II.

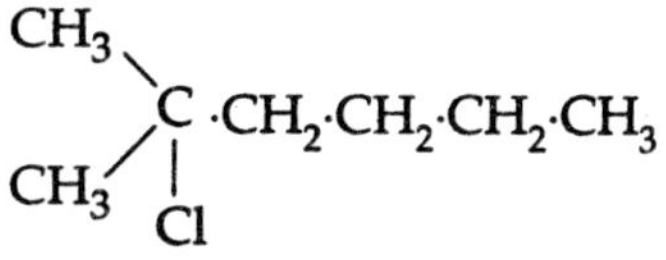

(I)

or

(II)

(iv) The elimination reactions by (I) and (II) yield;

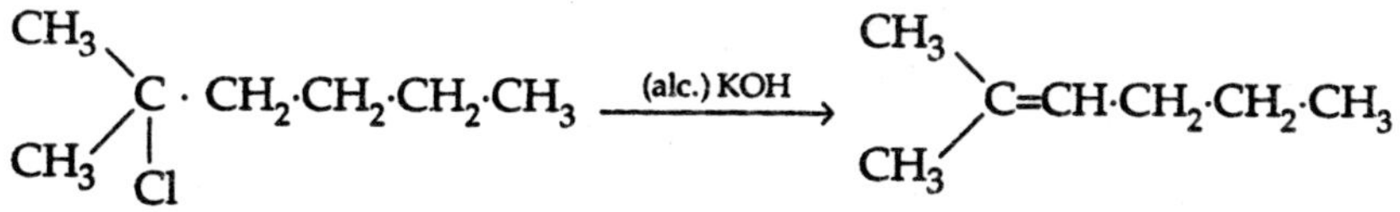

(I) — Major product

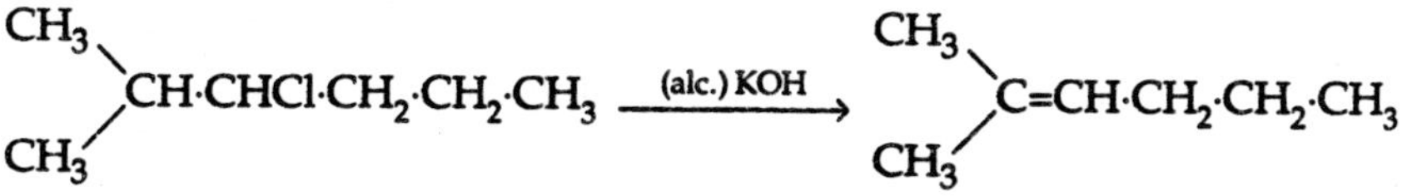

(II) — Major product

Since, in both the cases, major product formed is same as given and thus, (A) has either I or II structure.

Hence

(A) is

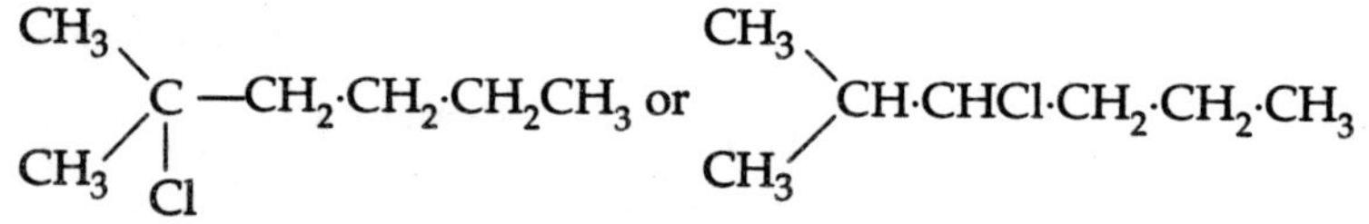

2-chloro 2-methyl hexane 3-chloro-2-methylhexane

(B)

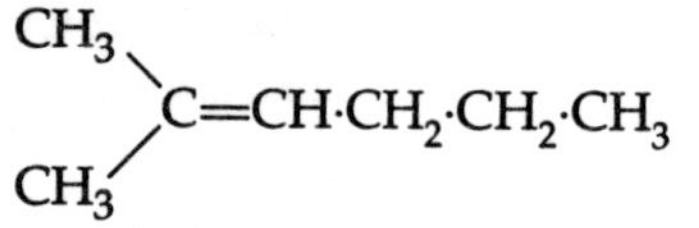

2-methylhex-2-ene

100. (i) Given, $C_4H_6 \xrightarrow{\text{Ozonolysis}} HCHO + \begin{array}{l} CHO \\ | \\ CHO \end{array}$

Methanal Glyoxal

(ii) Since, C_4H_6 on ozonolysis gives more than one products and thus, it may be alkadiene.

(iii) However, alkadiene on ozonolysis gives three molecules and thus, third one is also HCHO.

(iv) Keeping in view of above facts C_4H_6 is

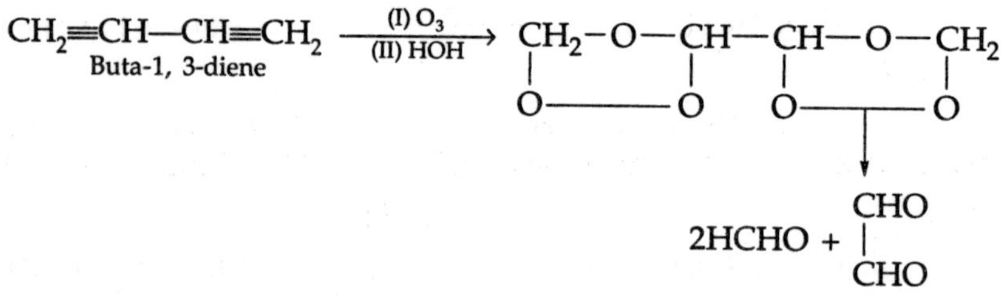

101. (i) Given,

$C_4H_6 \xrightarrow{\text{Ozonolysis}} HCHO + CO_2 + CH_3CHO$

(ii) Since, 3 molecules are present and thus, rupture of C_4H_6 at two places occurs during ozonolysis so C_4H_6 has two double bonds.

(iii) Keeping in view of above facts C_4H_6 is

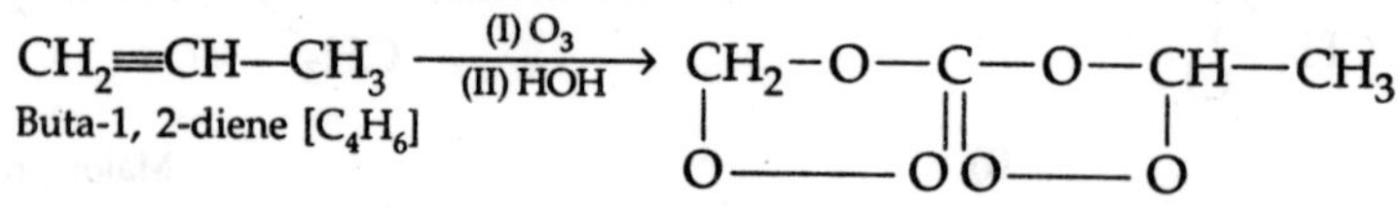

$HCHO + CO_2 + CH_3CHO$

Methanal Ethanal

102. (i) Compound (*A*) gives (*B*) on hydrogenation. Also (*A*) on ozonolysis gives HCHO and 2-ketopropanal. Thus (*A*) (C_5H_8) having two double bonds may be:

∴ *A* is

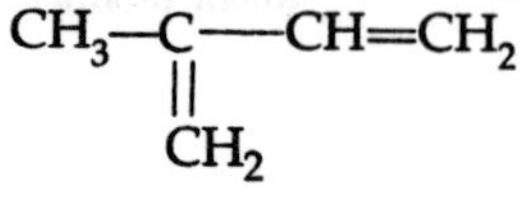

The reactions are

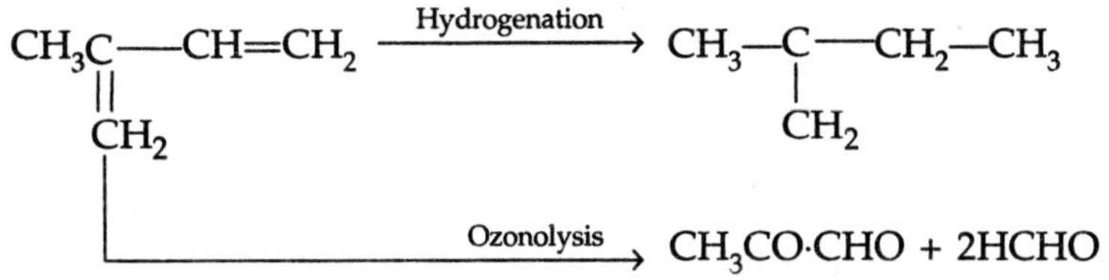
CH3C—CH=CH2
CH2
Hydrogenation
CH3—C—CH2—CH3
CH2
Ozonolysis
CH3CO·CHO + 2HCHO

103. (i) Compound (A) has formula C_6H_{12} and so it may be alkene or cycloalkane.
(ii) It gives positive test with Br_2 in CCl_4 and thus, it is alkene.
(iii) Oxidation of (*A*) with alkaline $KMnO_4$ yields only one salt of an acid and thus, (*A*) is symmetrical alkene, i.e., hex-3-ene.

$$CH_3{\cdot}CH_2{\cdot}CH{=}CH{\cdot}CH_2{\cdot}CH_3$$
(*A*)

Hence the reactions of the compound A are as follows:

(i) $CH_3{\cdot}CH_2{\cdot}CH{=}CH{\cdot}CH_2{\cdot}CH_3 \xrightarrow[\text{in } CCl_4]{Br_2} CH_3{\cdot}CH_2CHBr{\cdot}CHBr{\cdot}CH_2CH_3$

Hex-3-ene (*A*) — 3, 4-dibromohexane

(ii) $CH_3CH_2{\cdot}CH{=}CH{\cdot}CH_2{\cdot}CH_3 \xrightarrow{\text{alk. } KMnO_4} CH_3CH_2COOK$

(*A*) — (*B*) Pot. propanoate

104. Given $C_5H_{11}X \xrightarrow[-HX]{\text{KOH (alc.)}} CH_3{\cdot}CH_2{\cdot}CH_2CH{=\!=}CH_2$

Thus, possible structure of halide may be (I) or (II). However elimination of *HX*, according to Saytzeff rule, suggests that structure II is correct as it gives only pent-1-ene.

$CH_3{\cdot}CH_2{\cdot}CH_2{\cdot}CHX{\cdot}CH_3$ (I) $\xrightarrow{\text{(alc.) KOH}}$ $CH_3{\cdot}CH_2CH{=}CH{\cdot}CH_3$ Major, $CH_3{\cdot}CH_2{\cdot}CH_2{\cdot}CH{=}CH_2$ Minor

or

$CH_3CH_2{\cdot}CH_2CH_2{\cdot}CH_2X$ (II) $\xrightarrow{\text{(alc.) KOH}}$ $CH_3{\cdot}CH_2{\cdot}CH_2{\cdot}CH_2{=}CH_2$ pent-1-ene

Thus, the halide is $CH_3{\cdot}CH_2{\cdot}CH_2{\cdot}CH_2{\cdot}CH_2X$., i.e., 1-halopentane.

105. For empirical formula

C = 40.7/12 = 3.39	1
H = 5.1/1 = 5.1	1.5
O = 54.4/16 = 3.4	1.0

∴ Empirical formula of acid is $C_2H_3O_2$.

$$\frac{\text{wt. of Ag}}{\text{E. wt of Ag}} = \frac{\text{wt. of acid salt}}{\text{E. wt. of acid salt}}$$

∴ E. wt. of acid salt = (100 × 108)/65.1 = 165.89
∴ E. wt. of acid 165.89 – 108 + 1 = 58.89
∵ M. wt = 2 × V.D
∴ M. wt. of ethyl ester is 174 and thus, acid is dibasic because m. wt. of C_2H_5 group is only 29.

Thus Mol. wt. of acid = 58.89 × 2 = 117.78

∵ Empirical formula wt. = 59

∴ Mol. formula of acid = $(C_2H_3O_2)_2 = C_4H_6O_4$

Thus, acid (*A*) is $CH_3{\cdot}CH(COOH)_2$ the isomer of (*A*) may be $\begin{array}{l} CH_2COOH \\ | \\ CH_2COOH \end{array}$ or vice-versa

The two maybe distinguished by heating

(A) $CH_3CH(COOH)_2 \xrightarrow{\Delta} CH_3{\cdot}CH_2{\cdot}COOH + CO_2$

(Methyl malonic acid) (Propanoic acid)

(B) $\begin{array}{l} CH_2COOH \\ | \\ CH_2COOH \end{array} \xrightarrow{\Delta} \begin{array}{l} CH_2CO \\ | \\ CH_2CO \end{array} \!\!> O + H_2O$

(Succinic acid) (Succinic anhydride)

106. Given,

Addition of HBr in absence or in presence of O_2 gives same product and thus, the compound (*A*) is symmetrical alkene.

It C_6H_{12} does not docolourizes Br_2 water.

and on ozonolysis it does not give propanal. The compound A is

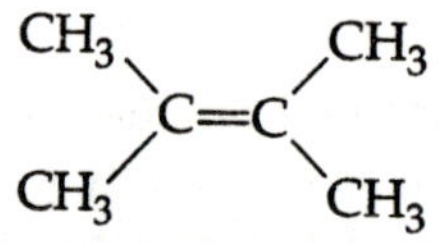
CH3
CH3
C=C
CH3
CH3

2, 3-methyl but-2-ene.

The reactions of this compound are as:

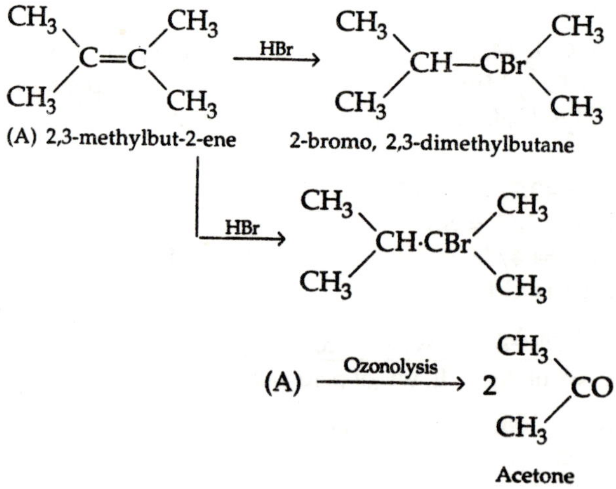
CH3
CH3
C=C
CH3
CH3
HBr
CH3
CH3
CH—CBr
CH3
CH3
(A) 2,3-methylbut-2-ene
2-bromo, 2,3-dimethylbutane
HBr
CH3
CH3
CH·CBr
CH3
CH3
(A)
Ozonolysis
2
CH3
CO
CH3
Acetone

107. Given, hydrocarbon (*A*) adds up one H_2 molecular and docolourizes Br_2/CCl_4 solution, so (*A*) is alkene.

Again (*A*) on hydrogeantion gives hexane so, (*A*) hexene C_6H_{12}.
and (*A*) on oxidation gives only one salt and thus, (*A*) is symmetrical alkene, i.e., hex-3-ene,
Thus the reactions of A are :

$$\underset{\text{(A) Hex-3-ene}}{CH_3{\cdot}CH_2{\cdot}CH{=}CH{\cdot}CH_2{\cdot}CH_3} \xrightarrow{Br_2/CCl_4} \underset{\text{3,4-diromobexane}}{CH_3{\cdot}CH_2{\cdot}CHBr{\cdot}CHBr{\cdot}CH_2{\cdot}CH_3}$$

$$\xrightarrow{H_2/Pt} \underset{\text{Hexane}}{CH_3{\cdot}CH_2{\cdot}CH_2{\cdot}CH_2{\cdot}CH_2{\cdot}CH_3}$$

$$\xrightarrow[\text{KMnO}_4\text{ alkaline}]{\text{Oxidation}} \underset{\text{Pot, propanoate}}{2CH_3{\cdot}CH_2{\cdot}COOK}$$

108. The reaction below shows the structures of the compound.

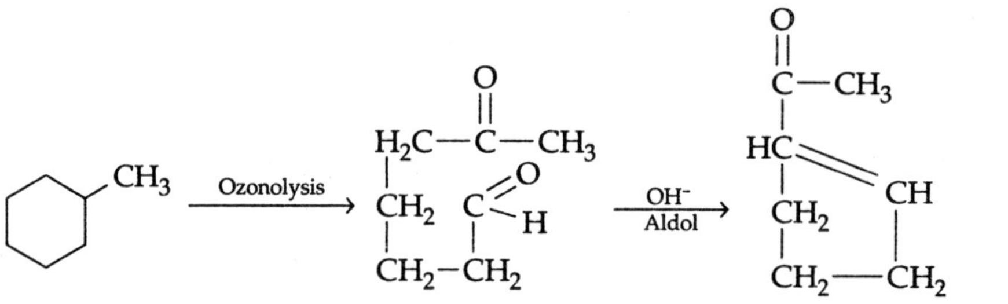

(A) (B) 1-acetyl cyclopentene

109. Given reactions are,

$$\underset{\substack{\text{1,1,2,2-tetrachloro}\\ \text{propane (m. wt. = 182)}}}{CHCl_2{-}CCl_2{\cdot}CH_3} \xrightarrow[\Delta]{\text{Zn dust}} \underset{\text{Propyne}}{CH{\equiv}C{-}CH_3} + 2ZnCl_3$$

$$\underset{\text{Propyne}}{CH{\equiv}C{-}CH_3} \xrightarrow[\text{Amm}]{Ag_2O} \underset{AgNO_3\ \text{Precipitate}}{C{\cdot}Ag{\equiv}C{\cdot}CH_3}$$

$\because$ 182 g compound (1,1,2,2-tetrachloropropane) gives = 147 g precipitate
$\therefore$ 36.4 g compound (1,2,3,3-tetrachloropropane) gives

$$= \frac{147 \times 36.4}{182}\text{ g}$$

$$= 29.4\text{ g}$$

$\therefore$ weight of precipitate = 29.4 g

110. Alkyne is terminal as it gives Ag salt. Let the alkyne be *R*—C≡CH and so formula of Ag salt is *R*—C≡C Ag

For 67.08 g Ag, wt. of silver salt = 100

$\therefore$ for 108 g; wt. of silver salt is = $\dfrac{100 \times 108}{67.08}$ = 161.00

$\therefore$ M. wt. of silver salt = 161.00

Note: If it would have acetylene then m.wt. would have been determined by 216 g Ag.

$\therefore$ $R—C{\equiv}C{\cdot}Ag = 161$

$R + 12 + 12 + 108 = 161$

$\therefore$ $R = 29$

or $C_nH_{2n+1} = 29$

$\therefore$ $n = 2$

$\therefore$ Alkyne is $CH_3CH_2C{\equiv}CH$
But-1-yne

111. 1. Given $C_5H_{11}X \xrightarrow[-HX]{KOH\,(alc.)} CH_3CH_2CH{=}CHCH_3$

2. Thus, possible structure of halide may be (I) or (II). However elimination of HX according to Saytzeff rule (the major product is one from which elimination of *H* occurs from less hydrogenated *C* atom) structure II is correct as it give only pent-2-ene.

$$\underset{(I)}{CH_3CH_2CH_2CHXCH_3} \xrightarrow{(alc.)\,KOH} \underset{major}{CH_3CH_2CH{=}CHCH_3}$$

$$+ \underset{minor}{CH_3CH_2CH_2CH{=}CH_2}$$

$$\underset{(II)}{CH_3CH_2CHXCH_2CH_3} \xrightarrow{(alc.)\,KOH} \underset{\text{only pent-2-ent}}{CH_3CH{=}CHCH_2CH_3}$$

Ans. Thus halide is $CH_3CH_2CHXCH_2CH_3$
3-halo pentane

112. (a) $\underset{\text{Pent-1-ene}}{CH_3CH_2CH_2CH{=}CH_2} + Br_2 \rightarrow \underset{\text{1,2-dibromopentane}}{CH_3CH_2CH_2CHBrCH_2Br}$

$\because$ 70 g C_5H_{10} requires 160 g Br_2

$\therefore$ 5 g C_5H_{10} requires $\frac{160 \times 5}{70}$ g Br_2 = 11.43 Br_2

(b) $\underset{\text{Pent-1-yne}}{CH_3CH_2C{\equiv}CH} + 2Br_2 \rightarrow \underset{\text{1,1,2,2-tetrabromopentane}}{CH_3CH_2CH_2CBr_2CHBr_2}$

$\because$ 68 g C_5H_8 requires 430 g Br_2

$\therefore$ 5 g C_5H_8 requires $\frac{320 \times 5}{68}$ g Br_2 = 23.53 g Br_2

(c) $CH_3CH_2CH_2CH_2CH_3 \xrightarrow{Br} C_5H_{11}Br + HBr$

$\because$ 72 g C_5H_{12} requires 160 g Br_2

$\therefore$ 5 g C_5H_{12} $\frac{160 \times 5}{72}$ g Br_2 = 11.11 g Br_2

113. The reaction for reduction is

$$C_6H_5—\underset{\underset{O}{\parallel}}{C}—\underset{\underset{O}{\parallel}}{C}— {}_6H_5 \xrightarrow{LiAlH_4} C_6H_5\underset{\underset{OH}{|}}{CH}\,\underset{\underset{OH}{|}}{CH}C_6H_5$$

Therefore two asymmetric carbon atoms in the product.

$\therefore$ $$a = 2^{n-1} = 2^{2-1} = 2$$

$$m = 2^{n/2-1} = 2^0 = 1$$

$\therefore$ No. of stereoisomers (atm) = 2 + 1 = 3

114. Here, V.D. = 36

$$\text{M. wt.} = 2 \times \text{V.D.} = 2 \times 36 = 72$$

The m. wt. suggests it to be alkane, i.e., C_nH_{2n+2}.

$\therefore$ $$12n + 2n + 2 = 72$$

or $$n = 5$$

$\therefore$ Alkane (A) is C_5H_{12}

Thus, (A) gives only one product on substitution, i.e. (A) is

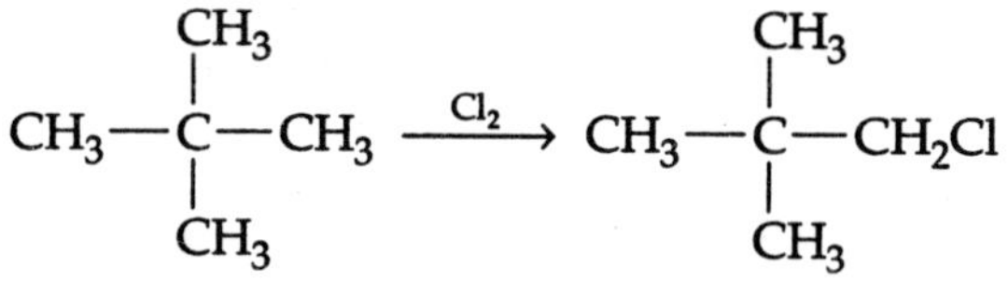

Hence, A is 2, 2-dimethylpropane (neo pentane).

115. As, Alkane is C_nH_{2n+2}; M. wt. = 86

$\therefore$ $$12n + 2n + 2 = 86$$

$\therefore$ $$n = 6$$

$\therefore$ Alkane is C_6H_{14}.

The alkane with this formula of bromination giving two products may be written as

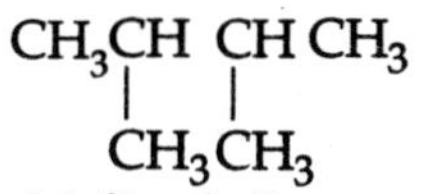

2,3-dimethylbutane

The reaction is:

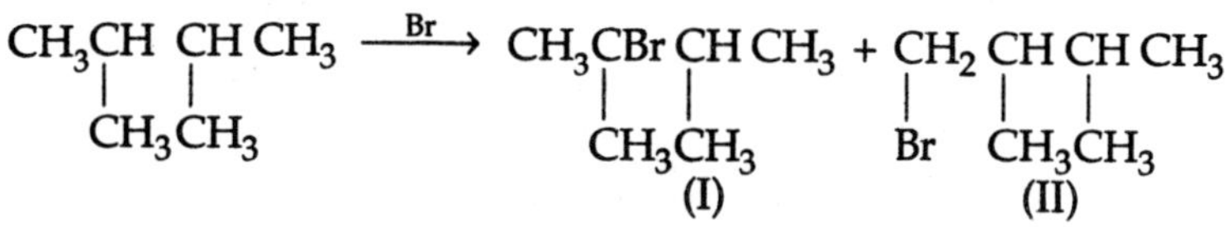

116. (i) When the molecule cannot be divided into equal parts, i.e., has no symmetry and n is the number of asymmetric carbon atoms, then

No. of optically active isomers, $a = 2^n$

Number of meso forms, $m = 0$

Total no. of recemic forms, $r = a/2$

Total no. of optical isomers, $= a + m$

$CH_3CHOHCHOHCOOH$

Here, asymmetric carbon atoms, $n = 2$

$$a = 2^2 = 4$$

$$m = 0$$

$$r = 0$$

Total optical isomers = 4 + 0 = 4

(ii) When the molecule can be divided into two equal halves (i.e., has symmetry) number of asymmetric carbon atoms is even, then

$$\text{Number of optically active isomers, } a = 2^{n-1}$$
$$\text{Number of meso forms, } m = 2^{(n/2)-1}$$
$$\text{Number of recemic forms, } r = a/2$$

∴ For COOHCHOHCHOHCOOH

Here, asymetric carbon atoms, $n = 2$

∴ $a = 2^{2-1} = 2$

$m = 2^{1-1} = 2^0 = 1$

$r = 1$

Total optical isomers = (a+m) = 2 + 1 = 3

117. Given, The compound (*B*) reacts with hydroxyl amine thus, (*B*) has carbonyl gp. (i.e., >C=O gp) either (>C=O) or aldehyde $(-\underset{\underset{H}{|}}{C}=O)$ gp.

Also (*B*) gives negative test with Tollen's reagent and thus, (*B*) is ketone.
(*B*) is also obtained by hydrolysis of (*A*) thus, both Cl atoms should be on same carbon atom, i.e., (*A*) must be gem dihalide as well as not on corners. So, only possibility of (*A*) is $CH_3 \cdot CCl_2 \cdot CH_2 \cdot CH_3$.
Hence we have,

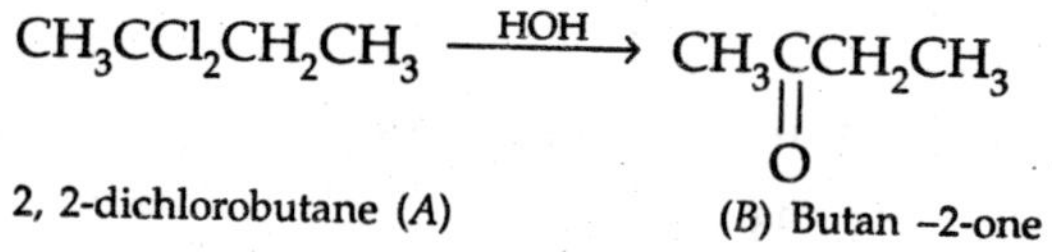
CH3CCl2CH2CH3
HOH
CH3CCH2CH3
O
2, 2-dichlorobutane (A)
(B) Butan -2-one

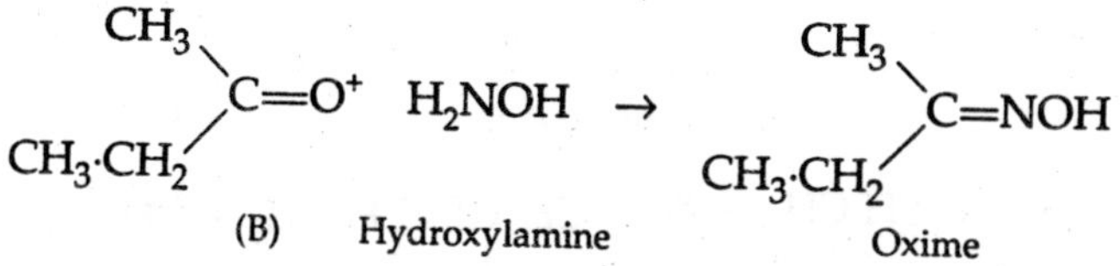
CH3
C=O+ H2NOH →
CH3·CH2
CH3
C=NOH
CH3·CH2
(B)
Hydroxylamine
Oxime

CH3
C=O
CH3·CH2
Tollen's regent
No reaction

118. Given compound (*A*) forms phenyl hydrozone and is thus a carbonyl compound, i.e., an aldehyde or ketone.
Also, (*A*) gives negative Tollen's reagent rest and thus, (*A*) is ketone.
and (*A*) also does not respond Iodoform test and thus, (*A*) is ketone.
Further on reduction it gives *n*-pentane and thus, (*A*) is straight chain compound.
Hence from the above facts and the formula of (*A*), is

$$C_5H_{10}O, (A) \text{ is } CH_3CH_2\underset{\underset{O}{\|}}{C}CH_2CH_3$$

Which is Pentan-3-one

So we have,

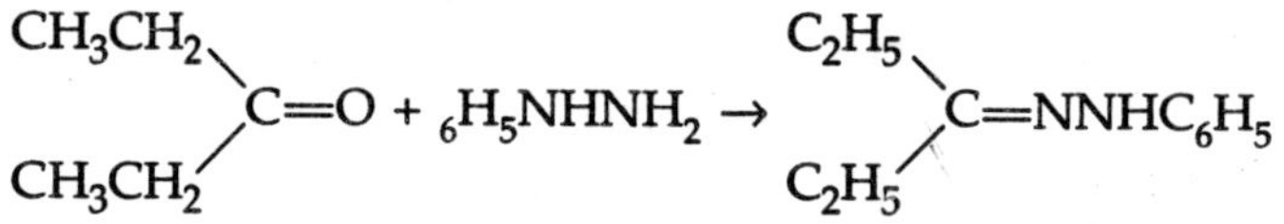

(A) Phenyl hydrazone

$$C_2H_5COC_2H_5 \xrightarrow[\text{regent}]{\text{Tollen's}} \text{No reaction}$$

$$C_2H_5COC_2H_5 \xrightarrow[\text{test}]{\text{Iodoform}} \text{No reaction}$$

$$\underset{\text{Pentan-3-one(a)}}{C_2H_5COC_2H_5} \xrightarrow{H} \underset{\text{n-Pentane}}{C_2H_5COC_2H_5}$$

119. Given,

Element	Percentage	Relative no. of atoms	Simplest ratio
C	69.77	5.76	5
H	11.63	11.63	10
O	19.20	1.2	1

∴ Empirical formula of compound is $C_5H_{10}O$.

∴ Empirical formula wt. = 86

Also Molecular wt. = 86

∴ Empirical formula of compound is $C_5H_{10}O$.

Again, Compound form bisulphite addition compound and thus, has carbonyl group. i.e., aldehyde or ketone.

and It does not reduce Fehling solution and thus, it is not aldehyde but it is ketone.

It also gives positive iodoform test and thus, it has

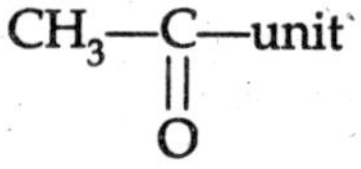

So we conclude that compound is

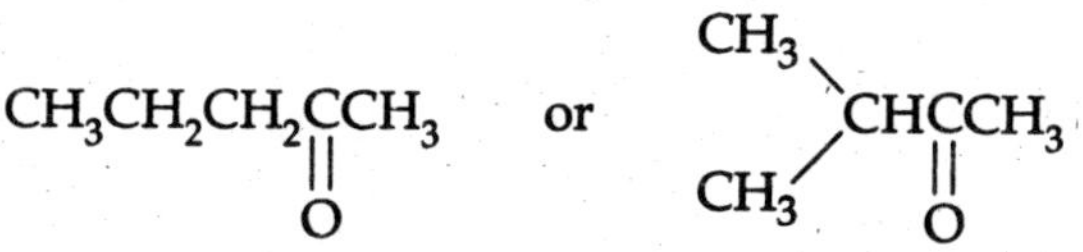

Pentan-2-one or 3-methylbutan-2-one

120. $C_4H_8O_2$ reacts with $NaHCO_3$ to liberate CO_2 and so it contains —COOH group, i.e., (*A*) is C_3H_7COOH.

Also, $$\underset{(A)}{C_3H_7COOH} \xrightarrow{NaOH + CaO} \underset{\text{Propane}}{C_3H_8} + Na_2CO_3$$

Thus, (*A*) may be

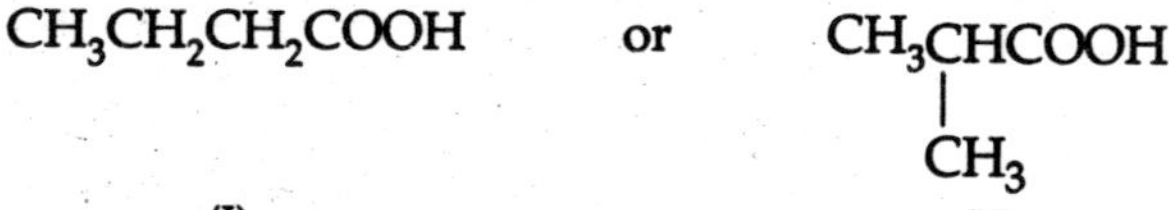

(I) Butanoic acid or (II) 2-methylpropanoic acid

Since both gives propane on decarboxylation
However, $C_8H_{14}O_4Ca$, i.e., Ca salt of acid on heating gives

Di isopropyl ketone

Di isopropyl ketone and thus, chain in acid is also not straight. Thus, structure (ii) is correct, i.e., (*A*) is

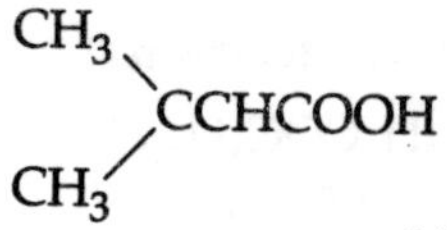

(2-methylpropanoic acid)

(*A*)

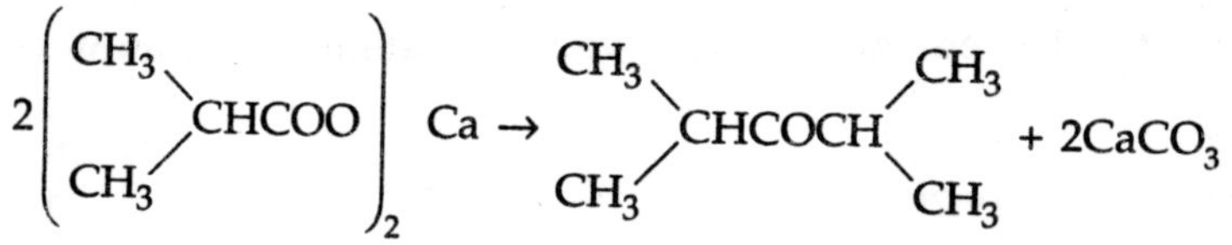

121. Given: A reacts with PCl_5 but does not react with Na thus, (*A*) may be carbonyl compound (not acid, not alcohol).
Also (*A*) undergoes haloform reaction it should have CH_3—CO—unit.
Buton Hydrogenation A gives 2-methylpentane

$$\therefore (A) \xrightarrow{H_2} \underset{\displaystyle CH_3}{CH_3\underset{|}{C}HCH_2HCH_2CH_2CH_3} \text{ (2-methylpentane)}$$

Also, During haloform reaction, acid salt is formed which is

$$CH_3-\underset{\underset{\displaystyle CH_3}{|}}{C}=CCOONa \text{ (3-methylbut-2-eoic acid salt)}$$

So from the above facts (*A*) is

$$(CH_3)_2C{=}CHCOCH_3 \text{ (4-methylpent-3-en-2-one)}$$

because:

$$\underset{(A)}{(CH_3)_2\,C{=}CH{\cdot}COCH_3} \xrightarrow{Na} \text{No reaction}$$

$$\underset{(A)}{(CH_3)_2C{=}CHCOCH_3} \xrightarrow{PCl_3} (CH_3)_2C{=}CHCCl_2CH_3$$

$$\underset{(A)}{(CH_3)_2C{=}CHCOCH_3} + X_2 + NaOH \longrightarrow (CH_3)_2C{=}CHCOONa + CHX_3$$

$$\xrightarrow{HOH} (CH_3)_2C{=}CHCOOH$$

122. Given, Two isomers (*A*) and (*B*) have formula CH_2O.

$$\text{V.D. of } (B) = 2 \times \text{V.D. of } (A)$$

(*A*) reduces fehling solution but does not react with $NaHCO_3$, so (*A*) is aldehyde.
(*B*) neither reduces Fehling solution, nor reacts with $NaHCO_3$. If (*A*) is HCHO then (B) is $C_2H_4O_2$ and so (*B*) must be an ester $HCOOH_3$.
Isomer of (*B*) reacts with $NaHCO_3$, then it is CH_3COOH, Thus,
A: HCHO (Formaldehyde)
B: $HCOOCH_3$ (Methyl methanoate)
The reactions are :

(i) HCHO $\xrightarrow{NaHCO_3}$ No reaction

HCHO $\xrightarrow{2CuO}$ $HCOOH + Cu_2O$

(ii) $CHOOCH_3$ $\xrightarrow{NaHCO_3}$ No reaction

$CHOOCH_3$ $\xrightarrow{2CuO}$ No reaction

(iii) $CH_3COOH \xrightarrow{NaHCO_3} CH_3COONa + CO_2 + H_2O$

123. The compound which gives a red ppt. with cuprous chloride must be an alkyne. So the hydrocarbon with three carbon atoms is propyne. Its halogen derivative is treated with a aqueous KOH with losses a water molecule to give an aldehyde

$$CH_3CH_2CHCl_2 \xrightarrow[\text{KOH}]{\text{alc.}} CH_3C{\equiv}CH$$

$CH_3CH_2CHCl_2 + KOH(aq) \rightarrow CH_3CH_2CH(OH)_2 \xrightarrow{-H_2O} CH_3CH_2CHO$ So '*X*' is propanal

124. This compound formed is due to Wurtz reaction because ether and sodium metal are used. So it is a combination of two molecules of alkyl halide.

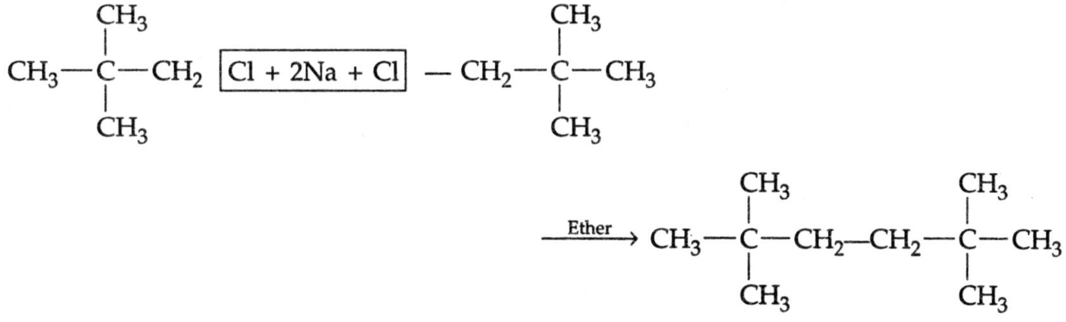

125. (a) 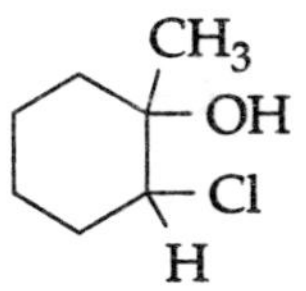 (b) 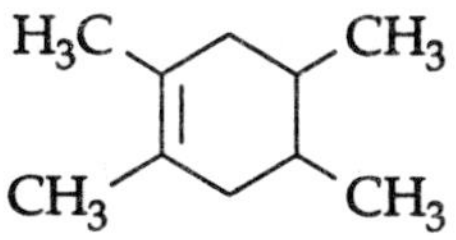

(c) 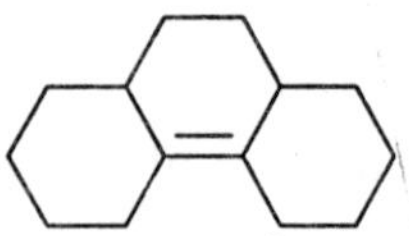

(d) 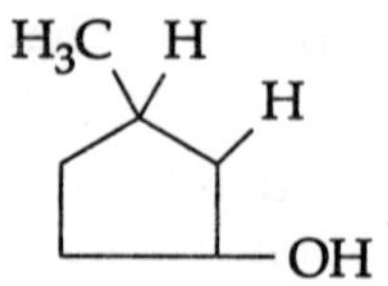

(e) 

(f) 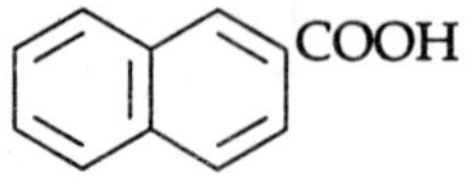

(g) 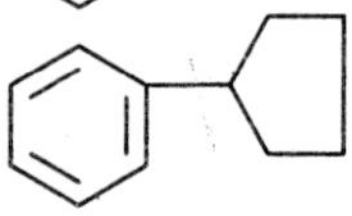

(h) 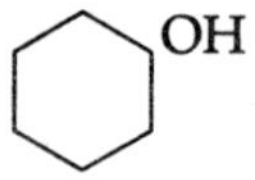

126. For empirical formula of *A*

Element	Percentage	Relative no. of atoms	Simplest ratio
C	90.6	7.7	1 or 4
H	9.4	8.4	1.25 or 5

∴ Empirical formula of *A* = C_4H_5
Empirical formula wt. of *A* = 53
Given, Mol. wt. of *A* = 106
∴ Molecular formula of *A* = C_8C_{10}
(*A*) forms (*B*) on oxidation which forms anhydride so (*B*) is dibasic acid and that two —COOH are at ortho positions. Also given data for empirical formula of (*B*) suggests its formula $C_8H_6O_4$.

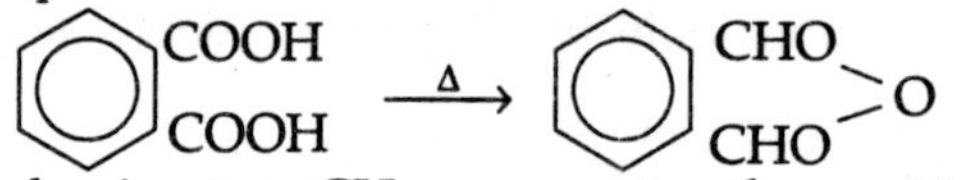
(Phthalic anhydride)

Thus, (*A*) is xylene having two CH_3 groups at ortho positions.

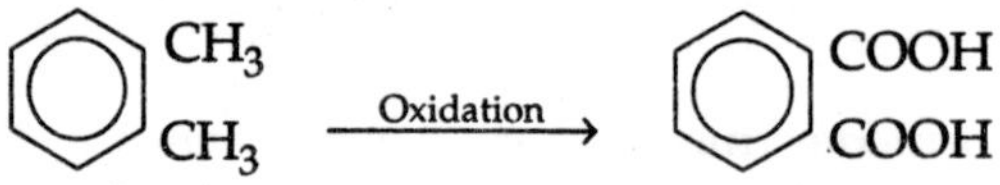
(A) (*o*-Xylene) (B) (Phthalic acid)

127. (i) It has two optically active forms

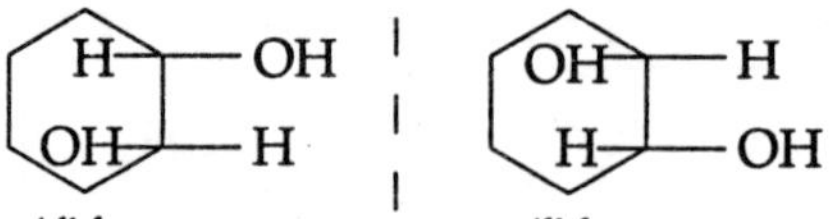
'*d*' form '*l*' form

(ii) It has four optically active forms:

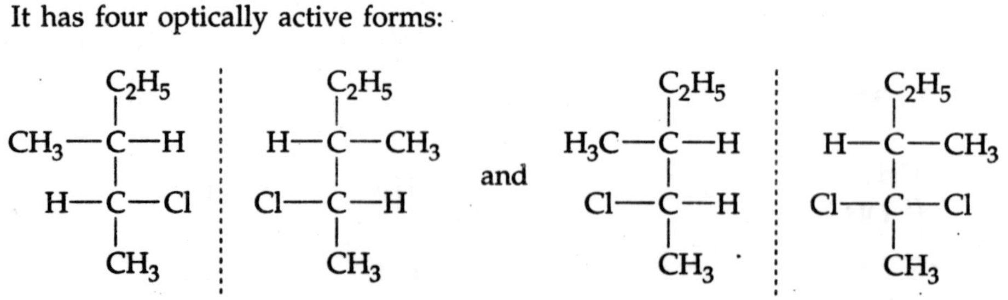
I '*d*' form II '*l*' form III '*d*' form IV '*l*' form

128. (i) (*A*) forms oxime and thus, contains carbonyl group.

(ii) (*A*) does not reduce Fehling solution but undergoes Cannizzaro reaction so, (*A*) is C_6H_5CHO.

$$\underset{\text{Benzaldehyde } (A)}{C_6H_5CHO} \xrightarrow{NH_2OH} C_6H_5CH{=}NOH \text{ oxime}$$

$$C_6H_5CHO \xrightarrow[\text{Canizzaro reaction}]{KOH_{(aq.)}} \underset{\text{Pot benzone}}{C_6H_5COOK} + \underset{\text{Benzyl alcohol}}{C_6H_5CH_2OH}$$

Note: C_6H_5CHO does not reduce Fehling solution but reduces amm. $AgNO_3$.

129.

Elements	Percentage	Relative No. of atoms	Simplest ratio
C	69.42	5.785	7
H	5.78	5.78	7
N	11.57	0.839	1
O	13.23	0.826	1

$\therefore$ Empirical formula of $A = C_7H_7NO$

Empirical formula weight = 121

Now Molecular weight = $2 \times 60.5 = 121$

$\therefore$ Molecular formula = C_7H_7NO

(*A*) gives NH_3 with caustic potash and thus, (*A*) should have —$CONH_2$ group.

(*A*) with molecular formula C_7H_7NO and group —$CONH_2$ can only be

CONH2

(A) Benzanide

The reactions of benzanide are:

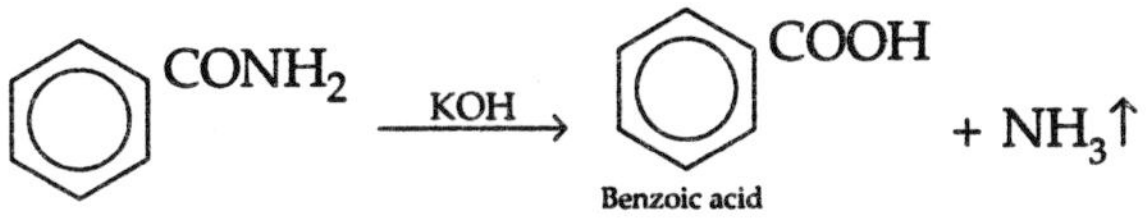
CONH2
KOH
COOH
+ NH3↑
Benzoic acid

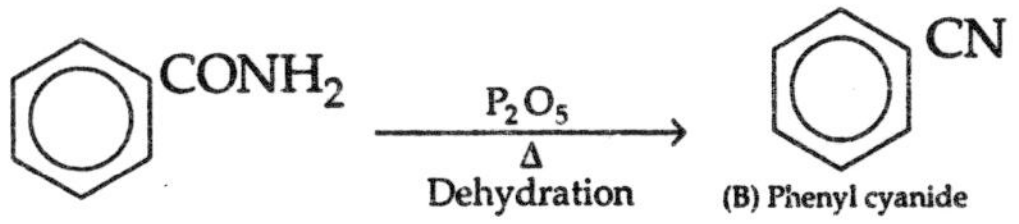
CONH2
P2O5
Δ
Dehydration
CN
(B) Phenyl cyanide

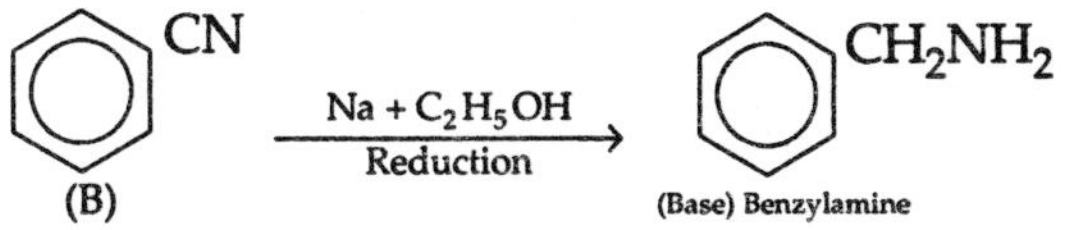
CN
(B)
Na + C2H5OH
Reduction
CH2NH2
(Base) Benzylamine

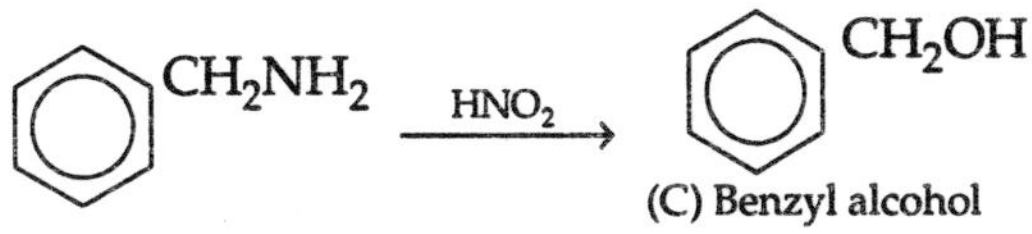
CH2NH2
HNO2
CH2OH
(C) Benzyl alcohol

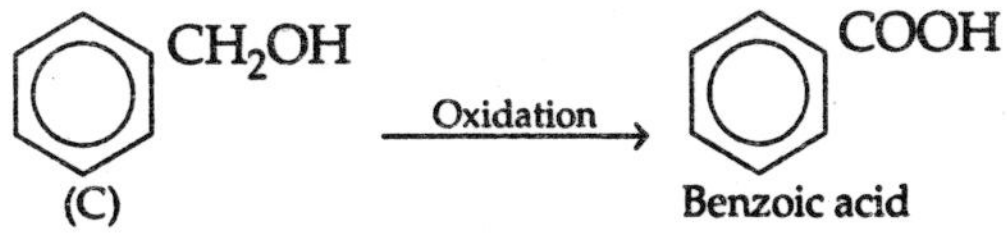
CH2OH
(C)
Oxidation
COOH
Benzoic acid

130. (i) The organic compound (A) releases N_2 with HNO_2 and thus, (A) is primary amine, i.e.,

$$\underset{(A)}{C_9H_{11}NH_2} \xrightarrow{HNO_2} \underset{(B)}{C_9H_{11}OH}$$

(ii) (A) It forms alcohol (B) with HNO_2 which is optically active and thus, keeping in view of molecular formula of (A) and (B) as well as the fact that (B) on oxidation gives dicarboxylic acid (i.e., two acid chain on C_6H_6 ring), (B) can be

$$C_6H_4\begin{matrix} \diagup CH_3 \\ \diagdown CHOHCH_3 \end{matrix} \xrightarrow{[O]} C_6H_4\begin{matrix} \diagup COOH \\ \diagdown COOH \end{matrix}$$

(iii) Since acid formed on oxidation of (B) forms anhydride and thus, two side chains are at ortho positions, i.e.,

COOH
COOH

The reactions are:

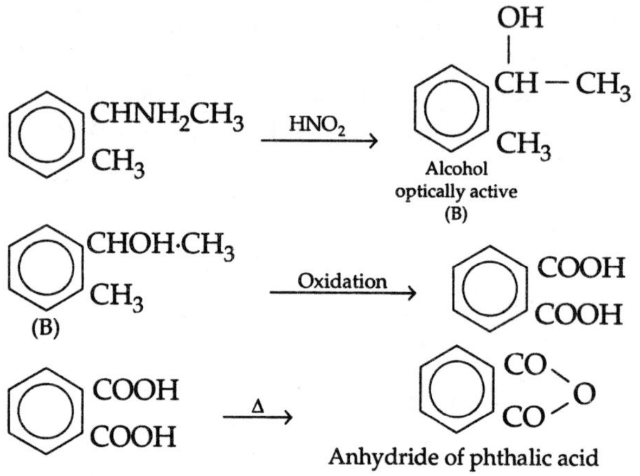
CHNH₂CH₃
CH₃
HNO₂
OH
CH – CH₃
CH₃
Alcohol
optically active
(B)
CHOH·CH₃
CH₃
(B)
Oxidation
COOH
COOH
COOH
COOH
Δ
CO
O
CO
Anhydride of phthalic acid

131. (i) (B) is monobasic acid and on decarboxylation gives C_6H_6 and so (B) is C_6H_5COOH

$$\underset{(B)}{C_6H_5COOH} \xrightarrow[\Delta]{Sodaline} C_6H_5$$

(ii) Since (A) on oxidation (by $KMnO_4$) forms a monobasic acid and thus, only one side chain must be present on C_6H_6 ring.

(iii) Keeping in view of these facts and molecular formula of (A) C_8H_{10}.

A is 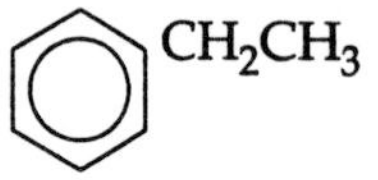
CH₂CH₃
(Ethyl benzene)

Thus, 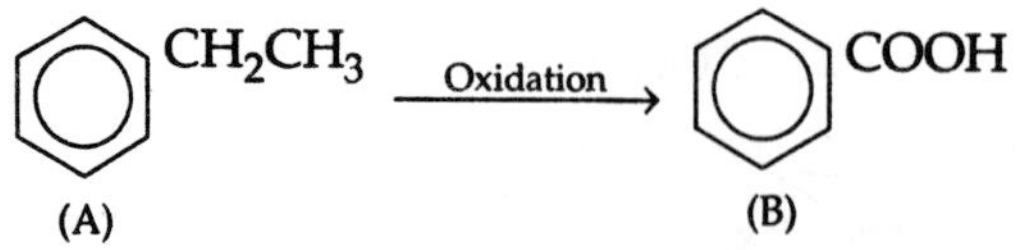
CH₂CH₃
Oxidation
COOH
(A)
(B)
(Benzoic acid)

132. The molecular formula of hydrocarbon (*A*) and step (3) suggest (*A*) to be an aromatic organic compound.
Step (2) suggests that (*A*) has a double bond in side chain. Thus (*A*) may be

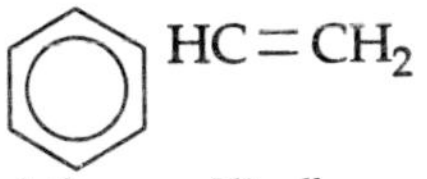

(A) Phenylethene or Vinylbenzene or Styene

Thus, the reactions involved are:

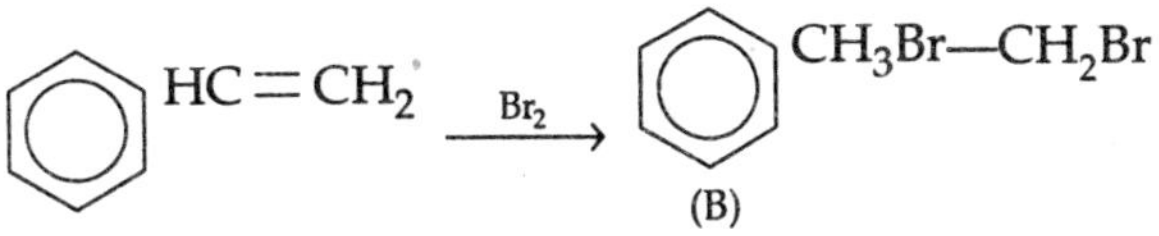

(B)

(C) Benzoic acid

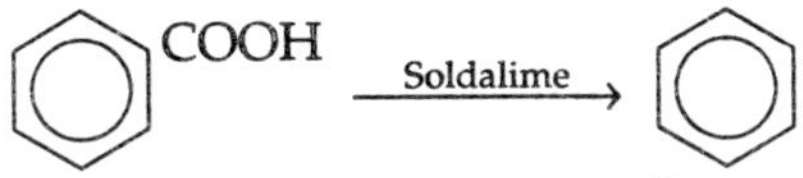

Benzene

133. The compound $C_2H_7NO_2$ on distillation loses water and forms compound (*B*) which on further distillation with P_2O_5 gives (*C*)

$$\underset{(A)}{C_2H_7NO_2} \xrightarrow[-H_2O]{\Delta} \underset{(B)}{C_2H_5NO} \xrightarrow[P_2O_5]{\Delta} \underset{(C)}{C_2H_3N}$$

(*C*) on hydrolysis gives (*D*) which on treating with P + Cl_2 followed with hydrolysis gives glycolic acid.

$$C_2H_3N \xrightarrow{HOH} (D) \xrightarrow[\text{(ii) HOH}]{\text{(i) P + Cl}_2} \underset{\text{Glycolic acid}}{\begin{array}{c} CH_2OH \\ | \\ COOH \end{array}}$$

All these facts can be explained if (*A*) is ammonium acetate., i.e., CH_3COONH_4.
The reactions of CH_3COONH_4 is summarized as:

$$\underset{\substack{\text{Ammonium acetate} \\ (A)}}{CH_3COONH_4} \xrightarrow[-H_2O]{\Delta} \underset{\substack{\text{Ethanamide} \\ (B)}}{CH_3CONH_2} \xrightarrow[P_2O_5]{\Delta} \underset{\substack{\text{Methyl cyanide} \\ (C)}}{CH_3CN}$$

$$\underset{\text{Glycolic}}{CH_2OH{\cdot}COOH} \xleftarrow{HOH} \underset{\text{(2-chloroacetic acid)}}{CH_2Cl{\cdot}COOH} \xleftarrow{P + CL_2} \underset{\text{(acetic acid)}}{CH_3COOH}$$

134. Both (*B*) and (*C*) of $C_8H_8O_2$ are monobasic acids, (*B*) on chlorination gives a compound having 3 Cl atoms whereas (*C*) on chlorination gives a compound having 2 Cl atom. Thus, (B) is CH_3 (C_6H_4) COOH and (C) is $C_6H_5CH_5COOH$.

The reactions are :

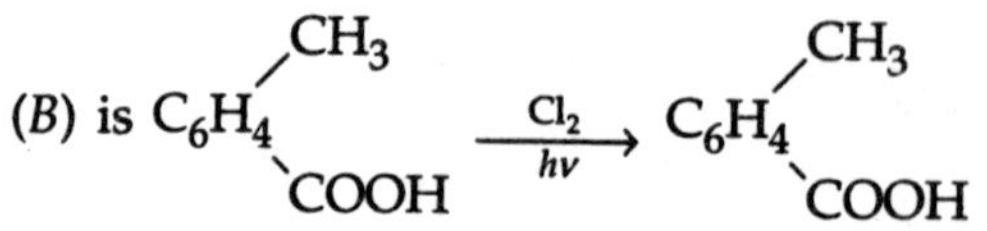

o-, m- and p- (B) o-, m- and p-(I)

$$(C)\text{ is } C_6H_5CH_2COOH \xrightarrow[hv]{Cl_2} C_6H_5CCl_2COOH$$

(C) (H)

Now, Acid (I) formed by (B) on hydrolysis gives dibasic acid (D), which given benzene with sodalime.

Thus,

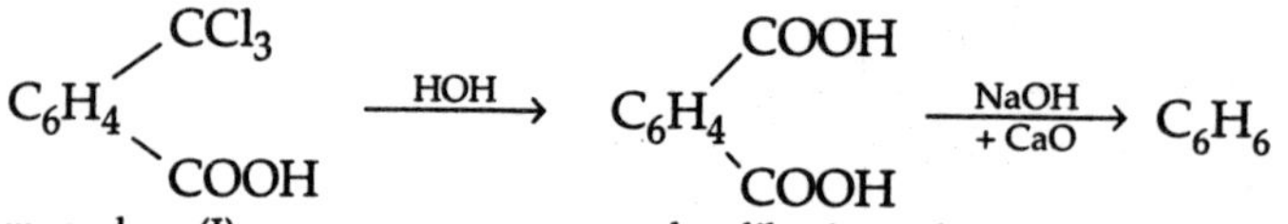

o-, m- and p- (I) o-, m- and p-dibasic acid

Compound (A) being neutral but gives C_6H_6 with sodalime and thus, (A) is

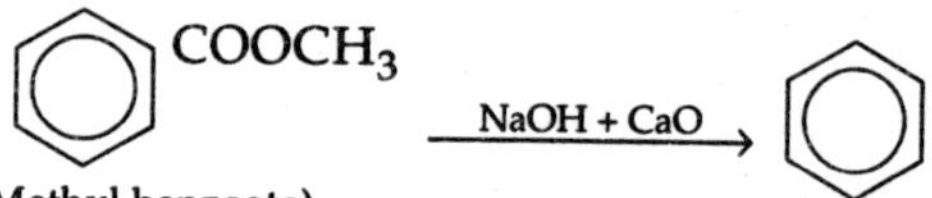

(A) (Methyl benzoate) Benzene

So, the reactions are:

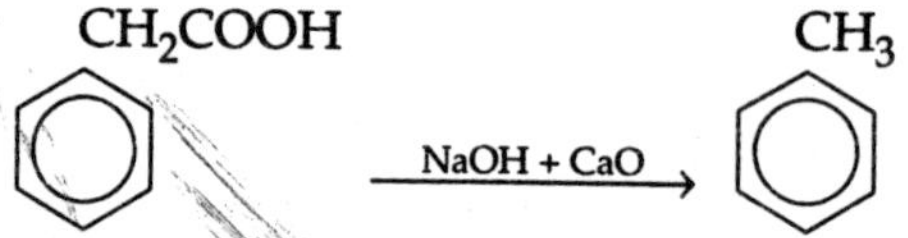

(C) Phenyl acetic acid (Toluene)

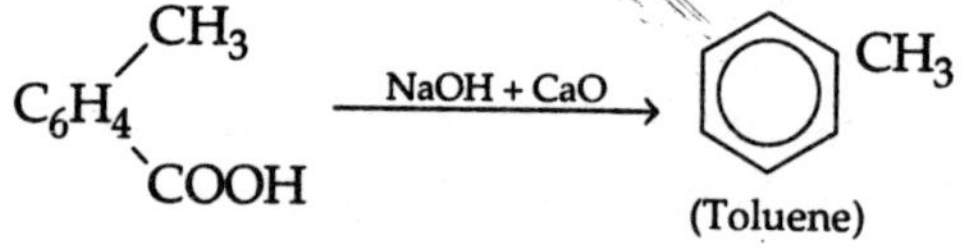

(Toluene)

135. (a) Molecular formula of (A) suggests that it is either

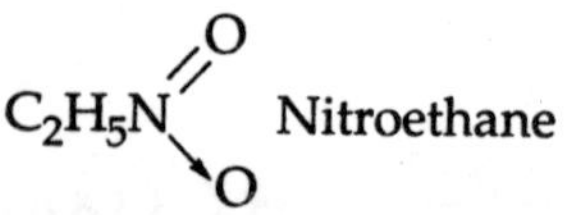

$C_2H_5O—N=O$ Ethyl nitrite

(b) Since (A) undergoes hydrolysis to give and alcohol (B) and an acid, it is ethyl nitrite $C_2H_5O—N{=}O$.

The Reactions are:

$$\underset{\text{Ethyl nitrite }(A)}{C_2H_5O\text{—}N\text{=}O} \xrightarrow[H^+]{H_2O} \underset{(B)}{C_2H_5OH} + HNO_2$$

Mol. wt. of (*B*) is = 46

$$C_2H_5O\text{—}N\text{=}O \xrightarrow{\text{Reduction}} C_2H_5OH + HNO_3$$

136. Both compounds (*A*) and (*B*) lose N_2 on treating with HNO_2 forming alcohols (*C*) and (*D*) respectively and thus, (*A*) and (*B*) both are primary amines.

$$\underset{A \text{ or } B}{C_4H_9NH_2} \xrightarrow{HNO_2} \underset{C \text{ or } D}{C_4H_9OH}$$

Now, (C) reacts with Lucas reagent immediately and so (C) is tertiary alcohol. The tertiary alcohol is of formula C_4H_9OH. Thus, A is 2 methyl propan-2-amine and (C) is 2-methyl propanol, i.e.

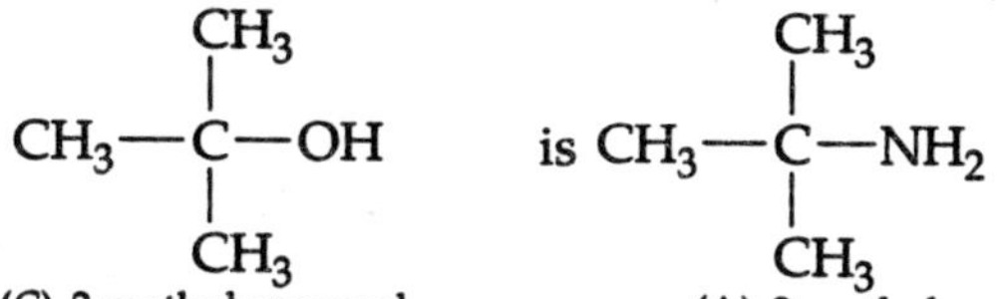

(C) 2-methylpropanol (A) 2-methylpropan-2-amine

(D) does not react with Lucas reagent and is easily oxidized and thus, it is primary alcohol (—CH_2OH). Complete methylation of (*B*) does not give on decomposition butene-1. Thus, (*B*) and (*D*) do not have straight chain. Therefore, (*B*) is 2-methyl propan-1-amine and (D) is 2-methyl propanol.

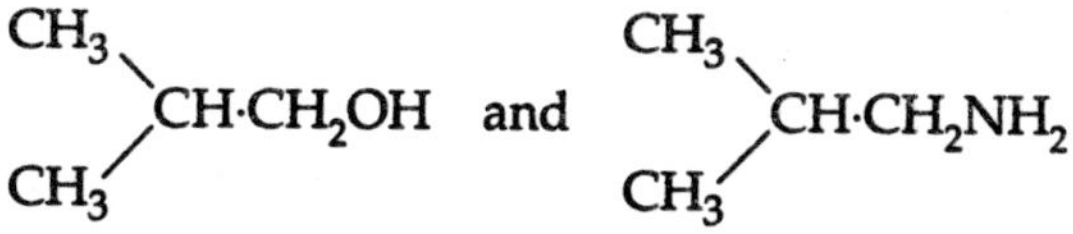

(*D*) 2-methylpropanol (*B*) 2-methylpropan-1-amine

The reactions are therefore:

$$\underset{(A)}{(CH_3)_3C\cdot NH_2} \xrightarrow{HNO_2} \underset{(C)}{(CH_3)_3COH} \xrightarrow[\text{reagent}]{\text{Lucas}} \text{Immediate ppt.}$$

$$\underset{(B)}{(CH_3)_2CHCH_2NH_2} \xrightarrow{HNO_2} \underset{(D)}{(CH_3)_2CHCH_2OH}$$

$$\xrightarrow[\text{reagent}]{\text{Lucas}} \text{No reaction in cold}$$

137. Compound C_3H_7NO i.e. (*A*) on hydrolysis gives an amine (*B*) and a carboxylic acid (*C*), (i.e. RNH_2 amine and *R*COOH).

Amine (*B*) reacts with benzene sulphonyl chloride and the product is insoluble in aqueous solution of NaOH and therefore, amine (*B*) is secondary amine, i.e.,

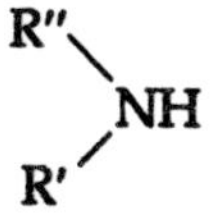

Acid (C) reacts with Tollen's reagent and thus. It may be formic acid HCOOH.
Above findings suggest that (*A*) C_3H_7NO can be

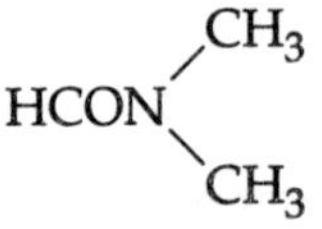

(N, N'-dimethylformamide)

The structure can be justified by the reactions given below:

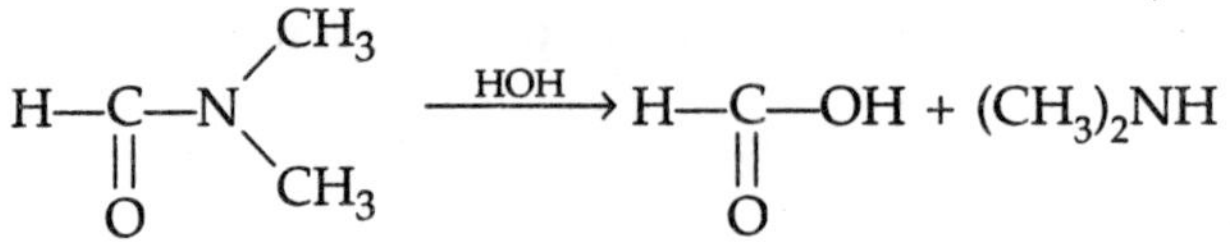

Formic acid (C) Secondary amine (B)

$$(CH_3)_2NH + C_6H_5SO_2Cl \longrightarrow (CH_3)_2N\ SO_2C_6H_5$$

Benzene Sulphonylchloride Sulphonamide Insoluble in NaOH (aq)

$$\underset{(C)}{HCOOH} + Ag_2) \longrightarrow \underset{\text{Silvermirror}}{2Ag} + H_2O + CO_2\uparrow$$

138. (i) For empirical formula of (*A*)

Element	Percentage	Relative no. of atoms	Simplest ratio
C	$\frac{12}{44} \times \frac{0.905}{0.45} \times 100 = 54.8$	4.58	4
H	$\frac{2}{18} \times \frac{0.185}{0.45} \times 100 = 54.8$	4.58	4
Cl	$\frac{35.5}{143.5} \times \frac{0.574}{0.350} \times 100 = 40.57$	1.3	1

∴ Empirical formula of (*A*) = C_4H_4Cl
and Empirical formula weight = 87.5
Molecular weight of (*A*) =2 × 87.5 = 175

∴ Molecular formula of (*A*) is $C_8H_8Cl_2$.

(ii) (*A*) on hydrolysis gives (*B*) which on mild reduction gives optically active compound (C). (C) shows haloform reaction. Thus, (C) should have CH_3CO— unit.

(iii) Keeping in view of above facts, we have

$$\underset{(A)}{C_6H_5CCl_2{\cdot}CH_3} \xrightarrow[Ca(OH)_2]{HOH} \underset{(B)\ \text{Acetophenone}}{C_6H_5COCH_3} \xrightarrow{\text{Reduction}} \underset{\text{optically active (C)}}{C_6H_5CHOH{\cdot}CH_3}$$

$$\underset{\substack{(C)\\ \text{1-Phenylethanol}}}{C_6H_5CHOHCH_5} \xrightarrow[\text{Iodoform test}]{I_2 + NaOH} \underset{\substack{(D)\ \text{Sod. benzone}\\ \text{Soluble in hot water}}}{C_6H_5COONa} + \underset{\text{Iodoform}}{CH_3}$$

139. (a) The empirical formula of compound (*A*) can be derived as

Element	Percentage	Relative No. of atoms	Simplest ratio
C	81.6	$\frac{81.6}{12} = 6.8$	$\frac{6.8}{0.97} = 7$
H	4.8	$\frac{4.8}{1} = 4.8$	$\frac{4.8}{0.97} = 5$
N	13.6	$\frac{13.6}{14} = 0.97$	$\frac{0.97}{0.97} = 1$

Empirical formula is C_7H_5N; Empirical formula wt. = 103
Mol. wt. 2 × vapour density = 2 × 51.5 = 103
Molecular formula of (*A*) = C_7H_5N

(b) The probable compound (*A*) can be 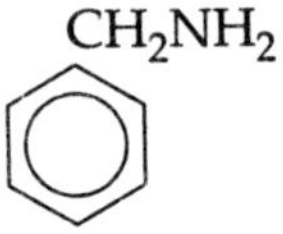 i.e., Phenyl cyanide

(c) The reactions are summarized below:

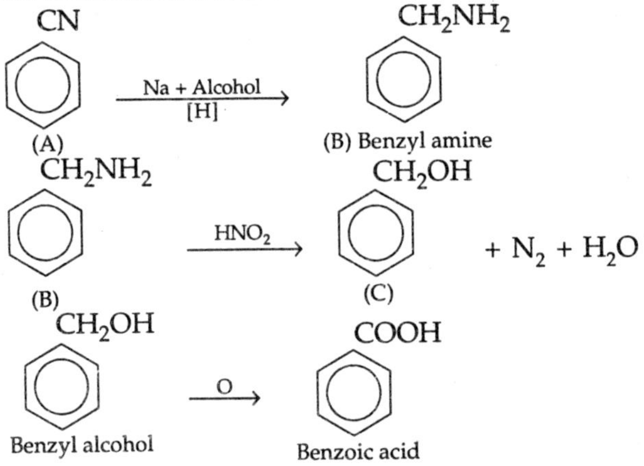

140. (a) $C_6H_5CH_3$

(b) 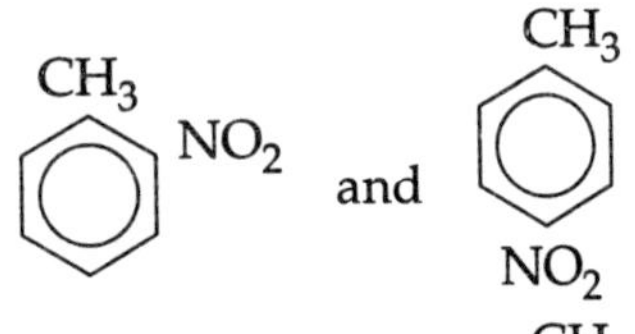

(c) 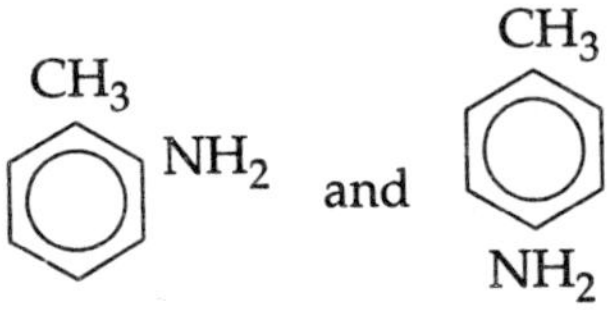

(d) 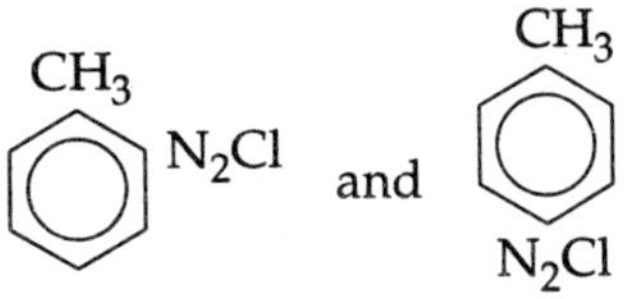

(e) 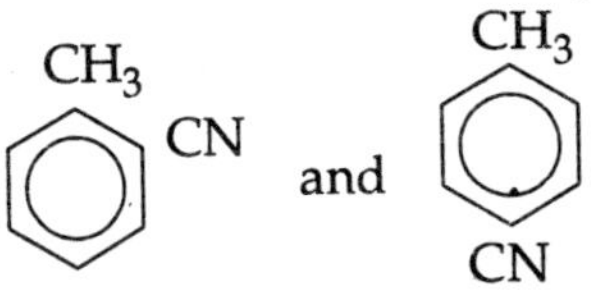

Chapter 14 : Practice Papers

1. Shielding and large ionic radius minimize electrostatic attraction.
2. F has only one negative oxidation state (–).
3. Be^{2+} now has electrons in the first energy level only.
4. $O=C=O$

   Oxidation is more electronegative than carbon, resulting in polar bonding. Because there are no unshared pairs of electrons for carbon, a linear molecule results.

5. 

   There are three molecules listed that exhibit resonance:

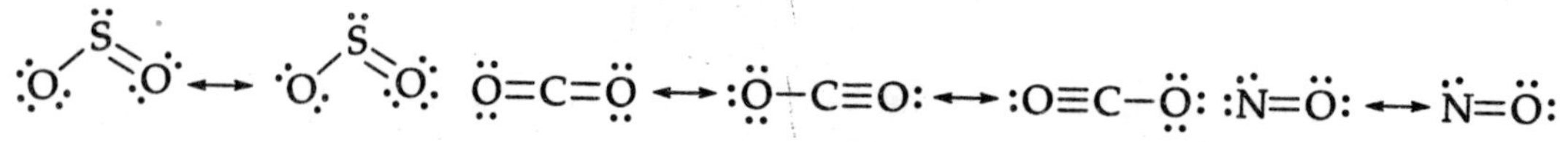

6. $$\text{bond order} = \frac{\text{number of bonding electrons} - \text{number of antibonding electrons}}{2}$$

   or bond order = (total valence electrons – non bonding electron) × $\frac{1}{2}$

   $$= (11 - 6) \times \frac{1}{2} = 2\frac{1}{2}$$

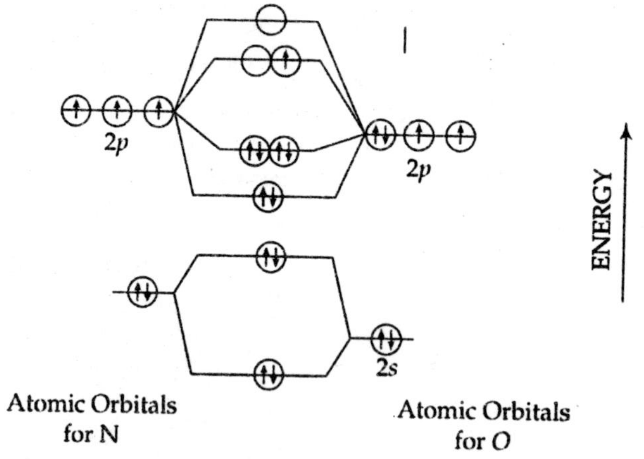

7. 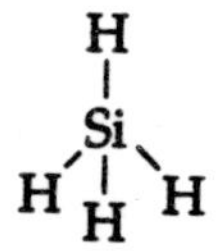

   Silicon, in order to bond four hydrogen atoms to itself, must exhibit sp^3 by hybridization.

8. Two of the four valence electrons would go to the σ_{2s} bonding orbital, and the other two would go the σ^*_{2s} antibonding orbital. The electron configuration would be $(\sigma_{2s})^2(\sigma^*_{2s})$.

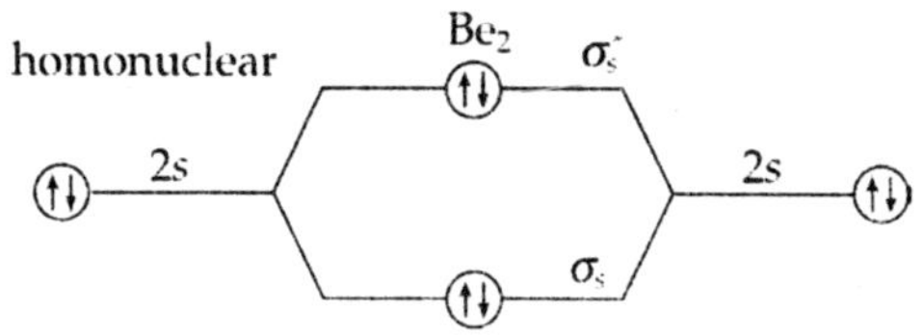

9. "Normal means 1 atom (760 mm Hg) pressure. Boiling occurs at a temperature at which the substance's vapor pressure becomes equal to the pressure above its surface. On this phase diagram, at 1 atm pressure, there is no intercept on a line separating the liquid phase from the gas phase. In other words, carbon dioxide cannot be liquified at 1 atm pressure. It is in the liquid form only under very high pressure. At 1.0 atm, pressure, solid CO_2 will sublime—that is, go directly to the gas phase.
10. The critical point is the point at which the liquid— gas curve ends at a point at which the temperature and pressure have their critical values. Critical temperature is the temperature above which the liquid state of a substance no longer exists. Critical pressure is the pressure at the critical temperature.
11. All three phases are in equilibrium at the triple point. The solid CO_2 sublimes if warmed at any pressure below 5.1 atm. Above 5.1 atm, the solid melts if warmed.
12. The greater the disorder of the system, the larger the entropy. There is an increase in the number of molecules and this greater disorder.
13. Entropy increases upon expansion. The molecules under 1.0 atm of pressure are more free to move around — less constricted.
14. Sublimation means the change from the ordered sold phase to the random gas phase.
15. There are three molecules on the left for every two on the right. Things are becoming more ordered on the right.
16. The system is in equilibrium. The rate of forward reaction equals the rate of the reverse reaction. No one particular side is becoming more (or less) ordered than the order. No additional stress is becoming placed on the system.
17. 8.00 g of oxygen atoms represent 0.500 mole.

$$\frac{8.00 \text{ g O}}{0.500 \text{ mole O}} \times \frac{2 \text{ moles O}}{1 \text{ mole X}} \times \frac{44.0 \text{ g X}}{8.00 \text{ g O}} = 176 \text{ g X/mole X}$$

18. Remember that in an endothermic process, energy is being absorbed. All endothermic changes are defined with a + sign Going from the liquid to the gaseous phase requires energy and thus is endothermic.
19. Remember Boyle's law: As the volume decreases (at constant temperature), the pressure increases.
20. The atomic radius decreases because increasing effective nuclear charge and electrosatic attraction. There are more protons and electrons, so electrons are needed to create a complete shell; thus, there is an increase in electronegativity.

21. Refer to the table entitled "Geometry and Hybridization Patterns," in text books.
22. NH_3 is a neutral ligand; the bromide and the chloride ion both have a –1 charge. Cobalt would have to have +3 charge in this compound for the complex compound to be electrically neutral.

Naming Complex Compounds

1. Name the cation before the anion.
2. Name the ligands before the metal ion.
3. For the ligands:
   (a) Add –o the root name fo the anion—In this case, bromo.
   (b) use aquo for H_2O, amine for NH3, carbonyl for Co, nitrosyl for NO.
   (c) Use the name of the ligand for other neutral ligands.
   (d) Name the ligands alphabetically.
4. Use prefixes to denote the number of ligands of the same kind present.
   (a) mono-, di-, tri-, penta-, hexa- (here, pentaamine).
   (b) bis-, tris-, tetrais for complex ligands.
5. Use Roman numerals to designate the oxidation state of the metal ion (cobalt (III)].
6. Add –ate as a suffix to the metal if the complex ion is an anion.

23. Hydrogen bonding is a very strong *inter* molecular force that occurs between an H atom of one molecule that bonded to either a fluorine, an oxygen, or a nitrogen atom. In choice (D), the hydrogens are bonded to carbon, not to F, O, or N.
24. The molarity of a solution multiplies by its volume equals the number of moles of solute. In this case, 450 mL pf 0.35 M $Al(NO_3)_3$ can be shown as

$$\frac{0.35 \text{ mole Al }(NO_3)_3}{1 \text{ }\cancel{\text{liter solution}}} \times \frac{0.45 \text{ }\cancel{\text{liter solution}}}{1} = 0.16 \text{ mole Al }(NO_3)_3$$

$Al(NO_3)_3$ is completely soluble, so there wold be three times the number of moles of nitrate ions present in the solution because

$$Al(NO_3)_3 \rightarrow Al^{3+} (aq) + 3O_3^-(aq)$$

Therefore, the number of moles of nitrate ions in the original solution would 0.16 × 3 = 0.48.

The number of moles of nitrate ions needs to be brought up to 0.77 because the volume did not change (it remainder at 0.45 liter).

$$\frac{1.7 \text{ moles } NO_3^-}{1 \text{ }\cancel{\text{liter solution}}} \times \frac{0.45 \text{ }\cancel{\text{liter soluion}}}{1} = 0.77 \text{ mole of } NO_3^- \text{ in final solution}$$

The solution begins with 0.48 mole of nitrate ions and must end up with 0.77 moles of nitrate ions; therefore, the solution needs an additional 0.29 mole of nitrate ions:

$$(0.77 - 0.48) = 0.29 \text{ mole } NO_3^- \text{ needed}$$

Calcium nitrate, $Ca(NO_3)_2$, produces 2 moles of nitrate ins in solution for each mole of solid calcium nitrate added to the solution. Therefore, because 0.29 mole of NO_3^- is needed, you will need 0.29/2 = 0.15 mole of solid $Ca(NO_3)_2$.

25. In examining the rate expression, note that B is first-order, so the rate is directly proportional to the concentration of the reactant. Holding $[A]$ constant, and doubling $[B]$ would double the rate.

26. The term $-d[\mathrm{B}]/dt$ represents the rate of decrease in the concentration of B as time elapses. For every mole of B that is lost on the reactant side, $\frac{1}{2} \times 4$, or 2 moles of D are gained on the product side over the same amount of time(dt).

27. The rate constant is independent of the concentration of the reactant. However, k depends on two factors:
    - The nature of the reaction. "Fast" reactions typically have large rate constants.
    - The temperature. Usually k increases with an increase in temperature.

    With all other variables held constant, choice (D) would reduce the rate of molecular collisions, but increasing the volume is analogous to decreasing the concentration.

28. The overall order of the reaction is the sum of the orders of the individual reactants. Here, $[A]^2[B]^1 = 2 + 1 = 3$. For reactions with an overall order of 3, the rate is proportional to the cube of the concentration of the reactants. Reducing the volume by 1/3 effectively triples their concentration: $3^{(\text{conc})3(\text{order})} = 27$.

29. *Step 1:* Write the balanced equation in equilibrium:

$$2SO_3(g) \leftrightarrow 2\ SO_2(g) + O_2(g)$$

Step 2: Write the equilibrium expression:

$$K_c = \frac{[SO_2]^2\ [O_2]}{[SO_3]^2}$$

Step 3: Create a chart showing initial and final concentrations.

Species	*Initial Concentration*	*Final Concentration*
SO_3	1.20 M	1.20 M – 0.20 M = 1.00 M
SO_2	0 M	0.20 M
O_2	0 M	0.10 M

Step 4: Substitute the final equilibrium concentrations into the equilibrium expression.

$$K_e = \frac{[SO_2]^2\ [O_2]}{[SO_2]^2} = \frac{(0.20)^2\ (0.10)}{(1.00)} = 4.0 \times 10^{-3}$$

30. Remember, $\log\,[H^+] = -pH$, so $[H^+] = 10^{-pH}$.

31. When $\Delta G° = 0$,

$$T = \frac{[(-600.) + (-400.)\ [1170.]]}{[(0.0700) + (0.2000)] + [0.10000]}$$
$$= 1.00 \times 10^3\ \mathrm{K}$$

32. $k = \dfrac{0.693}{t_{1/2}} = \dfrac{0.693}{6930\ \mathrm{yr}} = 1.00 \times 10^{-4}\ \mathrm{y^{r-1}}$

33. The functional group of a salt is

$$\overset{\displaystyle O}{\overset{||}{-C}}-O^{-}\cdot M^{+} \text{ where, M = metal}$$

The name of this compound is potassium propioate.

34. The functional group of a ketone is

$$-\overset{\displaystyle O}{\overset{||}{C}}-$$

The name of this ketone is methyl ethyl ketone.

35. The functional group of an amide is

$$-\overset{\displaystyle O}{\overset{||}{C}}-N\begin{smallmatrix}/\\ \backslash\end{smallmatrix}$$

The name of this amide is acetamide.

36. The functional group of an amine is $-\overset{|}{N}-$

The name of this amine is dimethylamine.

37. The functional group of a thiol is

—S—H

The name of this thiol is ethanethiol.

38. Do this problem by using the factor-lavel method m.t. stands for metric tons, 1 m.t. = 103 kg).

$$\frac{6.0\times 10^{2}\ \text{m.t. alloy}}{1}\times\frac{12.709\ \text{m.t. Cu}}{100.\ \text{m.t. alloy}}\times\frac{159.15\ \text{m.t. Cu}_2\text{S}}{127.09\ \text{m.t. Cu}}$$

$$=\frac{100\ \text{m.t. ore}}{15.915\ \text{m.t. Cu}_2\text{S}} = 6.0\times 10^{2}\ \text{m.t. ore}$$

39. Begin this problem by balancing the reaction.

$$2\ C_2H_6(g) + 7\ O_2(g) \rightarrow 4CO_2(g) + 6H_2O(g)$$

Because $\Delta H^{\circ} = \Sigma\Delta H^{\circ}_{f\ \text{products}} - \Sigma\Delta H^{\circ}_{f\ \text{reactants}}$, you can substitute at this point.

$$\Delta H^{\circ} = [4(-392.9)+6(-250.8)-2(-83.6)]$$

However, remember that the question calls for the answer per mole of C_2H_6. Thus, =–2909.2 because the balanced equation is written for 2 moles of C_2H_6, simply divide –2909.2 by 2 and you get the answer –1454.6.

40. Review the postulates of the kinetic molecular theory of gases.

41. A principal quantum number of 4 tells you that you are in the fourth energy level. The fourth energy level contains electrons in the *s*, *p*, *d* and of orbitals. Counting the maximum numbers of electrons available in each of the four types of sublevels — 2 in the *s*, 6 in the *p*, 10 in the *d*, and 14 in the *f* — yields a total of 32.

42. When presented with a generic formula, such AB_2, the best way to answer the question is to use familiar examples that satisfy the conditions of the questions. H_2O would satisfy AB_2 as a bent molecule, and CO_2 would itself AB_2 as a linear molecule. In CO_2, a linear molecule, the two dipoles cancel each other, resulting in a nonpolar molecule. However, in H_2O, which also satisfies the AB_2 requirement, the two dipoles do not cancel each other out and result in a net dipole moment and a bent molecule. For both CO_2 and H_2O, we have data on ionization energy, electron affinity, electronegativity, and bond energy, but these ar eof no use, by themselves in determining the geometry of the species.

43. Water molecules are neutral. Hydroxide ions (OH^-) have a – 1 charge. The overall charge of the complex is +1. Zinc would have top have a +2 charge in order for the complex to end up with a + 1 charge. If you let x the charge of the zinc ion, then

$$+1 = x + 30 + 1(-1)$$
$$x = +2$$

44. To do this problem, we use the Clausius-Clapeyron equation:

$$\log\frac{P_2}{P_1} = \frac{\Delta H_{vap}}{2.303\,R}\left(\frac{T_2 - T_1}{T_2 T_1}\right)$$

where

P_1(132 mm Hg) is the vapor pressure of the liquid at T_1(327 K).
$P_2(x)$ is the vapor pressure of the liquid at T_2 (310 K).
R is a universal gas constant: 8.314 joules/(mole · K).

Although the problem does not require you to solve the equation, it is presented below. Substituting the values of the problem into the equation gives

$$\log\frac{x}{132\text{ mm Hg}} = \frac{4.33\times10^4\ \text{J/}\cancel{\text{mole}}}{2.303\times8.314\ \cancel{\text{J/(mole}\cdot\text{K)}}}\left(\frac{310\ \cancel{K} - 327\ \cancel{K}}{310\ \cancel{K}\cdot 327\ \cancel{K}}\right)$$

Simplifying this problem gives you

$$\log\frac{x}{32} = -0.379$$

Solving for x yield x = 55.2 mm Hg.

Note that the question tells you there are 2.00 moles of the compound. This information is irrelevant to solving the problem since equilibrium vapor pressure is independent of the amount of compound.

45. This problem can be solved by using the factor-label method. 100.g solution – 10.g solute = 90g solvent (H_2O)

$$\frac{10.\ \cancel{\text{g } C_5H_2O}}{90.\ \cancel{\text{g } H_2O}} \times \frac{1000\ \cancel{\text{g } H_2O}}{1\text{ kg } H_2O} \times \frac{1\text{ mole } C_6H_2O}{90.\ \cancel{\text{g } C_6H_2O}} \approx 1.2m$$

46. The Al^{3+} has been diluted tenfold:

$$\frac{1.0\text{ M}\times1\ \cancel{\text{cm}^2}}{10.\ \cancel{\text{cm}^2}} = 0.10\text{ M}$$

47. $\frac{\text{rate}_2}{\text{rate}_1} = \frac{(\text{new conc.})}{(\text{old conc.})} = \frac{8}{1} = \left(\frac{2 \cdot \text{old conc.}}{\text{old conc.}}\right) = 2^x$

$$8 = 2^x$$
$$x = 3$$

48. All intermediate mechanisms must add up to yield the original, overall balanced equation.

$$NO(g) + O_2(g) \rightarrow \cancel{NO_3(g)}$$
$$\cancel{NO_3(g)} + NO(g) \rightarrow 2NO_2(g)$$
$$\overline{2NO(g) + O_2(g) \rightarrow 2NO_2(g)}$$

49. Remember that $K_a \times K_b = 10^{-14}$. Therefore,

$$K_b \text{ Lac}^- = \frac{10^{-14}}{1.5 \times 10^{-5}} = 6.7 \times 10^{-10}$$

50. Use the relation ship gram-equivalent acid = gram-equivalent base. Solve for the gram-equivalent of acid.

$$\frac{0.700 \cancel{\text{ liter and}}}{1} \times \frac{3.0 \text{ gram} - \text{equiv. acid}}{1 \cancel{\text{ liter}}} = 2.1 \text{ gram} - \text{equivalent of acid}$$

At neutralization, the gram-equivalents of acid = the gram-equivalents of base. Therefore,

$$\frac{2.1 \cancel{\text{ gram equiv. NaOH}}}{1} \times \frac{40.00 \text{ g NaOH}}{1 \text{ gram equiv. NaOH}} = 84 \text{ g NaOH}$$

51. ΔG° represents the free energy at standard conditions: 298K and 1 atm pressure. ΔG represents the free energy at nonstandard conditions. In this problems, we have the non-standard condition of 373K. In order to solve for the free energy fo this reaction, you must use the following equation:

$$\Delta G = \Delta G^\circ + 2.303 \text{ RT} \log Q_p$$

where the constant $R = 8.314 \text{ J} \cdot \text{K}^{-1} \cdot \text{mole}^{-1}$ and Q_p is called the reaction quotient. The reaction quotient has the same form as the equilibrium constant K_p.

Step 1: Write a balanced equation.

$$Q_p = [CO_2(g)] = 1.00$$

Step 2: Determine the value of Q_p, the reaction quotient.

$$Q_p = [CO_2(g)] = 1.00$$

Step 3: Substitute into the equation

$$\Delta G = \Delta G^\circ + 2.303 \, RT \log Q_p$$
$$= 130{,}240 \text{ J mol}^{-1} + 2.303(8.314 \text{ J} \cdot \text{K}^{-1} \cdot \text{mole}^{-1}) (373 \text{ K})(\log 1.00)$$
$$= 130{,}240 \text{ J mol}^{-1}$$
$$= 130.240 \text{ kJ mol}^{-1}$$

52. The position is a particle with the same mass the electron but the opposite charge. The net effect is to change a proton to a neutron. Begin by writing the nuclear equation.

$$^{239}_{94}Pu \rightarrow ^{0}_{1}e + ^{A}_{Z}X$$

Remember that the total of the A and Z values must be the same on both sides of the equation.
Solve for the Z value of X: $Z + 1 = 94$, so $Z = 93$.
Solve for the A value of X: $A + 0 = 239$, so $A = 239$.

Therefore, you have $^{239}_{93}X$ or $^{239}_{93}Np$.

53. Ni(s) | Ni^{2+}(aq) | | Ag^{+}(aq) | Ag(s)
anode (oxidation | | cathode (reduction)

$$Ni(s) + Ni^{2+}(aq) + 2^{e-} \| 2\ Ag^{+}(aq) + 2e^{-} \rightarrow 2\ Ag(s)$$

By convention, in he representation of the cell, the anode is reprsented on the left and the cathode on the right. The anode is the electrode at which oxidation occurs (AN OX), and the cathode is the electrode at which reduction takes place (RED CAT). The single vertical lines (|) indicate contact between the electrode and solution. The double vertical line (||) represent the porous partition, or salt bridge, between the two solutions in the two half-cells. The ion concentration or pressures of a gas are enclosed in parentheses.
Take the two equations that decoded the standard cell notation and include the $E°$ reduction and the $E°$ oxidation voltages:
(Change the sign!)

$$\text{ox: } Ni(s) \rightarrow Ni^{2+}(aq) + 2e^{-}$$
$$\text{red: } 2\ Ag^{+}(aq) + 2e^{-} \rightarrow 2\ Ag(s)$$

$$E^{\circ}_{ox} = +0.25 \text{ volts}$$

$$E^{\circ}_{red} + 0.80 \text{ volt}$$

$$E^{\circ}_{cell} + 1.05 \text{ volts}$$

By definition, E°_{cell} voltages that are positive indicate a spontaneous reaction.

54. Begin by writing as much of an equation as you can:

$$S_8(s) + M(s) \rightarrow M_aS_b(s)$$

From the information provided, you can determine that the residue, $M_aS_b(s)$, weighed 21.00 grams (62.00 – 41.00) and that the metal M weighed 14.00 grams 55.00 – 41.00). According to the Law of Conversion of Mass, the sulfur that reacted with the metal must have weighed 7.00 grams (21.00 – 14.00). You can now set up a proportion that relates the grams of S_8 and M to their respective equivalent weights.

$$\frac{7.00 \text{ grams sulfur } (S_s)}{16.0 \text{ grams/equiv. sulfur}} \times \frac{14.00 \text{ grams M}}{x \text{ grams/equiv.}}$$

Solving for x, you obtain 32.00 grams/eqiuv. For metal M. From this information, it would seem reasonable that the unknown metal is copper, forming the compound CuS. Copper, with a +2 valence has an equivalent weight of 31.78.

55. The law of Dulong and Petit states that

$$(\text{molar mass}) \times (\text{specific heat}) \approx 25 \text{ J mole}^{-1} \text{ K}^{-1}$$

Substituting the given information into this relationship yields

$$(x \text{ g mole}) \times 0.052 \text{ cal/g} \cdot °\text{C} \approx 25 \text{ J mole}^{-1} \text{ K}^{-1}$$

Because you know that 4.184 joules = 1 calories, convert the 25 joules to calories so that units can cancel.

$$(x \text{ g mole}^{-1}) \times 0.052 \text{ cal g}^{-1} \cdot \text{K}^{-1} = 6.4 \text{ cal mole}^{-1} \cdot \text{K}^{-1}$$

Solving for x yields ~124.

56. $\text{Density} = \dfrac{\text{mass}}{\text{V}} = \dfrac{\text{P}\cdot\text{M}}{\text{R}\cdot\text{T}} = \dfrac{\cancel{\text{atm}}\cdot \text{g}/\cancel{\text{mole}}}{\text{L}\cdot\cancel{\text{atm}\cdot\text{mole}^{-1}\text{K}^{-1}\cdot\text{K}}} = \dfrac{\text{g}}{\text{L}}$

Many students would choose answer choice (d). The reason molecular mass is the wrong answer as the question did not state the condition when comparing one gas to another. For example, 1 mole of hydrogen gas at 50 K in a 1.0-liter container might have a higher density that 0.01 mole of uranium hexafluoride (UF_6 at 200 K in a 20-liter container, even though the UF6 has a greater molecular mass.

57. Element A keys out to be Al, which, begin a metal in Group IIIA, would have a +3 charge. Element B would keys out as sulfur, nonmetal with a charge of –2, giving the formula Al_2S_3, or A_2B_3.

58. If you missed this question, go back to the rules for naming coordination compounds.

The (III) indicates that the central positive ion should have a +3 charge. In adding up the charges for the ligand you get:

$$\begin{array}{l} 2H_2O = 0 \\ 2NH_3 = 0 \\ 2(Br^-) = -2 \\ \underline{1(Cr^{+?}) = ?} \\ +1 \text{ (overall charge of compound)} \end{array}$$

Therefore, the Cr ion must have a charge of +3; choice (D) also yields a+3 charge for Cr and yet does not confirm to standard methods of writing complex ion formulas.

59. Raoult's law states that the partial pressure of a solvent over a solution, P_1 is given by the vapor pressure of the pure solvent, P_1° items the mole fraction of the solvent in the solution X_1.

$$P_1 = X_1 P_1^\circ$$

A dilution in vapor pressure is directly to the concentration (measured as mole fraction) of the solute present.

60. In dilution problems, we use the formula $M_1V_1 = M_2V_2$; therefore, it is necessary to determine the molarity of the initial solution first.

$$\frac{1.50 \text{ g } \cancel{\text{solution}}}{1 \text{ m}\cancel{\text{L solution}}} \times \frac{1000 \text{ mL solution}}{1 \text{ liter solution}} \times \frac{36.45 \text{ g HCl}}{100 \text{ g } \cancel{\text{solution}}} \times \frac{1 \text{ mole HCl}}{36.45 \text{ } \cancel{\text{g HCl}}}$$

Next we use the relationship $M_1V_1 = M_2V_2$:

(15.0 M)(x liters) = (5.0 M)(9.0 liters)

$$x = 3.0 \text{ liters}$$

61. In examining the balanced equation, note that for each mole N_2O_5 gas the decomposes, $\frac{1}{2}$ mole of O_2 gas is formed. Therefore, the rate of formation of oxygen gas should be half the rate of decomposition if the N_2O_5.

62. Begin by writing the equilibrium equation.

$$PbCl_2 \rightleftarrows Pb^{2+}(aq) + 2Cl^-(aq)$$

Next, write the equilibrium expression.

$$K_{sp} = [Pb^{2+}][Cl^-]^2$$

In reference to the chloride ion concentration, rewrite the expression for $[Cl^-]$:

$$[Cl^-] = \left(\frac{K_{sp}}{[Pb^{2+}]}\right)^{\frac{1}{2}}$$

At any value greater than this expression, $PbCl_2(s)$ will precipitate, removing $Cl^-(aq)$ from solution.

63. Use the equation $N_aV_a = N_bV_b$. Solve the equation for N_a.

$$N_a = \frac{N_b V_b}{V_s} = \frac{0.50\ \text{N} \times 50.0\ \text{mL}}{25.0\ \text{mL}} = 1.0\ \text{N}$$

64. This problem can be done using the factor-label method.

$$\frac{1.0\ \text{gram equiv. } HC_2H_3O_2}{1\ \text{liter vinegar}} \times \frac{60.00\ \text{g } C_2H_3O_2}{1\ \text{gram equiv. } HC_2H_3O_2/\text{liter of vinegar}}$$

65. This problem can be solved using the factor-label method.

$$\text{wt.\%} = \frac{\text{parts } HC_2H_3O_2}{\text{solution}} \times 100\%$$

$$= \frac{60.\ \text{g } HC_2H_3O_2}{1000.\ \text{mL solution} \cdot 1.0\ \text{g/mL}} \times 100\% = 6.0\%$$

66. Heat needs to be absorbed when a solid melts; therefore, the reaction is endothermic $\Delta H > 0$. When a solid metals and becomes a liquid, it is becoming more disordered, $\Delta S > 0$.

67. The equation concerns the effect of changing standard conditions of a cell to nonstandard conditions. The calculate the voltage of a cell under nonstandard conditions use the Nernst equation

$$E = E° - \frac{0.0591}{n}\log Q = E° - \frac{0.0591}{2}\log\frac{[Zn^{2+}]}{[Ag^+]^2}$$

where $E°$ represents the cell voltage under standard conditions, E represents the cell voltage under nonstandard conditions, n represents the number of moles of electrons passing through the cell, and Q represents the reaction quotient.

Choices (D) and (E) would have the effect of increasing the cell voltage. Choices (A) and (B) would have no effect on the cell voltage.

68. To solve this problem, use the equation

$$\log \frac{x_0}{x} = \frac{k \cdot t}{2.30}$$

with the corresponding half-life $t_{1/2} = 0.693/k$, where x_0 is the number of original radioactive nuclei and x represents the number of radioactive nuclei at time t. k represents the first-order rate constant. Substituting into equation yield

$$\log \frac{x_0}{x} = \frac{(0.693/6.93 \cancel{\text{years}})(11.5 \cancel{\text{years}})}{2.30} = 0.5$$

$$\frac{x_0}{x} \approx 3$$

$$\frac{x}{x_0} = \frac{1}{3} \times 100\% \approx 33\% \text{ that remain uncreated}$$

69. Begin by writing the equations which define the equilibrium constants.

$AgCl(s) \rightarrow Ag^+(aq) + Cl^-(aq)$ $\quad K_{sp1} = 1.6 \times 10^{-10}\ mol^2 \cdot L^{-2}$

$AgI(s) \rightarrow Ag^+(aq) + I^-(aq)$ $\quad K_{sp2} = 8.0 \times 10^{-17}\ mol^2 \cdot L^{-2}$

The K_{eq} is needed for the following equation:

$$AgCl(s) + I^-(aq) \rightarrow AgI(s) + Cl^-(aq) \qquad K_{eq} = \frac{[Cl^-]}{[I^-]}$$

$$\frac{1.6 \times 10^{-10} \cancel{\text{mole}^2}}{\cancel{L^2}} \times \frac{\cancel{L^2}}{8.0 \times 10^{-17} \cancel{\text{mole}^2}} = 2.0 \times 10^6$$

70. Bond breaking (ΔH_1) = H—H + F—F = 430.5K.J mol^{-1} + 137.9K.J $mole^{-1}$
Bond forming (ΔH_2) = 2 (H—F) = 2(–564.3K.J. mol^{-1}) = –1128.6 K.J mol^{-1}
$\Delta H° = \Delta H_1 + \Delta H_2$ = 568.4 K.J. mol^{-1} + (–1128.6 K.J. mol^{-1}) = –560.2 K.J. mol^{-1}

71. Whether you can answer this question depends in whether you are acquainted with what is known as the Maxwell-Boltzmann distribution. The distribution describes the way that molecular speeds or energies are shaerd among the molecules of a ages. If you missed this question, examine the following figure and refer to your textbook for a complete description of the Maxwell-Bioltzmann distribution.

72. Because all choices have 18 electrons in their valence shell, you should pick the species with the fewest protons in the nucleus; this would result in the weakest electrostatic attraction. That species is sulfur.

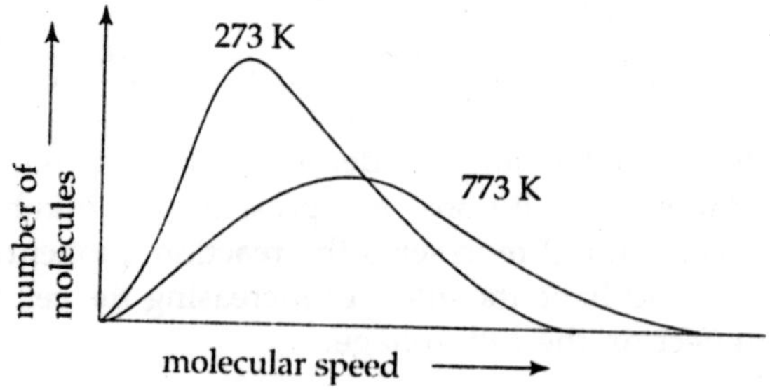

73. The central metal ion from only two bonds to ligands, so the coordination number is 2.

74. The body-centered cubic cell looks like this:

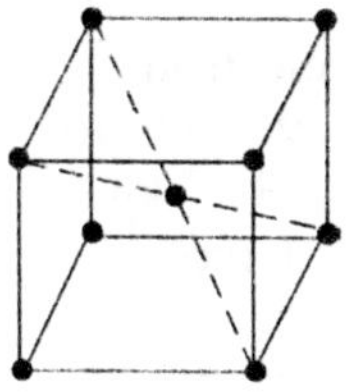

The formula that relates the atomic radius (r) to the length of one edge of the cube(s) for a body-centered cubic cell is $4r = s\sqrt{3}$.

$$S = \frac{4(0.255 \text{ nm})}{\sqrt{3}} = 0.60 \text{ nm}$$

75. H_3PO_4 is a triprotic acid; that is there are 3 moles of H^+ ions produced for each mole of H_3PO_4 that completely ionizes. Normality is the number of equivalents per liter. Assuming complete or 100% ionization, a 1-molar HCl solution is 1 normal. A 1-molar H_2SO_4 solution is 2 normal, and a 1-molar solution of H_3PO_4 is 4 normal.

In order to use the concept of normality, one must know the reaction involved. Phosphoric acid, H_3PO4 undergoes three simultaneous ionizations:

$$H_3PO_4 + H_2O \rightleftarrows H_3O^+ + H_2PO_4^- \qquad K_{a1} = 5.9 \times 10^{-3}$$
$$H_2PO_4^- + H_2O \rightleftarrows H_3O^+ + HPO_4^{2-} \qquad K_{a2} = 6.2 \times 10^{-8}$$
$$HPO_4^{2-} + H_2O \rightleftarrows H_3O^+ + PO_4^{3-} \qquad K_{a3} = 4.8 \times 10^{-13}$$

Because 1 mole of H_3PO_4 weights 97.995 grams, 1 equivalent of H_3PO_4 would weigh $\frac{1}{3}$ as much, or 32.665 grams. Given this relationship, it is now possible to do this problem by using factor-label techniques.

$$\frac{100.0 \text{ mL}}{1} \times \frac{1 \text{ liter}}{1000 \text{ mL}} \times \frac{0.100 \text{ equiv}}{1 \text{ liter}} \times \frac{32.665 \text{ g}}{1 \text{ equiv}} = 0.327 \text{ g}$$

Chapter 15 : Test Paper I

1. Mass of electron = 9.108×10^{-31} kg.
so,

Mass of one mole of electron
$= 6.023 \times 10^{23} \times 9.108 \times 10^{-31}$
$= 6.023 \times 9.108 \times 10^{-8}$ kg

Since $6.023 \times 9.108 \times 10^{-8}$ kg. of electron = 1 mol

$$\therefore \quad \text{kg of electron} = \frac{1}{9.108 \times 6.023 \times 10^{-8}}$$

or
$$= \frac{1}{6.023 \times 9.108} \times 10^{8} \text{ moles.}$$

2. From gas equation
at constant pressure (i.e. one atm)

$$\frac{V_1}{T_1} = \frac{V_2}{T_2}$$

$$\therefore \quad \frac{22.4}{270} = \frac{V_2}{373}$$

or
$$V_2 = \frac{373}{270} \times 22.4 = 30.6 \text{ Liter.}$$

3. 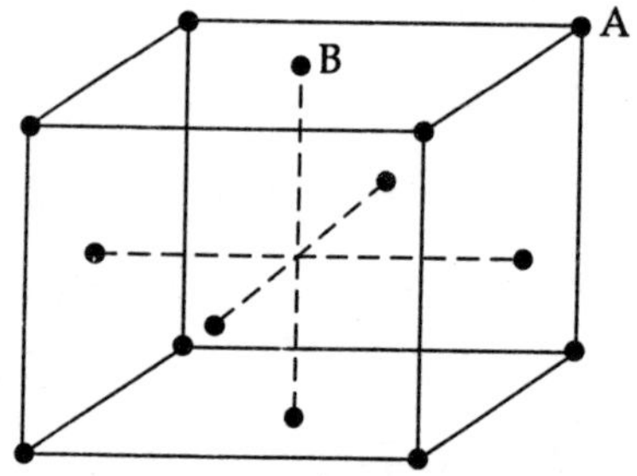

Eight corners of cube is occupied by atom A.
$\therefore$ Number of particle/atom of A

$$= \frac{N_{corner}}{8} = \frac{8}{8} = 1$$

Six face of cube is occupied by atom B

$$\therefore \text{ No of particle/atom of } B = \frac{N_{face}}{2} = \frac{6}{2} = 3$$

Thus, A × By can be
A_1B_3 or AB_3
so, (a) is correct Ans.

4. Nuclephile are negative charged centres which attacks the most electrophilic site first. Among aldehyde and a ketone, aldehydes are ore electrophilic as in ketones the δ^+ change on carbonyl carbone is decreased by +1 effect of both alkyl groups

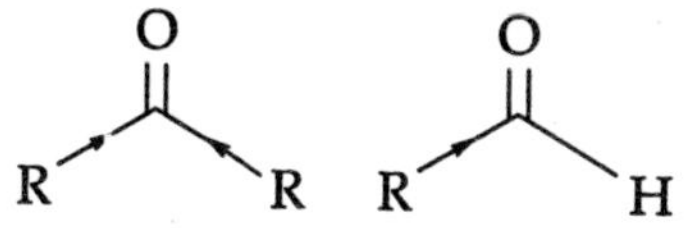

More over in tetrahedral intermediate aldehyds have less steric repulsion than ketone and aldehydes increases the –ve change on oxygen less in comparison to ketones.

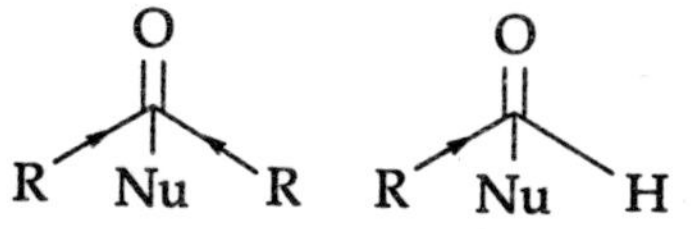

Based on the above the order of reactivity is (II) > (I) > (III)

5. We can see it in two steps i.e., 60° each rotation anticlockwise along the axis $C_2 - C_3$. Finally we will get Gauche conformation from anti conformation.

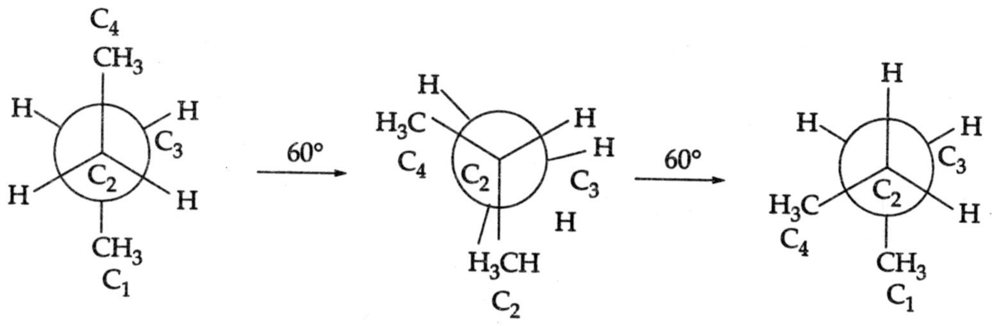

6. Change in enthalpy = $\Delta U + \Delta n R \Delta T$

Here,

$$\Delta n = 1 \text{ mol}$$
$$R = .0821 \text{ lit-atm/k/mol.}$$
$$\Delta T = 245 - 95 = 150 \text{ k}$$
$$\Delta U = 30.0 \text{ L atm.}$$
$$\therefore \quad \Delta H = 30.0 + 1 \times 0.0821 \times 150$$
$$= 42.31 \text{ L atm.}$$

7. $POCl_3$ is a dehydrating agent which removes one molecule of water from it and gives cyanide.

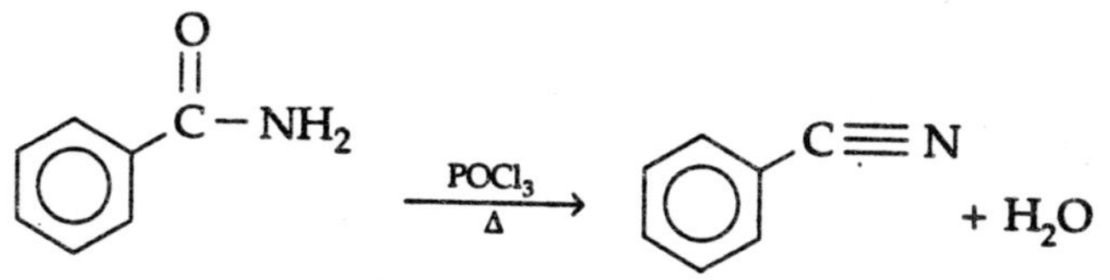

8. Z-phenyl propene on protonation gives 3° carbocation which further undergo hydration, the reaction follows the trend as

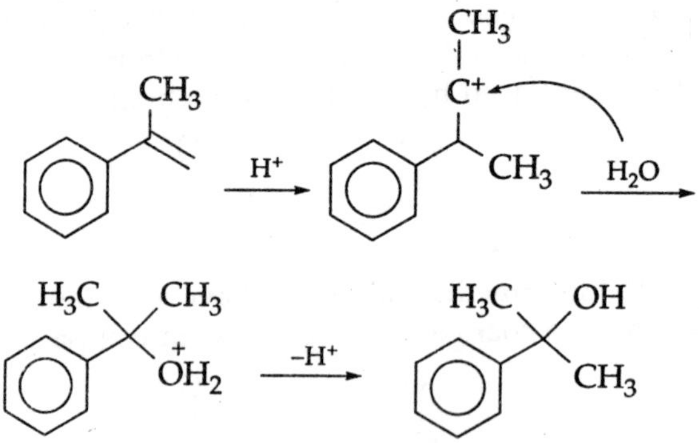
CH3
CH3
C+
H+
CH3
H2O
H3C
CH3
+
OH2
–H+
H3C
OH
CH3

9. The group which contains lone pair of electron are generally ring activatory except halogens. The phenyl ring having H – N < group is activated while another one is deactivated due to $-\overset{\overset{\displaystyle O}{||}}{C}-$, so electrophilic aromatic bromination will occur at para position with respect to H – N < group in activated ring.

10. MnO_4^- and HCl may be represented as follows

$$2MnO_4^-(aq)\ 16H^+ + Cl^- \rightarrow 2Mn^{2+}(aq) + 8H_2O(l) + 5Cl_2(g)$$

Above reaction represents in electrochemical cell.

$$PlCl_2(latm)/Cl^-_{(aq)} \,||\, MnO_4^-(aq)/Mn^{2+}\ (aq)\ E^\circ_{cell} = E^\circ_R - E^\circ_L = 1.51 - 1.40 = 0.11\ V$$

Since E°_{cell} is +ve the value of free energy change will be –ve it means this reaction is feasible MnO_4^- ion will not oxidize Fe^{2+} to Fe^{3+} but Cl^- to Cl_2 also. Therefore for quantitative, estimation of aqueous $Fe(NO_3)_2$ it is not a suitable reagent.

11. Cell reaction:

$$Zn + Fe^{2+} \rightarrow Zn^{2+} + Fe$$

$$E = E^\circ = \frac{0.0591}{2}\log\frac{10^{-2}}{10^{-3}}$$

$$\therefore \qquad E^\circ = 0.2905 + \frac{0.0591}{2} = 0.32$$

or

$$0.32 = \frac{0.0591}{2}\log K_{eq}$$

$$\therefore \qquad K_{eq} = 10^{\frac{0.32}{0.0295}}$$

12. $(NH_4)_2\ Cr_2\ O_7 \xrightarrow{\Delta} N_2\uparrow + Cr_2O_3 + 4H_2O$

$NH_4NO_2 \xrightarrow{\Delta} N_2\uparrow + 2H_2O$

13. $X^- + H_2O \rightarrow HX + OH^-$

$$K_h = \frac{K_w}{K_a} = 10^{-9}$$

$K_h = C\alpha^2 = 10^{-9}$

$\therefore\ \alpha^2 = 10^{-8} \Rightarrow \alpha = 10^{-4}$

% degree of dissociation= $10^{-4} \times 100 = 0.01\%$

14. Since

$$P = \frac{1}{3} \times \frac{m}{v} \times c^2 c = \text{(rms velocity}$$

$$\Rightarrow \qquad \frac{\sqrt{3PV}}{m} = c$$

For 1 molecule $PV = kT \Rightarrow c = \dfrac{\sqrt{3kT}}{m}$

$$\therefore \qquad KE = \frac{3}{2}\,kT = 2KE$$

$$= 3kT \quad \Rightarrow \quad c = \frac{\sqrt{2KE}}{m}$$

15. Electronic configuration for the compound $H_g[Co(SCN_4)]$

Co^{+2} =

3d						4s		4p		
↿⇂	↿⇂	↿	↿	↿						

In this complex Co^{2+} ion have 3 unpaired electrons ($n = 3$) so spin only magnetic moment $\mu_s = \sqrt{n(n+2)}$ will be

$\sqrt{3(3+2)}$ i.e. $\sqrt{15}$ BM

16. Given,

$$N_2O_4(g) \rightleftarrows 2NO_2(g)$$

At equilibrium $(1 - \alpha)$ mole $\qquad 2\alpha$ mole

Total number of moles = $1 - \alpha + 2\alpha = 1 + \alpha$ Now at equilibrium P is total pressure

$$P\,N_2O_4 = \frac{(1-\alpha)}{(1-\alpha)} \cdot P$$

$$P\,NO_2 = \frac{(2\alpha)}{(1+\alpha)} \cdot P$$

or
$$K_P = \frac{(P\,NO_2)^2}{(P\,N_2O_4)} = \frac{\left(\dfrac{(2\alpha)}{(1+\alpha)} \cdot P\right)^2}{\dfrac{(1-\alpha)}{(1+\alpha)}} \cdot P$$

or

$$K_p = \frac{\dfrac{4\alpha^2}{(1+\alpha)(1+\alpha)} \cdot P^2}{\dfrac{1-\alpha}{(1+\alpha)} \cdot P}$$

$$K_p = \frac{4\alpha^2}{(1+\alpha)(1+\alpha)} \cdot P$$

Since K_p is constant finally it will not affected by pressure change because as pressure doubled by reducing volume by half of its initial α will be lowered.

17. We have by electronic configuration,

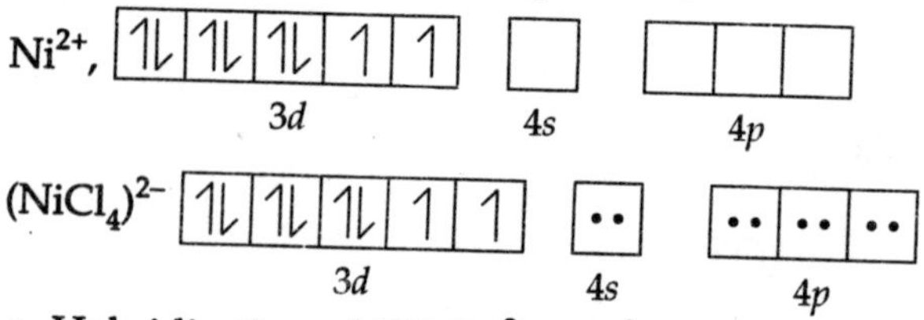

∴ Hybridisation of $[NiCl_4]^{2-} = sp^3$

Shape of $[NiCl_4]^{2-}$ = Tetrahedral

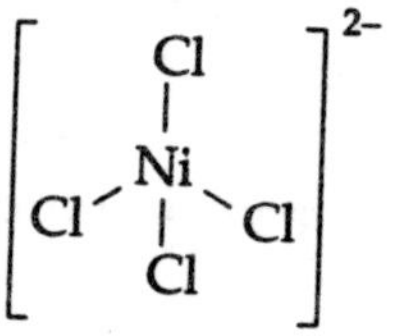

sp^3, tetrahedral

18. In the given reaction

$N_2(g) + 3H_2(g) \rightarrow 2NH_3$

The progress of reaction decreases the concentration of both $N_2(g)$ and $H_2(g)$ but the concentration of NH_3 will increase. Hence for reactants concentration we will use negative sign and for product positive sign

$$\text{rate} = \frac{d[N_2]}{dt} = \frac{-1d[H_2]}{3dt} = +\frac{1}{2}\frac{d[NH_3]}{dt}$$

So, (a) is correct Ans.

19. The reaction are

$$MgCl_2 + 2Na\,HCO_3 \rightarrow \underset{\text{soluble}}{Mg\,(HCO_3)_2} + 2NaCl$$

$$Mg\,(HOC_3)_2 \xrightarrow{\Delta} \underset{\text{(white)}}{MgCO_3\downarrow} + H_2O + CO_2$$

Ans (b)

20. According to Bohr' theory the radius of H-Like atom is $r_H = 0.529\,\frac{n^2}{Z}$ Å

For hydrogen $n = 1$ and Z = 1. Therefore $r_H = 0.529$ Å
For B^{3+} : Z = 4 and $n = 2$. Therefore,

$$r_{Be^{3+}} = \frac{0.529 \times 2^2}{4} = 0.529 \text{ Å}$$

∴ Ans (b) is correct

21. Acidic strength of any compound depends upon *pKa* value. If *pKa* value of any compound is less it is more acidic. *pKa* value of carboxylic group is less than *pKa* of $-\overset{+}{N}H_3(y)$ in amino acid and $-\overset{+}{N}H_3(z)$ will have comparatively less *pKa* than $-NH_3(z)$ due to –I effect of carboxylane group.

22. For the reaction

$$\begin{array}{ccc} N_2SO_4 & \rightleftharpoons 2Na^+ & + SO_4^{2-} \\ (0.004 - x) & 2x & x \end{array}$$

Since both the solutions are isotonic

$$0.004 + 2x = 0.01$$

∴ $$x = 3 \times 10^{-3}$$

$$\text{Percentage dissociation} = 3 \times \frac{10^{-3}}{0.004} \times 100 = 75\%$$

23. Solubility in aqueous medium depends upon electrolytic nature. Na_2S strong electrolytes, among ZnS and CuS first one is more soluble in aqueous medium because its k_{sp} value is more than k_{sp} value of CuS.
Hence correct order of solubility is

$$Na_2S > ZnS > CuS$$

24. In the extractive metallurgy of magnesium fused oxide or fused $MgCl_2$ salt is electrolysed because of its reactive nature.

$$MgCl_2 \rightarrow Mg^{2+} + 2Cl^-$$

At cathode $Mg^{2+} + e^- \rightarrow Mg$

At anode $2Cl^- \rightarrow Cl_2 + e^-$

25. We have, $\underset{X}{Cl_2} + H_2 \rightarrow HOCl + HCl$

$$HCl + AgNO_3 \rightarrow \underset{\text{white ppt.}}{AgCl} + HNO_3$$

$$2HCl + Mg \rightarrow MgCl_2 + \underset{Y}{H_2}$$

26. We have, $\Delta G = \Delta H - T\Delta S$
At equilibrium $\Delta G = 0$

$$T = \frac{\Delta H}{\Delta S} = \frac{30 \times 10^3}{75} = 400 \text{ K}$$

27. Enthalpy is isothermal and reversible press is H = E + PV and enthalpy change in the same process is

$$\Delta H = \Delta E + \Delta(PV)$$

or $\Delta H = \Delta E = nR\Delta T$

$\Delta T = 0$

$\Delta E = 0$

$\therefore$ $\Delta H = 0$

So no enthalpy change will happen

28. Alkynes get partially reduced by active metal in liquid ammonia takes place through trans vinylic anion which ultimately produces trans alkene.

29. From the given data we can deduce that

$$t_{1/2} = \frac{40}{2} = 20 \text{ min}$$

The rate of reaction when [X] is 0.01 M = $k\,[X] = \frac{0.693}{20} \times 0.01 = 3.47 \times 10^{-4}$ M min^{-1}

30. We have, $2KMnO_4 + 2\,KOH \rightarrow 2K_2\,MnO_4 + H_2O + O$

$2K_2\,MnO_4 + 2H_2O \rightarrow 2MnO_2 + 4\,KOH + 2O$

$$2\,KMnO_4 + H_2O \xrightarrow{\text{alkaline}} 2MnO_2 + 2KOH + 3[O]$$

$$KI + [O] \rightarrow KIO_3$$

$$\therefore\ 2KMnO_4 + KI + H_2O \rightarrow 2KOH + 2MnO_2 + KIO_3$$

31. Monochlorinated products of 2-methyl butane are

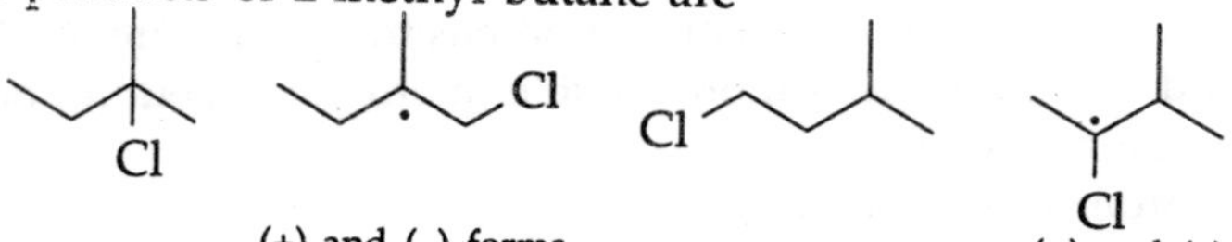

(+) and (–) forms (+) and (–)

Therefore a total of four chiral compounds are obtained

32. The reaction is:

$$FeO + SiO_2 \rightarrow FeSiO_3$$

33. All the species have 14 electrons distributed in molecular orbitals, this

Bond order = (10 – 4)\2 = 3

34. The structure are $H_2S_2O_5$,

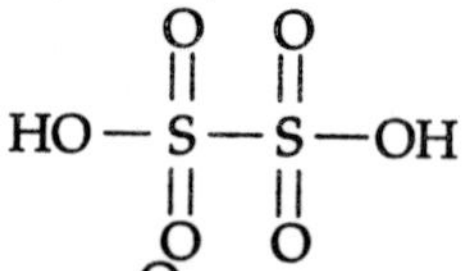

$H_2S_2O_5$,

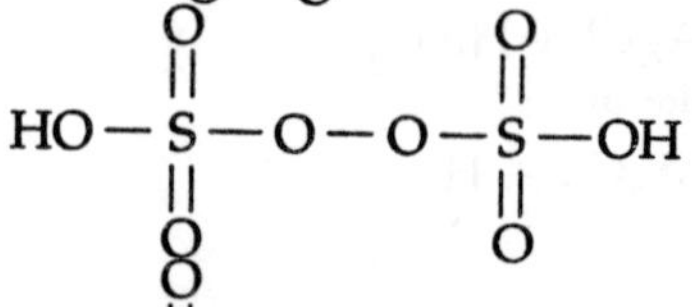

$H_2S_2O_3$,

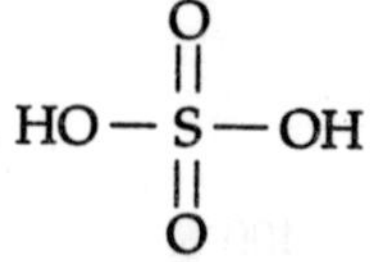

$H_2S_4O_6$,

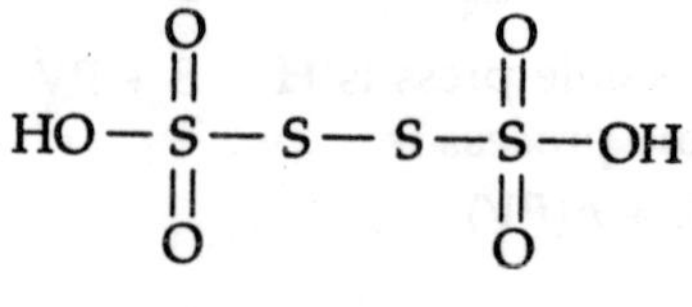

Test Paper 2

1. The structure is

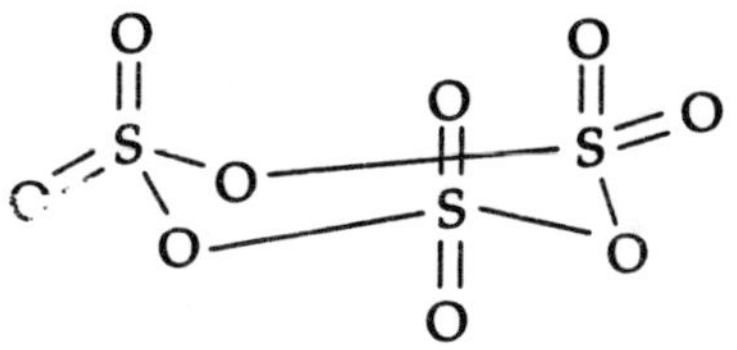

Thus these are no S-S bonds in S_3O_9

2. Since gram equivalent of oxalic acid in 250 ml = $\frac{6.3}{63}$ = 0.1

Grain equivalent per litre = 0.1 $\frac{1000}{250}$ = 0.4 N

∴ $V_1N_1 = V_2N_2$

[number of grain equivalent × volume = constant]

$$V_2 = \frac{V_1 N_1}{N_2} V_2 = \frac{0.4 \times 10}{0.1} = \mathbf{40\ ml}$$

3. From the structure of AB as shown in figure Number of A atoms per unit cell

= 8(1/8) + 4(1/2) = 3

Number of B atoms per unit cell

= 12(1/3) + 1 = 4

Hence, the stoichiometry of solid is A_3B_4

4. Higher the oxidation number higher the acidity.

5. Given acidic compound contains four acidic hydrogen. Our of these four the order of acidic strength is —COOH > —OH (NO_2 substituted phenolic group) > —OH (1 phenolic) > alkynic H(≡CH). Two mole of NH_2^- ion will take out two mole of the most acidic hydrogen from given compound. The product will contain $—COO^-$ ion with O^- ion.

6. Reduction of dichromatic solution can be done by the acidified H_2S or SO_2. It runs acidified dichromate solution green. But the H_2S gas has rotten egg like small where as SO_2 has irritating smell H_2S is evolved by the action of acid upon sulphide whereas SO_2 can be obtained by the action of acid upon sulphide.

7. As S_N2 reaction at an asymmetric carbon always gives a single stenoisomer due to the pressure of two or more asymmetric centres.

8. The reaction is:

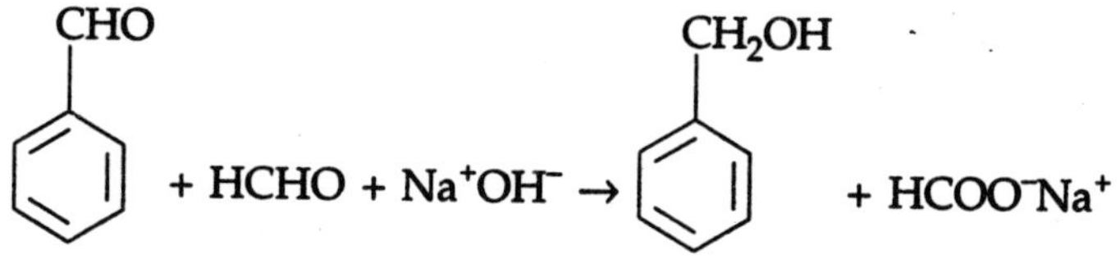
CHO
+ HCHO + Na⁺OH⁻ →
CH₂OH
+ HCOO⁻Na⁺

9. In this oxidation state of Mn is +7, hence configuration of Mn^{+7} = [Ar]$3d^0$ $4s^0$
10. The atomic radius increases gradually hence ionization potential decreases.
11. Hybridisation of N in NH_3 is sp^3
Hybridisation of Pt in $[Ptcl_4]^{2-}$ is dsp^2
Hybridisation of P in PCl_5 is sp^3d
Hybridisation of B in BCl_3 is sp^2
12. NO_3^- and CO_3^{2-} each have 32 numebr of electrons whereas ClO_3^- and SO_3 have 42 and 40 electrons respectively. Hybridization of N in $NO_3^- = \frac{1}{2}$[NOM + VEN – ION]

$$= \frac{1}{2}[0 + 5 + 1] = 3 \text{ or } sp^2$$

Hybridization of S inSO_3

$$= \frac{1}{2}[\text{NOM + VEN – ION}] = \frac{1}{2}[0 + 6 - 0]$$

$$= 3 \text{ or } sp^2$$

NO_3^- and CO_3^{2-} both have same sp^2 hybridized central atom hence isostructural

13. $\overset{sp^1}{H_2C}=\overset{sp^2}{CH}-\overset{sp}{C}\equiv\overset{sp}{CH}$

—— left to right ——→

14. 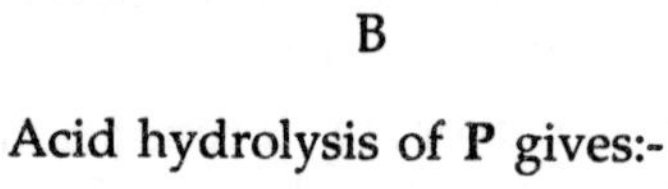

A

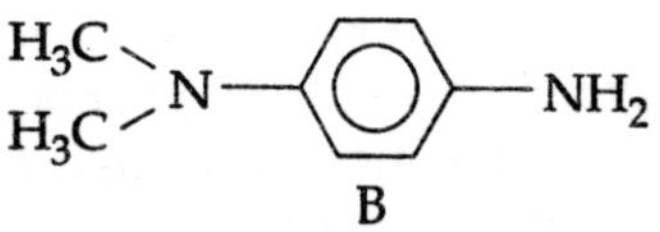

B

15. Acid hydrolysis of **P** gives:-

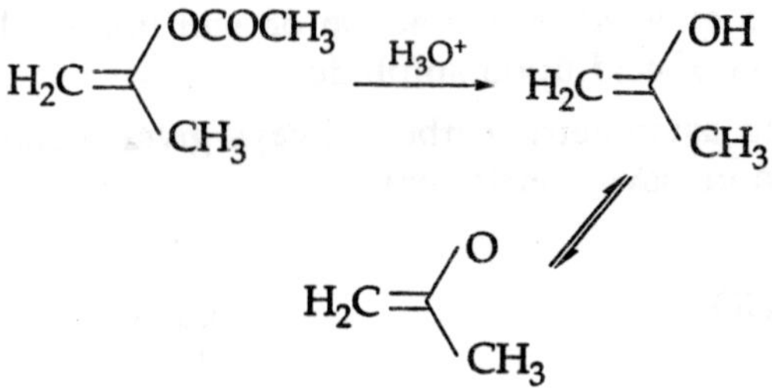

Acid hydrolysis of **Q** gives :-

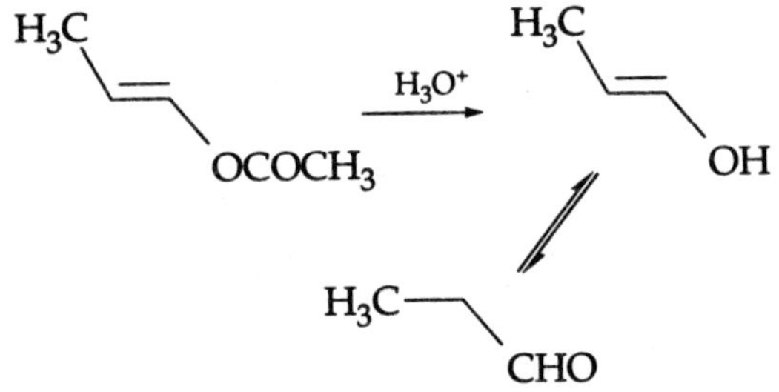
H3C
H3O+
OCOCH3
H3C
OH
H3C
CHO

Ketone and aldehyde can be distinguished by Fehlings solution. It is because aldehyde reduces **Fehings solution** whereas ketone is non-reducing in nature.

16. $C_2H_5O^-$ ion take out proton form phenol converting the phenol in to phenoxide ion. Alkyl iodine is prone to nucleophile attack $C_2H_5O^-$ is stronger and better nucleophile than $C_6H_5\,O^-$ (phenoxide) ion. It is due to the higher electron density over oxygen. But in the case of $C_2H_5O^-$ (phenoxide) ion negative change is delocalised over the ring system. That is why $C_2H_5O^-$ ion (nucleophile) will attack at the ethyl iodide to give diethyl ether (Willamson's synthesis).

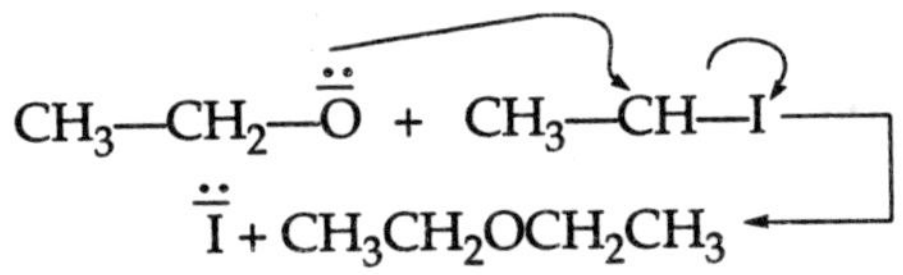
CH3—CH2—Ö + CH3—CH—I
Ï + CH3CH2OCH2CH3

17. In H_3BO_3 the central boron atom is electron-deficient. Boric acid is a Lewis acid with one vacant *p*-orbital. There is no d-orbital of suitable energy in boron atom, that is why it can accommodate only one additional electrons pair in its outermost shell

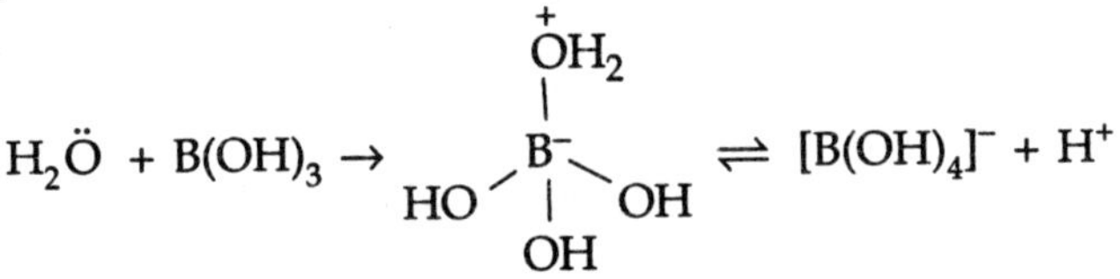
H2Ö + B(OH)3 →
OH2
B−
HO
OH
OH
⇌ [B(OH)4]− + H+

18. $K = \frac{2.303}{1}\log\frac{[A]_0}{[A]_1} = \frac{2.303}{2\times10^4}\log\frac{800}{50}$

$= 1.386 \times 10^4\ s^{-1}$

19. 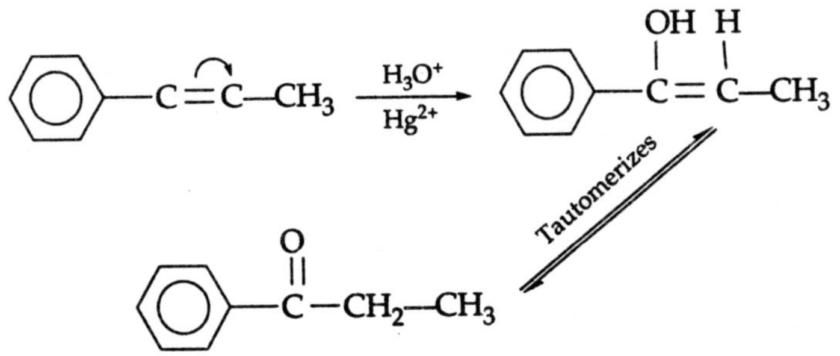
C=C—CH3
H3O+
Hg2+
OH H
C=C—CH3
Tautomerizes
O
C—CH2—CH3

20. 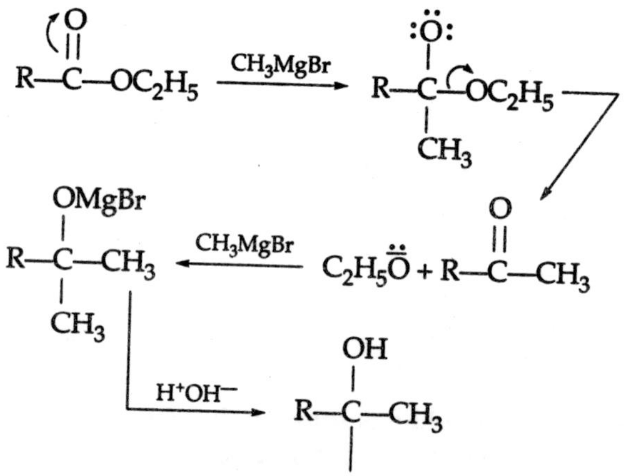
CH3MgBr
CH3MgBr
H+OH–

21. Hgs will precipitate first because the value of Ksp of HgS is lowest among FeS, MgS, ZnS and HgS.

22. $2Au + 4CN^- + H_2O + \frac{1}{2}O_2 \rightarrow 2[Au(CN)_2]^- + OH^-$

$$2[Au(CN)_2]^- \ Zn \rightarrow [Zn(CN)_4]^{2-} + 2Au$$

23. 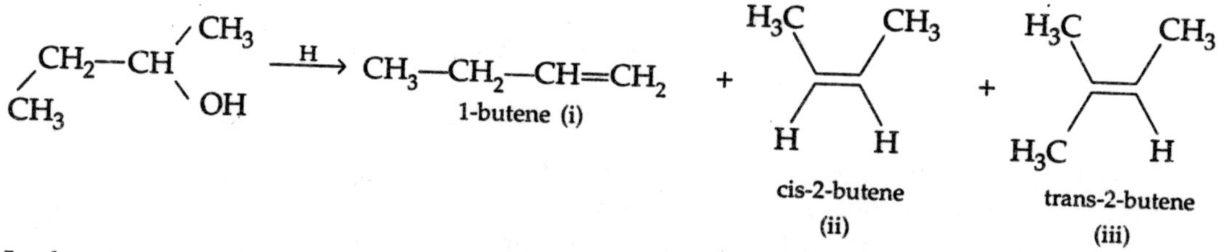
1-butene (i)
cis-2-butene
(ii)
trans-2-butene
(iii)

24. In the given reaction "X" is a free radical and major product is that which is more stable. Among given option (B) is more stable. Among given option (B) is more stable because it contain tertiary free radical.

25. Teminal alkynes are acidic due to acidic hydrogen where as central alkynes are not due to absence of replaceable acidic hydrogen.
In butyne – 1 acidic hydrogen is present

$$H{-}C{\equiv}C{-}CH_2{-}CH_3$$
$$\uparrow$$

Unsaturation at one terminal is the cause of acidic character of hydrogen

While is (butyne-2) $H_3C{-}C{\equiv}C{-}CH_3$ acidic hydrogen is not present acid this unsaturation is further satisfied by two methyl group from both end hence less reactive. Due to above reason butyne-I gives red ppt, with ammonical Cu_2Cl_2 solutions. While butyne 2 does not give these ppt

$$2CH_3{-}CH_2C{\equiv}CH + Cu_2Cl_2 + 2NH_4OH$$

$$\rightarrow 2CH_3-CH_2-C{\equiv}C \cdot Cu\downarrow + 2NH_4Cl + 2H_2O$$

red ppt.

$$CH_3-C{\equiv}C-CH_3 + Cu_2Cl_2\ \ 2NH_4OH \rightarrow \text{No reaction}$$

26. $-NO_2$ and $-Cl$ groups are electron with drawing in nature which deactivates benzene ring $-NO_2$ is more electron withdrawing than $-Cl$ group. CH_3 group is electron pumping in nature which activates benzene ring. On the basis of activation and deactivation the order of reactivity in electrophilic substitution reaction follow the trend as

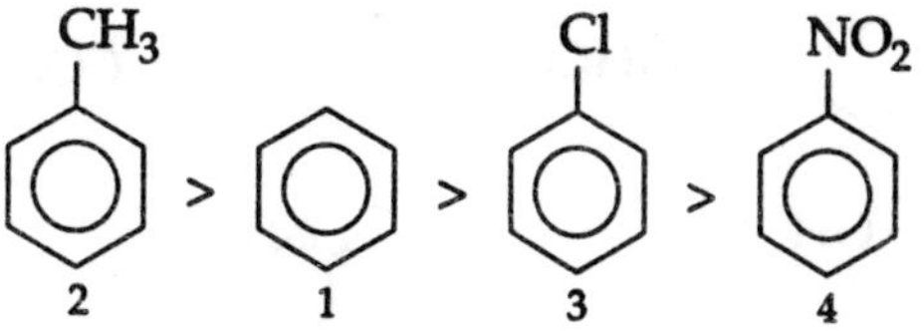
CH3
Cl
NO2
>
>
>

2 1 3 4

27. Primary alcohol on oxidation given aldehyde. So *A* is alcohol *B* is aldehyde.

$$\underset{(A)}{CH_3-CH_2-CH_2-OH} \xrightarrow{[o]} \underset{(B)}{CH_3-CH_2-CHO}$$

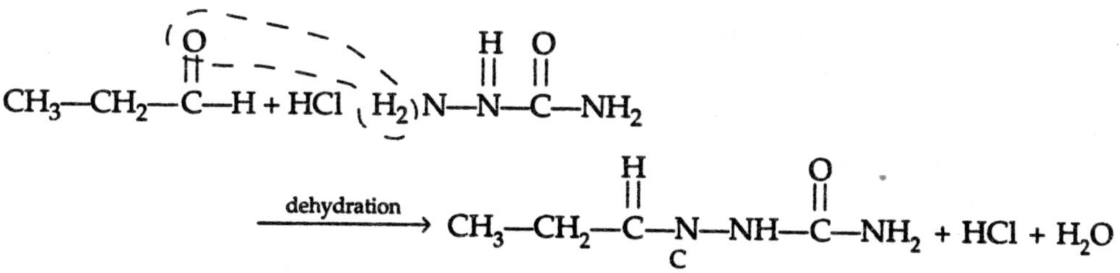

$$CH_3-CH_2-\overset{O}{\overset{\|}{C}}-H \xrightarrow{Ag_2O} 2Ag\downarrow + CH_3-CH_2-COOH$$

28. Dipole moment depends upon symmetry of molecule. If molecule is asymmetric dipole moment is more. If molecule is asymmetric dipole moment is more. If molecule is symmetric dipole moment is almost zero.

$CH_3-C{\equiv}C-CH_3$ is completely symmetric molecule hence its $\mu = 0$

29. Here, E° > O, ∴ reaction is spontaneous

30. $N^{-3} > O^{-2} > F^-$ because,

Greater negative charge in an isoelectronic species leads to the greater size.

31. Here, $-\ \alpha$ -H with respect to –OH is free

32. Use the fomula $-\Delta H_f = ZH_P - ZH_R$

33. The structures are:

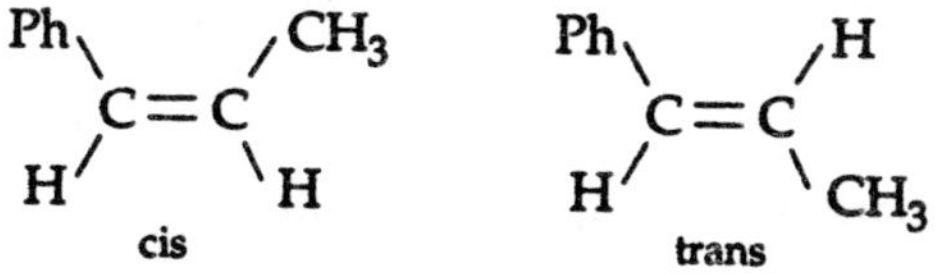
Ph
CH3
C=C
H
H
Ph
H
C=C
H
CH3

cis trans

34. Here, $K_p = 1.44 \times 10^{-5}$

$\therefore\ C = (0.082 \times 773)^{-2}$ mole lit^{-1}

Test Paper 3

2. Due to formation of chelated complex, the reaction moves in forward direction.
3. Due to formation of tetraminine zinc (II) complex $Zn^{+2} + NH_4OH \rightarrow [Zn(NH_3)_4]^{+2}$.
4. The reactions are:

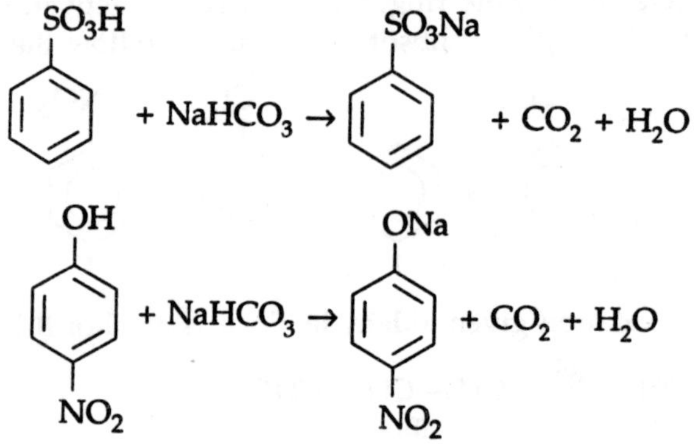

7. $N^+OCl^- \rightarrow$ Markonikov's Addition
8. $\Delta S_{(A \rightarrow B)} = \Delta S_{(A \rightarrow C)} + \Delta S_{(C \rightarrow D)} - \Delta S_{(B \rightarrow D)} = 50 + 30 - 20 = +60$ esu.
10. Isocyanide test/carbylamine reaction
11. The reactions which take place are:

$$Cu^{+2} + 2CN^- \rightarrow Cu(CN)_2$$
$$2Cu(CN)_2 \rightarrow 2CuCN + (CN)_2$$

14. $CO_2 + H_2O \rightleftharpoons H_2CO_3 \rightleftharpoons H^+ + HCO_3^- \rightleftharpoons 2H^+ + CO_3^{-2}$
15. Due to synergic bond formation between metal and CO, the bond order of CO decreases.
16. It is cumene hydroperoxide rearrangement reaction.

17. 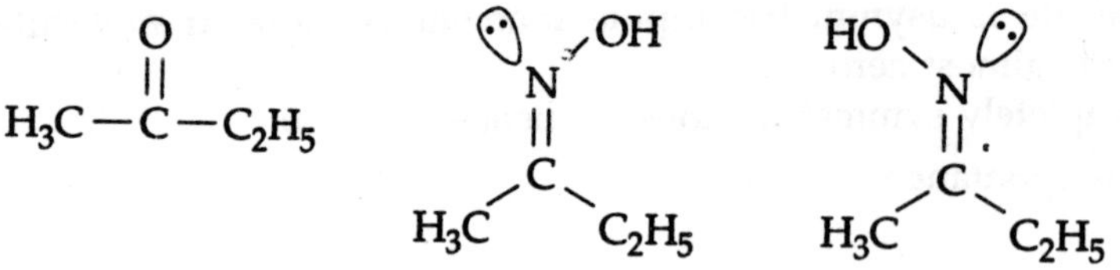

19. 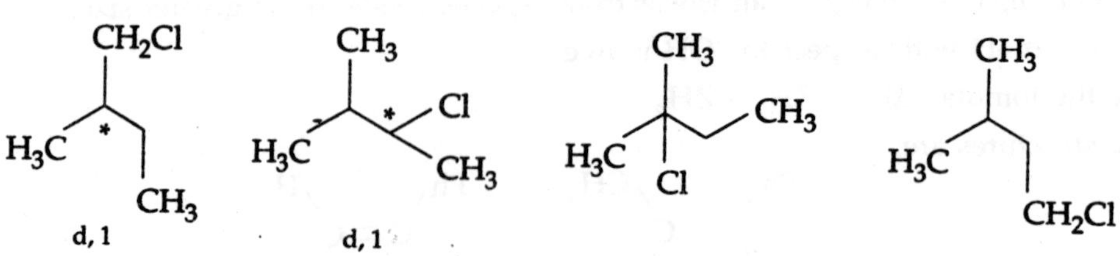

d, 1 d, 1

M → *d*, I cannot be separated by fractional distillation.

20. Test of Mg^{+2} ion
$Mg^{-2} + NH_4OH + Na_2HPO_4 \rightarrow Mg(NH_4)\ PO_4$

27. $E_{cell} = \frac{RT}{nF}$ in K

$$(0.8 - 0.05) = \frac{1}{2} \times \frac{0.0592}{2.303} \text{ in K}$$

$$\text{In K} = \frac{(0.8 - 0.05) \times 2 \times 2.303}{0.0592} = 58.38$$

28. On increasing concentration of NH_3, the concentration of H^+ ion decreases. Therefore, E_{red} increases.

33. $2C_6H_5OH \rightleftharpoons (C_6H_5OH)_2$

$$7 = 14 \times 0.8\left(\frac{2-\alpha}{2}\right)$$

$$\alpha = 0.75 = \mathbf{75\%}$$

34. $\Delta H = \Delta U + \Delta(PV)$

$$\Delta - \Delta U + VAP$$
$$\Delta U = \Delta H - VAP = -560 + 1 \times 30 \times 0.1$$
$$= -557$$

Absolute value = 557 kJ

35. The solubility of AgBr in prescribe of 10^{-7} molar $AgNO_3$ is 3×10^{-7} M.
Therefore $[Br^-] = 3 \times 10^{-4}\ m^3$, $[Ag^+] = 4 \times 10^{-4}\ m^3$ and $[NO_3^-] = 10^{-4}\ m^3$

Therefore $K_{mol} = K_{Br} + K_{Ag^+} + K_{NO_3^-} = 55\ Sm^{-1}$

36. $\rho = \frac{ZA}{NV}$

$$Z = \frac{\rho NV}{A} = \frac{2 \times 6 \times 10^{23} \times (5 \times 10^{-4})^3}{75}$$

$$n = 2$$

$$r = \frac{\sqrt{3}}{4}a = \frac{\sqrt{3}}{4} = 2.165\ \text{Å} = \mathbf{216.5\ pm}$$

Note: Answer may be 216 pm or 217 pm

TABLES

TABLE 1 Solubility Rules

Group	*Solubility*	*Exceptions*
Nitrates	NO_3	All soluble $Bi(NO_3)_3$, $Hg(NO_3)_2$
Chlorates ClO_3^-	All soluble	$Bi(ClO_3)_3$, $Hg(ClO_3)_2$
Perchlorates ClO_4^-	All soluble	$KClO_4^+$
Acetates $C_2H_3O_2^-$	All soluble	$AgC_2H_3O_3$
Alkali metal compounds		
Li, Na, K, Rb, Cs	All soluble	Some complex (ternary) alkali metal salts
Ammonium compounds NH_4^+	All soluble	none
Chlorides Cl^-	All soluble	none
Bromides Br^-		Hg_2^{2+}, Pb^{2+}, and $PbCl_2^*$,
Iodides I^-		$CrCl_3$, BiI_3, CuI, AuI, AuI_3, PtI_2, CrB_2^*, $CuBr_2^*$, $PtBr_4^+$, HgI_2^*
Sulfates SO_4^{2-}	Most are solubke	$SrSO_4$, $Sr_2(SO_4)_3$, $BaSO_4$, $PbSO_4$, $HgSO_4$, Hg_2SO_4, $CaSO_4^*$, $Ag_2SO_4^*$, $Cr_2(SO_4)_3^*$, $Fe_2(SO_4)_3^*$
Carbonates CO_3^{2-}	Group 1, $(NH_4)_2CO_3$, $CrCO_3$	All others
Phosphates PO_4^{3-}	Group 1, $(NH_4)_3PO_4$	All others
Sulfites SO_3^{2-}	Group 1. $(NH_4)_2SO_3$	All others
Hydroxides OH^-	Group 1, $Sr(OH)_2$, $Ba(OH)_2$, AuOH	All others, $Ca(OH)_2^*$
Sulfides S^{2-}	Group 1, $(NH_4)_2S$, SrS	All others, CaS*
Oxides O^{2-}	CaO, SrO, BaO	All others, Na_2O and K_2O decompose in H_2O

*Slightly soluble.

Solubility Rules (Abbreviated)

1. All common salts of the Group 1 (IA) elements and ammonium ion are soluble.
2. All common acetates and nitrates are soluble.
3. All binary compounds of Group 17 (VIIA) elements (other than F) with metals are soluble except those of silver, mercury (I), and lead.
4. All sulfates are soluble except those of barium, strontium, lead, calcium, silver and mercury (I).
5. Except for those in Rule 1, carbonates, hydroxides, oxides, sulfides, and phosphates are insoluble.

TABLE 2 Acid-Base Indicators

Range	*Indicator*	*Lower Color*	*Upper Color*
0.0–2.5	Methyl violet	Yellow-great	Violet
0.5–2.0	Malachite green HCl	Yellow	Blue
1.0–2.8	Thymol blue	Red	Yellow
1.2–4.0	Benzopurpurin	Violet	Red
1.3–3.0	Orange IV	Red	Yellow
1.5–2.6	Napthyol yellow S	Colorless	Yellow
2.1–2.8	*p*-Phenylazoaniline	Orange	Yellow
2.5–4.4	methyl orange	Red	Yellow
3.0–4.7	Bromophenol blue	Orange-yellow	Violet
3.0–5.0	Congo red	Blue	Red
3.5–6.3	Gallein	Orange	Red
3.8–5.4	Bromcresol green	Yellow	Blue
4.0–5.8	2, 5-Dinitrophenol	Colorless	Yellow
4.2–4.6	Ethyl orange	Colorless	Yellow
4.4–6.2	Methyl red	Salmon	Orange
4.5–8.3	Litmus	Red	Blue
4.8–6.2	Chlorphenol red	Yellow	Red
5.1–6.5	Propyl red	Pink	Yellow
5.4–6.8	Bromocresol purple	Green-yellow	Violet
5.6–7.2	Alizarin	Yellow	Red
6.0–7.6	Bromthymol blue	Yellow	Blue
6.0–7.6	Bromxylenol blue	Orange-yellow	Blue
6.4–8.2	Phenol red	Yellow	Red-violet
7.1–.8	Cresol red	Yellow	Violet
7.5–9.0	*m*-cresol purple	Yellow	Violet
8.1–9.5	Thymol blue	Yellow	Blue
8.3–10.0	Phenolphthalein	Colorless	dark pink
8.6–9.8	*o*-Cresolphthalein	Colorless	Pink
9.5–10.4	Thymolphthalein	Colorless	Blue
9.9–11.8	Alizarian yellow R	Yellow	Dark orange
10.6–13.4	Methyl blue	Blue	Pale violet
11.1–12.8	Acid fuchsin	Red	Colorless
11.4–13.0	Indigo carmine	Blue	Yellow
11.7–12.8	2, 4, 6-Trinitrotoluene	Colorless	Orange
12.0–14.0	Trinitrobenzene	Colorless	Orange

TABLE 3 Flame Tests for Elements

BLUE	
Azure	Lead, selenium, $CuCl_2$ (and other copper compounds when moistened with HCl); $CuBr_2$ appears azure blue, then is followed by green.
Light Blue	Arsenic and some of its compounds; slenium.
Blue-green	Phosphates moistened with sulfuric acid; B_2O_3.
GREEN	
Emerald Green	Copper compounds other than halides (when not moistened with HCl), thallium compounds.
Greenish-blue	$CuBr_2$; arsenic; lead; antimony.
Pure green	Thallium and tellurium compounds.
Yellow-green	Barium; possibly molybdenum; borates with (H_2SO_4).
Faint green	Antimony and ammonium compounds.
Whitish green	Zinc.
RED	
Carmine	Lithium compounds (masked by barium or sodium), are invisible when viewed through green glass, appear violet through cobalt glass.
Scarlet	Calcium compounds (masked by barium), appear greenish when viewed through cobalt glass and green through green glass.
Crimson	Strontium compounds (masked by barium), appear violet through cobalt glass, yellowish through green glass.
VIOLET	Potassium compounds other than silicates, phosphates and borates; rubidium and cesium are similar. Color is masked by lithium/and or sodium, appears purple-red through cobalt glass and bulish-green glass.
YELLOW	Sodium, even the smallest amount, invisible when viewed through cobalt glass.

TABLE 4 Qualitative Analysis of Cations

Flow Chart

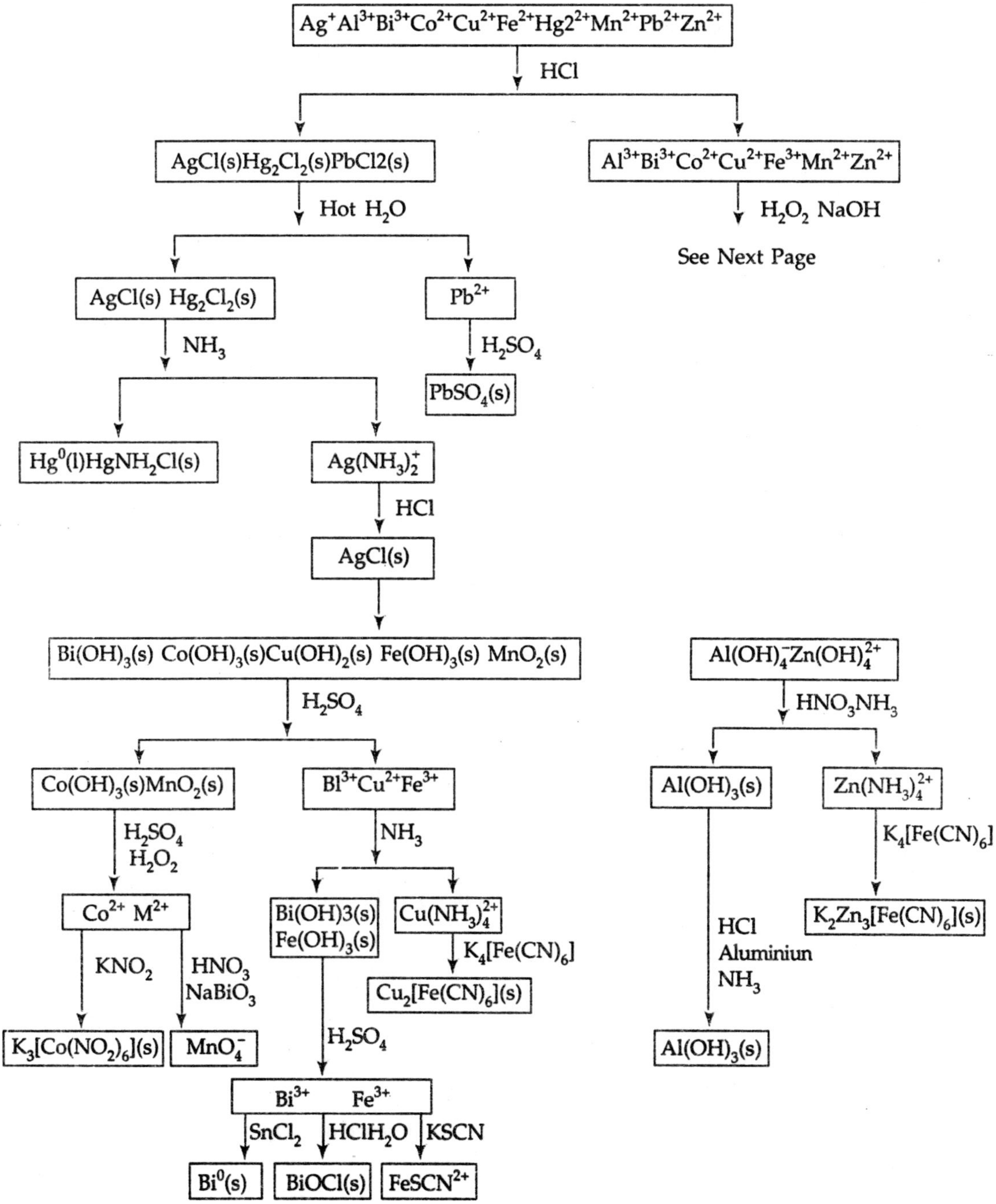

TABLE 5 Qualitative Analysis of Anions

Flow Chart

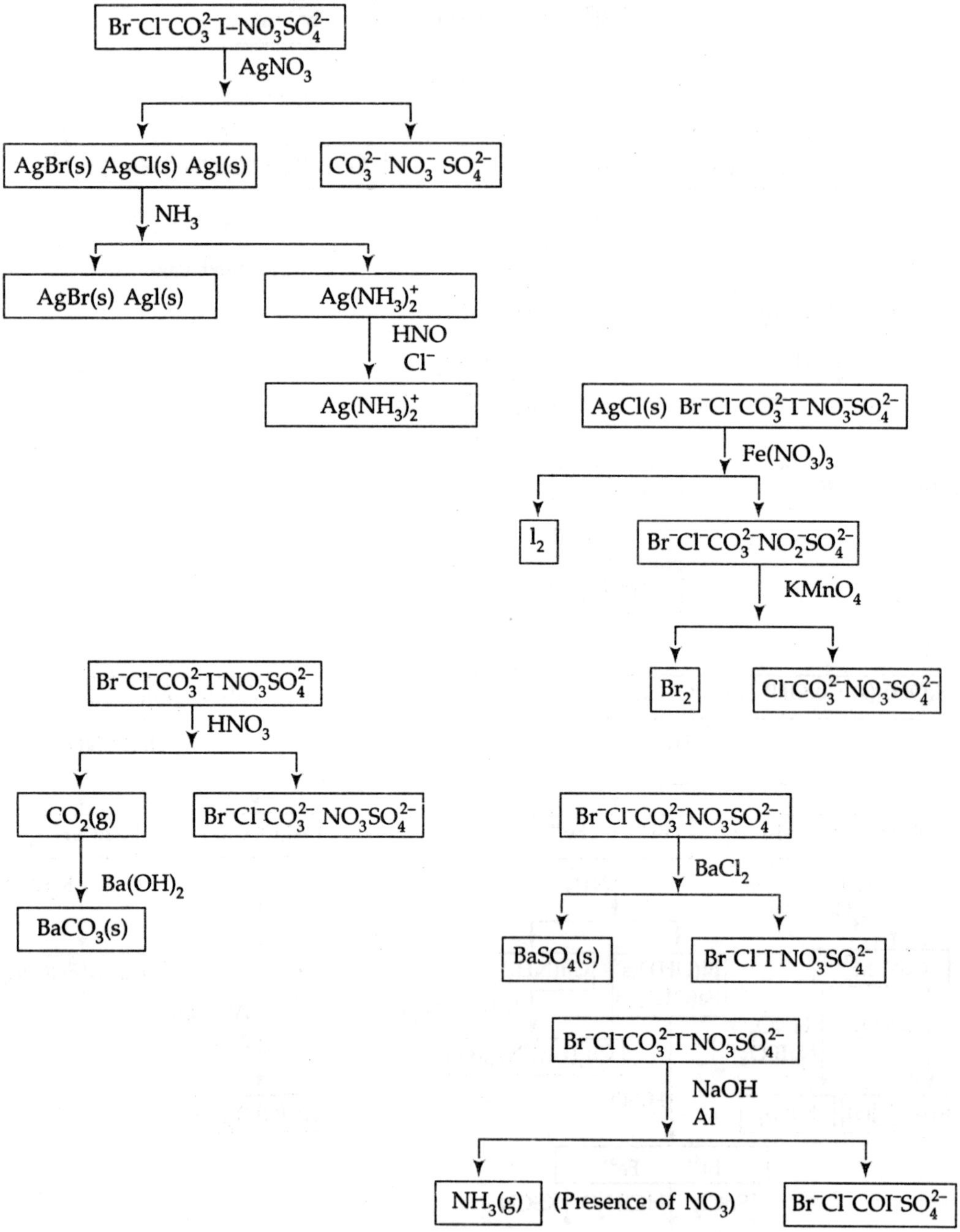

TABLE 6 Chemical Reactions

I. Metal Combining with Nonmetals

Metal + Nonmetal → Salt* (Metal Ion$^+$ + Nonmetal Ion$^-$)

Reaction of Alkali Metals and Alkaline Earth Metals

Group IA Metal (M)	*Combining Substance*	*Reaction*
All	Hydrogen	$2\ M(s) + H_2(g) \rightarrow 2\ MH(s)$
All	Halogen	$2\ M(s) + X_2 \rightarrow 2\ MX(s)$
Li	Nitrogen	$6\ Li(s) + N_2(g) \rightarrow 2\ Li_3N(s)$
All	Sulfur	$2\ M(s) + S(s) \rightarrow M_2S(s)$
Li	Oxygen	$4\ Li(s) + O_2(g) \rightarrow 2\ Li_2O(s)$
Na	Oxygen	$2\ Na(s) + O_2(g) \rightarrow Na_2O_2(s)$**
Na, K, Rb, Cs	Oxygen	$M(s) + O_2(g) \rightarrow MO_2(s)$
All	Water	$2\ M(s) + H_2O \rightarrow 2\ M+(aq) +$ $2\ OH^-(aq) + H_2(g)$

Group IIA Metal (M)	*Combining Substance*	*Reaction*
Ca, Sr, Ba	Hydrogen	$M(s) + H_2(g) \rightarrow MH_2(s)$
All	Halogens	$M(s) + X_2 \rightarrow MX_2(s)$
Mg, Ca, Sr, Ba	Nitrogen	$3\ M(s) + S(s) \rightarrow M_3N_2(s)$
Mg, Ca, Sr, Ba	Sulfur	$M(s) + S(s) \rightarrow MS(s)$
Be, Mg, Ca, Sr, Ba	Oxygen	$2\ M(s) + O_2(g) \rightarrow 2\ MO(s)$
Ba	Oxygen	$Ba(s) + O_2(g) \rightarrow BaO_2(s)$
Ca, Sr, Ba	Water	$M(s) + 2H_2O \rightarrow M^{2+}(aq) + 2OH^-(aq) + H_2(g)$
Mg	Water	$Mg(s) + H_2O(g) \rightarrow MgO(s) + H_2(g)$

M = symbol for the metal, *X* = any halogen

*By definition, a salt is an ionic compound made up of a cation other than H^+ and an anion other than OH^- or O^{2-}. However, in these examples a "salt" represents any ionic (metal-nonmetal) compound.

**Superoxides need special conditions.

1. Calcium metal is heater in the present of nitrogen gas:

   $$Ca + N_2 \rightarrow Ca_3N_2$$

2. Solid potassium is added to a flask of oxygen gas:

   $$K = O_2 \rightarrow KO_2$$

   This reaction is also true for rubidium and cesium. (Note: These are superoxides.)

3. A piece of solid magnesium, which is ignited, is added to water.

   $$Mg + H_2O \rightarrow MgO + H_2$$

II. Combustion

Substance + Oxygen Gas → Oxide of Element

The usual products are the oxides of the elements present in the original substance in their *higher* valence states. When N, Cl, Br, and I are present in the original compound, they are usually released as free elements, *not* as the oxides.

1. Solid copper (II) sulfide is heated strongly in oxygen gas:

   $$CuS + O_2 \rightarrow CuO + SO_2$$

   Or

   $$\rightarrow Cu_2O + SO_2$$

2. Carbon disulfide gas is burned in excess oxygen gas:

$$CS_2 + O_2 \rightarrow CO_2 + SO_2$$

3. Methanol is burned completely in air:

$$CH_3OH + O_2 \rightarrow CO_2 + H_2O$$

All alcohols (as well as hydrocarbons and carbohydrates) burn in oxygen gas to produce CO_2 and H_2O.

4. Silane is combusted in a stream of oxygen gas:

$$SiH_4 + O_2 \rightarrow SiO_2 + H_2O$$

III. Metallic Oxide + H_2O → Base (Metallic Hydroxide)

1. Lithium oxide is added to water:

$$Li_2O + H_2O \rightarrow Li^+ + OH^-$$

2. A solid piece of potassium oxide is dropped into cold water:

$$K_2O + H_2O \rightarrow K^+ + OH^-$$

IV. Nonmetallic Oxide + H_2O → Acid

1. Dinitrogen pentoxide is added to water:

$$N_2O_5 + H_2O \rightarrow H^+ + NO_3^-$$

2. Carbon dioxide gas is bubbled through water:

$$CO_2 + H_2O \rightarrow H_2CO_3 \text{ (carbonic acid)}$$

OR

$$\rightarrow H^+ + HCO_3^- \text{ (weak acid)}$$

3. Sulfur trioxide gas is bubbled through water:

$$SO_3 + H_2O \rightarrow H^+ + SO_4^{2-}$$

4. Phosphorus (V) oxytrichloride is added to water:

$$POCl_3 + H_2O \rightarrow H^+ + HPO_4^{2-} = Cl^-$$

V. Metallic Oxide + Acid → Salt + H_2O

1. Ferric oxide is added to hydrochloric acid:

$$Fe_2O_3 + H^+ \rightarrow Fe^{3+} + H_2O$$

2. Copper (II) oxide is added to nitric acid:

$$CuO + H^+ \rightarrow u^{2+} + H_2O$$

VI. Nonmetallic Oxide + base → Salt + H_2O

1. Carbon dioxide gas is bubbled through a solution of sodium hydroxide:

$$CO_2 + OH^- \rightarrow CO_3^{2-} + H_2O$$

2. Sulfur dioxide gas is bubbled through a solution of lithium hydroxide:

$$SO_2 + OH^- \rightarrow SO^{**} + H_2O$$

VII. Metallic Oxide + Nonmetallic Oxide → Salt

1. Magnesium oxide is heated strongly in carbon dioxide gas:

$$MgO + CO_2 \rightarrow MgCO_3$$

2. Calcium oxide is heated in an environment of sulfur trioxide gas:

$$CaO + SO_3 \rightarrow CaSO_4$$

VIII. Acid + Base → Salt + Water (Neutralization)

1. Hydrochloric acid is added to potassium hydroxide:

$$H^+ + OH^- \rightarrow H_2O$$

2. Equal volumes of 0.5 M sulfuric acid and 0.5 M sodium hydroxide are mixed:

$$HSO_4^- + OH^- \rightarrow SO_4^{2-} + H_2O$$

or

$$H^+ + OH^- \rightarrow H_2O$$

3. Hot nitric acid is added to solid sodium hydroxide:

$$H^+ + NaOH \rightarrow NOH + Na^+$$

IX. Acid + Metal → Salt + Hydrogen

1. Sulfuric acid is added to a solid strip of zinc:

$$H^+ + Zn \rightarrow H_2 + Zn^{2+}$$

2. A piece of magnesium is dropped into a beaker of 6 M hydrochloric acid:

$$Mg + H^+ \rightarrow Mg^{2+} + H_2$$

3. Calcium metal is added to a solution 4 M HCl:

$$Ca + H^+ \rightarrow Ca^{2+} + H_2$$

Some Common Departures

4. Lead shot is dropped into hot, concentrated sulfuric acid:

$$Pb + H^+SO_4^{2-} \rightarrow PbSO_4 + SO_2 + H_2O$$

5. Solid copper shavings are added to a concentrated nitric acid solution. (This reaction is well known and is covered quite extensively in textbooks. Note how it departs from the guideline.)

$$Cu + H^+ + NO_3^- \rightarrow Cu^{2+} + H_2O + NO_2$$

X. Base + Amphoteric Metal → Salt + Hydrogen Gas

Amphobetric metals (such as Al, Zn, Pb and Hg) have properties that may be intermediate between those of metals and those nonmetals. They will react with a base to from, a completed ion with oxygen. *This is rare problem.*

1. A piece of solid aluminuim is added to a 6 M solution of sodium hydroxide:

$$Al + OH^- \rightarrow AlO_3^{2-} + H_2$$

2. A solid piece of zinc is added to a 6 M solution of potassium hydroxide:

$$Zn + OH^- \rightarrow ZnO_2^{2-} + H_2$$

XI. Strong Acid + Salt of Weak Acid → Salt of Strong Acid + Weak Acid

1. Hydrochloric acid is added to potassium acetate:

$$H^+ + Ac^- \rightarrow HAc$$ (Do not dissociate a weak acid.)

2. A 9 M acid solution is added to a solution of potassium carbonate:

$$H^+ + CO_3^{2-} \rightarrow H_2CO_3$$

XII. Weak Acid + Weak Base → Conjugate Base + Conjugate Acid

1. A solution of ammonia is mixed with an equimolar solution of hydrofluoric acid:

$$NH_3 + HF \rightarrow NH_4^+ + F^-$$

2. Acetate acid is added to a solution of ammonia:

$$HC_2H_3O_2 + NH_3 \rightarrow C_2H_3O_2^- + NH_4^+$$

XIII. Weak Acid + Strong Base → Water + Conjugate Base

1. Hydrofluoric acid is added to a solution of sodium hydroxide:

$$HF + OH^- \rightarrow H_2O + F^-$$

2. A solution of vinegar (acetic acid) is titrated with lye (sodium hydroxide):

$$HC_2H_3O_2 + OH^- \rightarrow H_2O + C_2H_2O_2^-$$

XIV. Strong Acid + Weak Base + Conjugate Acid (Weak Acid)

1. Hydrochloric acid is added to a sodium of ammonia:

$$H^+ + NH_3 \rightarrow NH_4^+$$

2. Sulfuric acid is added to a solution of sodium fluoride:

$$H^+ + F^- \rightarrow HF$$

3. Hydrochloric acid is added to sodium carbonate:

$$H^+ + CO_3^{2-} \rightarrow HCO_3^-$$

or

$$\rightarrow H_2CO_3$$

or

$$\rightarrow CO_2 + H_2O$$

4. Sodium acetate is added to a weak solution of nitric acid:

$$H^+ + C_2H_3O_2^- \rightarrow HC_2H_3O_2$$

XV. Acid + Carbonate → Salt + CO_2 + Water

1. Hydrochloric acid is added to a sodium carbonate solution.

$$H^+ + CO_3^{2-} \rightarrow CO_2 + H_2O$$

2. Hydroiodic acid is mixed with solid calcium carbonate.

$$H^+ + CaCO_3 \rightarrow Ca^{2+} + CO_2 + H_2O$$

XVI. Acidic Anhydride + Base → Salt + Water

1. Dinitrogen pentoxide gas is bubbled through a solution of concentrated ammonia.

$$N_2O_5 + NH_3 \rightarrow NH_4^+ + NO_3^- + H_2O$$

2. Carbon dioxide gas is bubbled through a solution of lye.

$$CO_2 + OH^- \rightarrow CO_3^{2-} + H_2O$$

XVII. Acidic Anhydride + Basic Anhydride → Salt

1. Sulfur trioxide gas is heated in the presence of sodium oxide.

$$SO_3(g) + Na_2O(s) \rightarrow Na_2SO_4(s)$$

2. Carbon dioxide gas is heated in the presence of solid magnesium oxide.

$$CO_2(g) + MgO(s) \rightarrow MgCO_3(g)$$

XVIII. Basic Anhydride + Acid → Salt + Water

1. Solid copper (II) oxide is dropped into sulfuric acid.

$$Cu_2 + H^+ \rightarrow Cu^{2+} + H_2O$$

2. Chunks of magnesium oxide are dropped into hydrochloric acid.

$$MgO + H^+ \rightarrow Mg^{2+} + H_2O$$

XIX. Precipitation Reactions

These problems involve mixing two solutions. Each solution is a water solution of an ionic compound. From the mixture of the two solutions, at least one insoluble precipitate will for. The other ions present are probably soluble and are called spectatopr ions; they are not included in the net ionic equation. *You must kow your solubility rules to do these problems.*

1. A solution of silver nitrate is added to a solution of hydrochloric acid:

$$Ag^+ + Cl^- \rightarrow AgCl \text{ (All nitrates are soluble.)}$$

2. Silver nitrate is added to a solution of sodium chloride:

$$Ag^+ + CrO_4^{2-} \rightarrow Ag_2CrO_4$$

3. Lead nitrate is added to a solution of sodium chloride:

$$Pb^{2+} + Cl^- \rightarrow PbCl_2$$

4. Iron (III) nitrate is added to a strong sodium hydroxide solution:

$$Fe^{3+} + OH^- \rightarrow Fe(OH)_3$$

5. Strontium chloride is added to a solution of sodium sulfate:

$$Sr^{2+} + SO_4^{2-} \rightarrow SrSO_4$$

XX. Redox Reactions

1. A strip of zinc metal is added a solution of copper (II) nitrate:

$$Zn + Cu^{2+} \rightarrow Zn^{2+} + Cu$$

2. A piece of aluminium is dropped into a solution of lead chloride:

$$Al + Pb^{2+} \rightarrow Pb + Al^{3+}$$

3. Chlorine gas is bubbled through a solution of sodium iodide:

$$Cl_2 + I^- + I_2 + Cl^-$$

(4–20: Common Oxidation State Changes)

4. $Cl_2 \xrightarrow{\text{dilute basic sol'n}} ClO^-$
5. $Cl_2 \xrightarrow{\text{conc. basic sol'n}} ClO_2^-$
6. $Cl^- \rightarrow Cl_2$ (or other halide)
7. $Na \rightarrow Na^+$ (free metal)
8. $NO_2^- + NO_3^-$
9. $\underset{\text{(conc.)}}{HNO_3} \rightarrow NO_2$
10. $\underset{\text{(in acid sol'n.)}}{MNO_4^-} \rightarrow NO$
11. $SN^{2+} \rightarrow SN^{4+}$
12. $\underset{\text{(in acid sol'n.)}}{MnO_4^-} \rightarrow Mn^{2+}$
13. $\underset{\text{(in acid sol'n.)}}{MnO_2} \rightarrow Mn^{2+}$
14. $\underset{\text{(in necutral or basic sol'n.)}}{MnO_4^-} \rightarrow MnO_2$
15. $\underset{\text{(hot. conc.)}}{H_2SO_4} \rightarrow SO_2$
16. SO_3^{2-} or $SO_2(aq) \rightarrow SO_4^{2-}$
17. $Na_2O_2 \rightarrow NaOH$
18. $H_2O_2 \rightarrow H_2O + O_2$
19. $HClO_4 \rightarrow Cl^-$
20. $\underset{\text{(in acid)}}{CrO_7^{2-}} \rightarrow Cr^{3+}$

XXI. Synthesis Reactions (*A* + *B* → *AB*)

1. Solid calcium oxide is added to silicon dioxide and the mixture is heated strongly:

$$CaO + SiO_2 \rightarrow CaSiO_3$$

2. Hydrogen gas is mixed with oxygen gas and the mixture is sparked:

$$H_2 + O_2 \rightarrow H_2O$$

3. Boron trichloride gas and ammonia gas are mixed:

$$BCl_3 + NH_3 \rightarrow BCl_3NH_3$$

If this compound is not familiar, try drawing a Lewis diagram to see how it is put together.

XXII. Decomposition (*AB* → *A* + *B*)

1. A sample of calcium carbonate is heated:

$$CaCO_3 \rightarrow CaO + CO_2$$

2. Hydrogen peroxide is gently warmed:

$$H_2O_2 \rightarrow H_2O + O_2$$

3. Manganese dioxide (acting as a catalyst) is added to a solid sample of potassium chlorate and the mixture is then heated:

$$KClO_3 \xrightarrow{MnO_2} KCl + O_2$$

4. Solid ammonia hydroxide is heated:

$$Al(OH)_3 \rightarrow Al_2O_3 + H_2O$$

XXIII. Single Displacement (*A* + *BC* → *AC* + *B*)

1. Chlorine gas is bubbled through a strong solution of potassium bromide:

$$Cl_2 + Br^- \rightarrow Cl^- + Br_2$$

2. Powered lead is added to a warm solution of copper (II) sulfate:

$$Pb + Cu^{2+} + SO_4^{2-} \rightarrow PbSO_4 + Cu$$

3. Strontium turnings are added to a 4 M sulfuric acid solution:

$$Sr + H^+ + SO_4^{2-} \rightarrow SrSO_4 + H_2$$

4. Silver is added to a solution of hot hydrochloric acid:

$$Ag + H^+ + Cl^- \rightarrow AgCl + H_2$$

XXIV. Complex Ions

1. A concentrated solution of ammonia is added to solution of zinc nitrate:

$$ZN^2 + NH_3 \rightarrow Zn(NH_3)_2^{2+}$$

or

$$Zn^{2+} + NH_3 + H_2O \rightarrow ZN(OH)_2 + NH_4^+$$

The concentration of NH_3 determines the product.

2. A solution of iron (III) iodide is added to a solution of ammonium thiocyanate:

$$Fe^{3+} + SCN^- \rightarrow Fe(SCN)_6^{3-}$$

3. A solution of copper (II) nitrate is added to a strong solution of ammonia:

$$Cu^{2+} + NH_3 \rightarrow Cu(NH_3)_4^{2+}$$

4. A concentrated potassuim hydroxide solution is added to solid aluminium hydrozxide:

$$OH^- + Al(OH)_3 \rightarrow Al(OH)_4^-$$